Advances in

HEAT
TRANSFER

AUTHOR AND SUBJECT CUMULATIVE INDEX
Including Tables of Contents,
Volumes 1–31

Serial Editors

James P. Hartnett
Energy Resources Center
University of Illinois at Chicago
Chicago, Illinois

Thomas F. Irvine, Jr.
Department of Mechanical Engineering
State University of New York at Stony Brook
Stony Brook, New York

Serial Associate Editors

Young I. Cho
Department of Mechanical Engineering
Drexel University
Philadelphia, Pennsylvania

George A. Greene
Department of Advanced Technology
Brookhaven National Laboratory
Upton, New York

Volume 32

ACADEMIC PRESS
San Diego   London   Boston
New York   Sydney   Tokyo   Toronto

This book is printed on acid-free paper. ∞

Copyright © 1999 by Academic Press

ACADEMIC PRESS
525 B. Street, Suite 1900, San Diego, California 92101-4495, USA
http://www.apnet.com

Academic Press
24–28 Oval Road, London NW1 7DX, UK
http://www.hbuk.co.uk/ap/

International Standard Serial Number: 0065-2717
International Standard Book Number: 0-12-020032-5

Printed in the United States of America
99 00 01 02 03 BB 9 8 7 6 5 4 3 2 1

CONTENTS

Preface vii
Author Index 1
Subject Index 257
Appendix: Tables of Contents of Volumes 1–31 447

PREFACE

For over a third of a century, *Advances in Heat Transfer* has filled the information gap between regularly published journals and university-level textbooks. The series presents review articles on special topics of current interest. Each contribution starts from widely understood principles and brings the reader up to the forefront of the topic being addressed.

Now that 31 volumes have been published, it is becoming more difficult to locate a particular author citation or information on a given subject. Therefore, we have decided to issue the present volume containing cumulative subject and author indexes.

The cumulative subject index lists the *Advances in Heat Transfer* volume number and page number; for example: Bose-Einstein statistics, **12**:120. The cumulative author index is complicated by the fact that all volumes did not contain an author index. Those missing indices have been included in the present volume and an explanation of the numbering system is given below.

It is our aim in issuing this index volume that the information already published in volumes of *Advances in Heat Transfer* will become more readily available to the reader.

A Note to the Reader

Some volumes of *Advances in Heat Transfer* were published without an author index. The following index includes all published volumes in two different formats:

Entries for the already existing author indexes (Volumes 1–17, 27, and the unnumbered supplemental volume by R. Shah and A. London (1978) (denoted by **SUP**)), were organized in the following way: the author's name, the volume number (in boldface) and the volume page number where the author's work is referred. If an author's name was not cited in the text, the volume page number is followed by the chapter reference number in parentheses. This is then followed by a page number where the complete reference is listed. For example: Aamodt, R. E., **14**:97(186), 104.

Entries for the remaining volumes (Volumes 18–26, 28–31) that did not have an author index were organized in the following way: the author's name, the volume number (in boldface), the volume page number where the author's work is referred, and the reference number in parentheses. For example: Abbott, N. A. **24**:229(104).

Authors of the chapters are noted with "(Chapter Author)" following their names.

The Editors

ADVANCES IN HEAT TRANSFER

Volume 32

AUTHOR INDEX

Bold text indicates volume number. Supplementary volume published between volumes 14 and 15 is indicated by **SUP**. Numbers in parentheses indicate reference numbers.

A

Aakalu, N. G., **20**:244(192), 285(306), 308(192), 313(306)

Aamodt, R. E., **14**:97(186), 104

Abadzič, E. E., **7**:79, 81, 86; **11**:18, 49, 105(112), 114, 194; **26**:220(21), 227(21), 325(21)

Abakians, H., **19**:80(244, 245), 94(244, 245)

Abarbanel, S. S., **6**:95(156), 130; **7**:190, 214; **SUP**:73(196), 92(196), 167(196)

Abashechev, E. V., **31**:265(159), 329(159)

Abbot, I. H., **3**:64, 99

Abbote, D. E., **6**:109, 130

Abdalla, K. L., **4**:228

Abdel-Alim, A. H., **26**:10(1), 92(1)

Abdel-Khalik, A., **25**:259, 281, 286, 315

Abdel-Khalik, S. I., **18**:24(53), 81(53); **29**:167(122, 123, 124), 208(122, 123, 124)

Abdelsalam, M. A., **16**:95, 154, 181(17), 238

Abdel-Wahed, R. M., **19**:46(110), 88(110)

Abdollahian, D., **17**:42(144), 43(144), 44(144), 63

Abdulkadir, A., **10**:34(47), 37

Abel, W. T., **9**:118(8, 9), 177; **30**:236(1), 249(1)

Abelmessih, A. H., **15**:248, 280

Abeyata, C. N., **5**:295(43), 323

Abib, A. H., **28**:201(6), 204(6), 205(6), 208(6), 209(6), 219(1), 227(1), 228(6)

Abinante, M., **28**:15(21), 22(21), 71(21)

Ablott, D. E., **31**:167(73), 325(73)

Abolfadl, M. A., **29**:150(62), 180(164), 181(164), 194(62), 199(164), 205(62), 210(164)

Abou-Kassem, J. H., **23**:206(1), 267(1)

Abou-Sabe, A. H., **1**:382(44), 441

Abraham, F. F., **14**:282, 288(21), 300(5), 306, 310, 312(71), 314(74, 75), 317, 329, 338(133, 134, 135, 136), 339(5, 133, 134, 135, 136, 140), 341, 342, 343, 346

Abraham, G., **16**:9, 11, 23, 24, 26, 56

Abramouitch, G. N., **3**:80, 89(73), 93(73), 99

Abramov, A. I., **16**:110(68), 111(68), 114(68), 128(68), 131(68), 155

Abramovich, G. N., **13**:66, 114; **25**:131(202), 132(202), 150(202)

Abramowitz, M., **9**:374(42), 415; **SUP**:101(262); **20**:357, 360, 361, 364, 367, 370, 386

Abrams, M., **11**:412(198, 199), 415(198, 199), 416, 417, 418, 419, 439; **18**:29(64), 30(64), 33(64), 82(64); **19**:42(96), 87(96)

Abramson, P., **4**:213(98), 214(98), 227; **9**:259(104), 271

Abramzon, B., **26**:19(2), 21(3), 22(2), 23(3), 33(2), 67(4), 69(4), 71(4), 72(4), 79(4), 92(2, 3, 4)

Abriola, L. M., **23**:370(17), 461(17); **30**:93(2), 98(36), 111(54, 57, 58), 188(2), 190(36), 191(54, 57, 58)

Abuaf, N. (Chapter Author), **10**:167, 211, 212, 218; **14**:151(3), 242; **21**:280, 344; **23**:222(159), 274(159); **29**:17(14), 56(14)

Abu-Romia, M. M., **5**:254(9), 298(51), 308(51), 314(9), 316(9), 322, 323; **14**:139(1, 2), 144

Achari, P. J., **8**:57, 61, 90

Acharya, A., **15**:69(45), 136(45); **25**:272, 281, 315

Acharya, S., **20**:243(170), 307(170)

Achenbach, E., **8**:97(22), 100, 101, 110(33), 111(33), 159; **18**:34(69), 82(69), 91(22), 92(22), 95(22), 104(33), 105(33), 106(33), 120(54), 127(54), 158(22, 33), 159(54); **23**:336(252), 366(252)

Ackermann, G., **11**:202, 204, 252; **15**:232, 276

Ackermann, H., **16**:114(78), 155

Acloque, P., **11**:391(146), 438

Acosta, A. J., **25**:257, 286, 317
Acrivos, A., **2**:166, 170, 172, 269, 361(1), 390, 391(100), 394, 396; **4**:33, 63; **6**:61, 127; **7**:144(53), 161; **8**:36(198, 202), 89; **9**:296, 346; **15**:63(28), 64(28), 136(28), 153(12), 154(12), 155(12), 156(12), 162(12), 169(12), 170(12), 174(12), 175(12), 176(12), 185(55), 212(108), 215(108), 223(12), 224(55), 225(108); **18**:190(83, 84, 85), 191(83, 84, 85, 135), 193(83), 194(83), 195(84, 86), 196(85, 86), 197(85, 86), 201(93), 202(93), 203(93), 204(93), 237(83, 84, 85, 86), 238(93, 135); **23**:207(228), 277(228); **25**:257, 262, 263, 264, 265, 266, 267, 268, 270, 271, 279, 280, 281, 282, 283, 284, 286, 288, 289, 290, 293, 294, 315, 318; **26**:5(226), 6(226), 12(158), 19(226), 21(226), 34(5), 93(5), 100(158), 103(226)
Acton, L., **5**:295(40), 322
Adam, N. K., **9**:184, 185(3), 186, 188, 189(3), 198(3), 215, 268
Adam, O., **9**:134, 179
Adamek, T., **15**:287, 328
Adami, M., **20**:287(311), 288(311), 313(311)
Adams, B. B., **20**:285(308), 313(308)
Adams, C. C., **4**:98, 139
Adams, D., **25**:252, 293, 297, 302, 303, 304, 305, 306, 307, 308, 309, 311, 315
Adams, D. E., **4**:42, 64; **6**:460, 469(187), 475, 495; **9**:317(80), 347
Adams, F., **28**:124(103), 143(103)
Adams, J. A., **6**:312(101), 366; **9**:304(60, 61), 347
Adams, J. B., **10**:7, 8(14), 9, 21, 22(14), 24(32, 37, 41), 25(32, 37), 36
Adams, J. C., **1**:225, 264
Adams, J. H., **20**:250(232), 310(232)
Adams, J. M., **1**:239, 266; **4**:206(85), 226
Adams, J. S., **10**:270(102), 283
Adams, M. C., **1**:77, 121; **3**:61(62), 99
Adams, R. L., **19**:164(217, 218), 168(222), 171(217), 172(217), 190(217, 218, 222)
Adamski, J. A., **11**:412(191), 439; **30**:320(22), 323(26), 325(26), 345(26), 350(26), 412(26), 413(26), 417(26), 427(22, 26)
Adamson, T. C., Jr., **2**:213, 269
Adbelmessih, A. H., **22**:30(35), 154(35)
Additon, S. L., **29**:234(85, 86), 269(85), 320(86), 339(85, 86)

Addoms, J. M., **1**:423(123), 425, 426, 443
Addoms, J. N., **1**:213(23), 228, 239, 264, 265; **5**:66, 67, 125; **11**:105(58), 109(58), 110(58), 191; **20**:14(18), 15(18), 80(18)
Ade, P. A., **10**:34, 37, 53, 68, 81
Adeboyi, G. A., **21**:32(49), 52(49)
Adelberg, M., **4**:157, 158, 201, 224, 226
Adelman, M., **15**:162(22), 168(22), 223(22); **25**:260, 264, 286, 290, 317, 318, 319
Aden, M., **28**:124(1, 2), 138(1, 2)
Adenekan, A. E., **30**:185(162), 196(162)
Ades, M., **16**:12, 24, 39, 57
Adiutori, F. E., **18**:308(66a), 323(66a)
Adivarahan, P., **5**:155, 157, 248
Adler, P. M., **30**:94(11), 189(11)
Adomaitis, I.-E., **31**:160(56), 324(56)
Adomeit, G., **3**:217(63), 250
Adorni, N., **1**:379(33, 34, 35, 37), 381(40, 41), 385(34), 387(37), 389(34), 390(34), 401(34), 413(40, 41), 421(40, 41), 441, 441(40, 41)
Adrian, R. J., **21**:198(26), 237(26); **30**:255(3), 307(3)
Adrianov, V. N., **3**:201(23), 212, 215, 218(60), 242, 243, 244(60), 245(60), 249, 250, 251; **11**:390(143), 397(143), 438
Advani, G. H., **4**:90(28), 91(28), 92(28), 95(28), 109(28), 110(28), 138
Advani, S. G., **25**:298, 315
Aerov, G. E., **25**:73(120), 145(120)
Aerov, M. E., **14**:247; **19**:102(45), 185(45); **25**:223(110), 247(110)
Aeschlimann, R. W., **29**:271(141, 142), 342(141, 142)
Afanas'ev, Y. V., **28**:123(3), 138(3)
Afgan, N. (Chapter Author), **11**:1, 4(5), 19, 20(70), 21(70, 71, 72), 24(68), 25(68), 31(68, 71, 72, 73), 33(71), 34(72), 36(71), 37(7l), 38(68), 39(68), 40(68), 44(75), 45(75), 47, 49
Afgan, N. F., **23**:307(124, 125, 126), 359(124, 125, 126); **31**:160(66), 325(66)
Afgan, N. H., **16**:114(75), 117(75), 125, 128, 131, 155
Afrid, M., **20**:243(168), 307(168)
Afshari, B., **18**:34(68), 82(68)
Afzal, M., **3**:256(8), 260(8), 300
Agababov, S. G., **31**:202(107), 208(107), 212(107), 213(107), 327(107)
Agah, R., **22**:421(35), 422(35), 423(35),

424(35), 436(35)

Agar, J. N., **7**:90, 160; **12**:251(1), 270(1), 277

Agarwal, A. L., **17**:300, 305, 317

Agarwal, P. K., **23**:196(2), 206(2), 267(2)

Agarwal, S. C., **14**:158(59), 209, 243

Agbim, J. A., **25**:166(58), 199(58), 245(58)

Aggarwal, J. K., **5**:40, 53

Aggarwal, S. J., **22**:229(1), 347(1)

Aggarwal, S. K., **26**:13(11), 61(134), 89(6, 7, 8, 9, 10, 200), 93(6, 7, 8, 9, 10, 11), 99(134), 102(200)

Aggarwala, B. D., **SUP**:28(39), 62(76), 63(78), 64(76, 92), 67(76, 92), 68(130, 131), 72(167), 207(92), 208(39), 209(39), 225(78), 232(76, 92), 236(76, 78, 92), 237(92), 242(167), 248(92), 251(92), 259(76, 92), 371(131), 372(130), 373(131); **19**:278(63), 352(63)

Agnone, A., **6**:41(81), 65(81), 126

Agosta, C. C., **17**:133(140, 141), 156

Agrawal, D. K., **10**:21, 23, 36

Agrawal, H. C., **8**:24, 26, 32, 86, 87; **SUP**:174(350); **19**:39(90), 40(90), 87(90)

Agrawal, S., **10**:191, 217

Agrawal, S. K., *see* Aggarwal, S. K.

Agsten, R., **15**:51(84), 57(84)

Aguilo, M., **30**:396(162), 399(162), 434(162)

Aguirre-Puente, J., **19**:82(260), 94(260)

Agur, E. E., **28**:150(2), 227(2)

Agzamov, Sh. K., **31**:284(190, 191), 331(190, 191)

Aharoni, G., **26**:123(53), 126(53), 129(53), 150(53), 151(53), 210(53)

Ahern, J. E., **15**:50(82), 57(82)

Ahlers, G., **17**:102, 125, 127(126), 128, 129, 130, 131(122), 133, 149, 155, 156

Ahluwalia, K. S., **19**:144(154), 188(154)

Ahluwalia, M. S., **14**:151(12), 242

Ahmad, S. Y., **17**:30, 39(131), 60, 62

Ahmadzadeh, J., **26**:14(81), 96(81)

Ahmed, A. M., **11**:220, 223, 238, 258

Ahsmann, G., **14**:107, 116, 117(3, 4), 120, 121, 123, 144, 146

Ahuja, K. L., **8**:6, 24, 84, 87

Ahuja, S., **28**:268(96, 97, 98), 333(96, 97, 98)

Ai, D. K., **6**:53(100), 127

Aichi, K., **20**:357, 386

Aidoshin, G. T., **8**:36(206), 89

Aihara, T., **20**:186(18), 187(18, 19), 189(18), 202(81, 82, 83, 84), 300(18, 19), 303(81, 82, 83, 84); **23**:12(61, 62), 14(61, 62), 32(61, 62), 35(61), 39(61), 49(61), 53(61), 54(61), 58(61, 62), 60(61, 62), 89(61, 62), 92(61), 106(61), 109(62), 110(62), 120(62), 129(61, 62); **30**:405(175), 434(175)

Ainley, D. G., **9**:97, 98, 106

Ainola, L. Y., **8**:33, 88; **19**:202(58), 245(58)

Ainola, P. Ya., **25**:81(154), 147(154)

Ainshtein, V. G., **19**:119(100, 101), 187(100, 101)

Ainsworth, R. W., **23**:321(194, 200, 201), 363(194, 200, 201)

Aitkadi, A., **24**:109(90), 187(90)

Aitsam, A. L., **25**:81(146, 147, 148), 147(146, 147, 148)

Aitsam, A. M., **6**:407, 408(62), 409(62), 489

Ait'yan, S. K., **15**:291, 321

Aizawa, M., **10**:172(28), 215

Aizen, A. M., **15**:321, 323

Akagawa, K., **20**:110, 130; **23**:13(68), 15(68), 50(68), 106(68), 129(68)

Akai, S. I., **30**:323(25), 427(25)

Akao, F., **25**:81(130), 146(130)

Akbari, H., **18**:73(138), 86(138)

Akers, W. W., **9**:259(106, 107), 265(134), 271, 272; **15**:232(35), 277

Akfirat, J. C., **13**:6(26), 10(26, 35), 11(26), 12, 26(35), 41(26), 59, 60; **23**:5(5), 126(5); **26**:127(73), 132(73), 211(73)

Akharaz, A., **23**:253(267), 254(267), 255(267), 256(267), 260(267), 261(267), 278(267)

Akheizer, I., **8**:22(107), 86

Akilbayev, Z. S., **8**:103(29), 104(29), 105(29), 132(29), 133(29), 159; **11**:228(226), 262; **18**:98(29), 99(29), 130(29), 131(29), 158(29)

Akimenko, A. D., **23**:109(112), 110(112), 112(112), 132(112)

Akimenko, A. D. see above

Akin, G. A., **11**:105(57), 115(57), 191; **20**:15(26), 80(26)

Akinsete, V. A., **18**:50(98), 84(98)

Akiyama, M., **11**:14(46), 49; **19**:308(97), 313(97), 353(97); **28**:370(130), 421(130); **29**:151(69), 205(69)

Akiyoshi, M., **24**:6(21, 22), 36(21, 22)

Akiyoshi, R., **29**:131(14), 164(106), 203(14), 207(106)

Akmaev, R. A., **14**:265, 266, 267(38), 268(38), 269, 280

Akman, R. G., **6**:418(101), 491

Aksan, S. N., 21:88(51), 135(51)
Aksay, I. A., 24:102(11), 184(11)
Akselrod, L. S., 11:168(168), 170(168, 170), 196
Akselvoll, K., 25:353, 354, 358, 360, 392, 404, 408
Akyama, M., 15:244(65, 67), 279
Alad'ev, I. T., 4:206(84), 226; 6:551, 564
Aladyev, I. T., 1:244, 266, 416(108), 417(108), 443; 11:4(4), 47
Alamgir, M., 15:172(33), 224(33)
Al-Arabi, A., 30:280(64), 310(64)
Al-Arabi, M., 18:65(120), 85(120)
Al-Astrabadi, F. R., 20:268(274), 312(274)
Alavizadeh, N., 19:164(217, 218), 168(222), 171(217), 172(217), 190(217, 218, 222)
Alavyoon, F., 24:302(1), 318(1)
Al-Bahadili, H., 31:413(77), 429(77)
Alber, I. E., 6:29(38), 119, 124
Alber, J. E., 6:107(187), 132
Albergel, A., 19:78(190), 91(190)
Albers, J. A., 4:210, 211, 212(96), 213(96), 227; 9:259(105), 271
Albers, L. U., 2:420(13), 427(13), 450
Albert, R. V., 25:234(120), 248(120)
Albertson, M. L., 16:5, 9, 11, 31, 56
Alberty, G., 12:233(2), 277
Albini, F. A., 10:266, 283
Alboussiere, T., 30:333(57), 429(57)
Albright, D. E., 28:342(35), 416(35)
Alcock, J. F., 9:37, 106
Alder, B. J., 5:48(76), 54
Al-Dibouni, M. R., 23:243(83), 244(83), 257(83), 262(83), 270(83)
Aldoshin, G. T., 6:416(25), 418(152), 487, 493
Aldridge, K. D., 14:126, 147
Aleksandrov, A. A., 21:2(4, 8), 3(4, 8), 50(4), 51(8)
Aleksandrov, M. V., 26:314(86), 328(86)
Aleksashenko, A. A., 6:474(333), 502; SUP:12(19), 55(19), 137(19), 189(19); 25:11(39), 31(39), 141(39)
Aleksashenko, V. A., 6:474(333), 502; SUP:12(19), 55(19), 137(19), 189(19); 25:11(39), 31(39), 141(39)
Alekseenko, S. V., 26:123(70), 124(70), 211(70)
Alemasov, V. E., 25:4(19), 140(19)
Alescenkov, P. I., 1:435(140, 141), 444
Aleshin, I. V., 11:318(20), 435
Alexander, H., 30:358(118), 360(118), 432(118)

Al-Fariss, T. F., 23:193(3, 4), 196(4), 201(4), 202(4), 267(3, 4)
Al-Farris, T., 24:118(126), 189(126)
Alferov, N. S., 21:25(44), 52(44)
Algren, A. B., 11:219, 223, 237, 257
Alhaddad, A. A., 30:83(1), 87(1)
Al Hanai, W., 23:371(33), 462(33)
Ali, S., 25:272, 273, 318
Alievsky, M. Ya., 14:279(64), 280
Ali-Khishali, K. J., 29:73(34), 74(34), 75(34), 80(34), 81(34), 125(34)
Aliotta, C. F., 28:116(52), 141(52)
Alkidas, A. C., 23:341(277), 343(277, 287, 288, 289), 344(287), 367(277, 287, 288, 289)
Alkire, R. C., 12:233(3), 251(3), 270(3), 277
Allan, D. W., 15:286, 321
Allard, E. F., 14:321, 322, 344
Alleavitch, J., 7:238(41), 241, 242(41), 258(41), 260, 315, 320b, 320e
Allen, E., 23:262(5, 6), 267(5, 6)
Allen, J. F., 17:84(56), 95, 154
Allen, J. L., 23:321(200), 363(200)
Allen, J. T., 22:381(19), 384(19), 394(19), 435(19)
Allen, K. D., 28:351(78), 418(78)
Allen, L. S., 15:321
Allen, M. B., III, 30:101(38), 190(38)
Allen, M. D. (Chapter Author), 29:215, 218(18, 19), 219(18, 19, 24), 220(41), 221(27, 28, 29, 30, 31, 32, 33, 34, 35, 36, 37, 38, 39, 40, 41), 228(18), 230(41), 232(18), 233(18, 19), 244(18), 255(40), 264(126), 265(40, 41), 269(24, 40), 270(19, 40, 41), 271(28, 29, 30, 31), 276(41), 277(19, 24, 126), 278(40), 279(18, 19, 24), 280(28, 29, 31), 282(28, 29, 31, 41), 288(18), 290(40, 41), 293(29), 295(29), 297(18), 300(40), 302(36, 37), 305(19), 306(41), 309(29, 38), 310(32, 33, 34, 35, 36, 37, 40), 311(32, 33, 34, 35, 36, 37), 312(18, 24), 315(30, 40), 316(30, 40), 318(40), 319(19, 40), 322(19), 323(24, 40), 329(18), 335(18, 19, 24), 336(27, 28, 29, 30, 31, 32, 33, 34, 35, 36, 37, 38), 337(39, 40, 41), 341(126)
Allen, P. H. G., 14:119, 120, 124(5), 144
Allen, R. A., 5:304(63), 309(63), 323
Allen, R. G., 22:248(2), 250(2), 251(2), 252(2), 266(133), 268(133), 347(2), 353(133)
Allen, R. W., 6:525(34), 526, 528, 530(34), 531,

532, 562; **12**:93, 97, 113; **15**:105(133), 121(133), 126(133), 140(133)

Allen, T., **18**:73(139), 86(139)

Allen, W. G., **15**:287(81), 324

Allenman, J. E., **31**:458(1), 470(1)

Allingham, W. D., **4**:206, 227; **7**:254, 255, 317

Allison, C. M., **29**:215(1), 248(1), 334(1)

Allulli, S., **12**:233(2), 277

Allyn, P. A., **22**:222(108), 351(108)

Almazan, P., **31**:355(36), 426(36)

Al Naimi, A. E., **17**:68, 70(15), 153

Alpert, D., **5**:326(1), 505

Alsmeyer, H., **21**:327, 345

Alt, L. L., **28**:342(20), 416(20)

Al Taweel, A. M., **25**:259, 281, 286, 315

Altman, M., **1**:78, 121; **15**:287, 328

Altman, P. L., **4**:141

Altmayer, E. F., **18**:63(115, 116), 85(115, 116)

Altmos, D. A., **4**:228

Altshuler, V. S., **19**:110(71), 111(74), 186(71, 74)

Altunin, V. V., **21**:2(6), 3(6), 34(6), 38(6), 51(6)

Al Varado, D. A., **23**:193(7), 267(7)

Alvares, N. J., **10**:243(39), 282

Alves, G. E., **1**:357, 382(5), 440; **SUP**:47(55), 54(55); **15**:195(73), 224(73)

Alwang, W. G., **23**:280(8), 354(8)

Alziary de Roquefort, T., **21**:154, 156, 178

Alzofon, F. E., **3**:215(46), 216(46), 301

Amarasooriya, W. H., **29**:150(63), 180(165), 194(206, 207), 205(63), 210(165), 213(206, 207)

Amato, W. S., **11**:278, 280, 314; **15**:175(34), 176(34), 180(34), 224(34); **25**:259, 264, 265, 266, 268, 315

Amberg, G., **28**:253(57), 255(57), 262(70), 267(57, 70), 288(57), 289(57, 70), 290(70), 291(70), 292(70), 294(70), 295(70), 296(70), 331(57), 332(70), 334(121)

Ambirajan, A., **31**:347(34), 426(34)

Ambraziavičius, A. B., **8**:112(36), 159

Ambrok, G. S., **6**:474(255), 483(255), 498

Ambrose, J. H., **30**:121(83), 192(83)

Ambrose, T. W., **1**:357, 440

Amdur, I., **2**:326, 352; **3**:258, 264, 265(20), 293(92, 93), 300, 302

Amenitskii, A. N., **4**:206(84), 226

Ames, W. F., **15**:153(14), 223(14); **25**:262, 317

Ametistov, E. V., **18**:241(2), 249(2), 256(2), 310(2), 313(2), 320(2)

Ametistov, Y. V., **16**:216(27), 233, 234, 239; **17**:119(115), 146(166), 156, 157

Amir, A., **22**:220(179), 355(179)

Amundson, N. R., **15**:286, 301, 315, 327

Anand, D. K., **7**:227(14), 254, 256, 274, 276, 277, 296, 314, 317, 318

Anand, K. V., **28**:342(24), 416(24)

Anand, N. K., **30**:272(40), 309(40)

Ananier, E. P., **9**:259(109), 260, 271

Anantanarayanan, R., **11**:241, 242, 261

Anantharaman, T. R., **28**:26(48), 72(48)

Anastassakis, E., **10**:21, 23, 36

Anateg, R., **4**:82(20), 138

Ancudinov, V. B., **21**:8(16, 17, 18), 11(16), 12(16), 15(17), 21(17), 26(18), 27(18), 29(17, 18), 47(82), 48(82), 51(16, 17, 18), 53(82)

Andeen, B. R., **14**:186(115), 187(115), 188(115), 189(115), 191(115), 245; **19**:119(99), 187(99)

Andeen, G. B., **7**:227(16), 246, 314

Andereck, C. D., **17**:89(59), 154; **21**:178

Anderle, M., **29**:175(153), 210(153)

Andersen, J. W., **21**:319, 320, 321, 325, 329, 331, 344

Andersen, P. S., **16**:279, 297, 355, 356, 363, 365; **17**:37, 61

Anderson, A. C., **17**:69(26, 27), 71(26), 72(27), 153

Anderson, A. D., **1**:330, 354; **2**:134(19), 157, 162(31), 163(31), 170(31), 179, 182(31), 227, 230, 234, 235(19), 239(19), 247(19), 268, 269; **4**:35, 63; **6**:474(232), 480(232), 482, 483(232), 497

Anderson, A. M., **23**:280(21), 309(21), 310(21), 355(21)

Anderson, B. H., **5**:389, 510

Anderson, D., **10**:55(43), 82

Anderson, D. A., **24**:194, 268

Anderson, D. B., **2**:385(59), 395

Anderson, D. R., **18**:161(4), 175(4), 236(4)

Anderson, E. E. (Chapter Author), **11**:317, 361, 363, 365(99), 392(152), 395(152), 401(152), 402, 403, 404(173), 405(172), 406, 407(173), 423(173, 219a), 437, 438, 439, 440

Anderson, G. H., **1**:363(16), 377, 389, 390(16), 392, 427, 440, 441, 443; **10**:151, 165

Anderson, G. K., **28**:124(78), 142(78)

Anderson, H. E., **10**:267, 268(90), 269, 283

Anderson, I. E., **28**:11(15), 22(38), 71(15), 72(38)

Anderson, J. E., **4**:269(32), 284(32), 285, 286, 287, 288, 289, 314

Anderson, J. K., **17**:39(128), 62

Anderson, J. T., **5**:166(56), 248

Anderson, R. (Chapter Author), **18**:1, 18(36), 19(41, 42), 40(81), 41(81), 48(41), 53(102), 59(113, 117), 61(113), 62(113, 114), 63(113, 117), 64(36, 114), 65(36), 71(135), 72(135), 81(36, 41, 42), 83(81), 84(102), 85(113, 114, 117), 86(135); **20**:205(101), 304(101)

Anderson, R. E., **5**:381, 388, 509

Anderson, R. L., **28**:402(210), 424(210)

Anderson, R. N., **14**:41(95), 42(95), 102

Anderson, R. P., **29**:136(22), 145(38), 147(38), 168(38), 180(38), 183(172), 190(197), 203(22), 204(38), 211(172), 212(197)

Anderson, R. R., **22**:278(4), 347(4)

Anderson, S. L., **25**:294, 315

Anderson, T. B., **23**:374(38), 377(38), 462(38); **25**:195(98), 247(98)

Anderson, T. J., **28**:344(59), 417(59)

Anderson, T. M., **20**:246(208), 247(208), 248(208), 309(208); **23**:65(85), 100(85), 130(85)

Anderssen, R. S., **19**:197(29, 30), 244(29, 30)

Andersson, B. A., **27**:187(39)

Andersson, H. I., **25**:270, 315

Andersson, J., **29**:249(101), 340(101)

Ando, T., **17**:140(158), 157; **28**:20(36), 21(36), 22(36), 71(36)

Andoe, W. V., **26**:19(242), 104(242)

Andracchio, C. A., **23**:298(83), 357(83)

Andracchio, C. R., **4**:228; **17**:4, 8, 57

Andraka, C. E., **23**:282(31), 283(31), 355(31)

Andre, P., **23**:338(264), 366(264)

Andreescu, A., **9**:128(34), 178

Andreev, A. A., **13**:11(41), 60

Andreev, G. N., **3**:230, 251

Andreev, V. K., **17**:327(29), 329(29), 330(29), 336(29), 340

Andreopoulos, J., **20**:225(142), 306(142)

Andres, R. P., **14**:282, 314, 317(81), 341, 344; **26**:49(17), 93(17)

Andretta, A., **23**:292(54), 356(54)

Andrews, B. D., **25**:294, 317

Andrews, D. G., **23**:58(100), 60(100), 90(100), 91(100), 106(100), 131(100)

Andrews, G. E., **11**:220, 224, 235, 238, 258

Andrews, J. A., **20**:259(248, 250), 262(248), 311(248, 250)

Andrews, J. C., **5**:309(81), 324

Andrews, R. W., **28**:365(109), 420(109)

Andritsos, N., **20**:85, 130; **31**:462(38), 464(54), 472(38), 473(54)

Andritzky, H. K. M., **8**:99(26), 159

Andriyanov, P. A., **6**:474(307), 501

Andronikashvili, E. L., **17**:77(53), 154

Andros, F. E., **20**:283(301), 293(327), 294(301), 313(301)

Androulakis, J. G., **9**:394(71), 416

Andrussow, L., **2**:37, 106

Andvig, T. A., **9**:3(3), 66(3), 93(3), 106

Anfimov, N. A., **10**:53(34), 58(34), 64(34), 70(34), 71(34), 82

Angelini, S., **29**:153(78, 79, 81), 157(79), 206(78, 79, 81)

Angelino, H., **10**:185(76, 77), 205(76, 77), 216; **14**:152(35), 243; **15**:244(66), 247(79a), 279, 322

Angell, C. A., **22**:176(69), 350(69)

Angus, S., **17**:102, 155

Anisimov, S. I., **28**:123(4), 127(4), 138(4)

Annamalai, K., **26**:2(12), 84(12), 87(12), 93(12)

Annavarapu, S., **28**:1(2), 22(2), 64(2), 65(2), 66(2), 67(2), 68(2), 70(2)

Anno, J. N., **14**:139, 145

Ansari, M. A., **20**:205(105), 212(105), 304(105)

Anselmo, A. P. (Chapter Author), **30**:313, 328(30), 367(145), 368(145), 371(141, 145), 372(141, 145), 373(141), 374(141, 145), 375(141), 376(141), 377(141), 378(30, 141, 145), 379(141), 380(141), 381(141), 384(141, 145), 387(145), 392(141, 145), 394(145), 406(30, 141, 145), 407(30, 176), 408(30, 141, 145, 176, 177), 409(30), 410(177), 411(177), 412(145, 176, 177), 415(182), 416(182), 427(30), 433(141, 145), 434(176), 435(177, 182)

Antaki, P. J., **26**:81(13, 14, 15), 82(15), 93(13, 14, 15)

Anthony, M. L., **15**:321

Antipov, V. I., **17**:327(35), 335(38), 336(35), 337(35), 340

Anton, J. R., **10**:172, 177, 215

Antonetti, V. W., **20**:259(249), 261(252, 253), 262(255), 264(253), 267(253), 268(253,

273), 270(253), 311(249, 252, 253, 255), 312(273)

Antonia, R. A., **25**:383, 408

Antonir, I., **15**:242(53), 278

Antonishim, N. V., **14**:151(1), 152, 176(160, 183), 225, 236(183), 237(183), 242, 243, 246, 247

Antonishin, N. V., **19**:102(36), 116(36), 117(36), 122(113), 129(36, 113), 131(113), 132(36), 185(36), 187(113)

Antonopoulos-Domis, M., **25**:338, 361, 382, 399, 408

Antonyuk, K. S., **14**:176(119), 188(119), 191(119), 237(119), 245; **19**:160(193), 189(193)

Antuf'yev, V. M., **8**:94, 149, 157(17), 158, 159; **18**:88(8, 17), 147(17), 156(17), 158(8, 17); **31**:160(5), 182(5), 322(5)

Anzelius, A., **20**:171(18), 179(18)

Aoki, H., **7**:129, 161; **21**:165, 178

Aoki, S., **29**:160(94), 161(94), 207(94)

Aoki, T., **8**:232(4), 243(4), 282; **18**:246(8), 255(8), 257(8), 261(8), 265(59), 266(59), 270(8), 271(8), 272(8), 320(8), 322(59); **30**:202(58), 252(58); **31**:443(70), 474(70)

Aoyama, T., **28**:386(157), 387(157), 422(157)

Apelian, D., **23**:293(68), 357(68); **28**:1(2), 22(2), 23(41), 64(2, 41, 83), 65(2), 66(2), 67(2), 68(2), 70(2), 72(41), 73(83), 307(151, 152), 336(151, 152)

Apelt, C. J., **25**:81(160), 147(160)

Apfel, R. E., **10**:106(26, 27, 28), 107, 112, 113, 164

Apkarian, H., **24**:8(33), 14(33), 37(33)

Aplenc, A. J., **15**:52(88), 58(88)

Appelbaum, A., **28**:343(41, 42), 417(41, 42)

Appeldorn, J. K., **13**:219(67), 248(67), 265

Arai, A., **20**:75(65), 82(65)

Arai, H., **7**:129(39), 161; **21**:165, 178

Arai, N., **20**:75(65), 82(65)

Arai, Y., **30**:333(52), 349(101), 351(101), 378(101), 407(52), 408(52), 428(52), 431(101)

Arakawa, A., **14**:43, 102; **25**:371, 394, 408

Arakawa, Ch., **24**:247, 268

Arakawa, E. T., **28**:118(65), 119(65), 141(65)

Arakeri, V. H., **29**:130(8), 203(8)

Araki, T., **30**:379(147), 433(147)

Aral, B., **28**:193(31), 229(31)

Aravin, V. I., **24**:104(51), 186(51)

Arbarbanel, S. S., **1**:2(3), 49, 59, 121

Arbogast, T., **23**:371(29), 374(29), 462(29)

Archbold, E., **30**:261(17), 308(17)

Archer, C. T., **2**:352, 354

Ardon, K. H., **29**:254(107), 255(107), 340(107)

Arellano, F. E., **29**:221(25), 271(25), 309(25), 335(25)

Arenholz, E., **28**:102(37), 140(37)

Argumedo, A., **15**:77(61), 137(61)

Arie, M., **SUP**:164(342), 168(342), 190(342)

Arihara, N., **14**:16(43), 77, 87, 101, 103

Arinc, F., **24**:104(67), 187(67)

Aris, R., **13**:128(35), 202; **28**:402(199), 424(199)

Aritomi, M., **29**:160(94), 161(94), 207(94)

Arizumi, T., **11**:412(194, 195, 196), 439

Arkhipov, V. V., **17**:327(28, 29, 30), 328(28), 329, 330(28, 29), 331, 332(28), 335(28), 336(28, 29, 30), 340

Armaly, B. F., **9**:361(20), 364(33, 34), 382(20), 384(20), 385(56), 386(34), 387(33), 394(56), 397(56), 399(33), 414, 415, 416; **11**:376(134), 381, 438

Armand, A. A., **1**:375(24), 382, 390(23, 24), 441; **6**:474(272, 278, 303), 499, 500, 501, 551, 563

Armfield, S. W., **24**:300(35), 301(35), 320(35)

Armistead, R. A., Jr., **8**:293(20), 349

Arms, R. J., **17**:94, 154

Armstead, B. H., **7**:52(29), 84

Armstrong, B. H., **5**:281(25), 304(62), 309(62), 320(92), 322, 323, 324

Armstrong, D. R., **29**:136(22), 183(172), 190(197), 203(22), 211(172), 212(197)

Armstrong, R. C., **15**:62(7), 99(7), 100(7), 101(7), 135(7), 144(7), 149(7), 179(7), 223(7); **19**:251(10), 252(10), 295(10), 350(10); **23**:188(20), 189(20), 267(20); **24**:107(72), 108(72), 109(72), 187(72); **25**:253, 254, 255, 315; **28**:152(3), 227(3)

Arnaldos, J., **19**:102(42), 105(42), 107(42), 108(42), 180(42), 185(42); **25**:182(77), 184(79), 185(79), 186(79), 187(79), 189(86), 190(87, 88), 191(88), 192(88), 193(92), 194(92), 195(92), 200(79), 202(79), 204(88), 205(88), 222(88), 224(79), 227(88), 228(88), 234(138), 246(77, 79, 86, 87, 88, 92), 249(138)

Arnas, O. A., **SUP**:341(490), 342(490), 343(490), 344(490)

Arnaud, F. G., **22**:188(161), 354(161)

Arnaud, G., **30**:172(143), 195(143)

Arnberg, L., **28**:15(22), 71(22)

Arnett, C. D., **4**:150, 223

Arnitonov, R. V., **6**:416(90), 490

Arnold, J. N., **18**:17(29, 30), 80(29, 30)

Arnold, J. T., **20**:244(199), 309(199)

Aroeste, H., **5**:320(93), 324

Arola, R. A., **17**:175, 178(25), 314

Aronov, I. Z., **30**:212(16), 221(16), 223(16),
    250(16); **31**:200(102), 240(126), 294(197),
    304(197), 326(102), 328(126), 331(197)

Aronovich, F. D., **14**:220(157), 246

Aronson, R. B., **10**:278(110), 284

Arp, V. D., **17**:106, 138(147), 155, 319(1), 338

Arpaci, V. S. (Chapter Author), **4**:49(85), 64;
    5:336(25), 339(25), 341(25), 381, 382, 383,
    384, 386, 387(95, 96), 469, 506, 509, 516;
    6:474(265, 266), 499; **7**:178(41a), 212;
    9:216(40), 223(40), 269; **11**:322(37), 364,
    435, 437; **15**:211(105), 225(105); **21**:239,
    240(1, 2, 7, 8, 9, 10, 11, 12), 246(8, 12, 15,
    16, 17, 18, 19, 20, 21, 22, 25), 247(2, 25),
    256(17), 257(18, 22), 263(8, 11, 20, 21),
    267(39), 268(40), 273(19, 42, 43, 44),
    275(1, 2, 7, 8, 9, 10, 11, 12, 15, 16, 17, 18),
    276(19, 20, 21, 22, 25, 39, 40, 42, 43, 44);
    23:146(28), 185(28); **30**:1, 3(2, 3, 4, 5, 6, 7,
    8, 9), 9(13), 13(2, 3, 5, 6, 7, 8, 9), 25(11),
    28(10), 32(5), 34(12), 38(13), 39(13),
    42(12), 46(2, 3, 5, 6, 7, 8, 9), 53(14),
    61(14), 65(15), 72(15), 74(2, 3), 77(10),
    78(16), 84(16), 87(2), 88(3, 4, 5, 6, 7, 8, 9,
    10, 11, 12, 13, 14, 15, 16), 89(31)

Arshad, J., **16**:150(132), 156

Arts, T., **23**:280(26), 320(26, 167, 168), 355(26),
    361(167, 168)

Arturson, G., **22**:229(5), 239(107), 242(107),
    347(5), 351(107)

Artym, R. I., **1**:239, 265

Arunachalam, S. A., **26**:71(64), 95(64)

Arunachalam, Vr., **15**:91(82), 138(82)

Arvia, A. J., **12**:233(4), 251(4), 270(4), 277

Arvizu, D. E., **20**:227(149), 232(153, 159),
    233(153, 159), 234(159), 235(159),
    238(159), 242(153), 306(149, 153),
    307(159)

Asada, K., **7**:132(42), 161

Asahi, H., **28**:340(5), 415(5)

Asai, S., **28**:267(87), 316(163), 333(87),
    336(163)

Asako, Y., **20**:186(48), 187(48), 232(154),
    301(48), 306(154)

Asbedian, V. V., **5**:177(82), 248

Asch, V., **14**:130, 131, 144

Asfia, F. J., **29**:21(17), 23(17, 18, 19), 25(17),
    29(19), 30(19), 56(17, 18, 19)

Ash, R. L., **SUP**:79(225, 226), 156(226),
    157(226); **15**:321; **23**:301(101, 106),
    358(101, 106)

Ashar, S. T., **21**:172, 173, 180

Asher, U., **26**:304(82, 83), 328(82, 83)

Ashida, S., **17**:39(133), 62

Ashiwake, N., **20**:232(152), 242(152),
    306(152)

Ashmantas, L. A., **25**:4(7, 8), 114(7, 8, 180, 181,
    182, 184), 116(185), 129(7, 8), 140(7, 8),
    148(180, 181, 182), 149(184, 185)

Ashmantas, L.-V. A., **31**:160(51, 54, 58, 59),
    182(58), 298(58), 324(51, 54, 58, 59)

Ashton, G. D., **19**:334(146), 337(146), 338(146),
    355(146)

Ashton, R. A., **17**:100, 155

Ashworth, D. A., **23**:321(191, 195, 196),
    346(191), 362(191), 363(195, 196)

Asimacopoulos, P. J., **22**:321(150), 353(150)

Aso, S., **23**:293(70, 71), 357(70, 71)

Asomaning, S., **30**:201(2), 203(2), 208(2),
    249(2)

Assanis, D. N., **23**:343(291), 368(291)

Assens, G. E., **14**:6(24, 25), 11(24, 25), 100

Astarita, G., **2**:365(3), 394; **12**:103, 113;
    15:63(21), 64(21), 77(21), 92(86, 88),
    104(122), 111(122), 114(122), 121(122),
    135(21), 138(86, 88), 139(122), 144(9),
    193(9), 223(9); **18**:162(14), 163(14, 17),
    236(14, 17); **24**:107(71), 187(71); **25**:258,
    262, 286, 315, 318

Astfalk, G., **31**:401(55), 402(55), 427(55)

Astill, K. N., **5**:239, 251; **21**:158, 178

Astruc, J. M., **5**:412, 415, 416, 422, 513;
    11:105(68), 109, 111, 192; **16**:236

Atabek, H. B., **SUP**:71(160), 298(160)

Atanackovic, T. M., **24**:42(14), 98(14)

Atassi, H., **7**:193, 194(80), 214

Atherton, L. J., **30**:338(73), 339(73), 350(91),
    430(73, 91)

Atherton, R. W., **19**:202(56), 245(56)

Atkins, K. R., **17**:75, 102, 154

Atkins, W., **28**:413(230, 231), 425(230, 231)

Atkinson, **SUP**:69(138), 85(138), 437(138)

Atkinson, B., **SUP**:96(256), 97(257), 98(257), 168(257)

Atkinson, P. G., **14**:16(43), 80, 81(171), 82(171), 101, 104

Atkinson, W. H., **23**:280(15), 298(84, 85, 86, 87, 88), 306(84, 85, 86, 87, 120), 354(15), 357(84, 85, 86), 358(87, 88), 359(120)

Atreya, A., **17**:182, 188(36), 190, 300, 314

Atthey, D. R., **19**:25(48), 28(48), 85(48); **28**:77(5), 138(5)

Attridge, J. L., **6**:94(145), 129

Attwood, G. J., **29**:194(208), 213(208)

Atwell, N. P., **13**:80(35), 116

Aubert, J. H., **23**:193(8), 267(8)

Audunson, T., **11**:203(58), 206(58), 209(58), 216(58), 248(58), 254, 278(10), 279(10), 314

Aulds, D. D., **26**:220(12), 222(25), 236(12), 237(12), 242(12, 25), 244(12), 253(12, 25), 258(25), 267(12), 269(25), 291(12), 296(12), 300(12), 315(25), 321(12), 322(12), 325(12, 25)

Aulisio, C., **19**:160(194), 189(194)

Aung, W., **11**:290(24), 315; **18**:73(145), 86(145); **20**:185(23), 186(20, 22), 187(20, 21, 22, 23, 61, 62), 188(23), 190(22), 191(20, 22), 192(20, 22), 196(61, 62), 197(22, 61, 62), 199(61, 62), 200(61), 202(21), 211(62), 300(20, 21, 22, 23), 302(61, 62); **24**:2(2), 36(2); **31**:401(58), 428(58)

Auracher, H., **19**:81(252), 94(252); **23**:12(49), 14(49), 25(49), 128(49); **26**:123(61), 125(61), 127(61), 129(61), 130(61), 131(61), 133(61), 155(61), 211(61)

Auslander, D. M., **3**:35(24), 36(24), 98

Auson, D., **1**:394(69), 442

Austin, A. L., **14**:3(2), 100

Austin, J. M., **15**:70(50), 136(50)

Austin, L. G., **2**:138(22), 268

Auxer, W. L., **6**:41(89), 42(89), 69(89), 75(89), 82(89), 94(89), 95(89), 126

Avduevskii, V. S., **10**:53, 58(34), 64, 70, 71, 82

Avdyevsky, V. S., **25**:35(70a), 143(70a)

Avedisian, C. T., **26**:2(16), 48(16), 49(17), 93(16, 17); **28**:24(46), 72(46); **30**:173(148), 195(148)

Averin, E. K., **11**:105(61), 110, 191

Avery, W., **31**:436(52), 448(52), 460(52), 473(52)

Avizov, A. M., **6**:418(135), 492

Avkhimovich, B. M., **6**:418(112, 127), 491, 492

Avtokratowa, N. D., **13**:211(19), 264

Awberg, J. H., **11**:218, 221, 236, 256

Awberry, J. H., **3**:8, 11, 31

Awschalom, D. D., **17**:90, 99, 100, 154

Axford, R. A., **SUP**:354(495), 355(500, 501), 356(500, 501), 361(505), 363(500, 501, 505), 364(500, 501)

Aydelott, J. C., **4**:182(66), 183(66), 186(66), 195(66), 196(66), 225, 227, 228; **5**:351, 507

Ayers, J. D., **28**:22(38), 72(38)

Ayers, P., **10**:172, 175, 177(27), 215

Ayral, Y. S., **30**:96(28), 190(28)

Ayrton, W. E., **11**:202, 204, 209, 210, 252

Ayyaswamy, P. S. (Chapter Author), **22**:26(2), 59(3), 71(2), 137(2), 138(2), 139(2), 151(1), 152(1, 2, 3); **26**:1, 2(19, 20, 21, 193), 3(20, 193), 5(193), 6(192), 9(192), 16(192), 17(192), 19(20), 31(193), 32(52, 53, 216), 34(52, 192), 35(53), 36(18, 52, 91, 94, 95, 214, 215, 216, 217), 37(216), 39(216, 217), 40(94, 216), 41(94), 42(49, 50, 51, 215), 43(92, 215), 44(92), 45(21, 96), 59(52, 72, 73, 192), 60(72, 73, 105), 62(105), 63(73), 65(105), 71(19), 72(20, 93), 74(93), 80(92), 83(192), 93(18, 19, 20, 21), 95(49, 50, 51, 52, 53), 96(72, 73), 97(91, 92, 93, 94, 95, 96, 105), 101(192, 193), 102(214, 215, 216, 217)

Azar, M. Y., **19**:333(131), 334(131), 355(131)

Azarskov, V. M., **31**:160(35), 323(35)

Azevedo, L. F. A., **18**:15(100), 52(100), 53(100), 73(147, 149), 74(149), 84(100), 86(147, 149); **20**:186(57), 187(55, 57), 191(55), 301(55, 57); **26**:200(160), 216(160)

Aziz, K., **6**:474(257), 483(257), 499; **14**:21(55), 43, 48, 71(55), 72(55), 73(55), 101, 102; **20**:83, 84, 85, 88, 91, 93, 97, 99, 100, 101, 107, 108, 109, 110, 131; **23**:188(86), 192(86), 197(86), 270(86); **25**:253, 256, 316

Aziz, M., **26**:198(157), 199(157), 200(157), 216(157)

Aziz, M. J., **28**:2(1), 5(1), 94(1), 102(1)

Azizi, S., **31**:2(1), 5(1), 94(1), 102(1)

Azuma, T., **26**:142(95, 96, 97, 98, 99, 100), 144(95, 100), 157(95, 100), 158(95, 100), 212(95, 96), 213(97, 98, 99, 100)

B

Baba, T., **23**:284(36), 355(36)
Babcock, W. R., **28**:11(16), 71(16)
Babin, B. R., **20**:277(292), 313(292)
Babjack, S. J., **9**:417
Baboi, N. F., **14**:121, 130, 131, 134, 144
Babrov, H. J., **5**:301(57), 323; **9**:73(53), 76(53), 108
Babu, S. P., **19**:108(57), 186(57)
Babuska, I., **15**:321, 321(8); **24**:227, 268
Bach, G. R., **4**:241(7), 242(7), 313
Bach, J., **6**:334, 336(91), 366
Bach, P., **28**:39(64), 73(64)
Bache, T. C., **15**:325
Bachmann, R. C., **23**:285(44), 356(44)
Bachmat, Y., **13**:137, 202; **24**:104(64), 114(111, 116, 117), 187(64), 188(111, 116, 117)
Back, L. H., **2**:53, 54, 55, 56, 57, 60(59), 61(59), 62(59), 64(59), 65(59), 66(59), 68, 69(59), 70, 72, 73, 92, 94, 108; **3**:81(81), 84(81), 89(81), 93(82), 94, 95, 99; **4**:352, 446; **7**:339, 346, 354(14), 361(14), 377; **13**:5(21), 11(21), 59
Backmark, U., **28**:15(22), 71(22)
Backstorm, N., **28**:15(22), 71(22)
Bačlič, B. S. (Chapter Author), **20**:133, 173(20), 179(20); **24**:39, 48(18, 19), 99(18, 19); **26**:231(34), 234(34), 236(34), 246(34, 52), 251(34), 253(34, 52), 261(34), 269(34), 270(34, 52), 271(34), 274(52), 279(52), 281(34), 282(34), 283(34), 287(34), 291(34), 314(34, 52), 316(34, 52), 320(34), 322(34), 323(34), 326(34, 52)
Baddour, R. F., **14**:193, 245
Bader, H. J., **SUP**:80(231), 121(231), 149(231), 150(231)
Badger, W. L., **9**:265(139), 272
Badgwell, T. A., **28**:344(61), 345(61), 399(189), 400(189), 401(189), 402(61, 189, 209, 210, 211), 403(211), 404(211), 405(189), 418(61), 423(189), 424(209, 210, 211)
Badillo, E., **23**:343(291), 368(291)
Badrinarayan, M. A., **6**:29(48), 124
Bae, S., **9**:259(108), 260, 271
Baehr, H. D., **SUP**:57(67), 58(67), 59(67)
Baer, E., **9**:236(68), 237(68), 240(70), 243, 270
Baerg, A., **10**:201, 202, 203, 206, 209, 217; **14**:213, 216, 217, 218(146), 222(146), 226(146), 246

Baerg, C. A., **19**:122(114), 123(114), 129(114), 133(114), 134(114), 187(114)
Baeri, P., **28**:76(6), 138(6)
Bahl, S. K., **10**:255(56), 282
Bahrami, P. A., **20**:186(56), 187(56), 198(56), 301(56); **24**:3(17), 5(17), 36(17)
Bai, X., **22**:187(6), 347(6)
Baibikov, B. S., **25**:18(53), 45(71), 48(53), 75(122), 77(126), 81(152), 83(126), 142(53), 143(71), 145(122), 146(126), 147(152)
Baida, M. M., **6**:465, 495
Baijal, S. K., **23**:193(9), 267(9)
Bailes, P. J., **26**:14(22, 23), 93(22, 23)
Bailey, B. M., **5**:381, 509
Bailey, C. A., **16**:233
Bailey, C. D., **5**:347, 506
Bailey, K., **1**:10, 49
Bailey, R. V., **4**:191(69), 225
Bailey, T. E., **5**:390, 510
Bailie, R. C., **23**:215(265), 236(265), 238(265), 239(265), 240(265), 278(265)
Bailie, R. E., **31**:462(32), 472(32)
Baille, A., **11**:208, 220, 224, 240, 241, 243, 264
Baily, B., **5**:351(56), 507
Bainbridge, B. L., **23**:320(156), 361(156)
Baines, M., **29**:146(39, 44), 148(44), 171(138, 139), 172(44), 177(157), 178(157), 181(157), 204(39, 44), 209(138, 139), 210(157)
Baines, R. P., **23**:60(108), 102(108), 131(108)
Baines, W. D., **11**:232(170), 237, 260
Bainton, K. F., **7**:249, 317
Baird, M. H. I., **30**:242(3), 249(3)
Baird, R. M., **23**:217(223), 276(223)
Baish, J. W., **22**:26(2), 59(3), 71(2, 4), 137(2), 138(2), 139(2), 149(4), 151(1), 152(1, 2, 3, 4)
Bajorek, S. M., **18**:69(131), 70(131), 86(131)
Bajura, R. A., **16**:10(12), 11(12), 14(12), 16(12), 19(2), 27(12), 29(12), 32(12), 57; **23**:307(127), 359(127)
Bak, T. A., **12**:242(5), 277
Baker, B. S., **15**:321
Baker, C. B., **3**:281, 301
Baker, E., **20**:244(119, 120), 245(119), 305(119, 120)
Baker, H. D., **23**:288(51), 306(51), 356(51)
Baker, L., **11**:139(150), 195
Baker, L., Jr., **19**:80(225, 229), 93(225, 229);

29:1(1), 10(12), 55(1, 12), 188(189), 189(191, 192), 212(189, 191, 192), 271(137), 285(153), 288(153), 342(137), 343(153)

Baker, L. E., **22**:246(77), 350(77)

Baker, M., **10**:75(69), 83

Baker, M. J., **4**:50, 64

Baker, O., **1**:357, 373, 379(2), 382(2), 440; **5**:451, 452, 453, 515

Baker, P. E., **30**:154(113), 194(113)

Baker, R. P., **14**:177, 244

Bakewell, H. P., **8**:300, 301(51), 344(51), 345, 350

Bakhru, N., **11**:105(53), 109, 115, 191; **17**:7, 57; **20**:244(187), 308(187)

Bakhtiozin, R. A., **9**:119, 129(10), 177

Bakke, P., **13**:5(23), 11(23), 59

Bakker, C. A. P., **7**:116, 118, 160

Bakker, P. J., **25**:161(41, 42), 244(41, 42)

Baklastov, A. M., **9**:265(135), 272

Balakameswar, K., **SUP**:244(430)

Balakotaiah, V., **24**:102(13), 184(13)

Balakrishna, M., **23**:262(11), 267(11)

Balakrishnan, A., **10**:121(58), 165; **12**:138, 149, 150, 151, 152, 153, 154, 155, 178, 179, 180, 181, 182, 183, 184, 193; **19**:81(258), 94(258); **31**:389(43), 427(43)

Balakrishnan, X. X., **27**:183(35)

Balashov, V. V., **25**:114(184), 121(186, 187, 188, 189, 190, 191, 192), 129(188), 149(184, 186, 187, 188, 189, 190, 191, 192)

Balasubramaniam, R., **30**:364(129), 366(129), 432(129)

Balasubramaniam, T. A., **22**:10(2), 17(2), 381(18), 384(18), 394(18), 435(18)

Balcar, M., **23**:237(166), 238(166), 239(166), 240(166), 241(166), 242(166), 243(166), 246(166), 274(166)

Balcerzok, M. J., **8**:49, 89

Balcomb, J. D., **18**:10(7), 46(89), 67(125), 68(125), 79(7), 83(89), 85(125)

Bald, W. B., **17**:140, 141, 157; **22**:186(7), 221(7), 347(7)

Baldi, G., **30**:168(131), 195(131)

Baldridge, S., **10**:65(52), 67(52), 82

Baldwin, B. S., **3**:210(44), 250; **24**:234, 271

Baldwin, G. D., **11**:318(18, 19), 435

Baldwin, L. V., **7**:193, 194(79, 83), 200(79), 214; **11**:226, 259

Baldwin, R. B., **12**:206(6), 277

Balekjian, G., **5**:347, 353, 506; **30**:216(24), 250(24)

Bales, E., **23**:292(58), 356(58)

Bales, E. L., **29**:145(37), 171(37), 204(37)

Ball, E. F., **6**:418(103), 491

Ballance, J. O., **5**:40, 53

Ballou, C. O., **25**:161(39), 244(39)

Balunov, B. F., **21**:25(44), 52(44)

Balvanz, J. L., **18**:18(35), 81(35)

Balwanz, W. W., **5**:326(11), 505

Balzhiser, R. E., **5**:403, 511; **9**:218, 269; **11**:56(16), 84(16), 87(16), 90(16), 91(16), 105(16), 113(16), 148(16), 189

Bamford, C. H., **10**:243, 281; **17**:190, 315

Ban, V. S., **28**:351(67, 68), 366(114), 369(114), 418(67, 68), 420(114)

Banas, C. M., **19**:37(79), 86(79)

Banchero, J. T., **5**:67, 125, 432, 436, 512; **11**:105(62), 109(62), 110, 191; **20**:23(44), 81(44); **26**:10(66), 33(65), 95(65, 66)

Band, W., **14**:306, 310, 311(72), 342, 343

Bandrowski, J., **30**:279(58), 280(58), 309(58)

Bandyopadhyay, D., **28**:263(78), 332(78)

Banerjee, P. K., **19**:44(100), 87(100)

Banerjee, S., **17**:43(145), 63

Banez, B. T., **29**:271(141, 142), 342(141, 142)

Bang, K. H., **29**:180(159, 160), 183(175), 210(159, 160), 211(175)

Banijamali, S. H., **15**:96(95), 97(95), 138(95)

Bankoff, S. G., **1**:204, 206, 207(13), 208(18), 219, 220(41), 231, 232, 264, 265, 378, 441; **5**:82, 108, 126, 127; **6**:435(169, 170, 173), 437(169, 170, 173), 494; **8**:33, 88; **9**:262(123), 271; **10**:111, 112, 116, 121, 134, 136, 138, 139, 141, 162, 165, 166; **11**:9, 48, 95(51), 104, 105(118, 124, 133), 111, 114(118), 121(118), 191, 194, 411(183), 439; **15**:242(54), 254, 278, 280; **16**:64(7), 87, 153, 154; **17**:23, 59; **18**:245(5), 246(5), 320(5); **19**:2(2), 4(2), 5(2), 12(2), 14(2), 15(2), 16(2), 40(2), 83(2); **29**:140(32), 149(55, 56, 57), 159(93), 161(98), 162(93, 98), 168(126), 170(134), 172(126, 140, 142), 175(150), 185(177), 204(32), 205(55, 56, 57), 207(93, 98), 208(126), 209(134, 140, 142), 210(150), 211(177)

Banks, C. L., **21**:334, 344

Banks, W. H. H., **5**:164, 248

Bankston, C. A., **SUP**:18(32), 136(32), 146(330), 147(330), 148(330), 149(330), 150(330)

Bankvall, C. G., **24**:104(26), 185(26)

Bannister, J., **10**:203, 217

Bansal, R. K., **19**:160(195), 161(201), 189(195), 190(201)

Bansal, T. D., **11**:203, 206, 255

Barabash, P. A., **31**:278(180), 330(180)

Barakat, H. Z., **5**:341, 348, 394, 395, 396, 397, 398, 400, 506, 507, 511; **6**:474(249, 251), 483(249, 251), 484, 498

Barakat, N., **30**:296(98), 311(98)

Baram, J., **28**:15(20), 71(20)

Baranenko, V. I., **23**:9(23), 127(23)

Baranova, G. K., **28**:103(88), 142(88)

Baranovskii, V. O., **17**:319(9), 320(9), 327(9), 339

Baranovsky, I. V., **31**:160(15), 323(15)

Barazotti, A., **3**:79, 80, 99

Barbee, D. G., **SUP**:95(254)

Barber, R., **11**:425(233), 440

Barbone, R., **29**:163(104), 176(104), 207(104)

Barboza, M., **23**:193(12), 206(12), 226(12), 267(12); **25**:292, 308, 315

Barcatta, F. A., **7**:225(12), 314

Barcilon, A., **21**:154, 155, 178

Bar-Cohen, A., **20**:182(118), 183(17), 184(17, 118), 186(76), 187(76), 191(76), 194(76), 195(76), 202(76), 213(118), 217(123), 221 (118), 244(190, 199), 249(190), 250(190), 299(17), 302(76), 304(118), 305(123), 308 (190), 309(199); **23**:22(43), 46(67), 47(67), 48(67), 64(84), 77(84), 100(84), 121(67, 127), 122(127), 128(43), 129(67), 130(84), 132(127); **26**:170(123), 214(123); **30**:277 (53), 283(53), 309(53), 413(178), 435(178)

Bard, Y., **21**:328, 344

Barden, P., **9**:3(22), 66(22), 93(22), 95(22), 96(22), 107

Bardina, J., **25**:340, 341, 349, 361, 363, 381, 404, 408, 409

Bareiss, M., **19**:58(129, 130), 59(129, 130), 62(130), 63(130), 71(161), 72(161), 89(129, 130), 90(161); **24**:3(8, 11), 4(11), 6(11), 36(8, 11)

Barenghi, C. F., **17**:75(50), 87, 90(50), 91(50), 92, 93, 94, 102, 126, 127, 154, 155, 156

Bargeron, K. G., **SUP**:73(186), 92(186), 93(186), 96(186), 99(186), 300(186)

Barhorst, J. F., **28**:80(64), 141(64)

Bark, F. H., **12**:77(3), 111; **24**:302(1), 314(2), 318(1, 2)

Barkdoll, R. O., **5**:99, 127, 406(165), 512; **11**:105(90), 113(90), 193

Barker, G. E., **5**:67, 125, 432, 436, 512; **11**:105(62), 109(62), 110(62), 191

Barker, J. A., **14**:339(140), 346

Barker, J. J., **10**:171, 174, 215; **15**:285, 287, 321

Barker, V., **7**:309, 319

Barlett, R. N., **22**:222(108), 351(108)

Barlow, K. J., **28**:342(22), 416(22)

Barlow, R., **16**:329, 363

Barnacle, H. E., **9**:135, 156, 179

Barnard, A. J., **14**:321(93, 94), 323, 324(93, 94), 344

Barnard, D. A., **17**:37(123), 50, 62

Barnea, D. (Chapter Author), **20**:83, 85, 88, 93, 95, 99, 105, 108, 110, 111, 112, 121, 130, 132

Barnea, E., **21**:334, 344; **23**:207(13, 14, 15), 235(13, 14, 15), 267(13, 14, 15)

Barnes, D. K., **10**:118(50, 51, 52), 121(50, 51, 52), 165

Barnes, H., **24**:108(80), 187(80)

Barnes, H. A., **23**:190(16), 267(16); **25**:255, 315

Barnes, J. W., **6**:17(27), 18(27), 25(27), 27(27), 123

Barnett, D. O., **5**:389, 510

Barnett, P. G., **17**:30, 60

Baron, F., **25**:396, 409

Baron, J. R., **2**:53, 107; **8**:56, 90

Baronowsky, S., **29**:216(3), 258(3), 260(3), 262(3), 264(3), 334(3)

Barr, G., **20**:105, 130

Barrer, R. M., **15**:321

Barrère, J., **23**:409(70), 463(70)

Barrett, H. H., **30**:289(77), 310(77)

Barron, R. F., **5**:469, 470, 473, 516; **20**:209(114), 304(114); **26**:220(7, 12), 236(12), 237(12), 242(12), 244(12, 49), 253(12, 49), 267(12), 269(7), 291(12, 49), 296(12, 49), 300(12), 321(12), 322(12), 323(49), 324(7), 325(12), 326(49)

Barrow, H., **9**:73(55), 108; **SUP**:109(277); **19**:334(141), 336(141), 338(150), 355(141), 356(150); **26**:48(125), 98(125)

Barry, B., **23**:315(148), 360(148)

Barry, J. M., **14**:5(18), 100

Barry, R. E., **9**:218, 269; **11**:56(16), 84(16),

87(16), 90(16), 91(16), 105(16), 113(16), 148(16), 189
Barsanti, G., **26**:148(105), 151(105), 213(105)
Barsch, W. O. (Chapter Author), **7**:219
Barschdorff, D., **14**:337(130), 346
Barston, E. M., **14**:123, 144
Barstow, D., **16**:46, 57
Bart, G. C. J., **28**:306(148), 335(148)
Bartas, J. G., **2**:434(35), 451
Bartel, W. J., **14**:185(113, 114), 192, 193(121), 196, 197(126), 245; **19**:148(172, 173), 160(173), 189(172, 173)
Bartell, W. J., **10**:210, 218
Barthe, M. F., **28**:110(34), 140(34)
Bartholomew, R. N., **10**:200, 218
Bartko, F., **8**:281, 283
Bartlett, E. P., **2**:2(2), 105
Bartlit, J. R., **5**:356, 508
Bartman, A. B., **8**:40(211), 89
Bartnew, G. M., **13**:219(59), 265
Bartoli, B., **23**:292(54), 356(54)
Bartz, D. F., **31**:209(111), 327(111)
Bartz, D. R. (Chapter Author), **2**:1, 6(9), 7(9, 13, 14), 8(9), 9(9), 14, 17(9), 22(9), 23, 24(9), 27(9, 14), 29(9), 30(9), 32(9), 33(13), 34(9, 13), 35(13), 37(26, 27), 38(13, 26, 27), 40(9, 13), 62(14), 63(9), 95(79), 105, 106
Bartz, J. A., **7**:205, 217
Barua, A. K., **3**:256(8), 260(8), 300
Barulin, Y. D., **6**:551, 555(74), 563; **21**:21(35), 51(35)
Barzelay, M. E., **4**:238, 313
Basarow, I. P., **16**:68(18), 153
Basaskov, A. P., **10**:189, 190, 191, 198, 207, 208, 209, 216, 217, 218
Basel, M. D., **30**:154(116, 119), 155(116, 121), 156(116), 157(116), 184(160), 194(116, 119, 121), 196(160)
Baseman, R. J., **28**:103(59), 141(59)
Bashforth, F., **1**:225, 264
Basinlis, A., **20**:286(310), 287(310), 313(310)
Baskakov, A. P., **14**:151, 152(31, 32), 158(60), 160(20, 21, 80), 161(20, 21, 80), 166, 167, 169(80), 176(161), 209(142), 210(142), 221, 225, 242, 243, 244, 246, 247; **19**:100(21, 22, 23), 102(23), 103(21, 22), 116(22), 117(21), 118(21), 126(22), 127(22), 128(22), 164(210), 165(22), 169(210), 170(210), 171(22), 185(21, 22, 23), 190(210)

Baskov, V. L., **21**:34(53), 35(53), 52(53)
Bass, C. D., **11**:358(89), 359(89), 436
Bassanini, P., **7**:171, 172, 176, 178, 212
Bassett, C. E., **15**:101(114), 103(114), 139(114)
Bastaky, J., **22**:220(185), 355(185)
Bastin, J. A., **10**:2(7), 34, 35, 37, 44, 45, 46(20), 48, 53(31), 57(20), 62, 68(18, 31), 81
Basu, B., **19**:80(246), 94(246)
Basu, D. K., **14**:136, 144
Basu, P., **19**:164(209), 168(209), 170(209), 171(209), 190(209)
Basu, S., **29**:195(209), 213(209)
Basuilis, A., **7**:223(11), 224, 236, 250, 275(11), 314, 315
Bataill, J., **6**:424(157), 430(157), 493
Batanov, V. A., **28**:127(7), 138(7)
Batch, J. M., **17**:39(128), 62
Batchelor, G. K., **3**:35, 97; **4**:152, 223; **5**:177, 249; **6**:5, 59, 60, 122, 127; **18**:190(78, 80), 191(78, 80, 144), 192(79), 230(144), 237(78, 79, 80, 87), 239(144); **21**:195(21), 237(21), 267(38), 276(38); **23**:369(2), 407(2), 460(2); **24**:114(112), 188(112); **30**:4(17), 74(17), 88(17)
Batchelor, G. W., **8**:166, 167, 168, 170, 173, 176, 177, 179, 182, 184, 198, 226
Bateman, H., **1**:65, 121; **SUP**:2(4), 197(4), 223(4), 234(4), 247(4); **19**:258(11), 350(11); **24**:40(6), 98(6)
Bates, D. R., **4**:251, 252(15), 313
Bates, O. K., **18**:169(32), 170(32), 172(32), 218(32), 219(32), 236(32)
Bates, W. J., **6**:197, 364
Bathelt, A. G., **19**:45(106), 46(106, 107, 108), 47(106), 49(115), 54(108), 55(106, 108), 56(106, 108), 57(106, 108, 125), 88(106, 107, 108, 115), 89(125)
Bathkev, D. A., **20**:263(262), 311(262)
Bathla, P. S., **6**:474(330), 502
Batov, L. P., **6**:474(323), 502
Batsale, J.-C., **31**:2(13), 5(13), 89(13), 102(13)
Batt, R. G., **6**:36(66), 37(66), 51(66), 52(66), 125
Batten, J. J., **10**:270(102), 283
Battista, E., **19**:334(138), 336(138), 355(138)
Battya, P., **26**:50(24, 25), 52(197), 93(24), 94(25), 102(197)
Bau, H. H., **29**:38(28), 40(28), 56(28); **30**:116(68), 117(68), 122(68), 124(68, 85), 126(68, 85), 127(68, 85), 129(85), 130(85), 131(85), 192(68, 85)

Bauer, H. J., **23**:298(92), 358(92)

Bauer, R. C., **6**:25, 131, 132

Bauer, S., **25**:252, 292, 293, 316

Bauer, S. H., **14**:337(131), 346

Bauerle, D., **28**:102(37), 140(37)

Baughn, J. W., **23**:282(32), 324(208), 327(217, 218, 219), 330(208, 236, 237, 238), 336(254, 255, 256, 257, 258), 337(258), 355(32), 363(208), 364(217, 218, 219), 365(236, 237, 238), 366(254, 255, 256, 257, 258); **26**:132(78), 212(78)

Baule, B., **2**:339, 352

Baum, B. A., **28**:78(91), 85(91), 142(91)

Baum, E., **6**:20(31, 32, 33), 22, 23(33), 24(32), 36, 38, 39(32), 40(32), 101(161, 162, 163, 164), 102, 104, 105(164), 106(164), 107(163), 123, 130

Bauman, F. S., **18**:63(115, 116), 85(115, 116)

Baumeister, K. J., **5**:90, 126; **11**:58(23), 84(23, 42, 45, 46), 85(42), 86, 95(23, 45, 46), 96, 105(23, 98), 108, 109(23), 113, 118, 190, 191, 193

Baumgartl, J., **30**:334(58), 349(58), 351(96), 429(58), 431(96)

Bautista, P., **19**:317(107), 318(107), 354(107)

Baver, E., **5**:451, 453, 454, 515

Baveye, P., **23**:378(52), 463(52)

Baxley, A. L., **18**:162(16), 171(16), 179(16), 188(16), 236(16)

Baxter, C. A., **5**:159, 248

Baxter, C. R., **22**:229(1), 239(61), 242(61), 247(8), 347(1, 8), 349(61)

Baxter, D. C., **4**:334, 445; **19**:25(50), 85(50)

Bayazitoglu, Y., **19**:70(160), 71(160), 79(211), 90(160), 92(211); **20**:227(144), 229(144), 231(144), 306(144); **21**:246(16), 275(16); **24**:3(3, 10), 5(10), 36(3, 10); **26**:83(26), 94(26); **28**:18(29), 71(29)

Bayley, F. J., **4**:56, 64; **9**:3(9, 20), 29, 33, 34, 35, 36, 42, 45(8), 71, 77, 78, 84, 85, 86, 87, 96, 97(5, 20), 102, 106, 107; **15**:189(65), 224(65); **17**:50(156), 63

Bayly, J. G., **1**:398(81), 401, 429(81), 442

Bays, J. D., **5**:326(2), 505

Bazarov, I. P., **5**:262, 322

Bazett, H. C., **4**:88(25), 91(25), 138; **22**:20(5, 6, 7), 21(5), 23(5), 60(5, 6, 7), 61(6), 69(6), 70(6), 82(5), 152(5, 6, 7)

Bazley, N. W., **8**:248, 282

Bazzin, A. P., **8**:13, 85

Beach, H. L., **15**:321

Beacock, R. J., **23**:315(149), 360(149)

Beall, R. T., **5**:347, 506

Beam, J. E., **26**:162(116), 163(116), 170(116), 213(116); **30**:121(83), 192(83)

Beam, R. M., **24**:226, 275

Beaman, J. J., **22**:287(9), 288(9), 347(9)

Bear, J., **13**:178, 184, 203; **14**:3, 100; **20**:315(2), 349(2); **24**:104(53, 55, 60, 63, 64, 65, 66), 114(111, 116, 117), 186(53, 55, 60, 63), 187(64, 65, 66), 188(111, 116, 117); **30**:94(17), 95(17), 98(17), 189(17)

Beard, K. V., **21**:336, 337, 338, 339, 344; **26**:53(27), 94(27)

Bearden, J. A., **3**:255(5), 300

Beasley, D. E., **23**:338(266), 339(266), 340(268), 366(266), 367(268)

Beasley, W. H., **14**:269(58), 273(58), 277(58), 280

Beastall, D., **6**:29(42), 124

Beaton, W., **16**:70, 153

Beattie, J. R., **11**:365(116), 375(116), 425(229), 426(116, 229), 437, 440

Beattle, J. A., **28**:15(23), 71(23)

Beatty, K. O., Jr., **1**:232(54), 265; **9**:261(115), 271; **14**:136, 146, 151, 153, 208, 217, 242

Beaubouef, R. T., **16**:232; **20**:14(22), 15(22, 22), 16(22), 17(22), 80(22)

Beavers, G. S., **14**:5, 100; **SUP**:198(388), 199(388), 209(388), 210(388), 211(388); **19**:268(59), 272(59), 273(59), 274(59), 352(59); **23**:431(83), 464(83); **24**:112(100), 113(102), 188(100, 102)

Beck, D. F., **29**:164(107), 186(181), 190(198), 207(107), 211(181), 212(198), 268(134), 342(134)

Beck, J. L., **14**:14, 22, 24, 100; **24**:115(120), 189(120)

Beck, J. V., **6**:416(81, 83, 84), 490; **11**:408, 439; **15**:321, 327; **23**:301(105), 315(146), 320(156), 358(105), 360(146), 361(156)

Beck, W., **21**:171, 182

Becker, C., **14**:305, 327(36), 342

Becker, E., **SUP**:325(478); **17**:245, 316

Becker, E. B., **31**:336(22), 339(22), 340(22), 426(22)

Becker, E. W., **17**:127, 156

Becker, G. H., **7**:202, 216

Becker, K. M., **1**:422(120), 443; **5**:230(158), 233(162), 234, 238, 239, 240, 250, 251;

17:25(76), 29, 59, 60; **21**:160, 163, 164, 166, 168, 178; **29**:167(125), 208(125)

Becker, M., **7**:204, 217

Becker, R., **10**:92, 163; **11**:7, 48; **14**:287, 298, 304, 341

Becker, R. A., **28**:342(32), 416(32)

Beckermann, C., **28**:241(12), 242(12), 249(29), 252(12, 49, 51), 253(12), 255(12), 257(49), 258(49), 259(49), 260(49), 261(49), 262(12, 71), 263(49, 71, 81), 264(12, 49, 51, 82, 83, 84, 85, 86), 265(83), 266(12, 84), 267(49), 268(49, 81, 96, 97, 98), 269(12, 49, 82), 275(12), 276(12, 104), 285(49), 303(49, 81), 326(12), 327(49, 51, 84), 329(12), 330(29), 331(49, 51), 332(71, 81, 82, 83, 84), 333(85, 86, 96, 97, 98, 104); **30**:95(25), 108(44, 45, 51), 122(44, 45), 127(44, 51), 128(44), 129(44), 130(44), 131(44), 132(44, 45), 133(45), 139(51), 143(105), 144(105, 106), 145(106), 146(106), 147(106), 148(106), 149(106), 150(106), 151(105), 189(25), 190(44, 45), 191(51), 193(105, 106)

Beckers, H. L., **11**:203, 205, 209(38), 225, 253, 254; **SUP**:186(368)

Beckett, C. W., **2**:37(28), 38(28), 106

Beckett, P. M., **19**:80(243), 94(243)

Beckman, E. L., **4**:90(27), 91, 92, 93, 94(27), 95(33), 96, 138

Beckman, P., **9**:363(28), 415

Beckman, W. A., **5**:462, 515

Beckmann, J., **22**:220(10), 347(10)

Beckmann, R. B., **26**:19(106), 69(106), 97(106)

Beckmann, W., **4**:5, 15, 62; **11**:209, 263, 281, 282, 283, 314; **18**:167(52), 237(52); **30**:72(77), 91(77), 272(45), 273(45), 309(45)

Beckwith, I. E., **4**:318, 325, 330, 331, 334, 341, 342, 343, 344, 443, 444; **7**:202(124), 216; **8**:12, 85

Beder, E. C., **11**:358(89), 359, 436

Bedfair, S. M., **28**:343(43), 417(43)

Beecher, N., **1**:252(91), 256, 257, 258, 262, 266

Beek, O., **2**:353

Beer, H., **7**:320e(167), 320g; **10**:156(79), 165, 166; **19**:50(118), 51(118), 53(118), 57(118, 126), 58(118, 129, 130), 59(129, 130), 62(130), 63(130), 68(159), 69(159), 70(159), 71(161, 163), 72(161, 163), 88(118), 89(126, 129, 130), 90(159, 161,

163); **24**:3(8, 11), 4(11), 6(11), 36(8, 11)

Beggs, G. C., **26**:49(205), 102(205)

Beggs, H. D., **20**:84, 116, 120, 131

Behee, R. D., **26**:170(127), 214(127)

Behn, V. C., **2**:365(18), 394

Behning, F. P., **23**:345(296), 368(296)

Behrens, R. A., **25**:292, 296, 315

Behringer, R. P., **17**:125, 128, 129, 130, 131(122), 132, 133(137, 140, 141), 134, 135, 149, 156

Beilin, M. I., **10**:177, 216

Beitin, K. I., **20**:187(61, 62), 196(61, 62), 197(61, 62), 199(61, 62), 200(61, 62), 211(62), 302(61, 62)

Bejan, A. (Chapter Author), **15**:1, 11(34), 12(34), 14(34), 15(34), 16(34), 17(40), 19(40, 42), 20(40), 26(42), 28(42), 29(42), 31(42), 32(42), 33(42), 34(58), 35(58), 36(58), 37(58), 38(58), 40(61, 62), 42(62), 44(69), 47(72), 49(62), 56(34, 40, 42, 58), 57(61, 62, 69, 72), 58(91, 92, 93); **18**:15(94), 16(92), 19(41, 42), 48(41, 92), 50(97, 99), 51(97, 99), 68(129, 133), 70(129, 133), 81(41, 42), 84(92, 94, 97, 99), 85(129), 86(133, 134); **20**:315, 315(9), 316(9, 11, 13), 317(11), 329(46), 330(46), 333(11, 51), 334(51), 335(51), 337(57), 339(57), 340(57, 58), 341(58), 342(62), 343(62, 64), 347(11), 348(57, 58), 350(9, 11, 13), 351(46, 51), 352(57, 58, 62, 64); **21**:240(4, 5, 6), 275(4, 5, 6); **23**:369(4), 460(4); **24**:1, 8(32, 40), 10(40, 41, 42), 11(40), 14(40), 15(40), 16(40), 17(47), 18(48), 20(48), 22(48), 23(54), 24(54), 25(54), 26(54, 55), 27(55), 28(55), 29(55), 30(55, 66), 31(67), 32(67, 68), 33(68), 34(68), 37(32, 40, 41, 42, 47), 38(48, 54, 55, 66, 67), 104(70), 105(35), 106(42), 115(123), 179(70), 185(35), 186(42), 187(70), 189(123), 280(3), 299(50), 300(50), 302(50), 307(50), 318(3), 320(50); **28**:11(19), 71(19); **30**:111(60), 122(60), 191(60), 363(127), 432(127)

Belding, H. S., **4**:93, 138

Beletsky, G. S., **8**:94, 149, 157(17), 159; **18**:88(17), 147(17), 156(17), 158(17)

Beliakov, V. P., **17**:327(33), 334, 336, 340

Belichenko, N. P., **9**:103, 106

Belin, R. E., **1**:394(69), 442

Bell, D. W., **1**:410, 442

Bell, G. A., **9**:364(33), 386(58), 387(33), 394(58), 399(33), 401(58), 402(58), 415, 416

Bell, G. E., **19**:36(72), 86(72)

Bell, K. J., **5**:90, 126; **16**:148, 149, 156; **17**:4, 8, 57; **25**:252, 293, 297, 308, 315

Bell, N., **9**:85, 86, 96, 97(5), 106

Bell, R. P., **15**:321

Bell, S., **7**:190, 214

Bellemans, A., **13**:130(40), 164(40), 202; **15**:229(14), 230(14), 276

Belles, F. E., **29**:83(44), 125(44)

Bellet, D., **18**:171(119), 223(119), 225(119), 238(119)

Bellhouse, B. J., **8**:293(17), 349; **23**:337(259), 366(259)

Bellinghausen, R., **21**:5(12), 22(12), 23(12), 51(12)

Bellman, R., **3**:184, 249; **5**:121, 128, 434(203), 514

Bellmore, C. P., **21**:22(39), 23(39), 51(39)

Belostotskii, B. R., **6**:418(123), 492

Belotserkovski, O. M., **4**:334, 445

Belousov, V. P., **25**:64(111), 65(111, 112), 145(111, 112)

Belov, L. A., **23**:9(23), 127(23)

Belskaya, E. P., **18**:169(46), 171(46), 176(46), 228(46), 229(46), 237(46)

Beltaos, S., **26**:183(143), 215(143)

Belton, M. J. S., **8**:245, 282; **14**:253(14), 262, 264(14), 266, 267, 269(14), 273(14), 275, 279

Belyaev, N. M., **25**:4(13, 16, 17), 31(13), 33(13), 64(13), 96(13), 127(13), 140(13, 16, 17)

Belyaev, V. M., **21**:13(27), 14(27), 15(27), 29(27), 51(27)

Belyakov, I. I., **21**:30(46), 32(46), 52(46)

Belyakov, V. P., **17**:327(34), 336(34), 340

Belyayev, N. M., **19**:195(27), 243(27)

Ben-Amoz, M., **15**:321

Benard, C., **19**:62(144, 151), 63(144), 64(144), 65(144), 66(144, 151), 67(144), 89(144), 90(151)

Be'nard, H., **24**:174(152), 190(152)

Benchekchou, B., **23**:253(266, 267), 254(266, 267), 255(266, 267), 256(266, 267), 260(267), 261(266, 267), 278(266, 267)

Bendel, P., **30**:96(29), 190(29)

Bender, E., **SUP**:77(217), 97(258), 142(217), 146(217), 399(258)

Bendersky, D., **23**:314(137), 360(137)

Bendescu, I., **14**:231(167), 246; **19**:147(164), 189(164)

Bendet, J. S., **11**:34(74), 49

Bendiksen, K. H., **20**:100, 101, 102, 103, 104, 105, 106, 107, 130

Bendt, P., **18**:26(57), 82(57)

Benedick, W. B., **29**:96(62), 102(77), 126(62), 127(77), 165(117), 208(117)

Benedict, R. P., **23**:288(49), 289(49), 356(49)

Benedict, W. S., **2**:37(28), 38(28), 106

Benedikt, E. T., **4**:147(4), 222

Benek, J. A., **24**:218, 269, 274

Benenati, R. F., **23**:217(18), 267(18)

Bengis, M., **29**:216(3), 258(3), 260(3), 262(3), 264(3), 334(3)

Ben Haim, Y., **13**:67(6), 115

Benis, A. M., **23**:212(17), 267(17)

Benjamin, A. S., **21**:327, 330, 344

Benjamin, J. E., **16**:90, 91, 154

Benjamin, T. B., **20**:101, 102, 130; **21**:156, 178

Benke, R., **11**:218, 221, 236, 256

Benneman, K. H., **17**:103, 155

Bennethum, W. H., **23**:280(7), 354(7)

Bennett, A. W., **1**:245, 266, 421(115), 443; **17**:42, 62

Bennett, D. L., **16**:144, 156

Bennett, F. D., **6**:205, 220, 222(66), 364, 365; **28**:128(8), 138(8)

Bennett, G. S., **6**:211(61), 365

Bennett, S., **SUP**:73(196), 92(196), 167(196)

Bennett, T., **28**:24(43), 25(43), 26(43), 28(56), 29(43, 56), 31(56), 33(43), 35(43, 56), 36(43, 56), 37(43), 38(43), 56(56), 59(56), 72(43, 56)

Bennett, T. D. (Chapter Author), **28**:75, 103(9), 104(9), 105(9), 106(9), 107(9), 113(9), 114(9), 116(9), 117(9), 119(10), 120(10), 121(10), 122(10), 123(10), 138(9, 10)

Bennion, D., **12**:233(78), 237(78), 269(78), 280

Bennon, W. D., **28**:252(42, 43, 44, 45), 253(43, 45), 254(43, 45), 255(42, 43, 44, 45), 262(42, 43), 264(43), 266(42), 270(42, 43), 271(42), 272(101, 102), 273(101), 274(102), 275(42, 101, 102), 276(43, 45, 102), 282(43), 285(43), 286(128, 129, 131, 132), 287(43, 129, 131, 132), 288(131), 307(43), 311(43, 45), 319(43, 45), 326(43), 331(42, 43, 44, 45), 333(101, 102), 334(128), 335(129, 131, 132)

Benoit, J. R., **6**:196(30), 363
Benser, W. A., **2**:95, 108; **5**:365, 508
Bensmaili, A., **26**:123(69), 124(69), 148(69), 151(69), 154(69), 155(69), 172(130, 135), 177(130), 178(130), 181(135), 211(69), 214(130, 135)
Benson, S. V., **25**:356, 412
Benson, S. W., **22**:329(11), 347(11)
Bentley, R., **SUP**:75(201), 106(201), 305(201), 306(201), 309(201), 310(201)
Bentwich, M., **6**:474(207), 496; **26**:220(16), 255(16), 269(16), 325(16)
Benzing, H., **22**:18(28)
Benzinger, T. H., **4**:120(93), 140
Beran, M. J., **18**:190(82), 191(82), 198(90), 199(82), 237(82), 238(90)
Berenson, P. J., 67(18)(19), **5**:83, 84(19), 85(19), 88, 89, 103, 104, 105(62), 106(62), 107(62), 108(62), 111, 125, 127, 419, 420, 421, 424, 426, 428, 434(110), 436, 438, 510
Berenson, P. J., **1**:239, 265; **4**:98(41, 42), 100(41, 43), 139, 171, 225; **7**:320b(141), 320e; **9**:261(113), 271; **11**:9(18), 48, 56, 84(17), 87, 90, 91, 95(17), 96, 105(17, 48, 106), 112, 113, 189, 191, 193; **15**:244(70), 279; **16**:139(109), 141(109), 156; **17**:6, 57; **18**:245(7), 249(13), 255(7), 256(7), 257(7), 260(7), 270(7), 274(7), 290(7), 292(7), 293(7), 295(7), 298(7), 308(7), 313(7), 317(7), 320(7, 13)
Berenson, P. P., **16**:215, 238
Bereson, P. J., **20**:15(27), 80(27)
Beretta, G. P., **24**:8(39), 14(39), 37(39)
Berezovsky, A. A., **20**:217(122), 305(122)
Berg, B. V., **10**:207(123), 209, 217, 218; **14**:151(21), 160(21, 80), 161(21, 80), 166(21), 167, 169(80), 209, 210(142), 221(21), 242, 244, 246; **19**:100(22), 103(22), 116(22), 126(22), 127(22), 128(22), 165(22), 171(22), 185(22)
Berg, C. A., **15**:2(7), 54(7)
Berg, E. V., **29**:152(73), 206(73)
Berg, H. M., **20**:259(248, 250), 262(248), 311(248, 250)
Berg, J. C., **26**:14(39), 17(38), 18(38), 30(38, 39), 43(28), 94(28, 38, 39)
Bergelin, O. P., **8**:140(70, 71, 72, 73), 141(72), 142(72), 144(71), 148(72), 149(72), 151(72), 160; **18**:139(71, 72, 73), 140(72), 146(72), 147(72, 73), 159(71, 72, 73);

25:292, 293, 296, 302, 303, 304, 305, 306, 307, 309, 311, 315
Berger, D., **13**:126, 153(31), 161, 162, 169, 191, 201
Berger, J. L., **13**:261(124), 267
Berger, M., **24**:227, 269
Berger, S. A., **6**:6, 38, 98(12), 122, 125
Bergeron, K. D., **29**:188(188), 212(188), 217(13), 218(19), 219(19), 233(19), 234(73, 76), 270(19), 277(19), 279(19), 282(76), 288(76), 305(19), 313(76), 319(19), 322(19), 335(13, 19), 338(73), 339(76)
Berggaum, J. B., **9**:379(53), 386(53), 389(53), 394(53), 397(53), 416
Berghmans, J., **14**:133, 134, 144(11)
Bergles, A. E. (Chapter Author), **5**:441, 443, 445, 517; **9**:211(37), 269; **10**:149, 150, 152, 157, 165; **SUP**:76(211), 267(211, 446), 269(446), 381(446, 518), 382(446, 518), 393(521), 412(528), 413(528, 529); **15**:21(43, 44, 45, 47), 56(43, 44, 45, 47), 69(47), 136(47); **17**:2(3, 8), 34, 40(8), 46, 47, 56, 61, 63, 319, 339; **18**:257(39), 261(39), 295(39), 298(39), 321(39); **20**:203(92, 99), 204(92, 99), 205(92), 210(99), 212(99), 214(99), 216(99), 217(99), 244(184, 187, 189, 201, 204), 249(204), 250(216, 224, 226), 252(201), 253(201), 254(201), 255(201), 290(313), 303(92, 99), 308(184, 187, 189), 309(201, 204, 216), 310(224, 226), 314(313); **21**:125(117), 138(117); **23**:8(18), 9(27), 12(18, 35), 14(18, 35), 17(35), 26(52), 29(18), 30(18, 35), 32(18), 33(18, 35), 37(18), 39(18), 44(35), 46(18, 35), 47(18), 49(18, 35), 50(35), 53(18), 57(18), 58(18), 60(18), 78(18), 86(18), 90(18), 127(18, 27, 35), 128(52); **26**:123(58, 60), 125(58, 60), 127(58, 60, 72), 129(58, 60), 130(58, 60), 131(58, 72), 133(60), 149(58), 151(58), 155(58), 211(58, 60, 72); **30**:197, 198(4, 5, 6, 7, 8, 73), 207(55), 208(55), 216(55), 218(55), 221(55), 223(55), 224(55), 249(4, 5, 6, 7, 8), 252(55), 253(73); **31**:160(63), 278(63), 324(63), 456(68), 473(68)
Bergles, L. A., **20**:245(207), 246(207), 309(207)
Bergman, T. L., **18**:1(2), 79(2)
Bergmann, D., **18**:191(134), 198(134), 238(134)
Bergolin, D. P., **1**:385(52), 388(52), 441

Bergougnou, M. A., **23**:188(10), 253(10), 267(10)

Bergquist, T., **22**:347(12)

Bergsteinsson, P., **18**:328(23), 330(23), 362(23)

Beris, A. N., **23**:219(115, 217), 272(115), 276(217); **25**:292, 296, 305, 316, 319

Berkman, S., **28**:366(114), 369(114), 420(114)

Berkovich, S. Y., **6**:418(102), 491

Berkowitz, L., **1**:245, 266, 381(39), 411(39), 413(39), 441

Berl, W. G., **10**:221, 281

Berlad, A. L., **10**:261, 266(83), 269(75), 282, 283

Berlin, I. I. (Chapter Author), **11**:51, 115(187), 120(187), 122(187), 133(146), 144(146, 154, 155), 146(146, 154, 155), 151(155), 152(155), 156(155), 159(155), 170(154, 155), 172(155), 173(155), 176(155), 177(155), 179(155), 180(155), 182(155), 183(155), 185(155), 195, 196, 197; **18**:241, 258(53), 262(53), 311(68), 312(68), 316(70, 71), 322(53), 323(68, 70, 71); **25**:4(5), 33(5), 43(5), 52(5), 61(5), 92(5), 99(5), 106(5), 130(5), 139(5)

Berman, A. S., **14**:34(85), 72, 73(85), 102

Berman, M. (Chapter Author), **29**:59, 61(3, 4, 5, 6), 63(3, 4), 66(6), 67(3), 68(6), 69(3), 71(32), 72(3, 4, 5, 6), 84(6), 85(6), 102(77), 122(87), 123(3, 4, 5, 6), 125(32), 127(77, 87), 164(107), 165(114), 186(181), 189(193), 190(198), 194(205), 207(107), 208(114), 211(181), 212(193, 198), 213(205), 268(134), 304(166), 307(166), 342(134), 344(166)

Berman, N. S., **SUP**:92(249, 250, 251), 93(249), 96(249, 250, 251), 97(249), 166(249); **15**:92(90), 138(90)

Berman, R., **18**:161(12), 236(12)

Bernard, C., **22**:19(8), 59(8), 152(8); **28**:351(80), 418(80)

Bernardin, J. D., **30**:243(9), 249(9); **31**:459(2), 470(2)

Bernett, E. C., **10**:50, 51, 59, 68, 81

Bernhardt, E. C., **2**:358, 394

Bernhardt, E. C., ed., **18**:216(111), 238(111)

Bernis, A., **23**:338(263), 366(263)

Berns, M. W., **22**:248(13), 347(13)

Bernstein, F. V., **23**:342(283), 367(283)

Bernstein, J. T., **9**:405(97), 417

Berry, A. J., **2**:325, 355

Berry, C. J., **6**:98(159), 130

Berry, H. C., **SUP**:64(96), 357(96), 364(96)

Berry, R. A., **20**:232(158), 306(158); **23**:333(247), 365(247); **26**:170(127), 214(127)

Berry, V. J., **3**:120(23), 173; **11**:219, 222, 236, 257; **15**:104(119), 108(119), 121(119), 139(119)

Berry, V. J., Jr., **SUP**:185(366)

Bershader, D., **6**:244(75), 365

Berstein, B., **15**:62(10), 78, 135(10)

Bert, J. L., **15**:321

Berthoud, G. J., **29**:137(24), 151(66), 157(66), 170(133), 185(133), 203(24), 205(66), 209(133)

Bertin, H., **23**:371(33), 435(88), 462(33), 464(88)

Bertodano, M. L., **29**:231(58), 277(58, 149), 278(149), 338(58), 343(149)

Bertoletti, S., **1**:245(89), 266, 361(13, 14), 363(13), 378, 381(13, 39, 40, 41), 411(13, 39, 40, 41), 413(13, 39, 40, 41), 419(111), 421(13, 40, 41), 436(13), 440, 441, 443; **17**:28, 59

Bertoni, R., **17**:25(77), 59

Bertram, A., **2**:353

Bertram, M. H., **6**:79(134), 128

Bertrand, P. A., **28**:342(31), 416(31)

Bes, T., **SUP**:115(286), 117(286), 134(286), 174(286), 444(286)

Besant, R. W., **5**:178, 249

Besserman, D. L., **26**:123(63, 64), 124(63, 64), 211(63, 64)

Best, G., **19**:317(107), 318(107), 354(107)

Bestchastnov, S. P., **21**:39(64), 40(64), 52(64)

Betchov, R., **11**:208, 262

Bethe, H. A., **5**:286(22), 303(22), 322

Betts, K. R., **17**:119(112), 156

Betz, E., **22**:18(28)

Betzel, T., **19**:57(126), 71(163), 72(163), 89(126), 90(163)

Bevans, J. T., **12**:129(14), 138, 157, 169, 171, 174, 188, 192, 193

Bevans, R. S., **4**:108(53), 139

Beveridge, G. S. G., **23**:217(101), 271(101)

Bewilogua, L., **5**:412, 413, 513; **9**:3(10, 11), 82(10, 11), 83, 106; **11**:105(66), 111, 120(66), 121(66), 122(66), 191; **16**:114(78), 155, 233; **17**:6, 57, 140, 142, 146, 157

Beyer, E., **28**:124(1, 2), 138(1, 2)
Beyer, W. H., **31**:339(24), 426(24)
Bezrukova, E. E., **11**:360(93d), 437
Bhagat, P. M., **17**:202, 203, 315
Bharathan, D., **20**:90, 132
Bhat, G. N., **10**:204, 217
Bhat, M. S., **28**:267(95), 333(95)
Bhatnagar, P. L., **7**:171, 212
Bhatnagar, R. K., **25**:259, 269, 319
Bhatt, M. Siddhartha, **17**:265, 269(93), 316
Bhattacharya, D., **15**:323
Bhattacharya, S. C., **19**:174(226), 190(226)
Bhattacharyya, T. K., **SUP**:119(293), 136(319), 317(293, 473), 318(293)
Bhatti, M. S., **SUP**:160(340), 168(340), 192(340); **19**:248(2, 3), 268(37), 270(37), 277(2), 291(2), 292(2), 295(2), 297(2), 299(2), 306(2), 312(2), 322(3), 349(2, 3), 351(37)
Bhunia, S. K., **26**:137(89), 195(155), 196(155), 212(89), 215(155)
Bia, P., **14**:42, 47, 73, 76(96), 102
Bialokoz, J., **6**:538, 563; **7**:66(46), 85
Biancaniello, F. S., **28**:19(30, 31, 32), 71(30, 31, 32), 317(177), 337(177)
Biasi, L., **15**:243(58), 278; **17**:28, 59
Biberman, L. M., **5**:309(78), 310(82), 324
Bibier, C. A., **20**:232(155), 233(155), 242(155), 306(155)
Bibolaru, V., **10**:205, 217
Bicego, V., **24**:8(36), 14(36), 37(36)
Bickford, W. B., **15**:321, 321(18)
Bienert, W., **7**:213(95), 318
Bienkowski, G. K., **7**:174, 212
Bier, K., **16**:115, 117(80), 155, 171, 172, 173, 193(5), 215(5), 220, 221(5), 234, 238; **20**:23(43, 47), 81(43, 47)
Bierman, G. F., **12**:23(37), 74
Biggs, R. C., **9**:7, 106
Bijward, G., **31**:142(46), 158(46)
Bil, V. S., **13**:211(19), 264
Bilenas, J. A., **7**:320a, 320f
Billig, E., **11**:412(189), 439
Billman, G. W., **11**:219, 222, 236, 256
Bilson, E., **10**:27, 37, 57(47), 82
Binder, J. L., **29**:220(45, 52), 221(45, 46, 47, 48, 49, 50, 51, 52, 53, 54), 255(52), 270(52), 278(52), 282(52), 290(52), 310(52), 316(45), 337(45, 46, 47, 48, 49, 50, 51, 52, 53, 54)

Binder, K., **14**:339(137, 138), 346
Bingham, H., **22**:243(14), 347(14)
Binnie, A. M., **14**:333, 334(119, 120), 345
Biondo, P. P., **6**:57(101), 58(101), 62(101), 64(101), 65(101), 107(101), 127
Biot, M. A., **1**:79(36, 37, 38, 39, 40, 41), 82, 121; **8**:23(108, 109), 24, 25(125), 26, 31(111), 32(109, 111), 86, 87; **19**:197(31), 244(31); **24**:39(1, 2), 46(1, 2), 98(1, 2)
Bird, G. A., **6**:2(3), 122; **7**:198, 216; **28**:129(11), 138(11), 406(220), 407(220), 425(220); **31**:411(75), 429(75)
Bird, M. J., **29**:165(110), 207(110)
Bird, R. B., **1**:7(11), 49; **2**:38(35), 47(35), 106, 114(8), 268, 358, 368, 371, 374, 378, 379, 388, 394, 395, 397; **3**:192(17), 249, 254, 261, 262(3), 263, 267(3), 268(3, 22), 269(3), 282(3), 290(3), 291(3), 299(3), 300; **7**:10, 83; **11**:269(6a), 314; **12**:197(8, 9), 198(8, 9), 199(8), 206(9), 207(9), 208(9), 213(9), 219(9), 232(9), 244(9), 268(33), 269(8), 277, 278; **13**:7, 60, 126(33), 128(33), 131, 169(33), 201, 208(8), 209(8), 214(8), 219(52, 53), 236(105), 253, 264, 265, 266; **14**:194(124), 206(139), 245, 246; **SUP**:54(62), 109(275); **15**:13(37), 56(37), 62(5, 7, 9, 12), 76(59), 78(12), 79(68), 98(98), 99(7), 100(98), 101(7), 102(7), 103(7), 109(107), 110(107), 135(12), 137(59, 68), 138(98), 139(100, 107), 144(7), 149(7), 179(7), 223(7); **16**:147(126), 156; **17**:127, 156; **18**:216(110), 238(110); **19**:251(8, 10), 252(10), 295(10), 349(8), 350(10); **21**:240(3), 275(3); **23**:188(20), 189(19, 20, 21), 190(19), 195(21), 197(21), 198(226), 199(226), 200(226), 204(226), 205(226), 206(226), 218(226), 220(226), 221(226), 223(226), 226(226), 262(221b), 267(19, 20, 21), 276(221b, 226); **24**:107(72), 108(72, 76), 109(72), 187(72, 76); **25**:253, 254, 255, 315; **28**:152(3), 227(3), 267(88), 333(88), 353(93), 355(93), 356(93, 97), 358(93), 364(97), 419(93, 97); **29**:251(108), 293(108), 296(108), 340(108); **30**:364(130), 432(130); **31**:9(2), 102(2)
Birikh, R. V., **11**:297(28), 315; **30**:366(133), 432(133)
Biringen, S., **25**:382, 409

Birkebak, R. C. (Chapter Author), 1:34(31), 50; 2:403, 450; 3:81(83), 91(83), 99; 5:201, 249; 7:343(30, 31), 346, 354(46), 355(46), 360(30, 46), 363, 378; 10:1, 2(4, 5, 6), 12, 13, 16(22), 17, 19(24), 20, 21, 23(6, 29, 38, 39, 40), 24, 28, 29(40), 30(24, 38), 31(4, 6), 32(40), 34(4, 6, 47), 35, 36, 37, 52, 53(32, 37), 54(32, 37, 39), 69(32, 37), 70(37, 39), 75(39, 70, 71), 76(70), 77(39), 78(39, 70, 71), 80(39, 70), 81, 82, 83; 22:158(15), 347(15)

Birkhoff, G., 1:218, 264; 23:396(59), 463(59); 31:33(3), 37(3), 102(3)

Birmingham, B. W., 5:389(114), 510; 9:350(4), 352(4), 394(4), 414; 15:50(83), 57(83)

Birnbreier, H., 20:187(64), 195(64), 196(64), 302(64)

Birngruber, R., 22:318(16), 342(16), 347(16)

Birt, D. C. P., 9:244(73), 261(73), 270

Bischoff, K. B., 12:232(32), 278; 23:370(10), 427(10), 461(10)

Bishop, A. A., 5:102, 127; 7:50; 11:170(174), 173, 175, 176(174), 177(174), 185(174), 186(174), 196, 197; 25:81(158), 147(158)

Bishop, D. C., 28:343(45, 46), 417(45, 46)

Bishop, E. H., 11:285(19, 20), 286(20), 287(20), 288(19, 20, 2l), 315

Biskeborn, R. G., 20:183(5), 299(5)

Biton, M., 30:204(23), 250(23)

Bitsyutko, J. Ya., 6:474(213, 214), 478, 496

Biyikli, S., 19:164(219), 169(219), 171(219), 190(219)

Bizzell, G. D., 25:262, 315

Bjerketvedt, D., 29:98(69), 126(69)

Bjorklund, I. S., 5:233(171), 234, 235, 251; 11:203, 206, 219(229), 223(229), 262; 21:163, 178

Bjorklund, J. S., 5:166, 174, 248

Black, I. A., 9:401(83), 417

Black, S. A., 2:375(78), 396

Black, T. J., 16:363

Black, W. Z., 18:169(39), 170(39), 176(39), 230(39), 236(39)

Blacker, P. T., 1:394(70), 398(70), 442

Blackman, L. C. F., 21:102(79), 136(79)

Blackshear, P. L., Jr., 5:167, 168, 248; 10:225, 229, 243(39), 260, 281, 282; 17:175, 180, 182, 184(22), 186, 314

Blackwell, B. F., 16:280, 298, 349, 352, 355, 357, 363

Blackwell, J. H., 10:48, 81

Blair, C. K., 15:228(6), 276

Blair, M., 14:130, 144

Blais, N. C., 3:274, 301

Blake, G. K., 22:328(17), 329(17), 330(17), 333(17), 334(17), 336(17), 337(17), 338(17), 339(17), 340(17), 341(17), 342(17), 343(17), 347(17)

Blakely, A. D., 10:234(13), 281

Blalock, A., 4:118(77), 140

Blanchat, T. K., 29:220(41), 221(28, 32, 34, 35, 36, 37, 38, 39, 40, 41, 57), 222(57), 230(41), 255(40), 265(40, 41), 269(40), 270(40, 41), 271(28), 276(41), 278(40), 280(28), 282(28, 41), 290(40, 41), 300(40), 302(36, 37, 57), 306(41), 309(38), 310(32, 34, 35, 36, 37, 40), 311(32, 34, 35, 36, 37), 315(40), 316(40), 318(40), 319(40), 322(57), 323(40), 336(28, 32, 34, 35, 36, 37, 38), 337(39, 40, 41), 338(57)

Bland, M. E., 16:233

Blander, M., 10:105, 106, 164; 14:305(34), 309(69), 342, 343

Blankenship, V. D., 4:48, 49(83), 64

Blankenstein, E., 2:353

Blanks, R. F., 23:201(202), 202(202), 204(202), 205(202), 206(202), 216(202), 220(202), 275(202)

Blatt, T., 4:205(80), 206(80), 226

Blatz, W. J., 4:114(67), 140

Blau, W. J., 28:102(71), 141(71)

Blecher, S., 1:78, 121

Blenke, H., 14:176(94, 99), 178, 202(99), 203, 237(94, 132), 244, 245; 19:146(162), 148(175, 176), 188(162), 189(175, 176)

Bleviss, Z. O., 1:317, 318(56), 319, 322, 328, 354

Bligh, J., 22:360(1), 435(1)

Blinkov, G. N., 30:293(90), 296(99), 298(99), 311(90, 99)

Bliss, D. F., 30:320(22), 323(26), 325(26), 327(28), 335(28), 336(28), 345(26), 350(26), 351(28), 406(28), 412(26, 28), 413(26), 415(182), 416(182), 417(26, 28), 418(28), 420(28), 427(22, 26, 28), 435(182)

Bliss, F. E., Jr., 7:320b, 320e

Block, H., 15:63(22), 64(22), 135(22)

Block, J. A., 10:268, 283; 15:228(7), 270, 271(106), 276, 281; 17:192, 236, 315

Block, M. J., 30:43(18), 88(18)

Block, R., **20**:244(199), 309(199)

Blodgett, K. B., **2**:353

Bloem, J., **28**:365(113), 402(193), 420(113), 423(193)

Blom, J., **8**:299(43), 305, 324(43), 349

Blomquist, C. A., **29**:151(71), 206(71), 220(44), 221(44), 271(141, 142), 316(44), 323(44), 337(44), 342(141, 142)

Bloom, H., **12**:229(9a), 277

Bloom, M. H., **2**:266, 270

Blosch, E., **24**:222, 225, 269

Blottner, F. G., **2**:194(53), 230, 231, 233(53), 234(53), 235, 269, 270; **21**:279, 280, 320, 322, 325, 326, 327, 344; **24**:221, 269

Blottner, G. F., **2**:266, 270

Blue, R. E., **6**:244(72), 365

Blum, B., **22**:18(28)

Blum, W., **30**:361(122), 432(122)

Bluman, D. E., **9**:118(8, 9), 119, 177; **30**:236(1), 249(1)

Blumenkrantz, A. R., **15**:21(47), 56(47)

Blythe, P. A., **14**:332, 345

Board, S. J., **29**:132(15), 137(26), 140(26, 33, 34), 146(39, 44, 45), 148(44), 159(89), 172(44), 175(15), 185(26), 186(45), 203(15, 26), 204(33, 34, 39, 44, 45), 206(89)

Boardman, L. E., **3**:284, 301

Boarts, R. M., **9**:265(139), 272

Bobco, R. P., **4**:25, 63

Bobeck, G. E., **28**:8(9), 70(9)

Boberg, T., **30**:93(1), 188(1)

Bobianu, C., **28**:351(73), 418(73)

Bobrovich, G. I., **16**:127, 128, 129, 141(114), 155, 156, 235; **17**:7, 57; **20**:14(20), 15(20), 16(20), 80(20)

Boccio, J., **29**:84(49), 93(49), 94(49), 95(49), 100(49), 125(49)

Bochirol, I. J., **11**:105(107), 113, 193

Bochirol, L., **14**:120, 121(14, 15), 127, 128, 130(14, 15), 144

Bockris, J. O'M., **12**:207(10), 226(10), 229(10), 237(10), 250(10), 252(10), 255(10), 256(10), 277

Bodia, J. R., **11**:290(25), 315

Bodnarescu, M. V., **SUP**:115(285), 174(285)

Bodoia, J. R., **SUP**:73(172), 160(172), 161(172), 162(341), 163(172, 341), 165(172), 167(341), 168(341), 190(172), 191(172), 299(172), 399(172); **18**:73(142), 86(142);

19:268(52), 271(52), 273(52), 275(52), 351(52); **20**:186(24), 187(24), 300(24)

Bodunov, M. I., **31**:160(48, 61), 324(48, 61)

Bodvarsson, G. S., **20**:348(68), 352(68); **30**:140(97, 99), 141(97), 193(97, 99)

Boe, A., **20**:121, 122, 130

Boehler, J. P., **6**:38(72), 125

Boehm, R. F., **9**:364(31), 385(31), 386(31), 390(64, 65), 393(64, 65), 395(64), 415, 416; **11**:130, 131, 134(145), 135(145), 195; **15**:228(6), 267(100), 269, 272(113), 276, 280, 281; **18**:39(80), 40(80), 83(80); **26**:2(109), 97(109); **30**:168(129), 194(129)

Boehman, A. L., **23**:329(225), 330(225), 364(225)

Boehringer, J. C., **11**:376(132), 378, 388(132), 389(132), 390(132), 438

Boelter, L. M. K., **1**:357(1), 371(1), 440; **11**:202, 204, 248, 253; **SUP**:414(531), **15**:105(132), 140(132), 195(72, 73), 224(72, 73)

Boettinger, W. J., **28**:317(177, 178), 337(177, 178)

Boffé, N., **11**:425(228), 426(228), 440

Bogachev, V. A., **21**:16(31), 21(36, 37), 22(36), 51(31, 36, 37)

Bogdanavichyus, A. B., **25**:114(184), 149(184)

Bogdanoff, S. M., **7**:204(134), 217

Bogdanov, F. F., **31**:242(130), 328(130)

Bogdonoff, S. M., **6**:35(61), 80(135), 95(155), 125, 128, 130

Bogdzevich, A., **17**:116(106), 155

Boger, D. V., **19**:44(103), 45(103), 87(103); **24**:109(89), 187(89)

Boggs, J. H., **5**:242, 251; **18**:169(27), 170(27), 173(27), 218(27), 223(27), 224(27), 236(27)

Boggs, P. T., **19**:2(6), 83(6)

Bogomolov, Yu. N., **31**:282(186), 331(186)

Bogue, D. C., **2**:384, 385, 390, 391, 394, 395; **15**:62(11), 78(11), 135(11)

Boguslavskiy, N. M., **19**:111(75), 186(75)

Bohan, W., **5**:131, 246

Bohdansky, J., **7**:239, 250, 251, 271, 274(46), 275, 275(116), 289, 292, 315, 317, 318, 319

Böhm, U., **7**:98(13), 160; **23**:230(49), 231(49), 236(253), 252(253), 269(49), 278(253); **25**:291, 316

Bohn, M. S., **18**:9(6), 18(36), 53(101, 102), 54(101), 58(110, 111), 59(113), 60(110, 111), 61(113), 62(113, 114), 63(113),

Bohn, M. S. (*continued*)
  64(36, 114), 65(36), 66(121), 79(6), 81(36),
  84(101, 102, 110, 111), 85(113, 114, 121);
  20:205(100, 101), 210(100), 211(100),
  212(100), 303(100), 304(101)
Bohnert, M., **25**:353, 354, 358, 360, 362, 378,
  404, 409
Bohren, G. F., **23**:145(24), 185(24)
Boikov, G. P., **6**:474(328), 502
Boisdert, R. F., **19**:79(220), 93(220)
Boissin, J. C., **17**:323(21), 339
Bokor, J., **28**:117(27), 139(27)
Boldarev, A. M., **31**:160(9), 165(9), 182(9),
  322(9)
Boldmen, D. R., **6**:404, 488
Boldyrev, A. M., **25**:4(11), 140(11)
Boles, M. A., **11**:388(138), 438; **19**:38(86, 87),
  78(187), 87(86, 87), 91(187)
Boley, B. A., **8**:34, 88; **19**:14(27), 15(27), 16(27,
  30), 41(27), 44(27), 50(27), 84(27, 30);
  22:307(18), 308(18), 312(18), 347(18)
Boll, R. H., **5**:67, 125, 432, 436, 512;
  11:105(62), 109(62), 110(62), 191
Bollen, J. M., **28**:402(192), 423(192)
Bolling, G. F., **28**:231(1), 246(1, 23), 250(1),
  316(1, 23), 317(1), 329(1), 330(23)
Bollinger, A., **22**:225(19), 347(19)
Bologa, M. K., **14**:121, 122(36, 37), 127, 130,
  131, 132(55), 134(7, 55), 144, 145, 146;
  25:153(9), 162(9, 44), 163(44), 164(9, 44,
  47), 165(9, 44, 47), 166(49, 50, 51, 53),
  197(47), 201(44), 226(9), 243(9), 244(44,
  47), 245(49, 50, 51, 53); **31**:160(19),
  323(19)
Bolsamo, S. R., **10**:32, 37
Boltzmann, L., **19**:5(15), 84(15)
Bomberg, M., **23**:292(58), 356(58)
Bonacina, C., **22**:203(20), 347(20)
Bonch-Bruevich, A. M., **11**:318(20), 435
Bond, G., **4**:109(55), 110(55), 139
Bond, J. W., Jr., **5**:255(14), 259(14), 264(14),
  303(14), 304(14), 305(14), 310(14),
  319(14), 320(14), 322
Bond, L. A., **23**:330(235), 365(235)
Bondarchuk, T. P., **10**:203(107), 217
Bondi, A., **23**:193(100), 212(100), 229(100),
  271(100)
Boneberg, J., **28**:96(12), 138(12)
Bonhoeffer, K. F., **2**:330, 355
Bonilla, C. F., **1**:239, 265; **2**:38(36), 106;

  3:108(14), 172, 173; **4**:166, 224; **9**:251(79),
  270; **11**:278(8), 314; **16**:114(69, 70), 125,
  155, 233; **20**:14(21), 15(21), 16(21), 17(21),
  80(21)
Bonin, K. D., **30**:258(13), 307(13)
Bonjour, E., **11**:105(107), 113(107), 193; **14**:120,
  121(14, 15), 128, 129, 130(14, 15), 144
Bon Mardion, C., **17**:108(98), 109, 113, 114, 155
Bonnecaze, R. H., **20**:84, 85, 100, 102, 103, 130
Bonnett, W. S., **15**:213(110), 225(110)
Bonnier, J. J., **22**:279(209), 356(209), 426(38),
  436(38)
Bonyoucos, G. T., **31**:1(4), 102(4)
Booda, L. L., **4**:109(56, 57), 110(56), 139
Boom, R. W., **15**:44(70), 46(70), 57(70); **16**:233
Boon, S., **29**:68(17), 124(17)
Booney, N. M., **14**:160(81, 82), 161(81, 82),
  168(81, 82), 169(81, 82), 244
Booth, J. R., **2**:374(21), 395
Boothroyd, R. G., **9**:132, 167, 179
Boots, H. M. J., **28**:96(14), 139(14)
Booy, M. L., **28**:189(4), 228(4)
Booz, O., **21**:154, 179
Borde, I., **26**:19(2), 22(2), 33(2), 92(2)
Border, T. R., **14**:253(8), 279
Borell, G. J., **23**:301(108), 302(108), 343(108),
  358(108)
Borella, H. M., **23**:294(75), 357(75)
Borgers, T. R., **18**:73(138), 86(138)
Borghi, R., **29**:82(43), 125(43)
Bories, S., **14**:12(32), 42, 100, 102
Bories, S. A., **14**:3, 21, 72(9), 73(54), 74(9), 76,
  100, 101; **24**:103(19), 105(19), 185(19);
  30:169(137), 195(137)
Boris, J. P., **24**:234, 269; **29**:91(59), 126(59)
Borishansky, V. M., **1**:233, 238, 265; **5**:88, 89,
  90, 126, 419, 420, 422, 514; **7**:10, 83;
  11:55(2, 5, 6), 56(2, 7, 8, 9, 10, 11, 19, 20),
  57, 58, 59(7, 8, 9, 10, 11), 60(7, 8, 9, 10, 11,
  19, 20, 29), 95(50), 98(50), 101, 102(5),
  105(6, 19, 20, 29, 50, 54, 75, 76, 77),
  106(29), 107(50), 110, 111, 112, 117,
  118(29), 119, 120(29), 121(19, 29),
  122(29), 189, 190, 191, 192; **16**:163,
  181(14), 186, 187, 211, 226, 235, 237, 238;
  17:5, 57; **20**:14(20), 15(20), 16(20), 19(39),
  80(20), 81(39); **25**:63(107), 145(107);
  31:274(175), 330(175)
Borisov, V. V., **31**:165(72), 182(72), 186(72),
  325(72)

Borman, G. L., **26**:69(103), 97(103)

Born, M., **6**:135(1), 169, 179, 195, 219, 220, 292, 362; **11**:320(34), 322(40), 352(40), 401, 435; **23**:147(30), 185(30); **30**:295(92), 311(92)

Bornage, H., **6**:36(62), 125

Börner, H., **11**:261

Bornside, D. E., **30**:329(34), 350(92), 351(34), 358(34, 114), 428(34), 430(92), 432(114)

Borodulya, V. A., **19**:98(8), 100(28), 110(67), 111(28), 115(28), 116(28, 67), 117(28, 67), 118(28), 126(131), 129(28), 130(28), 135(28), 136(28), 137(28, 131, 144), 139(28), 140(131, 144), 144(28), 145(8, 131, 144, 158), 156(8, 187), 158(131), 159(8, 28, 144, 187), 160(187), 161(8, 144), 174(228), 184(8), 185(28), 186(67), 188(131, 144, 158), 189(187), 190(228); **25**:205(104, 105), 225(114), 234(114), 247(104, 105, 114)

Boronkov, V. P., **31**:282(186), 331(186)

Borovikov, V. P., **31**:278(179), 330(179)

Bortoli, R. A., **8**:135, 160; **18**:133(68), 134(68), 159(68)

Bosch, G., **23**:324(209), 363(209)

Bose, C. B., **28**:413(229), 425(229)

Bose, T. K., **19**:32(66), 37(66), 42(66), 86(66)

Bosnjakovič, F., **1**:216, 218, 264; **4**:259(23), 260(24), 262(23), 263(23), 264(24), 313; **11**:13, 48; **15**:8(27), 55(27)

Bosworth, R. C. L., **6**:497; **11**:202, 205, 211, 253

Bosworth, R. L. C., **30**:280(61), 310(61)

Boteler, W. C., **20**:187(37), 300(37)

Botnikov, Y. A., **14**:176(93), 178(93), 244

Botsaris, G. D., **11**:412(185), 439

Bott, M. H. P., **19**:73(169), 90(169)

Bott, T. J., **31**:449(79), 474(79)

Bott, T. R., **31**:447(3, 72), 450(4), 461(45), 470(3, 4), 472(45), 474(72)

Bottaro, A., **30**:371(143, 144), 382(143), 394(144), 433(143, 144)

Bottemanne, F. A., **11**:213, 215, 262

Botterill, J. S. M., **10**:184(70, 72), 185(72, 73, 74, 75), 192, 216; **14**:151(2, 4), 152(18, 33), 153, 154(33), 155(48, 55), 158(61, 63, 64), 160, 166, 169, 180(102), 193, 208, 217, 236, 242, 243, 244, 245, 247; **19**:98(2), 99(15, 17), 100(2, 15, 17, 24, 25, 31, 32), 101(17), 102(32, 35), 105(17, 32), 107(17, 32), 108(17, 32), 110(65, 68), 111(2, 68),

115(24, 68), 116(24, 68), 117(24, 68), 118(68), 129(24), 131(24), 137(25), 145(2), 154(2), 161(2), 164(213), 165(213), 169(213), 170(213), 171(213), 180(17, 32), 184(2), 185(15, 17, 24, 25, 31, 32, 35), 186(65,68), 190(213); **23**:188(22), 229(22), 233(22), 267(22); **25**:214(107), 229(107), 247(107), 252, 315

Bouabdallah, A., **21**:155, 178

Boucher, D., **25**:252, 292, 293, 316

Boucher, D. F., **SUP**:47(55), 54(55)

Boucheron, E. A., **19**:26(52), 27(52), 85(52)

Boudart, M., **14**:314, 344

Boulnois, J. L., **22**:279(21), 347(21)

Boulos, M. I., **23**:340(273), 367(273)

Bourdon, E. B. D., **28**:106(13), 139(13)

Bouré, J. A., **31**:112(1), 155(1)

Bourgeois, P., **19**:186(61)

Bourke, P. J., **7**:40(51), 42, 66, 67, 86

Bourne, D. E., **8**:36, 88; **SUP**:326(483)

Boussinesq, J., **11**:225, 258; **SUP**:69(136), 85(136); **17**:124, 156; **20**:353, 356, 359, 361, 362, 365, 370, 372, 373, 386; **24**:117(124), 189(124)

Boutette, M., **17**:181, 314

Boutron, P., **22**:177(22, 23), 347(22), 348(23)

Bouvier, J. E., **10**:107(35), 164

Bovenkirk, M. P., **9**:376(44), 415

Bowden, F. P., **24**:30(61), 38(61)

Bowden, M., **5**:227(144), 250

Bowell, E. L. C., **10**:2(7), 35

Bowen, B. D., **SUP**:349(493)

Bowen, W., **9**:132, 179

Bowersock, D. C., Jr., **5**:381, 509

Bowles, E. M., **29**:79(42), 80(42), 125(42)

Bowley, W. W., **15**:321; **16**:15, 16, 57

Bowling, K. M., **14**:180, 245

Bowlus, D. A., **8**:19(88), 86

Bowman, B. R., **7**:249, 317

Bowman, H. F., **16**:234; **22**:10(2), 17(2, 3), 26(10), 30(9), 152(9, 10), 158(24), 246(24), 342(24), 348(24), 381(18, 19), 384(18, 19), 394(18, 19), 435(18, 19)

Bowring, R. W., **17**:28, 60

Boyack, B. E., **12**:23(40), 74; **29**:244(96), 312(96), 320(96), 340(96)

Boyce, B. E., **9**:3(23), 107

Boyd, C. A., **3**:284, 301

Boyd, G. D., **30**:319(18), 427(18)

Boyd, L. W., **30**:222(10), 223(10), 250(10)

Boyer, H. E., **28**:7(6), 70(6)
Boyer, P. K., **28**:343(44), 417(44)
Boyer, R. F., **13**:210(14), 264
Boyko, L. D., **9**:259(109), 260(109), 261, 271
Boylan, D. E., **7**:204, 217
Boynton, F. P., **5**:295(43), 307(64), 308(64), 323
Bozowski, S., **17**:69(19), 70(19), 71(19), 72(19), 153
Braaten, M. E., **20**:243(169), 307(169); **24**:193, 196, 200, 201, 202, 210, 247, 249, 262, 263, 269, 273
Brabston, D. C., **26**:6(29, 30), 94(29, 30)
Bracewell, R., **30**:287(72), 310(72)
Brachel, H., **14**:160(75), 161(75), 188(75), 244
Bradburg, L. J. S., **11**:220, 224, 263
Bradbury, L. J. S., **18**:123(59), 159(59)
Braden, M., **15**:287(20), 321
Bradfield, W. S., **3**:55, 98; **5**:99, 127, 406(165), 512; **11**:56(14), 58, 60, 105(14, 78, 90), 111, 112, 113, 121(14), 189, 192, 193; **18**:257(36), 262(36), 304(36), 305(36), 321(36)
Bradley, A. B., **22**:248(220), 277(220), 279(220), 280(220), 356(220)
Bradley, D., **11**:220(137), 224(137), 235(137), 238(137), 258
Bradley, D. R., **18**:333(33, 38), 334(33, 38), 335(33), 362(33, 38); **21**:327, 344; **29**:71(27), 73(34), 74(34), 75(34), 80(34), 81(34), 121(27), 125(27, 34), 216(6), 268(134), 324(176), 325(176), 326(176), 327(176), 334(6), 342(134), 344(176)
Bradley, H. H., Jr., **11**:318(7), 376(7), 434
Bradley, R., **15**:287(81), 324
Bradshaw, A., **18**:330(29), 332(32), 333(32), 335(29), 362(29, 32)
Bradshaw, P., **6**:407, 488; **13**:80(35), 116; **16**:363; **21**:8(21), 51(21), 178
Bradshaw, R. D., **10**:172, 177, 215
Bradshaw, S. E., **28**:365(108), 420(108)
Brady, D. K., **12**:10(19), 12(19), 25(19), 73
Brady, J. F., **23**:370(23), 461(23); **24**:114(106), 188(106); **30**:111(61), 191(61)
Brag, K. N. C., **6**:95(157), 130
Bragina, O. N., **25**:28(57a), 142(57a)
Braginskii, S. I., **4**:268, 314
Brajuskovic, B., **23**:307(124, 125, 126), 359(124, 125, 126)
Brame, J. S. S., **17**:228, 316
Brand, H., **20**:323(18, 19), 328(37, 40), 350(18,

19), 351(37, 40)
Brand, R. S., **9**:302, 347
Brandon, C. A., **9**:125, 127, 167, 178
Brandt, A., **SUP**:73(192, 196), 92(196), 94(192), 161(192), 165(192), 166(192), 167(196); **24**:269
Branemark, P. I., **22**:226(25), 348(25)
Branson, D., **31**:458(21), 471(21)
Braren, B. E., **28**:116(52), 141(52)
Bras, G. H. P., **15**:266, 280
Brasen, D., **28**:343(41, 42), 417(41, 42)
Brash, A., **15**:63(26), 64(26), 135(26)
Brassington, D. J., **21**:26(45), 27(45), 52(45)
Brauer, A., **31**:242(129), 247(129), 328(129)
Brauer, F. E., **10**:109, 164
Brauer, H., **13**:16(46, 47), 60; **20**:377, 386
Brauerle, B., **28**:342(29), 343(29), 416(29)
Braumeister, K. J., **25**:12(46), 141(46)
Braun, W., **1**:287, 353; **14**:107(86a), 147
Braun, W. H., **5**:131, 207, 246; **8**:164, 226; **9**:296, 346
Brauner, N., **20**:110, 111, 112, 130
Braunschweig, S., **12**:24(41), 74
Brazelton, W. T., **10**:185, 200, 203(109), 208, 217, 218; **14**:151, 152(34, 37), 158(14, 34), 192(34), 193(34), 196(34), 201(34), 204(34), 208(141), 234, 242, 243, 246; **19**:128(137), 188(137)
Brdlick, P. M., **6**:312(100), 366; **9**:265(141), 272; **13**:15, 17(44, 45), 60; **26**:123(52), 126(52), 210(52)
Brea, F. M., **23**:193(23), 201(23), 202(23), 212(23), 213(23), 236(23, 253), 238(23), 240(23), 242(23), 246(23), 252(253), 267(23), 278(253)
Breeden, J. A., **18**:169(30), 170(30), 172(30), 209(30), 210(30), 236(30)
Breedijk, T., **28**:344(61), 345(61), 402(61), 418(61)
Breen, B. P., **11**:56(21), 58(21), 105(21), 108, 109, 111, 116, 190; **17**:147, 157
Breen, P. P., **5**:66, 67(13), 68(13), 69(13), 71(13), 125, 406, 428, 432, 436, 438, 439, 495, 512
Breene, R. G., **5**:278(23), 322
Breeze, J. C., **5**:295(39), 322
Breiland, W. C., **28**:402(212), 424(212)
Breinan, E. M., **19**:37(79), 86(79)
Breine, U., **22**:226(25), 348(25)
Breitner, M. C., **5**:177, 249
Bremert, O., **28**:102(62), 116(62), 141(62)

Bremhorst, K., **8**:299(41), 305, 349; **11**:240, 243, 261

Bremner, J. G. M., **2**:353

Brenner, H., **13**:170, 202; **18**:186(136), 191(136, 145), 239(136, 145); **23**:207(97), 208(97), 271(97), 370(23), 409(71), 461(23), 463(71); **25**:292, 317; **30**:94(11), 95(22), 100(22), 189(11, 22)

Brent, A. D., **28**:252(47), 253(47), 262(47), 264(47), 272(47), 331(47)

Brent, R., **30**:205(69), 208(69), 216(69), 221(69), 253(69)

Brent, R. P., **23**:418(77), 464(77)

Brentari, E. G., **5**:403(139), 404, 405, 406, 407(139), 408(139), 428, 440, 445, 511; **11**:105(104), 193; **17**:147, 148, 157

Brentau, E. G. F., **5**:469, 516

Breon, S. R., **17**:106(94), 107(94), 155

Brereton, C. M. H., **23**:340(271), 367(271)

Brereton, P. A., **11**:391(150), 411, 438

Breske, T. C., **31**:449(35), 472(35)

Bressler, R., **8**:107(31), 108, 159; **18**:101(31), 102(31), 158(31)

Bressler, R. G., **7**:308, 309, 319

Breton, J., **29**:102(76), 127(76)

Brett, J. R., **4**:95, 138

Breuer, N. M., **12**:242(11), 277

Breuer, S., **15**:321(21), 322

Brewer, D. F., **17**:105, 155

Brewster, D. B., **5**:241, 251

Brewster, M. Q., **23**:136(7), 154(40, 41), 184(7), 185(40, 41); **31**:334(1), 425(1)

Brickweede, F. G., **5**:406(159), 512

Bridgeman, K. D., **8**:248(23), 283

Bridgmann, P., **18**:216(104), 238(104)

Bridley, J., **21**:154, 155, 178

Briend, P., **23**:206(24), 218(24), 237(24), 239(24), 240(24), 242(24), 244(24), 267(24)

Briens, C. L., **23**:253(31, 32), 254(31, 32), 259(31, 32), 268(31, 32)

Brier, J. C., **11**:219, 222, 235, 237, 238, 257

Briggs, A. J., **29**:146(40), 204(40)

Briggs, C. W., **11**:425(232), 440

Briggs, D. C., **5**:241, 242(185), 243, 251

Briggs, D. G., **4**:38, 39, 64; **6**:474(242), 480(242), 482, 498; **9**:311(71), 347; **20**:186(75), 187(75), 201(75), 302(75)

Briggs, F., **10**:19, 21, 24(30), 36

Briggs, G. A., **12**:23(36), 74

Briggs, L. J., **10**:96, 97, 98(15), 164

Brigham, W. E., **14**:16(43), 92, 94(182), 97(193), 101, 104

Brignell, A. S., **26**:5(31), 20(32), 94(31, 32)

Brill, J. P., **20**:84, 115, 116, 119, 120, 121, 122, 123, 128, 131, 132

Briller, R., **9**:123, 124, 125, 167, 178

Brimacombe, J. K., **28**:306(145), 335(145)

Brindley, J., **4**:10, 62; **5**:250, 277(146); **8**:19, 86

Bringer, R. P., **6**:549, 551, 563; **7**:49, 84

Brinich, P. F., **3**:54, 98

Brink, J. C., **26**:19(110), 20(110), 23(110), 27(110), 97(110)

Brinkman, H. C., **2**:378, 394; **13**:219(62), 233, 265; **14**:5, 100; **SUP**:54(61), 109(61), 110(61), 129(61), 130(61); **23**:207(25), 268(25); **24**:110(93), 113(93), 114(93), 187(93)

Brinkmann, A., **13**:219(83), 266

Brinsfield, W. A., **19**:80(226), 93(226)

Broadbent, J. A., **19**:61(137), 62(137, 140), 89(137, 140)

Brock, J. R., **7**:180, 213

Brocklebank, M. P., **SUP**:97(257), 98(257), 168(257)

Brockmann, J. E., **29**:216(5, 6, 7, 8, 9, 11), 217(11), 221(25, 26, 27), 233(26, 67), 268(134), 269(5), 271(25), 276(148), 278(148), 303(26), 309(25, 67), 324(176), 325(176), 326(176), 327(176), 334(5, 6, 7, 8), 335(9, 11, 25), 336(26, 27), 338(67), 342(134), 343(148), 344(176)

Brodie, L. C., **17**:142(159), 143(162), 144(159), 145(162), 147(159), 149(162), 150(162), 151(178, 179), 157

Brodkey, R. S., **2**:361, 385, 388, 394, 395; **20**:107, 130; **25**:16(49), 21(49), 22(49), 23(49), 33(49), 142(49)

Brodnyan, J. G., **15**:182(45), 224(45)

Brodowicz, K., **6**:312(99), 366; **7**:66(46), 85; **9**:300(46), 301, 347; **30**:272(44), 309(44)

Broer, L. J. F., **11**:219, 223, 257

Brogan, J. J., **5**:390, 510; **15**:322

Brogan, T. R., **4**:310(69), 316

Brokaw, R. S., **2**:38, 106; **3**:269, 281, 300, 301

Bromley, L. A., **3**:276(52), 301; **4**:173, 174, 225; **5**:57, 60, 64, 65, 83, 95(44), 97(47), 98(47), 125, 126, 127, 428, 431, 432, 436, 464, 511; **9**:248(74), 261, 270, 271; **11**:73, 74, 75, 76, 83, 95(41), 96, 105(41), 109(41),

Bromley L. A. (*continued*)
110, 116, 123, 134, 138, 139, 140,
141(147), 142, 144, 161, 190, 195;
16:138(106), 141, 156; **18**:327(5), 337(49,
50, 51), 362(5), 363(49, 50, 51); **21**:88(55),
103(81), 136(55), 137(81); **23**:118(124),
132(124); **30**:39(19), 88(19), 246(63),
252(63)
Bromwich, T. J. I., **15**:285, 322
Brondyke, K. J., **29**:188(187), 212(187)
Bronson, J. P., **20**:232(158), 306(158)
Bronstein, E. M., **19**:202(65, 66), 207(66),
245(65, 66)
Brook, D. L., **24**:234, 269
Brooks, I., **8**:172, 173, 216, 217, 218, 219, 220,
221, 222, 227
Brooks, J. S., **17**:102, 110, 155
Brooks, N. H., **16**:9, 12, 16, 19, 27, 29, 57
Brooks, R. H., **29**:40(35), 57(35)
Brosens, P. J., **7**:275(110, 117), 297, 318, 319
Brosilow, C. B., **23**:217(18), 267(18)
Brost, J., **20**:183(8), 299(8)
Brost, R. A., **25**:375, 416
Brotz, W., **9**:127(28), 178; **14**:218(154),
223(154), 246
Broughton, J., **14**:151, 152, 242; **19**:108(53),
186(53)
Brounshtein, B. I., **26**:23(33), 94(33)
Brovkin, L. A., **6**:418(144), 493
Brow, J. E., **3**:285, 301
Brow, N. J., **17**:69(29), 153
Brown, A., **15**:286, 315, 322
Brown, A. H. O., **9**:66(15), 106
Brown, A. I., **24**:19(49), 38(49)
Brown, A. R., **9**:232(65), 270; **SUP**:412(525)
Brown, C. K., **7**:52(29), 84
Brown, D. D., **4**:202(77), 203(77), 226;
17:37(121), 62
Brown, D. J., **10**:229(17, 18), 281
Brown, D. R., **2**:358(63), 361, 395
Brown, D. W., **14**:97(186), 104
Brown, F., **31**:412(76), 429(76)
Brown, G., **15**:236, 262, 278
Brown, G. A., **8**:140(71, 72), 141(72), 142(72),
144(72), 148(72), 149(72), 151(72), 160;
18:139(72, 73), 140(72), 146(72), 147(72,
73), 159(72, 73); **25**:292, 293, 296, 302,
303, 304, 305, 306, 307, 309, 311, 315
Brown, G. M., **5**:191(104), 249; **SUP**:101(260),
170(260), 171(260)

Brown, G. W., **5**:3, 31(7), 44, 47, 48, 50
Brown, H., **12**:242(114), 282; **22**:219(104),
351(104)
Brown, J. C., **25**:294, 315
Brown, J. K., **10**:262, 283
Brown, L. E., **11**:105(108), 113, 193; **16**:134, 155
Brown, R. A., **28**:389(170), 390(172), 391(172),
392(172), 422(170, 172); **30**:313(5),
325(71), 329(34, 35), 337(71), 338(71, 73),
339(73), 347(71), 350(89, 90, 92), 351(34,
35), 358(34, 35, 113, 114), 360(119),
385(5, 71), 426(5), 428(34, 35), 429(71),
430(73, 89, 90, 92), 432(113, 114, 119)
Brown, R. A. S., **1**:376(29), 441; **3**:16(25), 32
Brown, R. E., **2**:326, 355
Brown, R. I., **17**:179, 314
Brown, T. W. F., **9**:3(14), 65, 71(14), 91(14),
93(13, 14), 95, 96, 97(14), 106
Brown, W. G., **18**:67(123), 72(123), 85(123)
Brown, W. H., **5**:390, 510
Brown, W. S., **7**:145, 161; **15**:287(97), 322, 325
Browne, F. L., **17**:186, 315
Brownell, D. H., **14**:78(162), 83(173), 84(162),
85(162), 88(173), 103, 104
Browning, B. L., **10**:226, 281
Browning, R., **3**:258, 271, 272, 297, 300
Broyer, E., **13**:261(127), 267
Bruckert, B., **29**:183(174), 211(174)
Bruggeman, D. A. G., **18**:181(67), 186(67),
228(67), 237(67)
Bruijn, P. J., **16**:83, 154
Bruines, J. J. P., **28**:96(14), 139(14)
Brun, E. A., **2**:322(24), 334(24), 337(24), 353;
6:38(72), 125; **7**:193, 194(80), 214; **10**:177,
216; **11**:219, 222, 226, 237, 240(210), 257,
262, 263
Brunco, D. P., **28**:76(55), 141(55)
Brun-Cottan, J. C., **18**:330(30), 335(30), 362(30)
Brundin, C. L., **7**:200, 201
Brundrett, E., **19**:334(144), 337(144), 355(144)
Brundrett, G. W., **10**:184(71), 216
Bruneau, C. H., **24**:247, 269
Brunello, G., **11**:219(110), 222(110), 237(110),
257
Brunjail, D., **23**:229(101a), 230(101a),
231(101a), 271(101a)
Brunn, P. O., **25**:292, 316; **26**:20(34), 94(34)
Brunner, M. J., **1**:273, 284, 353
Brunson, R. J., **18**:169(34), 170(34), 176(34),
223(34), 224(34), 225(34), 226(34),

236(34); **26**:19(242), 104(242)

Brunt, J. J., **9**:244(73), 261(73), 270

Bruschke, M. V., **25**:298, 315

Brusenback, R. A., **14**:152(36), 243

Brush, C. F., **2**:293, 353

Brusseau, M. L., **23**:370(14), 461(14)

Brutsaert, W., **14**:264(33), 280

Bruun, H. H., **8**:294(24), 349; **11**:240, 241, 242, 243, 261, 263

Bryan, R. L., **5**:99, 127

Bryant, W. A., **28**:342(17), 416(17)

Bryce, W. M., **29**:254(107), 255(107), 340(107)

Bryden, H., **18**:333(40), 362(40)

Bryne, J. T., **5**:99, 127; **11**:105(90), 113(90), 193

Bryson, J. O., **10**:265, 283

Brzustowski, T. A., **15**:27(55), 29(55), 56(55); **26**:71(141), 89(199), 99(141), 102(199)

Bubenchikov, A. M., **25**:96(174, 175), 148(174, 175)

Buchanan, D. J., **29**:158(88), 163(88), 175(151), 206(88), 210(151)

Buchanan, R. C., **20**:262(259), 311(259)

Buchanan, T. D., **23**:280(14), 304(14), 315(14), 322(14), 325(14), 354(14)

Buchberg, H., **18**:11(8), 16(21, 22), 79(8), 80(21, 22)

Buchlin, J.-M., **24**:104(66), 187(66); **30**:136(94), 193(94)

Buck, G. M., **23**:325(211), 364(211)

Buck, M., **29**:178(158), 180(169), 181(158), 210(158), 211(169)

Buckham, J. A., **14**:179(100), 188(100), 189(100), 191(100), 195(100), 245; **19**:113(89), 119(89), 148(89), 186(89)

Buckius, R. O., **31**:347(32, 33), 364(33), 367(32), 426(32, 33)

Buckle, E. R., **14**:306, 333(117), 342, 345

Buckley, H., **2**:424, 451

Buckmaster, J. D., **26**:61(35), 94(35)

Buddenberg, J. W., **3**:269, 300

Budenberg, J. W., **2**:38, 106

Budhia, H., **19**:12(19), 15(19), 84(19)

Budyko, M. I., **12**:25, 74

Buehler, E., **30**:314(10), 426(10)

Buell, J. C., **17**:137, 156; **25**:354, 355, 356, 415

Buelow, P. E. O., **24**:194, 272

Buetner, K., **11**:318(24, 25), 376(24, 25), 436

Buettner, K., **22**:11(4), 17(4)

Buevich, Yu. A., **25**:163(45), 244(45)

Buglaev, V. N., **31**:282(187), 331(187)

Bugrovsky, V. V., **25**:4(30, 32), 141(30, 32)

Buhler, R. D., **4**:283(48), 315

Bui, R. T., **23**:306(119), 359(119)

Bui, V. A., **29**:249(102), 340(102)

Bukhvotsov, A. P., **8**:24, 86

Bukreev, V. I., **25**:81(136, 153), 146(136), 147(153)

Bulanova, A. B., **11**:168(168), 170(168, 170), 196

Bulanova, L. B., **18**:264(57), 322(57)

Bulavin, P. E., **15**:285, 322

Buleev, N. I., **SUP**:355(496), 356(496)

Bulkley, R., **24**:109(85), 187(85)

Bull, D. C., **29**:101(72), 126(72)

Bull, M. K., **8**:293(15), 349

Buller, M. L., **20**:217(124), 220(124), 232(156), 238(156), 305(124), 306(156)

Bullister, E. T., **20**:232(166), 307(166)

Bullock, K. J., **8**:299(41), 305, 349; **11**:240, 243, 261

Bulthuis, K., **14**:256(22), 279

Bunche, C. M., **4**:116(71), 140

Bunde, R. E., **3**:284(69), 301

Bunditkul, S., **20**:328(34), 351(34)

Bundy, F. P., **9**:376(44), 415

Bundy, R. D., **4**:206(82), 226; **5**:131(14), 247; **19**:334(137), 336(137), 355(137)

Bune, A., **30**:351(96), 431(96)

Bunev, V. A., **29**:68(19), 124(19)

Buning, P. G., **24**:218, 269

Bunker, R. S., **23**:324(209), 363(209)

Bunkin, F. V., **28**:127(7), 138(7)

Bunn, R. L., **SUP**:393(521)

Bunow, B., **15**:291, 322

Burakov, B. A., **17**:8, 24(44), 58

Burbank, P. D., **6**:94(147), 129

Burch, D. E., **5**:279(24), 298(45), 322, 323

Burch, D. L., **12**:129(12), 138(12), 192

Burch, J. M., **6**:201, 364; **30**:261(17), 308(17)

Burcik, E. J., **23**:193(26, 27, 28, 29, 30), 206(26, 27, 28, 29, 30), 223(30), 226(26, 27, 28, 29, 30), 268(26, 27, 28, 29, 30)

Burdukov, A. P., **25**:4(10, 11), 33(10), 140(10, 11)

Burdukov, E. P., **31**:160(9), 165(9), 182(9), 322(9)

Buretta, R. J., **14**:34(85), 72, 73(85), 102

Burgens, J. M., **15**:236, 277

Burger, I. M., **3**:35, 97

Burger, J., **23**:219(68), 222(68), 270(68)

Bürger, M., **29**:152(73), 172(143), 173(143, 144, 145), 178(158), 180(169), 181(158, 171), 206(73), 209(143, 144, 145), 210(158), 211(169, 171)

Burgers, J. M., **12**:269(12), 277; **21**:185(1), 236(1); **25**:355, 409

Burgess, D. S., **29**:71(29), 125(29); **30**:76(20), 88(20)

Burgess, J. F., **20**:259(247), 311(247)

Burgess, R. W., **5**:345, 506

Burggraaf, A. J., **28**:343(52, 54, 56, 57), 417(52, 54, 56, 57)

Burggraf, F., **7**:344, 346, 352(38), 370(70), 371(38), 378, 379

Burggraf, O. R., **6**:60, 62(104, 105), 127, 416(82), 490

Burghardt, A., **SUP**:75(203)

Burghart, G. H., **6**:11(21), 30(21), 34(21), 123

Burgraff, O. R., **8**:11, 85

Burgraff, U. R., **3**:60, 81(84), 99

Burka, A. L., **6**:474(321, 322, 327), 502; **11**:408(178), 439

Burke, J. C., **5**:102, 127, 381, 445, 509, 514

Burke, J. P., **SUP**:92(249, 251), 93(249), 96(249, 251), 97(249), 166(249)

Burkel, W. E., **22**:250(225), 357(225)

Burkhalter, J. E., **21**:178

Burks, T., **29**:91(59), 126(59)

Burmeister, L. C., **23**:118(123), 132(123)

Burnet, C., **18**:330(30), 335(30), 362(30)

Burnett, F., **31**:246(153), 261(153), 329(153)

Burns, A. D., **25**:329, 384, 385, 409

Burns, M. A., **25**:235(139, 140, 141, 142, 143), 249(139, 140, 141, 142, 143)

Burns, P., **18**:67(126), 68(126), 85(126)

Burns, S. P., **31**:406(69), 428(69)

Burns, W. K., **7**:342(29), 350, 352(29), 359, 361, 378

Burow, S., **2**:363(13), 365, 394

Burriss, W. L., **4**:98, 100, 139

Burroughs, P. R., **19**:334(144), 337(144), 355(144)

Burru, I. G., **23**:253(31, 32), 254(31, 32), 259(31, 32), 268(31, 32)

Burton, A. C., **4**:88(25), 91(25), 138; **22**:20(11), 50(13), 81(12), 82(11, 12), 153(11, 12, 13)

Burton, J. A., **30**:331(42, 43), 428(42, 43)

Burton, J. J., **14**:288(22), 306, 341

Bury, P., **21**:2(9), 3(9), 51(9)

Busbee, B. L., **14**:43(102), 102

Busbridge, I. W., **3**:203(25), 249

Busbridge, L. W., **8**:34, 88

Buscall, R., **23**:262(32a), 268(32a)

Buscheck, T., **24**:103(24), 185(24)

Busey, F. L., **8**:94(4), 158; **18**:87(4), 157(4)

Bush, M. B., **23**:219(268), 278(268)

Bush, W. B., **1**:326, 327(59), 328(59), 335, 336, 338, 339, 340, 341, 344, 350, 354; **4**:335, 337, 446; **7**:197, 215; **8**:19, 86

Bushman, S. G., **28**:344(61), 345(61), 402(61), 418(61)

Bushnell, D. M., **31**:405(66), 428(66)

Buskirk, E. R., **4**:120, 140

Buss, H., **6**:14(25), 123

Busse, C. A., **7**:239, 248, 248(47, 62), 254, 265, 275(111), 289, 290, 291, 292, 315, 316, 318, 319, 320e(163), 320f

Busse, F. H., **14**:31, 33, 101; **17**:133, 156; **30**:29(21), 88(21), 362(124), 432(124)

Bussman, A., **4**:331, 444

Bussman, K., **4**:331, 445

Bussmann, M., **26**:78(180), 79(181), 101(180, 181)

Bussmann, P. J. T., **17**:207, 211, 222, 223, 225(69), 237, 246, 315, 316

Busulini, L., **13**:219(51), 265

Butenko, G. F., **1**:406(86), 442

Butler, B., **25**:223(111), 224(111), 247(111)

Butler, H. W., **8**:29, 87

Butler, J. C., **10**:57(46), 82

Butler, S. W., **28**:344(61), 345(61), 402(61), 418(61)

Butt, M. H. D., **10**:185(75), 216; **14**:151(18), 152(18), 153(18), 208(18), 217(18), 242

Butter, A. P., **17**:147, 157

Butterworth, D., **SUP**:35(43), 77(43), 148(43); **15**:229(11), 236(11), 276; **16**:142, 143(117), 147(128), 148(128), 156; **17**:2(9), 40(9), 56; **20**:250(217), 309(217)

Buttery, N. E., **29**:146(44), 148(44), 171(139), 172(44), 204(44), 209(139)

Butti, K., **18**:28(62), 82(62)

Büttner, K., **22**:233(27, 28), 234(28), 236(29), 237(26, 30), 239(30), 240(30), 241(30), 348(26, 27, 28, 29, 30)

Buxton, L. D., **29**:165(117), 208(117)

Buyco, E. H., **3**:124(31), 173

Buyevich, Y. A., **26**:114(35), 122(50), 202(35), 209(35), 210(50)

Buynyachenko, G. P., **6**:416(86), 490

Büyüktür, A. R., **3**:16(30), 19(30), 23(30, 39), 24(30), 32
Buznik, V. M., **31**:160(6), 322(6)
Byers, R. K., **29**:107(79), 127(79)
Byram, G. M., **10**:267, 283
Byrd, P. F., **8**:47(225), 89
Byrne, J. T., **5**:406(165), 512
Byrnes, W. R., **5**:381, 509
Bywater, R., **19**:81(253), 94(253)
Bywater, R. J., **25**:354, 355, 356, 415

C

Cabot, H., **25**:351, 353, 354, 360, 361, 378, 392, 396, 403, 404, 412, 416
Cabot, W., **25**:362, 416
Cade, C., **22**:421(31), 436(31)
Cadoret, R., **28**:351(80), 418(80)
Cahn, J. W., **28**:16(28), 71(28)
Cahn, R. P., **25**:189(84), 234(84, 130), 246(84), 248(130)
Cahoon, J. R., **28**:304(142), 335(142)
Cai, Z., **25**:252, 292, 293, 316
Cain, C. P., **22**:348(31), 386(21), 387(21), 404(21), 408(21), 435(21)
Cain, G. L., **10**:184(71), 185(75), 216; **14**:151(18), 152(18), 153(18), 208(18), 217(18), 242
Cairns, B. R., **28**:342(24), 416(24)
Cairns, D. N. H., **21**:26(45), 27(45), 52(45)
Cakl, J., **23**:193(33), 268(33)
Cala, G. C., **23**:294(75), 357(75)
Caldas, J., **10**:200, 218
Calder, P. H., **2**:52, 107
Calderbank, J. C., **11**:422(216), 440
Calderbank, P. H., **20**:372, 373, 374, 381, 386, 388; **25**:271, 317
Caldwell, D., **18**:328(14, 24, 25), 329(25), 330(25, 26), 333(39), 342(39), 347(14), 348(14), 349(14), 350(26), 359(26), 362(14, 24, 25, 26, 39)
Caldwell, D. R., **21**:151, 179
Caldwell, J., **SUP**:62(73), 322(73), 326(73)
Calehuff, G. L., **8**:305(57), 350
Calimbas, A. T., **7**:240, 271, 275(49), 315
Callahan, V., **26**:191(150, 151), 195(150, 151), 215(150, 151)
Callcott, T. A., **28**:118(65), 119(65), 141(65)
Callcott, T. G., **19**:108(52), 186(52)

Callen, H. B., **5**:261(16), 322; **14**:283(6), 341
Callen, N. S., **25**:342, 366, 415
Calmidi, V., **28**:391(178), 398(178), 399(178), 400(178), 423(178)
Caltagirone, J. P., **14**:46(110), 47(110), 72, 72(155), 73(155), 102, 103; **28**:370(127), 420(127)
Calus, W. F., **16**:82(30), 117(30, 82), 119, 120, 122, 125, 154, 155, 237
Calvert, S., **1**:389, 441
Caly, R., **5**:182, 251
Camarero, R., **24**:218, 272
Cambel, A. B., **4**:257(22), 258(22), 259(22), 260(22), 261(22), 262, 265(22), 266(22), 313
Camci, C., **23**:280(26), 320(26, 167, 168), 355(26), 361(167, 168)
Cameron, J. F., **1**:394(75), 397(75), 398(75), 442
Camia, F. M., **15**:322
Camm, J. C., **2**:127(16), 131(16), 134(16), 268
Camp, A. L., **29**:62(9), 108(81), 124(9), 127(81)
Camp, B., **31**:401(57), 402(57), 428(57)
Campbell, D. S., **23**:340(276), 341(276), 342(276), 367(276)
Campbell, D. T., **5**:135, 141, 149, 150(28), 247
Campbell, G. S., **31**:89(7), 102(7)
Campbell, J. A., **23**:337(261), 366(261)
Campbell, J. B., **4**:115(70), 140
Campbell, J. F., **13**:67(5), 114
Campbell, J. R., **10**:203, 217
Campbell, M. J., **10**:5(8), 19, 20, 24(27), 35, 36, 57(45), 73(59), 82
Campbell, P. M., **5**:36, 53
Campbell, W. B., **25**:294, 319
Campbell, W. D., **SUP**:70(148), 87(148), 160(148); **19**:268(36), 270(36), 351(36)
Campbell, W. F., **15**:287, 322
Campisano, S. U., **28**:76(6), 138(6)
Campo, A., **26**:188(148), 191(148), 215(148)
Campo, W. C., **20**:185(108), 205(108), 304(108)
Canada, G. S., **19**:98(7), 108(7), 137(7), 149(179), 160(7, 179), 184(7), 189(179)
Canavan, G. H., **8**:291(12), 344(12), 349
Candell, L. M., **5**:24, 51
Cane, K. L. D., **18**:16(24), 80(24)
Canjar, L. N., **14**:318(85), 321(85), 322(85), 323(85), 324(85), 344
Cann, G. L., **4**:283(48), 315
Cannon, C. N., **12**:5(13), 73
Cannon, J. N., **5**:242, 251

Canon, R. M., **18**:255(22), 257(22), 262(22), 321(22)

Canty, J. M., **5**:381, 509

Cao, Y., **23**:288(76), 295(76), 357(76)

Capatu, C., **9**:128(34), 178

Cape, J. A., **28**:352(90), 419(90)

Capes, C. E., **10**:203, 217

Capes, D. E., **19**:123(121), 125(121), 187(121)

Capp, S. P., **15**:188(57), 224(57)

Capriotti, D., **23**:326(212), 364(212)

Captieu, M., **9**:103, 106

Carachalios, C., **29**:173(144), 209(144)

Carberry, J. J., **23**:230(34, 187), 268(34), 275(187), 371(25), 461(25); **25**:273, 315, 318

Carbon, M. W., **17**:42(143), 44(143), 63

Carbonell, R. G., **18**:191(148, 149), 230(148), 239(148, 149); **23**:370(19, 20, 21), 372(34), 381(58), 386(34), 390(21, 34), 391(34), 398(34, 61, 62), 400(34), 404(34), 407(61), 409(34, 61, 62), 421(61), 422(34), 440(21), 442(34), 443(21), 447(90), 461(19, 20, 21), 462(34), 463(58, 61, 62), 464(90); **30**:95(19), 112(64), 189(19), 191(64); **31**:30(5), 32(6), 33(6), 34(6, 32c), 36(32c), 72(6), 76(32c), 80(32c), 82(6, 14, 15), 86(32c), 92(14, 15), 102(5, 6, 14, 15), 103(32c)

Carcassi, M., **29**:303(165), 304(165), 344(165)

Card, C. C. H., **SUP**:96(256), 97(257), 98(257), 168(257)

Carda, D. D., **31**:458(20), 471(20)

Carden, W. H., **7**:197, 215

Cardner, D. V., **9**:304(63), 347

Care, J. M., **14**:120, 144

Caren, R. P., **5**:477, 478, 479, 516; **9**:351(7), 363(27), 390(7, 27, 66), 393(54, 66), 406(100), 409(7), 411(107), 414, 415, 416, 417; **11**:400(162), 438

Caretto, L. S., **13**:67(10, 15), 102(15), 115; **SUP**:211(402); **19**:268(56), 271(56), 351(56)

Carey, G. F., **31**:336(22), 339(22), 340(22), 426(22)

Carey, P. G., **28**:96(63), 98(105), 141(63), 143(105)

Carey, V. P., **19**:79(200), 92(200); **20**:203(88), 303(88); **23**:12(74), 14(74), 53(74, 75), 56(74), 57(75), 58(74), 61(74), 87(74), 90(74), 106(74), 130(74, 75); **28**:127(15),

139(15); **29**:39(32), 57(32); **30**:117(70), 192(70)

Carichner, G. E., **6**:69(120), 71, 128

Carlen, R. A., **20**:285(306), 313(306)

Carleson, T. E., **26**:14(39), 26(151), 27(151), 30(39), 94(39), 99(151)

Carley, C. T., **11**:288(21), 315

Carlier, C. C., **14**:314, 343

Carlomango, G. M., **23**:328(224), 364(224)

Carlson, A. B., **31**:411(75), 429(75)

Carlson, B. G., **23**:139(14), 140(14), 184(14)

Carlson, D. J., **28**:11(17), 71(17)

Carlson, G. A., **7**:274, 294, 318, 320a, 320f, 348, 355, 379; **SUP**:73(178, 179), 211(178), 212(178, 179); **19**:268(58), 271(58), 351(58)

Carlson, L. D., **4**:92(30), 138

Carlson, L. W., **SUP**:229(421)

Carlson, O., **30**:272(37), 308(37)

Carlson, R., **9**:117(5), 129(5), 177

Carlson, R. D., **17**:321(13), 339

Carlson, R. O., **20**:259(247), 311(247)

Carlson, R. W., **30**:236(50), 252(50)

Carlson, W. D., **6**:474(252), 483(252), 484, 498

Carlson, W. O., **4**:152, 223; **8**:167, 168, 172, 176, 177(24), 178(24), 183(24), 184, 192, 195, 226; **11**:266, 314; **18**:13(19b), 14(19b), 80(19b)

Carman, E. H., **2**:297(65), 354; **3**:273(37), 300

Carman, M. F., **10**:57(46), 82

Carman, P. C., **23**:188(35), 196(35), 268(35); **24**:104(45), 186(45); **25**:294, 315

Carmi, A., **5**:166, 248

Carnahan, B., **30**:272(41), 309(41)

Carne, E. B., **11**:213, 214, 255

Carnesale, A., **7**:238(36, 39, 40), 241(36, 39, 40), 242(36, 39, 40), 315

Carolan, M., **28**:343(51), 417(51)

Carpenter, J. R., **20**:186(75), 187(75), 201(75), 302(75)

Carper, H. J., **7**:348, 358(54), 361(54), 379

Carper, H. J., Jr., **26**:114(31), 203(176, 177), 209(31), 217(176, 177)

Carra, S., **26**:45(36), 94(36)

Carreau, P., **24**:109(90), 187(90)

Carreau, P. J., **15**:62(12, 13), 78(12, 13), 79(13), 86(13), 135(12, 13)

Carrier, G. F., **8**:34, 88, 170, 226

Carrier, W. D., III, **10**:55(43), 56, 82

Carrier, W. H., **8**:94(4), 158; **18**:87(4), 157(4)

Carriere, P., **6**:87, 129

Carroll, D. E., **29**:234(73, 76), 276(147), 282(76), 288(76), 313(76), 329(147), 338(73), 339(76), 343(147)

Carruthers, J. R., **11**:318(4), 412(4), 434; **30**:367(134, 135), 369(135), 396(134), 397(134), 432(134), 433(135)

Carslaw, G., **25**:13(47), 141(47)

Carslaw, H. S., **1**:52, 54(1), 60(1), 62(1), 63(1), 66(1), 74, 99(1), 120, 211(22), 264; **3**:122(26), 126(33), 132(33), 173; **5**:40, 54, 336(26), 347, 506; **6**:489; **8**:3, 53(10), 75(10), 78(10), 84; **10**:48, 81; **11**:12(31), 13, 48, 322(36), 378(36), 383(36), 435; **SUP**:119(290); **15**:285, 287, 289, 313, 322; **18**:169(55), 237(55); **19**:4(12), 5(12), 6(12), 8(12), 12(12), 14(12), 18(12), 41(12), 84(12), 192(8), 193(8), 243(8); **20**:354, 355, 356, 387; **22**:159(33), 184(33), 185(33), 251(33), 271(33), 348(32, 33); **23**:317(150), 360(150), 432(86), 464(86); **24**:56(20), 99(20); **29**:138(27), 204(27)

Carslon, W. O., **6**:312(97), 366

Carson, W. W., **5**:40(69). 45, 53

Carstens, M. K., **6**:405, 406(60), 407(60), 408(60), 409(60), 410(60), 489

Carter, C. B., **28**:103(87), 142(87)

Carter, J. C., **29**:234(85, 86), 269(85), 320(86), 339(85, 86)

Carter, L. L., **31**:402(62), 428(62)

Cartigny, J. D., **23**:154(37, 42), 185(37, 42)

Cartwright, R., **30**:333(56), 429(56)

Caruso, R., **30**:358(109, 110), 431(109, 110)

Carvalho, R. D. M., **23**:26(52), 128(52)

Carver, D. B., **23**:313(135), 314(135), 360(135)

Carver, J. R., **6**:549(67), 551, 553(67), 563; **7**:50

Casagrande, I., **1**:367(17), 379(33, 34, 35, 36, 38), 385(34), 389(34, 36), 390(34, 36), 393(36), 401(34), 440, 441

Casal, J., **19**:102(42), 105(42), 107(42), 108(42), 180(42), 185(42); **25**:182(76, 77), 183(76), 184(76, 79), 185(79), 186(76, 79), 187(79), 189(86), 190(87, 88), 191(88), 192(88), 193(92), 194(92), 195(92), 200(76, 79), 202(79), 204(88), 205(88), 222(88), 224(79), 227(88), 228(88), 234(138), 246(76, 77, 79, 86, 87, 88, 92), 249(138)

Casamatta, G., **15**:322

Casarella, M. J., **SUP**:64(93), 75(93)

Casassa, G., **24**:30(65), 38(65)

Case, K. M., **5**:310(84), 324; **8**:70, 90; **9**:363(26), 415

Case, R. B., **22**:30(14), 153(14)

Cashwell, E. D., **5**:3, 50

Caskey, J. A., **13**:211(23), 264; **18**:169(42), 170(42), 173(42), 214(42), 217(42), 220(42), 236(42)

Caspi, S., **17**:115, 155

Cass, A., **31**:89(7), 102(7)

Cassidy, W. A., **10**:19(25), 24(25), 36

Cassulo, J. C., **29**:20(16), 56(16)

Castellana, F. S., **22**:30(14), 153(14)

Castelli, V. J., **18**:328(12, 13), 362(12, 13)

Casterline, J. E., **17**:39(129), 62

Castle, J. N., **21**:327, 344

Castleden, J. A., **15**:287, 289, 323

Castro, I. P., **11**:220, 224, 263; **18**:123(59), 159(59)

Caswell, B., **13**:262(132), 267; **23**:262(35a), 268(35a)

Catalano, A. P., **28**:342(25), 416(25)

Cates, M. R., **23**:294(75), 357(75)

Cates, R. E., **12**:3(9), 13, 14(9), 18(23), 73, 74

Catipovic, N. M., **19**:98(6), 99(6), 100(30), 136(30), 137(30), 139(30), 145(30), 184(6), 185(30); **23**:340(272), 367(272)

Catton, I., **17**:137, 156; **18**:11(8, 9), 17(29, 30), 79(8, 9), 80(29, 30); **19**:79(214, 215), 80(226), 92(214, 215), 93(226); **21**:327, 344; **29**:9(11), 47(43), 48(43), 49(45), 55(11), 57(43, 45), 130(8), 203(8); **30**:94(15), 118(73), 120(73), 134(93), 135(73), 136(93), 155(122), 189(15), 192(73), 193(93), 194(122)

Cavallini, A., **9**:261(110), 271; **15**:48(75), 57(75)

Cavendish, J. C., **15**:287(78), 324

Cavers, S. D., **6**:515(28), 562

Ceaglske, N. H., **13**:121, 122, 124, 185, 190, 201

Cebeci, T., **13**:79(29), 116; **15**:189(58), 224(58)

Cecil, E. A., **12**:3(11), 14(11), 15(11), 59(11, 62), 67, 68, 69(62), 73, 75

Celata, G. P., **26**:45(37), 94(37)

Celia, M. A., **30**:95(26), 189(26)

Cen, K., **19**:164(216), 170(216), 171(216), 174(216), 190(216)

Centolanzi, F. J., **6**:77(133), 128

Cercignani, C., **7**:171(28, 29), 172, 172(28, 29), 175, 176(29), 178(29), 181, 212

Cermak, J., **28**:376(144), 421(144)

Cermak, J. E., **SUP**:170(345), 177(345)

Cernescu, V., **14**:231(167), 246; **19**:147(164), 189(164)

Cerny, R., **26**:83(26), 94(26); **28**:18(29), 71(29), 76(86), 142(86)

Cess, R. D. (Chapter Author), **1**:1, 36(32), 50, 115, 122, 292, 294(31), 353; **2**:400, 450; **3**:219(69), 230, 232, 233(85), 236, 251; **4**:28, 63, 224, 228, 313; **5**:82, 96, 99, 126, 127, 145, 148, 149, 254(3), 255(3), 298(54), 306(3), 312(3), 313(3), 316(3), 321, 323; **6**:474(196, 210), 475, 476, 495, 496; **8**:11, 22, 71, 73, 74, 85, 86, 91, 229, 250(30), 252(30), 253(33, 35), 256(30, 35), 258(33), 259(33), 260, 263(39), 264(30), 267(32), 276(48), 278(49), 279(49), 283; **9**:294(18), 346, 362(21), 363(21), 364(21), 365(21), 377(21), 379(21), 396(21), 398(21), 415; **11**:67, 80, 82(33), 127, 128, 129, 131, 134, 143, 144(143), 190, 195, 326(47), 329(47), 390(141), 397(141), 399(141), 403(47), 435, 438; **12**:119, 120, 143(4), 177, 178, 192, 193; **14**:265(37), 280; **SUP**:69(141), 161(141), 177(355), 179(360), 180(360), 185(360, 363); **16**:136, 156; **17**:242(79), 316; **19**:268(43), 270(43), 351(43); **20**:353, 359, 360, 361, 362, 387; **23**:136(3), 184(3)

Cetas, T. C., **22**:30(23, 49), 153(23), 154(49), 367(11), 373(11), 421(34), 435(11), 436(34)

Cevantes, J., **19**:74(171), 91(171)

Chaboki, A., **20**:232(161), 237(161), 238(161), 239(161), 240(161), 307(161)

Chahine, M. T., **7**:191, 192(73), 214

Chakalev, K. N., **6**:376(6), 414(6), 415(6), 463(6), 464(6), 467(6), 486; **25**:11(37), 32(37), 141(37)

Chakma, A., **24**:104(28), 185(28)

Chakrabarti, A., **20**:329(44), 351(44)

Chakrabarti, U. K., **15**:287, 322

Chalfant, A. I., **6**:551, 564

Challis, L. J., **5**:487(269), 517

Challis, L. S., **17**:68, 69(17, 21, 32), 70(13, 21, 25, 32), 71(21), 72(13), 153

Chalmers, B., **22**:291(34), 348(34); **28**:79(44), 140(44)

Chamberlain, A. T., **29**:130(3), 202(3)

Chambers, J. E., **18**:169(30), 170(30), 172(30), 209(30), 210(30), 236(30)

Chambers, J. T., **23**:285(44), 356(44)

Chambers, L. G., **19**:198(46), 244(46)

Chambre, P. L., **1**:2(2), 49, 58, 60(13), 121; **2**:166, 170, 172, 269; **6**:460, 468, 474(200), 477, 495, 496; **7**:164(2), 165(2), 168, 187, 188, 190(2), 203(2), 211; **8**:36(198, 201, 202), 38(203), 89

Champagne, F. H., **11**:200(61), 240(81), 241, 242, 255; **21**:185(3), 204(3), 207(3), 236(3)

Chamra, L. M., **31**:456(78), 474(78)

Chan, B. K. C., **14**:5, 100

Chan, C. F., **29**:64(12), 124(12)

Chan, C. K., **9**:388(59), 391(59), 394(72), 398(72), 416; **23**:161(44), 186(44)

Chan, C. L., **28**:123(16), 139(16), 267(94), 333(94)

Chan, E. K. Y., **22**:421(36, 37), 428(37), 432(40), 433(40), 436(36, 37, 40)

Chan, G. K. C., **16**:53, 57

Chan, H. W., **2**:25, 106; **7**:343(32), 344, 361(32), 378

Chan, R. C., **18**:327(5), 362(5)

Chan, S. H., **12**:177, 193; **19**:38(85), 42(85), 43(85), 87(85); **29**:309(170), 344(170); **30**:243(9), 249(9), 335(65), 429(65); **31**:294(203), 299(203), 332(203), 459(2), 470(2)

Chan, S. K., **14**:33, 101

Chan, Y. T., **30**:335(69), 336(69), 347(69), 368(137), 429(69), 433(137)

Chandler, R. D., **SUP**:76(212), 171(212)

Chandler, T. R. D., **8**:248(23), 283; **14**:256(23), 279

Chandna, R. C., **11**:203, 206, 255

Chandra, B., **14**:126, 144

Chandra, R., **15**:327

Chandra, S., **28**:24(46), 72(46)

Chandran, R., **19**:137(145), 145(183), 160(183), 188(145), 189(183)

Chandrasekhar, B. S., **5**:326(10), 505

Chandrasekhar, S., **1**:5(10), 7(10), 49, 288, 353; **3**:176(1), 178(1), 180(1), 183(1), 184(1), 185(1), 186(1), 194(1), 200(1), 201(1), 203(1), 204(1), 206, 207, 215(1), 248; **5**:13(15), 40, 51, 53, 130(1d), 131, 204, 205(116), 228, 236, 246, 249, 250, 251, 304, 323; **8**:34, 88; **11**:325(43), 326(43), 337(43), 435; **15**:206(92), 207(92), 208(92), 211(102, 103), 225(92, 102, 103); **17**:133, 156; **21**:150, 151, 157, 178; **23**:136(5), 139(5), 145(5), 184(5); **24**:174(154), 190(154); **28**:108(17),

139(17); **30**:50(22), 88(22), 369(138),
433(138)

Chandrasekhara, B. C., **23**:217(36), 268(36)

Chandrupatla, A. R., **SUP**:216(407, 407a),
218(407), 219(407), 222(407a), 402(407);
19:261(23), 262(23), 264(23), 265(23),
268(23), 269(23), 271(23), 273(23),
276(23), 277(23), 279(23), 281(23, 81, 82),
283(23), 285(23, 81, 82), 286(23), 287(23),
288(23), 291(23), 293(23), 295(23),
296(23), 297(23), 299(23), 300(23),
303(23), 304(23), 305(23), 306(23),
312(23), 350(23), 352(81), 353(82)

Chaney, R. E., **30**:318(13), 427(13)

Chang, A., **30**:51(50), 52(50), 89(50)

Chang, C. C., **1**:300(44), 353; **SUP**:71(160),
298(160)

Chang, C. J., **30**:350(89), 430(89)

Chang, C.-N., **23**:206(45), 226(45), 229(45),
269(45)

Chang, C. T., **19**:81(254), 94(254); **26**:123(59),
124(59), 172(59), 178(59), 211(59)

Chang, C. W., **28**:318(190), 337(190)

Chang, H. C., **18**:191(146, 147), 239(146, 147)

Chang, I. D., **29**:172(141), 209(141)

Chang, I-Dee., **14**:63, 64, 68(135), 69(148), 70,
103

Chang, K. C., **24**:263, 264, 274

Chang, K. I., **9**:265(147), 272; **15**:62(14),
135(14)

Chang, L. C., **18**:69(132), 70(132), 86(132)

Chang, L. S., **26**:14(39), 17(38), 18(38), 30(38,
39), 94(38, 39)

Chang, M. J., **20**:232(157), 306(157)

Chang, P. Y., **24**:215, 265, 266, 267, 269, 273

Chang, R. H., **29**:39(33), 40(33), 41(33), 44(33),
57(33)

Chang, S. L., **31**:392(49), 427(49)

Chang, T. H., **26**:43(40), 94(40)

Chang, T. Y., **19**:81(254), 94(254)

Chang, W. J., **20**:336(52, 53), 351(52, 53)

Chang, W. S., **26**:123(65), 125(65), 211(65)

Chang, Y., **20**:248(210), 309(210)

Chang, Y. A., **25**:292, 296, 316

Chang, Y. P., **5**:82, 83(32), 88, 89, 110, 126, 409,
411, 419, 420, 422, 428, 513, 514;
11:84(47), 91, 95, 96(47), 105(91), 191,
193, 408, 439; **SUP**:266(442), 367(442),
368(442), 369(442); **18**:164(25), 236(25),
249(12), 320(12)

Changal Raju, D., **SUP**:249(433)

Channapragada, R. S., **6**:17, 123, 416(93), 490

Chao, A., **19**:122(110), 123(110), 124(110),
129(110), 130(110), 131(110), 180(110),
187(110)

Chao, B. T., **2**:443(41), 451; **6**:474(204, 224),
476, 477, 496, 497; **8**:19, 86; **9**:134, 135,
144, 179, 227, 269; **15**:242, 243, 244, 245,
249(59), 278; **19**:125(125, 126), 187(125,
126); **21**:335, 337, 346; **26**:10(244), 23(41,
42), 28(41), 30(41), 94(41, 42), 104(244)

Chao, P. K. B., **18**:19(39), 81(39)

Chapman, A. J., **9**:262(122), 271; **14**:138, 145;
26:231(32), 247(32), 248(32), 253(32),
261(32), 268(32), 269(32), 281(32),
282(32), 291(32), 297(32), 300(32),
316(32), 325(32)

Chapman, D., **2**:269; **3**:50, 51(56), 53, 59(56), 98

Chapman, D. R., **6**:5, 7, 13, 33, 39(22), 84,
89(7), 122, 123; **8**:22, 86; **25**:326, 331, 333,
397, 409

Chapman, R. C., **5**:91, 126; **11**:105(94), 113(94),
193

Chapman, S., **3**:287(86), 302; **4**:273(35), 314;
14:269(61), 280; **28**:406(221), 425(221)

Chapman, T. W., **12**:229(13), 237(13), 238(14),
248(14), 251(14), 261(13, 14), 266(13),
269(14), 270(14), 277, 278

Charach, Ch., **19**:30(59), 86(59)

Characklis, W. G., **31**:435(5), 436(5), 470(5)

Charette, A., **23**:306(119), 359(119)

Charette, M. P., **10**:7, 8, 9, 36

Chariton, J., **2**:353

Charles, M. E., **SUP**:325(480), 326(483)

Charlier, J. P., **28**:402(197), 404(197), 424(197)

Charlson, G. S., **17**:131, 156

Charm, S. E., **2**:374, 394; **18**:169(33), 170(33),
172(33), 212(33), 236(33)

Charney, J., **5**:210(128), 214, 250

Charny, C. K. (Chapter Author), **22**:19, 26(18),
30(15, 16), 71(18, 19), 127(17), 134(16),
151(19), 152(19), 153(15, 16, 17, 18, 19)

Charpakov, P. V., **8**:67, 90

Charters, W. M. S., **19**:334(145), 337(145),
355(145)

Charvonia, D. A., **1**:385(53), 388(53), 441

Charwat, A. F. (Chapter Author), **6**:1, 11(21), 23,
30(21), 33, 34(21), 35(53), 41(83), 44,
51(53), 68(83), 115, 69(115), 70(115), 73,
75(130), 76(83), 77(130), 78, 79(83),

Charwat, A. F. (*continued*)
  80(130), 81(130), 82, 123, 124, 126, 127, 128; 7:206, 217
Chase, M. W., Jr., 29:188(186), 212(186)
Chastain, J. W., 1:384(48), 394(48, 72), 398(48, 72), 441, 442
Chasteen, A. J., 14:3(3), 100
Chato, J. C. (Chapter Author), 6:474(243), 480(243), 483(243), 498; 7:304, 319; 9:73(16, 17), 75, 79, 107, 253(83), 270; SUP:414(536); 22:1, 1(8, 9, 10), 10(5), 13(7), 17(5, 6, 7, 8, 9, 10), 19(54), 30(54), 84(20), 87(20), 88(20), 91(20), 92(20), 94(20), 104(20), 105(20), 106(20), 109(20), 111(20), 138(20), 139(20), 153(20), 155(54), 158(35, 37), 276(36), 342(35), 348(35, 36, 37); 26:220(15), 234(15), 254(15), 269(15), 305(15), 325(15)
Chatterjee, A., 19:156(186), 179(186), 189(186)
Chatterjee, S., 28:344(61), 345(61), 402(61), 418(61)
Chatto, J. C., 20:248(211), 249(211), 309(211)
Chau, W. C., 14:63, 103
Chaudhury, Z. H., 20:252(237), 310(237); 26:114(30), 209(30)
Chauveteau, G., 23:219(170), 222(37), 225(37, 170), 226(199, 200), 228(199, 200), 268(37), 274(170), 275(199, 200)
Chavarie, C., 15:322; 23:206(24), 218(24), 237(24), 239(24), 240(24), 242(24), 244(24), 267(24)
Chavent, G., 30:108(49), 191(49)
Chavez, S. A., 29:215(1), 248(1), 334(1)
Chawla, T. C., 19:79(203), 92(203); 29:45(40), 57(40), 175(148), 210(148)
Chechetkin, A. V., 10:202, 218
Cheeke, D., 17:68(7, 11), 71(33), 153
Cheema, L. S., 6:474(224), 477, 497
Cheeseman, K. J., 24:30(62), 38(62)
Cheesewright, R., 9:291, 346; 14:66, 103
Cheeti, S. K. R., 30:257(55), 278(55), 279(55), 280(55), 309(55)
Chekansky, V. V., 14:160(77), 161(77), 171(77), 172(77), 176(119), 188(119), 191(119), 237(119), 244, 245; 19:160(193), 189(193)
Chekryzhov, S. I., 25:134(207), 150(207)
Chelton, D. B., 15:45(71), 57(71)
Chen, A. J., 23:311(132), 314(132), 360(132)
Chen, C. F., 18:1(1), 79(1); 28:277(111), 280(111), 334(111)

Chen, C. J., 13:76, 78(25), 115; 16:53, 58
Chen, C. K., 24:127(133), 129(133), 135(136), 137(137), 139(137), 189(133, 136, 137); 25:259, 288, 317
Chen, C. P., 24:226, 227, 247, 269, 271
Chen, C. S., 13:126(32), 191(32), 201
Chen, C. T., 16:55, 58; 18:330(28, 29), 333(28), 334(59), 335(29, 59), 340(59), 341(59), 342(59), 348(28), 350(28), 359(28), 362(28, 29), 363(59)
Chen, C. Y., SUP:291(462); 25:252, 292, 294, 316
Chen, D. T. W., SUP:68(133), 370(133), 371(133)
Chen, F., 28:277(111), 280(111), 282(117), 334(111, 117)
Chen, F. F., 4:256(21), 313; 19:50(116), 88(116); 28:133(18), 139(18)
Chen, G., 28:135(19), 139(19)
Chen, H. K., 14:16(43), 101
Chen, H. S., 18:190(85), 191(85), 196(85), 197(85), 237(85)
Chen, H. T., 24:127(133), 129(133), 135(136), 137(137), 139(137), 189(133, 136, 137)
Chen, J. C., 3:204(30), 239, 241, 249, 251; 5:441, 442, 443, 445, 514; 9:378(48), 379(48), 415; 10:159, 160, 166; 14:177(96), 226(96, 162), 228, 244, 246; SUP:81(235), 123(235), 124(235); 16:143, 144, 156, 180, 181, 238; 19:122(112), 128(134), 129(112), 131(112), 137(145), 145(183), 149(180), 150(169), 153(169), 154(169), 160(183), 164(215, 219), 168(215), 169(219), 171(215, 219), 173(223), 187(112), 188(134, 145), 189(169, 180, 183), 190(215, 219, 223); 21:279, 282, 299, 300, 306, 307, 313, 314, 345; 23:136(10), 161(10), 184(10); 24:199, 215, 271; 25:153(6), 242(6)
Chen, J. L. S., 25:262, 316
Chen, K., 28:376(146), 421(146)
Chen, K.-H., 24:199, 273
Chen, K. L., 19:173(223), 190(223)
Chen, K. S., 18:36(75), 83(75)
Chen, K. W., SUP:361(508), 363(508), 364(508)
Chen, L., 26:23(42), 94(42)
Chen, L. T., 20:24(53), 27(53), 81(53)
Chen, L. W., 26:53(252, 253), 104(252, 253)
Chen, M.-H., 24:193, 202, 214, 227, 230, 231, 263, 264, 269, 274; 28:308(154), 336(154)

Chen, M. M., **3**:108(13), 173; **5**:469, 516; **9**:250(78), 251, 252, 257, 270; **19**:80(229), 93(229), 125(125, 126), 187(125, 126); **21**:67(10), 134(10); **22**:10(12), 13(11), 17(11, 12), 26(22), 30(21), 45(22), 46(22), 48(22), 49(22), 51(22), 58(22), 85(22), 104(22), 106(22), 122(22), 139(22), 153(21, 22), 245(38), 348(38)

Chen, R.-Y., **SUP**:74(197), 94(197), 98(197), 99(197), 165(197), 166(197), 168(197), 169(197)

Chen, S. C., **20**:248(211), 249(211), 309(211)

Chen, S. F., **18**:327(5), 362(5)

Chen, S. J., **18**:189(141), 239(141); **23**:12(70, 71), 14(70, 71), 51(70, 71), 129(70, 71), 320(162), 361(162); **26**:105(2), 208(2)

Chen, T.-F., **26**:137(87), 212(87)

Chen, T. S., **7**:191(74), 192(74), 214, 331(3), 332, 338, 346, 353, 353(3), 354, 377; **11**:420(213), 440; **SUP**:269(447)

Chen, T. Y. W., **15**:156(15), 157(16), 163(15), 164(15), 165(15), 223(15, 16)

Chen, W. C. V., **15**:322

Chen, W. H., **10**:211(140, 141), 218; **15**:287, 322

Chen, W. S., **15**:287, 322

Chen, W. Y., **8**:291, 349

Chen, X., **28**:135(20), 139(20); **29**:157(108), 164(108), 174(108), 180(168), 181(108), 182(168), 191(200), 200(108, 168), 201(168), 207(108), 211(168), 212(200)

Chen, Y. L., **18**:36(73), 38(78), 39(78), 83(73, 78)

Chen, Y.-M., **20**:378, 387

Chen, Y. S., **24**:269

Chen, Y. T., **30**:168(129), 194(129)

Chen, Z.-X., **23**:371(28), 373(28), 462(28)

Cheney, A. G., **25**:166(57), 245(57)

Cheney, A. J., **8**:135, 160; **11**:219, 222, 236(98), 256; **18**:133(69), 134(69), 159(69)

Cheng, A. L., **7**:197, 216

Cheng, C., **14**:311(72), 343

Cheng, C. J., **11**:202, 205, 219, 222, 236, 241, 242, 253

Cheng, G. T., **15**:267, 268(102), 269, 280

Cheng, H. K., **2**:266, 270; **7**:197, 215, 216

Cheng, H. M., **SUP**:205(396); **19**:278(61), 352(61)

Cheng, H. S., **31**:112(34, 35, 36, 37, 38), 119(34, 35, 36, 37, 38), 120(35, 36, 37, 38), 121(34), 128(34), 137(34, 38), 153(34, 35, 36, 37), 157(34, 35, 36, 37, 38)

Cheng, K. C., **SUP**:58(69), 62(71), 63(71, 77), 65(71, 106, 107, 109, 110, 111), 80(228, 229), 83(239), 109(228), 110(228), 111(228), 129(239), 130(239), 131(239), 156(336), 157(229), 176(229), 177(357), 178(336, 357, 358), 184(229), 205(107), 208(107), 225(107), 262(69, 106, 107), 263(69, 106, 107), 273(111), 325(110), 328(110), 329(110), 333(110), 346(109), 348(109), 350(109, 111); **19**:77(180), 78(182), 91(180, 182), 308(97), 313(97), 353(97); **28**:370(125, 130), 420(125), 421(130)

Cheng, P. (Chapter Author), **3**:28(47), 32; **8**:14, 86; **14**:1, 17(47), 30, 31(69, 70), 38, 39(92), 40(92), 47, 48(111), 49, 50(47), 55(126, 127, 128, 129), 56(126, 127, 128, 129), 57(126, 127, 128, 129), 58(126, 127, 128, 129), 59(126, 127, 128, 129), 60(126, 127, 128, 129), 61, 62(133, 134, 135, 137, 138), 63(133, 134, 135), 64, 65(137, 138), 66, 68(133, 134, 135), 69(148, 149), 70(149), 71(151, 152, 153), 75(126, 127, 128, 129), 101, 102, 103; **20**:315(5, 6), 316(5, 6), 348(6), 349(5, 6); **24**:103(21, 22), 105(21, 22, 32), 122(130), 185(21, 22, 32), 189(130); **30**:93, 93(8), 94(16), 103(16), 104(16), 105(16), 108(16, 46, 47), 111(62), 137(46), 138(46), 139(46), 142(100), 147(100), 148(109), 149(109), 177(47), 178(47), 179(47), 180(47), 181(47), 182(159), 189(8, 16), 190(46), 191(47, 62), 193(100, 109), 196(159)

Cheng, R., **6**:38(73), 125

Cheng, R. T., **SUP**:73(194), 165(194), 166(194), 169(194)

Cheng, S., **18**:257(47), 262(47), 267(47), 268(47), 269(47), 322(47); **24**:8(31), 37(31)

Cheng, S. C., **17**:38, 62; **18**:179(76), 180(76), 228(76), 237(76)

Cheng, S. I., **2**:213, 269

Chenoweth, D. R., **24**:300(33), 320(33)

Chenoweth, J. M., **1**:373, 441; **5**:451, 454, 515; **30**:250(11); **31**:449(35), 465(7), 466(6, 7), 468(6), 470(6, 7), 472(35)

Chentsov, A. K., **6**:418(102), 491

Cheong, W.-F., **22**:277(40), 319(206), 348(40), 356(206, 217)

Cheremisinoff, M. P., **25**:189(83), 234(83), 246(83)

Cherepanov, G. P., **25**:163(46), 244(46)

Cherkasov, V. I., **6**:418(122), 492

Chermak, I., **25**:4(28), 141(28)

Cherney, W., **19**:11(18), 23(18), 50(18), 51(18), 54(18), 84(18)

Chernobyl'skii, I. I., **20**:42(58), 43(58), 49(58), 82(58)

Chernousov, S. V., **25**:73(120), 145(120)

Chernov, V. D., **14**:234, 235(176), 247

Chernykh, L. F., **15**:321

Cherrington, D. C., **19**:113(92), 121(92), 187(92)

Cherrington, R. D., **19**:113(93), 187(93)

Chesnokov, B. V., **19**:110(70), 111(70), 186(70)

Chesshire, G., **24**:218, 269

Chetty, A. S., **25**:235(143), 249(143)

Chetyrin, V. F., **25**:33(63), 77(128), 84(169), 92(169), 121(189), 142(63), 146(128), 148(169), 149(189)

Cheung, F. B., **19**:4(10, 11), 73(10, 11), 74(11, 173), 75(173), 77(10, 11, 179), 79(202, 217), 84(10, 11), 91(173, 179), 92(202), 93(217); **21**:300, 344; **30**:32(23), 33(23), 88(23)

Cheung, T. K., **19**:146(161), 147(161), 149(161), 150(161, 167), 151(161, 167), 152(161), 153(161), 154(161), 181(161), 188(161), 189(167)

Chez, R., **28**:341(15), 415(15)

Chhabra, P. S., **28**:250(30), 330(30)

Chhabra, R. P. (Chapter Author), **23**:187, 188(38), 190(38), 193(38, 40, 233), 194(38), 199(38, 243), 202(112, 171), 204(243), 205(243), 206(243), 208(39, 90, 91, 108, 109, 112, 171, 172), 210(112), 211(39, 112, 171), 212(108, 111, 112, 171), 214(108), 215(38, 233), 216(243), 217(243), 218(39), 219(104), 237(111, 233, 243), 238(111), 239(38a, 111, 233, 242), 240(38a, 111, 242), 241(243), 243(242, 243), 244(110, 242), 246(38b, 111), 247(110, 111), 248(111), 249(111), 250(110, 111), 251(111, 242), 262(41), 263(41), 264(41), 268(38a, 38, 38b, 39, 40, 41), 270(90, 91), 271(104, 108, 109, 110, 111, 112), 274(171, 172), 277(233, 242, 243); **25**:251, 253, 254, 293, 294, 295, 297, 298, 305, 310, 316, 319; **26**:2(43), 94(43)

Chi, J. W. H., **5**:445, 514; **11**:144(156, 157, 158, 159, 160), 145(156, 158, 159), 146(157, 158, 159, 160), 147(156), 162(157, 159, 160), 164(160), 170(156, 158, 159, 160), 173, 174, 177, 182(157, 159), 184(156), 196

Chi, M., **8**:49, 89; **SUP**:64(93), 75(93)

Chi, S. W., **3**:35(15), 36(15), 50, 56, 59, 60, 97; **7**:320b, 320e; **20**:273(286), 312(286)

Chiang, C. C., **28**:218(65), 230(65)

Chiang, C. H., **26**:75(44), 76(44, 45), 86(45), 87(45), 94(44, 45)

Chiang, F. P., **30**:257(9), 307(9)

Chiang, H. D., **23**:279(5), 288(5), 354(5)

Chiang, K. C., **28**:253(58), 304(58, 144), 331(58), 335(144)

Chiang, K. L., **28**:342(21), 416(21)

Chiang, T., **11**:203, 206, 254; **17**:321(13), 339

Chianta, M. A., **4**:129(110), 130(111), 135(110), 141; **22**:231(199), 240(197, 198, 199), 355(197, 198, 199)

Chichelli, M. T., **20**:14(21), 15(21), 16(21), 17(21), 80(21)

Chidiac, S. E., **28**:306(145), 335(145)

Chigier, N. A., **26**:2(46), 95(46)

Chiladakis, C., **4**:170(53), 225; **5**:462, 463, 464, 515

Childs, M. E., **6**:7(17, 18), 15(17), 123

Chilton, T. H., **7**:103, 125, 160, 161; **25**:293, 316

Chimah, B., **20**:185(23), 187(23), 188(23), 300(23)

Chin, D.-T., **26**:123(67), 124(67), 211(67); **28**:394(183), 423(183)

Chin, G. H., **3**:81, 99

Chin, J. H., **4**:153, 223; **5**:390, 391(119), 392, 393, 394, 395, 510; **7**:344, 346, 352, 370(70), 371(38), 378, 379

Ching, C. H., **28**:135(35, 66), 140(35), 141(66)

Ching, C. Y., **23**:320(176), 347(176), 362(176)

Ching, P. M., **4**:35, 63, 230, 273, 313

Chino, K., **15**:270(109), 281

Chinoda, Z., **24**:304(21), 319(21)

Chiou, C. S., **15**:63(33), 64(33), 83(33), 136(33)

Chiranjivi, C., **SUP**:217(410), 229(420), 244(430), 249(432), 256(437), 258(437), 264(440), 267(440)

Chiruvella, R. V., **28**:156(36), 171(7), 172(36), 173(7), 174(7), 175(7), 185(7), 195(5), 197(5), 198(5), 200(5), 201(6), 202(5), 204(6), 205(6), 207(5), 208(6), 209(6), 219(1, 8), 221(8), 222(8), 227(1), 228(5, 6, 7, 8), 229(36)

Chisholm, D., **31**:123(2), 134(2), 155(2)

Chitester, D. C., **19**:105(48), 108(48), 185(48)

Chitrangad, B., **18**:169(40), 170(40), 178(40), 220(40), 221(40), 236(40)

Chitsaz, B., **30**:261(19), 308(19)

Chitty, R., **17**:215, 315

Chiu, H. H., **2**:213, 269

Chiu, K. C., **28**:370(135), 421(135)

Chiu, S. C., **9**:363(26), 415

Chizhevskaya, S. N., **28**:84(36), 140(36); **30**:346(87), 430(87)

Chmiel, H., **22**:296(39), 348(39)

Chmielewski, C., **23**:193(42), 268(42); **25**:292, 308, 316

Cho, C. S. K., **23**:8(13), 12(13), 14(13), 25(13), 58(13), 60(13), 65(13), 78(13), 106(13), 126(13)

Cho, D. H., **19**:38(85), 42(85), 43(85), 87(85); **29**:136(23), 137(23), 147(49), 148(49), 183(172), 190(196, 197), 203(23), 205(49), 211(172), 212(196, 197), 309(170), 344(170); **30**:335(65), 429(65)

Cho, D. M., **21**:287, 300, 344

Cho, J., **29**:187(183), 211(183)

Cho, J. M., **22**:30(14), 153(14)

Cho, S. H., **19**:12(20), 13(20), 79(207), 84(20), 92(207); **29**:152(73), 178(158), 181(158), 206(73), 210(158)

Cho, Y. I. (Chapter Author), **15**:59, 63(16, 19, 34, 35), 64(16, 19, 34, 35), 71(19), 72(53), 73(16), 74(53, 55, 56, 57), 79(35), 80(35), 81(34), 90(78), 91(78), 94(34), 95(35, 162), 97(35), 109(19, 162), 110(140), 114(19), 118(19), 130(161), 131(78), 132(78), 135(16, 19), 136(34, 35), 137(53, 55, 56, 57), 138(78), 140(140), 141(161); **18**:169(29), 170(29), 172(29), 223(29), 225(29), 236(29); **19**:304(90), 305(90), 330(90), 332(129), 333(90, 129), 342(90), 353(90), 355(129); **24**:102(3), 184(3); **25**:253, 316; **28**:153(9), 228(9)

Choe, H., **16**:312(7), 314, 363

Choi, H., **9**:268; **SUP**:2(5), 139(5), 153(5), 190(5); **20**:244(188), 308(188)

Choi, H. K., **24**:313(14), 314(14), 319(14)

Choi, K. Y., **1**:234(60), 265; **4**:207(91), 227; **11**:273, 274(7), 314; **14**:128, 130, 132, 135, 137, 138, 144, 145; **15**:17(41), 56(41); **21**:105(85), 129(124), 137(85), 138(124)

Choi, K. J., **26**:85(47), 95(47)

Choi, K. Y., **23**:320(164), 361(164)

Choi, M., **19**:80(240), 93(240)

Choi, M. H., **13**:219(95), 266

Choi, T., **24**:199, 273

Choi, Y.-H., **24**:193, 272

Chojnowski, B., **16**:232, 233

Chollet, J. P., **25**:345, 378, 401, 409, 415

Chon, W. Y., **1**:232(54), 265; **14**:136, 146

Chong, J. P., **5**:130(2), 141, 149, 150, 155, 159(2), 246

Choplin, L., **24**:109(90), 187(90)

Chou, C. H., **19**:262(25), 263(25), 264(25), 265(25), 326(25), 330(25), 350(25); **25**:259, 319

Chou, C. T., **21**:20(34), 22(34), 25(34), 26(34), 51(34)

Chou, P. C., **15**:329

Choudary, D., **24**:199, 271

Choudhury, N. K. D., **15**:329

Chow, L. C., **26**:123(65, 66), 125(65, 66), 162(116), 163(116), 170(116), 211(65, 66), 213(116); **30**:121(83), 192(83)

Chow, L. W., **26**:148(107), 151(107), 154(107), 213(107)

Chow, R., **7**:197, 215

Chow, T. S., **15**:322

Chow, W., **31**:433(8), 447(8), 471(8)

Chow, W. L., **24**:199, 211, 272

Chowdbury, K., **26**:220(8), 324(8)

Choy, C. L., **18**:161(8), 164(26), 236(8, 26)

Chrisey, D. B., **28**:135(20a), 139(20a)

Christ, F. R., **23**:206(46), 226(46), 228(46), 269(46)

Christensen, D. O., **15**:325

Christensen, R. M., **9**:404(90), 417; **15**:78(66), 137(66)

Christenson, H., **5**:381, 510

Christenson, M. S., **28**:241(13), 272(102), 274(102), 275(102), 276(13, 102), 287(133), 329(13), 333(102), 335(133)

Christiaens, M., **17**:255, 262, 278, 316

Christiansen, E. B., **2**:367, 368, 370(15, 16), 373, 374, 383, 394, 395; **SUP**:73(174), 89(174), 93(174), 97(174), 98(174); **15**:63(37), 105(37), 136(37), 140(130), 197(86), 225(86)

Christiansen, J., **26**:304(82, 83), 328(82, 83)

Christiansen, W. H., **7**:200, 216

Christmann, L., **13**:218(33), 264

Christon, M. A., **31**:406(69), 428(69)

Christopher, P. J., **12**:3(6), 4(6), 11(6), 14(6), 42(6), 43(6), 44(6), 49, 51(6), 53(6), 68, 69(6), 73

Christopher, R. H., **23**:193(43), 198(43), 199(43), 200(43), 201(43), 268(43); **24**:110(94), 111(94), 112(94), 188(94)

Chrysler, G. M., **18**:73(147), 86(147); **20**:186(57), 187(57), 205(102, 104, 106), 207(102), 212(102, 104, 106), 301(57), 304(102, 104, 106)

Chu, B. T., **1**:276, 353; **3**:191(14), 249

Chu, C. C., **29**:151(68, 72), 205(68), 206(72)

Chu, C. M., **3**:207, 250

Chu, E. H. Y., **22**:353(141)

Chu, H. H. S., **18**:57(106), 84(106); **20**:217(125), 218(125), 219(125), 305(125); **30**:25(25), 88(25)

Chu, H. N., **8**:24, 26, 87

Chu, J. C., **16**:61(3), 153

Chu, N. C., **4**:152, 223; **9**:130, 132, 171, 172, 178

Chu, P. T., **9**:25, 35, 107

Chu, R. C., **20**:183(10, 16), 221(136), 262(260), 290(313), 299(10, 16), 305(136), 311(260), 314(313)

Chu, S. C., **6**:435(170), 437(170), 494

Chu, S. T., **2**:168, 269; **5**:167, 248

Chu, T. Y., **11**:300, 315; **30**:26(24), 27(24), 28(24, 45), 29(24, 45), 88(24), 89(45)

Chu, V. H., **16**:53, 57

Chu, W. T., **11**:241, 242, 261

Chuah, Y. K., **29**:39(32), 57(32); **30**:117(70), 192(70)

Chuang, C., **17**:70(24), 151, 153, 158

Chuang, Y. K., **19**:44(99), 87(99)

Chuanjiing, T., **23**:229(258), 278(258)

Chudnovskiy, A. F., **19**:193(11), 243(11)

Chue, R., **29**:87(54), 126(54)

Chuek, W., **19**:334(147), 337(147), 355(147)

Chui, G. K., **11**:360(91), 404(91), 436

Chui, W. K., **30**:356(104), 431(104)

Chukaev, A., **20**:227(146), 229(146), 230(146), 306(146)

Chun, M. H., **29**:259(122), 261(122), 262(122), 263(122), 264(122), 266(122), 341(122)

Chunchuzov, E. P., **14**:269(59), 274(59), 280

Chung, B. T. F., **10**:191, 203, 217; **19**:37(76), 39(76), 42(76), 81(254, 255), 86(76), 94(254, 255); **30**:339(81), 343(85), 430(81, 85)

Chung, C., **29**:259(123), 261(123), 262(123), 263(123), 264(123), 266(123), 341(123)

Chung, C. H., **28**:351(83), 418(83)

Chung, J. N., **26**:2(193), 3(193), 5(154, 193), 8(153, 154), 10(153), 15(54), 16(145), 22(152), 23(155), 24(155), 25(146, 155), 26(151, 155), 27(54, 151), 28(54), 29(54), 30(54), 31(143, 193), 32(52, 53), 34(52), 35(53), 36(52), 42(49, 50, 51), 43(40), 59(52), 82(48), 83(144), 94(40), 95(48, 49, 50, 51, 52, 53, 54), 99(143, 144, 145, 146, 151, 152, 153, 154), 100(155), 101(193); **30**:142(102), 147(102), 193(102)

Chung, K. S., **17**:51, 64

Chung, M., **30**:94(15), 134(93), 136(93), 155(122), 189(15), 193(93), 194(122)

Chung, M. K., **21**:198(26), 237(26)

Chung, P. M. (Chapter Author), **1**:330, 354; **2**:109, 134(19), 146, 152, 153, 157, 162(31), 163(31), 170(31), 172, 175, 176, 179, 182(31), 187, 189, 190(51), 192, 227, 230, 234, 235(19), 239(19), 247, 249(75), 255, 263, 267, 268, 269, 270; **6**:474(232), 480(232), 482, 483(232), 497; **8**:36(207), 42(207), 89; **9**:256(90), 270; **11**:219, 223, 237, 257; **19**:81(253), 94(253)

Chung, Y. C., **18**:171(77), 207(77), 208(77), 233(77), 237(77)

Chupp, R. E., **11**:365(110), 428(237, 238, 239, 239a), 429, 430(238, 239, 239a, 242), 431(242), 437, 440, 441; **23**:345(298), 368(298)

Chuprin, B. E., **25**:4(30, 32), 141(30, 32)

Churaev, N. V., **23**:187(43a), 188(43a), 193(43a), 268(43a)

Churchill, D. R., **5**:304(62), 309(62), 323

Churchill, R. V., **3**:125(32), 173; **15**:285, 322, 323

Churchill, S. W. (Chapter Author), **2**:16(21), 43, 106; **3**:201, 204(29), 249, 250; **4**:36, 37, 63; **6**:345, 346, 366, 474(234), 480(234), 483(234), 497; **7**:308, 319; **8**:167, 170, 172, 212, 224, 226; **9**:304(65, 66), 347, 378(47, 48), 379(48), 415; **11**:203, 206, 219, 222, 225, 235, 237, 238, 254, 257, 259; **13**:125, 164, 169, 201, 202; **SUP**:108(273), 128(306), 143(273), 148(306), 404(273, 306); **15**:185(53), 208(98), 209(98), 210(98), 224(53), 225(98); **18**:12(13, 18), 19(39), 57(106), 59(112), 60(112), 61(112),

63(112), 79(13, 18), 81(39), 84(106), 85(112), 162(71), 179(71), 237(71); **19**:20(38), 85(38); **20**:186(25), 187(25), 191(25), 195(79), 217(125), 218(125), 219(125), 300(25), 302(79), 305(125), 353, 367, 378, 380, 381, 382, 387, 388; **22**:186(41), 348(41); **23**:136(10), 161(10), 184(10); **24**:148(140), 189(140), 312(30, 43), 313(30, 31), 320(30, 31, 43); **25**:267, 316; **26**:131(77), 159(77), 212(77); **30**:25(25), 88(25)

Churgay, J. R., **5**:351(56), 507

Chuskin, P. I., **4**:334, 445

Chute, J., Jr., **10**:53(35), 74(35, 36), 82

Chyu, M. C., **20**:244(204), 249(204), 309(204); **23**:282(34), 355(34)

Chyu, W. J., **24**:218, 272

Chzhen-Yun, C., **SUP**:125(304), 145(304)

Ciborowski, J., **10**:177, 216

Ciccarelli, G., **29**:84(49), 93(49), 94(49), 95(49), 100(49), 125(49), 169(129), 176(129, 155, 156), 181(129), 183(173, 174), 185(173, 178, 179), 186(178, 179), 191(155), 208(129), 210(155, 156), 211(173, 174, 178, 179)

Cicchitti, A., **1**:410(96), 427(129), 442, 443

Ciccio, J. A., **20**:244(198), 308(198)

Cichelli, M. T., **1**:239, 265; **4**:166, 224; **16**:114(69), 125, 155, 233

Cielo, P., **28**:53(79), 73(79)

Çilesiz, I., **22**:287(42), 288(42), 289(42), 348(42)

Cimick, R. C., **9**:144(67), 147(67), 156(67), 179

Ciofalo, M. (Chapter Author), **25**:321, 323, 324, 331, 357, 367, 384, 385, 386, 387, 389, 390, 391, 392, 395, 402, 403, 409, 410

Cioulachtjian, S., **30**:137(96), 193(96)

Cipolla, J. W., **7**:179, 213

Cipriani, R., **17**:25(77), 59

Ciprios, G., **25**:193(94), 247(94)

Cirauqui, C., **29**:221(56), 276(56), 323(56), 338(56)

Çiray, C., **11**:240, 241, 242, 261, 263

Citakoglu, E., **9**:231(63), 264(63), 266, 269; **21**:82(37), 95(37), 135(37)

Cizsek, T. F., **30**:318(15), 427(15)

Claassen, L., **9**:296(35), 346

Clack, **18**:328(15), 329(15), 350(15), 362(15)

Clapier, R., **2**:324(28), 353

Clapp, J. T., **3**:8(4), 9(4), 31; **7**:119(32), 161; **11**:219(97), 222(97), 229(97), 231(97), 232(97), 236(97), 256

Clapp, R. M., **2**:384, 387, 394; **15**:105(129), 140(129)

Clark, C. A., **17**:133(141), 156

Clark, E. G., **7**:320b(143), 320e

Clark, H. B., **1**:209, 213, 264; **10**:119, 131, 132, 165; **11**:9(17), 48

Clark, J., **17**:66(1), 75, 153

Clark, J. A. (Chapter Author), **1**:195, 200(9), 226, 259, 263, 264; **4**:47, 48, 49(83, 85), 64, 161, 162, 169(39), 170(71), 171, 172(41), 173, 174, 206(88), 225, 226, 227; **5**:90, 91, 126, 127, 325, 335, 336(29), 339, 341, 347, 348, 349, 350, 351(52), 353(53), 354, 375(86), 376, 377(89), 378(41), 381, 382, 383, 384, 386, 387(95, 96), 389(41), 394, 395, 396, 398, 400, 401, 407(87), 413, 414, 422, 424(87), 425, 426(87), 428, 431, 432, 441, 459, 460, 461, 462, 463, 464, 465, 467(236), 468, 469, 506, 507, 508, 509, 511, 513, 516; **6**:474(251, 265, 266), 483(251), 484, 498, 499; **8**:300, 301(53), 350; **9**:256, 270; **10**:86, 163; **11**:60(27), 68(36), 71, 72(36), 95(36), 96, 105(27, 103), 106, 108(27), 112, 116, 118(27), 119(27, 36), 121(27), 122(27), 190, 193; **15**:242(59a), 278; **18**:257(37, 46), 260(37, 46), 299(37), 302(37), 321(37), 322(46); **23**:16(32), 127(32); **30**:42(37), 89(37)

Clark, R. A., **25**:341, 381, 410

Clark, R. J., **6**:211, 365

Clark, R. K., **10**:229(15), 281

Clark, S. H., **SUP**:64(94), 202(94), 203(94), 224(94), 225(94), **19**:307(91), 312(91), 353(91)

Clark, S. P., Jr., **10**:53(35), 54(36), 74(35, 36), 82

Clarke, A. M., **22**:254(43), 256(43), 257(43), 348(43), 350(86)

Clarke, I. H., **3**:35, 97

Clarke, J., **29**:87(54), 126(54)

Clarke, J. F., **2**:266, 270

Class, C. R., **5**:82, 126, 412(148), 512; **11**:105(80), 112, 121(79), 122(79), 192; **16**:234; **20**:24(51), 81(51)

Classen, W. A. P., **28**:402(193), 423(193)

Claudet, G., **17**:108(98), 109(98), 113(98), 114(98), 119(116), 120(117), 121(117), 122(117), 155, 156

Claus, J., **17**:218(67), 290, 294, 295, 301, 302(110), 315, 317

Clauser, F., **3**:35(25), 98

Clauser, F. H., **16**:251, 363

Clausing, A. M., **18**:33(66, 67), 35(67), 36(67), 82(66, 67)

Clausing, P., **2**:353

Clausius, R., **15**:8(26), 55(26)

Claussen, M., **21**:154, 178

Claxton, K. T., **10**:177(59), 178(59), 216; **31**:123(2), 134(2), 155(2)

Clay, W. G., **6**:36(63), 125

Clayton, J. T., **13**:126(32), 191(32), 201

Clayton, W. A., **23**:279(2), 304(2), 305(2, 111), 354(2), 359(111)

Clegg, P. E., **10**:34, 37, 44(18), 45(18), 46, 48(18), 53(18), 57, 62(18), 68(18), 81

Clegg, S. T., **22**:30(23), 153(23)

Clemans, J. E., **30**:415(181), 435(181)

Clement, B. W., **5**:481, 482, 483(265), 484(264), 485, 488, 492, 493, 494, 495, 497, 517; **11**:84(43), 93(43), 95(43), 97(43), 105(43), 112(43), 117(43), 190; **17**:74(40), 154

Clements, H. B., **10**:263, 264, 267(89), 283

Clements, L. D., **11**:105(105, 135), 113, 193, 195; **16**:106, 114(76), 134(76), 137(76), 154, 155

Clemmow, D. M., **3**:60, 99

Clerici, G. C., **17**:28(82), 59

Clery, M., **23**:294(74), 357(74)

Cleveland, R. G., **5**:451, 515

Clever, R. M., **11**:308, 315; **14**:31, 101; **17**:133, 156

Cliffe, K. A., **21**:156, 178

Clift, R., **19**:103(47), 185(47); **21**:334, 344; **25**:252, 316; **26**:2(55), 20(55), 23(55), 95(55); **28**:7(7), 70(7)

Clifton, D. G., **3**:270(30), 300

Climm, J., **30**:356(104), 431(104)

Cline, D., **5**:343(33), 506

Clipp, J. C., **18**:337(49, 51), 363(49, 51)

Clodfelter, R. G., **4**:162, 163, 168, 191, 210, 224, 227

Clomburg, L. A., **11**:420(209), 439

Clotan, V., **14**:231(167), 246; **19**:147(164), 189(164)

Clough, S. B., **2**:365(18), 394

Cloupeau, M., **14**:72(155), 73(155), 103

Clouston, M., **1**:430, 443

Clumpner, J. A., **14**:334(126), 335(126), 345

Clutter, D. W., **4**:331, 334, 444, 445

Clyne, T. W., **28**:16(25), 17(25), 71(25)

Coakley, W. T., **31**:461(45), 472(45)

Coantic, M., **8**:297, 300, 301(29), 314, 323, 349; **13**:86(46), 116

Coates, N. H., **9**:119, 177; **14**:185, 245

Coats, D. E., **6**:107(187), 132

Coats, K. H., **30**:154(114, 115), 194(114, 115)

Cobb, E. C., **5**:141, 144, 149, 150(31), 247

Cobble, M. H., **11**:318(21), 376(21), 384(21), 385(21), 387, 436, 438; **15**:286, 287, 289, 293, 299(124), 309, 310, 316, 323, 326

Cobble, M. M., **5**:177, 249

Cobbold, R. S. C., **22**:364(10), 367(10), 370(10), 373(10), 377(10), 382(10), 384(10), 386(10), 421(10), 435(10)

Cobine, J. D., **4**:248(20), 255(20), 313

Cobonpue, J., **13**:10(30), 12, 15, 17(30), 18, 19, 60

Cocci, A. A., **18**:163(24), 170(24), 177(24), 220(24), 221(24), 236(24)

Cochran, D. L., **20**:244(183), 308(183)

Cochran, J., **8**:12, 85

Cochran, J. R., **22**:187(208), 342(208), 356(208), 381(20), 384(20), 394(20), 435(20)

Cochran, T. H., **4**:182, 183, 186, 195, 196, 225

Cockayne, B., **30**:313(3), 315(3), 319(3, 20), 320(3), 322(3), 347(3), 426(3), 427(20)

Cockran, T. N., **17**:4, 8, 57

Cockrell, D. J., **13**:79(30), 116

Cockroft, J. O., **2**:353

Cocks, F. H., **22**:188(44, 102), 201(191), 348(44), 351(102), 355(191)

Coe, S. D., **30**:233(12), 250(12)

Coen, E., **11**:365(116), 375(116), 425, 426(116), 437

Coeuret, F., **26**:123(68, 69), 124(69), 125(68), 148(69), 151(69), 154(69), 155(69), 172(130, 135), 173(133, 136), 177(130, 133), 178(130, 133, 136), 181(135, 136), 211(68, 69), 214(130, 133, 135), 215(136)

Coffin, K. P., **3**:276, 301

Coger, R., **22**:177(45), 348(45)

Cogley, A. C., **8**:256, 283; **21**:246(14), 275(14)

Cognet, G., **21**:155, 178

Cohen, A. H., **25**:234(122, 123), 248(122, 123)

Cohen, A. J., **10**:19(25), 24(25), 36

Cohen, C. B., **4**:330, 331, 444; **6**:111, 131

Cohen, E. R., **10**:92, 164; **14**:242, 302(30), 306(61), 309(61), 343

Cohen, E. S., **3**:239, 244(93), 251; **5**:27, 28(32), 52; **27**:36(23)

Cohen, H., **9**:3(20), 77, 78, 84, 87, 96, 97(20), 107; **17**:50(156), 63

Cohen, H. I., **24**:314(2), 318(2)

Cohen, I. M., **3**:225, 251; **26**:60(105), 62(105), 65(105), 97(105)

Cohen, N. B., **4**:279, 314, 325, 330, 334, 341, 342, 343, 344, 444; **8**:12, 85

Cohen, S. (Chapter Author), **21**:141, 142, 179

Cohen, S. L., **20**:90, 98, 130

Cohen, Y., **23**:206(45, 46, 47), 216(47), 217(47), 226(44, 45, 46, 48), 228(44, 46), 229(45), 269(44, 45, 46, 47, 48)

Colak-Antic, P., **9**:322(84, 85), 348

Colberg, R. D., **25**:153(12), 195(12), 201(12), 232(12), 236(12), 243(12)

Colburn, A. P., **1**:433, 443; **2**:42, 107; **4**:54, 64; **7**:103, 160; **8**:140(70), 160; **SUP**:48(57), 53(57), 107(57), 108(57); **15**:189(63), 195(78), 224(63), 225(78), 232, 277; **16**:137, 147, 156; **18**:139(71), 159(71)

Colclough, C. D., **9**:3(21, 22), 66(21), 93(21, 22), 95, 96(21, 22), 107

Cole, E. H., **7**:339, 378

Cole, G. H., **23**:293(67), 295(67), 357(67)

Cole, G. S., **28**:231(1), 246(1, 23), 250(1), 316(1, 23), 317(1), 329(1), 330(23)

Cole, J., **11**:216, 225, 258; **20**:360, 387

Cole, J. A., **21**:143, 156, 159, 179

Cole, J. D., **4**:341, 446; **8**:7, 13, 84, 85

Cole, R. (Chapter Author), **10**:86, 155, 156, 165; **16**:75, 76, 88(39), 98, 100, 101, 104, 154, 188(19), 216(19), 238; **23**:16(33), 127(33); **30**:262(20), 308(20)

Cole, R. K., Jr., **29**:234(94), 340(94)

Cole, R. M., **23**:343(287), 344(287), 367(287)

Colella, P., **28**:129(21), 139(21)

Coleman, H. W., **16**:302, 306, 363; **23**:284(38), 314(139), 355(38), 360(139)

Coleman, T. A., **18**:50(98), 84(98)

Coles, D., **3**:52, 98; **5**:233, 235, 241(163), 251; **21**:147, 151, 152, 153, 155, 179

Coles, D. E., **2**:7, 44, 97, 98, 105; **4**:341, 342, 345, 446; **8**:12, 85; **16**:355, 356, 363

Colgate, S. A., **29**:130(9), 184(9), 203(9)

Collares Pereira, M., **18**:25(55), 81(55)

Collier, J. C., **17**:2(13), 26(13), 30(88), 40(13), 56, 60; **20**:250(221), 310(221)

Collier, J. G., **1**:245(87), 266, 390(58), 392(58), 410(58), 421(58, 115), 424(121), 425, 426, 428(58, 121), 441, 443; **9**:3(23), 107; **23**:7(11), 126(11); **31**:123(2), 134(2), 155(2)

Colling, K. M., **14**:329(108), 330(108), 345

Collings, E. W., **28**:24(45), 72(45)

Collins, D. G., **5**:37, 53

Collins, F. C., **14**:314, 316, 317(80), 344

Collins, G. J., **28**:343(44, 45, 46), 417(44, 45, 46)

Collins, I. K., **6**:94(147), 129

Collins, M., **2**:390, 391, 394; **SUP**:69(140), 161(140); **19**:268(42, 44), 269(42), 270(42, 44), 275(44), 277(42), 279(44), 351(42, 44)

Collins, M. W., **SUP**:148(331), 149(331); **25**:323, 331, 333, 340, 357, 361, 367, 382, 383, 385, 389, 391, 392, 394, 396, 397, 398, 399, 402, 410, 412, 413, 414, 418, 419

Collins, R., **20**:105, 130

Collins, R. E., **24**:104(49), 186(49)

Collins, R. S., **12**:18(30), 20(30), 21(30), 26(30), 74

Collis, D. C., **11**:202, 205, 208, 209, 216, 219, 223, 235, 237, 238, 248, 253, 257, 262

Collis, D. S., **8**:127(58), 128, 130(58), 160; **18**:123(57), 124(57), 127(57), 159(57)

Colner, D., **10**:275(105), 284

Colombo, G. V., **4**:228

Colton, C. K., **18**:208(97), 238(97)

Coltrin, M. E., **28**:366(116, 118), 382(116), 393(116, 181), 394(116), 420(116, 118), 423(181)

Coluzzi, B., **23**:292(54), 356(54)

Colver, C. P., **5**:414, 417, 418, 422, 438, 439, 513; **11**:105(69, 70, 105, 108, 135), 111(69, 70), 113, 116(69), 192, 193, 195; **16**:106, 114(76), 134, 137(76), 154, 155, 193, 220(21), 222(21), 223(21), 235, 236, 238

Colwell, R. E., **13**:219(43), 265

Combarnous, M., **14**:12(32), 100

Combarnous, M. A., **14**:3, 16(44), 21, 34, 42, 43, 47, 71, 72(9, 84, 155), 73(54, 84, 155), 74(9), 76, 100, 101, 102, 103; **24**:103(19), 105(19), 185(19)

Comings, E. W., **3**:8, 9, 31; **7**:119, 161; **11**:219, 222, 229(97), 231, 232, 236, 256; **13**:121, 175, 201

Comini, G., **19**:78(186), 91(186); **22**:203(20), 347(20)

Comiti, J., **23**:229(101a), 230(101a), 231(101a), 271(101a)

Commerford, G. E., **16**:232; **20**:14(22), 15(22, 22), 16(22), 17(22), 80(22)

Compton, R. N., **28**:113(90), 118(90), 119(90), 120(90), 142(90)

Comstock, D. F., **11**:318(14), 434

Comte-Bellot, G., **8**:296, 300, 301(50), 320, 342, 349, 350; **25**:325, 379, 381, 410

Concus, P., **15**:290, 323; **24**:201, 270

Condiff, D. W., **29**:146(46), 185(46), 204(46), 309(170), 344(170)

Condon, E. U., **11**:318(1), 321(1), 332(1), 434

Conel, J. E., **10**:11, 21, 22, 24(32, 33, 34), 36, 71(63), 73(63), 82

Conetti, G. S., **5**:168(73), 248

Coney, J. E. R., **SUP**:35(46), 73(46, 183), 77(46), 288(183), 299(183), 321(46); **21**:169, 182

Congelliere, J. T., **4**:146, 222, 228

Conlin, M. T., **21**:279, 282, 299, 300, 306, 307, 313, 314, 345

Conner, S., **24**:227, 271

Conner, W., **22**:421(34), 436(34)

Connolly, J. I., **17**:69(27), 72(27), 153

Consigny, H., **23**:320(163), 361(163)

Conti, R. J., **8**:14, 85

Conti, R. S., **29**:84(46), 125(46), 307(167), 344(167)

Continillo, G., **26**:89(56), 95(56)

Conway, E. C., **7**:223, 247, 273(7), 314

Conway, J. J., **28**:19(31), 71(31)

Cook, D. P., **28**:318(185), 337(185)

Cook, D. S., **20**:205(102), 207(102), 212(102), 304(102)

Cook, E. B., **29**:71(29), 125(29)

Cook, I., **29**:278(151), 343(151)

Cook, W. H., **1**:394(65), 398(65), 399, 420(112), 442, 443

Cook, W. J., **23**:318(151, 153), 361(151, 153)

Cooke, D. D., **23**:146(25), 185(25)

Cooper, A. L., **SUP**:70(156), 71(156), 240(156), 242(156)

Cooper, G. T., **10**:200, 201, 205, 209, 217; **14**:187, 217, 218(117), 225(117), 226(117), 227, 229, 245; **19**:134(141), 135(141), 180(141), 188(141)

Cooper, H. B., **11**:105(57), 115(57), 191

Cooper, I. S., **22**:17(13)

Cooper, J. L. B., **15**:323

Cooper, K. W., **10**:118(50), 165

Cooper, L. Y., **8**:19, 86

Cooper, M. G. (Chapter Author), **9**:359(16), 414; **10**:156, 166; **11**:17(57, 58), 19(57), 49; **16**:93, 94, 117(49), 154, 157, 159(1), 160, 161, 162(1), 163, 164, 165, 166, 167, 202, 211, 222, 237; **17**:4, 57; **20**:259(246), 311(246)

Cooper, T. E., **22**:10(36), 18(36); **23**:329(232), 365(232)

Cope, O., **22**:226(46), 348(46)

Cope, W. F., **3**:60, 99; **6**:94(145), 129; **31**:191(92), 326(92)

Copeland, D., **26**:172(132), 177(132), 178(132), 214(132)

Copeland, R. J., **18**:34(69a), 35(69a), 82(69a); **23**:12(40), 14(40), 21(40), 26(40), 27(40), 33(40), 36(40), 39(40), 58(40), 60(40), 67(40), 128(40)

Copes, C. E., **10**:171(17), 215

Copley, S. M., **28**:243(16), 277(16), 280(16), 329(16)

Coppage, J. E., **20**:133(3), 136(3), 155(3), 179(3)

Coppari, L. A., **15**:287, 328

Coppock, P. D., **15**:250, 280

Coppola, L., **23**:230(49), 231(49), 269(49); **25**:291, 316

Copus, E. R., **29**:268(134), 342(134)

Corapcioglu, M. Y., **24**:104(60, 63, 65), 186(60, 63), 187(65)

Corbett, R. J., **5**:356(69), 508

Corcoran, E., **25**:330, 410

Corcoran, W., **6**:515(27), 562

Corcoran, W. H., **7**:52(28), 84

Cordes, M. R., **28**:317(178), 337(178)

Cordoba, A., **23**:292(57), 356(57)

Core, T. C., **5**:445, 446, 447, 448, 449, 514

Corey, A. T., **24**:104(57), 186(57); **29**:40(35, 36), 57(35, 36)

Corey, T. C., **11**:170(185), 173, 175(185), 177(185), 197

Coriell, S. R., **19**:79(205, 220), 80(205), 92(205), 93(220); **28**:317(177, 178), 337(177, 178)

Corino, E. R., **25**:16(49), 21(49), 22(49), 23(49), 33(49), 142(49)

Corlett, R. C., **5**:21, 51; **10**:260, 282; **17**:194, 212, 214(62), 315; **30**:76(26, 27, 63), 77(27), 88(26), 89(27), 90(63)

Corlis, X. F., **28**:124(78), 142(78)

Cormack, K. H., **22**:224(84), 350(84)

Corman, J. C., **12**:103, 113; **15**:105(136), 107(136), 111(136), 114(136), 140(136)

Cornelissen, M. C. M., **21**:11(25), 51(25); **28**:306(148), 335(148)

Corneliussen, R. D., **18**:228(126), 238(126)

Cornell, D., **14**:4(13), 5, 100

Cornet, I., **15**:128(154), 141(154)

Cornish, R. J., **5**:239(180), 241, 251

Cornwell, K., **16**:205(28), 217, 239; **29**:39(29), 56(29)

Coronell, D. G., **28**:352(91), 402(218), 407(223), 408(223), 419(91), 425(218, 223)

Corpas, E. L., **4**:150(7), 222, 227, 228

Corpel, Ph. V., **23**:435(88), 464(88)

Corradini, M., **15**:270(107), 281; **21**:287, 300, 344

Corradini, M. L., **29**:136(21), 137(21), 139(21), 140(21), 147(21, 47), 149(52, 53), 151(68), 161(99), 164(105), 165(53, 117), 167(121), 168(21), 175(152, 154), 180(159, 160), 181(154, 170), 183(175), 188(170), 194(203), 203(21), 205(47, 52, 53, 68), 207(99, 105), 208(117, 121), 210(152, 154, 159, 160), 211(170, 175), 212(203), 233(69), 244(96), 312(96), 320(96), 338(69), 340(96)

Correa, S. M., **24**:199, 200, 201, 206, 210, 213, 274

Corrsin, S., **3**:43, 98; **11**:263; **21**:185(3), 200(29), 204(3, 32), 207(3), 208(32), 222(32), 223(32), 236(3), 237(29, 32), 267(37), 276(37); **25**:381, 410; **30**:4(28), 7(29), 74(28), 89(28, 29), 290(81), 310(81)

Corruccini, R. J., **9**:357, 358(12), 414

Corty, C., **1**:202(11), 203(11), 264; **10**:119, 165; **16**:64(6), 153; **20**:15(28), 80(28)

Cosart, W. P., **13**:5(22), 11(22), 59

Cosgrove, J. H., **7**:238(37, 38), 241(37, 38), 242(37, 38), 250, 252, 304, 304(37), 315

Cost, R. B., **5**:82, 126, 412(148), 512; **11**:105(79, 80), 112(79, 80), 121(79), 122(79), 192; **16**:234; **20**:24(51), 81(51)

Costa Ribeiro, J., **14**:119, 144

Costello, C. P., **1**:189(4), 190(4), 239, 240, 263, 265; **4**:206(85), 226, 227; **5**:515(230); **7**:260, 317; **11**:105(114), 114, 120(114), 121(114), 122(114), 194

Coston, R. M., **9**:384(55), 411(107), 413(108), 416, 417

Cotran, R. X., **22**:229(47), 348(47)

Cotta, R. M., **19**:281(75), 282(75), 284(75), 286(75), 294(75), 352(75)

Cotter, T. P., **7**:220, 227, 232, 233, 278, 284, 306, 313, 314; **20**:272(280), 283(302), 312(280), 313(302)

Cotton, K. C., **12**:5(13), 73

Cotton, M. A., **25**:81(133), 146(133)

Cotton, W. R., **25**:377, 419

Couderc, J. P., **10**:185(77), 205, 216; **14**:152(35), 243; **23**:235(50), 269(50)

Couette, M. M., **21**:143, 146, 161, 179

Coulaloglou, C. A., **25**:181(75), 182(75), 189(83, 85), 234(83, 85), 129), 246(75, 83, 85), 248(129)

Coulman, G. A., **30**:83(1), 87(1)

Coulson, J. M., **14**:130, 144; **15**:101(115), 102(115), 103(115), 104(115), 139(115)

Coulter, D. M., **11**:105(140), 195

Countryman, C. M., **10**:224, 265, 270, 271, 281, 283

Couper, J. R., **18**:179(58), 180(58), 237(58)

Couper, N. J., **28**:22(39), 72(39)

Courant, R., **8**:34(174), 54(238), 67(174), 88, 90; **29**:140(35), 144(35), 146(35), 204(35)

Courtney, W. G., **14**:306, 314(73), 342, 343

Courville, G. E., **23**:292(58), 356(58)

Coury, G. E., **11**:56(15), 59, 61(15), 95(15), 101, 104, 105(15), 113, 189

Cousteau, J. Y., **4**:109(58), 139

Couvillion, R. J., **19**:78(189), 91(189)

Covington, M. A., **19**:81(256), 94(256)

Covino, B. G., **4**:120(86), 140

Coward, H. F., **29**:64(13), 66(13), 69(13), 105(13), 124(13)

Cowin, M., **31**:246(153), 261(153), 329(153)

Cowley, C. W., **5**:423, 512; **11**:105(81), 112, 192

Cowling, T. G., **1**:287, 353; **3**:287(86), 302; **4**:273(35), 314; **28**:406(221), 425(221)

Cox, G., **17**:215, 315

Cox, H. W., **13**:219(90), 266

Cox, J. E., **10**:107(35), 164

Cox, R., **20**:217(128), 220(128), 305(128)

Cox, R. A., **18**:333(47), 336(47), 337(48), 363(47, 48)

Cox, R. G., **23**:370(23), 461(23); **26**:29(235), 103(235)

Cox, R. L., **11**:424, 440

Coxe, E. F., **5**:381, 390, 509, 510

Coyne, J. C., **20**:187(65), 196(65), 302(65)

Crabs, C. C., **4**:150(6), 211(96a), 222, 227

Crabtree, S. J., **23**:415(72), 463(72)

Craft, C., **19**:82(278), 95(278)

Craig, D. W., **23**:327(221), 364(221)

Craig, H. L., **15**:195(73), 225(73)

Craig, P. P., **17**:84(55), 104, 105(91), 106, 154, 155

Craig, S. E., Jr., **2**:367, 368, 370(15), 373, 394; **15**:63(37), 136(37), 197(86), 225(86)

Craik, A. D. D., **26**:202(167), 216(167)

Crain, R. W., **7**:249, 317

Craine, R. E., **19**:80(241), 93(241)

Cramer, E. R., **9**:132, 179

Cramer, K. R., **1**:298, 353; **4**:97(38), 100(38), 139

Crandall, J. H., **4**:53, 64

Crandall, S. H., **9**:324(93), 348

Crane, L. J., **9**:299(42), 347

Cranfield, R. R., **19**:99(16), 100(16), 102(16), 185(16)

Crank, J., **1**:102, 121; **10**:243(33), 281; **12**:278; **17**:190(45), 315; **19**:2(7), 24(47), 39(7), 41(92, 93), 83(7), 85(47), 87(92, 93); **22**:184(48), 349(48); **23**:379(56), 463(56)

Cravalho, E. G., **5**:479, 517; **9**:363(25, 27), 364(25), 390(27), 415; **19**:79(209), 92(209); **22**:12(14), 17(3, 14), 158(24), 199(180), 200(181), 202(182), 204(182), 205(182), 206(182), 207(182), 208(182), 209(182), 220(10), 246(24), 300(143), 304(183), 305(183), 306(183), 307(180, 183), 309(183), 311(183), 312(183), 321(57, 150), 322(151), 323(151), 324(151), 325(151), 326(151), 342(24), 347(10), 348(24), 349(57), 353(143, 150, 151), 355(180, 181, 182, 183)

Cravarolo, L., **1**:367(17), 379(33, 34, 35, 36), 385(34), 389(34, 36), 390(34, 36), 393(36), 394(73), 399(73), 401(34), 440, 441, 442

Crave, B., **2**:324(28), 353

Crawford, B. L., **5**:295(49), 323

Crawford, D. R., **6**:116(181c, 181d), 117(181c, 181d), 131

Crawford, H. R., **2**:385(87), 396; **15**:105(134), 107(134), 111(134), 114(134), 140(134)

Crawford, J. E., **9**:265(134), 272

Crawford, L. W., **11**:263

Crawford, M. E., **16**:312, 363; **20**:238(167), 307(167); **22**:188(62), 193(62), 194(62), 212(118), 349(62), 352(118)

Crawford, T., **20**:83, 132

Creager, M. O., **7**:189, 190(66), 200, 201(66), 214

Crease, J., **18**:333(42), 334(42), 335(42), 363(42)

Creekmore, H. S., **6**:94(146), 129

Creff, R., **23**:338(264), 366(264)

Cremers, C. J. (Chapter Author), **4**:248(13), 303(65), 305(65), 313, 315; **10**:2, 4, 5, 6, 12, 13, 16(22), 17, 19, 20, 21, 23(6, 26), 31(46), 34(4, 6), 35, 36, 39, 48, 52, 53(37), 54(37, 38, 39, 40, 41, 42), 57, 58(38, 41), 69(32, 37, 38, 39, 40, 41), 70(37, 38, 39, 40, 41), 71, 75(39, 71), 76(41, 70), 77, 78, 79(41), 80(38, 39, 70), 81, 82, 83; **19**:81(247, 248), 94(247, 248)

Cresci, R. J., **6**:41(79, 82), 65(79, 82), 126; **7**:331, 333, 334, 341, 353, 377

Cresswell, D. J., **9**:38, 70(88), 109

Criminale, W. O., **13**:213(25), 264

Crine, M., **30**:168(132), 195(132)

Crocco, L., **2**:41, 106; **3**:63(64b), 99; **4**:331, 444; **6**:5, 36, 85(6), 108, 109, 112, 122

Crochet, M. J., **25**:292, 296, 315; **28**:152(10), 228(10); **30**:338(74), 348(140), 350(74), 371(140), 382(140), 403(172), 404(172), 405(74), 430(74), 433(140), 434(172)

Crocker, G. H., **3**:21(37), 32

Cronenberg, A. W., **29**:136(19), 137(19), 139(19), 140(19), 175(147, 148), 203(19), 209(147), 210(148)

Croockewit, P., **21**:179

Crosbie, A. L., **11**:350, 351(71), 352, 376(134), 381(134), 392(153), 395(153), 398, 404(153), 436, 438

Cross, E. J., **6**:114, 131

Cross, M., **19**:26(54), 27(54), 85(54)

Cross, M. C., **17**:131(135), 149(135), 156

Crosser, O. K., **9**:259(106), 271; **18**:169(31, 36), 170(31, 36), 173(31, 36), 181(36), 184(31), 211(31), 212(31), 236(31, 36)

Crow, F. C., **31**:410(74), 428(74)

Crow, S. C., **8**:291(12), 344(12), 349; **21**:195(22), 237(22)

Crowe, C. T., **26**:88(57), 95(57); **28**:11(16), 71(16)

Crowley, A. B., **30**:350(93), 430(93)

Crowley, C. J., **15**:228(7), 270(106), 271(106), 276, 281

Crowther, D. J., **23**:326(215), 364(215)

Crowther, M. E., **19**:108(55), 186(55)

Crozier, R. D., **2**:374, 395

Cruddace, R. G., **9**:299(43), 347
Cryder, D. S., **16**:225(30), 227, 235, 239; **20**:14(19), 15(19), 80(19)
Csermerly, T. J., **6**:474(306), 501
Culkin, F., **18**:333(47), 336(47), 363(47)
Cummings, J. C., **29**:61(3), 62(9), 63(3), 67(3), 69(3), 72(3), 123(3), 124(9)
Cummings, K. N., **26**:130(76), 149(76), 151(76), 161(76), 212(76)
Cummings, R. D., **17**:146, 147, 157
Cummings, R. L., **4**:210, 214, 227
Cumo, M., **11**:18, 49; **17**:25, 59; **26**:45(37, 58), 94(37), 95(58)
Cunningham, T. M., **5**:477, 516
Cunnington, G. R. (Chapter Author), **9**:350, 351(7), 364(33), 367(38), 375(43), 386(58), 387(33), 388(59), 390(7, 63), 391(59), 394(58), 399(33), 401(58), 402(58, 85), 409(7), 414, 415, 416, 417; **20**:264(264), 311(264)
Cuomo, J. J., **28**:116(52), 141(52)
Cuomo, V., **23**:292(54), 356(54)
Curcio, L. A., Jr., **30**:232(13, 14, 54), 233(13), 234(13, 14, 54), 235(13), 250(13, 14), 252(54)
Curd, H. N., **8**:29, 87
Curievici, I., **9**:128, 129, 178
Curle, N., **6**:89, 108(143), 129; **8**:36(197), 89
Curr, R. M., **13**:67(10), 115; **SUP**:73(177), 209 (177), 211(177, 402), 212(177), 398(177); **19**:268(56, 57), 269(57), 271(56, 57), 272 (57), 273(57), 274(57), 276(57), 351(56, 57)
Curran, P. F., **12**:261(42a), 269(42a), 279
Curran, R. L., **11**:420(210, 211), 439, 440
Currie, I. G., **20**:186(26), 187(26), 300(26)
Currin, H. B., **1**:408(88), 411(88), 442
Curtis, B. J., **28**:351(87), 376(148, 149), 389(166), 419(87), 421(148, 149), 422(166)
Curtis, C. F., **1**:7(11), 49
Curtiss, C. F., **2**:38(35), 47(35), 106, 114(8), 115(10), 268; **3**:249, 254, 261, 262(3), 263, 267(3), 268(3), 269(3), 282(3, 63), 290(3), 291(3), 299(3), 300, 301; **7**:10, 83; **12**:197(8), 198(8), 199(8), 268(33), 269(9), 277, 278; **17**:127, 156; **28**:356(97), 364(97), 419(97)
Cushman, J. H., **23**:378(53), 463(53); **24**:114(115), 188(115)
Cussler, E. L., **12**:208(61), 213(61), 237(61), 242(61), 251(61), 261(61), 265(61),

266(61), 268(61), 269(61), 271(61), 280
Cybolski, R. J., **7**:193, 194(79), 200(79), 214
Cybulskis, P., **29**:329(182), 344(182)
Cygnarowicz, T. A., **7**:320b, 320e
Cyphers, J. A., **31**:242(128), 249(128), 328(128)
Cyr, M. A., **23**:280(15), 298(86, 87), 306(86, 87), 354(15), 357(86), 358(87)
Czekanski, J., **9**:34, 71, 106
Czerny, M., **11**:326(49), 362(49, 101), 435, 437; **12**:124, 192; **22**:420(28), 436(28)
Czikk, A. M., **20**:76(67), 82(67)
Czochralski, J., **30**:313(1), 426(1)

D

Dabbous, M. K., **23**:226(51), 269(51)
Dacev, A., **15**:323
Dacosta, M., **8**:53(236), 90
da Cunha, F. M., **21**:300, 345
Dagan, Z., **20**:251(234), 255(234), 256(234), 257(234), 310(234); **22**:19(24), 122(24), 131(24), 132(24), 151(24), 153(24); **23**:13(37, 46), 15(37, 46), 18(37), 19(37), 20(37), 22(37, 46), 26(37), 27(37, 46), 28(37), 34(37, 46), 38(46), 39(37, 46), 49(37, 46), 59(37, 46), 61(37, 46), 64(46), 65(37, 46), 77(37, 46), 79(46), 87(46), 90(46), 92(46), 100(37, 46), 101(37, 46), 128(37, 46); **26**:109(21), 113(21, 28), 114(33), 149(108), 151(108), 155(108), 172(108), 178(108), 209(21, 28, 33), 213(108)
Dagbjartsson, S., **7**:320c(146, 152), 320d, 320f
Daguenet, M., **12**:251(15a), 278
Dahlburg, J. P., **25**:362, 419
Dahlburg, R. B., **25**:362, 419
Dahlen, R. J., **11**:220, 264
Dahlhoff, B., **10**:210, 218
Dahlhoff, H., **14**:160(75), 161(75), 188(75), 244
Dahlkild, A. A., **24**:302(1), 318(1)
Dahm, W. K., **6**:40(75), 41(75), 65(75), 66(75), 75(75), 77(75), 125
Dai, Y. B., **16**:5(1), 9(1), 11(1), 31(1), 56
Daikoku, T., **16**:232, 233; **20**:75(66), 82(66), 232(152), 242(152), 306(152)
Daily, I. W., **6**:406, 407, 409(61), 410, 489
Daily, J. W., **5**:175, 177(82), 178, 181, 248; **10**:87(2), 163
Dake, J. M. K., **11**:318(31), 376(31), 435

Dakhno, V. N., **13**:11(41), 60
Dakin, J. T., **24**:265, 267, 269, 274
Dalbert, A. M., **20**:186(28), 187(27, 28), 300(27, 28)
D'Albon, G., **14**:231, 246; **19**:147(164), 189(164)
Dale, J. D., **15**:163(23), 164(23), 165(23), 223(23, 24)
Dalgleish, N., **6**:474(294), 500
Dalin, M. A., **14**:234(176), 235(176), 247
Dallaire, S., **28**:53(79), 73(79)
Dallavale, J. M., **18**:169(37), 170(37), 172(37), 236(37)
Dalla Valle, J. M., **2**:364, 396
Dalton, C., **SUP**:72(168), 199(168), 210(168), 213(168); **19**:268(51), 271(51), 272(51), 274(51), 278(51), 351(51)
Daly, J., **28**:376(153), 385(153), 422(153)
Dalzell, W. H., **23**:154(39), 168(39), 185(39)
Damkohler, G., **14**:318(86), 321, 322, 323(86), 324, 344
Danenberg, V. V., **6**:447(179), 448(180), 474(285, 317), 494, 500, 501
Danet, R., **22**:222(108), 351(108)
Daney, D. E., **21**:46(74), 47(74), 52(74)
Dang, K., **25**:345, 384, 410, 418
D'Angelo, N., **4**:251(17), 313
Danh, H. Q., **25**:383, 408
Daniel, A. J. P., **30**:266(30), 308(30)
Daniel, E. N., **25**:81(146, 147, 148), 147(146, 147, 148)
Danielejko, V. M., **15**:323
Daniels, D. J., **5**:91, 127; **11**:105(95), 113, 193; **30**:39(38), 89(38)
Daniels, L. C., **23**:320(187), 362(187)
Danielson, R. D., **20**:183(8), 299(8); **23**:22(43), 128(43)
Danilov, I. B., **5**:406(170), 513
Danilov, I. Yu., **31**:245(132), 328(132)
Danilov, Y. I., **6**:449(181), 453(181), 456(181), 457(181), 495; **25**:35(70a, 70b), 116(185), 143(70a, 70b), 149(185); **31**:160(51, 58), 182(58), 298(58), 324(51, 58)
Danilova, G. I., **31**:160(35), 323(35)
Danilova, G. N., **16**:215, 239; **20**:23(41), 81(41)
Danko, J. P., **19**:105(48), 108(48), 185(48)
Dankworth, D. C., **30**:112(65), 191(65)
Dannenberg, R. E., **7**:368, 379
Dannenburg, K., **7**:273(93), 317
D'Annibale, F., **26**:45(37), 94(37)

Danziger, W. J., **9**:121, 125, 178
Daoudi, S., **23**:225(52), 269(52)
Darcy, H., **14**:4, 100
Darcy, H. P. G., **24**:110(91), 114(91), 187(91)
Da Riva, I., **2**:247, 270
Darkoku, T., **20**:244(205), 309(205)
Darling, C. W. W., **SUP**:278(451), 357(451), 360(451)
Darling, R., **9**:3(22), 66(22), 93(22), 95(22), 96(22), 107
Darling, R. C., **22**:30(31, 32), 134(31, 32), 153(31, 32)
Darrieus, G., **15**:2(4), 8(28), 54(4), 55(28)
Darwish, S., **29**:234(71), 321(71), 338(71)
Daryanani, R. H., **13**:219(75), 234(75), 249, 265
Das, C. N., **10**:200, 218
Das, P., **28**:106(13), 139(13)
Da St. Jeronimo, M. A., **19**:186(62)
Datars, W. R., **11**:105(82), 112, 192
Date, A. W., **SUP**:64(99, 100, 101), 379(99, 100, 101), 380(99), 381(99, 100, 101), 382(99, 100, 101, 519), 383(100), 384(100, 520)
Dators, W. R., **5**:406(166), 512
Dauben, D. L., **23**:193(53), 204(53), 205(53), 206(53), 222(53), 226(53), 269(53)
Daughaday, H., **1**:79(41), 82, 121; **8**:24, 31(111), 32(111), 86
Daujotas, P. M., **18**:91(23), 92(23), 119(23), 120(23), 125(23), 126(23), 127(23), 158(23)
Daunt, J. G., **17**:82, 154
Davalath, J., **20**:227(144), 229(144), 231(144), 306(144)
Dave, R. M., **17**:300(108), 305(108), 317
Davenport, R., **18**:3(4), 6(4), 79(4)
Davenport, W. G., **21**:287, 300, 345
Davey, A., **9**:324, 348; **21**:149, 153, 165, 179
Davey, G., **16**:92, 93, 97, 98, 99, 102, 154, 233
David, M. M., **1**:425, 427, 443
Davidson, J. F., **1**:360, 363(11), 440; **10**:169(1, 5), 170(13), 171, 174, 176, 214, 215; **14**:154(42), 243; **19**:115(96), 116(96), 138(146), 140(146), 160(196), 187(96), 188(146), 189(196); **20**:100, 101, 105, 130, 131; **23**:188(54), 233(54), 269(54); **25**:152(4), 168(4), 181(4), 242(4)
Davidson, M. R., **15**:323
Davidson, N., **5**:259(14), 265(15), 267(15), 322
Davidson, P. A., **28**:306(149), 335(149)
Davidzon, M. I., **13**:2(16), 59

Davies, A. R., **28**:152(10), 228(10)

Davies, C. A., **29**:188(186), 212(186)

Davies, C. B., **24**:215, 270

Davies, C. N., **9**:126, 178

Davies, D. R., **5**:149, 152, 159, 247, 248; **8**:36, 88

Davies, E. A., **16**:70, 153

Davies, G. A., **9**:227(58, 59), 229(58, 59), 231, 269

Davies, J. T., **9**:184, 185, 198(4), 199(4), 202(4), 268

Davies, M. R. D., **23**:321(200), 363(200)

Davies, O. H., **23**:222(119), 226(119), 272(119)

Davies, P. O., **18**:123(58), 159(58)

Davies, P. O. A. L., **8**:128(59, 60), 160, 294(23), 349; **11**:208(121), 219, 223, 235, 238, 240, 241, 242, 261, 268

Davies, R. M., **20**:100, 101, 130

Davies, R. T., **7**:197, 215

Davies, T. H., **9**:2, 3, 4, 73, 74, 98, 107

Davies, T. W., **8**:295(26), 349

Davis, A. H., **4**:50, 54, 55, 59, 64; **8**:94(15), 128, 130(16), 159; **11**:200(2, 3), 202, 204, 209, 212, 213, 214, 218, 221, 252, 256; **18**:88(16), 124(16), 127(16), 158(16)

Davis, A. S., **3**:64(65), 71(65), 99

Davis, B. M., **6**:98(159), 130

Davis, D. H., **5**:40, 53

Davis, D. L., **16**:214(24), 216(24), 233, 238; **24**:218, 270

Davis, E. J., **1**:425, 427, 443; **10**:151, 165; **11**:15, 16, 49; **SUP**:12(22), 135(312), 188(22)

Davis, E. S., **SUP**:291(461)

Davis, G. F., **25**:234(137), 235(137), 249(137)

Davis, J. R., **26**:137(87), 212(87)

Davis, L. P., **20**:186(29), 187(29), 300(29)

Davis, L. R., **16**:10, 11, 15, 16, 57

Davis, M., **19**:80(242), 94(242)

Davis, M. R., **11**:235, 238, 240, 261

Davis, R. S., **16**:364

Davis, S. H., **14**:25(59), 101; **19**:79(218), 80(224), 93(218, 224); **28**:41(71), 73(71)

Davis, S. H., Jr., **9**:265(134), 272

Davis, T. H., **5**:242, 244, 251

Davison, B., **3**:206, 249; **23**:136(2), 184(2)

Davison, W. F., **5**:453, 515

Davy, A., **5**:233, 251

Dawson, D. B., **14**:333, 334(118), 335, 345

Dawson, J. R., **10**:19, 20, 21, 23(26, 29, 40), 24, 28, 29(40), 31, 36, 53(32), 54(32), 69(32), 81

Dawson, P. R., **28**:252(40), 330(40)

Day, R., **4**:93(31), 94(31), 96(31), 138

Day, R. S., **5**:66, 67, 125; **11**:105(58), 109(58), 110(58), 191; **20**:14(18), 15(18), 80(18)

Day, W. J., **14**:140, 141, 145, 146

Deak, G., **5**:130(3), 177(3), 179(3), 180, 246

Dean, C. W., **10**:158, 161, 166

Dean, D. S., **20**:290(314), 314(314)

Dean, O. C., **14**:217, 218(151), 222(151), 246; **19**:134(142), 188(142)

Deans, H. A., **9**:259(106), 271

Deans, S. R., **30**:285(70), 295(70), 310(70)

Deardorff, J. W., **25**:334, 337, 342, 343, 346, 347, 360, 361, 362, 370, 371, 372, 373, 374, 375, 376, 378, 379, 401, 402, 404, 405, 411, 417

Dearing, D. L., **9**:405(96), 406(96), 407(96), 417

Deaver, F. K., **11**:203, 206, 248, 254

Deavours, C. A., **SUP**:174(351, 352)

DeBartoli, R. A., **1**:245, 266

de Boer, I. H., **1**:137, 184

de Boer, J. H., **2**:353

Debolt, H., **4**:283(49), 315

De Bortoli, R. A., **1**:405(85), 409(85), 411(85), 442

Debrule, P. M., **12**:102, 103, 104, 108, 113

Debrule, R. M., **15**:104(126), 121(126), 140(126)

Debrun, J. L., **28**:110(34), 140(34)

Debrus, S., **30**:256(4, 6), 307(4, 6)

Debs, R. J., **14**:140, 146

Debye, P., **11**:354, 436

Dec, J. E., **30**:28(10), 77(10), 78(16), 83(30), 84(16), 88(10, 16), 89(30, 31)

Deck, B. D., **30**:241(15), 250(15)

Decker, D., **19**:112(78), 186(78)

Decker, N. A., **19**:100(27), 136(27), 138(27), 139(27), 173(224), 174(224), 175(224), 176(224), 177(224), 181(224), 185(27), 190(224)

Deckwer, W.-D., **23**:253(214, 266), 254(195, 214, 231, 266), 255(214, 266), 256(214, 266), 258(195, 214, 231), 259(214), 261(266), 275(195), 276(214), 277(231), 278(266)

de Coursin, D. V., **3**:55, 98

Deeds, R. S., **15**:272(113), 281

Deev, V. I., **16**:233; **17**:327(28, 29, 30), 328, 329(29), 330(28, 29, 30), 331(30), 332(30), 335(28), 336(28, 29, 30), 340; **21**:37(61),

Deev, V. I. (*continued*)
   39(61), 40(61), 41(61), 52(61)
De Farias, F. N., **6**:435(172), 467, 494
Defay, R., **13**:130, 164, 202; **15**:229(14),
   230(14), 276; **31**:50(8), 102(8)
Defenbaugh, D. M., **26**:203(177), 217(177)
deForge Deelman, A. S., **9**:3(23), 107
Degaleesan, T. E., **15**:250(85), 280
de G. Allen, D. N., **1**:119, 122
Degiovanni, A., **15**:323; **31**:2(1, 10, 13), 5(1, 13),
   89(13), 94(1), 102(1, 10, 13)
de Goede, J., **15**:287(82), 290(82), 324
DeGraaf, J. G. A., **18**:13(19a), 14(19a),
   80(19a)
de Graff, J. G. A., **8**:176, 227
de Groh, H. C., III, **28**:268(96, 97, 98), 333(96,
   97, 98)
de Groot, J., **28**:402(198), 424(198)
de Groot, S. R., **1**:182(26), 184; **8**:26(134),
   29(134), 87; **12**:206(16), 264(16), 271(16),
   278; **15**:11(31), 55(31); **23**:397(60),
   463(60); **31**:86(9), 102(9)
de Groot, W. F., **17**:185(39), 314
DeHaan, H. J., **14**:118(26), 123(26), 145
DeHaan, J. R., **5**:82, 126, 412(148), 512;
   11:105(79, 80), 112(79, 80), 121(79),
   122(79), 192; **16**:234; **20**:24(51), 81(51)
de Haart, L. G. J., **28**:343(52), 417(52)
de Haas, N., **3**:274, 276(41, 53, 57), 282(41, 53),
   299(99), 300, 301, 302
DeHoffmann, F., **5**:310(84), 324
Dehority, G. L., **11**:318(7), 376(7), 434
Deiber, J. A., **23**:219(55), 269(55)
Deinken, H. P., **10**:177(43), 216
de Iribarne, A. P., **8**:298(35), 300, 301(54), 349,
   350
Deissler, R., **21**:212(34), 213(34), 229(34, 36),
   237(34, 36)
Deissler, R. G., **2**:42, 107, 386, 395; **3**:35, 97,
   211(47), 212, 239, 241(47), 250; **5**:28(38),
   32(38), 52; **6**:404, 488, 504(10, 11), 506,
   512, 515, 524, 528, 534(24), 537, 544(24),
   561, 562, 563; **7**:52, 52(31), 84, 85, 103,
   160, 170(19), 185, 211; **11**:332(57), 364,
   436; **12**:92, 97, 101, 112; **15**:104(120),
   108(120), 121(120), 139(120);
   19:324(113), 354(113); **21**:172, 179,
   185(6), 236(6)
De Kee, D., **26**:2(43), 94(43)
Dekker, K., **23**:418(76), 436(76), 464(76)

DeKoning, W. L., **28**:402(215, 217), 424(215),
   425(217)
Delage, S. L., **28**:340(8), 415(8)
de La Harpe, A., **1**:404(84), 409(91), 411(84),
   427(84), 434, 442, 443
De Land, E. C., **5**:132(18), 134(18), 135, 140, 247
De La Rue, R. M., **18**:187(70), 237(70)
Delattre, M., **4**:146, 222
Del Casal, E., **9**:296(34), 304(59), 343(34), 346,
   347
DeLepeleire, G., **17**:171, 241(77), 255, 262, 278,
   285, 314, 316, 317
Delery, J., **6**:30(50), 31(51), 51(96), 65(96),
   66(96), 113(51), 124, 127
Delgado, M., **18**:25(55), 81(55)
Delgado Domingos, J., **6**:434(164), 494
Delhaye, J. M., **11**:44, 49
Dell, F. R., **17**:37(123), 50(123), 62
Delleur, J., **11**:219, 223, 241, 242, 258, 261
Dell'oca, C. J., **28**:341(14), 415(14)
Delmas, H., **15**:244(66), 247(79a), 279
Del Notario, P. P., **10**:261(73), 272(73), 283
Delsing, J., **17**:241, 242, 316
DeLuca, L., **23**:328(224), 364(224)
Demetri, E. P., **9**:199(14), 200(14), 201(14),
   203(14), 268; **26**:220(19), 221(19),
   236(19), 243(19), 249(19), 250(19),
   253(19), 269(19), 281(19), 300(19),
   302(19), 325(19)
Demetriades, A., **6**:36(64, 65), 125
Demirdzic, I., **24**:199, 270
De Moraes, F. F., **20**:105, 130
Demos, G., **31**:410(74), 428(74)
Demuren, A. D., **24**:247, 268
Denega, A. I., **10**:177(51), 216
de Never, N., **15**:250, 280
Dengler, C. E., **1**:382(43), 423(123), 425, 426,
   441, 443
Denham, M. K., **29**:155(84, 85), 156(84, 85),
   165(84, 85), 206(84, 85)
Denington, R. J., **4**:214, 227
Denison, M. R., **2**:46, 107; **3**:35, 98; **6**:20(31,
   32), 22, 24(32), 36(32), 38(32), 39(32),
   40(32), 101(163), 104, 107(163), 123, 130
Denisov, S. V., **25**:76(124), 83(124), 145(124)
Denisov, Y. H., **29**:91(60), 126(60)
Denisov, Y. P., **11**:60, 189; **18**:257(34, 35),
   259(34), 261(34, 35), 264(34), 265(34),
   295(34), 298(34), 299(34), 300(34),
   301(34), 304(34), 305(34), 321(34, 35)

Denk, E. G., **11**:412(185), 439

Denloye, A. L. O., **14**:158(63, 64), 160, 244

Denloye, A. O. O., **19**:100(24, 25), 102(35), 115(24), 116(24), 117(24), 127(132), 129(24), 131(24), 137(25), 185(24, 25, 35), 188(132)

Denn, M. M., **15**:77(63), 137(63), 179(38), 207(95), 213(113), 224(38), 225(95, 113); **25**:262, 316

Denner, H. D., **17**:69(23), 70(23), 71(23), 153

Denney, D. A., **2**:361, 395

Denning, R. S., **29**:244(96), 312(96), 320(96), 329(182), 340(96), 344(182)

Dennis, S. C. R., **8**:70, 90; **11**:226, 259; **SUP**:186(369), 213(405)

Dennis, S. R. C., **20**:377, 387

Denny, V. E., **9**:258(98), 265(133), 270, 272; **12**:136(24), 186(24), 193; **15**:232, 277; **26**:55(97), 97(97)

Denny, W. E., **SUP**:73(193), 92(193), 94(193), 165(193)

den Otter, J. L., **13**:229(99), 266

den Ouden, C., **9**:46, 47, 48(104), 107, 110; **23**:329(230), 365(230)

Denson, C. D., **25**:292, 296, 315; **28**:190(11), 228(11)

Dent, D. C., **19**:123(118, 119), 125(118, 119), 187(118, 119)

Dent, J. C., **9**:262(125), 271; **11**:235, 238, 260

Denton, E. B., **2**:386(57), 395; **7**:88(2), 91(2), 98(2), 108(2), 160; **12**:233(62), 251(62), 270(62), 280; **15**:128(151), 141(151)

Denton, W. H., **5**:327(17), 331, 333(17), 343(17), 505; **7**:40(51), 42, 66(51), 67(51), 86

Depew, C. A. (Chapter Author), **9**:113, 122, 127(18, 20), 129, 130, 132, 161, 169, 172(19), 178, 179; **12**:7(17), 73

Derby, J. J., **30**:338(73, 75), 339(73, 75), 350(90, 91), 356(105), 394(105), 395(105), 430(73, 75, 90, 91), 431(105)

de Reuck, K. M., **17**:102(78), 155

Derevianko, N. F., **8**:298(36), 349

Dergarabedian, P., **1**:218, 264; **11**:13(41), 48

de Ris, J., **10**:260, 267(87), 282, 283; **30**:71(53), 76(32), 77(32), 89(32), 90(53)

Derjaguin, B. V., **14**:305, 306, 342

Derrick, G. H., **18**:185(132), 186(132), 191(132), 238(132)

Desai, A., **19**:108(54), 186(54)

Desai, H. S., **11**:105(127), 115(127), 194

Desai, M., **14**:158(61), 243; **19**:110(68), 111(68), 115(68), 116(68), 117(68), 118(68), 186(68); **29**:131(13), 203(13)

Desai, P. V., **19**:160(195), 161(201), 189(195), 190(201)

De Santos, J. M., **30**:94(12), 189(12)

Desaussure, V. A., **18**:337(49, 51), 363(49, 51)

Desbiolles, J. L., **28**:263(77), 327(77), 332(77)

de Schor, B. B., **23**:314(138), 360(138)

Deshmukh, K. G., **14**:136, 146

Deshmukh, R., **28**:267(90), 333(90)

De Silva, C. N., **1**:23, 50; **3**:230, 235, 251

Desmon, L. G., **2**:42(42), 107; **6**:538, 563

Desmond, L. G., **5**:368, 508

Desnoyers, J. E., **18**:337(53), 363(53)

DeSoete, G. G., **29**:68(16), 124(16)

DeSoto, S., **5**:313(86), 324; **6**:474(324), 502

DeStefano, S., **23**:292(54), 356(54)

Detlof, A. A., **1**:141(14), 184

Detra, R. W., **4**:334(44), 445

Deutsch, T. F., **28**:342(37), 416(37)

De Vahl Davis, G., **8**:169, 173, 193, 194(29), 226; **SUP**:77(213), 185(213); **21**:161, 167, 181, 182; **24**:192, 195, 270

DeValk, J. P. J. M. M., **25**:375, 416

Deverall, J. E., **7**:230, 231, 235, 238, 247, 247(23), 267, 268, 270, 274, 275(23), 306, 314, 316, 317, 318, 320a, 320e

Deverall, L. I., **6**:416(93), 490

Devienne, F. M. (Chapter Author), **2**:271, 284(26), 317(20), 319(22), 322, 324(28), 334, 337, 353; **7**:168, 211

Devoino, A. N., **6**:417(97), 490; **25**:73(120), 145(120)

Devonshire, A. F., **2**:346, 353, 354

DeVos, J. C., **2**:426, 451

DeVries, A. L., **22**:196(114), 220(179), 352(114), 355(179)

DeVries, D. A., **13**:125, 133, 201

de Vries, K. J., **28**:343(52, 54), 417(52, 54)

Devyatov, B. N., **6**:474(290), 500

DeWaard, H., **28**:402(217), 425(217)

de Waele, A., **24**:108(83), 187(83)

Dewar, M. J. S., **21**:102(79), 136(79)

Dewey, C. F., Jr. (Chapter Author), **4**:317, 318, 326(4), 330, 345, 354(4), 443, 444; **6**:22, 36(67), 37(67), 41(67, 83), 68(83, 115), 69(115), 70(115), 73(83), 76(83), 78(83), 79(83), 82(83), 123, 125, 126, 127; **7**:200, 201, 216; **8**:12, 85

Dewey, W. A., **15**:322

Dewey, W. C., **22**:361(8), 435(8)

De Wiest, R. J. M., **24**:104(52, 54), 186(52, 54)

DeWitt, D. P., **11**:424(226), 440; **19**:144(156), 188(156); **23**:343(290), 367(290)

DeWitt, J. R., **SUP**:216(409)

DeWitt, K. J., **26**:8(157), 20(157), 22(157), 27(156), 100(156, 157)

De Wolf, M., **28**:124(102), 143(102)

Dey, K. N., **25**:257, 289, 290, 291, 316

Dey, N. C., **23**:193(9), 267(9)

DeYoung, S. H., **15**:198(87), 199(87), 200(87), 202(87), 203(87), 204(87), 220(87), 225(87)

Dhanak, A. M., **6**:61(106), 127

Dharmadhikari, R. V., **23**:201(56), 204(56), 205(56), 206(56), 226(56), 269(56); **24**:110(96), 111(96), 112(96), 188(96)

Dharmadurai, G., **17**:68(12), 153

Dhawan, S., **3**:53, 98

Dhir, V. K. (Chapter Author), **11**:105(52), 109, 116, 191; **16**:130, 155; **17**:12(53), 14, 58; **19**:80(230, 231, 232), 93(230, 231, 232); **20**:250(230), 310(230); **21**:327, 344; **29**:1, 21(17), 23(17, 18), 25(17), 41(37), 43(38), 45(38), 46(41), 47(43), 48(41, 43), 50(37, 47), 56(17, 18), 57(37, 38, 41, 43, 47); **30**:93(6), 94(6), 112(63), 134(92), 135(92), 151(112), 152(6), 175(152), 176(152), 189(6), 191(63), 193(92), 194(112), 196(152)

Diab, S., **30**:296(98), 311(98)

Diaconis, N. S., **2**:156, 269; **3**:54, 98

Diaguila, A. J., **9**:3(36), 9, 38, 71(36), 91(36), 93, 94, 95, 101, 107

Diamond, A. E., **18**:337(50), 363(50)

Diao, Q. Z., **28**:304(140, 141, 143), 335(140, 141, 143)

Diaper, M. P., **22**:304(164), 354(164)

Diaz, F., **30**:396(162), 399(162), 434(162)

Diaz, L. A., **19**:43(98), 87(98)

DiBartolomeo, D., **23**:292(55), 356(55)

DiCicco, D. A., **11**:105(64), 111, 191

Dick, J. J., **15**:328

Dickey, J. B., Jr., **12**:13, 18(23), 74

Dickins, B. G., **2**:325, 327(97), 353, 355

Dickinson, D. R., **12**:41, 75

Dickinson, N. L., **6**:551, 564; **7**:49, 84

Dickinson, R. E., **14**:253(10), 256(10), 279; **19**:82(275), 95(275)

Dickman, T., **11**:241, 243, 262

Dickson, P. F., **9**:399(77), 416

Diekerman, P. J., **1**:315, 316(52), 354

Dieleman, J., **28**:103(101), 143(101)

Dielsi, G. J., **23**:327(220), 364(220)

Dienemann, W., **25**:283, 316

Diep, G. B., **11**:240(210), 262

Dieperink, G. W., **9**:46, 47, 48(104), 107, 110

Diery, W., **26**:227(31), 228(31), 325(31)

Diesselhorst, T., **17**:8, 9, 58

Dietrick, J. R., **10**:109(36), 164

Dietsche, C., **19**:79(218, 219), 93(218, 219)

Dietsche, W., **17**:72(35, 36), 153, 154

Dietz, A. G. H., **11**:365(119), 374(119), 437

Dietz, P. W., **14**:127(27), 145

di Federico, I., **2**:380, 395

DiFelice, R., **25**:199(100), 247(100)

Dilawari, A. H., **28**:376(153, 154), 385(153, 154), 422(153, 154)

Dillard, D. S., **9**:360(19), 414

Diller, K. R. (Chapter Author), **22**:157, 162(53, 54), 163(53), 165(53), 166(54), 169(54), 171(54), 173(54), 175(54), 176(54), 178(55), 179(55), 180(55), 181(55), 182(55), 186(51, 52), 187(91, 208), 188(62, 93), 189(93), 190(95), 191(90, 93), 192(95), 193(62), 194(62), 196(96), 198(96), 199(96), 212(118), 215(96), 228(82), 229(1), 231(175), 239(61, 176, 205), 240(49), 242(61), 243(60), 287(42), 288(42), 289(42), 296(94, 95), 297(94), 299(92, 94), 301(92), 302(92), 303(92), 304(92), 320(90), 321(57), 326(56), 328(17), 329(17, 58), 330(17), 333(17), 334(17), 336(17), 337(17), 338(17), 339(17), 340(17), 341(17), 342(17, 208), 343(17), 347(1, 17), 348(42), 349(49, 50, 51, 52, 53, 54, 55, 56, 57, 58, 59, 60, 61, 62), 350(82, 90), 351(91, 92, 93, 94, 95, 96), 352(118), 354(168, 175, 176), 356(205, 208, 217), 381(20), 384(20), 394(20), 435(20)

Diller, T. E. (Chapter Author), **23**:279, 280(18), 282(31), 283(31), 296(78, 79, 80, 81), 297(79, 80), 298(81), 301(108), 302(108), 326(18), 333(249, 251), 334(251), 335(251), 337(18), 340(275, 276), 341(79, 276), 342(276), 343(79, 108), 347(302), 348(302), 349(302, 303, 304), 352(81), 354(18), 355(31), 357(78, 79, 80, 81),

358(108), 365(249), 366(251), 367(275, 276), 368(302, 303, 304)

Dillon, R. C., **4**:228

Dimaczek, G., **25**:390, 411

Dimant, Y. (Chapter Author), **12**:77, 77(4), 78(4), 79, 80(4), 82(4), 91(25), 93(25), 96(25), 99, 111, 112; **15**:121(144), 140(144); **24**:102(2), 184(2)

Di Marco, P., **26**:137(84), 148(84), 151(84), 152(84), 154(84), 167(84), 212(84)

Dimenna, A. R., **31**:106(3), 138(3), 155(3)

Dimic, M., **15**:245, 279

Dimitriads, K. P., **24**:199, 271

Dimopoulos, H. G., **7**:117, 120, 142(30), 161; **12**:233(18), 251(18), 270(18), 278

Dingee, D. A., **1**:384(48), 394(48, 72), 398(48, 72), 441, 442

Dingman, S. E., **29**:108(81), 127(81)

Dingus, C. A., **23**:327(219), 364(219)

Dingwall, A. G. F., **15**:290, 323

Dinh, N., **29**:249(101), 340(101)

Dinh, T. N., **29**:249(102), 340(102)

Dinulescu, H., **9**:128(35, 36), 129, 178

Di Piazza, P., **29**:191(200), 212(200)

DiPippo, R., **15**:2(11), 55(11)

Dipprey, D. F., **6**:525(37), 526, 562; **31**:205(110), 209(110), 211(110), 327(110)

DiPrima, R. C., **5**:231, 232(155), 233, 236, 237, 250, 251; **21**:149, 150, 153, 154, 157, 160, 179, 180, 183

Disimile, P. J., **26**:200(161), 216(161)

Dismukes, J. P., **28**:376(149), 421(149)

Dittman, F. W., **28**:376(143), 421(143)

Dittmer, D. S., **4**:141

Dittus, F. W., **15**:105(132), 140(132)

Divova, G. V., **31**:160(42), 324(42)

Dix, G. E., **17**:30(92), 60

Dixon, A. G., **23**:217(57), 269(57)

Dixon, J. C., **7**:223(11), 224, 236, 250, 275(11), 313, 315

Dixon-Lewis, G., **29**:77(37), 125(37)

Djukic, Dj., **19**:199(47), 244(47); **24**:41(10), 98(10)

Djuric, M., **31**:343(28), 426(28)

Djursing, D., **17**:29(86), 60

Dlott, D. D., **28**:77(22), 139(22)

Dmitriev, B. A., **9**:3(27), 107

Dobashin, P. A., **20**:264(267), 312(267)

Dobbe, C., **29**:218(17), 335(17)

Dobbins, R. A., **14**:325(101, 102), 344

Doberstein, S. C., **8**:140(72), 141(72), 142(72), 148(72), 149(72), 160; **18**:139(73), 147(73), 159(73); **25**:292, 296, 302, 303, 304, 305, 306, 307, 309, 311, 315

Dobry, R., **7**:116, 118, 160

Dobryakov, T. S., **31**:160(10), 322(10)

Dock, E. H., **2**:354

Dodge, C., **25**:153(6), 242(6)

Dodge, D. W., **2**:358(25), 382(26), 383(26), 384, 385(26), 387, 391, 395; **15**:88(75), 137(75), 190(68), 220(68), 224(68); **19**:325(119), 354(119)

Dodson, A. G., **19**:268(32), 317(32), 350(32)

Doering, W., **14**:287, 298, 304, 341

Doetsch, G., **2**:244(68), 270; **8**:3(13), 84

Doherty, J. A., **19**:113(87), 121(107), 155(190), 156(190), 157(87, 190), 158(87), 186(87), 187(107), 189(190)

Doi, H., **21**:166, 174, 180

Doig, I. D., **9**:137, 145, 156, 179

Dokuchaev, N. F., **1**:134, 184

Dolejs, V., **23**:215(168), 218(167), 274(167, 168)

Dolgova, I. M., **15**:326

Dolidovich, A. F. (Chapter Author), **19**:148(170), 150(166), 151(166), 153(170), 154(166, 170), 178(232), 189(166, 170), 190(232); **25**:151

Dolležhal, N. A., **1**:435(140, 141), 444

Dollfus, A., **10**:19(28), 36

Dolton, T. A., **1**:60, 63, 64, 67, 121

Domin, G., **7**:31(15), 32, 33(15), 84

Dominguez, J. G., **23**:193(58), 269(58)

Dominguez, J. S., **23**:226(262), 227(262), 278(262)

Domoto, G. A., **9**:364(31), 385, 386(31, 35), 390(57, 65), 391(68), 393(65, 68), 407(57), 415, 416; **19**:30(60), 31(62), 32(62), 86(60, 62)

Donaghey, L. F., **28**:376(145), 421(145)

Donaldson, A. B., **15**:327

Donaldson, C., **3**:60, 99

Donaldson, C. D., **26**:183(144), 215(144)

Donaldson, I. G., **9**:3(28), 63, 64, 107; **14**:36(87, 88, 89), 52(123), 53(123), 54(123), 78, 102, 103, 104; **15**:323

Donaldson, I. S., **6**:51(95), 52(95), 126

Donaldson, J. O., **4**:153(24), 223; **5**:390, 510

Donaldson, R., **21**:55(1), 133(1)

Donatelli, A. A., **25**:261, 267, 268, 269, 272, 317

Donato, M., **29**:101(73), 102(73), 126(73)

Donelly, R. J., **5**:228, 236, 237, 238, 250, 251

Dong, M., **31**:1(11a), 40(11a), 41(11a), 102(11a)

Dongara, J. J., **26**:304(84), 328(84)

Donizeau, A., **26**:173(133), 177(133), 178(133), 214(133)

Donnadieu, G. A., **10**:173, 175, 177(44), 215, 216

Donnelly, J. A., **17**:102(83), 155

Donnelly, R. J.(Chapter Author), **8**:26(135), 87; **17**:65, 75(49, 50), 82, 85, 86, 87(50), 88, 90(50), 91(50, 64), 92(50), 93(50), 94(50), 100(64), 101(64), 102(50, 82), 110, 126(82), 127(82), 129(131), 154, 155, 156; **21**:149, 151, 153, 154, 155, 158, 161, 179, 182, 183

Donnelly, V. M., **28**:343(41, 42), 417(41, 42)

Donoughe, P. L., **1**:33(28), 35(28), 36, 37(28), 50; **4**:331(25), 444

Donovan, J., **16**:101, 154

Dooley, D. A., **2**:46, 107, 213, 269

Doorly, D. J., **23**:321(190, 192, 193), 346(190), 362(190), 363(192, 193)

Doorly, J. E., **23**:321(197, 198, 199, 200), 363(197, 198, 199, 200)

Doornink, D. G., **11**:408(179), 409, 417(179), 439; **17**:204(57), 219(57), 315

Doraiswamy, L. K., **24**:103(16), 185(16)

Dore, F. I., **3**:35, 97

Dorf, C. J., **5**:327(18), 505

Dorfman, A., **8**:74(275), 75, 91

Dorfman, A. Sh., **25**:31(60), 142(60)

Dorfman, L. A., **5**:130, 131, 132, 133, 149, 150, 152, 153, 154, 161, 162, 163, 164, 166, 167, 246, 247; **21**:143, 179

Döring, W., **10**:92, 163; **11**:7, 48

Dorner, S., **7**:320c, 320f

Dorodnitsyn, A. A., **1**:109, 122; **3**:35, 97; **6**:116, 131

Dorokhov, A. R., **21**:107(93), 137(93)

Doroshchuk, V. E., **1**:244(85), 266, 416(108), 417(108), 443; **17**:319(4, 5, 6, 8, 9, 10, 11, 12), 320(4, 5, 6, 8, 9, 10, 11, 12), 327(4, 5, 6, 8, 9, 10, 11, 12), 333, 339

Dörr, J., **11**:402(167), 438

Dorrance, W. H., **2**:111(3), 268; **3**:35, 97

Dorsey, N. E., **18**:327(3), 362(3)

Dosch, F. P., **11**:203, 205, 209, 253

Dosdogru, G. A., **13**:5, 6(20), 10(20), 41(20), 59

Doshi, M. R., **SUP**:107(272)

Dotson, J. M., **25**:161(40), 165(40), 244(40)

Doty, R. T., **24**:278(20), 286(20), 302(20), 319(20)

Dougall, R. S., **5**:102, 127; **11**:144(165), 146(165), 151, 152, 155, 170(165), 172, 173, 174(165), 177, 185(165), 196; **20**:10(13), 80(13)

Dougherti, J. E., **10**:177(43), 216

Doughman, E., **5**:178, 181, 191, 192, 193, 194, 199, 249

Doughty, C., **30**:120(77), 121(77), 192(77)

Doughty, D. L., **2**:45(48), 107; **7**:58, 85

Douglas, E. C., **15**:290, 323

Douglas, J., **23**:371(29), 374(29), 462(29)

Douglas, W. J. M., **11**:225, 259; **23**:7(10), 126(10), 320(170, 172), 337(262), 361(170), 362(172), 366(262); **26**:132(80), 212(80)

Doul, M. J., **12**:229(9a), 277

Dow, J. E., **11**:2(2), 47

Dow, W. M., **10**:181, 192, 199, 203, 204, 205, 206, 216; **14**:151(11), 242

Dowden, J. M., **19**:80(242), 94(242); **28**:111(30), 139(30)

Dowling, M. F., **29**:167(122, 124), 208(122, 124)

Downes, G. L., **15**:286, 322

Downey, J. A., **22**:30(31, 32), 134(31, 32), 153(31, 32)

Downey, J. R., Jr., **29**:188(186), 212(186)

Downing, S., **26**:123(59), 124(59), 172(59), 178(59), 211(59)

Downs, S. J., **23**:7(9), 126(9); **26**:106(6), 136(6), 208(6)

Downs, W. F., **14**:54(124), 103

Dowrick, G., **15**:328

Dowson, D. A., **31**:265(164), 329(164)

Doyle, E. F., **5**:102, 127

Drake, D. B., **2**:374, 397

Drake, D. G., **6**:474(208), 496; **8**:12, 85

Drake, E. M., **10**:54(36), 74(36), 82

Drake, R. M., Jr., **1**:3(8), 9(8), 49, 132, 184; **2**:23(23), 106, 365, 395, 403(2), 405(2), 407(2), 411, 412(2), 417(2), 450; **3**:197, 198(21), 212(21), 249; **4**:153, 223, 270(33), 314; **6**:211(57), 364; **7**:58, 85, 193, 202, 209, 214, 216, 218; **11**:225, 258, 266(5), 278(5), 280(3), 314, 322(39), 410(39), 413(39), 421(39), 435; **13**:222(98), 243(98), 266; **15**:100(105), 139(105); **17**:225(70), 315; **23**:285(41), 317(41), 356(41); **24**:19(51), 38(51), 41(12), 98(12)

Drake, W. B., **SUP**:109(276)

Drakhlin, E., **8**:164, 226

Dransfeld, K., **5**:487(269), 517

Dransfield, K., **17**:68(13), 70(13), 72(13), 153

Draper, J. W., **5**:423, 512; **11**:105(65), 109(65), 111(65), 191; **16**:234

Draper, R., **7**:60, 61, 76, 76(44), 77, 78, 79, 80, 81, 82(44), 85

Drawin, H. W., **4**:245, 251(18), 313

Drayer, D. E., **5**:406(156), 512

Drayson, S. R., **14**:277(19), 279

Drazin, P. G., **19**:82(276), 95(276); **25**:324, 417

Drbohlav, R., **10**:169(4), 215

Dregalin, A. F., **25**:4(18, 19), 131(203), 132(203, 206), 140(18, 19), 150(203, 206)

Dreher, K. D., **23**:222(59), 269(59)

Dreitser, G. A. (Chapter Author), **6**:367, 449(181, 182, 183, 184, 193), 453(181, 182, 183, 184, 193), 456(181, 182, 183, 184), 457(181, 182, 183, 184), 495; **23**:330(239), 365(239); **25**:1, 3(1), 4(3, 4, 5, 6, 7, 8, 9, 10, 15), 12(1, 3, 44), 18(53), 22(3), 31(1, 44), 33(3, 4, 5, 9, 10, 44, 45, 63, 64, 65, 66, 67, 68, 69, 70), 35(70a, 70b), 43(3, 4, 5), 44(66), 45(66, 71), 48(53), 50(4), 52(3, 4, 5, 68, 70, 89, 90, 91, 92, 93), 58(94), 61(5, 102, 103, 104, 105, 106), 64(3), 70(94), 74(1), 75(122), 77(126, 127, 128), 83(126, 127), 84(167, 168, 169), 92(3, 4, 5, 104, 106, 169), 96(178), 99(3, 4, 5), 106(3, 4, 5), 108(3), 111(4), 114(7, 8, 180, 181, 182, 183, 184), 116(185), 120(3), 121(186, 187, 188, 189, 190, 191, 192), 129(7, 8, 188), 130(3, 5, 198, 199, 200), 131(201, 205), 132(201), 133(1, 198, 199, 200, 205), 139(1, 3, 4, 5), 140(6, 7, 8, 9, 10, 15), 141(44, 45), 142(53, 63, 64, 65, 66, 67, 68), 143(69, 70a, 70b, 70, 71), 144(89, 90, 91, 92, 93, 94, 102, 103, 104), 145(105, 106, 122), 146(126, 127, 128), 148(167, 168, 169, 178, 180, 181, 182, 183), 149(184, 185, 186, 187, 188, 189, 190, 191, 192, 198, 199, 200), 150(201, 205); **30**:212(16), 221(16), 223(16), 250(16); **31**:159, 159(1), 160(1, 12, 29, 51, 54, 57, 58, 59, 71), 161(1), 177(1), 182(1, 58), 183(12), 187(12, 87), 200(102), 201(103, 104, 105, 106), 205(104, 105, 106), 207(106), 216(114), 217(105), 221(105, 106), 222(105, 106), 224(105, 106), 228(118), 232(105, 120), 242(127), 243(12), 247(1), 250(144), 252(144), 253(12, 144), 254(148), 262(1), 263(1, 114, 157), 265(158, 159, 160), 266(29, 57, 114, 168, 169, 170, 171, 172), 270(29, 57, 114, 158, 170, 171, 172), 271(57, 173), 279(57, 172, 181, 182, 183, 184), 281(196), 282(183, 188), 283(189), 284(157, 190, 191, 192, 193, 194), 287(57, 183, 188, 195), 292(57, 183, 188), 294(57, 194, 197, 198, 199, 200), 298(58), 301(57), 304(197), 305(57, 209, 210, 211, 212, 213, 214), 322(1, 12), 323(29), 324(51, 54, 57, 58, 59), 325(71), 326(87, 102, 103, 104, 105), 327(106, 114, 118, 120, 121, 122), 328 (127, 144), 329(148, 157, 158, 159, 160), 330(168, 169, 170, 171, 172, 173, 181, 182, 183, 184), 331(188, 189, 190, 191, 192, 193, 194, 195, 196, 197, 198, 199, 200), 332(209, 210, 211, 212, 213, 214)

Drell, I. L., **29**:83(44), 125(44)

Drenker, G., **18**:42(85), 44(85), 83(85)

Dresner, L., **17**:122(118), 123, 152, 156, 158

Drew, A. A., **28**:257(67), 332(67)

Drew, T. B., **3**:16, 32; **5**:56, 102, 104, 125; **7**:125(36), 161; **11**:55(1), 105(1), 110(1), 189; **SUP**:2(3), 101(3), 105(3), 171(3), 172(3), **15**:232, 277; **16**:137, 147, 156

Dreyfus, R. W., **28**:111(49, 50), 115(50), 116(50), 124(77), 140(49, 50), 142(77)

Dreyster, G. A., *see* Dreitser, G. A.

Dring, R. P., **6**:312(103, 104), 366, 474(258, 264), 480(258, 264), 483(258), 499; **9**:311(72), 317(81), 328(99, 100), 330(99), 347, 348

Drinkenburg, A. A. H., **10**:192, 217

Drolen, B. L., **23**:154(38), 155(43), 161(43), 166(43), 167(43), 168(38, 43), 185(38), 186(43)

Dropkin, D., **5**:166, 206, 207, 248, 249; **11**:298, 299, 300, 315; **14**:130, 144; **16**:234; **20**:24(50), 81(50); **30**:28(44), 29(44), 89(44)

Drougge, G., **6**:95(150), 129

Drufes, H., **4**:284(52), 315

Drumheller, D. S., **29**:160(97), 207(97)

Drummond, J. E., **18**:49(93), 84(93); **25**:293, 316

Dryden, H. L., **11**:232(169), 237, 260; **SUP**:2(4), 54(60), 197(4), 223(4), 234(4), 247(4); **19**:258(11), 350(11); **24**:40(6), 98(6)

Dryer, F. L., **26**:77(116, 250), 98(116), 104(250)

Dsyubenko, B. V., *see* Dzyubenko, B. V.

Dube, S. N., **25**:51(82, 83), 143(82, 83)

Dubis, A., **SUP**:75(203)

Dubke, M., **28**:318(184, 193, 194), 337(184), 338(193, 194)

Dubner, A. D., **28**:343(49), 417(49)

Dubnitsky, V. N., **1**:178, 184

DuBois, E. F., **4**:67(6, 8), 137; **22**:20(29), 21(29), 153(29)

Dubreuil, B., **28**:110(34), 140(34)

Dubroff, W., **28**:8(8, 10), 12(8), 13(8), 17(8), 18(8), 70(8, 10)

Dubrovich, E. M., **6**:424(156), 493

Dubrovina, E. N., **7**:57, 58, 60, 85

Dubrovsky, E. V., **31**:159(1), 160(1, 14), 161(1), 177(1, 14), 182(1), 236(14), 247(1), 261(154, 155, 156), 262(1), 263(1), 279(184), 322(1), 323(14), 329(154, 155, 156), 330(184)

Duclos, T. G., **20**:217(124), 220(124), 305(124)

Duda, J. L., **SUP**:73(186), 92(186), 93(186), 96(186, 255), 99(186), 300(186); **19**:19(35), 22(46), 28(46), 85(35, 46); **23**:193(61, 254), 206(60), 226(61, 103, 254), 269(60, 61), 271(103), 278(254)

Duda, P. M., **29**:163(101), 175(153), 207(101), 210(153)

Dudis, J. J., **14**:264(34, 35, 36), 265(34, 35, 36), 266(34, 35, 36), 267, 280

Dudukovič, M. P., **30**:339(79), 430(79)

Duffey, R. B., **23**:108(110), 109(110), 110(110), 132(110)

Duffield, P. L., **21**:328, 345

Duffin, J. H., **11**:420(214), 440

Duffin, R. J., **8**:33(156), 87

Dufton, A. F., **4**:67, 137

Dugue, J., **18**:25(55), 81(55)

Duignan, M. R., Jr., **21**:329, 330, 331, 333, 344

Duke, E. E., **23**:33(63), 129(63)

Duke, F. R., **12**:229(18a), 251(18a), 278

Duke, W. M., **5**:327(15), 331, 333(15), 505

Dukler, A. E., **1**:376, 385(52), 387, 388(52), 389, 390(26), 392, 441; **5**:451, 515; **9**:262(128), 271; **20**:83, 84, 85, 88, 89, 91, 92, 93, 95, 97, 99, 100, 108, 110, 111, 112, 115, 120, 121, 130, 131, 132

Dukler, A. I., **11**:56(14), 59, 61(15), 95(15), 101, 104, 105(15), 113, 189

Dukler, D. A., **1**:369, 376, 380(19), 387, 390(19), 393, 440

Dukowicz, J. K., **26**:88(59), 95(59)

Dulkin, I. N., **31**:160(20), 323(20)

Dullenkopf, K., **23**:298(93), 358(93)

Dullforce, T. A., **29**:158(88), 163(88), 206(88)

Dullien, F. A. L., **13**:187, 203; **23**:196(62), 206(62), 215(62, 165), 269(62), 274(165); **24**:104(58), 105(31), 185(31), 186(58); **25**:292, 316; **31**:1(11a), 40(11a), 41(11a), 102(11a)

Dulnev, G. N., **6**:418(122), 492; **SUP**:193(380)

Duman, J. G., **22**:196(115), 352(115)

Dumitrescu, D. T., **20**:100, 101, 105, 130

DuMoré, J. M., **SUP**:120(294), 121(294)

Duncan, D., **20**:83, 132

Dunham, W. H., **6**:73(125), 128

Dunkle, R. V., **9**:391(67), 416; **12**:124, 129(14), 138(14), 157, 171, 174, 188, 192, 193

Dunkler, W. L., **5**:233(172), 236, 251

Dunlop, **18**:328(22), 329(22), 362(22)

Dunlop, P. J., **12**:206(6), 277

Dunn, D. W., **5**:233, 251; **21**:188(19), 227(19), 237(19)

Dunn, J. S. C., **1**:376(27), 441

Dunn, M. G., **23**:280(28), 319(28), 320(28, 177, 178, 179, 180, 181, 182, 183, 184, 185, 186), 345(298), 355(28), 362(177, 178, 179, 180, 181, 182, 183, 184, 185, 186), 368(298)

Dunn, P. D., **17**:292, 295, 317

Dunn, S. T., **5**:26, 51

Dunning, W. J., **14**:282, 286(11), 307, 309, 341, 343

Dunn-Rankin, D., **26**:86(147), 99(147)

Dunskus, T., **5**:109, 128; **11**:105(109), 113, 193

Dunsky, V. D., **14**:155, 243

Dunwoody, J., **20**:377, 387

Dunwoody, N. T., **SUP**:248(434), 249(434), 251(434)

Du Pleiss, J. P., **24**:114(118), 188(118)

Dupre, A., **9**:188(6), 268

Dupre, G., **29**:100(71), 126(71)

Dupret, F., **30**:313(7), 338(7, 74), 350(7, 74), 358(7), 403(172, 173), 404(7, 172), 405(74, 173), 406(7), 426(7), 430(74), 434(172, 173)

Dupuis, C. S., **14**:329(108), 330(108), 345

Durand, F., **28**:318(183), 337(183)

Durfee, R. L., **14**:129(65, 66), 130, 133, 134, 146

Durgin, F. H., **2**:53, 107

Duric, M. D., **6**:396(33), 487

During, E. H., **3**:81, 99

Durlofsky, L., **24**:114(106), 188(106)

Durpe, G. D., **25**:234(126, 127, 128), 248(126, 127, 128)

Durst, F., **8**:298, 342(64), 349, 350; **23**:206(63, 64, 65, 93), 219(63, 64, 65), 222(63, 64, 65, 93), 223(93), 225(63, 64, 65, 93), 269(63, 64, 65), 271(93); **25**:292, 316

Dury, T., **SUP**:85(246)

Dushin, A. N., **25**:4(32), 141(32)

Dushman, S., **2**:353

Dussan, B. K., **22**:222(63), 349(63)

Dussan, B. V., **SUP**:123(301)

Dussan, E. B., V, **28**:41(71, 72), 73(71, 72)

Dussourd, J. L., **9**:143, 179

Dutt, G. S., **17**:265, 277, 317

Dutta, A. (Chapter Author), **18**:161, 209(98), 224(98, 120), 234(98, 129), 238(98, 120, 129); **25**:254, 316

Dutta, P., **30**:385(152), 386(152), 387(152), 396(152), 400(152), 401(152), 402(152), 433(152)

Duty, R. L., **11**:412(186), 439

Duval, X., **2**:353

Duvaut, G., **19**:40(91), 87(91)

Duzy, C., **28**:133(23), 139(23)

Dvin, Yu. P., **15**:323

Dwight, H. B., **5**:12(13), 51

Dwivedi, P. N., **25**:298, 319

Dwyer, H. A., **24**:213, 227, 270; **26**:2(60), 45(63), 72(61, 62), 75(161), 95(60, 61, 62, 63), 100(161)

Dwyer, O. E., **3**:106(6), 123, 124(29), 130, 171(28), 172, 173; **8**:149, 151(72), 160; **10**:159, 160, 166; **SUP**:64(96), 290(460), 294(463), 357(96), 364(96); **18**:147(81), 159(81)

Dyad'kin, Yu. D., **14**:97, 104

Dyadyakin, B. V., **6**:540(62), 542(62), 543(62), 563; **21**:21(35), 51(35)

Dyban, E. P., **8**:122(56), 123, 160; **11**:229(173), 260; **13**:43, 60; **18**:116(53), 118(53), 128(53), 159(53); **31**:160(39), 323(39)

Dybbs, A., **13**:164(51), 202; **14**:76, 103; **23**:305(114), 359(114)

Dyer, I., **12**:22, 74

Dyer, J. R., **11**:203, 206, 254; **20**:186(30, 31), 187(30, 31), 300(30, 31)

Dykshoorn, P., **20**:232(165), 233(165), 236(165), 237(165), 238(165), 256(165), 307(165)

Dzakowic, G. S., **7**:264, 265, 317; **10**:152, 153, 165; **11**:17(56), 49

Dziubinski, M., **23**:192(134), 197(134), 198(134), 206(134), 214(134), 218(133, 136), 220(133, 136), 221(133), 272(133, 134), 273(136)

Džolev, M., **26**:246(55), 327(55)

Dzung, L. S., **SUP**:120(298), 136(316), 185(364)

Dzyubenko, B. V., **25**:4(7, 8), 114(7, 8, 180, 181, 182, 183, 184), 116(185), 121(191, 192), 129(7, 8), 140(7, 8), 148(180, 181, 182, 183), 149(184, 185, 191, 192); **31**:159(1), 160(1, 51, 54, 58, 59), 161(1), 177(1), 182(1, 58), 247(1), 262(1), 263(1), 298(58), 322(1), 324(51, 54, 58, 59)

E

Eady, E. T., **5**:215, 250

Eagles, P. M., **21**:153, 161, 179

Earl, D. E., **17**:176, 177, 181, 314

Early, R. A., **3**:275(46), 276(46), 301

East, L. F., **6**:73(116, 128), 127, 128; **31**:172(79), 325(79)

Easterday, O. T., **30**:108(46), 137(46), 138(46), 139(46), 190(46)

Eastman, G. Y., **7**:235(20), 248(71), 273(20), 274(20), 275(71, 113), 314, 316, 318; **20**:273(281), 312(281)

Eastman, P. C., **5**:406(166), 512; **11**:105(82), 112, 192

Eastwood, J., **23**:217(66), 269(66)

Eatough, M. J., **29**:166(118), 208(118)

Ebaugh, F. G., Jr., **4**:121(101), 140

Ebel, J., **19**:82(265), 94(265)

Eberhart, R. C., **4**:295, 315; **22**:1(33), 14(33), 18(33), 30(25), 153(25), 158(193), 290(152), 353(152), 355(193)

Eberly, D. K., **7**:202, 204, 216

Ebersole, T. E., **19**:37(82), 87(82)

Ebert, C. H. V., **10**:246(41), 282

Ebert, E. E., **25**:376, 411

Ebert, W. A., **4**:274(37), 314

Eccleston, W., **28**:342(22), 416(22)

Echenique, P. M., **28**:118(82), 119(82), 120(82), 142(82)

Echigo, R., **17**:7(36), 57
Eck, G., **19**:80(227), 93(227)
Eckart, C., **5**:216(138), 250; **18**:333(36), 334(36), 335(36), 362(36)
Eckelmann, H., **25**:381, 385, 386, 413
Ecker, G., **4**:295(62), 315
Eckert, E. R. G. (Chapter Author), **1**:3(8), 9(8), 34(31), 47(37), 49, 50, 132, 184; **2**:17(22), 23(23), 39, 40, 41, 45, 106, 107, 207, 211, 269, 403, 405, 407(2), 411, 412(2), 417(2), 420(13), 427(13), 437(39), 439(39), 441(40), 443(40), 445(45, 46), 448(46), 449(46), 450, 451, 452; **3**:35(31), 60, 75(71), 81(83), 91(83), 98, 99, 100, 197, 198(21), 212(21), 249, 250, 275, 301; **4**:9, 16, 17, 31, 41, 42, 44, 51, 55, 59, 62, 63, 64, 223, 224, 229, 269(32), 270(33), 274, 284(32), 285, 286, 287, 288, 289, 293(58), 294(58), 300(64), 301(64), 303(64), 304(66), 315, 331, 444; **5**:38(55), 53, 177, 178, 249; **6**:40(76), 41(76), 77(76), 83(136a), 126, 128, 211, 312(95, 96, 97, 98), 364, 366, 474(227), 480(227), 483, 497, 525(34), 526, 528, 530(34), 531, 532, 562; **7**:55, 56, 57, 85, 322, 340(24), 341(27), 342(28), 343(30, 31), 346, 348, 350, 353(42), 354(44, 46), 355(46), 359(42), 360(30, 46), 361(30, 61, 62, 63), 363, 364, 365, 366, 367(62, 63), 368(24), 370, 371, 371(28, 71, 72), 372, 373, 373(28), 374, 375(71), 377, 378, 379; **8**:35(182, 183), 88, 117(46), 119(49, 50), 120(49, 50), 121(49, 50), 159, 160, 167, 168, 172, 176, 177(24), 178(24), 183(24), 184, 192, 195, 226; **9**:11, 12, 37, 38, 91(106), 97, 107, 110, 317(77), 322(83), 333, 347, 348; **11**:219, 220, 222, 224, 257, 263, 266, 278, 280, 314, 322(39), 410(39), 413(39), 421(39), 435; **12**:93, 97, 113, 129, 133(19), 192; **13**:222(98), 243(98), 266; **SUP**:18(27), 62(27), 198(386, 387), 227(419), 229(422), 264(27, 419), 266(27, 419), 411(524), 413(524); **15**:21(46), 56(46), 100(105), 105(133), 121(133), 126(133), 139(105), 140(133), 189(64), 224(64); **17**:225(70), 315; **18**:13(19b), 14(19b), 80(19b), 112(44), 113(44, 48), 114(44, 49), 115(48, 49), 156(85), 158(44), 159(48, 49, 85); **19**:334(139, 140), 336(139, 140), 338(139, 140), 355(139, 140); **23**:285(41), 288(45), 317(41), 356(41, 45); **24**:19(51), 38(51), 41(12), 98(12); **30**:169(135), 195(135), 272(37), 308(37)
Eckert, J. S., **15**:264(97), 280
Eckhardt, G., **11**:360(93b), 437
Eckrich, T. L., **20**:150(15), 179(15)
Economos, C., **1**:78, 121; **2**:265(94), 270
Econonmos, C., **29**:84(49), 93(49), 94(49), 95(49), 100(49), 125(49)
Eddington, A. S., **3**:204, 206(31), 249
Eddington, R. I., **16**:70, 71, 77, 79, 153
Eddy, K. C., **1**:394(68), 398(68), 442
Ede, A. J. (Chapter Author), **4**:1, 18, 33, 61, 63, 64; **9**:288, 346; **11**:105(88), 113, 121, 192, 213, 214, 255; **15**:144(1), 149(1), 223(1)
Eden, H. F., **5**:227(144), 250
Edeskuty, F. J., **5**:351(55), 389(55), 507
Edgar, T. F., **28**:344(61), 345(61), 399(189), 400(189), 401(189), 402(61, 189, 209, 210, 211), 403(211), 404(211), 405(189), 418(61), 423(189), 424(209, 210, 211)
Edholm, O. G., **4**:94(34), 138; **22**:81(12), 82(12), 153(12)
Edie, D. D., **23**:193(67), 269(67)
Edinger, J. E., **12**:10(19), 12, 25, 73
Edkie, R. G., **14**:136, 145, 146
Edmond, A., **2**:322(24), 334(24), 337(24), 353
Edmonds, D. K., **5**:451, 515
Edward, D. H., **29**:98(66), 126(66)
Edwards, A., **3**:13, 23, 32
Edwards, D. A., **23**:409(71), 463(71); **30**:95(22), 100(22), 189(22)
Edwards, D. K. (Chapter Author), **5**:293(35), 294(35), 301(58), 308(67), 313(36), 322, 323, 324, 469, 516; **8**:241, 243(8, 9), 244, 245, 256, 257, 266, 282; **9**:394(75), 395(75), 416; **11**:344(64, 64a), 348, 400(164), 436, 438; **12**:116, 117(2, 3), 118, 120, 124(5), 129, 131(3), 132, 136(24), 137, 138, 139, 140, 141, 142(3), 146, 147, 149, 150, 151, 152, 153, 154, 155, 156, 157, 164, 165, 176, 177, 178, 179, 180, 181, 182, 183, 184, 185, 186, 192, 193; **14**:252(5), 254(5), 279; **18**:11(8), 16(21, 22), 17(29, 30), 79(8), 80(21, 22, 29, 30); **27**:183(33, 34, 35); **31**:389(42, 43), 427(42, 43)
Edwards, D. O., **17**:105, 155
Edwards, F. J., **31**:261(152), 329(152)

Edwards, J. P., **9**:97, 107

Edwards, M. F., **15**:136(42); **23**:193(23),
    201(23), 202(23), 212(23), 213(23),
    236(23), 238(23), 240(23), 242(23),
    246(23), 267(23)

Edwards, O. J., **11**:358(86), 359(86), 436

Efimov, V. A., **20**:250(225), 310(225)

Efthimiadis, T., **29**:171(136), 209(136)

Egbert, R. B., **12**:160, 193

Egen, R. A., **1**:384, 394(48), 398(48), 441

Eggers, D. F., **5**:295(49), 323

Eggington, A., **14**:298(29), 342

Eggink, H., **6**:29(42), 124

Eggleton, A. E. J., **2**:353

Egolpopoulos, F. N., **29**:77(38, 40), 78(38),
    125(38, 40)

Egorov, V. S., **6**:418(134), 492

Eguchi, M., **30**:392(157), 393(157), 396(163,
    164), 399(163, 164), 434(157, 163, 164)

Ehert, L., **SUP**:154(335)

Ehlers, R. C., **2**:53, 56, 69, 71, 94, 108;
    4:206(86), 226; **5**:458, 459, 515; **6**:404(58),
    488

Ehrenburg, D. O., **15**:286, 323

Ehrich, F. F., **26**:120(42), 210(42)

Ehrlich, D. J., **28**:342(32, 37), 343(38, 39),
    416(32, 37, 38), 417(39)

Ehrlich, L. W., **5**:40(72), 54

Eichhorm, K., **6**:474(235), 480(235), 482, 497

Eichhorn, R., **4**:13, 15, 26, 27, 37, 62, 63;
    6:474(241), 480(241), 482(241), 498; **9**:54,
    110, 140, 147, 179, 294(17), 311(68), 346,
    347; **SUP**:2(9); **15**:185(51), 224(51);
    17:8(40), 9(46), 10, 14, 57, 58; **23**:74(88,
    89), 93(89), 94(88), 95(89), 130(88),
    131(89); **24**:146(139), 189(139)

Eid, J. C., **20**:262(255), 311(255)

Eide, S. A., **18**:330(26), 350(26), 359(26),
    362(26)

Eidel'man, E. D., **28**:102(24), 139(24)

Eidsath, A., **23**:398(62), 409(62), 463(62)

Eiermann, K., **13**:211(17), 264; **18**:161(1, 3),
    214(3), 219(3), 226(3), 236(1, 3)

Eigenson, L. S., **4**:21(31, 32), 59(32), 63;
    11:213, 214, 215

Einarsson, A., **7**:133(44), 161

Einarsson, J. G., **16**:205(28), 217(28), 239

Einarsson, S. S., **14**:3(4), 41(4), 100

Einshtein, V. G., **10**:174(40, 41), 175(40), 177,
    181(41), 198, 200, 201(154), 202(155),
    203, 204(41), 205(116), 206, 210(139),
    215, 217, 218; **14**:151(6), 152, 155, 160(71,
    72), 161(71, 72), 163(71, 72), 164(71, 72),
    166(71, 72), 167(71, 72), 168(71, 72),
    169(71, 72), 170(71, 72), 171(71, 72),
    172(71, 72), 176(71, 72, 106, 107, 163),
    180(106, 107), 181(109), 187, 188(106,
    107), 189, 190(106, 107), 191(106, 107),
    195(125), 203(133, 134), 206(71, 72, 133,
    134, 138), 220(157), 226, 228(163, 164),
    237(6, 106, 107, 109, 163), 242, 244, 245,
    246; **19**:98(5), 100(5), 102(5), 103(5),
    111(5), 113(90), 116(5), 128(5), 129(5),
    146(163), 149(177, 178), 154(5), 159(191),
    160(192), 173(5), 184(5), 187(90),
    188(163), 189(177, 178, 191, 192);
    25:153(7), 242(7)

Einstein, T. H., **1**:18, 19, 21, 49; **3**:243, 244(99),
    245(98), 246, 251

Eiseman, P., **24**:213, 227, 270

Eisenberg, A. A., **16**:114(70), 155

Eisenberg, L., **22**:20(6), 60(6), 61(6), 69(6),
    70(6), 152(6)

Eisenberg, M., **4**:16(25), 63; **7**:89(6), 96(6),
    98(6, 11), 130, 160, 161; **12**:233(105),
    251(105, 117, 118), 270(105), 281, 282;
    21:170, 171, 172, 174, 179

Eisenklam, P., **26**:71(64), 95(64)

Eissenberg, D. M., **2**:385, 395

Eisworth, E. A., **23**:343(286), 367(286)

Eklind, O., **17**:29(86), 60

Eklund, T. I., **14**:325(101, 102), 344

Ekman, V., **18**:339(55), 363(55)

Ekman, V. W., **13**:93, 116

Ekonomov, S. P., **10**:53(34), 58(34), 64(34),
    70(34), 71(34), 82

Elata, C., **12**:79, 111; **23**:219(68), 222(68),
    270(68)

Elberly, D. K., **7**:203

Elbert, D. D., **5**:205(116), 228(149), 249, 250;
    21:150, 151, 178

Elbirli, B., **15**:79(69), 137(69); **23**:218(69),
    270(69); **28**:150(12), 228(12)

Elder, J. P., **12**:233(9), 251(19), 270(19), 278

Elder, J. W., **6**:474(335), 502; **8**:168, 173(27),
    176, 184, 185(27), 186(27), 187(27),
    188(27), 189(27), 190(27), 191(27), 193,
    195, 226, 227; **11**:266, 314; **14**:3, 15(6),
    37(91), 38(91), 43, 71, 73(91), 100, 102;
    30:27(33), 89(33)

Eldridge, J. M., **20**:274(287), 312(287)

El'Ehwany, A. A. M., **7**:331, 333(6), 335, 353, 377

Elenbaas, W., **4**:17, 20, 63; **8**:161, 226; **11**:202, 205, 213, 214, 253, 278, 279, 280, 290, 291, 292, 310(22), 314, 315; **18**:72(136), 75(136), 86(136); **20**:186(32), 187(32), 300(32); **24**:270; **30**:272(36), 275(51), 308(36), 309(51)

Elgar, E. C., **5**:233(167), 238, 239(179), 251; **21**:156, 157, 180

El-Ghandoor, H., **30**:296(98), 311(98)

Elgin, D. R., **9**:413(108), 417

Elgobashi, S., **21**:185(7), 188(15, 16, 17), 199(15), 236(7, 15, 16, 17)

Eliason, M. A., **3**:270, 298(97), 300, 302

Eligenhausen, H., **11**:360(93a), 437

Elison, B., **26**:123(55), 125(55), 129(55), 148(55), 151(55), 210(55)

Elizalde, J. C., **5**:458(228), 515

El-Kaddah, N., **28**:318(182, 191, 196), 337(182), 338(191, 196)

El-Kaissy, M. M., **23**:208(70), 270(70)

Elkouh, A. F., **15**:2(12), 55(12)

Eller, E., **11**:402(167), 438

Ellerbrock, H., **9**:3(33), 107

Ellerbrock, H. H., **11**:170(183), 173, 174(183), 177(183), 185(183), 197

Ellerbrook, H. H., **5**:447, 514

Ellinwood, J. W. E., **7**:206, 207, 217, 218; **8**:11, 85

Ellion, M. E., **1**:219, 222(40), 223, 264; **5**:74, 75, 96, 104, 126

Elliot, C. M., **19**:14(28), 16(28), 24(28), 25(28), 28(28), 40(28), 84(28)

Elliot, D. F., **17**:29(84), 60

Elliot, E., **7**:206, 217

Elliot, R., **19**:79(196), 92(196)

Elliott, D. E., **10**:189(71), 216

Elliott, D. G., **2**:7(14), 9, 27, 62(14), 105

Elliott, E. R., **10**:267(89), 283

Elliott, E. S., **31**:468(58), 473(58)

Elliott, J. H., **15**:92(93), 97(93), 138(93)

Elliott, J. K., **28**:399(189), 400(189), 401(189), 402(189, 209, 210), 405(189), 423(189), 424(209, 210)

Elliott, T. C., **10**:278(111), 284

Ellis, C. D., **20**:182(4), 299(4); **23**:32(60), 129(60)

Ellis, D., **5**:133(22), 149, 155, 156, 247

Ellis, W. E., **26**:231(35), 236(35), 248(35), 249(35), 253(35), 261(35), 269(35), 281(35), 282(35), 284(35), 285(35), 286(35), 291(35), 296(35), 299(35), 300(35), 303(35), 304(35), 316(35), 320(35), 326(35)

Ellison, G. N., **15**:323; **20**:221(137), 305(137)

Ellison, P. G., **29**:20(16), 56(16)

Ellison, T. H., **13**:96, 99, 116

Ellsworth, J., **29**:101(72), 126(72)

Elmas, M., **10**:211, 218

El Masri, M. A., **23**:58(108), 60(108), 102(108), 131(108)

El-Masry, N. A., **28**:343(43), 417(43)

Elmer, T., **24**:27(57), 38(57)

Elperin, I. T., **6**:418(126), 492

Elphimov, G. I., **6**:539, 563

El-Refaee, M. M., **18**:65(120), 85(120)

El-Riedy, M. K., **30**:280(64), 310(64)

Elrod, H. G., **6**:418(100), 474(274), 491, 499

Elrod, J. C., **23**:319(154), 361(154)

Elrod, W. C., **5**:413, 441, 513

El-Saiedi, A. F. I., **9**:134, 144, 179

Elsasser, W. M., **4**:67, 138

Elsayed, M., **18**:198(91), 238(91)

El-Sayed, M. S., **23**:215(165), 274(165)

El-Sayed, Y. M., **15**:52(87, 88), 58(87)

Elsgolts, L. E., **19**:202(55), 245(55)

El-Shaarawi, M. A. I., **SUP**:35(46), 73(46, 183), 77(46), 288(183), 299(183), 321(46); **21**:179

El-Shayeb, M. M., **23**:230(232), 232(232), 277(232)

ElSherbiny, S. M., **18**:13(19), 14(19), 65(119), 80(19), 85(119)

El-Shirbini, A. A., **17**:31(101), 61

Elsner, R. W., **4**:92(30), 138

Elterman, V. M., **14**:266(40), 280

El-Wakil, M. M., **9**:258(97), 270; **18**:13(19d), 14(19d), 17(31), 25(54), 80(19d, 31), 81(54)

Elwert, G., **4**:250, 259, 313

Elzinga, E. R., **26**:10(66), 33(65), 95(65, 66)

Emara, A. A., **29**:7(6), 8(9), 9(6), 27(9), 30(9), 33(9), 55(6, 9); **30**:32(59), 33(59), 90(59)

Ember, G., **3**:286, 288, 301

Emde, F., **3**:134(39), 174; **5**:12(14), 51

Emelianov, D. S., **10**:177(53), 216

Emeliyanov, A. I., **31**:160(46), 324(46)

Emel'janov, I. Ja., **1**:435(140, 141), 444

Emergy, K. A., **28**:343(44, 45, 46), 417(44, 45, 46)

Emerman, S. H., **19**:73(168), 82(266), 90(168), 95(266); **24**:7(28), 37(28)

Emery, A., **4**:152, 223; **6**:42(90), 126

Emery, A. F., **5**:40(69), 45, 53; **8**:176, 195, 227; **SUP**:77(214), 148(214); **15**:163(24), 165(24), 223(24); **18**:18(37), 81(37); **19**:317(110), 324(110, 117), 339(110), 354(110, 117)

Emery, J. C., **6**:95(149), 129

Emmerman, P. J., **30**:288(75), 299(75), 310(75)

Emmerson, G. S., **1**:409(89), 442

Emmet, R. T., **18**:333(46), 363(46)

Emmons, H. W., **4**:331, 444; **9**:206, 269; **10**:221, 243, 245(1), 250, 255, 257, 269, 281, 282, 283; **17**:182, 188(36), 190, 194, 210, 213, 300, 314, 315; **30**:58(34), 59(34), 60(34), 89(34)

Emms, P. W., **28**:282(122), 334(122)

Emswiler, J. E., **18**:67(122), 85(122)

Enazen, V. A., **6**:418(132), 492

Encranez, T., **8**:231(3), 282

Ende, W., **1**:216, 223, 264; **4**:178, 225; **11**:14, 15, 49

Enders, J. H., **4**:150(6a), 222

Endo, H., **10**:269(94), 283

Endo, S., **20**:278(295), 313(295)

Endo, Y., **7**:181(50), 213

Endoh, K., **11**:231, 232, 263; **16**:234

Ene, H. I., **24**:104(62), 186(62)

Engel, A. H. H., **28**:286(129, 130), 287(129, 130), 335(129, 130)

Engel, R. K., **20**:186(33, 34), 187(33, 34), 190(33), 300(33, 34)

Engelhorn, H. R., **16**:171(6), 172, 173, 193(6), 215(6), 220, 221(6), 234, 238; **20**:23(43), 81(43)

Engelman, F., **11**:318(12), 434

Engl, N. L., **28**:402(196), 424(196)

Englberger, W., **19**:79(222), 93(222)

English, D., **1**:394(70), 398(70), 442

Englund, D. R., **23**:280(6), 305(6), 306(118), 354(6), 359(118)

Engquist, B., **24**:221, 227, 230, 270

Enjalbert, M., **10**:185(76, 77), 205(76, 77), 216; **14**:152(35), 243

Enloe, C. L., **28**:135(25), 139(25)

Ennos, A. E., **30**:261(17), 308(17)

Enomoto, Y., **23**:314(140), 360(140)

Enya, S., **21**:84(42), 100(42), 135(42)

Epanov, Yu. G., **14**:154(38), 243; **19**:100(29), 136(29), 137(29), 139(29), 140(29), 145(29), 148(170), 153(170), 154(170), 157(29), 159(29), 161(29), 185(29), 189(170)

Epick, E. V., **8**:122(56), 123, 160; **18**:116(53), 118(53), 128(53), 159(53)

Epick, E. Y., **11**:229(173), 260

Epifanov, V. M., **31**:160(49), 324(49)

Eppich, H. M., **23**:299(95), 358(95)

Epstein, A. H., **23**:288(76), 295(76, 77), 346(299, 300, 301), 347(300), 357(76, 77), 368(299, 300, 301)

Epstein, E., **29**:189(194), 212(194)

Epstein, E. E., **14**:277(19), 279

Epstein, H. H., **1**:394(72), 398(72), 442

Epstein, M., **5**:381, 388, 509; **15**:323; **19**:4(10, 11), 73(10, 11), 74(11, 173), 75(173), 77(10, 11), 80(228), 84(10, 11), 91(173), 93(228); **21**:287, 300, 344; **28**:28(54), 36(54), 56(54), 72(54); **29**:175(149), 189(192), 190(195), 191(200), 210(149), 212(192, 195, 200), 250(103), 340(103)

Epstein, N., **SUP**:66(112), 346(112), 349(112), 360(503); **23**:196(73), 217(66), 253(72), 255(71), 269(66), 270(71, 72, 73); **30**:199(17), 201(18), 204(17), 207(18), 209(19), 212(19), 213(19), 233(19), 250(17, 18, 19), 252(67); **31**:434(11), 441(12), 442(12), 471(11, 12)

Epstein, P., **2**:353; **10**:40, 80

Epstein, S. G., **29**:130(4), 203(4)

Erb, R. A., **21**:102(80), 103(80, 82), 136(80), 137(82)

Erbeck, R., **30**:262(21), 263(21), 289(80), 290(80), 308(21), 310(80)

Erbland, P. J., **23**:280(12), 298(12), 305(12), 354(12)

Erdelyi, A., **8**:7(30), 13, 19(96), 20, 84, 85, 86

Erdman, C. A., **29**:168(127), 169(127), 174(127), 208(127), 273(145), 343(145)

Erdmann, C., **31**:462(61), 473(61)

Erdmann, S. F., **6**:202, 364

Erdos, J., **2**:231(65), 233(65), 269; **6**:87, 89(140), 90, 91(140), 129

Eremenko, E. V., **6**:399, 401, 488

Ergun, S., **10**:170, 215; **19**:100(20), 101(20), 185(20); **23**:213(74), 235(74), 270(74); **24**:112(97), 188(97); **25**:223(108, 109), 247(108, 109)

Ericksen, J. L., **13**:213(25), 221(96), 264, 266

Erik, S., **9**:262(126), 263(126), 271

Eriksen, V. L., **7**:342(28), 350, 370, 371, 371(28), 372, 373, 373(28), 374, 378

Erikson, G. F., **20**:272(280), 312(280)

Eriskine, W., Jr., **20**:84, 85, 100, 102, 103, 130

Erk, S., **4**:152(14), 208(14), 223; **22**:161(83), 350(83)

Erlebacher, G., **25**:362, 405, 411, 418

Erlich, R., **23**:415(72), 463(72)

Erlich, S. W., **11**:278(8), 314

Ermakov, G. V., **10**:104, 164

Ermolchik, V. N., **6**:474(319), 501

Ernst, D. M., **7**:241, 248, 265, 266, 275(71), 285, 286, 287, 288, 316

Ernst, R., **14**:151, 208, 217, 242

Ernst, W. D., **5**:177(82), 248

Ero, M. I. O., **16**:53, 57

Erohin, N. F., **21**:47(80), 53(80)

Eroschenko, V. M., **21**:16(31), 21(36, 37), 22(36), 34(54), 51(31, 36, 37), 52(54); **31**:160(47), 324(47)

Errico, M., **26**:137(85), 212(85)

Ersan, G. S., **11**:412(185), 439

Ershaghi, I., **23**:193(75), 270(75)

Eryou, N. D., **11**:407(171), 424(171), 439

Eryu, O., **28**:135(26), 139(26)

Eschenroeder, A. Q., **1**:316(55), 349(55), 354

Escudier, M. P., **7**:348, 355, 379

Eshghy, S., **4**:49, 64; **11**:213, 214, 255

Esikov, V. I., **17**:328(36), 340

Esmail, M. N., **26**:8(67), 95(67)

Espefalt, R., **29**:234(90), 339(90)

Espina, P. I., **28**:19(30, 31), 71(30, 31)

Esseghir, M., **28**:150(14, 15), 151(59), 154(14, 57), 155(13), 166(59), 168(59), 169(59), 170(59), 171(7, 59), 173(7), 174(7, 13), 175(7), 184(13, 14, 15), 185(7, 14, 15), 186(15), 187(15), 188(59), 189(59), 190(57), 198(57), 205(57), 206(57), 208(22), 211(22), 228(7, 13, 14, 15, 22), 230(57, 59)

Essenhigh, R. H., **11**:390(139), 438

Essig, R. H., **11**:228(156), 259

Estes, R. C., **15**:286, 323

Etemad, G. A., **5**:166(57), 248; **11**:203, 205, 253

Ettinger, H., **17**:68(7, 11), 153

Eubank, C. C., **15**:195(79), 196(79), 225(79)

Eubank, O. C., **SUP**:412(526)

Eucken, A., **2**:298(119), 338(119), 353, 355; **9**:205, 268

Eurenius, K., **22**:229(64, 144), 349(64), 353(144)

Eustis, R. H., **3**:93(87), 94(87), 100; **13**:11(36, 40), 60

Evans, D. C. B., **24**:30(62), 38(62)

Evans, D. G., **4**:200, 201, 205, 226

Evans, G. H., **14**:16(42), 58, 69, 75(42), 76(42), 101; **28**:370(128, 129), 371(129), 389(163), 391(163), 392(179, 180), 393(179, 181), 420(128), 421(129), 422(163), 423(179, 180, 181)

Evans, G. W., **1**:72, 75(25), 121; **19**:13(26), 19(26), 84(26)

Evans, H. L., **4**:331, 444; **20**:378, 387; **26**:121(45), 210(45)

Evans, J. W., **28**:318(185, 189, 195), 337(185, 189), 338(195)

Evans, L. B., **6**:474(254), 483(254), 484, 498; **22**:186(41), 348(41); **23**:139(12), 184(12)

Evans, N. A., **29**:146(42), 204(42)

Evans, P. V., **28**:28(55), 36(55), 56(55), 72(55)

Evans, R., **15**:2(9), 8(9), 55(9)

Evans, R. B., **15**:52(87), 58(87)

Evans-Lutterodt, K., **7**:26(9), 37, 39, 40(20), 42, 66, 67, 67(9), 83, 84

Evdokimov, E. D., **25**:84(167), 148(167)

Evdokimov, V. D., **25**:52(91), 61(102, 103, 104, 105, 106), 92(104, 106), 144(91, 102, 103, 104), 145(105, 106)

Everaarts, D. H., **9**:49, 102, 107

Everett, C. J., **5**:3, 50

Everett, D. H., **15**:229(14), 230(14), 276; **24**:104(47), 186(47)

Everett, D. J., **19**:125(127), 188(127)

Everly, D., **19**:149(181), 189(181)

Eversteyn, F. C., **28**:365(110), 366(110), 367(110), 368(110), 420(110)

Evgrafov, M. A., **8**:41(213), 89

Eyering, H., **22**:319(79), 350(79)

Eyring, H., **2**:361(40), 395; **9**:218(46), 269; **12**:242(26), 278; **15**:79(70), 137(70)

Eyssa, Y. M., **15**:48(77), 57(77)

F

Faber, O. C., Jr., **4**:164, 165, 224

Fadali, D. A., **23**:230(232), 232(232), 277(232)

Fadden, W., **15**:52(90), 58(90)

Faeth, G. M., **26**:2(68), 54(68), 71(68), 73(68), 88(68), 96(68)

Fafurin, A. V., **25**:51(76), 61(95, 96, 98), 81(143), 143(76), 144(95, 96, 98), 146(143)

Fagan, W. S., **2**:316(108), 355

Fage, A., **2**:172, 269; **3**:14, 32; **5**:239(181), 241, 251; **8**:100, 101, 159; **18**:94(27), 95(27), 158(27)

Faggiani, S., **26**:106(12), 148(105, 106), 151(105, 106), 154(106), 155(106), 156(106), 208(12), 213(105, 106)

Faghri, A., **25**:51(86), 144(86); **26**:201(164, 165, 166), 202(165), 203(166, 168, 169, 170, 171, 172, 173, 174, 175), 216(164, 165, 166, 168, 169, 170, 171, 172, 173, 174), 217(175); **28**:255(65), 276(65, 105), 332(65), 333(105); **30**:93(3), 188(3)

Faghri, M., **SUP**:135(313); **20**:187(70), 232(154), 302(70), 306(154); **25**:51(86), 144(86); **30**:169(135), 195(135)

Fagrell, B., **22**:225(65), 349(65)

Fahidy, T. Z., **21**:175, 181

Fahy, G. M., **22**:176(68, 69), 177(70), 178(70), 188(66), 196(67), 312(70), 349(66, 67, 68), 350(69, 70)

Faig, J., **29**:221(56), 276(56), 323(56), 338(56)

Fair, J. R., **21**:319, 320, 321, 325, 329, 331, 344

Fair, R., **15**:264(98), 266, 280

Fairbank, H. A., **17**:69(28), 153

Fairbanks, D., **3**:299(98), 302

Fairbanks, D. F., **10**:186, 187, 188(82), 189(81, 82), 191, 205(82), 206(82), 216; **14**:151, 169, 207(140), 216(15), 217(15), 218(15), 222(15), 242, 246; **19**:123(116), 129(116), 130(116), 134(143), 173(143), 187(116), 188(143)

Fairclough, J. M., **11**:220, 223, 258

Faktorovič, L. E., **1**:416(106, 107), 417(107), 443

Falacy, S. F., **29**:69(20), 124(20)

Falii, V. A., **25**:4(13), 31(13), 33(13), 64(13), 96(13), 127(13), 140(13)

Falii, V. F., **25**:64(109, 110), 96(110), 145(109, 110)

Falkner, V. M., **2**:172, 269; **3**:14, 32; **4**:336, 446; **26**:111(26), 120(26), 209(26)

Fallis, B., **3**:55, 98

Fallon, R. H., **22**:227(71), 350(71)

Fallon, R. J., **3**:270(33), 271(33), 300

Falls, A. H., **23**:193(76), 270(76)

Falta, R. W., **30**:102(40), 154(117), 157(123), 159(123), 181(157), 185(40, 117), 190(40), 194(117, 123), 196(157)

Faltzin, A. E., **31**:464(13), 471(13)

Fan, C., **30**:108(44, 45), 122(44, 45), 127(44), 128(44), 129(44), 130(44), 131(44), 132(44, 45), 133(45), 190(44, 45)

Fan, L. H., **16**:5, 9, 12, 13, 14, 16, 19, 27, 28, 29, 51, 56, 57

Fan, L. N., **13**:67(8), 115

Fan, L. S., **19**:105(48), 108(48), 185(48); **23**:188(77), 193(259), 207(116), 229(259), 236(259), 250(259), 253(77), 270(77), 272(116), 278(259)

Fan, L. T., **10**:196(89), 203(89), 217; **SUP**:77(218), 183(362), 190(218), 191(218), 192(218), 193(218); **22**:19(26), 30(26), 153(26)

Fan, S. K., **23**:193(61), 226(61), 269(61)

Fand, R. M., **3**:28, 32; **8**:130(62), 160; **11**:203, 206, 219, 220, 223, 224, 226, 233, 248, 249, 250, 254, 258, 259, 263; **18**:127(61), 159(61); **23**:196(78), 216(79), 270(78, 79); **30**:242(20), 250(20)

Fang, L. J., **19**:79(202), 92(202); **20**:232(157), 306(157)

Fang, P. C., **11**:438

Fang, Q. T., **19**:79(220), 93(220)

Fanger, P. O., **22**:8(16), 17(16)

Fann, W. S., **28**:117(27), 139(27)

Fanning, M. W., **15**:270(106), 271(106), 281

Fano, L., **2**:37(28), 38(28), 106

Fappo, M., **2**:353

Farag, H. A., **25**:259, 281, 286, 315

Farag, I., **31**:395(50), 427(50)

Farbar, L., **9**:114, 116, 122, 125, 127(1, 18, 20), 129, 177, 178

Farber, E. A., **5**:65, 66, 67, 104, 125; **11**:2, 47, 105(71), 109, 111, 191; **18**:257(33), 258(33), 259(33), 260(33), 264(33), 293(33), 306(33), 321(33); **20**:15(29), 80(29)

Farber, M., **2**:326, 327, 332, 355

Farbor, E. A., **17**:23(74), 59

Farello, G. E., **26**:45(37), 94(37)

Farges, J., **14**:330, 345

Fargie, D., **SUP**:70(152), 87(152)

Farhadieh, R., **19**:80(225, 229), 93(225, 229)

Farhat, C., **31**:407(70), 428(70)

Farkas, L., **14**:287, 298, 341
Farley, F. J. M., **14**:296(26), 297(26), 314, 342
Farmer, C. L., **29**:132(15), 175(15), 203(15)
Farmer, J. T. (Chapter Author), **31**:333, 335(19), 336(19, 20, 21), 340(19), 395(52, 53), 399(19, 20, 21), 419(19), 420(53), 426(19, 20, 21), 427(52, 53)
Farmer, W. S., **5**:74, 126
Farnia, K., **SUP**:73(181), 88(181), 89(181), 164(181), 298(181), **19**:268(53), 271(53), 351(53)
Farouk, B., **23**:293(68), 357(68); **28**:307(151, 152), 336(151, 152), 376(147, 152), 379(152), 381(152), 383(152), 384(152), 421(147), 422(152)
Farouq, S. M., **14**:77(160), 78, 82, 83(160), 103
Farouq Ali, S. M., **23**:193(106, 107, 248), 206(1), 267(1), 271(106, 107), 277(248)
Farr, W. W., **24**:102(13), 184(13)
Farran, R. A., **7**:244, 245, 316
Farrant, J., **22**:353(141)
Farrell, P. V., **30**:295(93), 296(96), 299(93), 300(96), 311(93, 96)
Farrelo, G. E., **11**:18(61), 49
Farrow, R. F. C., **28**:366(117), 420(117)
Fartushnov, A. V., **31**:271(173), 330(173)
Fasano, A., **22**:203(20), 347(20)
Fasel, H., **21**:154, 179
Fastovskii, V. G., **1**:239, 265
Fastovsky, V. G., **31**:160(3), 245(134), 322(3), 328(134)
Fateeva, L. A., **15**:323
Fatica, N., **9**:204, 233(21), 236, 238, 243, 268
Fatika, N., **21**:100(78), 136(78)
Fatt, I., **15**:321
Fauchet, P. M., **28**:102(28), 139(28)
Faulkner, S., **20**:367, 387
Fauske, H. K., **1**:410(94), 442; **17**:23, 59; **29**:134(18), 137(25), 139(18, 29), 140(18), 147(49), 148(49, 50), 175(148), 189(194), 190(195), 191(200), 203(18, 25), 204(29), 205(49, 50), 210(148), 212(194, 195, 200)
Faust, A. S., **10**:119, 165; **11**:105(60), 110(60), 191
Faust, C. R., **14**:6(26), 78, 79(169), 83(169), 88, 100, 104; **30**:98(35), 99(35), 190(35)
Favier, J. J., **30**:332(48, 49), 428(48, 49)
Favre, A., **6**:36(62), 125; **13**:86(46), 116
Faw, R. E., **29**:10(12), 55(12)
Fawkes, M. J., **26**:202(167), 216(167)

Faxen, H., **15**:73(54), 137(54)
Fay, J., **3**:34(4), 97
Fay, J. A., **1**:284(17), 344(17), 346(17), 349, 353, 354; **2**:49, 107, 139, 196, 198, 199(24), 200(24), 201(24), 268; **4**:271, 278, 279, 281, 314; **7**:196, 197, 214; **16**:5(3), 8(3), 56
Fearn, R. F., **5**:390, 510
Fedders, H., **16**:211, 216, 226, 235, 238
Feddosev, D. V., **8**:49(228), 89
Feder, J., **9**:202, 268; **14**:304(31), 307(31, 64), 308(64), 309(31, 64), 310, 314, 316, 317, 319, 320(31), 342, 343
Fedorov, E. P., **21**:48(83), 53(83)
Fedorov, V. B., **6**:418(134), 492; **28**:127(7), 138(7)
Fedorov, V. I., **25**:4(33), 141(33)
Fedorov, V. N., **25**:127(194), 149(194)
Fedorova, L. E., **6**:418(134), 492
Fedorovich, E. D., **9**:222(51), 269
Fedotkin, I. M., **15**:323; **31**:160(11), 322(11)
Fedotov, Yu. P., **6**:539, 563
Fedyaevsky, K. K., **6**:410(72), 411(71, 72), 489
Fedyakin, N. N., **1**:137, 184
Fehlner, F., **18**:20(49), 81(49)
Feider, M., **4**:116(71), 140
Feigelson, E. M., **14**:255, 279
Feigenbaum, M. J., **25**:356, 411
Feigl, G., **30**:378(146), 433(146)
Feiler, C. E., **2**:4(5), 86(5), 105
Feldman, A., **15**:232(34), 277
Feldman, E. E., **SUP**:73(184), 77(184), 333(184), 335(487), 337(184, 487), 338(487), 339(184, 487), 340(487)
Feldman, E. M., **23**:318(151), 361(151)
Feldman, K. T., Jr., **7**:223(8, 9), 235(21, 22), 236(8), 246, 273(8, 9), 274, 303, 314
Feldman, R. D., **28**:342(33), 416(33)
Feldmanis, C. J., **4**:205, 226
Feldshtein, Ya. M., **25**:127(196), 149(196)
Feldstein, N., **28**:341(10), 415(10)
Felenbok, P., **4**:245, 313
Felicelli, S. D., **28**:262(72), 303(72, 138, 139), 309(156, 157), 332(72), 335(138, 139), 336(156, 157)
Felske, J. D., **12**:143, 193; **17**:218, 236(68), 315; **23**:146(26), 185(26)
Fench, E. J., **7**:100, 160
Fendell, F. E., **2**:255, 263, 263(89), 270; **5**:131, 246; **8**:12, 85
Feneberg, W., **4**:268, 314

Feng, S. K., **14**:311(72), 343

Fenner, R. T., **13**:261(119), 266; **28**:150(16, 17), 156(17), 228(16, 17)

Fenske, R. M., **10**:203, 217

Fenster, S. K., **5**:375(86), 381, 508, 510

Fenstermacher, P. R., **21**:147, 153, 154, 155, 179, 182

Ferell, H., **5**:242, 251

Ferguson, C. R., **21**:263(29, 30), 276(29, 30)

Ferguson, E. T., **17**:166, 314

Ferguson, M. G., **8**:248(25), 283

Fernandes, R. C., **20**:84, 85, 92, 95, 97, 99, 108, 109, 110, 130, 131

Fernandes, R. T., **30**:151(110), 193(110)

Fernandez, F. L., **8**:12(46), 85

Fernandez, R. T., **29**:48(44), 57(44)

Fernandez-Pello, A. C., **26**:61(176), 70(177), 71(69), 77(116), 96(69), 98(116), 101(176, 177)

Ferrand, L. A., **30**:95(26), 96(27), 189(26, 27)

Ferrell, J. K., **7**:238(39), 241(39, 41), 242, 258, 260, 315, 320b, 320e

Ferrer, J., **23**:193(29), 206(29), 226(29), 268(29)

Ferretti, O. A., **23**:370(11), 461(11)

Ferri, A., **7**:197, 215

Ferriera de Sousa, M., **14**:119(70), 120(70), 146

Ferris, D. H., **6**:404, 488

Ferriso, C. C., **5**:295(40, 43), 307(64), 322, 323; **12**:129, 137, 193

Ferriss, D. H., **13**:80(35), 116

Ferron, J. R., **3**:286, 288(72), 301; **10**:173, 215

Ferry, J. D., **13**:251(112), 266; **18**:214(124), 238(124)

Fersiger, J. H., **8**:70, 90

Ferubko, N. E., **21**:39(65), 52(65)

Ferziger, J. H., **5**:28(41), 52; **18**:34(68), 82(68); **21**:185(5), 236(5); **24**:218, 275; **25**:333, 334, 335, 337, 340, 341, 342, 343, 349, 353, 354, 358, 360, 361, 362, 363, 367, 369, 370, 374, 375, 378, 380, 381, 382, 386, 397, 401, 404, 408, 409, 410, 411, 414, 415, 416

Feshbach, H., **3**:156(43), 174; **8**:14(77), 33(77), 45(77), 86; **24**:40(7), 98(7)

Fetter, A. L., **17**:88, 154

Fetting, F., **10**:193, 202, 203, 205, 206, 217; **14**:218(155, 156), 222(155, 156), 246

Feucht, D. L., **15**:330

Feuerstein, G., **16**:236

Feurstein, G., **31**:247(139, 140), 328(139, 140)

Feustel, J., **18**:3(4), 6(4), 79(4)

Feyerherm, A. M., **4**:104(50), 106(50), 139

Feynman, R. P., **17**:91, 97, 154

Fick, A., **12**:204(20), 278

Fick, J. L., **9**:402(84), 417

Fiedler, B., **25**:374, 412

Fiedler, H. E., **13**:86(47), 116; **29**:7(7), 55(7); **30**:32(35), 89(35)

Fieg, G., **29**:31(24), 56(24)

Field, R. J., **23**:329(232), 365(232)

Fielder, G., **10**:34, 37, 44(18), 45(18), 48(18), 53(18), 62(18), 68(18), 81

Fieschi, R., **14**:119, 145

Fife, P. C., **8**:13, 85

Figari, A., **24**:8(36), 14(36), 37(36)

Figliola, R. S., **18**:71(135), 72(135), 86(135); **23**:338(266), 339(266), 340(267, 268), 366(266, 267), 367(268); **28**:11(15), 71(15)

Figueiredo, O., **SUP**:326(483)

Filatov, I. A., **20**:24(54), 27(54), 81(54)

Filatov, V. V., **31**:160(33), 323(33)

Filbey, G. L., **13**:213(25), 264

Filimonov, S. S., **3**:219(68), 230(68), 250; **11**:390(142), 397(142), 438; **15**:323; **25**:27(56), 142(56); **31**:193(97), 326(97)

Filin, N. V., **25**:92(173), 148(173)

Filin, V. A., **15**:328

Filipchuk, V. E., **26**:252(63, 64), 253(63, 64), 261(63), 281(64), 316(64), 317(64), 318(63, 64), 327(63, 64)

Filipovic, J., **23**:123(128), 132(128)

Filippov, A. G., **1**:435(141), 444

Filippov, G. A., **31**:160(55), 324(55)

Filippov, M. V., **25**:153(13, 14, 15), 154(14, 16), 155(14, 15, 16, 17), 156(14), 157(13, 15, 16, 17, 19, 20, 21, 22, 23), 158(17, 19, 20), 163(14), 196(14, 15, 16, 17), 197(20), 198(20), 200(13, 14, 15, 16, 17, 19), 201(14, 15, 19, 21), 203(14, 15, 19, 21), 228(15, 16, 17), 237(15), 243(13, 14, 15, 16, 17, 19, 20, 21, 22, 23)

Filippovsky, N. F., **10**:208, 218; **14**:151(21), 152(32), 160(21), 161(21), 166(21), 221(21), 242, 243, 247; **19**:100(22), 103(22), 116(22), 126(22), 127(22), 128(22), 164(210), 165(22), 169(210), 170(210), 171(22), 185(22), 190(210)

Fillipov, L. P., **11**:332(58), 362(58), 436

Fillipovskii, N. F., *see* Filippovsky, N. F.

Filonenko, G. K., **25**:48(73), 143(73)

Finalborgo, A. C., **20**:14(19), 15(19), 80(19)
Findikakis, A., **25**:382, 412
Findlay, A. J., **31**:109(45), 110(45), 158(45)
Findlay, J. A., **20**:96, 107, 108, 132
Fine, R. A., **18**:333(45), 363(45)
Fineschi, F., **29**:303(165), 304(165), 344(165)
Finfrock, C., **29**:84(49), 93(49), 94(49), 95(49), 100(49), 125(49), 259(118), 261(118, 119, 121), 262(119), 263(118, 119, 121), 264(118, 119, 121), 265(118, 119, 121), 282(118), 328(181), 341(118, 119, 121), 344(181)
Fink, J., **16**:144, 156
Finke, B. R., **28**:111(29, 30), 139(29, 30)
Finke, M., **28**:111(30), 139(30)
Finkelnburg, W., **1**:283(15), 284(15), 353; **4**:232(3), 233(3), 234(3), 237(4), 240(4), 245, 248(3), 260, 293(3), 313
Finkelstein, A. B., **7**:279, 319
Finkelstein, Y., **15**:229(12), 230(12), 235(12), 237(12), 276
Finlay, I. C., **6**:474(294, 304), 500, 501
Finlayson, B. A., **8**:2, 24(8), 25(126), 32(8, 126, 155), 84, 87; **SUP**:67(123); **19**:195(26), 198(36, 37), 243(26), 244(36, 37); **25**:260, 268, 319; **28**:405(219), 425(219)
Finlayson, R. D., **28**:149(55), 230(55)
Finn, R. K., **7**:116, 118, 160; **12**:256(21), 278
Finston, M., **4**:22, 59, 63, 64; **9**:286(2), 346
Finzi, S., **1**:433, 443
Fiori, M. P., **17**:47, 63
Fiorini, A., **22**:421(33), 436(33)
Firsov, V. P., **31**:265(159), 329(159)
Firsova, E. V., **16**:235
Firstenberg, H., **1**:411(99), 442
Fischer, E. C., **18**:328(13), 362(13)
Fischer, F., **13**:211(20), 264
Fischer, J., **11**:203, 205, 209, 253
Fischman, G. S., **28**:351(82), 418(82)
Fisenko, V. V., **23**:9(23), 127(23)
Fishbein, G. A., **26**:8(187), 21(3, 189), 22(189), 23(3, 33), 92(3), 94(33), 101(187, 189)
Fishenden, J., **9**:296(32), 346
Fishenden, M., **11**:300, 315; **18**:167(51), 237(51); **30**:280(60), 309(60)
Fisher, D. H., **15**:94(92), 97(92), 138(92)
Fisher, D. J., **28**:262(68), 263(68), 288(68), 301(68), 332(68)
Fisher, D. P., **SUP**:115(288), 118(288), 135(288)
Fisher, E. M., **18**:59(113), 61(113), 62(113, 114), 63(113), 64(114), 85(113, 114)
Fisher, J. C., **1**:208, 264; **10**:111, 165
Fisher, K. M., **28**:299(137), 335(137)
Fisher, M. J., **8**:128(59), 160; **11**:208(121), 219, 223, 241, 242, 258; **18**:123(58), 159(58)
Fisher, R., **14**:155(56), 157, 243
Fisher, R. J., **19**:164(205), 171(205), 173(205), 190(205)
Fisher, S. A., **9**:3(114), 110
Fisov, V. P., **31**:265(160), 329(160)
Fitch, E. C., **5**:242, 251
Fitch, J. S., **30**:163(124), 165(124), 166(124), 194(124)
Fitzgerald, J. M., **15**:323
Fitzgerald, T. J., **19**:98(6), 99(6), 100(30), 136(30), 137(30), 139(30), 145(30), 184(6), 185(30); **23**:340(272), 367(272); **25**:234(132, 133), 235(132, 133), 248(132, 133)
Fitzjarrald, D. E., **30**:28(36), 29(36), 89(36)
Fitzroy, N. D., **20**:202(85), 303(85)
Fiveland, W. A., **23**:139(16), 140(16), 177(51), 184(16), 186(51)
Fix, G. J., **26**:89(7), 93(7)
Flack, R. D., **18**:50(95, 96), 51(95, 96), 84(95, 96); **20**:217(126, 133), 219(126, 133), 305(126, 133); **30**:281(66, 67), 310(66, 67)
Flage, R. A., **4**:228
Flamand, J. C., **11**:18(66), 31(66), 49
Flamant, G., **19**:164(212), 171(212), 190(212)
Flanagan, M., **29**:91(59), 126(59)
Fleck, J. A., Jr., **5**:36, 52
Fleischman, G. L., **7**:320b, 320e
Fleischmann, M., **12**:242(22), 278
Fleming, D. P., **SUP**:72(166), 199(166), 210(166), 213(166), 240(166), 241(166), 242(166); **19**:268(50), 271(50), 272(50), 274(50), 351(50)
Fleming, R., **15**:287, 289, 323
Fleming, R. B., **5**:356(67), 508
Fleming, R. H., **18**:339(54), 363(54)
Flemings, M. C., **22**:291(72), 350(72); **28**:65(84), 74(84), 231(2, 3, 4, 5, 6), 232(7), 238(7), 245(21), 251(2, 3, 4, 6, 36, 37), 266(5, 21, 36, 37), 267(21), 299(7), 304(4), 309(36, 158), 329(2, 3, 4, 5, 6, 7), 330(21, 36, 37), 336(158)
Flemmer, R. L. C., **21**:334, 344
Flemming, H., **31**:436(14), 459(15), 461(15), 471(14, 15)

Fletcher, D. D., **7**:350, 354, 354(44), 355(45), 358(45), 361(45), 375, 378

Fletcher, D. F. (Chapter Author), **29**:129, 132(16), 145(38), 147(16, 38), 149(54), 150(54, 58, 59), 152(74, 75, 76), 153(77), 155(84, 85), 156(84, 85, 86), 159(16, 92), 164(16), 165(84, 85, 115), 168(38), 180(38, 161, 162, 163), 181(162), 186(182), 189(190), 194(208), 203(16), 204(38), 205(54, 58, 59), 206(74, 75, 76, 77, 84, 85, 86), 207(92), 208(115), 210(161, 162, 163), 211(182), 212(190), 213(208)

Fletcher, L. S., **9**:390(61), 391, 416; **11**:290(24), 315; **18**:73(145), 86(145); **20**:186(22), 187(22), 190(22), 191(22), 192(22), 197(22), 260(251), 261(251), 262(251, 257), 263(257), 265(257), 266(270), 300(22), 311(251, 257), 312(270); **24**:17(46), 37(46); **30**:258(11), 260(11), 269(11), 270(11), 272(11, 40), 274(50), 276(50), 277(50, 52), 278(52), 281(52, 59), 282(52), 284(52), 285(52), 307(11), 309(40, 50, 52, 59)

Fletcher, N. H., **22**:177(73), 350(73)

Flint, E., **17**:143, 151, 157

Flood, H., **9**:202, 268; **14**:321, 322(95), 323, 324(95), 325, 344

Flood, S. C., **28**:306(149), 308(155), 335(149), 336(155)

Flores, A. F., **19**:281(76), 285(76), 294(76), 352(76)

Florinsky, B. V., **1**:435(140), 444

Florschuetz, L. W., **9**:227, 269; **15**:242, 243, 278; **16**:90, 154, 232; **18**:295(62), 298(62), 300(62), 322(62); **23**:333(247, 248), 365(247, 248); **26**:170(126, 127), 214(126, 127)

Flory, K., **10**:109, 164; **29**:166(119), 208(119)

Flower, J. F., **21**:179, 180

Flügge-Lotz, I., **4**:334, 445; **7**:197, 215

Fluker, B. J., **15**:322

Flynn, G. P., **3**:255(6), 256(6, 8), 260(8), 300

Flynn, T. M., **5**:423, 512; **11**:105(65), 109, 111, 191; **16**:234; **18**:258(49), 260(49), 322(49)

Focke, W. W., **26**:255(77), 328(77)

Fodemski, T. R., **25**:357, 383, 385, 391, 399, 410, 412

Fofonoff, N. P., **18**:333(34), 340(60), 341(60), 342(34, 56, 60), 362(34), 363(56, 60)

Fogelberg, C. V., **9**:408(104), 417

Foglia, J. J., **1**:394(72), 398(72), 442

Fohr, J. P., **30**:172(143), 195(143)

Fokin, B. S., **11**:56(7, 8, 9, 10, 11), 58(7, 8, 9, 10, 11), 59(7, 8, 9, 10, 11), 60(7, 8, 9, 10, 11, 29), 95(50), 98(50), 101, 105(75, 76, 77, 101), 106(29), 107(50), 108(29), 112, 117, 118(29), 119, 120(24), 121(29), 122(29), 189, 191, 192, 193

Folinsbee, J. T., **17**:69(26), 71(26), 153

Folkner, V. M., **8**:100, 101, 159; **18**:94(27), 95(27), 158(27)

Folland, G. B., **15**:323

Fomin, A. V., **25**:61(99), 144(99)

Fomin, N. A., **30**:283(69), 286(69), 291(87), 292(87, 89), 293(87, 90), 296(97, 99), 298(99), 299(97), 310(69), 311(87, 89, 90, 97, 99)

Fomin, S. A., **24**:8(31), 37(31)

Fonad, M. G., **7**:98(12), 160

Fonarev, A. S., **3**:225, 251

Foner, S., **15**:43(64), 57(64)

Fontana, D. M., **11**:19(67), 31(67), 49

Fontana, M. H., **29**:234(85, 86), 269(85), 320(86), 339(85, 86)

Foord, J. S., **28**:343(48), 417(48)

Foote, J. R., **4**:23, 63

Foraboschi, F. P., **2**:380, 395

Foraboschi, P., **6**:474(286), 500

Forbes, S. W., **29**:266(131), 342(131)

Forchheimer, P., **24**:110(92), 112(92), 114(92), 187(92)

Forchheimer, P. H., **14**:4, 100

Ford, G. W., **29**:120(86), 127(86)

Forest, A. E., **23**:315(148), 320(188, 189), 360(148), 362(188, 189)

Forgó, L., **12**:49, 75

Fornasieri, E., **15**:48(75), 57(75)

Forney, L. J., **16**:5(3), 8(3), 56

Forrest, G., **13**:219(76), 243(76), 265

Forrester, A. T., **7**:225(12), 314

Forsberg, C. H., **9**:364(35), 386(35), 390(57), 409(57), 415, 416

Forsch, C. J., **11**:357(80), 436

Forsdyke, E. J., **13**:119(1), 200

Forshey, D. R., **14**:306(41), 342

Forslund, R. P., **4**:213(98), 214(98), 227; **9**:259(104), 271; **11**:144(162, 166), 146(162, 166), 147(166), 148(166), 164, 167(166), 168(166), 169(166), 170(162, 166), 173, 177, 184, 196; **31**:137(4), 155(4)

Forster, D., **6**:474(244), 483(244), 498
Forster, H. K., **4**:178, 225; **5**:409, 411, 413, 414, 440, 467, 513; **11**:13(39), 14, 15, 48; **15**:242(55), 278; **16**:181(12), 238
Forster, H. K. E., **1**:195, 217, 218(32), 221(32), 230, 259, 263, 264, 265
Forster, K. E., **16**:181(11), 238
Forster, R. E., **22**:20(6), 60(6), 61(6), 69(6), 70(6), 152(6)
Forster, V. T., **12**:3(6), 4(6), 11(6), 14(6), 42(6), 43(6), 44(6), 49, 51(6), 53(6), 68, 69(6), 73
Forstrom, R. J., **4**:274(37), 314; **9**:300(47), 301, 347
Forsyth, M., **22**:196(132), 352(132)
Forsyth, P. A., **30**:102(40), 175(151), 185(40, 163), 190(40), 196(151, 163)
Forsyth, T. H., **13**:219(94), 266
Fort, C., **23**:326(213), 364(213)
Fort, E. F., **23**:343(290), 367(290)
Forteskie, P., **31**:245(136), 328(136)
Forth, C. I. P., **23**:321(200), 363(200)
Fortini, A., **2**:53, 56, 69, 71, 94, 108
Fortmann, C. M., **28**:342(35), 416(35)
Foscolo, P. U., **23**:196(80), 270(80); **25**:199(99, 100), 230(99), 247(99, 100)
Foster, C. V., **9**:5, 11, 29, 30, 107
Foster, K. R., **22**:26(2), 59(3), 71(2), 137(2), 138(2), 139(2), 151(1), 152(1, 2, 3)
Foster, P. J., **31**:390(44), 427(44)
Foster, W. R., **2**:375(96), 396
Foti, J. J., **20**:244(186), 308(186)
Fotiadis, D. I., **28**:375(139), 389(171), 390(171, 173), 391(171), 398(171), 421(139), 422(171), 423(173)
Fountain, J. A., **10**:48, 51, 52, 56, 57, 81
Fouske, H. K., **15**:243(58), 278
Foust, A. S., **1**:202(11), 203(11), 264; **5**:74, 126; **16**:64(6), 153; **20**:15(28), 80(28)
Foust, O. J., **21**:71(128), 139(128)
Fowkes, N. D., **8**:13, 85
Fowle, A. A., **5**:381, 509; **11**:318(14), 434
Fowler, A. C., **28**:282(122), 334(122)
Fowler, A. G., **SUP**:62(76), 64(76), 67(76), 232(76), 236(76), 259(76)
Fowler, A. J., **24**:30(66), 38(66)
Fowles, G. R., **29**:187(184), 212(184)
Fowlis, W. W., **5**:210(132), 213, 214, 216(132), 250; **24**:290(15), 319(15)
Fox, D. G., **16**:9, 57
Fox, G. C., **30**:356(106), 357(106), 431(106);

31:400(54), 402(54), 403(54), 405(54), 427(54)
Fox, G. R., **9**:405(97), 417
Fox, H., **8**:67, 68, 90
Fox, H. W., **9**:190, 227(8), 229(8), 268
Fox, J., **5**:143, 247; **6**:76, 128
Fox, J. E., Jr., **2**:385(59), 395
Fox, J. W., **3**:258, 271, 272, 297, 300
Fox, L., **19**:9(17), 16(17), 39(17), 54(17), 84(17)
Fox, L. R., **20**:292(318), 293(318), 294(318), 314(318)
Fox, P. F., **15**:327
Fox, R. H., **4**:94(34), 138
Foxon, C. T., **28**:340(4), 415(4)
Foyle, R. T., **9**:11, 30, 107
Fraas, A. P., **15**:26(52), 56(52)
Fraas, L. M., **28**:352(90), 419(90)
Fradkov, A. B., **5**:406(170), 513
Fragstein, C., **9**:362(23), 415
France, D. M., **17**:321, 339
France, J., **31**:247(141), 328(141)
Franchett, M. E., **26**:182(140), 183(140), 184(140), 185(140), 215(140)
Francis, A., **19**:80(241), 93(241)
Francis, J. E., **11**:365(120, 121, 122), 366(121), 371(120, 121, 122), 374(120), 437
Francis, P., **29**:130(10), 203(10)
Franck, J. P., **14**:325, 344
Francois, D., **23**:294(72, 73, 74), 357(72, 73, 74)
Françon, M., **6**:174, 195, 197(37), 199, 363, 364; **30**:255(1), 256(4), 257(1), 307(1, 4)
Frandsen, W. H., **10**:244(40), 266, 269(99), 282, 283
Frank, B., **8**:36(205), 89
Frank, S., **7**:236, 236(26), 270(26), 297, 298, 300, 301, 302, 314
Frankel, L. E., **8**:12, 85
Frankel, N. A., **9**:262(123), 271
Frankel, W. V., **8**:24, 86
Frankenfield, T. R., **19**:164(215), 168(215), 171(215), 190(215)
Frank-Kamenetsky, D. A., **3**:183, 248, 284(66), 288(66), 301; **8**:36, 89; **21**:180
Frank-Kamenickÿ, D. A., **4**:54, 64
Frankl, F., **3**:35, 60, 97
Franklin, J. L., **20**:262(258), 311(258)
Franklin, R. E., **11**:219, 222, 225, 237, 257
Franks, D. E., **11**:170(171), 196
Franks, F., **22**:177(75), 187(74), 350(74, 75)
Fransen, J. W. M., **9**:102, 107

Fransen, P., **28**:343(54), 417(54)

Frantisak, F., **8**:298(35), 300(54), 301(54), 349, 350

Frantz, B., **29**:21(17), 23(17), 25(17), 56(17)

Franzeck, U. K., **22**:225(19), 347(19)

Fraser, D. B., **28**:340(3), 415(3)

Fraser, R. G. J., **2**:353

Frazier, G., **3**:287(83), 288, 301

Frea, W. J., **4**:206, 227; **17**:50(157), 52, 63

Freche, J. C., **9**:3(36), 9, 38, 71(36), 91(36), 93, 94, 95, 101, 107

Frederick, D., **19**:41(94), 87(94)

Frederick, N. V., **17**:150, 157

Frederickson, A. G., **18**:177(57), 237(57)

Frederking, T. H. K., **4**:173, 174, 225; **5**:67, 91, 125, 126, 127, 428, 429, 430, 431, 432, 436, 481, 482, 483(265), 484(264), 485, 488, 492, 493, 494, 495(273), 497, 501, 512, 517; **11**:60(26), 68(37), 70, 71, 72, 84(43), 93, 95, 96, 97(43), 105(37, 43, 94, 95), 109(26), 110, 111, 112, 113, 116, 117, 118(26), 119(36), 121, 190, 193; **17**:70(24), 74(40), 108(95, 97), 110(95), 111(95), 112, 115, 117(107), 151(186), 153, 154, 155, 158; **30**:39(38), 42(37), 89(37, 38)

Fredrickson, A. G., **2**:358, 360, 361, 363, 388, 395; **13**:236(105), 266

Fredriksson, H., **28**:262(70), 267(70), 289(70), 290(70), 291(70), 292(70), 294(70), 295(70), 296(70), 332(70)

Freeborn, R. D. J., **SUP**:120(295)

Freeby, W. A., **14**:179(100), 188(100), 189(100), 191(100), 195(100), 245; **19**:113(89), 119(89), 148(89), 186(89)

Freeman, N. C., **2**:270

Freeman, S. M., **5**:347, 506

Frei, A. M., **7**:98(13), 160

Freilich, M. H., **21**:180

Frenkel, J., **9**:190(9), 192, 199(9), 268; **10**:92, 164; **11**:7, 10, 48; **14**:285(9), 287(9), 293, 298, 304, 306, 307, 341; **28**:16(27), 71(27)

Frenkel, L. I., **14**:205(137), 206(137), 245

Frenken, H., **13**:35, 60

Frenkiel, F. N., **8**:291, 349

Frenzen, P., **5**:205, 249

Frey, F., **14**:318(84), 323(84), 324, 344

Frick, C. W., **11**:202, 204, 252

Fricke, A. L., **13**:211(23), 264; **18**:169(41, 42, 43), 170(41, 42, 43), 173(42, 43), 214(42, 43), 215(43), 217(42, 43), 218(43), 220(42,

43), 236(41, 42, 43)

Fricke, H., **18**:181(64), 183(64), 193(64), 211(64), 237(64)

Frid, W., **29**:151(72), 206(72), 270(136), 342(136)

Fridlender, N. A., **6**:418(109), 491

Fried, E., **5**:467, 511; **9**:359(17), 414; **20**:266(271), 268(271), 312(271)

Fried, J. J., **14**:16(44), 101; **24**:104(56), 186(56)

Fried, N., **5**:403, 405, 411, 440, 511

Friedel, P., **28**:342(23), 416(23)

Friedland, A. J., **3**:106(5), 172

Friedman, A., **6**:474(209), 496; **8**:13, 85

Friedman, M. D., **8**:47(225), 89

Friedman, R., **5**:100, 127, 445, 514; **11**:170(172), 174(172), 182(172), 183(172), 196, 318(8), 376(8), 434

Friedman, S., **18**:208(97), 238(97)

Friedman, S. J., **13**:18, 26(50), 60

Friedmann, M., **SUP**:73(187), 92(187), 93(187), 94(187), 95(187), 98(187), 439(187)

Friedrich, C. R., **26**:221(22), 325(22)

Friedrich, R., **9**:3(37), 91(37), 93, 94, 107; **20**:327(33), 351(33); **25**:380, 391, 406, 412, 417

Friedrichs, K. O., **8**:53(238), 90; **29**:140(35), 144(35), 146(35), 204(35)

Friehe, C. A., **11**:240, 241, 242, 261

Friend, P. S., **2**:386, 387, 396; **15**:104(128), 105(128), 107(128), 140(128); **19**:338(153), 356(153)

Friend, W. L., **2**:386, 395; **7**:103, 160; **12**:88, 101, 112

Fries, P., **9**:3(38), 107

Frieser, R. G., **20**:244(191), 308(191)

Frigaard, I. A., **28**:69(86, 87), 74(86, 87)

Frind, E. O., **30**:181(158), 196(158)

Frisch, H. L., **14**:314, 343

Frisch, U., **30**:7(39), 89(39)

Frischmuth, R. W., Jr., **14**:190(175), 234, 246

Frish, S., **2**:353

Fritsch, C. A., **4**:30, 63; **7**:66(47), 85; **21**:37(57), 38(57), 52(57)

Fritz, J. C., **10**:173, 215

Fritz, J. E., **20**:182(3), 290(315), 291(315, 317), 292(315), 299(3), 314(315, 317)

Fritz, J. J., **5**:406(168), 512

Fritz, R., **7**:320c(146), 320f

Fritz, W., **1**:216, 221, 223, 264; **4**:178, 181, 225; **10**:207, 217; **11**:14, 15, 49; **20**:6(6), 9(10),

Fritz, W. (*continued*)
  15(10), 79(6), 80(10)
Fritzsche, A. F., **8**:157(83), 160; **18**:156(82),
  159(82)
Frivik, P. E., **19**:78(186), 91(186)
Froes, F. H., **28**:8(9), 70(9)
Froese, C., **18**:342(56), 363(56)
Fröhlich, G., **29**:163(102), 172(143), 173(143),
  175(153), 207(102), 209(143), 210(153)
Froishteter, G. B., **13**:219(74), 265
Froman, N., **8**:19(95), 86; **SUP**:171(346)
Froman, P. O., **8**:19(95), 86; **SUP**:171(346)
Froment, G. F., **23**:370(10), 427(10), 461(10)
Fromm, J., **6**:2, 122
Frössling, N., **3**:20, 32; **8**:119, 120, 121, 160;
  18:113(47), 114(47), 116(47), 159(47);
  20:383, 387; **26**:53(70), 96(70)
Frost, A. A., **2**:123(11), 268
Frost, D. L., **29**:163(104), 169(129), 176(104,
  129, 155, 156), 181(129), 183(173, 174),
  185(173, 178, 179), 186(178, 179),
  191(155), 207(104), 208(129), 210(155,
  156), 211(173, 174, 178, 179)
Frost, S., **11**:17(56), 49
Frost, W., **10**:152, 153, 165
Frurie, D. J., **29**:188(186), 212(186)
Fruth, H., **15**:52(89), 58(89)
Fruttuoso, G. A., **29**:234(93), 340(93)
Fry, C. J., **29**:146(41), 204(41)
Frysinger, G. R., **7**:275(113), 318
Frysinger, T. C., **4**:182(66), 183(66), 186(66),
  195(66), 196(66), 225, 228
Fryxell, D., **10**:55, 82
Fu, B.-I., **18**:58(108), 59(108), 84(108)
Fu, Z., **19**:164(217), 171(217), 172(217),
  190(217)
Fuchs, N. A., **25**:292, 293, 317
Fugii, N., **10**:64(50, 51), 67(51, 53), 82
Fugitt, C. E., **22**:330(76), 331(76), 332(76),
  350(76)
Fujie, K., **20**:75(65), 82(65)
Fujii, E., **28**:376(142), 421(142)
Fujii, F., **6**:551, 555(82), 564
Fujii, K., **18**:12(18), 79(18)
Fujii, M., **11**:213(75), 215(75), 255; **15**:166(25,
  26, 27), 167(25, 26, 27, 28), 223(25, 26,
  27), 224(28); **16**:232; **18**:73(144), 86(144);
  20:186(46), 187(46), 191(46), 209(113),
  217(121), 244(197), 301(46), 304(113),
  305(121), 308(197)

Fujii, T., **4**:10, 16, 51, 57, 61, 62, 64; **9**:299(44),
  302, 347; **11**:213, 215, 255, 293, 296, 298,
  299, 303, 315; **15**:166(25, 26, 27), 167(25,
  26, 27, 28), 189(66), 223(25, 26, 27),
  224(24, 66); **18**:73(143, 144), 86(143, 144);
  20:186(44, 45, 46), 187(43, 44, 45, 46),
  191(43, 44, 45, 46), 209(113), 217(121),
  301(43, 44, 45, 46), 304(113), 305(121);
  21:105(86, 89, 90), 106(91, 92), 109(98),
  113(98), 115(100), 124(113), 137(86, 89,
  90, 91, 92, 98, 100), 138(113); **25**:51(84),
  143(84), 286, 316; **28**:251(37), 266(37),
  330(37); **30**:77(40), 89(40)
Fujikake, J., **20**:277(293), 313(293)
Fujikawa, S., **21**:79(24), 134(24)
Fujimoto, H., **28**:36(63), 39(63), 73(63)
Fujisaki, M., **27**:181(32), 186(32); **31**:391(48),
  413(48), 418(48), 427(48)
Fujita, H., **1**:101, 121
Fujita, J., **17**:50(161), 63
Fujita, K., **30**:323(25), 427(25)
Fujita, S., **28**:342(36), 416(36)
Fujita, T., **11**:281(18), 315
Fujita, Y. (Chapter Author), **SUP**:18(31),
  83(243), 136(31); **16**:166(3), 170(3, 4),
  171(3, 4), 173(3, 4), 174(3, 4), 187(3),
  188(3), 189(3), 191(20), 193(3, 4, 20),
  194(20), 203(3), 209(3), 214(3), 215(3),
  219(20), 220(3, 4, 20), 221(3, 4), 222(3, 4),
  226, 232, 233, 235, 236, 238; **20**:1, 5(4),
  13(14), 16(35), 17(35), 18(35), 20(40),
  22(40), 27(55), 33(55), 76(70), 79(4),
  80(14), 81(35, 40, 55), 82(70)
Fujitani, K., **25**:358, 416
Fujiwara, H., **30**:329(33), 428(33)
Fujiwara, T., **30**:329(33), 428(33)
Fukai, J., **28**:36(57, 58), 37(57), 39(57), 40(57),
  41(57), 42(57), 43(57, 58), 44(57), 45(57),
  46(57), 47(57, 77), 48(57, 77, 78), 49(77),
  50(77), 51(77), 52(78), 53(77, 78), 54(57,
  58), 72(57, 58), 73(77, 78)
Fukni, S., **11**:68(38), 69, 70(38), 83, 105(38),
  109, 112, 121(38), 122(38), 123, 190;
  21:163, 182
Fukuchi, T., **27**:ix(11, 13, 14), 146(28, 29),
  147(28)
Fukuda, A., **7**:128(37), 161; **21**:181
Fukui, S., **5**:74, 91, 126, 233(166), 234,
  235(166), 251; **7**:129(38), 161
Fukusako, S., **SUP**:164(342), 168(342),

190(342); **19**:38(84), 43(97), 77(178), 79(213), 87(84, 97), 91(178), 92(213), 99(13), 185(13); **30**:151(111), 152(111), 194(111)

Fukushima, H., **21**:125(115), 138(115)

Fukushima, T., **20**:75(65), 82(65)

Fulford, G. D., **13**:120, 200

Fulk, M. M., **5**:474, 516; **9**:378(50), 381(50), 407(50), 415

Fulks, W. B., **14**:6(28), 100

Fuller, R. A., **19**:334(135), 335(135), 338(135), 355(135)

Fuller, R. E., **SUP**:73(195), 77(195), 300(195), 314(195), 321(195)

Fuller, T. R., **18**:169(43), 170(43), 173(43), 214(43), 215(43), 217(43), 218(43), 220(43), 236(43)

Fullington, F., **4**:92(30), 138

Fullmer, L. D., **30**:402(169), 434(169)

Fultz, D., **5**:131, 205, 206, 208, 210(129, 130, 131), 216(10), 218, 227(129), 228, 236, 246, 249, 250; **21**:149, 151, 179

Fumero, R., **22**:421(33), 436(33)

Funazaki, K., **27**:x(9), 174(9)

Funazu, M., **27**:viii(3)

Fung, Y. C., **SUP**:95(253), 96(253); **24**:102(9, 10), 184(9, 10)

Funk, J., **29**:98(68), 126(68)

Funk, J. E., **11**:365(117), 437

Fuoss, R. M., **12**:234(82), 235(82), 256(82), 259(82), 269(82), 280; **18**:328(16, 17), 329(17), 350(17), 362(16, 17)

Furber, B. N., **3**:13, 23, 32

Furmina, A. A., **30**:242(34), 251(34)

Furno, A. L., **29**:71(29), 125(29)

Furuhama, K., **16**:255, 312(12), 320, 363

Furuhama, S., **23**:314(140), 360(140)

Furukawa, Y., **23**:12(45), 14(45), 22(45), 31(45), 59(45), 61(45), 78(45), 85(45), 86(45), 93(45), 106(45), 128(45)

Furumoto, A. S., **14**:15(37), 100

Furzeland, R. M., **19**:21(41), 25(49), 26(41), 85(41, 49)

Fusegi, T., **24**:310(4), 311(4, 5), 318(4, 5)

Futagami, K., **SUP**:413(530); **15**:215(115), 225(115)

Futagawa, M., **21**:174, 180

G

Gabe, D. R., **21**:143, 180

Gabel, R. M., **9**:3(39, 40, 41), 91(39, 40, 41), 95, 97, 107, 108

Gabis, D. H., **25**:235(143), 249(143)

Gabitto, J. F., **24**:102(13), 184(13)

Gabor, D., **6**:200, 364

Gabor, J. D. (Chapter Author), **10**:185, 186, 216; **14**:149, 151, 152(28), 153(28), 154(40, 41, 44), 158, 159(66), 208, 217, 232(168), 234, 242, 243, 244, 246; **19**:98(3, 4), 111(3, 4), 112(3), 145(3), 154(3), 161(3, 4), 174(227), 184(3, 4), 190(227); **25**:164(48), 205(48), 214(48), 229(117), 244(48), 248(117); **29**:20(16), 56(16)

Gabour, L. A., **26**:109(23), 110(23), 111(23), 117(36), 118(23), 139(23), 141(23), 149(36), 151(36), 152(23), 153(23), 155(36, 114, 115), 193(114, 115), 194(114), 195(36, 114), 197(36), 198(36), 209(23), 210(36), 213(114, 115)

Gacesa, M., **9**:261(117), 271

Gadd, G. E., **6**:29(39), 35(57), 36, 94(145), 95(151, 154, 157), 108(154, 167), 124, 129, 130

Gaddis, E. S., **16**:144(121), 156

Gadgil, A. J., **18**:63(115, 116), 85(115, 116); **19**:62(149), 64(149), 65(149), 66(149), 90(149)

Gador, A. Y., **23**:298(90), 358(90)

Gaede, W., **2**:353

Gaertner, R. F., **16**:232; **17**:3, 16, 56; **18**:247(10, 11), 259(10), 320(10, 11); **20**:14(15), 15(15), 80(15)

Gage, K. S., **28**:370(136), 421(136)

Gageaut, C., **23**:294(72, 73, 74), 357(72, 73, 74)

Gagge, A. P., **4**:67(9), 83, 88(25), 91(25), 104(47), 110, 111(63), 118(78), 138, 139, 140; **22**:20(27), 21(27), 28(27), 82(27), 153(27)

Gaggioli, R. A., **15**:2(12), 52(85), 54(6), 55(12), 58(85)

Gago, J., **24**:227, 268

Gaitonde, N. Y., **23**:193(82), 220(82), 270(82)

Galagan, V. V., **31**:281(185), 331(185)

Galamba, D., **19**:80(232), 93(232)

Galanin, A. N., **1**:435(140), 444

Galante, S. R. (Chapter Author), **20**:353

Galaup, J. P., **11**:44(76), 49

Gale, E. H., Jr., **20**:271(277), 312(277)

Galegar, W. C., **1**:358, 440

Galershtein, D. M. (Chapter Author), **14**:149, 154(38), 160(79), 161(79), 167(79), 169(79), 176(119, 178), 188(119), 191(119), 232(79), 233(79, 173), 235(173, 178), 237(79, 119), 243, 244, 245, 246, 247; **19**:98(3), 100(29), 111(3), 112(3), 136(29), 137(29), 139(29), 140(29), 145(3, 29), 154(3), 157(29), 159(29), 160(193), 161(3, 29), 184(3), 185(29), 189(193); **25**:164(48), 205(48), 214(48), 244(48)

Galichskii, A. K., **6**:418(132), 492

Galili, N., **13**:219(73, 78), 221(78), 265

Galin, N. M., **31**:196(101), 202(107), 208(107), 212(107), 213(107), 242(131), 326(101), 327(107), 328(131)

Galitseisky, B. M., **6**:449(181, 182, 183, 184), 453(181, 182, 183, 184), 456(181, 182, 183, 184), 457(181, 182, 183, 184), 495; **25**:4(10, 12), 31(12), 33(10), 35(70a, 70b), 140(10, 12), 143(70a, 70b); **30**:83(41), 89(41); **31**:160(21), 182(21), 323(21)

Galjee, F. W. B. M., **17**:47, 63

Gall, T. L., **28**:7(6), 70(6)

Gallagher, J. S., **31**:115(6), 117(6), 155(6)

Gallagher, L. W., **4**:153(24), 223

Galli, A. F., **9**:119(9a), 177

Galloway, L. R., **SUP**:360(503)

Galloway, T. R., **10**:208, 217; **11**:220, 224, 237, 262; **19**:143(152), 188(152)

Gal-Or, B., **8**:29, 87; **23**:193(81), 208(81), 270(81)

Galowin, L. S., **7**:309, 319

Galvin, G. J., **28**:80(31), 139(31)

Gamayunov, N. I., **8**:66, 90

Gambill, W. R., **4**:206, 226; **5**:131(14), 247; **17**:2(2, 6), 32, 56, 61; **19**:334(137), 336(137), 355(137); **20**:250(214, 223), 309(214), 310(223)

Gambolati, G., **23**:418(75), 464(75)

Gammel, G., **7**:320c, 320f

Gamo, K., **28**:343(50), 417(50)

Gamson, B. W., **10**:200, 201, 218; **14**:214, 226(147), 228, 246

Gan, Y. P., **23**:12(49), 14(49), 25(49), 128(49)

Gandelsman, A. F., **6**:390(15, 19), 405(15, 19), 463

Gandin, C.-A., **28**:263(79, 80), 332(79, 80)

Ganesan, S., **28**:253(55, 56), 255(55), 256(55, 56), 258(55), 267(94), 331(55, 56), 333(94)

Gangal, M. K., **SUP**:68(130, 131), 72(167), 240(426), 241(426), 242(167, 426), 371(131), 372(130), 373(131, 426), 398(426)

Ganguli, A., **29**:161(98), 162(98), 207(98)

Ganic, E., **11**:19, 28(69), 37(69), 38(69), 49

Ganley, J. T., **21**:158, 178

Gannett, H. J., **16**:134(99), 155

Ganoulis, J., **25**:292, 316

Gantmacher, F. R., **8**:64(249), 90

Ganzha, V. L. (Chapter Author), **19**:98(8), 99(14), 100(14, 28), 102(14, 37), 103(14), 107(14), 108(14), 110(14, 37, 67), 111(14, 28, 37), 114(14, 37), 115(28), 116(28, 37, 67), 117(28, 37, 67), 118(14, 28), 126(131), 129(28), 130(28), 135(28), 136(14, 28), 137(28, 131, 144), 139(28), 140(131, 144, 147), 142(147, 149), 143(147), 144(28, 147), 145(8, 131, 144, 147, 158), 156(8, 187), 158(131), 159(8, 28, 144, 187), 160(14, 187), 161(8, 14, 144), 162(14, 147), 180(14, 147), 184(8), 185(14, 28, 37), 186(67), 188(131, 144, 147, 149, 158), 189(187); **25**:151, 152(1), 204(1, 102), 205(1, 103, 104, 105), 225(114), 226(115, 116), 227(1), 230(1), 231(1), 234(114), 238(1), 242(1), 247(102, 103, 104, 105, 114), 248(115, 116)

Gao, H., **17**:131(137), 132(137), 133(137), 134(137), 135(137), 156

Gao, S., **25**:361, 394, 412

Garandet, J. P., **30**:332(49), 333(57), 428(49), 429(57)

Garber, H. J., **20**:6(7), 79(7)

Garbuny, M., **11**:356(79), 358(79), 436

Garcia, A., **19**:334(147), 337(147), 355(147)

Garcia, H., **31**:464(13), 471(13)

Garcia, J., **23**:314(138), 360(138)

Gardiner, C. P., **6**:17(27), 18(27), 25(27), 27(27), 123

Gardiner, S. R. M., **21**:168, 180

Gardiner, W. C., **30**:295(94), 311(94)

Gardner, R., **5**:381, 509

Gardner, R. F., **11**:10(20), 48

Gardner, W., **13**:121, 201

Gardon, R., **11**:318(2), 326, 360(91), 365(115), 375, 376(131), 378(131), 382(131), 384, 385, 391(2, 149, 151), 397(52), 404(52, 91), 411, 434, 435, 436, 437, 438; **13**:6(26),

10(26, 30, 35), 11, 12, 15, 17(30), 18, 19, 26(35), 41(26), 60; **23**:5(5), 126(5), 300(98, 99, 100), 301(98), 302(98), 358(98, 99, 100); **26**:127(73), 132(73), 211(73)

Garfunkel, J. H., **11**:425(232), 440

Garg, D. R., **10**:243(36), 282

Garg, N. S., **25**:259, 286, 316

Garg, S. K., **14**:78(162, 165), 83, 84(162), 85(162), 88, 90, 92(180), 93(180), 103, 104; **24**:103(20), 185(20)

Garg, V. K., **26**:122(48), 210(48)

Garimella, S. V., **26**:123(62), 125(62), 132(62), 211(62); **28**:316(162), 336(162)

Garipay, R. R., **10**:73(65, 66), 82, 83

Garlick, G. F. J., **10**:19(28), 36

Garloin, E. L., **5**:326(12), 505

Garon, A. M., **30**:28(42), 29(42), 30(42), 31(42), 89(42)

Garret-Price, B. A., **31**:433(16), 471(16)

Garrett, P. H., **22**:367(15), 370(15), 373(15), 384(15), 386(15), 405(15), 412(15), 435(15)

Garrett, W. D., **9**:183(2), 199(2), 268

Garrett, W. E., **31**:456(17), 471(17)

Garribba, S., **17**:28(82), 59

Garris, C. A., **10**:253, 254(48), 255(48), 256(48), 257, 282

Garrison, C. J., **6**:411(74), 489

Garrison, J. D., **18**:20(48), 81(48)

Garside, J., **23**:243(83), 244(83), 257(83), 262(83), 270(83)

Garsuch, P. D., **2**:156(29), 269

Garth, J. C., **15**:287, 290, 323

Gartling, D. K., **24**:102(12), 184(12), 225, 226, 270

Gartner, J. R., **6**:474(295), 500

Gartrell, H. E., **4**:51(92), 64

Gartshore, I. S., **11**:220, 224, 238, 258

Gartside, G., **19**:123(119), 125(119), 187(119)

Gasith, M., **12**:233(23), 256(23), 278

Gaskell, D. R., **28**:255(61), 301(136), 302(61, 136), 303(136), 304(136), 306(136), 308(136), 319(61), 320(61), 331(61), 335(136)

Gaskill, H. S., **2**:386(57), 395; **7**:88(2), 91(2), 98(2), 108(2), 160; **12**:233(62), 251(62), 270(62), 280; **15**:128(151), 141(151)

Gaskins, F. H., **15**:182(45), 224(45)

Gaspari, G. P., **17**:28(81), 59

Gasser, R. D., **29**:234(76), 282(76), 288(76), 313(76), 339(76)

Gasterstaedt, J., **9**:135, 179

Gates, C. W., Jr., **1**:206(14), 264

Gates, J. W., **6**:201(46), 364

Gatos, H. C., **30**:402(170), 434(170)

Gatowski, J. A., **23**:341(277), 343(277), 367(277)

Gau, C., **19**:60(134), 62(146), 63(146), 67(146, 155), 68(155), 79(221), 89(134, 146), 90(155), 93(221)

Gaudet, J. P., **19**:78(190), 91(190)

Gaugler, R. S., **7**:219, 313; **20**:272(279), 312(279)

Gaul, W. W., **18**:26(57), 82(57)

Gauntner, J. W., **26**:127(74), 211(74)

Gautherie, M., **22**:11(17), 17(17)

Gavalas, G. R., **28**:343(53, 58), 417(53, 58)

Gaviglio, J., **6**:36(62), 125

Gavis, J., **13**:219(45), 265

Gavrilakis, S., **25**:330, 382, 412

Gavrilov, N. M., **14**:267(48), 269(56), 271, 276(48, 56), 277, 278(56), 280

Gavrilov, Y. A., **SUP**:193(380)

Gay, R. R., **15**:270, 281

Gaydon, F. A., **SUP**:346(491)

Gayhart, E. L., **6**:202, 364

Gazda, I. W., **30**:28(79), 91(79)

Gazley, C., **3**:23, 32; **5**:233(168), 251

Gazley, C. J. R., **21**:163, 168, 180

Geake, J. E., **10**:19(28), 36

Gear, A. E., **10**:46(20), 57(20), 81

Gebhart, B. (Chapter Author), **2**:407, 450; **4**:32, 39, 40, 41, 42, 63, 64; **5**:20, 51, 473, 516; **6**:211, 312(103, 104), 365, 366, 460, 469(187), 474(231, 239, 240, 258, 264), 475, 480(231, 239, 240, 258, 264), 481, 482, 483(258), 495, 497, 498, 499; **9**:273, 283(1), 286(1), 289, 290(12), 292(12, 15), 297(36), 299(45), 300(45, 48), 301, 304(55, 67), 305(45), 310(67), 311(70, 72, 73), 312(75, 76), 317(80, 81), 320(73), 324, 326(36, 95), 327, 328, 330(98, 99), 332, 333, 334, 335(90), 339(102), 342(102), 343, 346, 347, 348; **11**:203, 206, 208, 209(58), 216(58), 226, 248, 254, 259, 278, 279, 314; **14**:60(130), 66, 103; **15**:144(2), 223(2); **16**:1, 24, 47, 54, 55, 57; **18**:65(117a), 85(117a), 325, 330(1), 331(1), 332(1), 350(1), 359(1), 361(1); **19**:79(200), 92(200); **20**:203(94, 96, 97), 303(94, 96, 97), 316(12), 350(12); **24**:196, 217, 226,

Gebhart, B. (*continued*)
  274; **28**:239(9), 329(9), 407(222),
  425(222); **30**:280(65), 310(65), 362(125),
  413(125), 432(125)
Gebhart, E., **20**:203(95), 303(95)
Geddes, L. A., **22**:17(18), 246(77), 350(77)
Gee, D. L., **31**:406(68), 428(68)
Gee, J. M., **28**:343(40), 417(40)
Gee, R. E., **2**:377(33), 395; **13**:219(65), 265
Geel, P. J. Van, **30**:181(156), 196(156)
Geeraets, W. J., **22**:254(43), 256(43), 257(43),
  348(43), 350(86)
Geery, E. L., **5**:359, 360, 508
Geesey, G. G., **31**:445(18), 471(18)
Geffcken, W., **11**:326, 332(55), 344(65), 345,
  349, 363, 391, 435, 436
Gehrke, E., **2**:293, 353
Geiger, G. E., **19**:79(206), 92(206)
Geiger, G. P., **28**:20(37), 21(37), 72(37)
Geiger, G. T., **19**:71(162), 72(162), 90(162);
  20:377, 383, 388
Geis, T., **5**:132, 135, 247
Geising, J., **5**:133(22), 149, 155, 156, 247
Geist, G. A., **28**:76(108, 109), 143(108, 109)
Gelb, G. H., **5**:206, 249
Gel'd, P. V., **28**:78(91), 85(91), 142(91)
Geldart, D., **19**:99(9), 100(9), 105(9), 184(9);
  25:152(3), 242(3)
Geldart, G., **19**:99(16), 100(16), 102(16),
  185(16)
Gelder, D., **19**:37(81), 86(81)
Geller, A. S., **23**:320(165), 361(165);
  29:115(85), 127(85)
Geller, H. S., **17**:265, 277, 291(100), 294, 317
Geller, J. T., **30**:111(56), 191(56)
Geller, M. A., **14**:152(27), 242
Gelmont, B. L., **14**:122, 145
Gelperin, L. G., **10**:207(123), 217
Gelperin, N. I., **10**:174(40, 41), 175, 177,
  181(41), 198, 200, 201, 202, 203, 204(41),
  205(116), 206, 210(139), 215, 217, 218;
  14:151(6), 152, 155, 160(71, 72), 161(71,
  72), 163, 164, 166(71, 72), 167(71, 72),
  168(71, 72), 169(71, 72), 170(71, 72),
  172(71, 72), 176(71, 72, 106, 107, 163),
  180, 181, 188(106, 107), 190(106, 107),
  191(106, 107), 195, 203, 206, 206(138),
  220, 226, 228(164), 237(106, 107, 109,
  163), 244, 245, 246, 247; **19**:98(5), 100(5),
  102(5), 103(5), 111(5), 113(90), 116(5),

  119(101), 128(5), 129(5), 146(163),
  149(177, 178), 154(5), 159(191), 160(192),
  173(5), 184(5), 187(90, 101), 188(163),
  189(177, 178, 191, 192); **25**:153(7), 242(7)
Genereaux, R. P., **25**:293, 316
Genetti, W. E., **10**:209, 210(136), 218;
  14:177(97), 185(113), 192(121), 193(121),
  196, 197(126), 198(127), 199(127, 128),
  200(129), 201(129), 202, 204, 206,
  207(97), 210, 227, 228(122), 229, 244, 245;
  19:120(102), 122(102), 129(102),
  130(102), 131(102), 148(102, 171, 172,
  173, 174), 149(181, 182), 153(171),
  160(173), 187(102), 189(171, 172, 173,
  174, 181, 182)
Genin, L. G., **25**:48(72), 82(72), 122(72),
  143(72)
Génot, J., **9**:3(42, 44), 71, 79, 80, 91(44), 96,
  97(42, 44), 103, 108
Gentile, D., **17**:74(39), 123, 149, 152(182), 154,
  158
Gentry, C. C., **15**:170(30), 171(30), 224(30);
  25:280, 281, 286, 316
Genzel, L., **11**:326, 332(50), 358(81), 362(49,
  50, 100, 101), 397(50), 435, 436, 437
Geohegan, D. B., **28**:135(32, 33), 139(32),
  140(33)
George, A. H., **19**:164(214), 169(214), 170(214),
  190(214); **23**:299(94), 358(94)
George, C. W., **10**:234(13), 281
George, P. M., **10**:267(89), 283
George, W. K., **15**:92(90), 138(92), 188(57),
  224(57); **23**:317(152), 318(152), 320(185),
  345(298), 361(152), 362(185), 368(298)
Georgi, W., **13**:248(107), 266
Georgius, H. K., **5**:381, 388, 509
Geraghty, J. J., **12**:18(30), 20(30), 21(30), 26(30),
  74
Gerak, A., **20**:296(323), 314(323)
Gerasimov, S. V., **25**:81(143, 150), 146(143),
  147(150)
Gergau, A. A., **25**:127(196), 149(196)
Gerlach, L., **29**:84(49), 93(49), 94(49), 95(49),
  100(49), 125(49)
Germano, M., **25**:337, 351, 352, 353, 354, 360,
  361, 378, 392, 396, 404, 412
Gernier, E. M., **8**:49(228), 89
Gerrard, J. E., **2**:376, 378, 395; **13**:219(67, 68),
  248, 265
Gerrekens, P., **19**:81(257), 94(257)

Gerretsen, J. C. R., **9**:102, 110

Gerrish, R. W., **15**:324

Gersh, S. Ya., **26**:225(27), 226(27), 227(27), 325(27)

Gershuni, G. Z., **1**:295, 298, 353; **8**:168, 173, 212, 226, 227; **11**:297, 315; **20**:331(48), 332(50), 351(48, 50)

Gerstaman, J., **14**:137, 145

Gersten, K., **3**:30, 32

Gerstmann, J., **9**:255, 270

Gerz, T., **21**:185(7), 236(7)

Gessner, F. B., **19**:317(110), 324(110, 117), 339(110), 354(110, 117)

Getner, J. L., **28**:351(80), 418(80)

Getty, R. C., **11**:170(171), 196

Getz, R. J., **18**:326(2), 328(2), 329(2), 350(2), 361(2)

Geurts, B. J., **25**:364, 412, 419

Geuzens, P., **25**:176(71), 246(71)

Geva, M., **28**:343(41, 42), 417(41, 42)

Geyer, J. C., **12**:10(19), 12(19), 25(19), 73

Geyling, F. R., **30**:348(140), 371(140), 382(140), 433(140)

Ghaddar, N. K., **20**:243(174), 307(174)

Ghaffari, A., **15**:236, 277

Ghajar, A. J., **19**:333(131), 334(131), 355(131)

Ghaly, M. A., **16**:148, 149, 156

Gherson, P., **29**:171(137), 209(137)

Ghia, K. N., **24**:192, 244, 270

Ghia, U., **24**:192, 244, 270

Ghiaasiaan, S. M., **19**:80(226), 93(226)

Ghil, M., **19**:82(278), 95(278)

Ghosh, S. K., **15**:323

Ghosh, U. K. (Chapter Author), **25**:251, 257, 275, 276, 277, 278, 284, 285, 289, 290, 291, 306, 310, 316

Giakoumakis, S., **19**:78(190), 91(190)

Giamei, A. F., **28**:243(15, 16), 244(15, 16), 277(16), 280(15, 16), 303(15), 329(15, 16)

Gianetto, A., **24**:104(28), 185(28)

Giarratano, P. J., **5**:403, 404, 428, 440, 445, 511; **11**:175, 177(177), 185, 187(177), 197; **17**:150, 157, 323, 328(26), 333(26), 334, 335(26), 340

Giat, M., **6**:424(157), 430(157), 493

Gibbard, I. J., **31**:453(42), 472(42)

Gibbons, J. H., **17**:32(105), 61

Gibbs, J. W., **14**:287, 341

Gibeling, H. J., **30**:335(69), 336(69), 347(69), 368(137), 429(69), 433(137)

Gibert, T., **28**:110(34), 140(34)

Gibilaro, L. G., **23**:196(80), 270(80); **25**:199(99, 100), 230(99), 247(99, 100)

Gibson, A. H., **11**:218, 221, 236, 256

Gibson, J., **20**:83, 132

Gibson, M. M., **13**:81(43), 116

Gibson, R. D., **19**:74(174), 75(174), 91(174)

Gibson, T. S., **6**:73(123), 128

Gibson, W., **2**:266, 270

Gidaspow, D., **SUP**:28(38), 64(38), 202(38), 208(38), 209(38), 214(38), 215(38), 216(38), 218(38); **15**:321, 327; **19**:289(85), 291(85), 353(85)

Giddins, J. C., **12**:242(24, 26), 278

Gido, R. G., **29**:234(74, 79), 339(74, 79)

Giedt, W. H., **3**:12, 31; **7**:197, 202, 215, 216; **8**:121, 160; **11**:231, 232, 236, 259; **12**:129, 138(17, 27), 147(17, 27), 192, 193; **18**:116(50), 159(50); **23**:282(30), 283(30), 285(44), 355(30), 356(44)

Giegerich, W., **11**:391(147), 438

Gielisee, P. J., **11**:412(191), 439

Gier, H. L., **2**:53(59), 54(59), 55(59), 56(59), 57(59), 60(59), 61(59), 62(59), 64(59), 65(59), 66(59), 68(59), 69(59), 70(59), 72(59), 73(59), 94(59), 108

Gierasch, K. J., **14**:253(15), 279

Gierasch, P., **8**:281, 283

Gierasch, P. J., **14**:268(49), 280

Giere, A. C., **15**:323

Gieseke, J. A., **29**:329(182), 344(182)

Giess, E. A., **28**:341(15), 415(15)

Gifford, W. E., **9**:3(43), 82, 108

Gigliotti, M. F. X., **11**:318(5), 412(5), 413(5), 417(5), 434

Gigov, V. K., **25**:96(177), 148(177)

Gijbels, R., **28**:124(102, 103), 143(102, 103)

Gijsbers, G. H. M., **22**:279(209), 356(209), 426(38), 436(38)

Gilbert, C. S., **1**:201(10), 264; **10**:96(14), 97(14), 98(14), 99(14), 164

Gilbert, D. W., **29**:216(11), 217(11), 315(172), 323(172), 335(11), 344(172)

Gilbert, M., **30**:64(69), 90(69)

Gilbert, S. L., **28**:351(67), 418(67)

Gilbert, W., **16**:46(20), 57

Gilbertson, M. A., **29**:152(74, 75), 206(74, 75)

Gilcrest, A. S., **5**:477, 478, 479, 516; **9**:402(84), 406(100), 417

Gileadi, E., **12**:226(27a), 250(27), 251(27), 278

Giles, G. E., **19**:37(80), 86(80); **28**:76(110), 143(110)

Giles, M. B., **23**:346(300), 347(300), 368(300)

Gilfanov, K. Kh., **25**:61(98), 144(98)

Gilgenbach, R. M., **28**:135(25, 35, 66), 139(25), 140(35), 141(66)

Giling, L. J., **28**:369(120, 124), 373(120), 420(120, 124)

Gill, A. E., **8**:168, 171, 173, 176, 193, 216, 226; **9**:292, 324, 346, 348; **14**:66, 103; **18**:48(91), 84(91)

Gill, J., **30**:26(43), 89(43)

Gill, L. E., **1**:377(31), 385(31), 387, 389(31), 441; **7**:40(51), 42, 66(51), 67(51), 86

Gill, W. N., **2**:374, 378, 380, 395; **3**:167, 169, 170, 174; **6**:447, 494; **9**:296(34), 304(59), 343, 346, 347; **13**:164(54), 202; **SUP**:12(22), 188(22)

Gille, J., **8**:247, 282; **12**:180, 193

Gille, J. C., **14**:269(55), 270(55), 276(55), 277, 280

Gillen, A. G., **28**:64(83), 73(83)

Gilles, H. L., **16**:144(122), 156

Gilles, S. E., **8**:248, 249, 256(36), 273, 283; **21**:246(14), 275(14)

Gilli, M., **2**:322, 327, 353

Gilliland, E. R., **13**:120, 201; **16**:225(30), 227(30), 236, 239

Gillis, J., **SUP**:69(139), 73(187, 192, 196), 85(139), 92(187, 196), 93(187), 94(187, 192), 95(187), 98(187), 161(192), 165(192), 166(192), 167(196)

Gillis, J. C., **16**:265, 320, 321, 322, 324, 363

Gilmore, C. H., **16**:181(15), 238

Gilmore, F. R., **5**:309(77), 324

Gilmore, G. D., **18**:216(112), 238(112)

Gilmour, C. H., **5**:440, 517

Gilpin, R. R., **19**:75(175), 76(175), 77(180), 91(175, 180)

Gilsinn, D., **10**:275(105), 284

Ginevskii, A. S., **6**:410(72), 411(71, 72), 489

Gingrich, W. K., **24**:196, 217, 226, 274

Ginkhovskiy, A. V., **21**:30(46), 32(46), 52(46)

Ginn, R. E., **2**:361, 395

Ginn, R. F., **15**:182(47), 224(47)

Ginoux, J. J., **6**:33, 67, 73, 80, 87(185), 92(144), 93(144), 95(144), 96(144), 124, 127, 128, 129, 131

Ginsberg, T., **16**:12, 24, 39, 57; **29**:84(49), 93(49), 94(49), 95(49), 100(49), 125(49), 195(209), 213(209), 233(62, 70), 234(81), 259(118), 261(118, 119, 120, 121), 262(119), 263(118, 119, 120, 121), 264(118, 119, 120, 121), 265(118, 119, 121), 269(135), 274(146), 277(62), 282(118, 152), 328(181), 338(62, 70), 339(81), 341(118, 119, 120, 121), 342(135), 343(146, 152), 344(181)

Ginsburg, A. S., **14**:255, 279

Ginwala, K., **4**:205, 206, 226; **7**:241, 316; **9**:258, 262(103), 271

Ginzburg, A. S., **13**:120, 200

Ginzburg, V. M., **11**:423(221, 222, 223), 440

Giouanelli, R. G., **3**:206, 249

Giovannelli, A. D., **20**:296(322), 314(322)

Gipouloux, O., **23**:409(70), 463(70)

Giradkar, J. R., **25**:161(43), 162(43), 184(43), 200(43), 244(43)

Gishler, P. E., **10**:201(104), 202(104), 203, 206(104), 209(104), 217; **14**:213(146), 216(146), 217(146), 218(146), 222(146), 226(146), 246; **19**:122(114), 123(114), 129(114), 133(114), 134(114), 187(114)

Githinji, P. M., **20**:24(49), 27(49), 81(49)

Gitman, G. M., **14**:152(30), 243

Gittleman, J. I., **17**:69(19), 70(19), 71(19), 72(19), 153

Gittus, J., **22**:196(78), 350(78)

Glaberson, W. I., **17**:75(46), 89(59), 102, 154

Gladden, H. J., **23**:306(117), 319(155), 345(297), 351(297), 359(117), 361(155), 368(297)

Gladkov, A. A., **8**:13, 85

Glagoleva, N. N., **28**:84(36), 140(36); **30**:346(87), 430(87)

Glansdorff, P., **8**:26, 27(139), 29(142), 87, 87(137, 139); **24**:40(3), 46(3), 98(3)

Glaser, D. A., **1**:214, 264

Glaser, H., **13**:2, 18, 59; **SUP**:82(237), 154(237), 203(237)

Glaser, P. E., **9**:351(6), 352(6), 390(6), 414; **10**:42, 43, 45, 48, 50, 58(22), 59(22), 64, 68, 81

Glasov, V. M., **30**:346(87), 430(87)

Glass, B., **10**:25, 36

Glass, D. H., **14**:162, 244; **19**:112(82), 186(82)

Glass, I. I., **14**:336, 345

Glass, R. J., **30**:96(31), 190(31)

Glassen, L. K., **5**:293(36), 294(36), 301(36), 322; **8**:245(10), 256(10), 257(10), 266(10), 282; **31**:389(42), 427(42)

Glasser, O., **22**:4(19), 17(19)

Glassford, A. P. M., **9**:399(78), 404(78), 416

Glassman, I., **10**:267, 283

Glasstone, S., **2**:355; **22**:319(79), 350(79)

Glatz, G. A., **6**:474(250), 483(250), 498

Glatzmaier, G. C., **23**:370(12), 372(12), 411(12), 427(12), 438(12), 439(12), 442(12), 443(12), 461(12)

Glauert, M. B., **13**:2, 5(17), 11(17), 59

Glawe, G. E., **11**:219, 222, 257

Glazov, V. N., **28**:84(36), 140(36)

Glazunov, Yu. T., **19**:198(38, 39, 40, 41, 42), 244(38, 39, 40, 41, 42); **24**:42(15), 99(15)

Gleadow, P., **31**:440(19), 471(19)

Glebov, G. A., **25**:35(70a), 131(203), 132(203), 143(70a), 150(203)

Glenn, D. C., **29**:170(132), 209(132)

Glenn, L. A., **6**:474(324), 502

Glick, H. S., **6**:109, 130

Glicksman, L. R., **11**:407(171), 420(215), 423(215), 424(171), 439, 440; **12**:50, 61, 75; **14**:186(115), 187, 187(115), 188(115), 189(115), 191(115), 245; **18**:55(105), 56(105), 84(105); **19**:100(27), 112(78), 119(99), 136(27), 138(27), 139(27), 173(224), 174(224), 175(224), 176(224), 177(224), 181(224), 185(27), 186(78), 187(99), 190(224); **21**:96(76, 77), 97(76, 77), 125(114), 136(76, 77), 138(114)

Glicksman, M. E., **19**:79(205, 220), 80(205), 92(205), 93(220)

Glikman, B. F., **25**:4(21, 23), 140(21, 23)

Globe, S., **5**:207, 249; **11**:300, 315; **30**:28(44), 29(44), 89(44)

Gloersen, P., **1**:269(4), 352

Gluck, D. F., **2**:372(71), 396; **5**:351(58), 381, 507; **15**:196(84), 217(84), 219(84), 225(84)

Gluekler, E. L., **29**:1(1), 55(1)

Glushchenko, L. F., **17**:32(103, 104), 61

Glushko, G. S., **6**:398, 399, 488; **13**:80(34), 81(34), 83, 92, 116

Gnielinski, V., **13**:2(1), 6(1), 8, 9(1), 10(1), 11(1), 15, 16, 17(1), 25(1), 41, 42(1), 59; **18**:128(64), 159(64)

Gnilitsky, G. V., **18**:171(56), 176(56), 177(56), 212(56), 237(56)

Gnoffo, P. A., **23**:320(159), 361(159); **24**:213, 270

Gobin, D., **19**:62(144, 149, 151), 63(144), 64(144, 149), 65(144, 149), 66(144, 149, 151), 67(144), 89(144), 90(149, 151)

Gobis, L., **28**:343(46), 417(46)

Goblirsch, G., **19**:111(76), 121(76), 122(76), 186(76)

Gochernaur, J. E., **23**:319(154), 361(154)

Godbee, H. W., **11**:360(96), 437

Godbole, P. N., **SUP**:213(404)

Godefroy, J. C., **23**:280(9, 10), 294(72, 73, 74), 354(9, 10), 357(72, 73, 74)

Godfrey, J. C., **23**:192(84), 270(84); **25**:256, 316; **26**:2(71), 96(71)

Goel, I., **19**:113(91), 119(91), 120(91), 150(166), 151(166), 154(166), 187(91), 189(166)

Goel, S., **20**:203(89), 243(172), 303(89), 307(172)

Goela, J. S., **28**:342(18), 348(18), 350(18), 416(18)

Gogarty, W. B., **23**:222(59), 227(85), 269(59), 270(85)

Gogate, D. V. A., **11**:105(127), 115(127), 194

Goglia, G. L., **9**:199(13), 200, 201, 268

Gogolin, A. A., **31**:160(35), 323(35)

Gogonin, I. I., **16**:141(114), 156; **17**:7(33), 57; **21**:107(93), 137(93)

Gogos, C., **28**:146(62), 149(62), 150(62), 153(62), 184(62), 193(31), 223(62), 229(31), 230(62)

Gogos, C. G., **13**:219(70), 261(124), 265, 267

Gogos, G., **26**:56(101, 102), 57(101), 58(101), 59(72, 73), 60(72, 73), 63(73), 96(72, 73), 97(101, 102)

Goin, K. L., **6**:29(45), 124

Golan, L. P., **19**:113(92, 93), 121(92), 145(184), 187(92, 93), 189(184); **25**:233(118), 234(118), 248(118)

Gold, T., **10**:5(8), 19, 21, 24(27), 27, 36, 37, 57, 82

Goldberg, M. B., **16**:53, 57

Goldberg, N., **20**:244(196), 308(196)

Goldberg, P., **SUP**:139(320)

Goldemberg, J., **17**:179, 314

Golden, J. O., **19**:22(44), 85(44)

Goldenberg, H., **15**:324

Goldfarb, E. M., **19**:191(3), 242(3)

Goldman, K., **1**:411, 414(104), 442, 443; **4**:293(55), 315; **7**:30, 53, 84, 85

Goldman, R. F., **4**:93(33), 94(33), 95(33), 96(33), 138, 139

Goldmann, K., **6**:506, 515, 544, 558, 559(64), 562, 563

Goldobin, J. M., **14**:151(21), 160(21), 161(21), 166(21), 221(21), 242; **19**:100(22), 103(22), 116(22), 126(22), 127(22), 128(22), 164(210), 165(22), 169(210), 170(210), 171(22), 185(22), 190(210)

Goldshtik, M. A., **31**:160(30), 323(30)

Goldsmith, A., **11**:354(76), 436

Goldsmith, H. L., **20**:105, 131

Goldsmith, N., **28**:366(114), 369(114), 420(114)

Goldstein, C. A., **13**:219(48), 265

Goldstein, G. A., **SUP**:325(481)

Goldstein, R. J. (Chapter Author), **4**:16, 17, 38, 39, 41, 42, 63, 64, 157, 224; **6**:221, 365, 474(227, 242), 480(227, 242), 483, 497, 498; **7**:321, 331(3), 332, 337, 338, 340(24), 341(27), 342(28), 346, 348, 350, 353(3, 42), 354, 354(44), 356, 357, 359(42), 361(61, 62, 63), 364, 365, 366, 367, 367(62, 63), 368(24), 370, 371, 371(28, 71, 72), 372, 373, 373(28), 374, 375(71), 377, 378, 379; **8**:298(37), 349; **9**:311(71), 317(77), 347; **11**:300, 315, 422(218), 440; **SUP**:209(401), 210(401), 211(401); **18**:12(17), 26(58, 59, 60, 61), 27(61), 28(60), 79(17), 82(58, 59, 60, 61); **19**:47(113), 55(113), 56(113), 57(113), 88(113), 258(14), 259(14), 268(14), 350(14); **23**:279(5), 288(5, 45), 329(231), 354(5), 356(45), 365(231); **26**:182(140), 183(140), 184(140), 185(140), 215(140); **29**:4(2), 5(2), 9(10), 10(2), 55(2, 10); **30**:26(24), 27(24), 28(24, 42, 45, 46), 29(24, 42, 45, 46), 30(42), 31(42), 88(24), 89(42, 45, 46), 258(10), 278(54), 280(54, 63), 307(10), 309(54), 310(63)

Goldstein, R. W., **18**:156(85), 159(85)

Goldstein, S., **5**:149, 247; **6**:23, 123, 312(96), 366, 397(41), 488; **SUP**:69(138), 85(138)

Goldston, C. S., **22**:201(191), 355(191)

Golem, P. J., **15**:27(55), 29, 56(55)

Golfer, C., **17**:300, 317

Golik, R. J., **6**:109(174, 176), 110(174), 112(174), 113(174), 131

Golike, R. C., **2**:295, 327, 337, 355

Golitsyn, G. S., **14**:253(12), 266(42, 43), 268(51, 53), 269(54, 59), 270(51, 54), 271(51), 274(59), 276(42, 43), 279, 280

Golladay, R. L., **7**:358, 379

Gollub, J. P., **21**:147, 153, 154, 155, 179, 180, 182; **25**:356, 412

Golos, S., **SUP**:83(240)

Golosov, A. C., **8**:36(206), 89

Golosov, A. S., **6**:416(25), 487

Golovanov, L. B., **9**:402(86), 417

Golovastikov, P. P., **6**:418(102), 491

Golovnya, V. N., **25**:73(120), 145(120)

Golub, C. H., **14**:102

Golub, G. H., **23**:418(74), 464(74); **24**:201, 270

Golubev, A. V., **31**:278(180), 330(180)

Gomelauri, V. I., **9**:258(99), 271; **31**:203(108, 109), 208(108), 210(108, 109, 112), 304(207), 327(108, 109, 112), 332(207)

Gomez, A. V., **6**:23, 123

Gomez, R. D., **29**:221(25), 271(25), 309(25), 335(25)

Gomf, G. E., **2**:6(11), 105

Gomi, T., **26**:123(61), 125(61), 127(61), 129(61), 130(61), 131(61), 133(61), 155(61), 211(61)

Gomon, V. I., **30**:212(16), 221(16), 223(16), 250(16); **31**:200(102), 238(125), 240(126), 294(197), 304(197), 326(102), 327(125), 328(126), 331(197)

Gomory, R. E., **31**:404(64), 428(64)

Goncharov, E. I., **6**:418(133, 145), 492, 493

Gong, L., **30**:172(144), 195(144)

Gonsales, H., **1**:394(71), 398(71), 442

Gonzalez, F., **21**:287, 300, 344

Goobie, G. J., **18**:169(44), 170(44), 178(44), 220(44), 221(44), 233(44), 236(44)

Good, R. J., **14**:338(132), 346

Goodell, H., **4**:121(103), 141

Goodier, J. N., **SUP**:62(72), 67(72), 197(72), 264(72), 272(72)

Gooding, C. H., **23**:193(67), 269(67)

Goodling, J. S., **16**:234; **17**:74(41), 119(41), 154; **19**:42(95), 87(95); **20**:182(4), 252(241), 257(241), 258(241), 299(4), 310(241); **23**:32(60), 129(60)

Goodman, J. W., **30**:265(29), 295(29), 308(29)

Goodman, T. R. (Chapter Author), **1**:1, 56(17), 61, 68, 71(21), 72, 74(24), 75(24), 78, 86, 89(45), 95(45), 96, 108, 110, 113(60), 121, 122; **6**:474(197, 198), 475, 476, 480, 495, 496; **8**:2(9), 4, 24(9), 84; **10**:212, 218; **15**:324; **19**:2(3), 21(3), 34(3, 69), 35(3), 39(3, 69), 83(3), 86(69), 195(23, 24), 243(23, 24); **31**:107(5), 113(5), 138(5), 143(5), 145(5), 155(5)

Goodrich, J., **24**:306(42), 320(42)

Goodrich, L. E., **19**:27(55), 85(55)

Goodson, R. E., **11**:420(213), 439, 440

Goodwin, F. K., **6**:116(183), 117(183), 118(183), 131

Goodwin, G., **2**:247, 249(75), 270; **7**:189, 190(66), 200, 201(66), 214

Goodwin, J., **17**:128, 156

Goodwin, J. W., **23**:262(32a), 268(32a)

Goody, R. M., **3**:205(32), 249; **5**:254(12), 268(12), 283(12), 288(12), 291(12), 300(12), 301(12), 313(12), 322; **8**:238(6), 242(6), 245, 247, 248, 249(6), 250(6), 256, 273, 281, 282; **12**:7, 8, 73, 127(10), 134, 177(10), 180, 192, 193; **14**:253(13, 14, 15), 254(13, 14, 15), 255(13, 14, 15), 256(13, 14, 15), 262, 263(13, 14, 15, 31), 264(13, 14, 15), 266, 267, 268(49), 269(13, 14, 15), 273(13, 14, 15), 275, 279, 280; **21**:250(28), 276(28)

Goodykoontz, J. H., **5**:449, 515; **11**:170(184), 173(184), 174(184), 177(184), 185(184), 197

Goose, J., **15**:324

Gopalakrishna, S., **28**:151(18), 160(18), 215(18), 216(18), 217(18), 228(18)

Goranson, C. B., **24**:103(23), 185(23)

Gorbatchev, S. P., **17**:327(33, 34), 334(33, 34), 336(33, 34), 340

Gorbis, Z. R., **9**:119, 129(10), 167, 177, 180

Gorbunov, A. A., **28**:103(88), 142(88)

Gordeev, Yu. V., **17**:327(28), 328(28), 330(28), 332(28), 335(28), 336(28), 340

Gordon, **18**:328(18), 329(18), 350(18), 362(18)

Gordon, A. R., **SUP**:197(384), 209(384), 210(384); **19**:258(16), 259(16), 268(16), 350(16)

Gordon, H. S., **11**:202, 205, 253

Gordon, R., **30**:299(101), 311(101)

Gordon, R. G., **22**:153(28)

Gordon, R. J., **15**:63(33), 64(33), 69(46), 83(33), 136(33, 46)

Gordon, S., **29**:62(11), 124(11)

Gordon, S. A., **23**:315(142), 360(142)

Gordov, A. N., **6**:418(98, 99), 490; **19**:193(13), 243(13)

Gordy, L. H., **11**:105(113), 114, 194

Gorelik, A. G., **10**:194, 217

Gorelik, S. S., **22**:196(80), 350(80)

Goren, S. L., **7**:133, 161; **15**:236, 278; **25**:252, 319

Gorenflo, D., **16**:115(80), 117(80), 155, 171(5), 172(5), 173(5), 193(5), 215(5), 220(5), 221(5), 234, 238; **20**:23(43), 81(43)

Gorenflo, K., **20**:23(42, 47), 81(42, 47)

Gorkow, L. P., **30**:50(47), 89(47)

Gorla, R. S. R., **25**:261, 280, 290, 291, 316, 318

Gorland, S. H., **4**:211(96a), 227

Gorman, V. I., *see* Gomon, V. I.

Gorodov, A. K., **16**:235, 236; **17**:22(69), 59

Goroff, I. R., **1**:288(28), 353; **5**:207, 249

Gorokhov, V. V., **20**:24(54), 27(54), 81(54)

Goroshko, V. D., **19**:101(33), 185(33); **25**:157(18), 243(18)

Gorring, R L., **7**:308, 319

Gorring, R. L., **13**:164, 202; **18**:162(71), 179(71), 237(71)

Gorshkov, V. V., **31**:160(17), 323(17)

Gort, C., **11**:302(37), 315

Gorter, C. J., **17**:94(67), 154

Goryainov, L. A., **3**:243, 251

Gose, E. E., **9**:236(68), 237, 270

Goshayeshi, A., **19**:164(217, 218), 168(222), 171(217), 172(217), 190(217, 218, 222)

Gosman, A. D., **9**:99, 100, 108; **12**:237(40), 251(40), 279; **13**:12(43), 60, 67(9, 15), 102(15), 115, 236, 237(104), 266; **24**:193, 194, 199, 270

Gosman, A. O., **7**:132(43), 161

Gosse, J., **11**:208, 219, 223, 231, 232, 235, 237, 238, 239, 257; **22**:161(81), 350(81)

Gosting, L. J., **12**:206(6), 277; **18**:328(21, 22), 329(21, 22), 350(21, 22), 362(21, 22)

Goto, M., **21**:115(100), 137(100)

Gotsis, A. D., **28**:190(19), 193(31), 228(19), 229(31)

Gottfried, B. S., **30**:353(102), 354(102), 431(102)

Gottifredi, J. C., **19**:281(76), 285(76), 294(76), 352(76)

Gottlieb, D., **15**:321(21), 322

Gottlieb, J. J., **24**:213, 270

Gottzmann, C. F., **20**:76(67, 68, 69), 82(67, 68, 69)

Götz, J., **16**:171(7), 172, 173, 193(7), 215, 220, 221(7), 234, 238

Gough, M. J., **30**:227(21, 22), 228(21, 22), 250(21, 22); **31**:453(42), 472(42)

Goukhman, A., **11**:218, 221, 236, 256

Goulard, M., **1**:2, 6(7), 9, 21, 49; **3**:215(61), 225, 250, 251; **5**:28(37), 52

Goulard, R., **1**:2, 6(7), 9, 21, 49, 345, 346(76), 354; **3**:192, 213(58), 215(58, 61), 216(58), 217(58), 218(58), 219(58), 225, 249, 250, 251; **5**:28(37), 52; **11**:426(236), 440; **30**:288(75), 299(75), 310(75)

Goulard, R. J., **2**:136(20), 146, 268

Gould, T. L., **14**:98(198), 105

Gourrier, S., **28**:342(23), 416(23)

Gouse, S. W., Jr., **17**:30(95), 60

Gouy, M., **15**:2(1), 54(1)

Govier, G. W., **1**:376, 441; **20**:107, 108, 131; **23**:188(86), 192(86), 197(86), 270(86); **25**:253, 256, 316

Govinda Rao, N. S., **SUP**:70(149), 87(149)

Gowen, R. A., **31**:191(94), 211(94), 326(94)

Gozum, D., **21**:246(15), 275(15)

Gräber, H., **SUP**:84(244), 159(244), 296(244), 364(244)

Grace, J. R., **19**:99(12), 100(12), 103(47), 111(73), 118(73), 185(12, 47), 186(73); **21**:334, 344; **23**:340(270, 271), 367(270, 271); **25**:252, 316; **26**:2(55, 74), 12(75), 20(55), 23(55), 95(55), 96(74, 75); **28**:7(7), 70(7)

Grachev, N. S., **25**:51(78), 143(78); **31**:265(162), 329(162)

Graczyk, Cz., **6**:474(289), 500

Grad, H., **7**:170(20), 211

Gradon, L., **19**:79(201), 92(201)

Graetz, L., **13**:234, 266; **SUP**:2(1), 74(1), 79(1), 99(1), 125(1), 169(1)

Graf, A., **18**:67(124), 68(124), 85(124)

Graf, J., **25**:377, 412

Grafton, P. E., **15**:232, 277

Graham, C., **21**:82(34), 83(34), 91(34), 95(34), 96(34), 97(34), 135(34)

Graham, D. J., **11**:217, 228, 259

Graham, K. J., **29**:188(185), 212(185)

Graham, R. W., **1**:211, 216, 224, 225(21), 231, 264; **2**:95, 108; **4**:201, 206(86), 226; **5**:100, 127, 355(63), 360, 361, 362, 365, 368, 370, 372, 373, 374, 445, 446, 447, 448, 449, 450, 451, 458, 459, 507, 508; **6**:404, 488, 550(88), 564; **7**:53(35), 85; **10**:128, 148 (61), 149, 165; **11**:17(59), 18(59), 49, 170 (172, 180), 173(180), 174(180), 175(180), 177(180), 182, 183(172), 185(180), 186(180), 196, 197; **23**:344(294), 368(294)

Grand, D., **25**:346, 392, 406, 417

Grange, J. M., **6**:109(175), 119, 131

Grant, M. A., **14**:95(183, 184), 96(183, 184), 104

Grant, N. J., **28**:10(12), 11(12, 13), 12(13), 13(13), 14(13), 15(13, 20, 21), 17(12, 13), 18(12, 13), 19(12, 34), 20(36), 21(36), 22(21, 36), 23(12), 71(12, 13, 20, 21, 34, 36)

Granville, P. S., **2**:392, 395; **12**:87, 112

Grassi, W., **26**:106(12), 137(84, 90), 148(84, 90, 105, 106), 151(84, 90, 105, 106), 152(84), 154(84, 106), 155(106), 156(106), 167(84), 208(12), 212(84, 90), 213(105, 106)

Grassmann, P., **7**:116, 118, 120, 160; **11**:105(110, 111), 113, 114, 193; **15**:254, 280

Graumann, D. W., **7**:320b(141), 320e

Gravas, N., **21**:158, 180

Graves, D. B., **28**:382(155), 397(187), 398(187), 402(155), 404(155), 405(155), 422(155), 423(187)

Graves, D. J., **25**:235(139, 140, 141), 249(139, 140, 141)

Graves, R. S., **20**:264(266), 312(266)

Gray, L. D., **5**:293(34), 322

Gray, V. E., **10**:280(118), 284

Gray, V. H., **7**:223, 314

Gray, W. G., **13**:137, 144, 148, 149, 157, 158, 202; **14**:6(23), 7(23), 8(23), 10(23), 11(31), 27(23), 100; **23**:377(50), 378(51), 381(57), 463(50, 51, 57); **24**:114(114), 188(114); **28**:255(62, 63), 257(62, 63), 258(62, 63), 332(62, 63); **30**:95(20), 96(30), 98(20), 189(20), 190(30)

Grayson, J., **22**:30(35), 154(35)

Grcar, J. F., **28**:402(212), 424(212)

Greaves, M., **23**:193(87), 270(87)

Greber, I., **6**:95(156), 130

Greebler, P., **9**:369(39), 415

Green, A., **6**:197(32), 364

Green, A. E., **19**:268(30), 330(30), 350(30)

Green, D. M., **22**:228(82), 350(82)

Green, J. R., **14**:333, 334(119), 345

Green, J. T., **18**:237(87)

Green, J. W., **1**:103, 122

Green, L. C., **22**:231(200), 240(200), 356(200)

Green, M. J., **19**:81(256), 94(256)

Green, N. W., **10**:109(40), 164

Green, S. J., **1**:245(86), 266, 421, 443

Green, S. V., **1**:405(85), 409(85), 411(85), 442

Green, T., **15**:211(106), 225(106); **20**:323(20), 350(20)

Green, W. J., **17**:32, 61

Greenberg, H. J., **11**:425(230), 440

Greenberg, I.., **4**:67(5), 137

Greenberg, R. A., **6**:57(101), 58(101), 62(101), 64(101), 65(101), 107(101), 127

Greene, A. O., **20**:243(178), 308(178)

Greene, G. A. (Chapter Author), **21**:277, 279, 280, 281, 282, 287, 293, 299, 300, 301, 306, 307, 312, 313, 314, 316, 329, 330, 331, 333, 344, 345; **29**:16(13), 17(14), 56(13, 14)

Greene, H. L., **15**:196(82), 205(91), 225(82, 91)

Greene, L. C., **4**:119(81), 121(81), 123(81), 124(81), 125(81, 108), 127(81), 140, 141

Greene, N. D., **4**:206, 226

Greenkorn, R. A., **13**:164, 202; **20**:315(4), 349(4); **23**:188(88), 206(88), 270(88), 274(177); **24**:104(59), 105(33), 185(33), 186(59)

Greenshields, J. B., **14**:306(41), 342

Greenspan, H. P., **24**:286(6), 318(6)

Greenwood, W., **10**:55(43), 82

Greer, A. L., **28**:28(55), 36(55), 56(55), 72(55)

Greer, S. E., **2**:38(36), 106

Gref, H., **13**:35, 54, 60

Gregary, H. S., **2**:354

Gregg, J. L., **1**:33(29), 40, 41, 48, 50; **2**:427(28), 451; **3**:108(11), 173; **4**:12, 15, 17, 21, 23, 29, 42(18), 62, 63; **5**:132(16), 134(16), 144, 145, 247; **6**:474, 476, 480(228), 481, 495, 497; **9**:248, 249, 251(76), 261(114), 270, 271, 286(3), 287(5), 303(54), 325(5), 346, 347; **11**:213, 214, 247, 255, 261, 294, 315; **15**:185(51), 224(51); **20**:191(78), 302(78); **21**:67(9), 134(9); **24**:146(139), 189(139); **28**:369(123), 420(123); **30**:34(81), 91(81), 272(42), 309(42)

Gregor, L. C., **20**:244(193), 308(193)

Gregorash, D. A., **29**:271(141, 142), 342(141, 142)

Gregorig, R., **8**:94, 99(26), 115(44), 157(19), 159; **18**:88(19), 156(19), 158(19); **21**:116(101), 137(101)

Gregory, D. R., **23**:193(89, 237, 257), 204(89), 205(89), 218(237), 220(237), 270(89), 277(237), 278(257)

Gregory, G. A., **20**:83, 84, 85, 88, 91, 93, 97, 99, 100, 101, 109, 110, 113, 131

Gregory, H. S., **2**:353, 354

Gregory, J. J., **29**:234(78, 79), 328(78), 329(78), 339(78, 79)

Gregory, N., **5**:149, 156, 182, 247

Greif, R., **1**:195, 230, 259, 263; **5**:310(83), 324, 409, 411, 413, 414, 440, 513; **7**:177, 212; **11**:344(63), 348, 404(168), 436, 438; **12**:139, 193; **15**:128(154), 141(154); **18**:65(118), 69(118, 130), 70(118, 130), 85(118, 130); **19**:41(94), 80(240), 87(94), 93(240); **20**:187(35), 300(35); **21**:182; **24**:227, 270; **28**:370(128, 129), 371(129), 389(163), 391(163), 392(179, 180), 393(179), 420(128), 421(129), 422(163), 423(179, 180)

Greifinger, P. S., **1**:283(14), 284(14), 353

Greig, D., **18**:212(103), 222(103), 238(103)

Greiling, P. T., **30**:320(21), 427(21)

Grenier, P., **19**:186(61)

Grens, E. A., II, **12**:233(3, 28), 251(3, 28), 270(3, 28), 277, 278

Gresho, P. M., **24**:226, 227, 270; **30**:350(91), 430(91)

Greskovich, E. J., **20**:84, 85, 100, 102, 103, 109, 113, 130, 131

Gretsov, V. K., **2**:68, 108

Grevstad, P. E., **4**:210(93), 214(93), 227

Grewal, N. S. (Chapter Author), **14**:149; **19**:98(3), 108(58), 110(66), 111(3, 76, 77), 112(3, 79, 80), 113(79, 80), 117(80, 98), 118(98), 119(79, 104), 120(79), 121(76), 122(76, 79), 129(98), 145(3, 77), 146(159, 161), 147(159, 161), 149(159, 161), 150(77, 159, 161, 167), 151(159, 161, 167), 152(159, 161), 153(159, 161), 154(3, 77, 159, 161), 157(188, 189), 159(77, 188, 189), 160(77, 79, 188, 189, 197, 200), 161(3, 77, 200, 202), 172(79), 174(202, 227), 178(232), 180(79, 80, 98), 181(161, 189, 200), 184(3), 186(58, 66, 76, 77, 79, 80), 187(98, 104), 188(159, 161), 189(167, 188, 189), 190(197, 200, 202, 227, 232); **25**:164(48), 205(48), 214(48), 225(113), 244(48), 247(113)

Grey, J. D., **6**:95(152), 129

Gribanov, Y. I., **SUP**:355(496), 356(496)

Gribbon, P. W. F., **26**:202(167), 216(167)

Grief, R., **8**:253, 283; **16**:181(11), 238

Griem, H. R., **4**:234(5), 242(5), 245, 313; **5**:264(19), 277(19), 281(19), 303(19), 322

Grier, J. T., **12**:129(14), 138(14), 192

Griesmeyer, J. M., **29**:216(10), 335(10)

Griffel, D. H., **19**:82(276), 95(276)

Griffi, E. P., **19**:37(82), 87(82)

Griffin, R. N., **2**:164, 269

Griffis, C. L., **18**:179(58), 180(58), 237(58)

Griffith, M. V., **4**:135, 136, 141; **15**:324

Griffith, P., **1**:202(12), 210, 213, 218, 219, 221, 233, 264, 265, 359, 361, 390(59), 392, 406, 407(87), 440, 441, 442; **5**:419(194), 420, 514; **7**:37, 40, 84; **9**:204(19), 206(19), 207, 210, 231(64), 232(64), 233(64), 241, 255, 268, 270; **10**:122, 123, 125, 126, 127, 128, 130, 131, 132(59), 133, 152, 165; **11**:9(16), 10, 11(22, 23), 14(22, 23, 43, 49), 15(49), 16, 48, 49; **15**:271, 281; **17**:9, 32, 58, 61; **20**:15(30), 80(30), 100, 102, 110, 111, 131, 132; **21**:80(31), 134(31)

Griffith, R. M., **13**:233, 266; **26**:22(76), 96(76); **28**:149(20), 228(20)

Griffith, R. O., **29**:218(22), 221(29, 30, 31, 32, 33, 40), 234(22), 244(22), 253(109), 255(40), 265(40), 266(132), 269(40), 270(40), 271(29, 30, 31), 275(22), 278(40), 280(22, 29, 31), 281(22), 282(29, 31), 290(40), 293(29), 295(22, 29), 296(22), 297(22), 300(40), 301(22), 306(22), 307(22), 308(22), 309(29), 310(22, 32, 33, 40), 311(22, 32, 33), 315(30, 40), 316(30, 40), 318(40), 319(22, 40), 321(22), 323(40), 335(22), 336(29, 30, 31, 32, 33), 337(40), 340(109), 342(132)

Griffiths, E., **3**:8, 11, 31; **4**:50, 54, 55, 59, 64; **11**:213, 214, 218, 221, 236, 255, 256

Griffiths, S. K., **26**:29(77, 78), 30(77), 96(77, 78)

Griggs, D. T., **14**:97(195), 105

Grigor'ev, B. A., **11**:376(132a), 378(132a), 381(132a), 438

Grigor'ev, L. N., **16**:66, 116, 117, 125, 153, 155

Grigor'ev, V. A., *see* Grigoriev, V. A.

Grigoriev, L. V., **31**:282(186), 331(186)

Grigoriev, M. M., **25**:92(170, 171), 148(170, 171)

Grigoriev, V. A., **16**:233, 234; **17**:146, 157, 335(38), 340; **18**:241(2), 249(2), 256(2, 26, 27), 257(26, 27), 263(26, 27), 274(26, 27), 290(26, 27), 291(26), 294(27), 298(27), 310(2), 313(2), 320(2), 321(26, 27); **31**:160(38), 323(38)

Grigoriev, V. S., **21**:8(16), 11(16), 12(16), 32(50), 33(50), 51(16), 52(50)

Grigoriev, V. V., **16**:216, 239

Grigoriev, Y. M., **6**:474(205), 479, 496

Grigoropoulos, C. P. (Chapter Author), **28**:75, 80(113, 114), 81(114), 82(114), 83(114), 84(114), 85(114), 86(114), 87(113, 115), 88(114), 89(115), 90(115), 93(113), 94(113), 96(117), 97(117), 99(117), 100(117), 101(117), 102(117), 103(9), 104(9), 105(9), 106(9), 107(9), 113(9), 114(9), 116(9), 117(9), 119(10), 120(10), 121(10), 122(10), 123(10), 126(39), 130(39), 131(39), 132(39), 133(38), 134(39), 138(9, 10), 140(38, 39), 143(113, 114), 144(115, 117)

Grigoryants, A. N., **1**:435(140), 444

Grigull, U. (Chapter Author), **2**:374, 395; **4**:152(14), 208(14), 223; **6**:134, 150(11), 221, 312(87), 316(87), 331(90), 340(92), 363, 365, 366; **7**:55(37), 79, 81, 85, 86; **9**:11(106), 91(106), 110, 207(30), 262(126), 263(126), 269, 271; **11**:18, 49, 105(112), 114, 194, 281, 282, 283, 315, 363(105), 422(217), 423(217), 437, 440; **SUP**:21(34), 76(34), 102(34), 107(34), 119(34), 126(34), 128(34); **16**:225(29), 227(29), 236, 239; **17**:9(48), 58; **21**:92(64, 65), 136(64, 65); **22**:161(83), 350(83)

Grillaud, G., **21**:154, 156, 178

Grilly, E. R., **2**:327, 330, 354; **3**:273, 300

Grimado, P. B., **15**:324

Grimes, R., **29**:130(3), 202(3)

Grimison, E. D., **8**:94, 148, 149, 158; **11**:241, 242, 260; **15**:23(49), 56(49); **18**:88(9), 109(9), 146(9), 147(9), 158(9)

Grimley, T. A., **23**:89(101), 131(101)

Grimmble, R. E., **8**:135(68), 160; **18**:133(68), 134(68), 159(68)

Grimmet, E. S., **14**:177(97), 192(121), 193(121), 204(97), 206(97), 207(97), 229(97), 244, 245; **19**:120(102), 122(102), 129(102), 130(102), 131(102), 148(102, 172), 187(102), 189(172)

Grimmett, E. S., **10**:209(134), 210(136), 218

Grimshaw, C. A., **29**:234(81), 339(81)

Grinberg, G. A., **15**:287, 324; **19**:193(14), 243(14)

Grinchenkov, V. T., **8**:35, 88

Grindrod, K. J., **23**:321(191), 346(191), 362(191)

Gringarten, A. C., **14**:97, 104

Grishin, M. A., **10**:177, 216

Griskey, R. G., **13**:219(93, 95), 266; **15**:62(4), 128(153), 134(4), 141(153); **19**:281(71), 284(71), 294(71), 352(71); **23**:193(89, 237), 204(89), 205(89), 218(237), 220(237), 270(89), 277(237)

Grizzle, T. A., **9**:125(27), 127, 178

Gröber, H., **4**:152, 208, 223; **9**:262, 263, 271; **22**:161(83), 350(83)

Groden, C. M., **6**:497

Groehn, H. G., **23**:336(253), 366(253)

Groeneveld, D. C., **17**:39, 43(145), 48, 62, 63; **18**:257(47), 262(47), 267(47), 268(47), 269(47), 322(47)

Groenig, H., **23**:320(157), 361(157)

Grohse, E. W., **1**:394(74), 442

Groll, M., **7**:320c(152), 320d(153, 158, 159), 320e(160, 161, 162), 320f; **30**:117(71), 118(71), 192(71)

Grolmes, M. A., **19**:80(228), 93(228); **29**:147(49), 148(49), 175(147), 205(49), 209(147), 259(110), 341(110)

Gromeka, I. S., **6**:396, 487

Gromyko, A. G., **6**:416(90), 490

Groninger, G. L., **11**:430(240, 242, 243), 431(242), 440

Groothuis, H., **1**:427, 443

Grosbie, A. L., **6**:474(325, 329), 502

Grosh, R. J., **1**:12, 13(15), 14(15), 15, 16(15, 16), 17, 18(16), 19(16), 21, 22(16), 23, 26, 33, 49, 50, 429, 432, 443; **3**:203(24), 205(34), 207, 211(24), 219(34), 226, 229(77), 230, 231, 232(77), 233(77), 249, 250, 251; **4**:30, 63; **5**:27, 28(33), 52, 73, 141(32), 142, 143(32), 247; **7**:66(47), 85; **8**:34(167), 88; **9**:296, 346; **11**:67(32), 80, 82, 190, 390(140), 397, 400, 401(165), 438; **20**:353, 359, 360, 361, 387; **21**:37(57), 38(57), 52(57)

Groshev, A. I., **25**:51(79), 143(79)

Grosjean, C. C., **SUP**:156(338)

Gross, A., **21**:180

Gross, B., **14**:119, 145

Gross, D., **10**:235(25), 258, 265, 268, 280(117, 118), 281, 282, 283, 284

Gross, E. P., **7**:171, 172, 212

Gross, J. F. (Chapter Author), **3**:35(22), 98; **4**:317, 318, 326(4), 347, 354(4), 443, 446; **8**:12, 85

Gross, M. J., **14**:122, 126(34, 35), 145

Grossberg, P., **5**:241, 251

Grosse, E., **26**:304(84), 328(84)

Grosser, K. A., **30**:112(64), 191(64)

Grossman, G., **15**:271(111), 281

Grossman, L., **19**:82(268), 95(268)

Grossman, L. M., **1**:402(83), 404(83), 426, 442

Grosu, F. P., **14**:122(36, 37), 127, 132(55), 134(55), 145, 146; **31**:160(19), 323(19)

Grothaus, M. G., **31**:461(40), 472(40)

Grötzbach, G., **25**:326, 330, 333, 344, 368, 379, 380, 398, 399, 412

Grove, A. S., **7**:144(53), 161

Grove, F. J., **11**:358(84, 85), 436

Grover, C. P., **30**:256(6), 307(6)

Grover, G. M., **7**:220, 222, 227(3), 234, 246, 247, 247(61), 274(61), 313, 316; **20**:272(280), 312(280)

Grover, G. P., **30**:256(4), 307(4)

Grover, P. D., **26**:49(172), 50(170, 171), 100(170, 171, 172)

Grover, S. S., **30**:216(24), 250(24)

Groves, S. H., **28**:389(170), 422(170)

Grozev, G. T., **25**:158(25, 26), 159(25, 26), 200(25, 26), 236(25), 243(25, 26)

Grube, G. W., **29**:172(141), 209(141)

Gruber, C. E., **31**:458(20), 471(20)

Gruber, R. P., **4**:227, 228

Grubin, H. L., **30**:335(69), 336(69), 347(69), 368(137), 429(69), 433(137)

Grujin, S., **26**:246(52), 253(52), 270(52), 274(52), 279(52), 314(52), 316(52), 326(52)

Grumer, J., **30**:76(20), 88(20)

Grummer, M., **10**:181(64, 65), 189, 203, 216

Grummitt, W. E., **2**:355

Grunblatt, L., **12**:82, 86(13), 117

Grundy, I., **19**:338(152), 356(152)

Grunfest, I. J., **2**:378, 395; **13**:256, 266

Gruszczynski, J. S., **4**:280, 314; **5**:309(79), 324

Gryaznov, I. D., **31**:160(49), 324(49)

Gryte, C. C., **15**:326; **30**:96(28), 190(28)

Gu, C. B., **26**:123(66), 125(66), 148(107), 151(107), 154(107), 162(116), 163(116), 170(116), 211(66), 213(107, 116)

Gu, D., **23**:208(89a), 245(89a), 246(89a), 270(89a)

Guay, K. P., **29**:166(118, 120), 208(118, 120)

Gubin, V. E., **SUP**:70(147), 87(147)

Guenette, G. R., **23**:288(76), 295(76, 77), 346(299, 300), 347(300), 357(76, 77), 368(299, 300)

Guenigault, R., **19**:60(135), 89(135)
Guenther, M. E., **2**:75(69), 95(69), 108
Guenther, R. B., **14**:6(28), 100
Guernsey, R., **17**:143(160), 151(160), 157
Guerrieri, R., **28**:402(208), 424(208)
Guerrieri, S. A., **1**:426, 443
Guerry, D., **22**:251(85), 350(85)
Guevara, F. A., **7**:280, 319
Guggenheim, E. A., **12**:200, 202(30), 226(29, 30), 259(30), 261(29, 30), 268(29, 30), 269, 271, 278
Guglielmini, G., **23**:9(24), 127(24)
Guiglian, C., **14**:152(35), 243
Guillemin, V., Jr., **4**:75, 138
Guirao, C. M., **29**:96(62, 63), 98(67), 100(67), 126(62, 63, 67)
Guiraud, J. P., **3**:225, 251
Gukhman, A. A., **6**:390(15, 19), 405, 486; **11**:168, 170(168, 170), 196
Gulaev, V. A., **11**:18(63), 19(63), 31(63), 49
Gulglion, C., **10**:185(77), 216
Gulin, N. P., **25**:81(155), 147(155)
Gullino, P. M., **22**:11(25), 18(25)
Gumley, P., **31**:247(141), 248(143), 250(143), 328(141, 143)
Gummalam, S., **23**:208(90, 91), 270(90, 91)
Gundappa, M., **23**:340(275, 276), 341(276), 342(276), 367(275, 276)
Gunn, D. J., **SUP**:278(451), 357(451), 360(451)
Gunn, G. C., **8**:19, 86
Gunness, R. C., Jr., **9**:334, 348
Gunter, A. Y., **25**:295, 316
Gunther, F. C., **1**:219, 226, 231, 264, 265
Guo, K.-F., **27**:167(30)
Guo, K.-H., **27**:ix(15), 187(36, 37, 38), 193(37, 38)
Guo, Z.-Y., **30**:274(48, 49), 275(48), 276(48), 309(48, 49)
Gupalo, Iu. M., **26**:21(79), 96(79)
Gupalo, Yu. P., **25**:163(46), 244(46)
Gupta, A., **28**:103(59), 141(59)
Gupta, A. S., **1**:293, 353; **20**:329(44), 351(44); **25**:192(91), 246(91)
Gupta, C. L., **6**:474(256), 483(256), 498
Gupta, J. P., **13**:125, 169, 201; **19**:62(152), 67(152, 153), 90(152, 153); **28**:251(34, 35), 330(34, 35)
Gupta, K. P., **30**:328(30), 371(141), 372(141), 373(141), 374(141), 375(141), 376(141), 377(141), 378(30, 141), 379(141), 380(141), 381(141), 384(141), 392(141), 406(30, 141), 407(30), 408(30, 141), 409(30), 427(30), 433(141)
Gupta, M., **28**:150(21), 166(21), 208(22), 211(22), 228(21, 22)
Gupta, M. K., **12**:88, 102, 103, 105, 106, 107(34), 108, 112, 113; **15**:104(123), 111(123), 121(123), 139(123)
Gupta, M. L., **6**:474(256), 483(256), 498
Gupta, O. P., **25**:257, 286, 317
Gupta, R. C., **2**:390, 395; **SUP**:70(150), 160(150); **19**:268(34), 270(34), 351(34)
Gupta, R. K., **23**:206(92), 219(92, 104), 223(92), 270(92), 271(104)
Gupta, R. P., **31**:371(37), 426(37)
Gupta, R. S., **19**:41(93), 87(93)
Gupta, S. C., **8**:26(132), 30(150), 31(150), 32, 87; **SUP**:67(120)
Gupta, S. N., **25**:257, 276, 284, 285, 289, 290, 291, 293, 297, 298, 300, 304, 305, 306, 309, 310, 316, 318
Gupta, V. P., **14**:33, 34(82), 72(82), 101
Gurney, G. B., **15**:92(90), 138(90)
Gurtcheff, G. A., **19**:61(138), 89(138)
Gurtin, M. E., **8**:33, 88; **18**:163(18), 236(18)
Gurtman, G. A., **15**:325
Gurukul, S. M. K. A., **26**:255(72, 73), 304(72, 73), 305(72, 73), 327(72, 73)
Gurvich, A. M., **3**:218(65), 250
Gusev, **31**:458(39), 472(39)
Gusev, J. A., **6**:474(206), 496
Guseva, I. N., **11**:423(221, 222, 223), 440
Gustavsson, V., **29**:234(91), 321(91), 340(91)
Gutfinger, C. (Chapter Author), **10**:167, 211(140, 141, 142), 212, 218; **13**:261(127), 267; **14**:151(3), 242; **23**:222(159), 274(159)
Guthrie, R. I. L., **28**:128(41), 140(41)
Gutierrez-Miravete, E., **28**:10(12), 11(12, 13), 12(13), 13(13), 14(13), 15(13), 17(12, 13), 18(12, 13), 19(12), 23(12), 64(82), 71(12, 13), 73(82)
Gutowski, T. G., **25**:252, 292, 293, 316
Guy, A. C., **19**:37(81), 86(81)
Guyton, Arthur C., **4**:141
Gvozdenac, D. D., **20**:173(20), 179(20); **24**:48(19), 99(19); **26**:231(34), 234(34), 236(34), 246(34), 251(34), 253(34), 261(34), 269(34), 270(34), 271(34), 281(34), 282(34), 283(34), 287(34),

291(34), 314(34), 316(34), 320(34),
322(34), 323(34), 326(34)
Gyarmathy, G., **14**:333, 334(121), 345
Gyorog, D. A., **9**:390(61), 391(61), 416

H

Ha, M. Y., **28**:307(150), 336(150)
Haaland, S. E., **14**:69(150), 103
Haar, L., **31**:115(6), 117(6), 155(6)
Haas, F. C., **5**:230, 233, 235(161), 251; **21**:180
Haas, J. C., **21**:273(44), 276(44)
Haas, R., **23**:206(63, 64, 65, 93), 219(63, 64,
65), 222(63, 64, 65, 93, 94, 150, 151),
223(93), 225(63, 64, 65, 93, 94, 150, 151),
269(63, 64, 65), 271(93, 94), 273(150, 151)
Haase, R., **12**:224, 259(30a), 261(30a), 269(30a),
278; **16**:65, 153
Haasen, P., **30**:358(118), 360(118), 432(118)
Habchi, S., **20**:243(170), 307(170)
Habchi, S. D., **24**:199, 275
Habdas, E. P., **20**:76(67), 82(67); **30**:219(76),
253(76)
Haben, R. L., **17**:117, 155
Haber, F., **6**:197(35), 364
Haber, S., **15**:324
Haberman, W. L., **15**:250, 280; **26**:49(80), 96(80)
Habetler, G., **9**:73(53), 76(53), 108
Habib, I. S., **8**:253, 283; **11**:344(63, 67), 345(67),
348, 349, 412(202, 203), 415(202, 203),
416, 436, 439; **19**:37(77, 78), 38(77),
86(77, 78)
Habip, L. M., **4**:146, 148, 222
Hackenberg, C. M., **26**:79(112, 113), 98(112, 113)
Hacker, D. S., **3**:71, 99
Hackworth, J. H., **SUP**:201(389), 202(389)
Hadfield, M. G., **25**:377, 412
Hadid, A., **29**:149(57), 205(57)
Haenseler, F., **5**:494, 517
Hafer, C. A., **10**:31, 281
Haffner, H., **20**:23(46), 81(46)
Hagemeyer, W. A., **10**:71(63), 73(63, 64, 65), 82,
83
Hagen, G., **SUP**:78(221)
Hagen, S. L., **SUP**:66(113), 350(113)
Hager, J. M., **23**:296(78, 79, 80, 81, 82), 297(79,
80), 298(81), 341(79), 343(79), 347(302),
348(302), 349(302), 352(81, 82), 357(78,
79, 80, 81, 82), 368(302)

Hager, N. E., Jr., **11**:360(97), 437; **23**:292(53,
61), 356(53, 61)
Hager, W. R., **14**:202, 245
Hagmann, M. J., **22**:30(15, 42), 153(15), 154(42)
Hahn, G., **6**:265(79), 365
Hahn, S. J., **2**:361, 395
Hahne, E., **16**:236; **17**:8, 9(48), 58
Hahne, E. W. P., **9**:91, 108; **21**:3(7), 42(7), 43(7),
44(7), 51(7)
Hahnemann, H., **9**:41, 108
Hahnemann, H. W., **SUP**:154(335)
Haigh, C. P., **16**:232, 233, 235, 236
Haimes, R., **23**:346(300), 347(300), 368(300)
Haines, K. A., **6**:200(45), 364
Hains, F. D., **1**:314, 316(51), 354
Haise, F. W., Jr., **4**:150(6a), 222
Hajicek, D. R., **19**:160(197), 190(197)
Haji-Sheikh, A., 40(68)(70)(71), **5**:44, 53, 54
Haji-Sheikh, A., **6**:415(79), 489; **7**:337, 338,
356, 357, 360, 361(62, 63), 364, 365, 366,
367(62, 63), 377, 379; **SUP**:66(114),
232(423), 236(423), 237(423), 262(114),
264(423, 441), 265(441), 266(423, 441),
267(114), 268(114); **31**:334(9), 344(31),
425(9), 426(31)
Hakkinen, R. J., **6**:95(156), 130
Hala, E., **16**:61(4), 153
Halajian, J. D., **10**:73(58), 82
Halbritter, J., **17**:68(8), 153
Haldie, P. H., **5**:453, 515
Hale, B. N., **14**:306, 343
Hale, D. V., **9**:404(88), 417
Hale, F. J., **4**:307, 309, 310, 316
Hale, G. M., **22**:426(39), 436(39)
Hale, N. W., Jr., **19**:45(105), 62(142), 63(142),
88(105), 89(142)
Haley, P. J., **28**:41(69), 73(69)
Halin, J., **31**:138(44), 157(44)
Hall, A. L., **4**:215, 218, 227
Hall, D. M., **11**:13(34), 48
Hall, H. E., **17**:88, 91, 154
Hall, I. S., **11**:13(33), 48
Hall, J. A., **29**:260(114), 263(114), 264(114),
265(114), 341(114)
Hall, N. A., **20**:89, 131
Hall, R. G. N., **6**:201(46), 364
Hall, R. S., **29**:137(26), 140(26), 185(26), 203(26)
Hall, R. W., **29**:137(26), 140(26, 33, 34), 146(39,
44, 45), 148(44), 152(74), 153(77),
159(89), 172(44), 185(26), 186(45),

Hall, R. W. (*continued*)
  203(26), 204(33, 34, 39, 44, 45), 206(74, 77, 89)
Hall, S. F., **29**:73(34), 74(34), 75(34), 80(34), 81(34), 125(34)
Hall, V., **31**:245(136), 328(136)
Hall, W. B. (Chapter Author), **3**:41, 42, 98, 106, 172; **6**:550, 551, 563; **7**:1, 24, 31, 35, 50, 52(12), 54, 66, 66(52), 68, 70(53), 72, 72(53), 83, 84, 86, 249, 275(74, 114), 317, 318; **SUP**:83(242), 85(242, 245), **21**:1(2), 4(2), 10(23), 13(2), 14(23), 16(23), 30(2), 32(49), 50(2), 51(23), 52(49); **25**:28(57), 142(57)
Hall, W. D., **26**:8(124), 12(124), 44(124), 98(124)
Haller, H. C., **7**:274, 274(102), 318
Halliwell, N. A., **15**:92(91), 138(91)
Hallman, T. M., **3**:126(34), 132(34), 173; **9**:169, 180; **12**:96(29), 113; **SUP**:18(29), 125(29), 129(29), 135(29), 136(29), 145(29), 179(29), 188(29), 414(532); **15**:100(102), 120(142), 121(142), 139(102), 140(142)
Hallock, K. R., **14**:134(47), 145
Hall-Taylor, N. S., **17**:2(7), 42, 56; **20**:250(215), 309(215)
Halmos, A. L., **24**:19(52), 38(52)
Halsey, G. H., **14**:118, 148
Halteman, E. K., **15**:324
Ham, A. W., **22**:224(84), 350(84)
Ham, W. T., Jr., **22**:251(85), 254(43), 256(43), 257(43), 348(43), 350(85, 86)
Hama, F. R., **6**:35(59a, 59b, 59c), 44, 50(59a, 59b, 59c), 51(59a, 59b, 59c), 52(59a, 59b, 59c), 59, 105(59a, 59b, 59c), 125; **11**:241, 242, 263; **17**:94, 154
Hamadah, T., **20**:243(173), 307(173)
Hamaker, H. C., **3**:204(28), 249
Hamano, Y., **10**:67(53), 82
Hamby, R. K., **25**:153(12), 195(12, 96), 201(12), 232(12), 236(12), 243(12), 247(96)
Hamed, A. M., **30**:296(98), 311(98)
Hamielec, A. E., **23**:208(161), 274(161); **25**:295, 296, 317; **26**:8(124), 10(1), 12(124), 19(241), 44(124), 53(245), 92(1), 98(124), 104(241, 245)
Hamill, T. D., **5**:90, 126; **11**:84(46), 86, 95(46), 105(98), 191, 193; **25**:12(46), 141(46)
Hamilton, D. C., **2**:411, 412(10), 417(10), 450; **9**:73(47), 76(47), 108

Hamilton, J. C., **11**:420(205), 439
Hamilton, R. L., **18**:169(31, 36), 170(31, 36), 173(31, 36), 181(36), 184(31), 211(31), 212(31), 236(31, 36)
Hamilton, R. M., **6**:526, 562; **7**:108, 109, 160
Hammecke, K., **8**:146, 148, 149, 160; **18**:145(78), 146(78), 147(78), 159(78)
Hammel, E. F., Jr., **5**:327(14), 505; **17**:84, 104, 105(91), 106(91), 154, 155
Hammel, R. L., **4**:228
Hammerling, P., **2**:127(15), 131(15), 134(15), 268
Hammersley, J. M., **5**:2, 3, 11(2), 12(2), 13, 47, 50; **31**:334(5), 425(5)
Hammersley, R. J., **29**:220(55), 221(55, 56), 234(85), 269(85), 276(56), 316(173), 318(173), 323(56), 338(55, 56), 339(85), 344(173)
Hammitt, F. G., **9**:3(48), 25, 35, 107, 108; **10**:87(2), 163; **19**:113(92), 121(92), 187(92)
Hammond, M. B., Jr., **9**:404(89), 417
Hammond, M. L., **28**:344(62), 365(62), 418(62)
Hammond, R. B., **28**:80(31), 139(31)
Hammond, W. E., **5**:244, 251
Hammond, W. H., **4**:94(34), 138
Hampel, V. E., **7**:236(32), 240, 275, 294, 296, 315
Hampson, H., **9**:265(138), 272; **21**:102(79), 136(79)
Han, A. J., **22**:320(87), 350(87)
Han, B. S., **8**:70, 90; **SUP**:249(435), 250(435), 251(435)
Han, C., **11**:11(22, 23), 14(22, 23), 15, 16, 48
Han, C. D., **28**:152(23), 228(23)
Han, C.-Y., **10**:132, 166
Han, J. C., **31**:259(150, 151), 329(150, 151)
Han, J. H., **28**:351(83), 418(83)
Han, J. T., **8**:195, 227
Han, L. S., **5**:469, 470, 473, 516; **11**:228(156), 259; **SUP**:62(70), 70(154, 155, 156), 71(154, 156, 158), 77(155), 163(154), 168(154), 192(155), 197(70), 198(154, 158), 199(154, 158), 200(158), 205(70), 208(70, 398), 210(154, 158), 213(158), 223(70), 224(70), 234(70), 236(70), 240(156, 158), 241(158), 242(156, 158); **19**:258(12), 268(45, 47, 49), 270(45, 47), 271(45, 49), 272(49), 274(49), 350(12),

351(45, 47, 49)

Han, S. H., **29**:149(55, 56), 205(55, 56)

Han, T., **24**:227, 270

Hanamura, K., **30**:154(120), 194(120)

Hanasaki, K., **23**:109(114, 115, 116), 110(114, 115, 116), 113(114, 115, 116), 114(116), 115(115, 116), 132(114, 115, 116)

Hanby, V. I., **30**:83(48), 89(48)

Hand, J. H., **15**:63(24), 64(24), 135(24)

Hand, J. W., **22**:290(88), 350(88)

Handler, R. A., **25**:359, 415

Hands, D., **18**:161(7), 166(7), 167(7), 169(7, 28), 170(28), 173(28), 174(28), 213(7), 217(28), 218(28), 219(7), 220(116), 236(7, 28), 238(116)

Handscomb, D. C., **5**:2, 3, 11(2), 12(2), 13, 47, 50; **31**:334(5), 425(5)

Hanel, R. A., **8**:281, 283

Hanesian, D., **11**:213, 215, 255; **14**:133(63), 146

Hanin, M., **24**:315(7), 318(7)

Hanjalič, K., **8**:289, 348; **25**:367, 412

Hanke, C. C., **9**:407(102), 417

Hankel, R., **3**:109(16), 173

Hankey, W. L., **6**:114, 131; **26**:201(164, 165, 166), 202(165), 203(166, 168, 172, 173, 174, 175), 216(164, 165, 166, 168, 172, 173, 174), 217(175)

Hanks, R. V., **3**:255(6), 256(6), 260(6), 300

Hanks, R. W., **2**:374, 383, 395; **15**:196(90), 205(90), 225(90)

Hanna, M. G., **22**:353(141)

Hanna, M. R., **23**:198(142, 146), 206(142, 146), 216(95), 222(95), 223(142), 226(95, 142, 146), 228(142), 229(95, 142, 146), 271(95, 146), 273(142, 146)

Hanna, O. T., **9**:304(62), 347; **SUP**:50(58), 51(58, 59), 126(58), 181(59), 420(58); **15**:122(148), 141(148)

Hanna, S. R., **12**:24(44), 74

Hannah, D. M., **5**:167, 248

Hannemann, R. J., **20**:183(15, 17), 184(17), 292(318), 293(318), 294(318), 299(15, 17), 314(318); **21**:88(56), 136(56); **23**:298(89), 358(89)

Hannes, H., **6**:205, 211, 275, 320, 322(89), 364, 365, 366

Hanratty, T. J., **2**:388, 396; **7**:88, 108, 110, 110(25), 112, 114, 117, 120, 134, 134(46a), 140, 142(30), 160, 161; **8**:293(18, 19), 299(40), 305, 349; **9**:156, 180; **12**:233(18,

71, 86, 110), 251(18, 71, 86, 110), 270(18, 71, 86, 110), 271(71), 278, 280, 281, 282; **SUP**:414(534); **15**:122(146), 128(146), 140(146), 141(146), 195(75, 76, 77), 225(75, 76, 77); **20**:85, 90, 98, 130; **23**:337(261), 366(261)

Hanrew, W. L., **6**:406(61), 407(61), 409(61), 410(61), 489

Hansen, A. G., **9**:296, 346; **15**:153(13), 185(13), 223(13); **25**:262, 318

Hansen, C. F., **2**:123(14), 133(14), 268; **3**:275(46), 276(46), 301

Hansen, D., **13**:211(16), 264; **18**:161(2), 219(114), 220(117), 236(2), 238(114, 117)

Hansen, G., **6**:211, 365

Hansen, J. E., **23**:147(31), 185(31)

Hansen, J. S., **24**:213, 270

Hansen, L. W., **22**:30(31), 134(31), 153(31)

Hansen, P. B., **31**:408(71), 428(71)

Hansinger, S. H., **5**:177, 249

Hanson, A. R., **6**:40(75), 41(75), 65(75), 66(75), 75(75), 77(75), 125

Hanson, C., **26**:2(71), 96(71)

Hanson, F. B., **4**:334, 445

Hanson, J. P., **7**:235, 284, 285, 314

Hanson, W. B., **5**:406(169), 512

Hansson Mild, K., **15**:287, 326

Hantman, R., **8**:172, 213, 214, 215, 216, 218, 222, 227

Hanzawa, T., **28**:369(122), 420(122)

Hapke, B., **10**:2(1), 7(1), 9, 10, 19(25), 21(31), 24(25), 35, 36

Happel, J., **5**:308(73), 309(73), 323; **23**:207(96, 97), 208(97), 271(96, 97); **25**:292, 293, 297, 316, 317

Happel, O., **16**:114(73), 117(73), 155

Haq, S., **24**:167(150), 170(150), 171(150), 190(150)

Haquet, Y., **31**:453(25), 471(25)

Hara, T., **4**:29, 63

Harada, I., **24**:315(8), 319(8)

Harada, J., **23**:9(20), 109(20, 117), 110(20, 117), 113(20, 117), 115(20, 117), 119(20), 127(20), 132(117)

Harada, K., **17**:50(161), 63

Harakas, N. K., **14**:151, 153, 208, 217, 242

Haramura, Y., **17**:12(54), 15, 16(54, 57), 18(54, 57), 58; **23**:58(107), 60(107), 63(81), 96(81), 98(81), 100(81), 102(107), 130(81), 131(107)

Harasgama, S. P., **23**:320(171), 361(171)

Harbaugh, W. F., **7**:223, 223(10), 240(51), 248(10, 65), 273(10), 274(51), 275(65), 314, 315, 316

Harbour, P. J., **7**:204, 217

Harchenkov, S. V., **21**:34(54), 52(54)

Hardee, H. C., **14**:98(196, 197), 105; **29**:38(27), 40(27), 56(27); **30**:134(91), 135(91), 193(91)

Harder, R. L., **4**:283(48), 315

Hardie, A. D. K., **17**:177, 314

Hardy, J., **31**:410(74), 428(74)

Hardy, J. D., **4**:67(6, 8), 69(12, 13), 70, 71(13), 75(12, 17), 81(18), 82(19), 104(47), 116(73), 117(73), 119(80, 83), 121(97, 99), 123(99), 125(108), 137, 138, 139, 140, 141; **11**:318(26), 376(26), 435; **22**:20(29, 30), 21(29, 30), 30(57), 82(30), 134(57), 153(29, 30), 155(57)

Hare, E. F., **9**:231(62), 269

Harel, O., **25**:194(95), 234(131), 247(95), 248(131)

Hargis, J. H., **5**:356, 363, 364, 508

Hargis, P. J., Jr., **28**:343(40), 417(40)

Harith, M. A., **28**:112(79), 142(79)

Harkee, J. E., **11**:170(185), 173(185), 175(185), 177(185), 197

Harkee, J. F., **5**:445, 446, 447, 448, 449, 514

Harker, J. H., **26**:14(81), 96(81)

Harkins, W. D., **15**:232(34), 277

Harkness, A. L., **3**:293(93), 302

Harleman, D. R. F., **11**:318(31), 376(31), 435; **12**:12, 74; **13**:108, 109, 117

Harlow, F. H., **13**:80(31), 94(31), 116; **14**:97, 104; **24**:194, 270, 315(9), 319(9); **29**:184(176), 211(176)

Harman, B. N., **16**:234

Harman, G. G., **11**:105(113), 114, 194

Harmathy, T. Z., **4**:190, 225; **20**:108, 131

Harmon, J. C., **30**:222(10), 223(10), 250(10)

Harned, H. S., **12**:234(31), 235(31), 236(31), 237(31), 240(31), 256(31), 259(31), 269(31), 278

Harnon, B. N., **5**:405, 511

Harper, E. Y., **2**:75, 76(68), 88(68), 89(68), 90(68), 91(68), 92, 93(68), 108; **4**:153(24), 223; **5**:390, 391(119, 120), 392, 393, 394, 395, 510; **29**:172(141), 209(141)

Harper, J. C., **SUP**:80(231), 121(231), 149(231), 150(231)

Harper, J. F., **26**:2(82), 10(82, 83), 96(82, 83)

Harper, J. M., **28**:149(24), 184(24), 213(24), 228(24)

Harpole, G. M., **26**:37(84), 66(84), 96(84)

Harr, M. E., **24**:104(50), 186(50)

Harrington, R. E., **23**:193(98), 223(98), 271(98)

Harrington, S. A., **7**:206, 218

Harriott, P., **7**:108, 109, 160; **15**:267, 268, 269, 270(103), 280

Harris, J. A., **18**:44(87), 45(87), 83(87)

Harris, L. P., **1**:351, 354

Harris, W. W., **23**:274(177)

Harris, Y. G., **21**:185(3), 204(3), 207(3), 236(3)

Harrison, D., **10**:169(1, 2), 170, 171, 174, 214, 215; **14**:154(42), 160(81), 161(81), 162, 168(81), 169, 243, 244; **19**:108(56), 112(82), 115(96), 116(96), 138(146), 140(146), 174(226), 186(56, 82), 187(96), 188(146), 190(226); **20**:105, 130; **23**:188(54), 233(54), 269(54); **25**:152(4), 168(4), 181(4), 242(4)

Harrison, H. L., **6**:474(295), 500

Harrison, I., **28**:106(13), 139(13)

Harrison, J. P., **17**:68, 153

Harrison, R. F., **28**:124(78), 142(78)

Harrison, S. F., **14**:306(51), 342

Harrison, W. B., **20**:187(37), 300(37)

Harrison, W. C., **29**:71(33), 125(33)

Harrje, D. T., **2**:4(3, 4), 86(3), 105

Harrop, R., **3**:60, 99

Harshvardhan, **14**:265(37), 280

Hart, G. K., **15**:272(114), 281

Hart, J., **11**:297, 315

Hart, J. E., **18**:12(12), 79(12)

Hart, W. F., **21**:320, 321, 325, 345

Harteck, P., **3**:257, 258, 271, 291, 300

Harten, A., **24**:234, 240, 270

Hartland, S., **26**:2(212), 102(212)

Hartly, J. G., **19**:78(189), 91(189)

Hartmann, J., **1**:269, 300, 352

Hartmann, U., **19**:79(197, 198), 92(197, 198); **22**:209(89), 210(89), 211(89), 214(89), 215(89), 217(89), 218(89), 219(89), 293(224), 350(89), 357(224)

Hartnett, J., **5**:177, 249

Hartnett, J. P. (Chapter Author), **1**:34(31), 47(37), 50; **2**:207, 211, 269; **3**:35(22), 81, 91, 98, 99, 108(12), 173; **4**:331, 445; **5**:132(18), 134(18), 134(25), 135, 140, 149, 152, 201,

247, 248, 249; **7**:168, 191(5), 211, 343, 346, 351(30), 354(46), 355(46), 360(30, 46), 361(30), 363, 378; **8**:2(1, 2, 3, 4, 5, 6, 7), 34, 88, 163, 226; **9**:20, 32, 33, 35, 40, 108, 109, 250(77), 251(77), 261, 270, 271; **11**:363(107), 437; **12**:88(21), 97, 102(21), 112, 113; **15**:59, 63(19, 34, 35), 64(19, 34, 35), 71(19, 52), 72(53), 74(53, 55, 56, 57), 79(35), 80(35), 81(34), 85(52, 72), 88(74), 90(77, 78), 91(78), 94(34), 95(35, 162), 97(35), 102(77), 104(77, 118, 121, 123), 108(121), 109(19, 162), 110(140), 111(123), 114(19), 118(19, 52, 72), 119(72), 121(121, 123), 130(161), 131(74, 78), 132(74, 78), 135(19), 136(34, 35), 137(53, 55, 56, 57, 74), 138(77, 78), 139(118, 121, 123), 140(140), 141(161, 162); **18**:169(29), 170(29), 172(29), 223(29), 225(29), 236(29); **19**:247, 264(28, 29), 267(28, 29), 268(29), 304(90), 305(90), 311(104), 314(104), 315(104), 316(104), 317(29, 108), 319(29, 104, 108), 320(29), 321(104), 322(104), 324(114), 325(118), 326(28, 121, 122), 327(122, 123), 328(122), 329(122, 125), 330(90, 121, 127), 331(28), 332(28, 128, 129), 333(28, 90, 128, 129, 130), 334(148), 337(148), 338(148), 339(148), 340(148), 341(122), 342(90, 128, 130), 343(128), 344(130), 350(28, 29), 353(90), 354(104, 108, 114, 118, 121, 122, 123, 125), 355(127, 128, 129, 130), 356(148); **24**:102(3, 5), 174(5), 184(3, 5); **25**:253, 260, 279, 280, 281, 316, 318; **28**:153(9), 228(9); **30**:34(54), 38(54), 39(54), 90(54)

Hartop, R., **20**:263(262), 311(262)

Hartree, D. R., **4**:331, 445; **8**:21, 86

Hartunian, R. A., **2**:164, 190, 269

Haruaki, F., **10**:177, 216

Harvey, A. H., **23**:271(99)

Harvey, E. N., **10**:118(51, 52, 53), 121(50, 51, 52, 53), 165

Harwell, W., **7**:320a, 320f

Hasan, A. R., **20**:100, 102, 104, 131

Hasan, M. M., **17**:8(40), 9, 14(56), 57, 58

Hasan, M. Z., **17**:8, 22, 57, 59; **23**:74(90), 94(90), 131(90)

Hasatani, M., **11**:399(161), 438; **19**:310(94), 312(94), 353(94)

Hasegawa, S., **6**:551, 555(82), 564; **7**:66(48), 85;

9:9, 10(52), 21(51, 52), 23, 36, 37, 99, 108; **SUP**:18(31), 83(243), 136(31); **17**:7, 57; **21**:38(62), 52(62)

Haselden, G. G., **1**:427(128), 443; **5**:406(158), 512

Haseler, L. E., **26**:255(71), 305(71), 327(71)

Hashim, Z., **18**:237(88)

Hashimoto, K., **26**:45(224), 103(224)

Hashimoto, Y., **26**:14(219), 103(219)

Haskin, W. J., **7**:235(25), 238, 271, 293, 314, 320b, 320e

Hasoon, M. A., **21**:158, 181

Hasper, A., **28**:394(185), 395(185), 423(185)

Hassager, O., **15**:62(7), 99(7), 100(7), 101(7), 135(7), 144(7), 149(7), 179(7), 223(7); **19**:251(10), 252(10), 295(10), 350(10); **23**:188(20), 189(20), 267(20); **24**:107(72), 108(72), 109(72), 187(72); **25**:253, 254, 255, 315; **28**:39(64), 73(64)

Hassan, H., **11**:235, 238, 260

Hassan, K., **11**:298, 299, 315

Hassan, K. E., **9**:256(91), 270

Hassanizadeh, M., **24**:114(114), 188(114); **28**:255(62, 63), 257(62, 63), 258(62, 63), 332(62, 63)

Hassanizadeh, S. M., **23**:378(51), 463(51); **30**:95(20), 98(20), 189(20)

Hassell, H. L., **23**:193(100), 212(100), 229(100), 271(100)

Hassenzahl, W. V., **17**:74(39), 123, 149, 152(182), 154, 158

Hassid, A., **1**:379(33, 34, 35, 36), 385(34), 389(34, 36), 390(34, 36), 393(36), 394(73), 399(73), 401(34), 441, 442

Hassid, S., **13**:81(40), 116

Hassler, J. C., **18**:169(41), 170(41), 236(41)

Hasson, D., **15**:229(9, 10), 233, 234(37), 235(9, 37), 239(9), 276, 277; **30**:204(23), 250(23); **31**:440(22), 441(22), 458(21), 471(21, 22)

Hastings, C., **31**:440(19), 471(19)

Hastings, R. C., **6**:29(41), 124

Hata, K., **17**:22(70), 24(70), 59

Hatami, R., **25**:81(149), 147(149)

Hatamiya, S., **21**:73(19), 74(19), 79(19), 83(41, 41), 89(41, 41), 90(41, 41), 134(19), 135(41, 41)

Hatano, M., **11**:105(99), 114(99), 193

Hatano, Y., **23**:13(30), 15(30), 16(30), 17(30), 23(30), 24(30), 25(30), 27(30), 127(30)

Hatch, J. E., **7**:338, 344, 359, 368(11), 377

Hatch, M. R., **5**:451, 455, 456, 515

Hatch, W. H., **25**:168(65), 169(65), 170(65), 175(65), 176(65), 177(65), 228(65), 232(65), 233(65), 245(65)

Hatcher, J. D., **13**:129(37), 202; **15**:183(49), 224(49); **25**:259, 281, 317

Hatfield, D. W., **17**:70(24), 153

Hatsopoulos, G. N., **21**:271(41), 276(41)

Hatta, N., **23**:9(20), 109(20, 114, 115, 116, 117, 118, 119), 110(20, 114, 115, 116, 117, 118, 119), 113(20, 114, 115, 116, 117), 114(116), 115(20, 115, 116, 117), 116(118, 119), 119(20), 127(20), 132(114, 115, 116, 117, 118, 119)

Hatton, A. P., **5**:191(103), 249; **11**:13, 48, 203, 206, 210, 220, 223, 235, 238, 248, 254; **SUP**:177(356), 306(469), 310(469), 316(469), **19**:338(151, 152), 340(151), 356(151, 152)

Hattori, M., **31**:334(12), 425(12)

Hattori, S., **23**:284(36), 355(36)

Haucke, H. J., **17**:126(123), 156

Hauf, T., **25**:376, 401, 417

Hauf, W. (Chapter Author), **6**:134, 244(76), 312(87), 316(87), 331(90), 365, 366; **11**:281, 282, 283, 315, 422(217), 423(217), 440

Haugen, R. L., **6**:61(106), 127

Haugh, J. J., **29**:299(157), 302(157), 343(157)

Haughey, D. P., **23**:217(101), 271(101)

Hauks, K. H., **25**:51(85), 143(85)

Haumann, W., **5**:326(6), 505

Haung, Z. Q., **28**:135(119), 144(119)

Haupimann, E. G., **21**:22(38), 23(38), 51(38)

Haupin, W. E., **23**:292(59), 356(59)

Haurwitz, B., **4**:67, 138

Hause, A., **23**:320(180), 362(180)

Hause, C. D., **6**:211(61), 365

Hausen, A. G., **4**:33, 63

Hausen, H., **20**:133(2), 134(2), 148(11), 149(2), 179(2, 11); **26**:220(10), 238(10), 246(10), 253(10), 269(10), 274(10), 304(10), 315(10), 324(10)

Hausenblas, H., **13**:219(35), 264

Hauser, D. R., **9**:407(103), 417

Hauser, G. M., **29**:189(192), 212(192)

Hauser, J. J., **11**:105(110), 113, 193

Hauser, W. C., **5**:293(36), 301(36), 322; **8**:244(10), 256(10), 257(10), 266(10), 282; **31**:389(42), 427(42)

Havekotte, J. C., **7**:181(47), 213

Haviland, J. K., **5**:39, 40, 53; **7**:171, 174, 175, 212

Hawes, R. I., **9**:129, 178

Hawken, D. F., **24**:213, 270

Hawkins, G. A., **9**:350(1), 414; **11**:213(68), 214(68), 255; **SUP**:291(462)

Hawkins, T. D., **3**:109(16), 173

Hawley, M. C., **23**:201(202), 202(202), 204(202), 205(202), 206(202), 216(178, 202), 217(178), 220(202), 274(178), 275(202)

Haworth, D. R., **11**:139(150), 195

Hawthorn, R. D., **10**:188(82), 189(82), 205(82), 206(82), 216; **14**:207(140), 246

Hawthorne, R. D., **19**:123(116), 129(116), 130(116), 187(116)

Hay, N., **23**:280(23), 315(23), 325(23), 355(23)

Hay, R. D., **5**:365, 508

Hayasaka, H., **23**:161(46), 186(46); **27**:ix(6, 11), 146(28), 147(28)

Hayase, T., **24**:227, 270

Hayashi, M., **23**:293(70, 71), 357(70, 71)

Hayashi, T., **28**:390(175), 423(175)

Hayashi, Y., **19**:79(208), 92(208)

Hayasi, N., **6**:418(107), 491

Hayday, A. A., **5**:133, 136, 139, 141(23), 163, 247; **8**:19(88), 35(182, 183), 86, 88

Hayes, J., **18**:73(139), 86(139); **23**:300(96), 358(96)

Hayes, J. R., **23**:311(132), 314(132), 360(132)

Hayes, L. E., **3**:81(84), 99; **7**:344, 346, 352(35), 370(70), 371(38), 378, 379

Hayes, L. J., **22**:187(91), 188(62, 93), 189(93), 190(95), 191(90, 93), 192(95), 193(62), 194(62), 196(96), 198(96), 199(96), 215(96), 239(61), 242(61), 243(60), 277(219), 290(152), 296(94, 95), 297(94), 299(92, 94), 301(92), 302(92), 303(92), 304(92), 320(90), 328(17), 329(17, 58), 330(17), 333(17), 334(17), 336(17), 337(17), 338(17), 339(17), 340(17), 341(17), 342(17), 343(17), 347(17), 349(58, 59, 60, 61, 62), 350(90), 351(91, 92, 93, 94, 95, 96), 353(152), 356(219); **31**:401(58), 428(58)

Hayes, T. L., **22**:220(185), 355(185)

Hayes, W. D., **1**:334, 335, 341(66), 354; **4**:318, 443; **6**:105(165), 130; **7**:164(3), 187(3), 194(3), 203(3), 211

Haynes, F. D., **19**:334(146), 337(146), 338(146), 355(146)

Haynes, L. C., **28**:124(78), 142(78)

Hays, D. F., **8**:29, 87

Hays, G., **31**:451(23, 24, 36), 469(36), 471(23, 24), 472(36)

Hayward, G. L., **23**:340(274), 367(274)

Haywood, R. J., **26**:72(85, 182), 73(85, 182), 74(85), 79(86, 87), 80(86), 81(86), 96(85, 86, 87), 101(182)

Haywood, R. W., **15**:2(5), 54(5)

Hazell, T. D., **SUP**:35(43), 77(43), 148(43)

Hazet, P. R., **15**:122(148), 141(148)

Hazzah, A. S., **11**:408, 439

He, J., **26**:221(22), 325(22)

Head, R. R., **5**:381, 510

Healzer, J. M., **16**:302(28), 305, 309, 363, 364; **17**:42(144), 43(144), 44(144), 63

Hearne, L. F., **23**:304(110), 359(110)

Heaselet, M. A., **8**:264(40), 283

Heaslet, M. A., **2**:430(31), 451; **5**:28(35), 29, 52; **11**:344(61), 436; **27**:27(22)

Heasley, J. H., **15**:324

Heath, C. A., **7**:274(101), 318; **11**:105(114), 114, 120(114), 121(114), 122(114), 194

Heath, C. H., **29**:215(1), 248(1), 334(1)

Heatherly, E. R., **5**:471, 516

Heaton, H. S., **SUP**:35(44), 71(44), 77(44), 145(44), 193(44), 194(44), 287(44), 288(44), 297(44), 319(44), 321(44)

Heaviside, O., **15**:285, 286, 324

Hechanova, A. E., **23**:330(236), 365(236)

Hechler, K., **17**:321(17), 339

Hecht, E., **30**:261(18), 308(18)

Hechtel, L. M., **25**:377, 413

Hector, L. G., Jr., **28**:77(54), 141(54)

Hedberg, S., **11**:241, 243, 261

Hedden, K., **14**:218(153), 223(153), 246

Heddleson, C. F., **11**:228(156), 259

Hedgepeth, L. M., **4**:164, 165, 224

Hedstrom, B. O. A., **2**:381, 395

Heeren, J., **12**:11(21), 14(21), 74

Heertjes, P. M., **10**:172, 175, 177(46), 179, 215, 216; **25**:161(41, 42), 181(73), 199(73), 244(41, 42), 246(73)

Hegazy, A., **20**:261(252, 254), 270(254), 311(252, 254)

Hegedus, L. L., **15**:287(78), 322, 324

Hegemier, G. A., **15**:325

Heggs, P. J. (Chapter Author), **20**:133, 136(7),

137(7), 155(7), 179(7)

Heighway, J. E., **9**:296(28), 346

Heiken, G., **10**:55(43), 82

Heilman, G., **11**:362(101), 437

Heilmann, R. H., **4**:54, 64

Heimburg, R. W., **20**:271(278), 312(278)

Heimenz, K., **20**:377, 387

Heimes, K. A., **15**:324

Hein, H., **21**:155, 182

Heinbockel, J. H., **SUP**:79(225)

Heindel, T. J., **26**:188(148), 191(148), 215(148)

Heinecke, E., **8**:145(77), 148, 149, 160; **18**:145(78), 146(78), 147(78), 159(78)

Heineman, J. B., **19**:334(136), 335(136), 355(136)

Heinisch, R. P., **31**:343(26), 426(26)

Heinrich, J. C., **28**:256(66), 262(72), 267(93, 95), 303(72, 138, 139), 309(156, 157), 332(66, 72), 333(93, 95), 335(138, 139), 336(156, 157)

Heisler, M. P., **22**:161(97), 166(97), 351(97); **26**:238(43), 253(43), 269(43), 326(43)

Heist, R. H., **14**:245, 327, 329(107, 108), 330(108)

Heitler, W., **5**:264(18), 265(18), 303(18), 322

Heitz, J., **28**:102(37), 140(37)

Heitz, W. L., **19**:45(104), 88(104)

Held, E. F. M. Van Der, *see* Van der Held, E. F. M.

Hell, H. L., **8**:140(71), 144(71), 160

Helland, K. N., **21**:185(4), 204(4), 206(4), 207(4), 236(4)

Hellawell, A., **28**:277(107, 108), 280(107, 108, 114), 282(118), 285(125, 126, 127), 303(108, 114), 310(107, 108), 333(107), 334(108, 114, 118, 125, 126, 127)

Heller, M. A., **14**:152(29), 243

Hellman, J. H., **17**:198, 315

Hellman, J. M. (Chapter Author), **10**:220, 261, 272, 283, 284

Hellman, S. K., **9**:73(53), 76(53), 108

Hellums, J. D., **4**:36, 37, 63; **6**:474(234, 257), 480(234), 483(234, 257), 497, 499; **8**:167, 226; **9**:304(63), 347; **20**:367, 387

Hellwedge, K. N., **18**:161(1), 236(1)

Hellwege, K. H., **13**:217(32), 218(32), 264

Hellwege, K. K., **18**:212(102), 222(102), 238(102)

Helvajian, H., **28**:118(53), 141(53)

Helvensteijn, B. P. M., **17**:106, 107, 155

Hemingway, B. S., **10**:44(17), 59(17), 60(16, 17, 49), 61, 63(17), 64(17, 49), 69(17), 70(17), 81, 82

Hempel, A., **6**:474(277), 499

Hench, J., **20**:108, 132

Hendal, W. P., **1**:427, 443

Henderson, F. M., **20**:91, 131

Hendler, E., **4**:121(97), 140

Hendricks, R. C., **4**:206(86), 226; **5**:100, 127, 355(63), 360, 361, 362, 368, 370, 372, 373, 374, 445, 446, 447, 448, 449, 450, 451, 458, 459, 507; **6**:550(88), 564; **7**:53, 85; **11**:58(23), 84(23, 45), 86, 95(23, 45), 96, 105(23), 108, 109(23), 113, 118, 170(172, 179, 180), 173, 174(172, 179, 180), 175(179), 177, 182(172), 183(172), 185(179, 180), 186, 190, 191, 196, 197; **23**:345(297), 351(297), 368(297)

Heng, K. T., **17**:38(126), 62

Hengstenberg, D., **10**:105(24), 106(24), 164

Henle, K. J., **22**:325(98), 351(98)

Henn, D. S., **25**:375, 418

Hennecke, D. K., **SUP**:58(68), 115(68), 116(68), 117(68), 118(68), 133(68), 134(68), 135(68)

Hennecke, F. W., **13**:2(5), 8(5), 11(5), 18(5), 19(5), 25(5), 32(5), 59

Hennig, J., **18**:212(101, 102), 222(101, 102), 238(101, 102)

Henning, C. D., **15**:43(65), 57(65)

Henrie, J. O., **29**:71(28), 125(28)

Henriques, F. C., **4**:119(82), 125, 140, 141; **22**:11(20), 17(20), 230(148), 231(148), 232(100), 313(99), 314(99, 148), 320(99), 329(100), 330(99), 331(99), 332(99), 336(99), 351(99, 100), 353(148)

Henry, H. R., **14**:40, 77, 102

Henry, J., **8**:297(33), 349

Henry, R. E., **29**:134(18), 139(18, 30), 140(18), 148(50), 159(91), 165(112, 113), 189(192), 203(18), 204(30), 205(50), 207(91), 208(112, 113), 212(192), 220(55), 221(55, 56), 234(85), 247(98), 266(98), 269(85), 271(98), 272(98), 276(56), 316(173), 318(173), 323(56), 338(55, 56), 339(85), 340(98), 344(173)

Henshaw, W. D., **24**:218, 269

Henwood, G. A., **25**:166(57), 245(57)

Henze, R. H., **18**:36(74), 83(74)

Heppel, J., **12**:138(28), 193

Herbert, D. M., **15**:213(109), 225(109)

Herbral, B., **17**:71(33), 153

Hering, R. G., **5**:141(32), 142, 143(32), 247; **9**:296, 346; **11**:408(179, 180), 409(181), 417(179), 439

Hering, W. S., **14**:253(8), 279

Herman, C. V., **26**:246(54), 327(54)

Herman, J. R., **14**:267(46), 280

Herman, R., **8**:26(135), 87

Hermann, M., **4**:283(49), 315

Hermann, R., **4**:51, 64; **11**:202, 204, 209, 253

Hernandez, E. J., **22**:30(25), 153(25)

Heroman, L. C., Jr., **5**:100, 127, 453, 515

Herrick, C. S., **28**:351(77), 418(77)

Herring, J. R., **25**:333, 340, 413; **30**:27(49), 89(49)

Herring, T. K., **7**:197, 215

Herrington, L. P., **4**:67(9), 111(63), 118(78), 138, 139, 140; **22**:20(27), 21(27), 28(27), 82(27), 153(27)

Herrmann, L. R., **23**:398(62), 409(62), 463(62)

Herschel, W., **22**:420(27), 436(27)

Herschel, W. H., **24**:109(85), 187(85)

Hersey, M. D., **13**:219(34), 264

Hershey, F. B., **4**:125(106), 141

Hertel, K. I., **25**:294, 319

Hertofilis, S. A., **6**:95(153), 129

Hertz, H. G., **14**:325, 344

Hertz, J., **5**:347, 506

Hertzberg, A., **12**:7(17), 73

Hertzberg, M., **29**:71(30), 84(46), 125(30, 46), 307(167), 344(167)

Hervey, G. R., **4**:94(34), 138

Herwijn, A., **17**:211, 216, 217, 226, 315

Herzberg, G., **5**:267(20, 21), 269(21), 322

Herzberger, M., **11**:374(126), 437

Herziger, G., **28**:124(1, 2), 138(1, 2)

Heselden, A. J., **10**:259(65), 282

Hess, C. F., **18**:36(74), 83(74)

Hess, D. W., **28**:344(59), 417(59)

Hess, H. L., **5**:355(61), 368, 369, 370, 371, 372, 374, 507; **7**:53, 85

Hess, J. L., **20**:367, 387

Hess, Kh. L., **6**:553, 564

Hess, P. D., **29**:188(187), 212(187)

Hess, R. C., **17**:323, 328(26), 333(26), 334(26), 335(26), 340

Hesse, G., **18**:255(16), 257(16), 260(16), 290(16), 306(16), 320(16)

Hesson, G. M., **1**:421(118), 443

Hesteness, M. R., **23**:418(73), 464(73)

Hetherington, H. J., **20**:186(36, 49), 187(36, 49), 300(36), 301(49)

Hetsroni, G., **15**:250(88, 89), 252(88), 253(89), 280, 324

Hewitt, G. F., **1**:377, 384(51), 385(31, 51), 387, 389(31), 390(58), 392(58), 410(58), 421(58), 427, 428(58), 441; **15**:229(11), 263(11), 276; **16**:145, 156; **17**:2(7, 9, 11, 14), 30, 34(112), 37(120), 40(9, 11, 14), 42(138, 140), 43(11, 140), 44(140), 45(11), 46, 47, 56, 60, 61, 62, 63, 321, 339; **20**:250(215, 217, 219, 222), 309(215, 217, 219), 310(222); **29**:266(133), 342(133); **31**:160(44), 324(44)

Hewitt, H. C., **9**:227(57), 269

Hewitt, J. M., **19**:82(262), 94(262)

Hews, D., **31**:245(137), 328(137)

Hey, A. J. G., **31**:406(68), 428(68)

Heyda, J. F., **SUP**:325(479)

Heynick, L. N., **22**:342(101), 351(101)

Heywood, N. I., **20**:109, 113, 131

Hibi, K., **25**:396, 416

Hibiya, T., **30**:392(157), 393(157), 396(163, 164), 399(163, 164), 434(157, 163, 164)

Hiby, J. W., **9**:127(28), 178

Hicken, E., **SUP**:57(67), 58(67), 59(67), 109(274), 128(274), 219(412, 413)

Hickerson, C. W., **10**:224(8), 253(8), 281

Hickman, H. J., **SUP**:80(232), 121(232), 159(232), 186(232)

Hickman, K., **15**:230, 231(15), 233, 276

Hickman, K. C. D., **5**:131, 246

Hickman, R., **19**:164(206), 170(206), 171(206), 190(206)

Hickman, R. G., **14**:156(58), 157(58), 243

Hickman, R. S., **7**:197, 215; **15**:232, 277

Hickox, C. E., **24**:102(12), 184(12)

Hicks, B. L., **5**:40, 53

Hicks, E., **29**:136(20), 143(20), 203(20)

Hidaka, A., **29**:165(116), 208(116)

Hidaka, S., **16**:166(3), 170(3), 171(3), 173(3), 174(4), 187(3), 188(3), 189(3), 193(3), 203(3), 209(3), 214(3), 215(3), 220(3), 221(3), 222(3), 232, 238; **20**:20(40), 22(40), 81(40)

Hidaka, Y., **30**:295(94), 311(94)

Hide, R., **5**:130(4), 131, 208, 210, 213, 214, 215, 216(132, 134), 219(141), 221, 226, 227, 246, 250

Hieber, C. A., **9**:324, 328(91), 332(91), 334(91, 92), 348; **11**:226, 259

Hiemenz, K., **18**:112(45), 159(45)

Higdon, J. J. L., **25**:256, 293, 317

Higeta, K., **21**:300, 345

Higgins, G. H., **14**:3(2), 100

Higuchi, Y., **30**:359(123), 361(123), 432(123)

Hihara, E., **29**:303(162), 343(162)

Hijikata, K., **21**:109(97, 99), 117(102, 103, 104, 105), 129(125), 130(126), 137(97, 99, 102), 138(103, 104, 105), 139(125, 126); **26**:44(88), 96(88); **30**:173(146), 195(146)

Hilal, M. A., **15**:44(70), 46(70), 48(76), 57(70, 76); **17**:119, 156; **23**:229(101a), 230(101a), 231(101a), 271(101a)

Hilbert, D., **8**:34(174), 67(174), 88

Hildebrand, B. P., **6**:200(45), 364

Hildebrand, F. B., **2**:422(19), 451; **8**:54(243), 90; **26**:269(78), 328(78)

Hildebrandt, G., **17**:323, 335(23), 340

Hildebrandt, W. H., **22**:188(44, 102), 348(44), 351(102)

Hilder, D. S. (Chapter Author), **16**:1

Hilditch, M. A., **23**:321(200, 201), 363(200, 201)

Hilgeroth, E., **13**:18, 26(52), 60

Hill, A. C., **15**:287, 324

Hill, D., **18**:67(126), 68(126), 85(126)

Hill, F. B., **19**:177(230), 190(230)

Hill, F. K., **3**:52, 61, 98

Hill, J. E., **15**:34(57), 56(57)

Hill, J. M., **19**:20(39), 21(39), 85(39)

Hill, P. G., **9**:199, 200(14), 201, 202(16), 203, 268; **14**:332(115), 333(115, 118), 334(115, 118), 335(118), 345

Hill, P. R., **6**:414, 418, 489

Hill, T., **4**:67, 75, 137

Hill, T. L., **14**:306, 342

Hillesland, K. E., **31**:402(62), 428(62)

Hillig, W. B., **28**:16(28), 71(28)

Hills, B. A., **18**:167(47), 237(47)

Hills, R. N., **17**:102(83), 155; **28**:252(41), 255(41), 288(41), 326(41), 331(41)

Hilpert, L., **8**:94(13), 128(13), 130(13), 159; **18**:88(13), 127(13), 158(13)

Hilpert, R., **3**:9, 11, 19, 31; **7**:119(33), 161; **11**:218, 221, 230, 233, 234, 235, 237, 238, 239, 256

Hilsenrath, J., **2**:37(28), 38(28), 106; **5**:309(76), 323; **11**:230(175, 177), 260

Hilton, R. M., **30**:320(22), 427(22)
Himeno, R., **25**:358, 416
Himmelblau, D. M., **12**:232(32), 278
Hinch, E. J., **12**:77(3), 111
Hinckley, R. B., **5**:475, 476, 516
Hindermann, J. D., **7**:241, 243, 259, 261, 303, 316
Hine, F., **7**:132(42), 161
Hine, R. J., **8**:248(20), 282
Hines, C. O., **14**:251(3), 268(50), 270(3, 50), 279, 280
Hines, F. F., **23**:293(62), 356(62)
Hines, F. L., **5**:390, 510
Hines, W. S., **5**:308(72), 323
Hingst, W. R., **23**:326(214), 364(214)
Hinkle, B. L., **9**:137, 179
Hintze, J. O., **30**:290(84), 311(84)
Hinze, J. O., **2**:72, 108; **6**:393(22), 394(22), 400(22), 487, 511(20), 562; **21**:195(23), 237(23); **25**:323, 327, 413; **31**:168(74), 325(74)
Hippensteele, S. A., **23**:298(83), 323(207), 324(207), 330(207), 331(240, 241, 242, 243), 332(243), 357(83), 363(207), 365(240, 241, 242, 243)
Hiquily, N., **19**:108(59), 186(59)
Hirai, E., **2**:373(47), 374, 395
Hiraku, K., **23**:109(117), 110(117), 113(117), 115(117), 132(117)
Hiramitsu, M., **18**:59(112), 60(112), 61(112), 63(112), 85(112)
Hirano, F., **1**:231(52), 265; **20**:4(2), 5(3), 7(3), 9(3), 10(3), 11(3), 15(3), 79(2, 3), 80(11)
Hirano, H., **11**:231(241), 232(241), 263; **28**:41(67), 73(67)
Hiraoka, S., **7**:122(35), 161; **SUP**:64(97, 98), 224(97), 237(97), 260(97, 98), 262(98)
Hirasaki, G. J., **23**:206(102), 226(102), 228(102), 271(102)
Hirasawa, S., **21**:117(102, 103, 104, 105), 137(102), 138(103, 104, 105); **28**:402(216), 425(216)
Hirata, A., **30**:358(115), 396(166), 400(166), 432(115), 434(166)
Hirata, H., **28**:316(166, 167), 336(166, 167)
Hirata, M., **3**:81(76), 82(76), 84(76), 87(76), 99; **6**:546(65), 563; **7**:34, 84, 346, 353(41), 378; **11**:281(18), 315; **13**:11(39), 60; **20**:252(238), 255(238), 310(238); **21**:4(10), 34(52), 35(52), 36(52), 37(59), 51(10),

52(52, 59); **23**:293(63), 356(63)
Hirata, S., **28**:342(36), 416(36)
Hirata, T., **19**:76(176), 91(176); **24**:6(25), 37(25)
Hirayama, H., **28**:340(5), 415(5)
Hirose, K., **19**:58(133), 59(133), 68(133), 70(133), 89(133); **24**:3(7), 36(7)
Hirose, T., **20**:370, 371, 380, 387; **25**:272, 273, 318
Hirsch, C., **24**:193, 226, 227, 234, 270
Hirsch, G., **15**:245, 246(75), 279
Hirsch, R. L., **22**:30(48), 36(48), 154(48)
Hirschfelder, J. O., **1**:7(11), 49; **2**:38, 47, 106, 114(8), 115(10), 253, 268, 270; **3**:190(13), 249, 254, 261, 262, 263, 267, 268(22), 269(3), 270(29), 279, 282, 290, 291, 298(97), 299, 300, 301, 302; **7**:10, 83; **12**:197(8), 198(8), 199(8), 268(33), 269(8), 277, 278; **17**:127, 156; **28**:356(97), 364(97), 419(97)
Hirschhorn, H. J., **11**:354(76), 436
Hirst, E. A., **16**:9, 10, 11, 12, 19, 23, 26, 27, 28, 32, 46, 57
Hirt, C. W., **28**:41(65), 73(65)
Hirth, J. P., **10**:92(92), 164; **14**:307(64, 65, 66), 308(64), 309(65, 66), 343
Hishida, M., **SUP**:69(145), 73(145), 160(145); **19**:268(38), 270(38), 351(38)
Hishio, S., **29**:164(106), 207(106)
Hitchcock, J. E., **3**:212, 243, 246, 250; **5**:28(34), 52; **6**:312(105), 366; **SUP**:190(376), 191(376), 194(376); **19**:308(92), 312(92), 353(92); **23**:319(154), 361(154)
Hitchman, M. L., **28**:351(87), 389(166), 402(191, 194), 404(191), 419(87), 422(166), 423(191), 424(194)
Hitz, J. A., **6**:41(83), 68(83), 115, 69(115), 70(115), 73(83), 76(83), 78(83), 79(83), 82(83), 126, 127
Hixon, C. W., **SUP**:42(50), 43(50), 209(50), 210(50), 211(50), **19**:258(15), 259(15), 268(15), 272(15), 274(15), 334(142), 336(142), 340(142), 350(15), 355(142)
Hjellming, L. N., **30**:333(54), 351(54), 371(142), 390(54), 429(54), 433(142)
Hlavacek, B., **23**:206(24), 218(24), 237(24), 239(24), 240(24), 242(24), 244(24), 267(24)
Ho, C. C., **13**:211(16), 264; **18**:161(2), 219(114), 220(117), 236(2), 238(114, 117)
Ho, C.-J., **19**:58(131), 59(131), 60(131),

62(145), 63(145), 65(145), 66(145), 67(154), 89(131, 145), 90(154); **24**:3(9), 36(9)

Ho, C. K., **30**:175(150), 195(150)

Ho, C.-T., **28**:213(37), 229(37)

Ho, C. Y., **21**:164, 166, 180; **24**:27(56), 38(56)

Ho, H. T., **7**:197, 215

Ho, J.-R. (Chapter Author), **28**:75, 96(117), 97(117), 99(117), 100(117), 101(117), 102(117), 126(39), 130(39), 131(39), 132(39), 133(38), 134(39), 140(38, 39), 144(117)

Ho, L. Y., **21**:151, 181

Ho, T. C., **19**:179(235), 190(235)

Hoagland, L. C., **19**:324(116), 354(116)

Hobart, H., **23**:298(85), 306(85), 357(85)

Hobbs, P. B., **22**:177(103), 351(103)

Hochberg, K., **28**:402(207), 424(207)

Hochstim, A. R., **1**:334(68), 354

Hocking, L. M., **28**:41(74), 73(74)

Hodge, J. K., **23**:311(132), 314(132), 360(132)

Hodge, S. A., **29**:234(94), 340(94)

Hodges, K. I., **29**:194(208), 213(208)

Hodges, R. R., **14**:256(21), 259, 269(58), 273(58), 275, 277(58), 279, 280

Hodgkiess, T., **31**:459(43), 472(43)

Hodgkins, W. R., **19**:2(5), 83(5)

Hodgson, J. P., **8**:248(20), 282

Hodgson, J. W., **11**:220, 223, 237, 258

Hodredge, R. M., **17**:119(113), 156

Hoe, Y. L., **1**:382(42), 441

Hoehne, J. C., **16**:114(72), 155, 235; **20**:16(36), 17(36), 81(36)

Hoelscher, H. E., **15**:250(85), 280

Hofeldt, D. L., **30**:295(93), 299(93), 311(93)

Hoff, H. E., **22**:17(18)

Hoffman, D. P., **12**:14(25), 74

Hoffman, E. J., **9**:258(101), 271

Hoffman, H. W., **SUP**:361(506), 363(506)

Hoffman, M. A., **7**:320a, 320f; **23**:327(217, 218, 219), 330(238), 336(256, 257, 258), 337(258), 364(217, 218, 219), 365(238), 366(256, 257, 258)

Hoffman, T. W., **23**:305(116), 359(116)

Hoffmann, E. G., **18**:255(17), 257(17), 261(17), 320(17)

Hofmann, A., **17**:146, 157

Hofmann, D., **30**:320(23), 427(23)

Hogan, J. S., **8**:231(3), 281, 282, 283

Hogan, W. T., **1**:349, 354

Hoge, H. J., **2**:37(28), 38(28), 106; **5**:406(159), 512

Hogge, E. A., **12**:233(34), 251(34), 270(34), 278

Hogge, M., **19**:81(257), 94(257)

Hoglund, B. M., **1**:189(4), 190(4), 240, 263

Hoglund, R. F., **28**:11(17), 71(17)

Hoh, C. Y., **5**:233(169), 235(169), 251

Hoheisel, W., **28**:118(40), 119(40), 120(40), 140(40)

Hohenberg, P. C., **17**:131(135), 149(135), 156; **20**:323(18), 350(18)

Hohenemser, K., **13**:211(24), 264

Hohmann, H., **29**:140(31), 156(87), 165(31, 112, 113), 173(145), 178(31, 158), 181(158), 204(31), 206(87), 208(112, 113), 209(145), 210(158)

Hohnstreiter, G. F., **9**:144(67), 147(67), 156(67), 179

Holanda, R., **23**:298(83), 357(83)

Holben, C. D., **9**:408(104), 417

Holborn, E., **16**:179, 238

Holbrook, J. A., **26**:11(89, 90), 97(89, 90)

Holcombe, A. H., **5**:327(18), 505

Holder, G. W., **6**:29(39), 35(57), 36, 124, 195, 363

Holdredge, R. M., **5**:481(263), 487, 488, 489, 490, 491, 494, 495(263), 517; **11**:105(138), 195

Holland, E., **9**:129(37), 178

Hollands, K. G. T. (Chapter Author), **11**:266; **18**:12(14), 13(19), 14(19), 15(20), 16(23, 24, 25, 26), 17(25, 28, 32), 65(119), 73(141), 74(141), 79(14), 80(19, 20, 23, 24, 25, 26, 28, 32), 85(119), 86(141); **20**:186(50), 187(50), 301(50); **23**:333(244), 365(244)

Holleman, J., **28**:394(185), 395(185), 423(185)

Höller, H., **25**:376, 401, 417

Hollingsworth, D. K., **23**:329(225, 226, 227), 330(225), 364(225, 226, 227)

Hollman, T. M., **6**:524(31), 562

Holloway, F. A. L., **15**:264(96), 267(96), 280

Holloway, P. F., **6**:94(146), 129

Hollwegen, D. J., **9**:365(36), 372(36), 415

Hollworth, B. R., **23**:293(67), 295(67), 338(265), 357(67), 366(265)

Holly, L., **12**:11(21), 14(21), 74

Holm, F. W., **4**:90(28), 91(28), 92(28), 95(28), 109(28), 110(28), 138; **7**:310, 319

Holman, J. P., **4**:51, 64; **10**:179(60), 193(86), 216; **20**:255(242), 310(242); **23**:13(41), 15(41), 21(41), 26(41), 27(41), 30(41), 37(41), 39(41), 109(41), 110(41), 117(41), 128(41); **30**:271(34), 308(34)

Holman, K. I., **21**:172, 173, 180

Holman, T. P., **23**:285(39), 355(39)

Holmes, D. B., **SUP**:197(383), 198(383); **19**:258(13), 259(13), 268(13), 350(13)

Holmes, D. G., **24**:227, 271

Holmes, D. S., **17**:152(184), 158

Holmes, K. R., **22**:10(12), 13(11), 17(11, 12), 26(22), 30(21), 45(22), 46(22), 48(22), 49(22), 51(22), 58(22), 85(22), 104(22), 106(22), 122(22), 139(22), 153(21, 22)

Holmes, R. E., **9**:262(122), 271; **14**:138, 139, 145

Holmes, T., **14**:158(65), 244

Holst, P. H., **14**:43, 48, 102

Holt, J. L., **23**:320(179), 362(179)

Holt, J. S. C., **11**:278(12), 279(12), 305, 314

Holt, M., **6**:109(171), 116, 117(181a, 181b, 181c), 118, 130, 131

Holten, D. C., **5**:471, 472, 516

Holtz, R. E., **10**:158, 159(83, 84), 166

Holweg, J., **25**:292, 316

Holzwarth, H., **9**:3(54), 11, 91(54), 108

Homann, F., **20**:369, 387; **26**:109(13), 208(13)

Hommert, P. J., **11**:428(238), 429, 430(238, 241, 241a, 243), 440, 441

Homsy, G. M., **14**:16(41), 19(49), 20(41), 22, 34(86), 67, 69(149), 70(149), 75(41), 101, 102, 103; **19**:202(56), 245(56); **23**:207(270), 208(70), 270(70), 278(270), 399(68), 463(68); **28**:334(121)

Honda, H., **21**:105(89, 90), 106(91, 92), 120(106, 107), 121(110), 122(106, 107, 111), 137(89, 90, 91, 92), 138(106, 107, 110, 111)

Honda, M., **6**:109, 130

Hong, J. T., **30**:83(62), 90(62)

Hong, S.-A., **23**:206(60), 226(103), 269(60), 271(103)

Hong, S. K., **4**:90(26), 92(26, 36), 120(86, 87), 138, 140

Hong, S. W., **SUP**:76(211), 267(211, 446), 269(446), 381(446, 518), 382(446, 518)

Hongisto, O., **29**:20(15), 56(15)

Hongo, T., **21**:174, 180

Honig, C. C., **21**:179

Honig, E. P., **12**:242(35), 243(35), 256(36, 44),

267(44), 278, 279

Hoogendoorn, C. J., **11**:219(113), 223(113), 257; **18**:12(15), 13(15, 19c), 14(19c), 16(27), 18(38), 79(15), 80(19c, 27), 81(38); **21**:11(25), 51(25); **23**:329(229, 230), 364(229), 365(230); **26**:166(120), 214 (120); **28**:356(96), 363(98), 364(98), 394 (96, 185), 395(185), 419(96, 98), 423(185)

Hoogervorst, C. J. P., **15**:287, 290, 324

Hooker, W. J., **8**:248(21), 283

Hooper, F. C., **10**:121(58), 165; **15**:248(80a), 280

Hooper, G., **29**:101(72), 126(72)

Hooper, M. S., **SUP**:62(74), 322(74), 324(74), 341(74)

Hoori, Z., **10**:179(62), 216

Hoory, S., **31**:462(27), 471(27)

Hope, A. B., **12**:247(37), 279

Hopenfeld, J., **11**:68(37), 70, 72(37), 105(37), 111, 121, 190

Hopkins, J. A., **28**:282(119), 334(119)

Hopper, J. R., **2**:365(48), 395; **19**:179(235), 190(235)

Hopper, P., **28**:413(230, 231), 425(230, 231)

Höppner, G., **13**:26(51), 60

Horai, K., **10**:43, 44(15), 45, 60(15), 62, 64, 65(15, 52), 67, 81

Horgan, C. O., **15**:321(83), 324

Hori, E., **7**:150, 161

Hori, Y., **19**:74(172), 76(172), 77(172), 91(172)

Horigome, T., **25**:51(87), 144(87)

Horiuti, K., **25**:344, 350, 360, 384, 404, 405, 413, 419

Horkowitz, J. P., **23**:415(72), 463(72)

Horkowitz, K. O., **23**:415(72), 463(72)

Horlor, M. L., **1**:394(69), 442

Hornbaker, D. R., **23**:279(4), 285(43), 309(4), 354(4), 356(43)

Hornbeck, R. W., **SUP**:73(173, 178), 77(216), 87(173), 88(173, 216), 93(173), 97(173), 98(173), 139(216), 140(216), 141(216), 142(216), 143(216), 144(216), 145(216), 146(216), 147(216), 211(178), 212(178), 336(173); **19**:268(58), 271(58), 351(58)

Hornbecker, M. F., **28**:243(16), 277(16), 280(16), 329(16)

Horne, R. A., **18**:327(7), 330(27), 338(27), 339(27), 362(7, 27)

Horne, R. N., **14**:43, 44, 45(101), 46(104, 105), 47(104, 105), 48(111), 49(101), 50(101, 117, 118), 72, 99, 102, 105

Horning, W. A., **1**:218(34), 264

Horsfall, F., **18**:169(28), 170(28), 173(28), 174(28), 217(28), 218(28), 236(28)

Horstman, C. C., **7**:206, 208(152), 218

Hortmann, M., **24**:247, 271

Horton, F. G., **23**:315(149), 320(189), 360(149), 362(189)

Horton, G. K., **4**:135, 136, 141; **15**:324

Horton, S. F., **20**:205(107), 208(107), 210(107), 304(107)

Horváth, C. D., **26**:231(33), 250(33, 58), 253(58), 261(33, 58), 269(33), 281(33, 58), 316(58), 320(58), 326(33), 327(58)

Horvath, J. L., **20**:183(5), 299(5)

Horvath, L., **20**:296(323), 314(323)

Horvath, S. M., **22**:153(28)

Horvath, T. J., **23**:322(203), 325(203), 363(203)

Horvay, G., **8**:53(236), 90; **15**:324, 325

Hoshigaki, H., **4**:278, 279, 314

Hoshikawa, K., **28**:316(165, 166), 336(165, 166)

Hoshino, T., **26**:142(95, 96, 97, 98, 99), 144(95), 157(95), 158(95), 212(95, 96), 213(97, 98, 99)

Hoshizaki, H., **5**:309(81), 324; **7**:197, 216

Hoskin, N. E., **5**:173, 248

Hosler, E. K., **5**:514(204)

Hosler, E. R., **5**:88, 90, 112, 113, 126

Hosler, I. R., **11**:56, 57(12), 91, 95, 96(12), 105(12), 112, 121(12), 122(12), 189

Hosni, M. H., **23**:284(38), 314(139), 355(38), 360(139)

Hossager, O., **28**:152(3), 227(3)

Hottel, H. C., **1**:2, 7(5), 13, 49; **2**:407, 411, 414, 450; **3**:212(56), 215(56), 239, 242, 244(93), 250, 251; **5**:20, 21, 27, 28(32), 51, 52, 254(1), 255(1), 306(1), 307, 308, 309, 311(1), 313(1), 321, 473, 516; **9**:362(22), 363(22), 364(22), 365(22), 415; **10**:269, 283; **11**:323(42), 325(42), 376(129), 420(209), 435, 437, 439; **12**:129, 133(18, 21), 136, 160, 162, 192, 193; **17**:204(56), 218, 315; **23**:139(12), 154(39), 168(39), 184(12), 185(39); **27**:14(20), 36(23, 25); **30**:338(76), 430(76)

Houchin, W. R., **23**:9(26), 127(26)

Houf, W. G., **28**:402(212), 424(212)

Hougen, O. A., **1**:433, 443; **4**:54, 64; **13**:121, 122, 124, 185, 190, 201; **15**:189(63), 224(63); **16**:147, 156

Houghten, F. C., **4**:104(46), 139

Houghton, G. L., **2**:371(74), 372(74), 396; **15**:101(116), 102(116), 139(116)

Houghton, W. T., **2**:385(76), 391(75, 76), 396

Hoult, D. P., **12**:23(37), 74; **16**:5, 8, 56

Hours, R., **1**:398(78), 442

Houtman, C., **28**:397(187), 398(187), 423(187)

Hovestreijdt, J., **16**:132, 155

Howard, J. H., **14**:3(2), 14(34), 100

Howard, J. N., **12**:129(12), 138, 192

Howard, L. N., **5**:209(127), 250; **14**:33, 44(106), 101, 102; **24**:286(6), 318(6)

Howard, R. G., **12**:87, 112

Howarth, A., **2**:343, 354

Howarth, C. R., **31**:390(44), 427(44)

Howarth, L., **5**:159, 160, 161(49), 248; **15**:15(39), 56(39); **24**:315(10), 319(10)

Howe, J. T., **1**:23, 50

Howe, R. T., **23**:281(29), 355(29)

Howe, W. C., **19**:160(194), 189(194)

Howell, B. J., **4**:120(86), 140

Howell, J. R. (Chapter Author), **1**:15, 16(17), 49; **3**:211(49), 250; **4**:169, 225; **5**:2, 24, 25, 26, 28(42), 29, 30, 32, 33, 33(44), 34(46), 34(47), 34(48), 34(49), 35, 36, 51, 52, 313(87), 324; **8**:34, 88; **11**:11(25), 12(25), 48, 319, 321(33), 329(33), 335(33), 338(33), 339(33), 435; **12**:124, 168(9), 192; **23**:135(1), 145(1), 148(1), 149(1), 150(1), 151(1), 161(45), 164(1), 184(1), 186(45); **27**:23(21), 36(24); **30**:338(77), 339(77), 430(77); **31**:333, 334(3, 6, 17), 336(20, 21), 341(17), 395(51, 52), 399(20, 21), 425(3, 6, 17), 426(20, 21), 427(51, 52)

Howes, F. A., **31**:13(11b), 102(11b)

Hoyt, J. W., **12**:77(1), 111; **15**:63(25), 64(25), 135(25); **25**:286, 317; **26**:137(86), 212(86)

Hozawa, M., **30**:358(115), 432(115)

Hozumi, M., **2**:380(77), 396

Hrycak, P., **15**:62(4), 134(4); **19**:281(71), 284(71), 294(71), 352(71); **20**:255(243), 310(243); **26**:106(8), 127(74), 132(79), 136(8), 208(8), 211(74), 212(79)

Hsia, H. M., **8**:66, 90

Hsiao, J. S., **19**:81(254, 255), 94(254, 255)

Hsieh, A. C. L., **4**:92(30), 138

Hsieh, C. K., **18**:22(51, 52), 23(51), 81(51, 52); **23**:142(21), 185(21)

Hsieh, T., **12**:42(53), 44(53), 45, 49, 50(53), 75

Hsieh, T. C., **12**:139, 193

Hsu, C. C., **8**:19(90), 86; **24**:218, 269, 271

Hsu, C. J., **8**:56, 90; **SUP**:81(233), 118(289), 120(297), 121(233, 297, 300), 122(297), 125(305), 126(305), 132(300), 133(300, 311), 134(311), 135(311), 136(305), 183(311), 186(297), 187(297), 262(439), 309(470), 318(474), 364(513); **20**:353, 360, 361, 370, 387; **23**:198(143, 144), 203(144), 206(143, 144), 226(143, 144), 228(143), 229(143, 144), 273(143, 144)

Hsu, C. T., **8**:248, 282; **14**:69, 70(149), 71(151, 152, 153), 103; **30**:111(62), 191(62)

Hsu, F. T., **22**:19(26), 30(26), 153(26)

Hsu, H. C., **31**:325(75)

Hsu, I-Yu, **6**:549, 563

Hsu, N. T., **3**:16(23), 32; **6**:515(28), 562

Hsu, P.-F., **31**:395(51, 53), 420(53), 427(51, 53)

Hsu, S. K., **7**:176, 212

Hsu, S. T., **11**:226, 259; **20**:15(31), 80(31)

Hsu, W. K., **19**:281(77), 285(77), 352(77)

Hsu, Y. Y., **1**:210, 211(20), 212(20), 213(20), 216, 224, 225(21), 231, 264, 425, 443; **4**:173, 201, 225, 226; **5**:67(20)(21), 75, 76, 78, 79, 81, 82, 100, 126, 127, 355(63), 360, 361, 362, 368, 370, 372, 373, 374, 428, 445, 446, 447, 448, 449, 450, 451, 464, 507, 512; **6**:550(88), 564; **7**:53(35), 85; **10**:128(62), 129, 130, 132, 133, 148, 152, 165; **11**:11, 17(59), 18, 48, 49, 55(4), 60(4), 95(25), 98(25), 99, 100(25), 104, 105(4), 112, 118, 120(4, 25), 121(25), 122(25), 170(172, 180), 173(180), 174(172, 180), 175(180), 177(180), 182(172), 183(172), 185(180), 186(180), 189, 190, 196, 197; **16**:138, 156; **23**:16(31), 18(31), 20(31), 127(31)

Hsui, A. T., **19**:82(261), 94(261)

Hu, K., **29**:171(137), 209(137)

Hu, M. H., **SUP**:67(129), 68(129), 265(443), 266(129, 442, 443), 275(129), 276(129, 443), 278(129, 443), 367(129, 442), 368(442, 443), 369(442, 443), 370(129)

Hu, R. Y. Z., **25**:260, 280, 281, 318

Hu, S., **21**:334, 335, 337, 339, 345

Hu, S. C., **15**:264(94), 265(94), 280

Hu, S. M., **15**:324; **28**:402(213), 424(213)

Hua, T. N., **23**:214(103a), 271(103a)

Huang, B., **23**:7(10), 126(10)

Huang, C. J., **SUP**:318(474)

Huang, G., **19**:164(216), 170(216), 171(216), 174(216), 190(216)

Huang, K., **11**:322(40), 352(40), 435

Huang, L.-J., **26**:36(18, 91, 94, 95), 40(94), 41(94), 43(92), 44(92), 45(96), 72(93), 74(93), 80(92), 93(18), 97(91, 92, 93, 94, 95, 96)

Huang, M.-J., **25**:259, 288, 317

Huang, P. G., **24**:227, 271

Huang, P. N. S., **19**:33(67), 37(67), 42(67), 86(67)

Huang, S. C., **19**:62(147), 63(147), 82(274), 89(147), 95(274)

Huang, W. C., **10**:260, 282

Huang, Y. S., **7**:307, 308, 319; **21**:72(13), 134(13)

Huang, Z. Q., **28**:135(118), 144(118)

Hubbard, D. W., **7**:108, 109, 160

Hubbard, G. L., **26**:55(97), 97(97)

Hubbard, H. A., **SUP**:65(104), 363(104), 364(104)

Hubbard, M. G., **20**:84, 85, 88, 91, 93, 97, 99, 100, 109, 110, 113, 130, 131

Hubbert, M. K., **14**:6, 100

Hubel, A., **22**:220(10), 347(10)

Huber, D. A., **16**:114(72), 155, 235; **20**:16(36), 17(36), 81(36)

Hubert, A., **30**:334(58), 349(58), 429(58)

Hubler, G. K., **28**:135(20a), 139(20a)

Huckaba, C. E., **22**:9(21), 17(21), 30(31, 32), 134(31, 32), 153(31, 32)

Hudders, R. S., **5**:327(18), 505

Hudson, F. L., **10**:208, 218

Hudson, J. D., **11**:226(151), 259

Hudson, J. L., **6**:435(169, 173), 437(169, 173), 494

Huebener, R. P., **17**:152(183), 158

Huesmann, K., **13**:29(53), 60

Hufen, J. H., **4**:331, 445

Huffman, D. R., **23**:145(24), 185(24)

Huge, E. S., **8**:94(11), 159; **18**:88(11), 158(11)

Huggins, C. E., **19**:79(209), 92(209); **22**:300(143), 353(143)

Hughart, D., **17**:160(1), 313

Hughes, D., **9**:265(146), 272

Hughes, E. D., **17**:42(143), 44(143), 63; **31**:106(7, 20), 138(7), 155(7), 156(20)

Hughes, J. A., **8**:94(2), 129(2), 130(2), 158; **11**:218, 221, 236, 256; **18**:87(2), 125(2), 127(2), 157(2)

Hughes, T. A., **1**:394(66), 398(66), 442

Hughes, T. P., **24**:30(61), 38(61)

Hughmark, G. A., **1**:367(18), 377, 383, 440; **15**: 122(147), 128(147), 141(147); **25**:298, 317

Huh, E., **28**:41(70), 73(70)

Huhtiniemi, I., **29**:233(69), 338(69)

Hui, T. -O., **16**:77, 78, 79, 111, 112, 154

Huige, N. J. J., **10**:192(90), 217

Huigol, R. R., **15**:77(64), 137(64)

Huilgol, R. R., **18**:163(21), 236(21)

Hulett, R. H., **7**:240, 271, 275(49), 315

Hull, H. L., **18**:139(72), 140(72), 146(72), 147(72), 159(72); **25**:292, 293, 315

Hulm, J. K., **5**:326(10), 505

Hultmark, E. B., **20**:183(5), 299(5)

Humble, L. V., **2**:42, 107; **6**:538, 563

Hume, J. R., **6**:474(281), 500

Hummel, R. L., **8**:293(21), 298(34, 35), 300(54), 301(54), 303(21), 349, 350; **15**:91(82), 138(82); **20**:15(34), 81(34); **26**:8(67, 196), 95(67), 101(196)

Humphrey, J. A. C., **18**:36(75), 38(77), 39(79), 41(83), 83(75, 77, 79, 83); **24**:227, 270; **28**:126(39), 130(39), 131(39), 132(39), 133(38), 134(39), 140(38, 39)

Humphrey, J. C., **5**:381, 509

Humphreys, C. G. R., **5**:453, 505

Humphreys, C. M., **4**:104(49), 105(49), 139

Humphreys, H. W., **9**:298(40), 347; **10**:250(45), 282

Humphreys, J. F., **5**:244, 251; **9**:73, 108; **SUP**:109(277)

Humphreys, J. S., **11**:217, 262

Humphreys, R. F., **9**:261(118), 271

Hundy, G. F., **11**:220(137), 224(137), 235(137), 238(137), 258

Hung, F. T., **6**:474(332), 502

Hunsbedt, A., **14**:97(187, 188, 189), 104

Hunt, A. W., Jr., **21**:96(76), 97(76), 136(76)

Hunt, B. G., **8**:281(50), 283

Hunt, G. R., **10**:11, 32, 36, 37

Hunt, H., **4**:90(27), 94(27), 138

Hunt, J. C. R., **28**:318(192), 338(192)

Hunt, J. D., **22**:219(104), 351(104); **28**:270(99, 100), 333(99, 100)

Hunt, J. R., **30**:111(55, 56), 154(55), 155(55), 156(55), 191(55, 56)

Hunt, L. P., **28**:351(74, 75, 76), 418(74, 75, 76)

Hunter, B. J., **9**:378(50), 381(50), 407(50), 415

Hunter, G. B., **28**:308(154), 336(154)

Hunter, J. C., **25**:361, 382, 413

Huntington, R. L., **1**:358(7), 440

Huntley, J. M., **30**:262(22), 308(22)

Huntley, S. C., **5**:389(115), 398, 510

Huntsinger, R. C., **10**:200, 218

Huppertz, H., **28**:402(196), 424(196)

Huppler, J. D., **15**:62(9), 78(9), 135(9)

Hurd, R. E., **2**:385(76), 391(76), 396

Hurd, S. E., **4**:153(24), 223; **5**:390, 391(119), 392, 393, 394, 395, 510

Hurlbut, F. C., **6**:38(73), 125; **7**:167, 168, 173, 211; **9**:358(15), 414

Hurle, D. T. J., **30**:313(3), 315(3), 319(3), 320(3), 322(3), 329(37), 330(40, 41), 331(40), 332(37), 333(55), 340(40), 347(3), 350(93), 390(156), 426(3), 428(37, 40, 41), 429(55), 430(93), 434(156)

Hurwicz, H., **11**:388(137), 438

Husain, A., **13**:126, 191, 201

Husain, S. Abid, **1**:397(77), 442

Husar, R. B., **9**:295(23), 345, 346, 348

Hush, D. R., **22**:419(26), 436(26)

Hussain, **30**:78(49a)

Hussain, A. K. M. F., **8**:291(13), 344(13), 349; **25**:381, 385, 386, 413; **26**:146(103), 213(103)

Hussaini, M. Y., **25**:362, 382, 405, 411, 416, 418

Hussey, R. G., **24**:294(47), 320(47)

Hustad, J. E., **29**:66(15), 68(15), 124(15)

Hustrup, R. C., **31**:209(111), 327(111)

Hutchinson, P., **17**:42(140, 141), 43(140), 44(140, 141), 62, 63

Hutton, J. F., **24**:108(80), 187(80)

Huyakorn, P. S., **30**:102(40), 185(40), 190(40)

Huyghe, J., **1**:384(49), 385(49), 388(49), 441

Huzarewicz, S., **23**:219(104), 271(104)

Hwang, B. K., Jr., **28**:190(11), 228(11)

Hwang, C. L., **10**:196(89), 203(89), 217; **SUP**:77(218), 183(362), 190(218, 375), 191(218, 375), 192(218, 375), 193(218); **22**:19(26), 30(26), 153(26)

Hwang, G. J., **SUP**:65(110), 120(296), 140(325), 141(325), 142(324), 143(324, 325), 146(324), 177(357), 178(357), 325(110), 328(110), 329(110), 333(110); **19**:308(97), 310(100), 313(97, 100), 353(97, 100); **28**:370(125, 130), 420(125), 421(130)

Hwang, N. M., **28**:351(84), 419(84)

Hwang, S. S., **16**:53, 57

Hwang, U. P., **20**:183(10), 244(195), 250(233), 251(233), 262(260), 299(10), 308(195), 310(233), 311(260)

Hwang, V. P., **20**:243(171), 307(171)
Hyde, M. A., **25**:272, 317
Hyman, C. R., **29**:234(94), 340(94)
Hyman, S. C., **11**:278, 314
Hynek, J. K., **31**:265(166), 330(166)
Hynek, S. J., **11**:170(186), 173, 177(186), 197
Hynynen, K., **22**:290(105), 351(105)
Hyun, J. M. (Chapter Author), **24**:277, 289(12), 290(12, 15), 291(12, 13), 292(13), 293(13), 294(11), 295(11), 306(18, 19), 307(18), 308(18), 309(18), 310(4), 311(4, 5), 313(14), 314(14), 315(16, 32), 318(4, 5), 319(11, 12, 13, 14, 15, 16, 18, 19), 320(32)

I

Iaglom, A. M., **8**:289(6), 291(6), 349
Ibele, W. E., **9**:253, 270; **15**:64(39), 136(39)
Ibl, N., **7**:88(2), 98, 116(28), 118(28), 120(28), 160; **12**:233(38), 251(38, 39), 270(38), 279
Ibragimov, M. H., **31**:317(215), 332(215)
Ibrahim, E. A., **16**:233
Ibrahim, M. Y., **31**:259(150), 329(150)
Ibriscu, M. M., **11**:318(7), 376(7), 434
Ibusuki, A., **7**:122(35), 161
Ichiraku, Y., **27**:196(42)
Iddan, G., **15**:238(46), 278
Idelchick, I. E., **31**:122(8), 134(8), 155(8)
Iden, B.-A., **23**:109(125), 110(125), 118(125), 132(125)
Ideriah, F. J. K., **24**:193, 194, 270
Ievlev, V. M., **14**:259(27), 263(27), 280; **25**:83(165), 114(180), 121(191), 147(165), 148(180), 149(191)
Ignatiev, M. P., **31**:254(148), 329(148)
Ignatiev, V. S., **25**:134(207), 150(207)
Iguchi, M., **25**:81(130, 132, 137, 138, 139, 157), 146(130, 132, 137, 138, 139), 147(157)
Ihrig, H. K., **6**:474(220), 479, 497
Iida, H. T., **5**:240, 241, 242, 251
Iida, T., **28**:128(41), 140(41)
Iida, Y., **18**:241(4), 257(4, 30), 263(4, 30), 267(4), 268(4), 269(4), 320(4), 321(30); **20**:33(56), 81(56)
Ikai, S., **25**:258, 283, 285, 319
Ikeda, M., **22**:219(184), 355(184)
Ikegawa, M., **28**:352(92), 419(92)

Ikezaki, E., **20**:387
Ikoku, C. U., **23**:212(105), 271(105)
Ikryannikov, N. P., **SUP**:203(394), 207(397); **21**:17(33), 20(33), 22(33), 51(33)
Ilchenko, A. I., **10**:208, 217; **14**:155(33, 49, 54), 243; **19**:164(204), 168(204), 171(204), 190(204)
Ilchenko, O. T., **6**:418(129, 136), 492
Ilegbusi, O. J., **28**:319(197, 198, 199), 338(197, 198, 199); **30**:333(56), 429(56)
Il'in, L. N., **6**:538, 541(48), 542, 563
Ilkovic, D., **7**:88(1), 160
Illingworth, C. R., **4**:33, 63; **8**:36, 88; **11**:226, 259
Ilyin, L. N., **8**:130(64), 160; **18**:127(63), 159(63)
Im, J. S., **28**:96(42), 140(42)
Imai, I., **8**:21, 22(97), 86
Imaishi, N., **30**:358(115), 432(115)
Imamura, T., **24**:6(26), 37(26); **26**:181(134), 182(134), 201(134), 202(134), 214(134)
Imas, Ia. A., **11**:318(20), 435
Imberger, J., **16**:9, 11, 56; **17**:207, 315; **24**:296(36), 299(36), 307(36), 312(36), 320(36)
Imura, H., **11**:296, 298, 299, 303, 315; **17**:50(159, 163), 51, 52, 63, 64; **30**:77(40), 89(40)
Inaba, H., **18**:12(16), 79(16); **19**:77(180), 78(188), 91(180, 188)
Inada, J., **21**:300, 345
Inada, S., **23**:8(19), 12(19, 34), 14(19, 34), 16(34), 28(19, 34), 29(19, 34, 55), 36(19, 34), 39(19, 34), 42(34), 46(34), 59(19, 34), 61(19, 34), 79(19), 80(19), 82(19, 55), 83(19), 84(19), 88(19), 93(19), 103(34), 106(19), 127(19, 34), 129(55); **26**:120(39), 121(39), 122(39), 123(39), 162(39, 117), 163(39, 117), 167(39), 169(39), 210(39), 214(117)
Inai, T., **11**:105(116), 114(116), 194; **17**:4(21), 6(21), 57
Inamura, H., **17**:19(61), 20, 58
Inayatov, A. Ya., **31**:236(123), 327(123)
Inayatov, I. A., **31**:296(206), 332(206)
Ince, E. L., **3**:174, 174(46)
Incropera, F. P. (Chapter Author), **18**:1(2), 79(2); **19**:144(156), 188(156), 300(88), 301(88), 311(102), 313(102), 353(88, 102); **20**:183(13, 14), 184(14), 221(14), 222(14, 138, 139, 140), 223(139), 224(138, 139,

140), 225(140), 226(139), 247(209), 256(138), 299(13, 14), 305(138), 306(139, 140), 309(209); **23**:1, 1(1, 2, 3), 7(3), 8(12), 10(12), 11(12), 13(12, 36, 69), 15(12, 36, 69), 18(12, 36), 29(12), 33(12), 43(12), 44(12), 46(12), 51(69), 53(73), 89(101), 90(102), 98(102), 99(102), 118(121), 120(121), 123(3, 128), 126(1, 2, 3, 12), 128(36), 129(69), 130(73), 131(101, 102), 132(121, 128); **26**:105(3), 106(9), 120(40), 121(46), 122(40, 47), 123(53, 54, 63, 64), 124(63, 64), 126(53, 54), 129(53, 54), 134(54), 135(54, 81, 83), 144(102), 145(46), 146(46), 147(40, 46, 102), 150(53, 54), 151(53, 54), 155(113), 161(54), 162(40, 47, 118), 163(47, 81, 118), 164(40, 102, 113), 165(46, 47, 102, 113), 166(113), 167(40, 47, 118), 168(40, 102), 169(47), 170(40, 113), 173(131, 137, 138), 177(131), 178(131), 182(137, 138), 187(47), 188(146, 148, 154), 189(47), 190(146), 191(148, 149), 192(154), 193(146), 204(178, 180), 208(3, 9), 210(40, 46, 47, 53, 54), 211(63, 64), 212(81, 83), 213(102, 113), 214(118, 131), 215(137, 138, 146, 148, 149, 154), 217(178, 180); **28**:231, 241(13), 243(17, 18), 244(17, 18), 245(18, 20), 246(18), 247(18), 248(18), 252(42, 43, 44, 45, 52, 53), 253(43, 45), 254(43, 45), 255(42, 43, 44, 45, 60, 61), 262(42, 43), 263(52, 53), 264(43), 266(42), 268(52, 53), 270(42, 43), 271(42), 272(101, 102), 273(101), 274(102), 275(42, 101, 102), 276(13, 43, 45, 102), 278(112, 113), 279(112, 113), 280(112, 113, 115), 281(113), 282(17, 18, 43, 112, 123), 283(18, 123), 284(18, 123), 285(18, 43, 52, 53, 123), 286(18, 128, 129, 130, 131, 132), 287(43, 129, 130, 131, 132, 133), 288(131), 290(134), 292(134), 298(134), 299(134), 300(134), 301(60, 136), 302(52, 53, 60, 61, 134, 136), 303(136), 304(136), 306(20, 136), 307(43), 308(136), 310(112, 159, 160), 311(43, 45, 159), 312(159, 160), 313(159), 314(159, 161), 315(160), 317(179), 319(43, 45, 61, 200, 201), 320(61, 134, 200), 321(201), 322(201), 323(200), 324(201), 325(134), 326(43), 327(52, 53), 329(13, 17), 330(18, 20), 331(42, 43, 44, 45, 52, 53, 60, 61),

333(101, 102), 334(112, 113, 115, 123, 128), 335(129, 130, 131, 132, 133, 134, 136), 336(159, 160, 161), 337(179), 338(200, 201), 370(132), 421(132); **30**:334(59), 429(59)

Incropera, T. A., **20**:183(13), 247(209), 299(13), 309(209); **23**:90(102), 98(102), 99(102), 131(102)

Ing, S. W., **28**:342(20), 416(20)

Inger, G. R., **2**:166, 169, 170(36), 175, 186, 219, 224, 225(62), 226(62), 246, 247, 251(70, 74), 269, 270; **8**:7, 84

Ingle, S. G., **25**:161(43), 162(43), 184(43), 200(43), 244(43)

Ingmanson, W. L., **25**:294, 317

Ingram, G. W., **23**:370(13), 461(13)

Ingrao, H. C., **10**:76, 78, 83

Inman, R. M., **2**:379, 395; **7**:185, 186, 187, 213; **SUP**:129(308)

Inocu, K., **11**:105(116), 114(116), 194

Inoue, A., **29**:159(93), 160(94), 161(94, 98), 162(93, 98, 100), 207(93, 94, 98, 100)

Inoue, H., **28**:343(55), 417(55)

Inoue, I., **25**:261, 269, 274, 275, 278, 318

Inoue, K., **17**:4(21), 6(21), 57

Inoue, M., **17**:21(64), 59

Inoue, N., **28**:316(167), 336(167)

Inoue, S., **17**:50(161), 63

Inoue, T., **23**:100(104), 131(104); **30**:396(165), 400(165), 434(165)

Inoue, Y., **28**:386(157), 387(157), 422(157)

Inouye, K., **6**:418(107), 491

Inouye, M., **2**:172, 173, 269

Inozemtseva, L. I., **25**:81(155), 147(155)

Interthal, W., **23**:206(64), 219(64), 222(64), 225(64), 269(64)

Inui, S., **28**:96(68), 141(68)

Inuzuka, T., **28**:343(47), 417(47)

Ioffe, I. V., **14**:122, 145

Ioselevich, V. A., **12**:80(11), 112; **15**:97(97), 138(97)

Iosff, D., **28**:387(159), 388(159), 389(159), 422(159)

Ip, B. M., **29**:167(124), 208(124)

Iqbal, M., **SUP**:28(39), 62(76), 63(78), 64(76, 92), 67(76, 92), 124(303), 207(92), 208(39), 209(39), 225(78), 232(76, 92), 236(76, 78, 92), 237(92), 248(92), 251(92), 259(76, 92); **19**:278(63), 352(63)

Iquchi, M., *see* Iguchi, M.

Ireland, P. T., **23**:323(204, 205, 206), 324(208, 210), 330(208), 363(204, 205, 206, 208, 210)

Irey, R. K., **5**:486, 487, 488, 517; **11**:105(137), 195; **16**:234; **17**:74(41), 119(41, 114), 154, 156

Iribarne, A., **7**:132, 161; **12**:237(40), 251(40), 279

Irmay, S., **24**:104(53), 186(53)

Irons, B. M., **SUP**:96(256)

Irvine, S. J. C., **28**:342(28, 34), 343(28), 416(28, 34)

Irvine, T. F., Jr., **1**:47(37), 50; **2**:441(40), 443(40), 451; **3**:275, 301; **4**:97(38), 100(38), 139; **8**:2(1, 2, 3, 4, 5, 6, 7), 84; **SUP**:18(27), 26(37), 62(27), 123(301), 198(386, 387), 227(419), 229(421, 422), 264(27, 419), 266(27, 419); **19**:248(4), 349(4); **21**:299, 301, 312, 329, 330, 331, 333, 344, 345; **24**:102(8), 184(8); **28**:153(25), 228(25); **31**:160(44), 324(44)

Irving, J. P., **2**:75(69), 95(69), 108

Isaacson, E., **1**:72(25), 75(25), 121; **19**:13(26), 19(26), 84(26)

Isachenko, V. P., **8**:140(73), 141(73), 148, 149, 160; **9**:237(69), 238(69), 270; **18**:139(74), 140(74), 146(74), 159(74); **19**:144(157), 188(157); **31**:160(45), 202(107), 208(107), 212(107), 213(107), 242(131), 275(177, 178), 281(178), 324(45), 327(107), 328(131), 330(177, 178)

Isaev, S. G., **25**:77(128), 146(128)

Isaji, A., **11**:220, 224, 237, 263

Isakeev, A. I., **31**:160(33), 323(33)

Isakoff, S. I., **11**:105(115), 114, 194

Isakson, V. E., **9**:408(104), 417

Isatayev, S. I., **8**:103(29), 104(29), 105(29), 132(29), 133(29), 159; **11**:228(226), 262; **18**:98(29), 99(29), 130(29), 131(29), 158(29)

Isayama, Y., **17**:42, 62

Isbin, H. S., **1**:357, 374, 375, 383(46), 394(68), 398(68), 410, 440, 441, 442; **15**:243(58), 278; **20**:15(33), 81(33)

Ischigai, S., **21**:29(47), 31(47), 52(47)

Isdale, J. D., **18**:327(4), 362(4); **31**:463(74), 474(74)

Iseler, G., **30**:415(182), 416(182), 435(182)

Isenberg, J., **15**:245, 246(78), 248(77), 249(77), 262(78), 279

Isermann, R., **6**:474(293), 500

Ishibashi, E., **20**:36(57), 42(57), 43(57), 81(57)

Ishida, M., **30**:359(123), 361(123), 432(123)

Ishida, N., **20**:277(293), 313(293)

Ishida, S., **20**:277(294), 283(304), 313(294, 304)

Ishigai, S., **11**:105(116), 114, 194; **17**:4(21), 6, 57; **18**:255(19), 257(19), 259(19), 262(19), 264(19), 265(19), 267(19), 268(19), 269(19), 270(19), 271(19), 272(19), 273(19), 274(19), 290(19), 313(19), 320(19); **23**:12(53), 14(53), 28(53), 29(53), 33(53), 36(53), 39(53), 58(53, 83), 59(93), 60(53, 83), 61(93), 63(83), 77(83, 93), 78(83, 93), 84(53), 85(83), 93(53, 83), 106(53, 93), 108(53), 109(53, 93, 113), 110(53, 93, 113), 111(53), 112(53, 93, 113), 115(53), 118(53, 122), 129(53), 130(83), 131(93), 132(113, 122); **26**:181(134), 182(134), 201(134), 202(134), 214(134)

Ishiguro, R., **20**:387; **21**:76(21), 77(21), 134(21)

Ishihara, M., **19**:76(176), 91(176)

Ishihara, T., **23**:193(175), 214(175), 216(175), 274(175); **25**:342, 361, 388, 402, 404, 413; **28**:345(64), 418(64)

Ishii, K., **17**:19(60), 20, 58; **20**:252(240), 310(240); **23**:8(15), 12(15), 14(15), 36(15), 39(15), 45(15), 58(15), 60(15), 78(15), 101(15), 102(15), 107(15), 126(15)

Ishii, M., **17**:37, 62; **29**:45(40), 57(40), 196(210), 213(210), 259(110, 129), 263(124), 270(124), 271(124, 138, 139, 140), 272(124, 139), 273(124, 129, 140, 144), 341(110, 124), 342(129, 138, 139, 140), 343(144); **31**:108(10), 109(13), 110(11), 111(10), 112(12), 156(10, 11, 12, 13)

Ishii, T., **23**:214(103a), 271(103a)

Ishimaru, A., **23**:154(36), 168(36), 185(36)

Ishimaru, M., **23**:12(62), 14(62), 32(62), 58(62), 60(62), 89(62), 109(62), 110(62), 120(62), 129(62)

Ishirawa, S., **6**:396(34), 487

Ishizawa, S., **SUP**:160(339), 168(339)

Ishizuka, M., **20**:187(66, 67), 302(66, 67)

Iskra, A. L., **6**:474(280), 500

Islam, I. R., **23**:193(107), 271(107)

Islam, K. M. M., **30**:175(153), 196(153)

Islam, M. R., **23**:193(106), 271(106); **24**:104(28), 185(28)

Ismael, J. O., **25**:81(133), 146(133)

Ismail, K. A. R., **19**:36(74), 86(74)

Isoda, H., **4**:215, 216, 217, 227

Isoda, Y., **26**:170(126), 214(126)

Isokrari, O. F., **14**:89, 90(178), 91(178), 104

Israeli, M., **14**:51, 102

Issa, R. I., **24**:199, 249, 270, 271

Isshiki, N., **16**:95, 96, 101, 127, 128, 154

Itamura, M. T., **30**:184(161), 196(161)

Ito, I., **26**:109(19), 113(19), 114(19, 34), 118(19, 34), 157(19), 171(19), 173(19), 174(19), 175(19), 177(19), 209(19, 34)

Ito, R., **7**:108(24), 110(24), 122(35), 128(37), 135, 160, 161; **15**:101(112), 102(112), 103(112), 139(112); **21**:181

Ito, S., **27**:196(42)

Ito, T., **11**:65, 67, 68, 76(30), 77(30), 78(30), 80, 82(30), 83(30), 126, 127, 131, 132, 134, 136(141), 190, 195; **16**:136, 155; **20**:64(61), 68(61), 70(62), 72(63), 76(70), 82(61, 62, 63, 70); **21**:37(58), 38(63), 52(58, 63)

Itoh, M., **11**:281, 315

Itsweire, E. C., **21**:185(4), 204(4), 206(4), 207(4), 236(4)

Ittner, W. B., **5**:326(9), 505

Iuanov, V. V., **14**:253(16), 279

Iuchi, M., **6**:69(119), 128

Ivakin, V. P., **6**:379(11), 380(11), 486; **25**:28(58), 142(58)

Ivanov, A. I., **21**:2(4), 3(4), 50(4)

Ivanov, D. G., **25**:158(24, 25, 26, 27, 28, 29, 30, 31, 32, 38), 159(24, 25, 26, 27, 32), 160(29), 161(28, 29, 38), 163(29), 165(28, 29, 38), 168(32), 200(25, 26, 27, 32), 201(32), 202(27), 236(25), 243(24, 25, 26, 27), 244(28, 29, 30, 31, 32, 38)

Ivanov, R. A., **6**:418(140), 493

Ivanov, V. L., **9**:95, 101, 103, 108; **31**:160(49), 324(49)

Ivanov, V. M., **25**:51(80, 81), 143(80, 81)

Ivanov, V. V., **6**:418(121, 146, 147, 148, 149, 150), 492, 493

Ivanova, A. V., **6**:526, 539, 562

Ivanovskii, M. N., **9**:241(71), 242, 255(71), 270; **21**:92(66), 136(66)

Ivanushkin, S. G., **25**:73(121), 145(121)

Ivashchenko, M. M., **6**:418(114, 115), 491; **25**:127(196), 149(196)

Ivashchenko, N. I., **6**:542, 563

Ivakevič, A. A., **1**:412, 443

Ives, D. J. G., **12**:225(41), 249(41), 250(41), 279

Ivey, C. M., **14**:5(18), 100

Ivey, G. N., **24**:300(17), 319(17)

Ivey, H. J., **1**:239, 265; **4**:206(83), 226; **5**:462, 515; **17**:23(73), 59

Ivey, R. K., **9**:358(14), 408(14), 414

Ivins, R. O., **29**:136(23), 137(23), 203(23)

Iwahori, K., **25**:73(113), 96(113), 145(113)

Iwamoto, M., **30**:334(63), 349(63), 429(63)

Iwasa, Y., **17**:151, 158

Iwasaki, H., **3**:269(23), 300

Iwatsu, R., **24**:306(18, 19), 307(18), 308(18), 309(18), 319(18, 19)

Iyengar, S. R. K., **26**:26(133), 99(133)

Iyer, K., **29**:171(137), 209(137)

Iyevlev, V. M., **31**:159(1), 160(1), 161(1), 177(1), 182(1), 247(1), 262(1), 263(1), 322(1)

Iynkaran, K., **18**:17(28), 80(28)

Izasimov, V. G., **25**:75(122), 145(122)

Izosimov, V. G., **6**:449(181, 182, 183, 184, 193), 453(181, 182, 183, 184, 193), 456(181, 182, 183, 184), 457(181, 182, 183, 184), 495

Izumi, M., **21**:91(61), 136(61)

Izumi, R., **26**:120(39), 121(39), 122(39), 123(39), 162(39), 163(39), 167(39), 169(39), 210(39)

J

Jablonic, R. M., **9**:71(117), 104(117), 111

Jackman, R. B., **28**:343(48), 417(48)

Jackson, G. W., **25**:252, 292, 317

Jackson, J., **17**:152(180), 157

Jackson, J. D., **6**:550(72), 551, 563; **7**:24(7), 31, 35, 37, 39, 40(20), 42, 50, 52, 54(12), 66, 66(12, 52), 67, 68, 68(7), 72, 83, 84, 86; **SUP**:83(242), 85(242); **15**:8(29), 55(29); **21**:1(2), 4(2), 10(23), 13(2), 14(23), 16(23), 30(2), 50(2), 51(23)

Jackson, J. M., **2**:343, 354

Jackson, K. A., **11**:318(4), 412(4), 434; **22**:219(104), 351(104); **28**:79(43, 44), 140(43, 44), 270(99, 100), 333(99, 100)

Jackson, K. W., **18**:169(39), 170(39), 176(39), 230(39), 236(39)

Jackson, M., **5**:26, 51

Jackson, R. G., **10**:170(16), 215; **23**:374(38), 377(38), 462(38); **25**:195(98), 247(98)

Jackson, R. G., **4**:228

Jackson, T., **4**:51, 55, 59, 64; **15**:189(64), 224(64)

Jackson, T. W., **9**:11, 12, 37, 38, 97, 107, 108; **15**:195(80), 225(80); **20**:187(37), 300(37)

Jacob, A., **10**:187, 189, 199, 202, 203, 205, 217

Jacob, M., **11**:300, 315; **18**:167(53), 237(53)

Jacobs, E. W., **18**:41(83), 83(83)

Jacobs, H., **29**:151(67), 155(83), 157(67), 174(146), 205(67), 206(83), 209(146)

Jacobs, H. R., **9**:258(96), 270; **11**:130, 131, 134(145), 135(145), 195; **15**:228(6), 245(74, 76a), 246, 248, 264, 267(100), 269, 272(113), 276, 279, 280, 281, 322, 325; **26**:1(98), 2(98), 19(98), 47(98, 99), 97(98, 99)

Jacobs, R. B., **5**:355, 375, 403, 451, 455, 456, 507, 511, 515

Jacobs, R. T., **1**:410, 442

Jacobsen, I. A., **22**:304(162), 354(162)

Jacobson, J. D., **5**:48(77), 54

Jaeckel, R., **2**:354

Jaeger, H. L., **9**:202(16), 268; **14**:332(115), 333, 334(115), 345

Jaeger, J. C., **1**:2(1), 49, 52, 54(1), 58, 60(1), 62(1), 63(1), 66(1), 74, 99(1), 120, 121, 211(22), 264; **3**:122(26), 126(33), 132(33), 173; **5**:40, 54, 336(26), 347, 506; **6**:489; **8**:3, 53(10), 75(10), 78(10), 84; **10**:40, 70, 71, 72, 78, 81, 82; **11**:12(31), 13, 48, 322(36), 378(36), 383(36), 435; **SUP**:119(290), **15**:285, 287, 289, 313, 322, 324; **18**:169(55), 237(55); **19**:4(12), 5(12), 6(12), 8(12), 12(12), 14(12), 18(12), 41(12), 84(12), 192(8), 193(8), 243(8); **20**:354, 355, 356, 358, 387; **22**:159(33), 184(33), 185(33), 251(33), 271(33), 348(33); **23**:317(150), 360(150), 432(86), 464(86); **24**:56(20), 99(20); **29**:138(27), 204(27)

Jaeger, J. S., **10**:48, 76, 81

Jaeger, R. C., **20**:182(4), 252(241), 257(241), 258(241), 299(4), 310(241); **23**:32(60), 129(60)

Jaffee, L. D., **10**:50(30), 51(30), 59(30), 68(30), 81

Jafri, I. H., **30**:408(177), 410(177), 411(177), 412(177), 435(177)

Jagadish, B. S., **21**:5(13), 51(13)

Jagannathan, P. S., **9**:361(20), 382(20), 384(20), 385(56), 394(56, 69, 72, 73), 397(56), 398(70, 72, 73), 414, 416

Jäger, K., **22**:225(19), 347(19)

Jagov, V. V., **16**:235, 236; **17**:22(69), 59

Jahn, M., **29**:5(3), 6(4), 7(3), 15(3), 17(3), 18(3), 19(3, 4), 24(3), 25(3), 26(3), 55(3, 4)

Jahne, B., **30**:402(167), 434(167)

Jahnke, E., **3**:134(39), 174; **5**:12(14), 51

Jahns, H. O., **23**:274(177)

Jain, A. C., **7**:204, 217

Jain, M. K., **2**:380, 395

Jain, P., **6**:396(39), 488

Jain, R. K., **22**:11(25), 18(25), 30(33), 84(33), 134(33), 154(33), 342(106), 351(106), 360(5), 435(5)

Jain, S. P., **6**:474(256), 483(256), 498

Jain, V. K., **30**:65(74), 90(74)

Jaiswal, A. K., **23**:202(112), 208(108, 109, 112), 210(112), 211(112), 212(108, 111, 112), 214(108), 237(111), 238(111), 239(111), 240(111), 244(110), 246(111), 247(110, 111), 248(111), 249(111), 250(110, 111), 251(111), 271(108, 109, 110, 111, 112)

Jaivin, G. I., **2**:75, 79, 86, 108

Jakeman, E., **30**:329(37), 332(37), 428(37)

Jakes, J., **20**:297(324), 314(324)

Jakob, M., **1**:187(1), 189(1), 194, 202(1), 208(1), 216, 226, 227(1), 259, 263; **2**:411, 412(8), 413(8), 450; **3**:95, 100, 272, 300; **4**:54, 64, 155, 156, 223; **8**:176, 195, 227; **9**:206, 220(26), 256(91), 258, 268, 269, 270, 350(1), 414; **10**:181, 184, 192, 203, 204, 206, 216; **11**:202, 204, 205, 213(68), 214(68), 253, 255; **14**:151(11), 242; **SUP**:101(261), 289(457); **18**:179(59), 180(59), 237(59); **20**:7(8), 9(10), 14(16), 15(10, 16), 24(48), 79(8), 80(10, 16), 81(48); **21**:80(29), 134(29); **26**:250(57), 327(57)

Jakobsson, J. D., **29**:49(45), 57(45)

Jakobsson, O. P., **22**:239(107), 242(107), 351(107)

Jalaluddin, A. K., **14**:135, 145

Jallouk, P. A., **9**:49, 50(64), 51(64), 53, 54(58, 64), 55(64), 56(64), 57(64), 60, 62(64), 67, 69(64), 70, 71, 103(64), 108, 109

Jaluria, Y. (Chapter Author), **18**:54(103), 65(117a), 84(103), 85(117a); **19**:62(152), 67(152, 153), 90(152, 153); **20**:203(89, 90, 91, 97, 98), 204(91), 243(172), 303(89, 90, 91, 97, 98), 307(172), 316(12), 350(12); **24**:115(122), 189(122); **28**:145, 150(21,

33), 151(18, 38, 58, 59), 156(36), 159(33), 160(18, 27, 33), 162(38), 163(38), 165(38), 166(21, 59), 168(59), 169(59), 170(59), 171(7, 59), 172(36), 173(7), 174(7), 175(7), 181(33), 185(7), 188(59), 189(59), 190(38, 58), 193(38, 58), 195(5), 197(5), 198(5), 200(5), 201(6), 202(5), 204(6), 205(6), 207(5), 208(6, 22), 209(6), 211(22, 44), 215(18), 216(18), 217(18), 219(1, 8), 221(8), 222(8), 227(1), 228(5, 6, 7, 8, 18, 21, 22, 26), 229(27, 33, 36, 38, 44), 230(58, 59), 239(9), 251(34, 35), 329(9), 330(34, 35); **30**:362(125), 413(125), 432(125)

Jamaluddin, A. S., **23**:139(17), 184(17)

James, D. D., **11**:13(32), 48, 203, 206(56), 210(56), 220(56), 223(56), 235, 238(56), 248(56), 254; **19**:334(143), 337(143), 339(143), 355(143); **23**:9(22), 30(22), 49(22), 127(22)

James, D. F., **23**:206(113), 219(115), 222(113, 114), 223(114), 225(113, 114), 272(113, 114, 115); **25**:252, 257, 286, 292, 317

James, E. H., **23**:7(9), 126(9); **26**:106(6), 136(6), 208(6)

James, G. B., **17**:147(172), 157

James, P. A., **SUP**:67(126), 249(126), 251(126), 280(126), 282(126)

James, P. W., **17**:42(141), 44(141), 63

James, W., **8**:128(61), 130(61), 160; **11**:218(96), 222(96), 256; **18**:123(60), 124(60), 127(60), 159(60)

Jameson, A., **24**:227, 234, 269, 271

Jameson, S. L., **15**:23(50), 56(50)

Jamieson, D. T., **18**:328(11), 362(11)

Jamil, M., **SUP**:65(109, 111), 273(111, 449), 274(449), 346(109), 348(109), 350(109, 111, 449), 352(449)

Jamin, J., **6**:195, 363

Jan, J. S., **17**:133(140), 156

Janeschitz-Kriegl, H., **13**:219(75), 234(75), 248(108), 249(75), 265, 266

Jang, J., **28**:285(125, 126), 334(125, 126)

Jang, J. Y., **20**:336(52, 53, 54), 351(52, 53), 352(54)

Janikowski, H. E., **18**:68(128), 70(128), 85(128)

Janour, Z., **20**:377, 387

Jansen, W., **4**:205(80), 206(80), 226

Janssen, E., **1**:244, 265; **2**:41, 106; **3**:35(33), 60, 98

Janssen, J. E., **4**:113(64), 139, 331, 444

Janssen, K., **14**:176(92), 244

Janssen, L. P. B. M., **28**:151(28), 229(28)

Janssen, R. J. A., **28**:306(148), 335(148)

Jansson, V. K., **5**:230, 250

Jantscher, H. N., **5**:149, 152, 248

Janz, G. J., **12**:225(41), 249(41), 250(41), 279

Japikse, D. (Chapter Author), **9**:1, 3(65), 5, 10(63), 12(59), 26, 27, 28, 29, 30, 31, 35, 36, 46, 48, 49, 50, 51(59, 64), 52, 53, 54(64), 55(64), 56, 57(64), 59, 62(60, 64), 65, 66, 67, 69(59, 64), 70(59, 64), 71, 91(65), 92(65), 100, 103, 104, 108, 109

Jaraiz-M, E., **25**:234(132, 133, 134, 135, 136), 235(132, 133, 134, 135), 248(132, 133, 134, 135), 249(136)

Jargon, J. R., **23**:212(255), 278(255)

Jarmai, L., **8**:66, 90

Jarre, G., **2**:243, 269

Jarzebski, A. B., **26**:234(38), 254(38), 326(38)

Jaskins, J. F., **5**:347, 506

Jasper, J. J., **16**:132, 155

Jassal, A. S., **28**:251(38), 275(38), 276(38), 330(38)

Jaupart, C., **28**:282(116), 334(116)

Jaussen, E., **1**:421(117), 443

Javandel, I., **30**:154(117), 181(157), 185(117), 194(117), 196(157)

Javanovic, G. N., **23**:340(272), 367(272)

Javdam, K., **25**:51(86), 144(86)

Javeri, V., **SUP**:21(35), 75(35, 207), 80(207), 103(207), 121(207), 142(207), 149(207), 151(207), 219(35), 252(35), **19**:281(66, 67, 68, 69), 283(66, 69), 284(67), 285(68), 291(66, 67, 68, 69), 292(68), 296(68), 299(68), 302(68), 306(68), 312(68), 352(66, 67, 68, 69)

Jaworski, W., **8**:76(279), 77(279), 91

Jay, K. M., **22**:222(108), 351(108)

Jayaraj, S., **26**:122(48), 210(48)

Jayaraman, K., **23**:193(42), 268(42); **25**:292, 308, 316

Jayaraman, R., **25**:81(145), 146(145)

Jean, R.-H., **23**:207(116), 272(116)

Jeans, A. H., **16**:329, 331, 363

Jefferson, T., **5**:390, 510

Jefferson, T. B., **2**:364, 395; **11**:203(44), 206(44), 248(44), 254; **18**:179(61), 180(61), 237(61)

Jefferys, B. S., **8**:19(94), 86

Jeffrey, D., **18**:190(81), 237(81)

Jeffrey, D. J., **15**:63(28), 64(28), 136(28)

Jeffrey, H., **18**:167(49), 237(49)

Jeffreys, H., **4**:152, 223

Jeffries, N. P., **7**:236(27), 247, 277(27), 314

Jeffs, A. T., **9**:405(93), 417

Jelinek, F., **20**:296(323), 314(323)

Jellison, G. E., Jr., **28**:76(46), 79(45, 67), 80(46), 99(111), 140(45, 46), 141(67), 143(111)

Jellyman, P. E., **11**:358(84), 436

Jeng, D. R., **6**:474(204), 476, 496; **8**:19, 86

Jeng, Y. N., **24**:215, 271

Jenkins, A. E., **3**:106, 124(30), 172, 173

Jenkins, L., **17**:143(160), 151(160), 157

Jennings, B. H., **4**:104(49), 105(49), 139

Jennings, R. R., **23**:222(117), 272(117)

Jensen, A., **17**:37(118), 61

Jensen, G. E., **2**:367, 370, 373, 394

Jensen, K. F., **28**:339(2), 344(2, 59), 350(2), 352(91), 358(2), 364(104, 105), 365(105), 369(2, 104, 105), 370(104, 105), 371(104), 372(104), 373(104, 105), 374(104), 375(2, 139, 140, 141), 376(2), 379(104), 382(155), 389(105, 171), 390(171, 173, 177), 391(171), 397(187, 188), 398(171, 187, 188), 402(155, 200, 201, 218), 404(155), 405(155, 200), 407(223), 408(223), 415(2), 417(59), 419(91, 104, 105), 421(139, 140, 141), 422(155, 171), 423(173, 177, 187, 188), 424(200, 201), 425(218, 223)

Jensen, M. K., **30**:198(8), 249(8)

Jensen, R. A., **16**:5(1), 9(1), 11(1), 31(1), 56

Jensen, R. D., **26**:61(100), 97(100)

Jenson, V. G., **2**:372, 396; **15**:64(38), 101(38), 102(38), 136(38), 196(85), 197(85), 225(85)

Jepson, G., **9**:127(29), 178

Jerauld, G. R., **25**:187(80), 198(80), 246(80)

Jerbens, R. H., **7**:125(36), 161

Jergel, M., **16**:233, 234; **17**:321, 339

Jerger, E. W., **4**:44, 46, 64

Jeschar, R., **13**:10(31), 17, 60; **23**:117(120), 132(120)

Jessen, C., **23**:320(157), 361(157)

Jessup, R. E., **23**:370(14), 461(14)

Jew, H., **9**:24, 109

Jia, H., **26**:56(101, 102), 57(101), 58(101), 97(101, 102)

Jiang, Y., **24**:226, 227, 247, 269, 271

Jiji, L. J., **26**:

Jiji, L. M., **19**:28(58), 29(58), 30(58, 61), 33(58, 61), 44(101), 78(101, 191, 192), 86(58, 61),

87(101), 91(191, 192); **20**:251(234), 255(234), 256(234), 257(234), 310(234); **22**:14(26, 37), 18(26, 37), 19(24, 34, 55, 56, 59, 60, 61, 62, 63), 24(41, 60), 26(60, 61), 33(61, 63), 45(34, 61), 69(56), 71(34, 55, 56, 59, 60, 61, 62, 63), 96(34, 55, 56, 59, 60, 61, 62, 63), 97(59), 99(59), 101(59), 103(34, 60), 104(59, 60), 105(59, 60), 106(60), 107(41, 59, 60), 108(34, 60), 109(34), 110(34, 60), 111(34), 113(34), 114(34), 116(34), 117(34), 118(34), 119(34), 120(34), 121(34), 122(24, 41, 60), 123(34, 41, 59, 60, 61), 124(34, 60, 61), 125(34, 59, 60, 61), 126(34, 60, 61), 128(61), 129(34, 60), 130(34, 60, 61), 131(24, 34, 55, 60), 132(24, 34, 41, 55, 60), 133(34, 55), 134(34, 55, 60), 135(55), 136(55), 137(55, 62, 63), 139(60, 62), 140(61, 62), 141(41), 142(61), 144(63), 146(63), 147(63), 150(63), 151(24, 55, 56), 152(41), 153(24), 154(34, 41), 155(55, 56, 59, 60, 61, 62, 63); **23**:13(37, 46), 15(37, 46), 16(32), 18(37), 19(37), 20(37), 22(37, 46), 26(37), 27(37, 46), 28(37), 34(37, 46), 38(46), 39(37, 46), 49(37, 46), 59(37, 46), 61(37, 46), 64(46), 65(37, 46), 77(37, 46), 79(46), 87(46), 90(46), 92(46), 100(37, 46), 101(37, 46), 127(32), 128(37, 46); **26**:109(21), 113(21, 28), 114(33), 149(108), 151(108), 155(108), 172(108), 178(108), 209(21, 28, 33), 213(108)

Jimbo, G., **19**:186(63)

Jimenez, J., **23**:292(56, 57), 356(56, 57)

Jin, J. D., **26**:69(103), 97(103)

Jischke, M. C., **8**:56, 90; **24**:278(20), 286(20), 302(20), 319(20)

Joanny, J. F., **28**:41(76), 73(76)

Joch, L., **15**:273(117), 281

Jochem, M., **22**:188(111), 352(111)

Jodlbauer, K., **11**:202, 204, 209, 253

Jodrey, W. S., **23**:161(47), 186(47)

Joffre, R. J., **20**:209(114), 304(114)

Jog, M. A., **26**:60(104, 105), 62(105), 65(105), 97(104, 105)

Joglekar, G., **19**:58(132), 59(132), 61(132), 89(132); **24**:3(6), 36(6)

Johannes, C., **17**:146, 157, 322, 323(21), 326, 335(18), 339

Johannessen, J. A., **16**:48, 57

Johanson, L. N., **10**:173, 215

John, J. E., **20**:264(263), 311(263)

John, R. R., **4**:283(49), 315

Johns, D. J., **22**:312(109), 351(109)

Johns, L. E., **26**:19(106), 69(106), 97(106)

Johnsen, G., **31**:112(9), 156(9)

Johnson, A. B., **23**:321(194), 363(194)

Johnson, A. E., **20**:217(127), 220(127), 305(127)

Johnson, A. I., **21**:172, 181; **26**:19(241), 104(241)

Johnson, B. M., **12**:41, 75

Johnson, C. A., **19**:126(130), 188(130)

Johnson, C. E., **20**:187(63), 195(63), 302(63)

Johnson, C. H., **14**:66, 103

Johnson, C. L., **9**:365(36), 372(36), 401(81), 415, 416; **22**:223(110), 228(110), 352(110)

Johnson, D. H., **18**:1(1), 79(1)

Johnson, D. S., **8**:299(44), 305, 349; **18**:327(7), 362(7)

Johnson, F. A., **18**:169(35), 170(35), 172(35), 182(35), 236(35)

Johnson, F. S., **10**:29, 37, 75(68), 83

Johnson, G. D., **7**:248, 316

Johnson, G. M., **10**:234(13), 281

Johnson, H. A., **1**:382(44), 441; **2**:41, 106; **4**:43(76), 64; **6**:460, 461(189), 467, 474(203), 477, 495, 496; **9**:317(79), 347

Johnson, H. L., **5**:406(168), 512

Johnson, H. R., **7**:320b, 320e

Johnson, I. E., **3**:54, 61, 98

Johnson, J. E., **20**:134(4), 179(4)

Johnson, K. D. B., **15**:228(2), 268, 276

Johnson, L. F., **30**:319(18), 427(18)

Johnson, M. A., **30**:356(106), 357(106), 431(106)

Johnson, M. M., **2**:383, 396

Johnson, M. W., **18**:339(54), 363(54)

Johnson, N. L., **2**:378(39), 395

Johnson, R. C., **5**:487(271), 517; **11**:219, 222, 257, 365(124), 368(124), 369, 370, 372, 437; **17**:68, 71(14), 72(14, 34), 153; **25**:294, 317

Johnson, R. D., **20**:296(321), 314(321)

Johnson, R. E., **26**:12(190, 194), 47(107), 48(107), 97(107), 101(190, 194)

Johnson, R. L., **14**:132, 134, 145

Johnson, R. R., **5**:304(62), 309(62), 320(92), 323, 324

Johnson, R. S., **19**:338(149), 356(149)

Johnson, S. M., **28**:243(16), 277(16), 280(16), 329(16)

Johnson, S. W., **10**:56(44), 82

Johnson, T. R., **11**:220, 223, 237, 258

Johnson, T. V., **10**:7, 8, 9, 35, 36

Johnson, V. J., **9**:415

Johnson, W. R., **22**:30(35), 154(35)

Johnson, W. W., **17**:114, 115, 155

Johnsson, F., **27**:187(39)

Johnston, G. H., **9**:3(66), 109

Johnston, H. L., **2**:327(57), 330(57), 354, 355; **3**:273, 300

Johnston, J. P., **16**:242, 265(13, 45), 319, 320(13, 45), 322(13), 329, 331, 363, 365; **31**:172(78), 174(78), 325(78)

Johnstone, H. F., **10**:199, 204, 205, 207, 217; **14**:211, 216, 217, 218(144), 227, 246; **19**:128(135), 134(135), 188(135)

Johnstone, R. K. M., **6**:211, 365; **9**:220(47), 269

Jolley, L. J., **19**:164(203), 190(203)

Jolls, K. R., **7**:134, 134(46a), 161

Jones, A., **28**:190(42), 229(42)

Jones, A. D. W., **30**:315(12), 363(128), 366(128), 367(128), 396(12), 398(12), 400(12), 427(12), 432(128)

Jones, A. S., **SUP**:115(287), 117(287), 183(361)

Jones, B. P., **10**:73(66), 83

Jones, C. A., **21**:180

Jones, D. C., **30**:280(63), 310(63)

Jones, D. M., **23**:206(122), 223(122), 272(122); **25**:292, 308, 317

Jones, D. R. H., **22**:219(112), 352(112)

Jones, G. W., **29**:64(13), 66(13), 69(13), 105(13), 124(13)

Jones, H., **17**:84(56), 154; **28**:23(40), 26(47), 72(40, 47)

Jones, I. P., **24**:192, 270; **25**:329, 384, 409

Jones, J. K., **31**:265(165), 330(165)

Jones, J. R., **8**:36(205), 89

Jones, K. A., **28**:351(70), 418(70)

Jones, L. T., **31**:89(7), 102(7)

Jones, M. C., **2**:352; **4**:170(53), 225; **5**:462, 463, 464, 515; **9**:364(32), 386(32), 399(77), 415, 416; **17**:114, 115, 138(147), 155, 157, 319(1), 323, 328(26), 333(26), 334(26), 335(26), 338, 340; **21**:46(74), 47(74), 52(74)

Jones, M. E., **28**:351(86), 419(86)

Jones, O. C., Jr., **15**:239, 278; **19**:323(111), 324(111), 330(111), 354(111); **21**:280, 344; **29**:17(14), 56(14); **31**:138(30), 157(30)

Jones, P. H., **14**:88(177), 104

Jones, R. L., **10**:24(41), 36
Jones, R. T., **11**:239, 261
Jones, R. W., **18**:46(89, 90), 83(89, 90)
Jones, S. E., **24**:41(13), 42(14), 98(13, 14)
Jones, T. B. (Chapter Author), **14**:107, 129, 132, 134(44, 45, 46, 47), 135, 136, 139(44, 45, 46, 47), 140(44, 45, 46, 47), 145; **21**:126(122), 138(122); **26**:2(108), 13(108), 97(108)
Jones, T. V., **23**:279(3), 315(3), 320(187), 323(204, 205, 206, 207), 324(207, 208, 210), 330(207, 208), 354(3), 362(187), 363(204, 205, 206, 207, 208, 210)
Jones, T. W., **23**:280(24), 309(24), 315(24), 317(24), 318(24), 319(24), 321(24), 322(24), 323(24), 325(24), 355(24)
Jones, W. M., **23**:222(118, 119, 120, 121), 226(118, 119, 120, 121), 272(118, 119, 120, 121)
Jones, W. P., **13**:81(39), 116; **15**:232, 277; **16**:347, 364; **21**:188(18), 237(18); **28**:319(202), 338(202); **30**:329(31), 427(31)
Jones, W. W., **17**:306(113), 317
Jones-Meehan, J., **31**:461(40), 472(40)
Jonke, A. A., **14**:232(168), 234(168), 246
Jonsson, V. K., **2**:421(14), 424(21), 427(14, 29), 430(21), 434(32), 436(32), 443(42), 444(42), 445(46), 448(46), 449(46), 450, 451, 452; **5**:25, 51; **7**:184, 184(53), 191(74), 192(74), 213; **SUP**:269(447), 325(482); **21**:150, 151, 152, 182
Joo, J. W., **28**:177(29), 184(40), 229(29, 40)
Joosen, C. J. J., **16**:138(108), 156
Jordan, A. S., **30**:348(140), 358(109, 110), 371(140), 382(140), 415(179, 180, 181), 431(109, 110), 433(140), 435(179, 180, 181)
Jordan, D. P. (Chapter Author), **5**:55; **11**:105(102), 193
Jordan, J. E., **3**:293(92), 302
Jordan, J. M., **6**:406(61), 407(61), 409(61), 410(61), 489
Jørgensen, F. E., **11**:241, 243, 262
Joseph, D. D., **6**:474(263), 499; **13**:219(37, 56), 264, 265; **14**:33, 34(81, 82), 72(81, 82), 101; **23**:431(83), 464(83); **24**:113(102), 188(102)
Joshi, D. D., **15**:69(47), 136(38)
Joshi, D. H., **24**:262, 263, 271

Joshi, J. B., **23**:237(157), 242(157), 274(157)
Joshi, M., **22**:226(25), 348(25)
Joshi, M. G., **28**:402(202), 424(202)
Joshi, R. K., **19**:142(150), 188(150)
Jost, W., **3**:283, 301; **12**:279
Joubert, P. N., **7**:346, 352(37), 378; **11**:220, 223, 237, 258
Joudi, K. A., **23**:9(22), 30(22), 49(22), 127(22)
Joukovsky, V., **11**:218(90), 221(90), 236(90), 256
Jouran, C., **24**:247, 269
Jovanovic, G. N., **19**:98(6), 99(6), 100(30), 136(30), 137(30), 139(30), 145(30), 184(6), 185(30)
Jovanovič, L., **11**:20(70), 21(70, 71, 72), 31(71, 72, 73), 33(71), 34(72), 36(71), 37(71), 44(75), 45(75), 49
Joy, P., **7**:320a, 320e
Joyce, A., **18**:25(55), 81(55)
Joyce, N., **17**:292(101), 295(101), 317
Judd, R. L., **16**:232
Judge, J. F., **7**:248(68), 275(68), 316
Jueland, A. C., **10**:177, 216
Jugel, R., **11**:363, 437
Juhasz, P., **28**:124(102), 143(102)
Jukoff, D., **2**:287, 355; **7**:190, 214
Julien, H. L., **16**:302, 363
Julien, H. P., **28**:15(23), 71(23)
June, R. R., **4**:50, 64
Jungnickel, H., **16**:114, 155
Junkhan, G. H., **SUP**:393(521); **30**:198(73), 253(73)
Jurmo, M. W., **30**:219(76), 253(76)
Jusionis, V. J., **9**:265(133), 272; **15**:232, 277
Juza, J., **28**:376(144), 421(144)

K

Kabel, R. L., **15**:195(76), 225(76)
Kabelitz, H. P., **6**:71(121, 122), 76(121, 122), 79(122), 80(122), 128
Kaberdian, E. G., **15**:128(155), 141(155)
Kabir, C. S., **20**:100, 102, 104, 131
Kacker, S. C., **7**:256, 257, 339(17, 19), 348, 350, 352(36), 355, 358, 362, 378, 379
Kaczmar, B. U., **23**:206(65), 219(65), 222(65), 225(65), 269(65)
Kada, H., **9**:156, 180
Kadaba, P. V., **19**:160(195), 161(201), 189(195), 190(201)

Kadaner, Y. S., **SUP**:81(236), 123(236), 124(236), 135(236)

Kadanoff, L. P., **3**:205(33), 249; **11**:318(9), 376(9), 388(9), 434

Kadar-Kallen, M. A., **30**:258(13), 307(13)

Kader, B. A., **6**:390(19), 405(19), 486, 513(25), 562; **SUP**:112(284), 115(284)

Kadi, F. J., **16**:143, 156

Kadle, D. S., **20**:232(163), 238(163), 240(163), 241(163), 307(163)

Kafarov, V. V., **25**:166(54), 245(54)

Kafoussias, N. G., **20**:336(55), 352(55); **24**:106(39), 186(39)

Kagan, J., **11**:7, 48

Kagan, Y., **10**:104, 105, 164

Kaganer, M. G., **9**:351(5), 352(5), 390(5), 399(5), 414

Kageyama, T., **23**:326(213), 364(213)

Kagii, I., **21**:29(47), 31(47), 52(47)

Kago, T., **23**:253(126), 254(126), 257(126), 261(126), 272(126)

Kaguei, S., **24**:102(14), 184(14)

Kahanovitz, A., **12**:79(7), 111

Kahawita, R., **19**:80(239), 93(239)

Kahl, G. D., **6**:205, 222(66), 364, 365

Kahl, W. H., **5**:356(66), 507

Kahn, H., **5**:2, 3, 50

Kahny, W.-H., **8**:291(11), 344(11), 349

Kahttab, A., **15**:189(58), 224(58)

Kaino, K., **21**:91(61), 136(61)

Kaisawan, P., **23**:327(221), 364(221)

Kaischew, R., **14**:287, 298, 341

Kaiser, T. M. V., **19**:77(181), 91(181)

Kaiser, W. K., **30**:318(14), 427(14)

Kaji, M., **23**:59(93), 61(93), 77(93), 78(93), 106(93), 109(93), 110(93), 112(93), 131(93)

Kajimoto, M., **29**:325(180), 326(180), 344(180)

Kajishima, T., **25**:384, 403, 413

Kak, A. C., **30**:287(73), 310(73)

Kakaç, S., **SUP**:77(215), 139(215), 145(215); **24**:104(67), 187(67); **25**:51(77, 88), 127(193), 143(77), 144(88), 149(193)

Kakarala, C. R., **6**:549(67), 551, 553(67), 563; **7**:50

Kakimoto, K., **30**:392(157), 393(157), 396(163, 164), 399(163, 164), 403(172), 404(172), 434(157, 163, 164, 172)

Kakota, H., **31**:388(40), 427(40)

Kalaba, R., **3**:184(10), 249

Kalcikhman, L. E., **3**:35, 91

Kale, D. D., **15**:92(83), 93(83), 105(83), 107(83), 114(83), 138(83), 179(39), 224(39); **23**:201(56), 204(56), 205(56), 206(56), 226(56), 269(56); **24**:110(96), 111(96), 112(96), 188(96)

Kale, S. R., **23**:307(127), 359(127)

Kalhori, B., **19**:58(128), 89(128)

Kalinin, E. K. (Chapter Author), **6**:367, 376(1), 387(14), 412(14), 416(96), 419(153), 449(181, 182, 183, 184, 193), 453(181, 182, 183, 184, 193), 456(181, 182, 183, 184, 193), 457(181, 182, 183, 184), 486, 490, 493, 495; **11**:51, 115, 120(187), 122(187), 133(146), 144(146, 154), 146(146, 154), 151, 152, 156, 170(154), 173, 176, 177, 182, 183, 195, 196, 197; **18**:241, 257(40), 258(40, 52, 53, 54a), 262(53), 263(52, 54a), 270(52), 310(40, 67), 311(67, 68), 312(67, 68), 316(54a, 70), 321(40), 322(52, 53, 54a), 323(67, 68, 70); **23**:330(239), 365(239); **25**:1, 3(1), 4(2, 3, 5), 10(36), 12(1, 3, 44), 22(3), 31(1, 44), 33(3, 5, 44, 64, 65, 68), 35(70a), 43(3, 5), 52(3, 5, 68, 89, 90, 91, 92, 93), 61(5, 103, 104, 106), 64(3), 74(1), 92(3, 5, 104, 106), 99(3, 5), 106(3, 5), 108(3), 120(3), 130(3, 5, 198, 199, 200), 133(1, 198, 199, 200), 139(1, 2, 3, 5), 141(36, 44), 142(64, 65, 68), 143(70a), 144(89, 90, 91, 92, 93, 103, 104), 145(106), 149(198, 199, 200); **31**:159, 159(1), 160(1, 12, 29, 57), 161(1), 165(72), 177(1), 182(1, 72), 183(12, 86), 186(72), 187(12, 87), 188(88, 89), 191(93, 95), 198(93, 95), 200(95), 201(103, 104, 105, 106), 205(104, 105, 106), 207(106), 208(95), 209(93), 217(105), 220(88), 221(105, 106), 222(105, 106), 224(105, 106), 225(117), 232(105, 120), 242(127), 243(12), 247(1), 250(144), 252(144), 253(12, 144), 262(1), 263(1), 265(158), 266(29, 57, 168, 169, 170), 270(29, 57, 158, 170), 271(57), 279(57, 181, 183, 184), 282(183, 188), 287(57, 183, 188), 292(57, 183, 188), 294(57), 301(57), 305(57), 322(1, 12), 323(29), 324(57), 325(72), 326(86, 87, 88, 89, 93, 95, 103, 104, 105), 327(106, 117, 120, 121, 122), 328(127, 144), 329(158), 330(168, 169, 170, 181, 183, 184), 331(188)

Kalinin, V. M., **25**:4(29), 141(29)

Kalish, R. L., **11**:213, 215, 255

Kalishevsky, L. L., **6**:409, 410, 489

Kallen, C., **19**:82(278), 95(278)

Kalman, H., **26**:47(126), 98(126)

Kaloni, P. N., **24**:106(40), 175(155), 177(155), 178(155), 186(40), 190(155)

Kalos, M. H., **31**:334(8), 381(8), 425(8)

Kaluarachchi, J. J., **30**:175(153), 196(153)

Kalvinskas, L. A., **5**:330, 404, 405, 409, 411, 412, 413, 420, 421, 440, 445, 511

Kalyano, Yu. M., **11**:412(193), 439

Kalyanov, B. I., **21**:47(80), 53(80)

Kalyon, D. M., **28**:150(43), 184(30), 190(19), 193(31), 228(19), 229(30, 31, 43)

Kamal, M. R., **13**:261(125), 267

Kamata, T., **23**:12(58, 59), 13(66), 14(58, 59), 15(66), 31(58, 59), 34(58, 59), 42(58, 59), 43(66), 58(58, 59), 60(58, 59), 87(58, 59), 91(58, 59), 92(59), 103(58, 59), 109(58, 59, 66), 110(58, 59, 66), 117(58, 59, 66), 129(58, 59, 66)

Kameda, T., **19**:127(133), 128(133), 188(133)

Kamenkovich, V. M., **14**:270(62), 280

Kamenshchikov, F. T., **31**:160(46), 324(46)

Kamins, T. I., **28**:342(21), 416(21)

Kaminski, D. A., **31**:343(29), 426(29)

Kamioka, Y., **17**:108(95), 110, 111, 151(186), 155, 158

Kamiuto, K., **23**:170(50), 186(50)

Kamm, R. F., **5**:326(2), 505

Kammerud, R. L., **18**:63(115, 116), 85(115, 116)

Kamotani, Y., **19**:309(98, 99), 310(101), 313(98, 99, 101), 353(98, 99, 101); **28**:370(126, 131), 420(126), 421(131); **30**:51(50), 52(50), 89(50)

Kanamaru, K., **SUP**:103(269)

Kancir, B., **26**:246(53), 253(53), 326(53)

Kandelaki, R. D., **31**:210(112), 327(112)

Kandlikar, S. G., **26**:255(76), 328(76)

Kandula, M., **23**:70(86), 95(86), 130(86)

Kane, J., **28**:402(191, 194), 404(191), 423(191), 424(194)

Kaneda, H., **27**:ix(15), 187(36, 37, 38), 193(37, 38)

Kaneko, T., **14**:21, 71, 72, 73, 73(55), 101

Kaneko, Y., **24**:6(25), 37(25)

Kanellopoulos, N. N., **23**:206(123), 272(123); **30**:136(95), 193(95)

Kanevets, G. E., **15**:323

Kang, B., **28**:54(81), 56(81), 57(81), 59(81), 60(81), 61(81), 62(81), 63(81), 73(81)

Kang, B. S., **4**:90(26), 92(26), 95(26, 36), 120(86), 138, 140

Kang, C. S., **11**:408, 439

Kang, D. H., **4**:90(26), 92(26), 95(26, 36), 138

Kang, W. K., **10**:171(17), 215; **19**:123(121), 125(121), 187(121)

Kang, Y., **23**:254(124), 261(124), 272(124), 329(233), 365(233)

Kannuluik, W. G., **2**:297(65), 354; **3**:273(37), 300

Kano, M., **25**:342, 361, 388, 402, 404, 413

Kanomori, H., **10**:43(15), 44(15), 45(15), 60(15), 62(15), 64(15, 50, 51), 65(15), 67, 81, 82

Kantayya, R. C., **18**:220(117), 238(117)

Kantha, L. L., **19**:82(272), 95(272)

Kantorovich, L. V., **8**:22(106), 28(106), 57(106), 86; **12**:96(30), 113; **19**:207(28), 227(28), 244(28)

Kantrowitz, A., **14**:314, 343

Kanury, A. M., **30**:57(52), 76(51), 77(51), 90(51, 52)

Kanve, M. V., **28**:213(37), 229(37)

Kao, H. C., **7**:112, 197; **8**:13(69), 14(69), 85

Kao, S., **26**:226(29), 227(29), 252(29), 254(29, 66), 262(66), 305(29, 66), 325(29), 327(66)

Kao, S.-H., **30**:25(11), 34(12), 42(12), 88(11, 12)

Kao, T. W., **11**:240(238), 263

Kapadia, P. D., **19**:80(242), 94(242); **28**:111(30), 139(30)

Kaparthi, R., **10**:173, 177(35), 179, 200, 205, 215, 217

Kaphengauz, N. L., **6**:558(97), 564

Kapila, A. K., **SUP**:69(144), 161(144)

Kapinos, V. M., **5**:177, 180, 191, 192, 193, 197(107), 198, 199, 249

Kapinos, V. N., **25**:127(197), 149(197)

Kapitza, P. L., **5**:484(266), 487(266), 517; **17**:67, 84, 153

Kaplan, C., **8**:184, 227

Kaplun, S., **2**:257(91), 270; **4**:341, 446; **8**:7, 13, 84, 85

Kappler, G., **9**:3(11), 82(11), 83, 106

Kaptilniy, A. G., **21**:8(18), 26(18), 27(18), 29(18), 51(18)

Kapur, D., **28**:397(188), 398(188), 423(188)

Kapur, J. N., **2**:390, 395

Karabelas, A. J., **31**:462(38), 464(54), 472(38), 473(54)

Karagounis, A., **5**:406(160), 512; **11**:105(111), 114, 193

Karakus, H. J., **5**:131(11), 246

Karam, N. H., **28**:343(43), 417(43)

Karch, G. E., **17**:181, 314

Kardas, A., **20**:150(13), 179(13)

Kardos, J. L., **25**:252, 292, 293, 297, 298, 308, 319

Karen, G., **28**:103(59), 141(59)

Karian, C., **28**:150(32), 229(32)

Karim, G. A., **29**:68(17), 124(17)

Karim, R. B., **19**:317(106), 318(106), 354(106)

Karimov, K. F., **31**:281(185), 331(185)

Karimov, R. V., **31**:160(18), 323(18)

Karioun, M., **23**:253(267), 254(267), 255(267), 256(267), 260(267), 261(267), 278(267)

Karker, Y. I., **6**:474(323), 502

Karkhu, V. A., **31**:278(179), 330(179)

Karki, K. C., **20**:202(86), 303(86); **24**:199, 211, 271

Karlekar, B. V., **2**:443(41), 451

Karlsson, S. K. F., **6**:411(73), 489

Karman, T., **6**:504, 561

Karman, T. von, *see* von Kármán, T.

Karni, J., **19**:248(4), 349(4); **24**:102(8), 184(8); **28**:153(25), 228(25)

Karniadakis, G. E., **20**:232(166), 243(175), 307(166, 175); **25**:394, 418

Karnik, U., **21**:204(31), 206(31), 207(31), 208(31), 209(31), 237(31)

Karnopp, D. C., **22**:287(113), 288(113), 352(113)

Karpin, E. B., **6**:418(118), 491

Karrow, A. M., **22**:290(163), 354(163)

Kartashov, E. M., **19**:191(6), 193(15), 195(19), 243(6, 15, 19)

Karwe, M. V., **28**:150(33), 151(18, 38, 58), 156(36), 159(33), 160(18, 33), 162(38), 163(38), 165(38), 172(36), 181(33), 190(38, 58), 193(38, 58), 195(5), 197(5), 198(5), 200(5, 34, 35), 202(5), 207(5), 215(18), 216(18), 217(18), 219(8), 221(8), 222(8), 228(5, 8, 18), 229(33, 34, 35, 36, 38), 230(58)

Kasahara, K., **23**:12(61), 14(61), 32(61), 35(61), 39(61), 49(61), 53(61), 54(61), 58(61), 60(61), 89(61), 92(61), 106(61), 129(61)

Kascheev, V. M., **15**:285, 322

Kase, S., **13**:260(115), 266

Kaser, F., **23**:222(125), 223(125), 224(125), 272(125)

Kashinsky, O. N., **25**:4(10), 33(10), 140(10)

Kasnov, M. L., **8**:34(178), 88

Kassemi, M., **18**:44(88), 83(88); **30**:339(82), 430(82)

Kassinos, A., **19**:21(40), 66(40), 85(40)

Kassner, J. L., **14**:321, 322, 344

Kassoy, D. R., **11**:226, 259; **14**:25, 31, 66, 101, 103; **24**:103(20), 185(20)

Kastell, D., **30**:258(11), 260(11), 269(11), 270(11), 272(11), 307(11)

Kastenberg, W. E., **29**:130(8), 203(8)

Kasturirangan, S., **10**:125, 127, 165

Kasuya, A., **28**:135(26), 139(26)

Katano, K., **30**:359(123), 361(123), 432(123)

Kataoka, H., **29**:160(94), 161(94), 207(94)

Kataoka, I., **29**:259(129), 273(129), 342(129)

Kataoka, K., **7**:128(37), 161; **21**:165, 166, 174, 180, 181

Katayama, K., **24**:3(5), 6(21, 22), 36(5, 21, 22)

Katayama, T., **26**:14(219), 103(219); **27**:ix(14), 167(30)

Katchalsky, A., **12**:261(42a), 269(42a), 279; **22**:287(154), 354(154)

Katgerman, L., **28**:308(155), 336(155)

Kathuria, D., **19**:112(83), 180(83), 186(83)

Katinas, V., **18**:107(38), 118(38), 119(38), 120(38), 121(38), 125(38), 126(38), 127(38), 158(38)

Katinas, V. J., **8**:113(38), 123(38), 124(38), 125(38), 129(38), 130(38), 159

Kato, E., **25**:81(144), 146(144)

Kato, H., **21**:37(59), 52(59); **28**:304(142), 335(142)

Kato, K., **28**:369(122), 420(122)

Kato, T., **25**:258, 283, 284, 285, 318

Kato, Y., **21**:106(92), 109(98), 113(98), 137(92, 98); **23**:253(126), 254(126), 257(126), 261(126), 272(126)

Katsma, K. R., **31**:106(7), 138(7), 155(7)

Katsnelson, B. D., **6**:474(211), 477, 496

Katsnelson, G. G., **6**:390(19), 405(19), 486

Katsuda, H., **21**:125(116), 138(116)

Katsuta, K., **9**:207(31), 210(31), 242, 269

Katsuta, M., **23**:12(48), 13(30), 14(48), 15(30), 16(30), 17(30), 23(30), 24(30), 25(30, 48), 27(30, 48), 37(48), 39(48), 58(48), 60(48), 63(79, 82), 64(48), 67(48), 75(79), 77(48), 78(48), 127(30), 128(48), 130(79, 82)

Kattawar, G. W., **5**:37, 53

Kattchee, N., **31**:250(145), 251(145), 252(145), 328(145)

Katto, Y. (Chapter Author), **14**:16(40), 19, 20(40), 71, 101; **17**:1, 4(22), 5(22), 6, 7, 8, 9, 11(54), 15, 16(54, 57), 18(54, 57), 19(59, 60), 20(59, 63), 21(59), 22(22), 23(22), 24(59), 26(78), 30(94), 31(99), 32, 33(99, 110a), 35(78), 37(114, 115, 116, 119), 39(132, 133), 40(132), 43(99), 45(150), 57, 58, 59, 60, 61; **18**:247(9), 248(9), 256(9), 257(9), 258(51), 259(9), 261(9, 51), 264(9), 270(9), 313(51), 320(9), 322(51); **20**:252(239, 240), 310(239, 240); **23**:8(14, 15, 16), 12(14, 15, 16, 44, 51, 57), 14(14, 15, 16, 44, 51, 57), 22(16, 44), 23(16), 24(16), 26(16, 51), 27(16), 29(14), 30(16, 57), 31(57), 33(14), 34(57), 35(14), 36(14, 15, 16, 44), 37(16), 38(14), 39(14, 15, 16, 44), 41(16), 45(15), 55(76), 58(14, 15, 44, 76, 106, 107), 59(16, 51, 57), 60(14, 15, 44, 76, 106, 107), 61(16, 51, 57), 62(14, 16, 44, 76), 63(14, 16, 44, 76, 81), 64(16), 67(14, 76), 68(14), 69(76), 70(76), 71(76), 74(16, 44, 76), 75(14, 16, 44, 76, 92), 77(16, 76), 78(14, 15, 16), 79(14, 16, 92), 85(14, 51), 86(14), 87(16), 90(16), 91(14), 93(14, 51), 95(76), 96(81), 97(92), 98(81), 100(81), 101(15, 16, 106), 102(15, 106, 107), 106(14, 16, 44), 107(15, 16, 44, 57), 126(14, 15), 127(16), 128(44, 51), 129(57), 130(76, 81), 131(92, 106, 107)

Katz, D. I., **31**:245(138), 328(138)

Katz, D. L., **5**:74, 126, 347, 353, 365(83), 506, 508; **9**:204, 233(21), 236, 238, 243, 268; **10**:109(43), 110, 164, 200, 218; **11**:50(60), 110(60), 191, 217, 228, 259; **14**:4(13), 5, 100; **21**:100(78), 136(78); **29**:130(7), 134(17), 203(7, 17); **30**:216(24), 250(24)

Katz, E., **6**:29(44), 124

Katz, H., **25**:153(8), 243(8)

Katz, I., **11**:358(88), 436

Katz, J. L., **10**:92, 105(24), 106(24), 164; **14**:305(35), 306(60, 61), 309(60, 61, 69), 323(97), 325, 326(104, 105, 106), 327(104, 105, 106), 328, 329, 330, 334, 339(104, 105, 106), 342, 343, 344, 345; **31**:458(28), 459(28), 471(28)

Katz, T., **31**:462(62), 473(62)

Katzoff, S., **7**:220, 223, 273, 276, 313

Kaufman, A., **9**:3(67), 97(67), 109

Kaufman, W. G., **12**:242(5), 277

Kauh, S., **24**:300(23), 301(23), 319(23)

Kautman, W. D., **20**:23(45), 81(45)

Kautsky, D. E., **16**:133, 134, 155

Kautzky, D. K., **11**:105(117), 114, 121(117), 122(117), 194

Kavanau, L. L., **6**:35(60), 125; **7**:193, 214

Kaviany, M. (Chapter Author), **6**:35(60), 125; **7**:193, 214; **23**:133, 136(9), 156(9), 161(9), 170(49), 172(49), 179(52), 184(9), 186(49, 52), 369(3), 370(3), 407(3), 443(3), 452(3), 460(3); **24**:104(69), 187(69); **30**:94(14), 95(14), 110(14), 111(14), 112(14), 154(120), 171(140), 172(140), 189(14), 194(120), 195(140)

Kawaguchi, S., **26**:44(88), 96(88)

Kawaguti, M., **6**:396(39), 488

Kawahara, M., **28**:41(67), 73(67)

Kawamata, M., **25**:81(144), 146(144)

Kawamura, H., **25**:73(113, 114), 81(161), 96(113, 114, 161), 145(113, 114), 147(161)

Kawamura, T., **23**:340(269), 341(269), 367(269)

Kawano, S., **21**:106(92), 137(92)

Kawarata, S., **30**:359(123), 361(123), 432(123)

Kawase, Y., **23**:201(127), 208(127, 128), 210(127), 214(128), 230(127, 128, 130, 130a, 132), 231(128), 237(131), 238(131), 239(131), 240(131), 241(131), 243(131), 246(131), 252(131), 262(129), 272(127, 128, 129, 130, 130a, 131, 132); **25**:260, 261, 269, 271, 272, 273, 274, 278, 279, 281, 286, 317

Kawaski, M., **27**:ix(9), 174(9)

Kaye, B. H., **5**:26, 51

Kaye, J., **5**:230(158), 233(162), 234, 238, 239, 240, 241, 250, 251; **11**:203, 206, 254; **21**:156, 157, 160, 163, 164, 166, 168, 178, 180

Kaye, W. H., **21**:163, 178, 180

Kaylor, R., **5**:131, 246

Kays, W. M. (Chapter Author), **2**:16(19), 25(19), 43, 105, 107; **3**:14(17), 23(17), 32, 81(79), 85(79), 86(79), 99, 106(9), 130, 138, 139, 171(37), 172, 174; **5**:166, 174, 233(171), 234, 235, 242, 248, 251, 356, 508; **6**:404, 488; **8**:94, 148, 157(18), 159; **11**:203, 206, 219(229), 223(229), 238, 263; **12**:40(48), 44(48), 45, 49, 50, 52(48), 54(48), 61, 74, 91(26), 92(26), 96, 109(26), 112; **13**:81(44), 116; **SUP**:2(6, 7, 12), 7(7), 9(7), 10(7),

13(7), 15(7), 18(7), 30(40), 31(6, 7, 41), 32(7, 41), 33(7, 41), 34(7), 35(44), 38(6, 7), 44(6), 53(6), 59(6), 64(94), 71(44), 77(40, 44), 78(7), 83(7), 86(7), 98(7), 119(7), 125(7, 40), 135(7), 136(7), 137(7), 138(40), 139(40), 143(40), 144(40), 145(7, 40, 44), 146(40), 148(7), 152(40), 153(332), 154(332), 169(332), 177(332), 185(332), 188(7), 193(44), 194(7, 44), 195(7), 202(94), 203(94), 214(7), 224(6, 7, 94), 225(6, 7, 94), 285(41), 287(44), 288(44), 289(41), 292(41), 297(44), 301(41), 302(41), 303(41), 306(41), 307(41), 309(41), 310(41), 311(41), 312(41), 313(41), 314(41), 319(44), 321(44), 366(6), 407(6), 409(6), 411(12), 414(12), 415(6, 7, 12), 416(12); **15**:23(48), 24(48), 29(48), 32(48), 56(48), 100(99), 121(99, 143), 122(99), 138(99), 139(99), 140(143); **16**:241, 242, 243, 245(35, 36, 37, 38), 247(30), 255(35, 36, 37, 38), 256(35, 36, 37, 38), 265(13, 45), 280(30), 299(32), 302(16, 28), 310(23), 312(7, 11, 18), 314(7), 320(13, 45), 322(13), 349(32), 356(35, 36, 37, 38), 363, 364, 365; **18**:88(18), 146(18), 156(18), 158(18); **19**:304(89), 307(91), 312(91), 353(89, 91); **20**:145(9), 179(9), 238(167), 307(167); **23**:284(37), 333(250), 355(37), 366(250); **26**:232(36), 237(36), 244(36), 248(36), 256(36), 261(36), 263(36), 279(36), 291(36), 305(36), 306(36), 322(36), 323(36), 326(36); **31**:160(2, 4), 322(2, 4)

Kazakevich, F. P., **8**:135, 137, 149, 160; **11**:219, 222, 237, 241, 242, 257; **18**:133(66), 134(66), 135(66), 147(66), 159(66)

Kazakova, E. A., **1**:239, 265; **10**:177(45, 51), 216

Kazenin, D. A., **11**:168(168), 170(168), 196

Kazimi, M. S., **21**:280, 345; **26**:52(210), 102(210)

Kazmierczak, M., **24**:304(21), 319(21)

Keairns, D. L., **14**:160(73, 74), 161(73, 74), 164, 244

Keane, M., **28**:231(5, 6), 251(6), 266(5), 329(5, 6)

Kear, B. H., **19**:37(79), 86(79); **28**:243(15), 244(15), 280(15), 303(15), 329(15)

Kearney, D. W., **12**:23(40), 74; **15**:58(91); **16**:280, 349, 363, 364

Kearsey, H. A., **17**:30(88), 37(120), 42(136, 138), 60, 62

Kearsley, E. A., **2**:378, 395; **13**:219(36), 264; **15**:62(10), 78(10), 135(10)

Kececioglu, I., **28**:253(54), 331(54)

Kech, P. H., **30**:318(14), 427(14)

Keck, J. C., **21**:263(29, 30), 276(29, 30)

Kee, R. J., **24**:213, 227, 270; **28**:366(116, 118), 382(116), 393(116, 181), 394(116), 420(116, 118), 423(181)

Keech, T. W., Jr., **23**:307(127), 359(127)

Keefer, T. N., **16**:6(4), 15(4), 16(4), 56

Keeling, K. B., Jr., **17**:6, 57

Keenan, J. H., **1**:199(7, 8), 201(7), 263, 264; **15**:2(3), 7, 26(24), 54(3), 55(24); **18**:343(62), 345(62), 363(62); **21**:271(41), 276(41)

Keesom, W. H., **2**:297, 354; **5**:332, 506; **17**:102, 155

Keever, S. L., **31**:461(40), 472(40)

Keey, R. B., **25**:260, 279, 317

Keys, R. K. F., **17**:42(138), 62

Kehat, E., **15**:228(5), 242(5), 276

Kehrt, L. G., **5**:177, 249

Kehtarnavaz, H., **19**:79(211), 92(211)

Keilin, V. E., **16**:233; **17**:325, 340; **31**:245(132), 328(132)

Keinonen, J., **24**:30(63), 38(63)

Keith, T. G., Jr., **23**:326(214), 364(214)

Keliher, T. J., **23**:304(110), 359(110)

Kelkar, A. S., **28**:390(176), 391(176), 408(176), 411(176), 423(176)

Kelkar, K. M., **24**:199, 271; **29**:25(20), 26(21), 56(20, 21)

Kell, G. S., **18**:334(57), 340(57), 341(57), 342(57), 363(57); **31**:115(6), 117(6), 155(6)

Kelleher, M. (Chapter Author), **16**:1

Keller, C. W., **9**:386(58), 394(58), 401(58), 402(58), 416

Keller, H. B., **6**:2, 122; **26**:6(30), 94(30)

Keller, J. B., **14**:44(108), 102; **25**:293, 317

Keller, J. O., **30**:78(16), 83(30), 84(16), 88(16), 89(30, 31), 291(86), 311(86)

Keller, J. R., **8**:13, 85

Keller, K. H., **18**:208(96), 211(96), 238(96); **22**:70(36), 71(36), 72(36), 76(36), 79(36), 80(36), 81(36), 82(36), 83(36), 85(36), 90(36), 100(36), 108(36), 109(36), 154(36)

Keller, R. J., **23**:222(125), 223(125), 224(125), 272(125)

Keller, W. E., **17**:75, 84(55), 102, 104, 105(91), 106(91), 127(47), 129, 154, 155

Kellett, B. S., **11**:391(148), 438

Kelley, A. J., **7**:258, 317

Kelley, M. J., **7**:223, 247, 273(7), 314

Kellogg, H. H., **12**:224(43), 241(43), 246(43), 247(43), 253(43), 259(43), 279

Kelly, G. E., **15**:34(57), 56(57)

Kelly, K. W., **30**:358(111, 112), 360(121), 431(111, 112), 432(121)

Kelly, M. J., **20**:266(271), 268(271), 312(271)

Kelly, P. S., **5**:320(92), 324

Kelly, R., **28**:106(51), 110(51), 111(47, 48, 49, 50), 115(50), 116(50, 52), 140(47, 48, 49, 50, 51), 141(52)

Kelman, R. B., **15**:324

Kelsey, F. J., **14**:78(168), 104

Keltner, N. R., **23**:301(103, 105), 315(146), 320(156), 358(103, 105), 360(146), 361(156)

Kelvin, W. T., Sir., **5**:129, 246

Kemblowski, Z., **23**:192(134), 197(134, 137), 198(134), 199(137), 201(137), 202(135, 137), 206(134, 137), 213(135), 214(134), 218(133, 136, 138), 220(133, 136), 221(133, 137), 272(133, 134), 273(135, 136, 137, 138); **24**:110(95), 111(95), 188(95)

Kemeny, C. A., **31**:242(128), 249(128), 328(128)

Kemeny, G. A., **9**:3(68), 109; **SUP**:414(533)

Kemink, R. G., **19**:47(112, 114), 48(114), 49(114), 58(114), 66(114), 88(112, 114)

Kemlelly, A. E., **11**:202, 204, 209(7), 218, 221, 252, 255

Kemme, J. E., **7**:230(17), 235, 238, 240, 247, 247(48), 260, 262, 263, 268, 269, 270, 271, 272, 274(23), 275(23), 305, 306, 314, 315, 316, 317

Kemp, N., **1**:335, 336, 338, 340, 343, 354

Kemp, N. H., **4**:278, 279, 281, 314, 318, 334, 444, 445; **7**:197, 214, 215

Kendall, M. G., **5**:10(10), 11(12), 50

Kenig, S., **13**:261(125), 267

Kenis, P. R., **15**:63(23, 36), 64(23, 36), 81(36), 135(23), 136(36)

Kennard, E. H., **2**:296, 325, 354; **3**:211(50), 250; **7**:169(10), 170, 211

Kennard, R. B., **6**:278, 365

Kennedy, E. D., **6**:14, 123

Kennedy, E. H., **10**:87(3), 107(32), 164, 169

Kennedy, G. C., **14**:97(195), 105

Kennedy, I. M., **26**:77(250), 104(250)

Kennedy, K. J., **20**:226(143), 306(143)

Kennedy, L. A., **15**:189(62), 224(62)

Kennel, W. E., **1**:213(23), 264; **11**:2(2), 47

Kenning, D. B. R., **10**:156, 166; **16**:70, 71, 72(23), 77, 79, 104(23), 144(23), 145(23), 153; **23**:329(228), 364(228); **29**:152(74, 75), 206(74, 75)

Kenrick, F. B., **1**:201, 264; **10**:96, 97, 98(14), 99, 164

Kenshalo, D. R., **4**:121(102), 123(102), 141

Kenty, **24**:265, 271

Kenty, C., **2**:354

Kerby, J. S., **20**:222(140), 224(140), 225(140), 306(140)

Kercher, D. M., **23**:320(169), 330(235), 361(169), 365(235); **26**:170(124), 214(124)

Kerhm, S., **10**:53(35), 74(35), 82

Kerimov, R. V., **9**:120, 129(12), 178

Kerjilian, G., **20**:185(108), 205(108), 304(108)

Kerker, M., **23**:146(25), 185(25)

Kerley, T. E., **29**:221(25), 271(25), 309(25), 335(25)

Kermode, R. I., **10**:210, 218; **14**:183, 184, 245; **SUP**:201(389), 202(389)

Kern, D. Q., **5**:131(11), 246; **30**:202(26), 216(25), 250(25, 26); **31**:160(13), 323(13), 441(29), 471(29)

Kern, W. I., **31**:453(30), 472(30)

Kerner, E. H., **18**:189(139), 239(139)

Kernerman, V. Sh., **19**:110(70), 111(70), 186(70)

Kerrebrock, J. L., **4**:290, 305, 306, 307, 309, 310, 315, 316

Kerrebrock, J. S., **1**:312, 354

Kervinen, J. A., **1**:244, 265

Kesavan, K., **29**:48(44), 57(44); **30**:151(110), 193(110)

Keshock, E. G., **4**:116, 145(5), 149(5), 162, 164, 175, 179, 180, 181, 182, 183(60), 184, 185, 186, 190, 199, 222, 225; **16**:97, 98, 154

Keshoeh, E. G., **11**:105(129), 115, 194

Kesselring, R. C., **11**:105(118, 122), 114, 121(118), 194; **18**:255(18), 257(18), 261(18), 270(18), 313(18), 320(18)

Kessler, D. P., **13**:164, 202

Kessler, S. W., **7**:275(114), 318; **20**:285(308), 313(308)

Kessler, T. J., **20**:187(61, 62), 196(61, 62), 197(61, 62), 199(61, 62), 200(61), 211(62), 302(61, 62)

Kesten, A. S., **11**:344(62), 345, 346(62), 347(62), 348, 436

Kester, R., **3**:50, 53, 98

Kestin, I., **3**:36, 44, 45, 98

Kestin, J. (Chapter Author), **2**:94(76, 77), 108; **3**:1, 16(30, 31, 32, 33, 34), 19(30, 33), 21(34), 23(30, 39), 24(30), 26, 30(50), 32, 256, 257(10, 12, 13), 260, 269(23), 300; **6**:511(23), 562; **7**:119, 161; **8**:122(54, 55), 160; **11**:229(167), 231, 232, 237, 240(210), 259, 260, 262; **15**:2(10, 11), 3(15, 16), 5(22), 11(33), 12(36), 55(10, 11, 15, 16, 22, 33), 56(36); **18**:116(52), 128(52), 159(52), 343(61), 344(61), 345(61), 346(61), 348(61), 349(61), 363(61); **26**:166(119), 196(156), 214(119), 216(156)

Keswani, K. K., **11**:220, 224, 226, 233, 248, 249, 250, 258, 259, 263

Ketelaar, J. A. A., **12**:256(36, 44, 45), 262(46), 267(44, 45), 279

Ketley, H. C., **19**:164(206), 170(206), 171(206), 190(206)

Ketley, M. C., **14**:156(58), 157(58), 243

Kettenring, K. N., **10**:172, 175, 177, 179, 215

Ketterson, J. B., **17**:103, 155

Kettleborough, C. F., **18**:73(146), 86(146); **20**:186(38), 187(38), 300(38)

Keulegan, G. H., **12**:233(47), 251(47), 270(47), 279

Keung, C. K. J., **13**:219(82), 243(82), 265; **19**:281(72), 284(72), 286(72), 294(72), 295(72), 296(72), 297(72), 302(72), 352(72)

Keuroghlian, P. S., **12**:78(5), 88(5), 102(5), 106(5), 111; **15**:104(124), 114(124), 140(124)

Kevorkian, J., **8**:13, 85

Kevrekidis, I. G., **30**:112(65), 191(65)

Keyes, F. G., **1**:199(8), 264

Keyes, J. J., Jr., **8**:293(20), 349

Keyhani, M., **20**:217(128), 220(128), 305(128)

Kezios, S. P., **1**:260, 261, 266; **5**:167, 170, 171, 172, 248; **9**:261, 271; **13**:11(37), 60

Khabakhpasheva, E. M., **8**:297(31), 349; **12**:103, 108, 113; **25**:81(162), 147(162)

Khabakhpasheva, Y. M., **15**:91(80), 92(80), 104(127), 107(80), 108(80), 138(80), 140(127)

Khabbag, G., **4**:205(80), 206(80), 226

Khader, M. S., **19**:42(95), 87(95)

Khair, K. R., **20**:333(51), 334(51), 335(51), 351(51)

Khait, M. Y., **6**:474(315), 501

Khajeh-Nouri, B., **21**:188(12), 196(12), 199(12), 236(12)

Khalatnikov, I. M., **5**:484, 488, 517; **17**:67, 153

Khalatov, A. A., **31**:160(34), 323(34)

Khalifa, H. E., **15**:2(11), 55(11)

Khalil, A., **17**:114, 140, 155, 157

Khalil, H. K., **11**:350, 352, 436

Khalil, Y. F., **29**:325(177), 326(177), 344(177)

Khalimov, A. A., **25**:92(170), 148(170)

Khamashta, M., **23**:193(139), 273(139)

Khambatta, F. B., **11**:412(191), 439

Khan, A. A., **20**:331(49), 351(49)

Khan, A. R., **16**:90, 154; **23**:242(139a), 262(139a), 273(139a)

Khan, M., **29**:253(106), 254(106), 340(106)

Khan, M. M. K., **23**:219(115), 272(115)

Khan, S. A., **6**:550(72), 551, 563; **7**:24(7), 35, 68(7), 83

Khanna, R., **9**:3(23), 107

Kharchenko, N. V., **14**:155, 243

Kharchenko, V. N., **6**:474(334), 502

Kharchenko, W. N., **3**:71, 74, 99

Khare, A. S., **23**:237(157), 242(157), 274(157)

Kharitonov, V. P., **6**:418(125), 492

Kharitonov, V. V., **20**:262(261), 270(261), 311(261)

Kharitonov, Y. V., **6**:474(285), 500

Khasaev, N. O., **31**:266(168, 169), 330(168, 169)

Khasanov, R. R., **14**:160(79), 161(79), 167, 169(79), 172(86), 176(177, 178), 232, 233(79, 86, 172, 173), 235(178), 237(79, 177), 244, 246, 247

Khatib, S. A. H., **14**:136, 146

Khatib-Rahbar, M., **29**:244(96), 312(96), 320(96), 340(96)

Khatry, A. K., **SUP**:28(39), 64(92), 67(92), 207(92), 208(39), 209(39), 232(92), 236(92), 237(92), 248(92), 251(92), 259(92); **19**:278(63), 352(63)

Kheifets, L. I., **31**:160(32), 323(32)

Khitrin, L. N., **10**:235(24), 281

Khomami, B., **25**:252, 292, 293, 297, 298, 308, 319

Khorev, V. I., **25**:73(120), 145(120)

Khosla, P. K., **SUP**:212(403)

Khristichenko, P. I., **6**:418(105), 491

Khrustalev, B. A., **3**:219(68), 230(68), 250; **11**:390(142), 397(142), 400(164a), 438; **25**:27(56), 142(56); **31**:193(97), 326(97)

Khudisko, V. V., **25**:51(78), 143(78)

Khursid, B. M., **18**:169(46), 171(46), 176(46), 228(46), 229(46), 237(46)

Khvastukhin, Yu. I., **14**:176, 176(91), 244

Khvostova, J. Y., **6**:474(311), 501

Kiang, C. S., **14**:298(29), 342

Kibel, I. A., **25**:5(35), 141(35)

Kichigin, A. M., **18**:257(43), 259(43, 55), 260(43), 265(43, 55), 266(43, 55), 321(43), 322(55)

Kida, M., **30**:333(52), 349(101), 351(101), 378(101), 407(52), 408(52), 428(52), 431(101)

Kidd, C. T., **23**:280(13), 299(97), 300(97), 311(13), 312(13), 313(13, 134, 135), 314(13, 135, 136), 315(143), 316(143), 342(284), 354(13), 358(97), 360(134, 135, 136, 143), 367(284)

Kidder, G. A., **3**:287(88), 302

Kieda, S., **28**:390(173), 423(173)

Kierkus, N., **11**:298, 299, 315

Kierkus, W. T., **6**:312(99), 366; **9**:300(46), 301, 347

Kiermasz, A., **28**:342(22), 416(22)

Kiewra, E. W., **20**:294(320), 314(320)

Kightley, J. R., **25**:329, 384, 409

Kihara, T., **3**:269(24), 300

Kihm, K. D. (Chapter Author), **30**:255, 257(55), 258(11), 260(11, 15), 269(11), 270(11), 271(35), 272(11), 274(50), 276(50), 277(50, 52), 278(52, 55), 279(55, 56), 280(55), 281(52, 59), 282(52), 284(52), 285(52), 286(71), 296(56), 297(56), 300(71), 301(71), 307(11, 15), 308(35), 309(50, 52, 55, 56, 59), 310(71)

Kikuchi, K., **21**:129(125), 130(126, 127), 139(125, 126, 127)

Kikuchi, R., **14**:306, 342

Kikuchi, T., **13**:164, 202

Kikukawa, H., **19**:108(54), 186(54)

Kilburn, R. F., **20**:232(156), 238(156), 306(156)

Kilgour, H., **11**:202, 204, 209, 210, 252

Kilkis, B., **24**:104(67), 187(67)

Killen, J. M., **23**:222(193), 224(193), 225(193), 275(193)

Killermann, F., **6**:190, 363

Killian, E. S., **9**:243(72), 270

Killian, W. R., **5**:356(69), 508

Kilsdonk, D., **29**:216(4), 258(4), 260(4), 262(4), 264(4), 334(4)

Kim, B. J., **29**:136(21), 137(21), 139(21), 140(21), 147(21), 168(21), 175(152), 203(21), 210(152)

Kim, B. K., **25**:282, 317

Kim, B. Y. K., **23**:196(78), 270(78)

Kim, C. B., **15**:171(31), 172(31), 218(31), 224(31); **25**:280, 317

Kim, C. H., **23**:254(140), 255(140), 257(140), 273(140)

Kim, C. S., **14**:124, 146

Kim, D., **14**:124(57), 146

Kim, D. S., **29**:172(143), 173(143, 145), 209(143, 145)

Kim, H., **29**:167(121), 208(121), 259(123), 261(123), 262(123), 263(123), 264(123), 266(123), 341(123)

Kim, H. J., **28**:96(42), 140(42)

Kim, H. T., **8**:300, 301(52), 303(52), 304(52, 56), 344(52), 350

Kim, I., **17**:70(24), 153

Kim, I. H., **28**:218(65), 230(65)

Kim, J., **24**:192, 271; **25**:342, 348, 367, 369, 370, 381, 382, 383, 385, 386, 392, 393, 394, 397, 402, 403, 404, 413, 415, 416; **28**:118(53), 141(53)

Kim, J. H., **23**:398(64), 415(64), 463(64); **30**:260(14), 274(50), 276(50), 277(50, 52), 278(52), 281(52, 59), 282(52), 284(52), 285(52), 307(14), 309(50, 52, 59); **31**:80(11c), 102(11c)

Kim, J. K., **16**:312(18), 364; **23**:12(61, 62), 14(61, 62), 32(61, 62), 35(61), 39(61), 49(61), 53(61), 54(61), 58(61, 62), 60(61, 62), 89(61, 62), 92(61), 106(61), 109(62), 110(62), 120(62), 129(61, 62)

Kim, J. S., **30**:71(53), 90(53)

Kim, K., **14**:124, 146; **28**:307(150), 336(150)

Kim, K. C., **28**:307(150), 336(150)

Kim, K. M., **28**:316(168), 317(171, 176), 336(168), 337(171, 176); **30**:333(51), 334(51), 402(170), 428(51), 434(170)

Kim, M., **14**:124(57), 146; **29**:259(123), 261(123), 262(123), 263(123), 264(123), 266(123), 341(123)

Kim, M. H., **29**:233(69), 259(127), 266(127), 269(127), 338(69), 342(127); **31**:304(208), 332(208)

Kim, N., **31**:462(31), 472(31)

Kim, N.-H., **30**:221(72), 223(27, 72), 250(27), 253(72); **31**:453(77), 456(77), 474(77)

Kim, P. K., **4**:90(26), 92(26), 95(26), 138

Kim, S., **29**:259(123), 261(123), 262(123), 263(123), 264(123), 266(123), 341(123)

Kim, S. B., **29**:259(127), 266(127), 269(127), 342(127)

Kim, S. D., **23**:254(124, 140), 255(140), 257(140), 261(124), 272(124), 273(140)

Kim, S. H., **30**:272(40), 309(40)

Kim, S. J., **28**:184(40), 229(40)

Kim, T. S., **13**:11(38), 60

Kim, W. S., **28**:77(54), 141(54)

Kim, Y. G., **23**:293(68), 357(68); **28**:307(151, 152), 336(151, 152)

Kim, Y. I., **17**:151, 158

Kim, Y. M., **24**:247, 269; **27**:ix(17, 18), 193(41)

Kimura, M., **19**:127(133), 128(133), 188(133)

Kimura, S., **30**:344(86), 430(86)

Kimzey, J. H., **4**:211, 218, 227

Kinbara, T., **10**:269, 283

Kincaide, W. C., **4**:103, 139

Kinchen, B., **23**:342(283), 367(283)

Kinder, H., **17**:72(35, 36), 153, 154

Kinder, W., **6**:195(27), 206, 211, 218, 224, 363, 364

King, C. D., **4**:114(65), 139

King, C. J., **3**:169, 174

King, C. Y., **6**:474(261), 483(261), 499

King, D. F., **19**:108(56), 115(96), 116(96), 138(146), 140(146), 186(56), 187(96), 188(146)

King, E. A., Jr., **10**:57, 82

King, E. C., **5**:109, 128

King, H. H., **6**:20(32), 22, 24(32), 36(32), 38(32), 39(32), 40(32), 123

King, J. G., **17**:228, 316

King, J. I. F., **5**:288(33), 322

King, L. V., **8**:94(1), 128(1), 158; **11**:208, 218, 221, 225, 235, 238, 239, 241, 242, 255, 256; **18**:87(1), 157(1); **20**:358, 359, 360, 388

King, N. K., **10**:265, 283

King, P. P., **7**:320a, 320e

King, R. D., **1**:384(51), 385(51), 441

King, T. R., **28**:124(78), 142(78)

King, W. J., **11**:202, 204, 213, 214, 225, 252

Kingery, W. D., **11**:360(93), 437

Kingrey, J., **25**:252, 292, 293, 316

Kingston, A. E., **4**:251(14, 15, 16), 252(15), 313

Kinney, G. F., **29**:188(185), 212(185)

Kinney, R. B., **7**:169, 211

Kinney, T. A., **30**:329(34, 35), 350(92), 351(34, 35), 358(34, 35, 113, 114), 428(34, 35), 430(92), 432(113, 114)

Kinoshita, M., **21**:37(58), 52(58)

Kinsey, J. L., **4**:110(60), 139

Kintner, R. C., **21**:334, 335, 337, 339, 345

Kinzie, P. A., **23**:288(50), 289(50), 356(50)

Kiper, A. M., **26**:170(122), 214(122)

Kipp, H. W., **23**:343(286), 367(286)

Kirakosyan, V. A., **14**:151(21), 160(21), 161(21), 166(21), 221(21), 242; **19**:100(22), 103(22), 116(22), 126(22), 127(22), 128(22), 165(22), 171(22), 185(22)

Kirano, F., **11**:12(27), 48

Kirby, D., **18**:259(56), 322(56)

Kirby, D. B., **20**:15(32), 80(32)

Kirby, G. I., **9**:129(37), 178

Kirby, G. J., **17**:30, 60

Kirchgässner, K., **5**:228, 250; **21**:150, 151, 180

Kirchgessner, T. A., **5**:363, 508; **6**:538, 563

Kirchhoff, G., **6**:2, 122; **24**:30(58), 38(58)

Kirchoff, R. H., **23**:301(104), 358(104)

Kirichenko, Y. A., **16**:235; **17**:140(154, 155), 146, 157

Kirillov, P. L., **21**:39(64), 40(64), 52(64); **25**:51(79), 143(79); **31**:265(162), 329(162)

Kirillov, V. V., **1**:141(14), 184; **6**:504(13), 523(13), 531(13), 539, 540(61), 541(61), 562, 563; **21**:14(29), 51(29)

Kirk, D. A., **4**:151, 223

Kirk, F. N., **6**:18, 123

Kirk, L. A., **10**:208, 218

Kirk, P. S., **8**:19, 86

Kirker, T. J., **23**:315(149), 360(149)

Kirko, I. M., **25**:153(13, 14), 154(14), 155(14), 156(14), 157(13), 163(14), 196(14), 200(13, 14), 201(14), 203(14), 243(13, 14)

Kirkpatrick, A. T., **18**:53(101), 54(101), 58(110, 111), 60(110, 111), 67(126), 68(126), 84(101, 110, 111), 85(126)

Kirkpatrick, J. R., **19**:37(80), 86(80)

Kirkpatrick, M. E., **4**:150, 223

Kirmse, D. W., **5**:48, 54

Kirowa-Eisner, E., **12**:226(27a), 278

Kirpichnikov, F. P., **31**:254(148), 329(148)

Kirsch, A. A., **25**:292, 293, 317

Kirschbaum, E., **16**:61(2), 153

Kirwan, D. J., **11**:412(185), 439
Kiryu, M., **26**:105(4), 208(4)
Kiselev, A. I., **8**:34(178), 88
Kiselev, I. G., **31**:160(33), 323(33)
Kiser, K. M., **15**:92(89), 138(89)
Kishinami, K., **11**:220, 224, 263; **19**:81(249), 94(249)
Kishore, J., **23**:192(152), 273(152)
Kisker, D. N., **28**:342(33), 416(33)
Kisker, D. W., **28**:342(19), 416(19)
Kiss, L. I., **23**:306(119), 359(119)
Kistiakowski, G. B., **2**:354
Kistler, A. L., **3**:43, 98; **6**:67, 127; **30**:290(81), 310(81)
Kitada, H., **26**:45(150), 99(150)
Kitahara, K., **14**:293, 341
Kitani, S., **26**:45(224), 103(224)
Kittel, C., **11**:321(35), 435
Kittl, J. A., **28**:76(55), 141(55)
Kitto, J. B., **17**:319, 338
Kitzinger, C., **4**:120(93), 140
Kivel, B., **1**:10, 49; **2**:127(15, 16), 131(15, 16), 134(15, 16), 268; **5**:320(94), 324
Kiwaki, J., **19**:310(94), 312(94), 353(94)
Kiwaki, Z., **17**:4(21), 57
Kiya, M., **SUP**:164(342), 168(342), 190(342)
Kjellström, B., **11**:241, 243, 261
Klages, J. R., **29**:259(118), 261(118, 121), 263(118, 121), 264(118, 121), 265(118, 121), 269(135), 282(118), 328(181), 341(118, 121), 342(135), 344(181)
Klamerus, E. W., **29**:219(24), 269(24), 277(24), 279(24), 312(24), 323(24), 335(24)
Klaren, D. G., **30**:236(28), 237(28), 238(28), 250(28); **31**:462(32), 472(32)
Klarsfeld, S., **14**:67, 103
Klassen, J., **10**:201(104), 202(104), 203, 206(104), 209(104), 217; **14**:213(146), 216(146), 217(146), 218(146), 222(146), 226(146), 246; **19**:122(114), 123(114), 129(114), 133(114), 134(114), 187(114)
Klassen, N. V., **28**:103(88), 142(88)
Klaus, E. E., **23**:193(61, 254), 206(60), 226(61, 103, 254), 269(60, 61), 271(103), 278(254)
Klebanoff, P. S., **2**:73, 108; **3**:43, 98; **7**:142(52), 153, 155, 161; **8**:291, 300, 301(48), 349, 350
Klee, A., **21**:336, 345
Kleijn, C. R., **28**:354(94), 356(96), 363(95, 98), 364(98), 394(95, 96, 185), 395(95, 185),

405(95), 419(94, 95, 96, 98), 423(185)
Klein, D. E., **23**:161(45), 186(45)
Klein, I., **13**:261(120), 266; **28**:149(63), 224(63), 230(63)
Klein, J., **1**:34(30), 50; **3**:132, 174; **7**:330, 377
Klein, J. D., **9**:378(49), 379(49), 415
Klein, J. S., **8**:22(102), 86; **SUP**:18(25), 21(25), 79(25), 101(25), 102(25), 119(25), 120(25), 125(25), 137(25), 170(25), 171(25), 177(354); **15**:100(101), 139(101)
Klein, M., **5**:309(76), 323
Klein, V., **3**:16, 32; **11**:218, 221, 236, 256
Kleiner, M. K., **22**:28(53), 154(53)
Kleinman, D. A., **11**:357(80), 436
Kleinstreuer, C., **25**:260, 270, 283, 319
Klell, M., **23**:343(292), 368(292)
Kleman, B., **22**:347(12)
Klemens, P. G., **9**:360(18), 361(18), 414; **11**:322(41), 352(41), 353(41), 355(41), 435; **24**:27(56), 38(56)
Klemm, A., **12**:229(50), 246(50), 269(48, 49), 279
Klems, J. H., **23**:292(55), 356(55)
Kleppe, J., **15**:177(35), 217(35), 224(35)
Klett, D. E., **9**:358(14), 408(14), 414
Klett, J. D., **26**:2(166), 20(166), 100(166)
Klevtsov, I. A., **23**:305(112), 359(112)
Klibanov, T. M., **3**:284, 288, 301
Klimenko, A. S., **15**:324
Klimenko, V. V., **17**:146(166), 157, 335(38), 340; **18**:256(28), 257(28), 258(28), 290(28), 291(28), 294(28), 298(28), 299(28), 308(28), 309(28), 310(28), 317(28), 321(28); **23**:9(25), 116(25), 127(25)
Klimov, N. N., **21**:39(65, 66), 41(66), 42(66), 52(65, 66)
Kline, D. E., **18**:220(118), 238(118)
Kline, J. F., **5**:351(58), 381, 449, 507, 515; **11**:170(184), 173(184), 174(184), 177(184), 185(184), 197
Kline, S. G., **3**:81(79), 85(79), 86(79), 99
Kline, S. J., **2**:16(19), 25(19), 105; **3**:14(17), 23(17), 32; **5**:93, 127; **8**:287(2), 297(30, 33), 300(52), 301(52), 303(52), 304(30, 52), 305(58), 344(52, 65), 348, 349, 350; **13**:79(30), 116; **SUP**:6(14), 69(14), 86(14), 87(14), 398(14); **16**:242, 244, 245(35, 36, 37, 38), 248, 255(35, 36, 37, 38), 256(35, 36, 37, 38), 271, 356(35, 36, 37, 38), 363,

364, 365; **23**:284(37), 355(37); **25**:21(55), 142(55); **31**:167(73), 172(78), 174(78, 80), 325(73, 78, 80)

Klineberg, J. M., **6**:109(175), 119(175), 131

Klinger, H. G., **22**:36(37, 38, 39), 39(37), 40(38), 43(39), 44(39), 45(39), 46(39), 59(39), 154(37, 38, 39)

Kloper, G., **12**:51(61), 75

Klotz, I. M., **12**:261(51), 262(51), 263(51), 279

Kluge, M. D., **28**:80(56), 141(56)

Klumel, G. S., **18**:171(56), 176(56), 177(56), 212(56), 237(56)

Klyukanov, A. A., **14**:131, 134(7), 144

Kmečko, I., **26**:237(41), 241(41), 246(55), 247(41), 248(41), 292(41), 294(41), 295(41), 296(41), 298(41), 299(41), 326(41), 327(55)

Kmetyk, L. N., **29**:218(21), 234(21, 95), 246(95), 335(21), 340(95)

Kmonicek, V., **6**:416(94), 490

Knapp, K. K., **7**:59, 60, 85

Knapp, R. M., **14**:89, 90(178, 179), 91(178, 179), 104

Knapp, R. T., **10**:87, 118, 119, 121, 163, 165

Knappe, W., **13**:211(21), 217(32), 218(32, 33), 264; **18**:161(6), 212(101, 102), 213(6), 222(101, 102), 236(6), 238(101, 102)

Knavin, A. A., **31**:245(133), 328(133)

Kneisel, K., **5**:141, 144, 149, 247

Knieper, P. J., **SUP**:183(362)

Knight, B. A., **4**:165, 166, 224

Knight, B. K., **7**:279, 319

Knight, C. A., **22**:196(114, 115), 352(114, 115)

Knight, C. J., **28**:111(57), 133(23), 139(23), 141(57)

Knight, L. H., **5**:390, 510

Knjazeva, G. D., **1**:435(141), 444

Knoche, K. F., **4**:259(23), 262(23), 263(23), 313

Knock, R. E., **20**:217(129), 218(129), 219(129), 305(129)

Knoebel, D. H., **17**:32(105), 61

Knoke, S., **25**:223(111), 224(111), 247(111)

Knoll, R. H., **4**:228

Knoner, R., **5**:412, 413, 513

Knöner, R., **9**:3(10, 11, 69), 82(10, 11, 69), 83, 106, 109; **11**:105(66), 111, 120(66), 121(66), 122(66), 191; **16**:114(78), 155, 233

Knopp, K., **8**:14(78), 17(78), 86

Knowles, C. P., **6**:211, 365; **9**:324(95), 326(95), 327, 328, 330(98), 332, 333, 348

Knowles, J. B., **29**:160(96), 162(96), 207(96)

Knowlton, T. M., **19**:110(69), 186(69)

Knox, A. L., **28**:370(132), 421(132)

Knudsen, J. G. (Chapter Author), **5**:365(83), 508; **7**:97(7), 98(7), 160; **11**:203, 206, 217, 228, 254, 259; **14**:192, 193, 196, 198, 199, 204, 210, 211(143), 226, 227, 228(122), 245, 246; **19**:135(236), 190(236); **30**:202(58), 203(29), 213(29), 216(24), 250(24), 251(29), 252(58); **31**:245(138), 294(201), 328(138), 431, 433(16), 443(70), 446(34), 447(33), 449(35), 451(24, 36), 456(37), 469(36), 471(16, 24), 472(33, 34, 35, 36, 37), 474(70)

Knudsen, M., **2**:297, 354; **18**:333(35), 336(35), 362(35)

Knudson, D. L., **29**:218(17, 18, 19), 219(18, 19), 228(18), 232(18), 233(18, 19), 244(18), 270(19), 277(19), 279(18, 19), 288(18), 297(18), 305(19), 312(18), 319(19), 322(19), 329(18), 335(17, 18, 19)

Knuth, E. L., **3**:63(64b), 99; **5**:348, 507; **6**:383(12), 474(12), 476, 486; **12**:120(6), 192

Knystautas, R. K., **29**:64(12), 86(51), 88(51), 89(51), 96(62, 63), 98(67), 100(67), 101(73), 102(73), 124(12), 125(51), 126(62, 63, 67, 73)

Ko, R. S., **18**:19(40), 48(40), 81(40)

Ko, S. Y., **23**:320(162), 361(162)

Koai, K., **30**:349(95), 350(95), 351(96), 358(111, 112), 360(121), 431(95, 96, 111, 112), 432(121)

Kobayashi, F., **20**:232(152), 242(152), 306(152)

Kobayashi, H., **17**:115, 116, 117, 118, 119, 155

Kobayashi, J., **28**:352(92), 419(92)

Kobayashi, K., **11**:139, 195

Kobayashi, M., **14**:152(37), 159, 208, 243, 244, 246

Kobayashi, M. H., **24**:213, 271

Kobayashi, N., **11**:412(194, 195, 196), 439; **30**:315(11), 329(38), 368(136), 371(11, 136), 427(11), 428(38), 433(136)

Kobayashi, S., **17**:152(181), 157; **30**:329(33), 428(33)

Kobayashi, T., **25**:342, 361, 388, 391, 393, 402, 404, 406, 413, 416; **30**:349(155), 390(155), 433(155)

Kobayasi, K., **18**:241(4), 257(4, 30), 263(4, 30), 267(4), 268(4), 269(4), 320(4), 321(30)

Kober, H., **8**:49(232), 89
Kobiyama, M., **27**:ix(5); **31**:409(73), 428(73)
Koblinger, K., **31**:334(10), 425(10)
Kocamustafaogullari, G., **19**:38(85), 42(85), 43(85), 87(85); **26**:123(59), 124(59), 172(59), 178(59), 211(59); **30**:335(65), 429(65); **31**:112(12), 156(12)
Koch, D. L., **23**:370(23), 461(23); **30**:111(61), 191(61)
Koch, F. A., **11**:220, 224, 237, 258
Koch, R., **31**:189(90), 190(90), 191(90), 196(90), 204(90), 220(90), 222(90), 223(90), 224(90), 225(90), 326(90)
Koch, W., **11**:202, 204, 209, 211, 213, 214, 252
Koch, W. B., **4**:21, 63, 104(49), 105(49), 139
Kochelaev, Y. S., **11**:133(146), 144(146, 154, 155), 146(146, 154, 155), 151(155), 152(155), 156(155), 159(155), 170(154, 155), 172(155), 173(155), 176(155), 177(155), 179(155), 180(155), 182(155), 183(155), 185(155), 195, 196
Kochenov, I. S., **6**:408, 409(66), 457, 489, 495; **25**:64(108, 109, 110), 96(110), 145(108, 109, 110)
Kochin, N. E., **25**:5(35), 141(35)
Kochubei, A. A., **25**:4(13), 31(13), 33(13), 64(13), 96(13), 127(13), 140(13)
Kochubey, A. A., **19**:195(27), 243(27)
Koczkur, E., **20**:187(68), 302(68)
Koency, J. E., **23**:305(113), 359(113)
Koenig, A. A., **18**:41(84), 42(85), 44(85, 86), 83(84, 85, 86)
Koenig, H. A., **15**:321
Koestel, A., **4**:214(98a), 227
Kofman, P. O., **14**:228, 228(164), 246
Kogan, A., **15**:228(3), 238(47), 239(47, 48), 242, 271(3, 48), 276, 278
Kogan, M. G., **6**:474(323), 502
Kogarko, S. M., **29**:96(61), 126(61)
Koh, B., **20**:264(263), 311(263)
Koh, J. C. Y., **1**:23, 50, 71, 121; **3**:108(12), 173, 230, 235, 251; **4**:331, 445; **5**:82, 126, 167, 168, 169, 248; **8**:34, 88; **9**:250, 251, 270; **11**:67(31), 68, 76, 77(31), 79, 80(31), 81(31), 82(31), 83(31), 84(31), 190; **15**:232, 277; **19**:324(114), 354(114); **30**:34(54), 38(54), 39(54), 90(54)
Koh, R. C. Y., **13**:67(8), 115
Kohara, M., **20**:183(7), 299(7)
Kohda, H., **28**:316(166), 336(166)

Kohing, F. C., **26**:105(1), 208(1)
Kohiro, K., **30**:320(24), 427(24)
Kohl, A. L., **15**:271(112), 281
Kohler, A., **30**:349(95), 350(95), 431(95)
Kohler, W., **18**:200(153), 239(153)
Kohlmayr, G. F., **6**:418(128), 492
Kohout, F. A., **14**:40, 77, 102
Koizumi, H., **23**:293(66), 349(66), 356(66)
Koizumi, T., **20**:277(293), 278(295), 313(293, 295)
Kokado, J., **23**:9(20), 109(20, 114, 115, 116, 117, 118), 110(20, 114, 115, 116, 117, 118), 113(20, 114, 115, 116, 117), 114(116), 115(20, 115, 116, 117), 116(118), 119(20), 127(20), 132(114, 115, 116, 117, 118)
Kokal, S., **20**:93, 100, 132
Kokini, J. L., **28**:154(41), 213(37), 229(37, 41)
Kokora, A., **28**:127(84), 142(84)
Kolar, A. K., **19**:100(29), 136(29), 137(29), 139(29), 140(29), 145(29), 157(29), 159(29), 161(29, 202), 174(202, 227), 185(29), 190(202, 227)
Kolar, M. J., **5**:389, 510
Kolb, E. D., **30**:331(43), 428(43)
Kolb, H., **20**:285(305), 313(305)
Kolbach, J. St., **31**:462(61), 473(61)
Kolbel, H., **21**:319, 320, 321, 325, 329, 331, 345
Kolesnichenko, A. F., **28**:316(164), 336(164)
Kolesnikov, P. M., **19**:197(32), 244(32)
Kolker, H., **30**:403(171), 434(171)
Kollera, M., **21**:92(64, 65), 136(64, 65)
Kollie, T. G., **20**:264(266), 312(266)
Kolmogorov, A. N., **3**:35, 97; **21**:265(35), 267(35), 276(35); **30**:4(55), 90(55)
Koln, H. H., **5**:365, 508
Kolodney, M. S., **22**:243(122), 244(122), 245(122), 246(122), 247(122, 123), 352(122, 123)
Kolovandin, B. A. (Chapter Author), **21**:185, 185(9, 10), 198(9, 10), 202(9, 10), 203(9, 10), 236(9, 10)
Kolozsi, J. J., **2**:53, 68, 69(57), 70, 107
Komai, T., **21**:166, 174, 180
Komatsu, T., **27**:181(32), 186(32); **31**:391(48), 413(48), 418(48), 427(48)
Komiya, S., **28**:390(175), 423(175)
Komoda, H., **6**:69(119), 128
Komori, T., **19**:79(208), 92(208)
Komoriya, T., **30**:151(111), 152(111), 194(111)

Komotori, K., **21**:300, 345; **26**:49(228, 229), 51(228, 229), 103(228, 229)

Kompaneets, A. S., **29**:140(36), 204(36)

Komzsik, L., **31**:403(63), 428(63)

Konagai, M., **28**:342(26), 416(26)

Konakov, P. K., **3**:219(68), 230, 250, 251

Kondera, H., **28**:97(58), 141(58)

Kondrashov, N. G., **6**:465(192), 495

Kondratenko, A. K., **21**:37(61), 39(61), 40(61), 41(61), 52(61)

Kondratiev, N. S., **6**:551, 564

Kondratyev, K. Ya., **14**:249(1), 279

Kondukov, N. B., **14**:205, 206(137), 245

Konev, S. V., **31**:160(40), 324(40)

Konicek, L., **18**:15(20), 80(20)

Konkov, A. S., **6**:551, 555(74), 563; **17**:319(4), 320(4), 327(4), 339; **21**:32(48), 52(48)

Kono, N., **19**:127(133), 128(133), 188(133)

Konopliv, N., **15**:325

Konopnicki, T. T., **18**:50(95), 51(95), 84(95); **30**:281(66), 310(66)

Konov, V. I., **28**:102(99), 103(88), 142(88), 143(99)

Konowalow, D. D., **3**:270(29), 300

Konsetov, V. V., **21**:279, 320, 321, 322, 324, 325, 326, 327, 328, 329, 331, 345

Konstanchuk, D. M., **16**:114(74), 155

Koo, V. H. S., **11**:412(188), 439

Kooijman, J. M., **SUP**:2(11)

Kooker, D. E., **21**:263(31), 276(31)

Koong, S., **5**:167, 168, 248

Koopman, R. P., **7**:236(32), 240, 275, 294, 296, 315

Kopal, Z., **7**:209, 218

Kopetsch, H., **30**:337(70), 351(70), 390(70), 429(70)

Köpf, U., **30**:256(5), 307(5)

Kopp, I. Z., **31**:266(170, 171), 270(170, 171), 330(170, 171)

Koppel, L. B., **6**:551, 564; **7**:50, 84; **10**:193, 203(109), 217; **14**:152(34), 158(34), 192(34), 193(34), 196(34), 201(34), 204(34), 243; **19**:128(137), 188(137)

Koppel, T. A., **25**:81(154), 147(154)

Körber, Ch., **19**:79(197, 198), 92(197, 198); **22**:188(111), 209(89), 210(89), 211(89), 212(117), 214(89), 215(89), 217(89), 218(89), 219(89), 220(10), 293(116, 224), 347(10), 350(89), 352(111, 116, 117), 357(224)

Korczak, K. Z., **20**:243(174), 307(174)

Korenberg, Yu. G., **14**:173(90), 176(90), 244; **19**:120(103), 187(103)

Korenevskiy, V. A., **23**:9(23), 127(23)

Korenic, B., **15**:3(14), 55(14)

Korger, M., **13**:10(33, 34), 12, 26(34), 41, 42(58), 43, 44, 60

Koriyama, K., **29**:151(70), 206(70)

Korjakin, Ju. I., **1**:435(141), 444

Korkegi, R. H., **3**:52, 98

Kormishin, V. A., **25**:4(32), 141(32)

Kornbau, R. W., **30**:222(30), 251(30)

Korner, M., **16**:66, 67, 99, 107, 117(14), 118, 123, 125, 126, 153, 155

Kornosky, R. M., **19**:105(48), 108(48), 185(48)

Korolev, A. L., **11**:144(155), 146(155), 151(155), 152(155), 156(155), 159(155), 170(155), 172(155), 173(155), 176(155), 177(155), 179(155), 180(155), 182(155), 183(155), 185(155), 195, 196; **31**:265(167), 330(167)

Korolev, V. N., **10**:208, 218; **19**:149(168), 150(168), 151(168), 153(168), 189(168)

Korolkov, B. P., **6**:474, 501; **20**:152(16), 179(16); **25**:4(26, 27), 140(26, 27)

Koroll, G. W., **29**:79(41, 42), 80(42), 125(41, 42)

Korotyanskaya, L. A., **10**:210(139), 218; **14**:160(72), 161(72), 164(72), 166(72), 167(72), 168(72), 169(72), 170(72), 171(72), 172(72), 176(72), 180(107), 181, 181(109), 188(107), 190(107), 191(107), 206(109), 237(109), 244, 245, 246; **19**:160(192), 189(192)

Korpela, S. A., **18**:49(93), 84(93)

Korshakov, A. P., **31**:242(130), 328(130)

Korst, H. H., **6**:5, 7, 15, 122, 123, 474(220), 479, 497

Korsunsky, V. I., **14**:205, 245

Korth, G. E., **29**:215(1), 248(1), 334(1)

Kortleven, A., **11**:219(113), 223(113), 257

Kosaka, S., **27**:181(32), 186(32); **31**:391(48), 413(48), 418(48), 427(48)

Koschmieder, E. L., **17**:131(136), 156; **21**:155, 178, 180

Koshkin, V. K., **6**:376(1), 449(181, 182, 183, 184), 453(181, 182, 183, 184), 456(181, 182, 183, 184), 457(181, 182, 183, 184), 486, 495; **11**:133, 144(146, 154, 155), 146(146, 154, 155), 151(155), 152(155), 156(155), 159(155), 170(154, 155), 172(155), 173(155), 176(155), 177(155),

Koshkin, V. K. (*continued*)
179(155), 180(155), 182(155), 183(155), 185(155), 195, 196; **18**:241(3), 294(3), 298(3), 300(3), 311(3), 320(3); **25**:4(2, 3), 10(36), 12(3), 22(3), 33(3), 35(70a, 70b), 43(3), 52(3), 64(3), 83(164), 92(3), 99(3), 106(3), 108(3), 120(3), 130(3), 139(2, 3), 141(36), 143(70a, 70b), 147(164); **31**:265(158), 270(158), 329(158)

Koshmarov, Y. A., **7**:206(141), 208, 209, 217

Kosky, P. G., **4**:170(53), 225; **5**:405, 462, 463, 464, 511, 515; **9**:229(61), 269; **10**:146, 165; **11**:105(120), 114(120), 194; **16**:114(79), 155, 234

Kosterin, S. I., **1**:398(79), 399, 442

Kostic, M. (Chapter Author), **19**:247, 264(29), 267(29), 268(29), 311(95), 313(95), 317(29, 108), 319(29, 108), 320(29), 325(118), 326(121), 327(123), 330(95, 121), 332(128), 333(128), 334(148), 337(148), 338(148), 339(148), 340(148), 342(128), 343(128), 350(29), 353(95), 354(108, 118, 121, 123), 355(128), 356(148); **24**:102(5), 174(5), 184(5)

Kostič, Ž., **8**:108(32), 109(32), 110(34), 159; **11**:220, 224, 237, 258; **18**:102(32), 103(32), 104(34), 105(34), 158(32, 34)

Kostin, M. D., **13**:169, 202; **15**:285(116), 287, 290, 326

Kostiouk, V. V. (Chapter Author), **18**:241, 316(70), 323(70); **25**:4(5, 10), 33(5, 10), 43(5), 52(5), 61(5), 92(5), 99(5), 106(5), 130(5), 139(5), 140(10)

Kostoglou, M., **31**:462(38), 472(38)

Kostyuk, A. G., **6**:418(118), 491

Kostyuk, V. V. (Chapter Author), **11**:51, 115(187), 120(187), 122(187), 133(146), 144(146, 154, 155), 146(146, 154, 155), 151(155), 152(155), 156(155), 159(155), 170(154), 172(155), 173(155), 176(155), 177(155), 179(155), 180(155), 182(155), 183(155), 185(155), 195, 196, 197

Kotchaphakde, P., **16**:233

Kothandaramaran, G., **28**:133(83), 142(83)

Kothari, A. K., **10**:174, 175, 177(38), 215

Kothari, J., **23**:12(70, 71), 14(70, 71), 51(70, 71), 129(70, 71); **26**:105(2), 208(2)

Kotlyar, I. V., **6**:474(319), 501

Kotorynski, W. P., **8**:15(79), 19, 68, 69(267), 70(267), 86, 90

Kottowski, H., **29**:165(112, 113), 208(112, 113)

Kottowski, H. M., **11**:4(4a), 47

Kou, S., **28**:251(36, 39), 266(36), 309(36), 330(36, 39)

Kouba, G. E., **20**:100, 101, 108, 131

Kourganoff, V., **1**:5, 8(9), 49; **3**:176(2), 178(2), 180(2), 181(2), 184(2), 186(2), 197(2), 201, 206, 207(2), 208(2), 248; **5**:13(16), 51, 313(85), 324; **8**:34, 88; **11**:325(44), 337(44), 435

Kourosh, S., **22**:212(118), 352(118)

Koval', Yu. D., **15**:324

Kovalenko, L. M., **31**:160(15), 323(15)

Kovalenko, V. F., **18**:337(52), 363(52)

Kovalev, I. A., **16**:233; **17**:325(27), 340

Kovalev, S. A., **18**:255(15, 20), 256(20), 257(20), 258(50), 261(50), 262(20), 264(58), 307(20), 315(69), 317(69), 320(15), 321(20), 322(50, 58), 323(69); **25**:48(72), 82(72), 122(72), 143(72)

Kovalev, S. D., **6**:416(95), 417(97), 490

Kovalnogov, N. N., **15**:328

Kovasznay, L. S. G., **30**:289(79), 310(79)

Kovelnogov, N. N., **25**:92(173), 148(173)

Kovensky, V. I., **19**:110(67), 116(67), 117(67), 174(228), 186(67), 190(228); **25**:225(114), 234(114), 247(114)

Koverianov, V. A., **6**:416(89), 490

Kovitz, A. A., **2**:213, 269

Kowal, M. G., **29**:167(122, 123), 208(122, 123)

Kowalski, C., **24**:104(28), 185(28)

Koyama, H., **24**:125(132), 127(132), 129(134), 130(134), 132(134), 134(134), 135(134), 189(132, 134); **25**:258, 262, 270, 318

Koyama, K., **SUP**:103(269)

Koyama, S., **21**:124(113), 138(113)

Koza, Y., **28**:366(115), 369(115), 389(168), 420(115), 422(168)

Kozachenko, L. S., **8**:94, 158; **18**:88(8), 158(8)

Kozai, H., **17**:51(163), 64

Kozanevich, Z. Y., **6**:474(297, 298, 299), 500

Kozdoba, L. A., **19**:193(17), 195(17, 20), 197(17), 243(17, 20)

Kozhinov, I. A., **9**:265(141), 272

Kozhosev, L. S., **20**:262(261), 270(261), 311(261)

Kozhukhar, I. A., **14**:122, 146; **31**:160(19), 323(19)

Kozicki, W., **19**:262(25, 26), 263(25), 264(25, 26), 265(25), 326(25), 330(25), 350(25,

26); **23**:187(141), 188(141), 196(148), 197(148), 198(141, 142, 143, 144, 145, 146, 147, 148, 149), 201(148), 203(144), 204(148), 206(142, 143, 144, 145, 146, 147, 148, 149), 216(95), 222(95), 223(142), 226(95, 142, 143, 144, 145, 146, 147, 148, 149), 228(142, 143), 229(95, 142, 143, 144, 145, 146, 147, 149), 271(95), 273(141, 142, 143, 144, 145, 146, 147, 148, 149); **24**:103(25), 105(38), 185(25), 186(38)

Koziol, J., **30**:328(30), 371(141), 372(141), 373(141), 374(141), 375(141), 376(141), 377(141), 378(30, 141), 379(141), 380(141), 381(141), 384(141), 392(141), 406(30, 141), 407(30, 176), 408(30, 141, 176), 409(30), 412(176), 427(30), 433(141), 434(176)

Kozlov, A. A., **25**:134(207), 150(207)

Kozlov, A. K., **31**:250(144), 252(144), 253(144), 327(121), 328(144)

Kozlov, B. K., **1**:358, 440

Kozlov, L. W., **3**:61, 99

Kozlov, S. M., **16**:235; **17**:140(154, 155), 146(168), 157

Kozlov, V. N., **19**:195(18), 243(18)

Kozlowski, H., **5**:178, 181, 249

Kozlu, H., **20**:243(176), 307(176)

Kozyrev, A., **16**:235

Kraabel, J. S., **18**:37(76), 83(76); **23**:336(254, 255), 366(254, 255)

Kraev, M. V., **31**:265(159), 329(159)

Krafft, G., **17**:138(148), 146, 157

Kraichman, M. B., **12**:233(34), 251(34), 270(34), 278

Kraichnan, R. H., **25**:345, 413

Kraitshev, S. G., **13**:2(4), 36, 37(4), 38(4), 39(4), 40(4), 59

Krajewski, B., **19**:199(52, 54), 200(52, 54), 202(52, 54), 244(52), 245(54)

Krajewski, N., **20**:183(8), 299(8)

Krajnovich, D. J., **28**:102(61), 103(9), 104(9), 105(9), 106(9), 107(9), 113(9), 114(9), 116(9), 117(9), 119(10), 120(10), 121(10), 122(10), 123(10), 138(9, 10), 141(60, 61)

Krakow, B., **5**:301(57), 323

Krall, K. M., **8**:119, 120(49), 121(49), 160; **11**:220, 224, 263; **18**:113(48), 115(48), 159(48)

Kramarenko, V. A., **11**:423(223), 440

Kramer, T. J. (Chapter Author), **9**:113, 132, 137, 138, 142, 143, 144, 150, 155, 156, 163, 170, 172, 179

Kramers, H., **11**:225, 258; **21**:179

Krane, M., **23**:305(114), 359(114)

Krane, M. J. M., **28**:245(20), 306(20), 330(20)

Krane, R. J., **20**:217(130), 220(130), 305(130); **26**:281(80), 297(80), 300(80), 328(80)

Krantz, P., **4**:111, 139

Krapez, J.-C., **28**:53(79), 73(79)

Kras, A., **29**:183(172), 211(172)

Krase, J. M., **12**:5(14), 7(14), 35(14), 73

Krasheninnikov, V. V., **6**:474(303, 313), 501

Krashtalev, P. A., **8**:103(29), 104(29), 105(29), 132(29), 133(29), 159; **18**:98(29), 99(29), 130(29), 131(29), 158(29)

Krashtalyev, P. A., **11**:228(226), 262

Krasin, A. K., **1**:435(140), 444

Krasnikova, O. K., **31**:192(96), 326(96)

Krasnoschekhov, E. A., **1**:414(105), 443; **6**:551, 552, 564; **7**:48(22), 49(22), 50, 84

Krasnov, S. I., **25**:73(116, 117, 118, 119), 145(116, 117, 118, 119)

Krasnov, S. N., **21**:44(72), 52(72)

Krasyakova, L. I., **1**:357, 384, 388(4), 390(4), 440

Krasyakova, L. Y., **21**:30(46), 32(46), 52(46)

Kratovil, M. T., **14**:199, 200(129, 130), 202, 245; **19**:148(171), 153(171), 189(171)

Kratzer, P. A., **12**:24(41), 74

Kraus, A. D., **20**:182(118), 184(118), 213(118), 221(118), 304(118); **30**:216(25), 250(25), 413(178), 435(178); **31**:160(13), 323(13)

Kraus, W., **4**:15, 63

Kraus, W. E., **16**:114(77), 155

Krause, F. R., **6**:40(75), 41(75), 65(75), 66(75), 75(75), 77(75), 125

Krause, W. B., **19**:146(160), 149(160), 188(160)

Kraushaar, R., **6**:199, 364

Kraushold, H., **5**:230(159), 250

Kraussold, H., **11**:266, 281, 283, 284, 314; **21**:160, 167, 180

Kravchenko, V. F., **15**:325; **19**:202(59), 245(59)

Kravtsov, S. F., **11**:203, 206, 255

Kraybill, R. R., **18**:171(109), 216(109), 228(109), 233(109), 238(109)

Kreatsoulas, J. S., **23**:299(95), 358(95)

Krebs, H.-U., **28**:102(62), 116(62), 141(62)

Kreid, D. K., **8**:298(37), 349; **SUP**:209(401), 210(401), 211(388); **19**:258(14), 259(14), 268(14), 350(14)

Kreider, J. F., **15**:48(81), 57(81); **18**:2(3), 6(3), 7(3), 9(3), 79(3)
Kreider, K. G., **23**:343(290), 367(290)
Kreider, M. B., **4**:123(104), 141
Krein, M. S., **29**:165(114), 208(114)
Kreith, F. (Chapter Author), **1**:226, 231, 265; **2**:372, 395, 411, 450; **5**:129, 130(2), 132(2), 133(22), 141, 144, 149, 150, 153(45), 155158, 159(2), 160, 161, 164166, 172, 177(3), 178, 179(3), 180182, 183(86), 184188, 190(101), 191194, 199, 227, 240, 241, 242, 246251, 339(28), 506; **6**:531, 562, 563; **11**:213, 214, 255; **SUP**:143(326); **15**:58(91); **18**:1, 2(3), 3(4), 6(3, 4), 7(3), 9(3, 6), 19(45), 28(63), 79(3, 4, 6), 81(45), 82(63); **19**:12(19), 15(19), 84(19); **21**:143, 180; **26**:2(109), 97(109)
Krekel, J., **13**:219(86), 256, 266
Kremer, A. M., **28**:389(171), 390(171), 391(171), 398(171), 422(171)
Kremlev, O. A., **31**:245(133), 328(133)
Kremmer, M., **14**:334, 334(125), 345
Kreplin, H. P., **25**:381, 385, 386, 413
Krettenauer, K., **25**:376, 413
Kretz, L. O., **23**:280(12), 298(12), 305(12), 354(12)
Kretzchmar, B., **16**:114(78), 155
Krey, R. U., **4**:241(7), 242(7), 313
Krichevsky, G. Ya., **14**:205, 245
Krischer, O., **13**:124, 201; **30**:168(133), 195(133)
Krishna, M. S., **5**:155, 157, 248
Krishna, P. M., **21**:337, 339, 345
Krishna, R., **15**:232, 276
Krishnamoorthy, G., **SUP**:74(198), 165(198), 166(198)
Krishnamurthy, M. V., **SUP**:317(473)
Krishnamurthy, S., **15**:287(163), 328
Krishnamurthy, V. V. G., **SUP**:172(349), 216(408), 242(428), 251(436), 271(448), 310(472)
Krishnamurti, R., **30**:29(56, 57, 58), 90(56, 57, 58), 362(126), 432(126)
Krishnamurty, K., **6**:69(126), 73(126), 128
Krishnamurty, V. V. G., **26**:220(6), 239(44), 241(44), 242(48), 246(48), 253(44, 48), 274(44, 48), 324(6), 326(44, 48)
Krishnan, K. S., **2**:426, 450, 451
Krishna Prasad, K. (Chapter Author), **17**:159, 162(12, 13), 173, 174(13), 178(13), 207,

222(69), 223(69), 225(69), 237, 241(77), 248, 269, 272(95), 275, 281, 296, 297, 313, 315, 316
Kristianpoller, N., **11**:358(88), 436
Kristichenko, P. I., **15**:325
Krist-Spit, C. E., **17**:291(99), 302(99), 317
Kritz, M. A., **10**:48, 81
Krivesko, A. A., **16**:95, 97(55), 114(71, 74), 116(71), 154, 155
Krivoshei, F. A., **15**:325
Křížek, F., **13**:10(33, 34), 12, 26(34), 44, 60
Krizek, T. J., **22**:226(119), 352(119)
Kroeger, P. G., **19**:78(183), 80(235), 91(183), 93(235)
Kroesser, F. W., **30**:71(53), 90(53)
Krog, J., **22**:60(52), 61(52), 70(52), 154(52)
Kröger, C., **11**:360(93a), 437
Kroger, D. G., **9**:218, 222, 255(42), 264(131), 269, 272
Krokhin, O. N., **28**:123(3), 138(3)
Krokhin, Yu. I., **31**:160(38), 323(38)
Krolenok, V. V., **31**:160(40), 324(40)
Kromann, G. B., **20**:292(318), 293(318), 294(318), 314(318)
Krome, H., **2**:353
Kronauer, R. E., **11**:241, 242, 261
Kröner, R., **17**:6(30), 57, 140(152), 142(152), 146(152), 157
Krongelb, S., **3**:287, 301
Kronig, R., **14**:107, 116, 117(3, 4), 120, 121, 123, 144, 146; **26**:19(110), 20(110), 23(110), 27(110), 97(110)
Kronin, I. V., **17**:37(122), 62
Krooger, P. G., **5**:356(70), 508
Krook, M., **3**:207, 210(41), 250
Kropschot, R. H., **5**:329(31), 343(31), 345, 346, 347, 474, 506; **9**:350(4), 352(4, 9), 378(50), 381(50), 394(4), 399(9), 404(9), 407(50), 414, 415; **15**:50(83), 57(83)
Krotikov, V. D., **10**:41, 71, 81
Krötzsch, P., **13**:2(2, 3, 5), 8(5), 11(5), 18, 19, 20, 21, 22, 25(5), 32(5), 48, 59
Kroujilin, G., **3**:16, 32
Krueger, A. W., **30**:228(31), 229(31), 251(31)
Krueger, E. R., **5**:236, 237, 251; **21**:180
Krueger, J., **29**:167(121), 208(121)
Krug, W., **6**:135(7), 195, 322, 363, 365
Kruger, C. H., **5**:255(13), 322
Kruger, Ch. H., **7**:170(22), 171(22), 211
Kruger, C. H., Jr., **8**:248, 282

Kruger, P., **14**:13(33), 97(33, 187, 188, 189, 193), 100, 104

Kruger, R. A., **5**:102, 127; **11**:144(163, 164), 146(163, 164), 159, 160(164), 162, 170(163, 164), 172, 177, 196

Kruglikov, V. Y., **10**:201(154), 218

Kruglikov, V. Ya., **19**:119(101), 187(101)

Krujilin, G., **11**:218, 221, 236(92), 256

Krummel, O., **18**:327(9), 328(9), 362(9)

Kruskal, M., **8**:28, 87

Kruzhilin, G. N., **8**:94(14), 116, 120, 121, 159; **9**:259(109), 260(109), 261, 271; **18**:88(14), 110(43), 114(43), 116(14), 158(14, 43); **25**:84(166), 147(166)

Krylov, V. I., **8**:22(106), 28(106), 57(106), 86; **12**:96(30); **19**:207(28), 227(28), 244(28)

Krylov, V. S., **26**:23(237), 104(237)

Krylovich, V. I., **8**:75, 76, 77(278), 91

Kryukov, A. P., **21**:73(16), 76(16), 134(16)

Kryukov, V. G., **25**:4(18, 19), 132(206), 140(18, 19), 150(206)

Kryukova, M. G., **6**:474(212), 478, 496

Ku, J. C., **23**:146(26), 185(26)

Kuang, Wey-Yen, **17**:69(18), 70(18), 71(18), 72(18), 153

Kuang, W.-Y., **5**:487(270), 517

Kubair, V. G., **15**:184(50), 185(50), 224(50)

Kuball, D., **12**:3(7), 14(7), 51(7), 59(7), 68(7), 69(7), 73

Kubanskii, P. N., **3**:28, 32

Kubelka, P., **22**:278(120, 121), 352(120, 121)

Kubie, J., **14**:151, 152, 242; **19**:188(148)

Kublbeck, K., **24**:289(22), 319(22)

Kubo, R., **23**:13(66), 15(66), 43(66), 109(66), 110(66), 117(66), 129(66)

Kubo, T., **30**:329(33), 428(33)

Kubota, H., **14**:232(169), 246

Kubota, T., **6**:14, 22, 33(52), 89(52), 91(52), 92(52), 123, 124; **8**:12(46), 85

Kucera, A., **19**:20(39), 21(39), 85(39)

Kuchta, J. M., **29**:71(29), 125(29)

Kudo, K. (Chapter Author), **23**:161(46), 186(46); **27**:ix(6, 7, 9, 10, 11, 13, 14, 15, 17, 18, 19), 46(6, 7), 146(28, 29), 147(28), 167(30), 174(9), 181(31, 32), 186(32), 187(36, 37, 38), 193(37, 38, 41); **31**:334(11, 12, 13, 14, 15, 16), 365(11), 388(39, 40), 391(48), 413(48), 418(48), 425(11, 12, 13, 14, 15, 16), 427(39, 40, 48)

Kudo, M., **4**:26(44), 63

Kudo, T., **29**:151(70), 165(116), 206(70), 208(116)

Kudryashev, L. I., **6**:376(8), 418(123), 463(8), 464, 466, 474(215, 216, 217), 478, 479, 480, 486, 492, 496

Kudryavtsev, A. P., **17**:22(68), 59

Kudryavtsev, E. V., **6**:376(6, 7), 414, 463, 464(6, 7), 467, 474(247), 480(247), 483(247), 486, 498; **25**:11(37), 32(37), 141(37)

Kuech, T. F., **28**:342(19), 375(141), 416(19), 421(141)

Kuehn, D. M., **6**:5(7), 7(7), 33(7), 84(7), 89(7), 122

Kuehn, T. H., **18**:18(35), 26(58, 59, 60, 61), 27(61), 28(60), 81(35), 82(58, 59, 60, 61)

Kuerten, H., **25**:364, 412

Kuerten, J. G. M., **25**:364, 419

Kuerti, G., **4**:318, 443

Kuethe, A. M., **3**:21(37), 32; **11**:229(172), 260

Kuga, O., **SUP**:102(268), 119(268, 291), 136(317)

Kuga, Y., **23**:154(36), 168(36), 185(36)

Kugath, D. A., **18**:41(84), 42(85), 44(85), 83(84, 85)

Kugushev, N. M., **1**:435(140), 444

Kuhlman, M. R., **29**:329(182), 344(182)

Kuhn, H., **6**:197(36), 364

Kuhns, P. W., **7**:200(121), 216

Kuhrt, F., **14**:290, 307, 341

Kuiken, H. K., **8**:12, 85; **9**:288, 346

Kuiper, A. E. T., **28**:402(198), 424(198)

Kukulka, D. J. (Chapter Author), **18**:325

Kukushkin, A. N., **31**:160(55), 324(55)

Kulacki, F. A., **24**:104(67), 187(67); **29**:4(2), 5(2), 7(6), 8(8, 9), 9(6, 10), 10(2, 12), 26(22), 27(9, 22), 28(22), 30(9, 22), 32(25), 33(9), 55(2, 6, 8, 9, 10, 12), 56(22, 25); **30**:32(59, 60), 33(59), 90(59, 60)

Kulakov, M. P., **28**:103(88), 142(88)

Kulhavy, J. T., **9**:3(70), 75(70), 109

Kulič, E., **26**:39(111), 98(111)

Kulicke, W.-M., **23**:222(94, 150, 151), 225(94, 150, 151), 271(94), 273(150, 151)

Kulkarni, A. K., **23**:302(109), 303(109), 304(109), 305(109), 358(109)

Kulkarni, B. D., **24**:103(16), 185(16)

Kulkarni, M. G., **18**:214(108, 121), 215(108), 226(121, 122, 123), 227(121), 229(121), 230(121), 233(108, 121), 238(108, 121, 122, 123)

Kuloor, N. R., **23**:262(11), 267(11)

Kumada, M., **13**:2(15), 10(15), 11(15), 59; **23**:293(63), 340(269), 341(269), 356(63), 367(269)

Kumada, T., **18**:169(38), 170(38), 172(38), 188(38), 189(38), 236(38); **20**:387

Kumagai, N., **27**:ix(14)

Kumagai, S., **4**:215, 216, 217, 227; **21**:125(115, 116), 138(115, 116); **23**:12(58, 59), 13(66), 14(58, 59), 15(66), 31(58, 59), 34(58, 59), 42(58, 59), 43(66), 58(58, 59), 60(58, 59), 87(58, 59), 91(58, 59), 92(59), 103(58, 59), 109(58, 59, 66), 110(58, 59, 66), 117(58, 59, 66), 129(58, 59, 66)

Kumagaya, T., **23**:58(94), 61(94), 77(94), 106(94), 131(94)

Kumar, A., **15**:250, 280

Kumar, I. J. (Chapter Author), **8**:1, 6, 29, 30(150), 31(150), 40, 42(213), 45, 46, 48, 56, 57, 61, 78(281, 282, 283), 84, 87, 89; **SUP**:171(347)

Kumar, M., **20**:150(14), 179(14)

Kumar, N. G., **14**:327(105), 328(105), 334(105), 339(105), 345

Kumar, R., **13**:219(49), 265

Kumar, R. K., **29**:66(14), 68(14), 71(33), 79(42), 80(42), 84(48), 124(14), 125(33, 42, 48), 303(163), 304(163), 307(169), 343(163), 344(169)

Kumar, S., **23**:139(18), 140(18), 153(35), 170(18), 185(18, 35), 192(152), 212(153), 213(153), 215(153), 230(153), 231(153), 236(153, 154), 238(153), 239(151a), 240(151a, 153), 242(153), 252(153, 154), 273(151a, 152, 153, 154); **25**:257, 260, 272, 274, 275, 276, 277, 278, 283, 284, 285, 289, 290, 291, 306, 310, 316, 317

Kume, S., **23**:293(63), 356(63)

Kumor, M. S., **12**:82, 112

Kumuduni, W. K. A., **28**:135(73), 142(73)

Kun, L. C., **SUP**:68(132), 374(132), 375(517)

Kunder, G. A., **12**:23(37), 74

Kundt, A., **2**:354

Kunes, J. J., **9**:3(71), 75, 109

Kung, C. T. V., **14**:337(131), 346

Kung, H. C., **10**:231, 281

Kunihiro, M., **17**:6, 57; **23**:8(14), 12(14), 14(14), 29(14), 33(14), 35(14), 36(14), 38(14), 39(14), 58(14), 60(14), 62(14), 63(14), 67(14), 68(14), 75(14), 78(14), 79(14),

85(14), 86(14), 91(14), 93(14), 106(14), 126(14)

Kunii, D., **10**:169(2, 6), 170(2), 171(19), 174(2, 19), 176(19), 179, 180(2), 181, 194, 195, 196(94), 197(94), 214, 215, 217; **14**:151(7, 23), 155(47), 157(47), 159, 242, 243, 244; **19**:99(10), 108(10, 51), 164(207), 173(207), 174(207), 175(207, 229), 176(207), 177(229), 184(10), 186(51), 190(207, 229); **23**:188(155), 233(155), 273(155); **25**:152(5), 201(5), 223(5), 225(5), 242(5)

Kunio, S., **6**:474(302), 501

Kunitomo, T., **27**:124(27)

Kuno, T., **18**:255(19), 257(19), 259(19), 262(19), 264(19), 265(19), 267(19), 268(19), 269(19), 270(19), 271(19), 272(19), 273(19), 274(19), 290(19), 313(19), 320(19)

Kunts, Kh. R., **6**:553, 564

Kunugi, M., **11**:420(206), 439

Kunz, H., **11**:424(226), 440

Kunz, H. R., **5**:355(61), 368, 369, 370, 371, 372, 374, 507; **7**:53, 85, 226(13), 236, 239, 241, 241(13), 243(13), 254(13), 274(13), 304(13), 314, 316

Kunze, H., **28**:124(2), 138(2)

Kunze, W., **13**:58, 60

Kuo, C. H., **23**:302(109), 303(109), 304(109), 305(109), 358(109)

Kuo, C. Y., **5**:240, 241, 242, 251

Kuo, H. L., **14**:30, 101

Kuo, M. C. T., **14**:97, 104

Kuo, Y. H., **8**:4, 84

Kuppusamy, T., **30**:98(34), 190(34)

Kupriyanova, A. V., **16**:215(26), 239

Kuraeva, I. V., **6**:551(84), 564

Kurata, C., **17**:8, 9, 58; **23**:58(106), 60(106), 101(106), 102(106), 131(106)

Kurayeva, I. V., **21**:34(53), 35(53), 39(65, 66), 41(66), 42(66), 52(53, 65, 66)

Kurganov, V. A., **21**:8(16, 17, 18), 11(16), 12(16, 24), 15(17), 21(17), 26(18), 27(18), 29(17, 18), 47(82), 48(82), 51(16, 17, 18, 24), 53(82)

Kuribayashi, T., **25**:81(131), 146(131)

Kurihara, H. M., **1**:230, 265; **20**:14(17), 15(17, 25), 80(17, 25); **21**:328, 345

Kuriwake, Y., **15**:101(112), 102(112), 103(112), 105(131), 139(112), 140(131)

Kurochkin, Yu. P., **14**:160(83), 161(83), 170, 177, 244

Kuroda, A., **31**:334(14, 15), 388(40), 391(48), 413(48), 418(48), 425(14, 15), 427(40, 48)

Kuroda, C., **25**:261, 269, 274, 275, 278, 318

Kurokawa, M., **19**:74(172), 76(172), 77(172), 91(172)

Kurose, T., **23**:12(48), 13(30), 14(48), 15(30), 16(30), 17(30), 23(30), 24(30), 25(30, 48), 27(30, 48), 37(48), 39(48), 58(48), 60(48), 64(48), 67(48), 77(48), 78(48), 127(30), 128(48)

Kurstedt, H. A., Jr., **17**:164, 314

Kurten, H., **28**:11(14), 71(14)

Kurtz, E. F., **4**:7(8), 53, 54(99), 62, 64; **9**:324(93), 348

Kurz, H. D., **13**:219(61), 265

Kurz, W., **28**:262(68), 263(68), 288(68), 301(68), 332(68)

Kusama, A., **27**:ix(6)

Kushnyrev, V. I., **31**:160(45), 324(45)

Kusko, A., **4**:283(49), 315

Kusnetschikowa, W. W., **13**:219(59), 265

Kussoy, M. I., **7**:206, 208(152), 218

Kusuda, H., **17**:50(159), 52, 63; **20**:55(59), 58(59, 60), 59(60), 82(59, 60); **23**:12(64), 13(64), 15(64), 35(64), 40(64), 55(77), 59(64, 77), 61(64, 77), 63(77), 69(77), 70(77), 74(77), 75(77), 77(77), 99(64), 100(64), 101(64), 106(64), 129(64), 130(77)

Kusumoto, Y., **28**:390(175), 423(175)

Kusy, R. P., **18**:228(126), 238(126)

Kutateladze, S. S., **1**:233, 239, 259, 265, 421, 443; **2**:25, 106; **3**:35(34), 36(34, 35, 36), 37(37), 45(42), 48(35, 36), 60, 62, 80(35, 61, 75), 95, 98, 99; **4**:167, 224; **5**:403, 404, 405, 406, 409, 411, 413, 415, 419, 420, 421, 422, 423, 511; **6**:506, 537, 562; **7**:10(5), 48, 50, 83, 331, 333(5), 334, 353, 355, 377; **11**:12, 13(37), 48, 73, 74, 83, 105(92), 158(92), 190, 193, 225, 259; **SUP**:224(414), 242(414); **15**:233, 240, 277; **16**:120, 141, 155, 156, 181(13), 235, 238; **17**:5, 7, 8, 10, 23(72), 24(44), 51(168), 57, 58, 59, 64, 157, 163; **18**:241(1), 249(1), 306(65), 316(65), 320(1), 323(65); **20**:250(227), 310(227); **21**:328, 329, 331, 345; **23**:82(96, 98), 83(96), 85(96), 131(96, 98); **25**:17(50), 63(107), 142(50), 145(107);

31:274(175, 176), 330(175, 176)

Kuvaki, Z., **11**:105(116), 114(116), 194

Kuwabara, S., **23**:207(156), 273(156); **25**:293, 295, 317

Kuwahara, H., **16**:232, 233; **20**:75(66), 82(66)

Kuwahara, K., **24**:306(18, 19), 307(18), 308(18), 309(18), 310(4), 311(4, 5), 318(4, 5), 319(18, 19)

Kuzay, T. M., **21**:168, 180

Kuzma-Kichta, Yu. A., **18**:264(58), 307(66), 322(58), 323(66)

Kuzmin, V. V., **25**:92(170, 171), 148(170, 171)

Kuzminov, V. A., **25**:4(4), 33(4), 43(4), 50(4), 52(4, 89, 90, 91, 92, 93), 92(4), 96(178), 99(4), 106(4), 111(4), 139(4), 144(89, 90, 91, 92, 93), 148(178); **31**:201(103, 104), 205(104), 242(127), 305(209, 210), 326(103, 104), 328(127), 332(209, 210)

Kuznetsov, E. S., **11**:350, 351, 436

Kuznetsov, E. V., **21**:17(32), 34(54), 51(32), 52(54)

Kuznetsov, I. A., **6**:418(106), 491

Kuznetsov, N. V., **8**:94, 114(39), 148, 149, 159; **18**:88(12), 109(39), 146(39), 147(39), 158(12, 39)

Kuznetsov, U. N., **6**:408, 409(66), 437, 457, 474(336, 337), 489, 495, 502

Kuznetsov, V. K., **6**:418(138), 492

Kuznetsov, Ye. S., **11**:344(60), 345, 436

Kuznetsov, Yu. N., **25**:11(42), 31(42), 33(62), 64(111), 65(111, 112), 81(163), 82(163), 141(42), 142(62), 145(111, 112), 147(163)

Kvasha, V. B., **14**:247

Kvasha, V. G., **25**:153(7), 242(7)

Kvasnyuk, S. V., **17**:327(29), 329(29), 330(29), 336(29), 340

Kvernvold, O., **14**:30(72), 31(72), 33(72), 101; **20**:330(47), 351(47)

Kvon, V. I., **6**:402, 488; **25**:81(142), 146(142)

Kwack, E. Y., **15**:63(34, 35), 64(34, 35), 79(35), 80(35), 81(34), 94(34), 95(35, 162), 97(35), 109(162), 136(34, 35), 141(162); **19**:264(28), 267(28), 295(87), 326(28), 331(28), 332(28, 129), 333(28, 129, 130), 342(130), 344(130), 350(28), 353(87), 355(129, 130); **25**:260, 280, 281, 318

Kwak, D., **25**:334, 380, 414

Kwak, J. C. T., **12**:256(45), 267(45), 279

Kwant, P. B., **13**:207(7), 264; **19**:281(73), 284(73), 294(73), 295(73), 308(73),

Kwant, P. B. (*continued*)
  312(73), 352(73)
Kwok, H. S., **28**:135(118, 119), 144(118, 119)
Kwon, T. H., **28**:150(21), 151(38), 154(57),
  162(38), 163(38), 165(38), 166(21),
  167(39), 177(29), 180(56), 183(56),
  184(40), 190(38, 57), 193(38), 198(57),
  201(39), 204(39), 205(57), 206(57),
  208(22), 211(22), 228(21, 22), 229(29, 38,
  39, 40), 230(56, 57)
Kyle, T. G., **14**:263(32), 280
Kymäläinen, D., **29**:20(15), 56(15)
Kyomen, S., **25**:81(134), 146(134)
Kyriakidas, I., **18**:36(72), 83(72)
Kyte, J. R., **11**:202, 205, 209(28), 213, 214, 253,
  278, 279, 280, 314

L

Laananen, D. H., **15**:38(60), 57(60)
Labeish, V. G., **31**:160(41), 324(41)
Labountzov, D. A., **5**:409, 411, 413, 440, 513
Labrecque, R. P., **25**:294, 317
Labuntsov, D. A., **11**:13(42), 15, 17, 48, 60(28),
  105(28), 117, 118, 190; **SUP**:79(224);
  16:235, 236; **17**:22(69), 59, 119(115), 156;
  21:46(75), 47(75), 52(75); **31**:273(174),
  274(174), 330(174)
Labuntzov, D. A., **21**:72(15), 73(16), 76(16),
  134(15, 16)
LaCava, A. I., **31**:464(13), 471(13)
Lacaze, A., **5**:412, 415, 416, 422, 512, 513;
  11:105(68, 84, 85), 109(68, 84), 111(68),
  112, 192
Lacey, P. M. C., **1**:245(87), 266, 390(58),
  392(58), 410(58), 421(58, 115), 428(58),
  441, 443; **17**:37(120), 42(136), 62
Lacey, R. E., **12**:242(52), 247(52), 279
Lacher, J. R., **2**:354
Lackey, D. L., **15**:236(41), 263, 267, 278
Laddha, G. S., **15**:250(85), 280
Ladeinde, F., **30**:329(36), 351(36), 352(36),
  428(36)
Ladenburg, R., **6**:195, 268(78), 274(78), 363,
  365
Ladiev, R. Y., **6**:474(297, 298, 299), 500
Lady, E. R., **5**:413, 441, 513
Laendle, K. W., **28**:342(20), 416(20)
Lafferanderie, J., **17**:120(117), 121(117),
  122(117), 156

LaFond, E., **18**:333(41), 335(41), 363(41)
Laganelli, A. L., **7**:340, 378
Lage, P. L. C., **26**:79(112, 113), 98(112, 113)
Lagendijk, J. J. W., **22**:87(40), 88(40), 154(40)
Lagerstrom, P. A., **4**:341, 446; **8**:7, 12, 84, 85
LaGraff, J. E., **23**:320(174), 321(191, 195, 196),
  346(191), 362(174, 191), 363(195, 196)
Lagun, I. M., **25**:61(101), 144(101)
Lagurie, C., **19**:108(59), 186(59)
Laheurte, J. P., **17**:69(31), 153
Lahey, F. J., **9**:302, 347
Lahey, R., **8**:304(56), 350
Lahey, R. T., Jr., **31**:130(14), 156(14)
Lai, C. H., **20**:348(68), 352(68); **30**:140(99),
  193(99)
Lai, F. C., **24**:163(148), 190(148)
Lai, L. S., **28**:154(41), 229(41)
Lai, W., **5**:230(157), 250
Laibowitz, R. B., **28**:103(59), 141(59)
Laidler, K., **22**:319(79), 350(79)
Laird, A. R. K., **14**:77, 103
Laity, R. W., **12**:229(18a), 251(18a), 253(55),
  259(53, 54), 260(55), 269(53, 54), 278, 279
Lakin, W. D., **SUP**:186(370)
Lakshmanan, S. M., **SUP**:197(385), 202(392),
  286(455); **19**:259(18), 262(18), 350(18)
Lakshmana Rao, N. S., **SUP**:41(47), 68(135),
  97(135), 168(135)
Lal, P., **23**:192(152), 273(152); **25**:260, 279, 291,
  317
Lalas, T., **11**:220, 224, 262
Lali, A. M., **23**:237(157), 242(157), 274(157)
Lall, P. S., **6**:474(301), 501
LaLonde, G. V., **19**:145(184), 189(184)
Lalude, O. A., **18**:16(22), 80(22)
Lam, A. C. C., **23**:196(78), 270(78)
Lam, S. H., **8**:11, 84
Lam, T. T., **11**:376(134), 381(134), 438
Lamaire, N. A., **3**:255(6), 256(6), 260(6), 300
Lamantia, F. P., **18**:234(128), 238(128)
Lamb, D. E., **7**:89, 160
Lamb, H., **1**:234, 236, 265; **5**:117, 128; **8**:96(21),
  100(21), 103(21), 159; **SUP**:285(454)
Lamb, L., **1**:284, 323(18), 326(18), 328(18), 353
Lamb, W., **10**:19(28), 36
Lambermont, J., **19**:199(50, 53), 200(50, 53),
  202(50, 53), 244(50), 245(53)
Lambert, G. A., **21**:287, 300, 344
Lambert, R. B., **5**:233, 251; **21**:153, 182
Lambert, R. H., **18**:41(84), 83(84)

Lambright, A. J., **21**:319, 320, 321, 325, 329, 331, 344

Lambrinos, G., **19**:82(260), 94(260)

Lamla, E., **6**:205, 211, 364

Lamm, O., **12**:242(56), 269(57), 279

Lamontagne, M., **28**:53(79), 73(79)

Lamp, Yu. Yu., **25**:81(154), 147(154)

Lamphrecht, R., **30**:396(161), 399(161), 434(161)

Lamvik, M., **23**:109(125), 110(125), 118(125), 132(125)

Lance, G. N., **5**:149, 247

Lancet, R. T., **4**:213, 214, 227; **9**:259(104), 271; **19**:334(134), 335(134), 355(134)

Landahl, H. D., **1**:55(5, 6), 110, 120

Landahl, M. T., **12**:77(3), 111

Landau, H. G., **1**:76, 77, 82, 121; **2**:344, 354; **8**:31, 32, 87; **19**:18(33), 24(33), 84(33)

Landau, L., **21**:154, 180, 181

Landau, L. D., **3**:191(15), 249; **30**:50(61), 90(61)

Landau, N., **25**:325, 327, 361, 414

Lande, D. P., **14**:14(34), 100

Lander, C. H., **11**:202, 204, 253

Landes, L. G., **11**:219, 222, 257

Landi, E., **28**:96(63), 141(63)

Landis, F., **5**:230(160), 251; **6**:474(253), 483(253), 498; **11**:281(15), 284(15), 315; **19**:22(43), 45(43), 85(43); **20**:186(73), 187(73), 198(73), 200(73), 302(73); **26**:123(59), 124(59), 172(59), 178(59), 211(59)

Landis, R. B., **26**:78(114), 98(114)

Landolt-Börnstein, Ç., **6**:284(83), 366; **11**:230, 260

Landram, C. S., **19**:80(237), 93(237)

Landrock, A. H., **10**:211, 218

Landrum, C. S., **23**:315(144), 360(144)

Lane, C. T., **5**:329(21), 331, 481, 506

Lane, K., **18**:214(107), 238(107)

Lang, K. W., **15**:321(127), 326

Lange-Hahn, R., **14**:119(87), 120(87), 124(87), 147

Langer, J. S., **17**:76, 77(51), 154

Langhaar, H. L., **SUP**:70(153), 77(153), 87(153), 145(153), 297(153); **15**:100(106), 139(106); **19**:268(46), 270(46), 351(46)

Langley, L. W., **23**:296(78, 79, 80), 297(79, 80), 341(79), 343(79), 357(78, 79, 80)

Langlois, W. E., **19**:79(216), 80(216), 93(216); **28**:317(169, 170, 171, 176), 336(169), 337(170, 171, 176); **30**:313(4), 315(4), 333(51), 334(51), 370(139), 371(142), 426(4), 428(51), 433(139, 142)

Langmuir, I., **2**:353, 354; **4**:14, 17, 63; **11**:202, 204, 209, 252, 266, 280, 314

Langseth, M. G., Jr., **10**:53, 54, 74(35), 82

Langston, L. S., **7**:226(13), 236, 239, 241, 243, 254, 274(13), 304(13, 155), 314, 316

Lankford, D. W., **18**:19(43), 48(43), 81(43)

Lanphear, R., **4**:109(55), 110(55), 139

Lanteri, S., **31**:407(70), 428(70)

Lantsman, F. P., **17**:319(4, 5, 6, 7, 8, 9, 11), 320(4, 5, 6, 7, 8, 9, 11), 327(4, 5, 6, 7, 8, 9, 11), 333(6, 8), 339

Lantz, E., **7**:274(101), 318

Lanzo, C. D., **5**:449, 450, 515

Lao, Y. J., **11**:56(16), 84(16), 87, 90, 91, 105(16), 113, 148(16), 189

Lapin, A., **5**:413, 451, 453, 454, 513, 515

Lapin, Yu. D., **9**:73, 95(55c), 97(72), 101(55b, 55c), 108, 109

Lapin, Yu. V., **3**:35, 98

Lapp, M. M., **5**:293(34), 322

Lapwood, E. R., **14**:18(48), 19, 22, 26, 33(48), 101; **24**:176(156), 190(156)

Lara-Urbaneja, P., **26**:66(115), 67(115), 78(115), 98(115)

Lardge, M. G. C., **14**:125, 133, 147

Lardner, T. J., **1**:64, 79(35), 81, 82, 121; **8**:86, 124(110); **19**:197(33), 198(33), 244(33)

Larimer, J. W., **19**:82(268), 95(268)

Larinoff, M. W., **12**:3(10), 14(10), 73

Larkin, B. K., **3**:204(29), 249; **9**:378(47), 415; **SUP**:101(264)

Larkin, B. S., **9**:3(73, 74), 76(73), 87, 88, 89, 109; **20**:297(325), 314(325)

Larsen, F. W., **9**:32, 33, 35(50), 40, 108, 109

Larsen, G. L., **12**:3(8), 11(8), 14(8), 73

Larsen, H., **28**:190(42), 229(42)

Larsen, M. A., **6**:474(282, 284), 500

Larsen, P. S., **5**:348, 349, 350, 351(53), 353(53), 354, 468, 469, 507, 516; **7**:178(41a), 212; **11**:364, 437; **21**:240(2), 246(17), 247(2), 256(17), 273(42), 275(2, 17), 276(42); **30**:9(13), 38(13), 39(13), 88(13)

Larsen, T. W., **6**:474(312), 501

Larson, B. C., **28**:80(64), 141(64)

Larson, D. E., **23**:322(202), 363(202)

Larson, D. W., **14**:98(197), 105

Larson, E. D., **19**:58(127), 62(141), 89(127, 141)

Larson, H. C., **1**:357(3), 374(3), 375(3), 383(46), 384, 395(47), 429, 440, 441

Larson, H. K., **6**:5(7), 7(7), 33(7), 39(74), 77, 78(74), 82(74), 84(7), 89(7), 122, 125

Larson, J. R., **7**:45, 62, 63, 85

Larson, R. E., **6**:40(75), 41(75), 65, 66, 75(75), 77(75), 125; **25**:256, 293, 317

Larson, R. G., **23**:212(158), 274(158); **25**:317

Larson, S. E., **20**:346(65, 66), 347(65, 66), 352(65, 66)

LaRue, J. C., **21**:188(17), 236(17); **23**:305(113), 359(113)

Lash, J. S., **28**:135(35), 140(35)

Lasher, L. E., **5**:310(83), 324

Lasheras, J. C., **26**:77(116), 98(116)

Lasseter, T. J., **14**:78, 83(164), 84(164), 104

Lasseux, D., **31**:1(12), 39(12), 40(12), 60(12), 102(12)

Lassiter, T. W., **6**:73(123), 128

Lathan, R. C., **26**:202(167), 216(167)

Lathrop, K. D., **23**:139(14), 140(14), 184(14)

Latko, R. J., **11**:397(158), 438

Lau, E., **6**:322, 365

Lau, K. H., **14**:30, 31(69, 70), 38(92), 39(92), 40(92), 49(116), 101, 102

Lau, K. S., **30**:278(54), 280(54), 309(54)

Lauber, T. S., **20**:186(39), 187(39), 300(39)

Laudise, R. A., **11**:318(3, 4), 412(4), 434

Laufer, G., **23**:222(159), 274(159)

Laufer, I., **3**:43, 98

Laufer, J., **7**:142(50, 51), 161, 200, 216; **8**:300, 301(46, 47), 350; **9**:126, 148, 178; **13**:82, 116; **25**:327, 379, 388, 414

Launder, B. E., **8**:288(4), 348; **13**:75, 76, 80(32), 81(22, 32, 39), 84, 91, 92, 115; **16**:248, 347, 364; **21**:185(11), 188(15, 16), 199(15), 236(11, 15, 16); **23**:327(219), 336(257), 364(219), 366(257); **24**:211, 227, 264, 270, 271; **25**:331, 351, 414; **28**:308(153), 319(153, 202), 336(153), 338(202); **30**:329(31), 427(31)

Laura, P. A., **8**:49, 89; **SUP**:64(93), 75(93)

Laurence, D., **25**:396, 409

Laurence, J. C., **7**:194(83), 214; **11**:219, 222, 226(144), 241, 242, 257, 259, 261

Laurence, R. L., **13**:219(41, 45, 48, 55), 221(55), 259, 264, 265

Lauriat, G., **18**:59(117), 63(117), 85(117)

Laurmann, J. A., **2**:354

Lauver, M. R., **3**:275(47), 301

Lauwerier, H. A., **SUP**:102(267)

Laverman, R. J., **26**:220(15), 234(15), 254(15), 269(15), 305(15), 325(15)

Lavernia, E. J., **28**:8(9), 10(12), 11(12, 13), 12(13), 13(13), 14(13), 15(13, 20, 21), 17(12, 13), 18(12, 13), 19(12, 34), 22(21), 23(12), 36(60, 61, 62), 53(60, 62), 54(61), 64(82), 69(60), 70(9), 71(12, 13, 20, 21, 34), 72(60, 61), 73(62, 82)

Laverty, W. F., **5**:102, 127, 445, 514; **11**:144(161), 146(161), 147(161), 148(161), 167, 168(161), 170(161), 173, 177, 181, 196

Lavin, M. L., **7**:171, 174, 175, 212

Law, C. K., **21**:263(32), 276(32); **26**:2(119), 36(122), 55(123), 61(117), 66(122), 77(118, 120, 240), 78(118, 174), 81(121), 82(138), 84(246), 98(117, 118, 119, 120, 121, 122, 123), 99(138), 100(174), 104(240, 246); **29**:77(38, 39, 40), 78(38), 125(38, 39, 40)

Law, H. K., **26**:77(120), 81(121), 98(120, 121)

Law, M., **10**:243, 259(65), 279(114), 281, 282, 284

Lawal, A., **19**:248(5), 281(83), 285(83), 317(109), 319(5, 109), 349(5), 353(83), 354(109); **28**:150(43), 229(43)

Lawler, J. F., **28**:102(71), 141(71)

Lawless, K. R., **28**:341(9), 415(9)

Lawley, A., **28**:1(2), 22(2), 23(41), 64(2, 41, 83), 65(2), 66(2), 67(2), 68(2), 70(2), 72(41), 73(83)

Lawrence, S. P. A., **29**:189(190), 212(190)

Lawrence, W. T., **6**:474(243), 480(243), 483(243), 498; **9**:73(17), 75, 107; **SUP**:414(536)

Lawson, D. I., **10**:221, 278(112), 281, 284

Lawson, M. L., **20**:328(34, 35), 351(34, 35)

Lawson, R., **22**:421(30), 436(30)

Lawther, K. R., **17**:30(93), 32, 60, 61

Lazareff, P., **2**:293, 354

Lazarenko, B. R., **14**:127, 132, 134, 146

Lazarids, L. J., **7**:275(115), 318

Lazaro, M., **25**:189(86), 246(86)

Lazarus, F., **1**:269, 300(2), 352

Lazgin, F. S., **20**:256(244), 311(244); **23**:13(65), 15(65), 37(65), 109(65), 110(65), 112(65), 129(65)

Leadon, B. M., **1**:317, 364; **3**:63(66), 64, 71, 75, 99

Leaf, G. K., **19**:79(203), 92(203); **24**:247, 274

Leal, L. G., **18**:163(23), 171(77), 201(23, 92), 202(23, 92), 203(23, 94, 95), 205(94), 206(92, 94, 95), 207(77), 208(77), 233(77), 236(23), 237(77), 238(92, 94, 95); **23**:219(198), 275(198)

Leam, A. J., **28**:402(195), 424(195)

Leary, B. G., **11**:241, 242, 261

Leary, P. A., **28**:116(52), 141(52)

Lebed', N. G., **SUP**:244(431)

Lebedev, P. D., **1**:124, 184; **9**:265(135), 272; **10**:174(40), 175(40), 215; **13**:120, 200

LeBlanc, M., **3**:287(85), 301

Lebofsky, L. A., **10**:7, 8, 9, 36

Lebon, G., **19**:199(50, 51, 53), 200(50, 51, 53), 202(50, 51, 53), 244(50, 51), 245(53)

LeCarpentier, G., **22**:278(149), 353(149)

Le Châtelier, H., **17**:181, 314

Lecjaks, Z., **23**:193(33), 237(166), 238(166), 239(166), 240(166), 241(166), 242(166), 243(166), 246(166), 268(33), 274(166)

Leckner, B., **12**:160, 193; **27**:187(39)

LeClair, B. P., **23**:208(160, 161), 274(160, 161); **25**:295, 296, 317; **26**:8(124), 12(124), 44(124), 98(124)

Lecomte, M., **30**:403(172), 404(172), 434(172)

Ledesma, V. L., **15**:267(102), 268(102), 269(102), 280

Ledoux, P., **3**:192(16), 249

Lee, C., **14**:124(56, 57), 146

Lee, C. F., **14**:337(129), 346

Lee, C. H., **26**:81(121), 98(121)

Lee, C.-J., **14**:124, 147

Lee, C. L., **24**:218, 271

Lee, C. O., **14**:124, 146

Lee, C. S., **13**:40, 60; **29**:259(122), 261(122), 262(122), 263(122), 264(122), 266(122), 341(122)

Lee, C. Y., **22**:220(185), 304(164), 354(164), 355(185)

Lee, D., **23**:327(218), 336(256, 257, 258), 337(258), 364(218), 366(256, 257, 258); **24**:215, 271

Lee, D. H., **17**:28, 59

Lee, D. N., **26**:45(220), 103(220)

Lee, D. T., **26**:127(74), 211(74)

Lee, D. Y., **4**:120, 140

Lee, H. H., **28**:340(6), 342(6), 345(6), 406(6), 415(6)

Lee, H. J., **26**:85(47), 95(47)

Lee, H. S., **15**:63(18), 64(18), 71(18), 135(18); **25**:282, 317

Lee, H. Y., **29**:259(127), 266(127), 269(127), 342(127)

Lee, I., **28**:118(65), 119(65), 141(65)

Lee, J., **9**:262(127), 263, 271

Lee, J. H. S. (Chapter Author), **29**:59, 64(12), 86(51), 87(54), 88(51), 89(51), 90(57), 91(57), 92(57), 96(62, 63, 64), 97(57), 98(67, 70), 99(57), 100(67, 71), 101(73), 102(73), 124(12), 125(51), 126(54, 57, 62, 63, 64, 67, 70, 71, 73), 185(178), 186(178), 211(178)

Lee, J. J., **19**:68(157), 90(157)

Lee, J. K., **14**:288(21), 339(140), 341, 346

Lee, J. M., **17**:108(95), 110(95), 111(95), 155

Lee, K., **26**:48(125), 98(125)

Lee, K. H., **15**:3(14), 55(14)

Lee, K.-J., **28**:317(169, 171), 336(169), 337(171)

Lee, K. W., **29**:329(182), 344(182)

Lee, K. Y., **21**:263(33), 276(33); **28**:351(83), 418(83)

Lee, L., **14**:139, 147; **16**:225(32), 227, 236, 239; **19**:79(212), 92(212); **26**:120(43), 165(43), 210(43)

Lee, M., **21**:280, 345

Lee, M. J., **25**:392

Lee, M. M., **4**:119(85), 140

Lee, M. S., **9**:231(64), 232(64), 233(64), 270

Lee, N., **20**:115, 132

Lee, P., **28**:397(188), 398(188), 423(188)

Lee, P. C. Y., **23**:377(50), 463(50)

Lee, P. H., **4**:120(87), 140

Lee, R., **24**:226, 227, 270

Lee, R. C., **22**:243(122), 244(122), 245(122), 246(122), 247(122, 123), 352(122, 123)

Lee, R. H. C., **5**:308(73), 309(73), 323

Lee, R. H. E., **12**:138(28), 193

Lee, R. L., **17**:117, 118, 123, 124, 156

Lee, R. Y., **29**:271(140), 273(140), 342(140)

Lee, S., **25**:361, 362, 403, 416

Lee, S. C., **23**:153(34), 185(34)

Lee, S. L. (Chapter Author), **10**:220, 221, 247, 248(43), 249, 250, 253, 254(48), 255(48, 50, 57), 256(48), 257(62), 260, 261(75), 262, 269(75), 272, 274(60), 281, 282, 283, 284; **17**:198, 213(50), 315

Lee, S. W., **28**:307(150), 336(150)

Lee, S. Y., **15**:153(14), 223(14); **25**:262, 317; **29**:47(42), 57(42)

Lee, T. G., **10**:280(118), 284

Lee, T.-L., **25**:261, 267, 268, 269, 317

Lee, T. S., **18**:21(50), 22(50), 23(50), 26(50), 81(50); **20**:187(40), 301(40); **24**:300(23), 301(23), 319(23)

Lee, T. Y., **23**:64(84), 77(84), 100(84), 130(84); **30**:173(147), 195(147)

Lee, W. C., **30**:142(102), 147(102), 172(144), 193(102), 195(144)

Lee, W. K., **25**:168(66), 171(66), 172(67), 173(66, 67, 68), 174(68, 69), 175(66, 68, 69), 176(66), 189(66), 196(66), 204(67, 68, 69), 228(67, 68, 69), 232(67, 68, 69), 245(66, 67, 68, 69)

Lee, W. Y., **18**:169(29), 170(29), 172(29), 223(29), 225(29), 236(29)

Lee, Y., **9**:3(76), 80, 82, 83, 109; **18**:49(93), 84(93); **31**:462(31), 472(31)

Lee, Y. D., **18**:228(127), 238(127)

Leefer, B. I., **7**:248(64), 275, 316

Lees, C. H., **18**:171(73), 237(73)

Lees, L., **2**:46, 47, 48, 49, 107, 111(1), 139, 148(1, 23), 149(1), 150(1, 23), 154(1), 157(1), 158, 159(23), 161(23), 201, 202(1, 23), 268; **3**:34(3), 36, 97; **4**:334, 335, 347, 445; **6**:5, 29(38), 33(52), 36, 85(6), 89(52), 91(52), 92(52), 108, 109, 110, 112, 113(172, 174), 119, 122, 124, 130, 131; **7**:171, 171(23), 172, 173, 178, 178(30), 180, 211, 212, 213

LeFevre, E. J., **4**:7, 18, 56, 62, 63, 64; **9**:210(35, 36), 232(36), 240, 244(36), 265(36), 269, 288, 346; **11**:213, 214, 255; **21**:79(27), 82(35, 36), 92(67), 93(67), 95(35, 36, 67), 96(67), 97(67), 134(27), 135(35, 36), 136(67)

Lefevre, M. R., **12**:23(31), 74

LeFoll, J., **15**:27(53), 56(53)

LeFur, B., **14**:34, 43, 71, 72(84), 76(84), 102

Leger, L., **28**:41(76), 73(76)

Leger, L. G., **11**:425(228), 426(228), 440

Le Goff, P., **23**:338(263), 366(263)

Legras, J., **8**:4, 84

leGrivès, E., **9**:3(42, 44), 71, 79, 80, 91(44), 96, 97(42, 44), 103, 108

Legros, J. C., **28**:370(133, 134), 421(133, 134)

Lehmann, G. L., **20**:227(147, 148), 228(147, 148), 229(147, 148), 230(147, 148), 237(147, 148), 306(147, 148)

Lehner, F. K., **23**:212(162), 274(162)

Lehner, J. R., **29**:233(63), 319(63), 338(63)

Lehnert, B., **1**:288, 293, 353

Lehninger, A. L., **4**:141

Lehongre, S., **17**:323, 339

Lehrer, J., **12**:79(7), 111

Lei, D. H., **23**:12(49), 14(49), 25(49), 128(49); **26**:123(60), 125(60), 127(60), 129(60), 130(60), 133(60), 211(60)

Leibo, S. P., **22**:353(141)

Leidenfrost, J. G., **5**:56, 90, 125

Leidenfrost, W., **1**:287, 353; **3**:257(10, 12), 260, 300; **14**:118(83, 84), 119, 124(83, 84), 147; **15**:3(13, 14), 55(13, 14); **19**:45(106), 46(106, 108), 47(106), 54(108), 55(106, 108), 56(106, 108), 57(106, 108), 68(158), 69(158), 88(106, 108), 90(158); **25**:51(85), 143(85)

Leiderer, P., **28**:96(12), 138(12)

Leigh, D., **4**:331, 444

Leighton, R. I., **25**:359, 415

Leiner, W. P., **12**:258(58), 279

Leinweber, G., **21**:300, 344

Leister, H.-J., **30**:349(95), 350(95), 382(150), 431(95), 433(150)

Leith, C. E., **25**:361, 414

Leith, E. N., **6**:200, 364

Lekach, S. V., **31**:112(34), 119(34), 121(34), 128(34), 137(34), 153(34), 157(34)

Lekhstner, E. N., **11**:423(221), 440

Lekvenshvili, N. N., **31**:304(207), 332(207)

Leland, J. L., **26**:123(65), 125(65), 211(65)

Lelchuk, V. L., **6**:539, 540(62), 542(62), 543(62), 563

Lelchyk, V. L., **25**:28(57a), 142(57a)

Lele, P. P., **22**:342(124), 352(124)

Leleev, N. S., **25**:4(34), 141(34)

Lemcoff, N. O., **23**:370(11), 461(11); **24**:102(15), 185(15)

Lemicus, G. P., **11**:105(136), 195

Lemieux, G. P., **5**:494, 495, 496, 497, 517

Lemlich, R., **4**:26, 58, 63; **10**:200, 218; **11**:203, 205, 206, 209, 253, 254; **SUP**:277(450), **30**:211(32), 242(32), 251(32)

Lemmon, H. E., **SUP**:73(174), 89(174), 93(174), 97(174), 98(174)

Lemons, D. E., **22**:14(26, 37), 18(26, 37), 19(34, 56, 60, 70), 24(41, 60), 26(60), 45(34), 69(56), 71(34, 56, 60, 70), 96(34, 56, 60, 70), 103(34, 60), 104(60), 105(60), 106(60), 107(41, 60), 108(34, 60), 109(34,

110(34, 60), 111(34), 113(34), 114(34), 116(34), 117(34), 118(34), 119(34), 120(34), 121(34), 122(41, 60), 123(34, 41, 60), 124(34, 60), 125(34, 60), 126(34, 60), 127(70), 129(34, 60), 130(34, 60), 131(34, 60), 132(34, 41, 60), 133(34), 134(34, 60), 138(70), 139(60), 141(41, 70), 142(70), 144(70), 145(70), 146(70), 151(56), 152(41), 154(34, 41), 155(56, 60, 70)

Lenard, M., **7**:197, 214, 215

Lenhard, R. J., **30**:98(33, 34), 190(33, 34)

Lennard-Jones, J. E., **2**:354

Lensesdey, A. G., **26**:252(65), 327(65)

Lenz, T. G., **18**:44(87), 45(87), 83(87)

Leonard, A., **3**:246(100), 251; **25**:333, 337, 414

Leonard, A. C., **5**:494, 495, 496, 497, 517; **11**:105(136, 140), 195; **17**:117(107), 119(112), 155, 156

Leonard, B. P., **24**:226, 239, 271; **31**:138(15), 156(15)

Leonard, E. F., **21**:328, 344

Leonard, R. A., **SUP**:277(450)

Leonardi, E., **21**:161, 167, 181

Leonberger, F. J., **28**:342(32), 416(32)

Leonhard, K. E., **5**:451, 453, 515; **11**:170(171), 196

Leonhardt, H., **7**:320c, 320f

Leonidopoulos, D. J., **16**:117(82), 120, 122, 125, 155

Leonova, G. M., **11**:168(168), 170(168, 169, 170), 196

Leont'ev, A. I. (Chapter Author), **2**:25, 106; **3**:33, 36(35, 36), 37(37), 48(35, 36), 60, 62, 76(71a), 80(35, 61, 75), 95(35, 36), 98, 99; **6**:379, 380(11), 486, 506, 537, 554, 562, 564; **7**:331, 333(5), 334, 353, 355, 377; **11**:12, 13, 48; **17**:10, 58; **25**:28(58), 142(58); **31**:265(163), 329(163)

Leontiev, A. S., **7**:48, 50, 84

Leontovich, A. M., **11**:423(220, 224), 440

Leopold, M. J., **5**:365, 508

Leovy, C. B., **14**:253(9), 266(41), 275(9), 279, 280

Lepere, J., **20**:244(206), 309(206)

Lepper, R., **4**:147(4), 222

Leppert, G. (Chapter Author), **1**:185, 189(4), 190(4), 240, 245, 246(90), 249(90), 254(90), 256(91), 257(90), 258(90), 259(91), 260(91), 261(91), 262(91), 263, 266; **3**:16, 26, 32; **5**:107, 127; **6**:539, 542(57), 543(57), 563; **7**:145(55), 161;

8:104, 105(30), 159; **9**:256(87), 270; **11**:219, 223, 228, 237, 257; **SUP**:414(535); **16**:128, 130, 155; **17**:2(15), 8, 10, 24(42), 56, 58; **18**:98(30), 99(30), 128(30), 158(30)

Lepple, F. K., **18**:333(43), 336(43), 363(43)

LeQuere, P., **18**:38(77), 83(77); **24**:310(24), 319(24)

Lerner, Y., **26**:47(126, 127), 98(126, 127)

Leroy, N. R., **5**:92, 93, 94, 95, 127; **11**:134(147), 138(147), 139(147), 140(147), 141(147), 142(147), 195

Lesage, J., **1**:245(89), 266, 361(13), 363(13), 378(13), 381(13, 39, 40, 41), 411(13, 39, 40, 41), 413(13, 39, 40, 41), 421(13, 40, 41), 436(13), 440, 441

Leschziner, M. A., **24**:199, 227, 262, 263, 271

Lese, H. K., **10**:210, 218; **14**:183, 184, 245

Lesieur, M., **25**:333, 344, 345, 346, 357, 361, 364, 378, 392, 401, 403, 406, 409, 414, 415, 416, 417

Leslie, D. C., **25**:333, 355, 361, 382, 383, 394, 411, 412, 413, 414, 418

Leslie, F. M., **9**:20, 21(78), 26, 27, 34, 39, 71(78), 109

Lessen, M., **21**:154, 155, 178

Lessman, R. C., **23**:333(246), 365(246)

Letan, R., **26**:19(128), 47(126, 127), 98(126, 127, 128)

Le Tournean, B. W., **1**:405(85), 409(85), 411(85), 442

Letourneau, B. W., **1**:245(86), 266

Leuenberger, H., **2**:411, 450

Leung, A., **17**:43(145), 63

Leung, E. Y., **3**:130, 138, 139, 171(37), 174; **21**:180

Leung, J. C., **29**:189(192), 212(192)

Leva, M., **10**:169(8), 181(65, 66), 189, 202, 203, 205, 206, 215, 216, 217; **14**:151(9, 45), 187, 242, 243, 245; **19**:102(46), 103(46), 143(46), 185(46); **25**:184(78), 246(78)

LeVan, M. D., **26**:11(89, 90), 12(129), 13(129), 97(89, 90), 98(129)

Levchenko, N. M., **16**:235; **17**:140(154), 146(168), 157

Levenson, L. L., **12**:129(14), 138(14), 192

Levenspiel, O., **10**:169(2), 170(2), 171(19), 172(25), 174(2, 19, 25), 175(25), 176(19), 177(25), 179, 180(2), 181(68), 182, 194, 195, 196(94), 197(94), 199, 203, 206, 214, 215, 217; **14**:151(7, 10, 23), 242; **19**:99(10),

Levenspiel, O. (*continued*)
  100(30), 108(10), 136(30), 137(30),
  139(30), 144(153), 145(30), 184(10),
  185(30), 188(153); **21**:334, 345;
  23:188(155), 233(155), 273(155);
  25:152(5), 201(5), 223(5), 225(5), 234(132,
  133, 134, 135, 136, 137), 235(132, 133,
  134, 135, 137), 242(5), 248(132, 133, 134,
  135), 249(136, 137)
Leventhal, E. L., **2**:434(34), 451
Leveque, M. A., **2**:366, 395
Lévêque, M. A., **SUP**:75(199), 105(199),
  106(199), 154(199), 171(199), 172(199),
  401(199); **15**:100(104), 139(104)
Leverett, M. C., **29**:37(26), 56(26)
Levi, C. G., **28**:15(24), 18(24), 71(24)
Levich, V. G., **20**:107, 131; **26**:12(130), 21(130),
  23(237), 43(130), 98(130), 104(237)
Levin, E. S., **31**:160(42), 279(182), 283(189),
  284(191), 324(42), 330(182), 331(189, 191)
Levin, R. L., **22**:26(18), 30(15, 16, 42), 71(18,
  19), 127(17), 134(16), 151(19), 152(19),
  153(15, 16, 17, 18, 19), 154(42), 291(126),
  352(125, 126)
Levin, V. S., **SUP**:70(147), 87(147)
Levin, Y. A., **6**:418(120), 492
Levinsky, E. S., **7**:197, 215
Levitan, L. L., **17**:319(4, 5, 6, 7, 8, 9, 11), 320(4,
  5, 6, 7, 8, 9, 11), 327(4, 5, 6, 7, 8, 9, 11),
  333(6, 8), 339
Levitch, B., **12**:233(59), 251(59), 270(59), 280
Levitch, V. G., **12**:88, 101, 112, 229(60),
  233(60), 239(60), 240(60), 241, 251(60),
  267(60), 270(60), 280
Levitin, P. S., **25**:11(41), 32(41), 141(41)
Levy, A., **6**:42(90), 126
Levy, A. M., **5**:390, 391(119), 392, 393, 394,
  395, 510
Levy, E. K., **7**:304, 305, 319; **20**:187(77),
  202(77), 302(77); **25**:153(6), 242(6);
  30:272(38), 308(38)
Levy, G. L., **23**:204(163), 205(163), 274(163)
Levy, R. H., **1**:269(5), 352
Levy, S., **1**:393, 428(130), 442, 443; **5**:100, 101,
  102, 127, 409, 410, 411, 413, 513; **8**:22, 86;
  11:170(178), 173(178), 175(178),
  177(178), 197; **15**:325; **17**:42, 43(144),
  44(144), 63; **19**:334(135), 335(135),
  338(135), 355(135); **26**:121(44), 210(44);
  29:259(128), 342(128)

Levy, S. E., **4**:118(77), 140
Lew, H. S., **SUP**:95(253), 96(253)
Lewis, B., **10**:247, 282; **29**:83(45), 125(45)
Lewis, D. J., **5**:110, 112, 121, 128
Lewis, E. E., **31**:334(7), 346(7), 357(7), 365(7),
  425(7)
Lewis, E. W., **4**:170, 171, 172, 173, 174, 224;
  5:375(88), 376, 416, 419, 423, 425, 426,
  427, 432, 433, 434, 435, 437, 508; **9**:256,
  270; **11**:105, 106, 107(93), 119(93),
  121(93), 193
Lewis, J. A., **8**:164, 166, 169, 170, 226
Lewis, J. B., **10**:177(59), 178(59), 216
Lewis, J. E., **6**:33, 89, 91(52), 92(52), 124, 129
Lewis, J. H., **7**:204, 217
Lewis, J. P., **5**:172(75, 76), 248, 449, 515;
  11:170(184), 173, 174(184), 177(184),
  185(184), 197, 220, 224, 258
Lewis, J. W., **21**:151, 181
Lewis, R. E., **14**:49(114), 102
Lewis, R. O., **30**:222(30), 251(30)
Lewis, W. K., **13**:120, 200
Lewkowicz, A. K., **15**:92(91), 138(91)
Lezak, D., **17**:151(178), 157
Lezec, H., **28**:343(49), 417(49)
Li, B. Q., **28**:318(185, 195), 337(185), 338(195)
Li, B.-X., **31**:334(15), 425(15)
Li, C. H., **21**:160, 181
Li, C. Y., **30**:53(14), 61(14), 88(14)
Li, H.-W., **18**:24(53), 81(53)
Li, J., **28**:103(87), 142(87)
Li, J.-Y., **30**:264(27), 265(27), 273(27),
  299(100), 300(100), 308(27), 311(100)
Li, K. W., **11**:203, 206, 263
Li, Q., **28**:245(19), 253(59), 255(19, 59), 305(19,
  59), 330(19), 331(59)
Li, Q. B., **30**:257(9), 307(9)
Li, T. M. C., **14**:54(124), 103
Li, T. Y., **4**:330, 347, 444, 446; **7**:204, 217; **8**:19,
  86
Li, X., **26**:74(183), 79(181), 101(181, 183)
Li, Y., **28**:410(227), 425(227)
Liang, S. F., **15**:212(108), 215(108), 225(108)
Liang, S.-J., **24**:269
Liang Wu, J., **15**:250(87), 280
Liao, N. S., **30**:83(62), 90(62)
Libby, P. A., **2**:242(66), 265, 269, 270;
  6:108(166), 130; **8**:67, 68, 90
Libby, R. A., **3**:35(17), 97
Libchaber, A., **17**:133, 156

Libet, B., **22**:20(7), 60(7), 152(7)

Librizzi, J., **7**:331, 333, 334, 341, 353, 377

Licht, S., **4**:141; **22**:352(127)

Licht, W., **21**:337, 339, 345

Lick, W., **3**:210(45), 250; **11**:404(169), 408(176), 438, 439

Liebenberg, D. H., **5**:351(55), 389(55), 507

Lieberman, J., **20**:203(95), 303(95)

Liebermann, R. W., **4**:241(7), 242(7), 283(49), 313, 315

Liebert, C. H., **23**:298(83), 310(129, 130, 131), 311(130, 131), 342(280, 281), 357(83), 359(129), 360(130, 131), 367(280, 281)

Lieblein, S., **2**:434(33), 451; **7**:274(102), 318; **9**:118, 177; **30**:236(43), 251(43)

Liedenfrost, W., **11**:397(158), 438

Lien, F.-S., **24**:262, 263, 271

Lienhard, J. H., **4**:167, 171, 199, 224, 225; **5**:113, 114, 115, 116(74), 117(74), 128, 403, 407, 419(143), 421, 425, 436, 439, 511; **9**:357(13), 358(13), 361(13), 384(13), 414; **11**:57, 58(22), 105(52, 53, 119), 109, 114, 115, 116, 190, 191, 194, 308, 315; **16**:130, 139(110), 155, 156, 228, 239; **17**:6, 7, 8(40), 9(46), 10, 12, 14(56), 22, 57, 58, 59

Lienhard, J. H., IV, **18**:67(127), 85(127), 249(14), 320(14); **20**:250(230, 231), 310(230, 231); **21**:328, 345; **23**:9(26), 74(88, 89, 90, 91), 75(91), 79(91), 93(89), 94(88, 90, 91), 95(89), 127(26), 130(88), 131(89, 90, 91); **29**:50(46), 57(46)

Lienhard, J. H., V, **18**:67(127), 85(127); **26**:106(10), 109(22, 23), 110(23), 111(23), 113(22), 114(22, 32), 115(22, 32), 116(22, 32), 117(32, 36), 118(22, 23, 32), 137(89), 139(23), 141(23), 144(22, 32), 149(22, 32, 36, 109, 110), 151(36, 110), 152(22, 23), 153(22, 23), 155(36, 114, 115), 156(22), 157(22), 158(22), 159(22), 193(114, 115), 194(114), 195(36, 114, 155), 197(36), 198(36), 201(163), 202(163), 203(163), 208(10), 209(22, 23, 32), 210(36), 212(89), 213(109, 110, 114, 115), 215(155), 216(163)

Licpmann, H. W., **3**:53, 98; **8**:36, 88; **13**:82, 116

Lieu, B. H., **7**:368, 379

Lieu, S. W., **8**:36(207), 42(207), 89

Liew, K. S., **25**:260, 264, 317

Liew, T. L., **11**:13(32), 48

Lifschitz, N., **25**:325, 327, 361, 414

Lifshitz, E. M., **3**:191(15), 192(15), 249; **21**:154, 181

Lightfoot, E. N., **2**:358(7), 368(7), 394; **3**:192(11), 249; **7**:108, 109, 160; **9**:304(64), 347; **11**:269(6a), 314; **12**:197(9), 198(9), 206(9), 207(9), 208(9, 61), 213(9, 61), 219(9), 232(9), 237(61), 242(61), 244(9), 251(61, 61a), 261(61), 265(61), 266(61), 268(61), 269(61), 271(61), 277, 280; **13**:7(27), 60, 126(33), 128(33), 131(33), 169(33), 201, 208(8), 209(8), 214(8), 264; **14**:28, 101, 194(124), 206(139), 245, 246; **SUP**:54(62); **15**:13(37), 56(37), 98(98), 100(98), 138(98); **16**:147(126), 156; **18**:216(110), 238(110); **20**:323(20), 350(20); **21**:240(3), 275(3); **23**:189(21), 195(21), 197(21), 267(21); **28**:267(88), 333(88), 353(93), 355(93), 356(93), 358(93), 419(93); **29**:251(108), 293(108), 296(108), 340(108); **30**:364(130), 432(130); **31**:9(2), 102(2)

Lighthill, M. J., **1**:37, 50; **2**:133, 172, 268, 269; **7**:146, 161; **8**:4, 22, 35, 36, 40, 42, 84, 86, 162(6), 226; **9**:9, 10, 11, 12, 15, 18, 19, 20, 22, 23, 27, 31, 36, 41, 42, 43, 46, 49(80), 99, 109

Ligrani, P. M., **16**:302, 310, 364; **23**:327(221), 364(221)

Lii, C. C., **11**:397(160), 401(160), 408, 438

Liijegren, J., **19**:125(125), 187(125)

Liiv, U. R., **25**:76(125), 81(146, 154), 83(125), 145(125), 147(146, 154)

Likov, V. V., **17**:325(27), 340

Liles, D. R., **31**:106(16), 138(16), 156(16)

Liley, P. E., **3**:254(2), 300; **11**:230(176), 260

Lilley, G. M., **8**:36, 88

Lilly, D. K., **25**:332, 334, 340, 341, 343, 353, 370, 414

Lim, C. J., **23**:340(270, 271), 367(270, 271)

Limpert, G. J. C., **30**:243(33), 251(33)

Lin, C. C., **9**:126(31), 178; **12**:88(24), 112

Lin, C. H., **14**:268(52), 280

Lin, C. K., **31**:172(78), 174(78), 325(78)

Lin, C. S., **2**:386, 395; **7**:88, 91, 98, 103, 108, 110, 160; **12**:233(62), 251(62), 270(62), 280; **15**:128(151), 141(151)

Lin, C. Y., **5**:230(160), 251

Lin, D. Y. T., **17**:19, 58

Lin, H.-T., **25**:259, 290, 317

Lin, J., **14**:306, 309, 342

Lin, J. S., **19**:125(126), 187(126)
Lin, N. N., **18**:69(133), 70(133), 86(133)
Lin, P., **28**:211(44), 229(44)
Lin, S., **19**:80(239), 82(259), 93(239), 94(259)
Lin, S. C., **1**:284, 323(18), 326(18), 328(18),
    353; **2**:131(17), 134(17), 268
Lin, S. H., **2**:379(103), 396, 427(30), 430(30),
    451; **4**:98(41, 42), 100(41, 43), 139; **7**:185,
    186, 213; **9**:265(140), 272; **13**:219(80),
    265; **SUP**:71(162), 72(162, 163), 77(162),
    87(162), 139(162), 145(162), 163(162),
    176(353), 180(353), 287(163), 288(163),
    298(163), 299(163), 336(163); **15**:232(20),
    277; **19**:268(48), 271(48), 281(77),
    285(77), 351(48), 352(77); **22**:256(128),
    352(128)
Lin, T., **19**:281(79, 80), 285(79, 80), 286(80),
    287(80), 300(79), 302(79), 352(79, 80)
Lin, T. F., **25**:51(85), 143(85)
Lin, Y. S., **28**:343(52, 54, 56, 57), 417(52, 54,
    56, 57)
Linan, A., **2**:247, 263, 270; **10**:243(37), 282
Liñán, A., **26**:61(131), 98(131)
Lindauer, G. C., **8**:56, 90
Lindberg, K., **17**:29(86), 60
Lindemoth, T. E., **15**:236(44), 272(44), 278
Linder, B., **3**:270(29), 300
Lindin, V. M., **10**:177, 216
Lindland, K. P., **29**:167(125), 208(125)
Lindley, R. A., **28**:135(35, 66), 140(35), 141(66)
Lindt, J. T., **28**:150(12, 45, 46), 228(12), 229(45,
    46)
Lindzen, R. S., **14**:269(57, 61), 273(57), 280
Linehan, J. H., **9**:258(97), 270; **19**:79(202),
    92(202); **21**:287, 300, 344
Linetskiy, V. N., **9**:256(92), 270
Ling, A. T., **21**:66(8), 134(8)
Ling, C. H., **10**:255, 282
Ling, Y., **23**:229(258), 278(258)
Lingineni, S., **30**:175(152), 176(152), 196(152)
Linhoff, B., **31**:431(73), 474(73)
Linke, W., **4**:54, 64; **11**:202, 204, 213, 214, 253;
    20:14(16), 15(16), 24(48), 80(16), 81(48)
Linnet, C., **17**:108(97), 111, 112, 155
Linsky, J. L., **10**:41, 47, 57, 71, 73(7), 76(73),
    78(73), 81, 83
Linsman, V. S., **31**:160(11), 322(11)
Linthorst, S. J. M., **18**:12(15), 13(15), 79(15)
Linton, W. H., Jr., **15**:128(150), 141(150)
Lion, K. S., **23**:288(52), 289(52), 290(52),
    356(52)
Lior, N., **18**:12(18), 59(112), 60(112), 61(112),
    63(112), 79(18), 85(112)
Li-Orlov, V. K., **6**:418(137), 492
Liou, K.-N., **23**:147(31), 185(31)
Liou, M.-S., **24**:218, 272
Liparulo, N. J., **29**:71(31), 125(31)
Lipinski, R. J., **29**:39(34), 40(34), 41(34),
    44(34), 48(34), 57(34), 324(176), 325(176),
    326(176), 327(176), 344(176); **30**:115(67),
    118(67), 119(67), 191(67)
Lipkin, M., **4**:119(80), 140
Lipkis, R. P., **SUP**:101(263)
Lipovski, G. J., **22**:411(24), 435(24)
Lippert, T. E., **20**:10(13), 80(13)
Lippmann, M. J., **14**:3(8), 16, 51, 100, 101, 102;
    24:103(23, 24), 185(23, 24); **30**:140(97),
    141(97), 193(97)
Lipsett, S. G., **10**:107(33), 164; **29**:130(5),
    203(5)
Lipska, A. E., **10**:229(14), 281
Lipton, J. M., **22**:360(3), 435(3)
Lira, I. H., **30**:303(106, 107), 304(107),
    305(107), 311(106, 107)
Liron, N., **SUP**:73(187), 92(187), 93(187),
    94(187), 95(187), 98(187)
Lis, J., **11**:281, 315
Lisovenko, A. T., **15**:321
List, E. J., **16**:9, 11, 56; **17**:207, 315
List, R., **9**:222(52), 225(52), 269
Litkoubi, B., **30**:339(81), 430(81)
Litsek, P. A., **24**:18(48), 20(48), 22(48), 26(55),
    27(55), 28(55), 29(55), 30(55), 38(48, 55)
Little, Arthur D., Inc., **5**:471, 472, 473, 516
Little, J. B., **30**:313(9), 426(9)
Little, N. C., **1**:288, 293(23), 353
Little, W. A., **5**:487(271), 517; **17**:68, 71(14),
    72(14), 153
Littles, J. W., **16**:233; **20**:24(52), 81(52)
Littrell, J. J., **30**:222(10), 223(10), 250(10)
Litvin, M., **31**:408(72), 428(72)
Liu, C., **11**:281, 284, 315; **29**:151(64), 205(64)
Liu, C. L., **19**:310(100), 313(100), 353(100)
Liu, C. Y., **2**:434(36), 451; **3**:257(10), 300;
    8:33(157), 87
Liu, D. D. S., **29**:69(25), 76(35), 124(25),
    125(35)
Liu, H., **28**:36(60, 61, 62), 53(60, 62), 54(61),
    69(60), 72(60, 61), 73(62)
Liu, J., **SUP**:73(182), 88(182), 89(182), 98(182),

163(182), 164(182), 167(182), 168(182), 287(182), 288(182), 298(182), 299(182), 301(182), 398(182); **24**:219, 274; **31**:391(46, 47), 427(46, 47)

Liu, N.-S., **30**:368(137), 433(137)

Liu, S. S., **21**:178

Liu, S. W., **2**:172(45, 46), 175, 176(45), 179(45), 247, 269, 270

Liu, T. C., **30**:287(74), 288(74), 310(74)

Liu, V. C., **9**:24, 109

Liu, X., **26**:109(22, 23), 110(23), 111(23), 113(22), 114(22, 32), 115(22, 32), 116(22, 32), 117(32, 36), 118(22, 23, 32), 139(23), 141(23), 144(22, 32), 149(22, 32, 36, 109, 110), 151(36, 110), 152(22, 23), 153(22, 23), 155(36), 156(22), 157(22), 158(22), 159(22), 195(36), 197(36), 198(36), 201(163), 202(163), 203(163), 209(22, 23, 32), 210(36), 213(109, 110), 216(163)

Liu, X. Q., **26**:77(240), 104(240)

Liu, Y., **28**:376(146), 421(146)

Liu, Y. A., **25**:153(12), 195(12, 96), 201(12), 232(12), 236(12), 243(12), 247(96)

Livingood, J. N. B., **1**:33(28), 35(28), 36, 37(28), 50; **5**:447, 514; **7**:322, 377; **9**:11, 97, 108; **11**:170(183), 173(183), 174(183), 177(183), 185(183), 197; **19**:334(133), 335(133), 355(133); **26**:106(8), 127(74), 136(8), 208(8), 211(74)

Livshits, M. N., **31**:282(187), 331(187)

Liyv, U. R., **6**:407, 408(62), 409(62, 63, 64, 65), 410(65), 489

Ljunggren, S., **12**:242(63), 280

Lloyd, A. J. P., **11**:17(57), 19(57), 49; **17**:4, 57

Lloyd, B. A., **15**:287(97), 325

Lloyd, J. R., **9**:345(107), 348; **11**:302, 315; **18**:69(131, 132), 70(131, 132), 86(131, 132); **19**:334(142), 336(142), 340(142), 355(142); **24**:303(26), 319(26); **30**:280(62), 310(62)

Lo, R. K., **1**:260, 261, 266

Lobb, R. K., **2**:16(18), 43, 105; **3**:53, 61, 98

Lobner, P., **29**:324(174), 344(174)

Lobov, I. V., **SUP**:244(431)

Lochiel, A. C., **20**:372, 373, 374, 381, 386, 388; **25**:271, 317

Lock, C. N., **11**:216, 228, 229(154), 259

Lock, G. S. H., **6**:474(250), 483(250), 498; **8**:19, 86; **9**:3, 5, 14, 29, 42, 44, 45, 54, 58, 59, 60, 102, 106, 109; **11**:302, 315; **SUP**:120(295);

18:19(40), 48(40), 81(40); **19**:77(181), 91(181)

Lockett, J. F., **25**:361, 382, 383, 389, 413, 414

Lockhart, R. W., **1**:370, 382, 441; **5**:451, 453, 454, 515

Lockwood, F. C., **9**:21(83), 24, 25, 26(90), 36, 52, 70, 99(43a), 100(43a), 108, 109

Lockwood, J. D., **15**:285, 325

Lockwood, R. W., **30**:76(63), 90(63)

Lodge, A. S., **13**:211, 212(26), 213(26), 264; **15**:79(67), 137(67)

Loeb, L. B., **1**:271(7), 272(7), 353; **2**:354

Loeb, S., **12**:242(52), 247(52), 279

Loeffler, A. L., Jr., **SUP**:65(103, 104), 355(103), 356(103), 357(103), 363(104), 364(104)

Loeffler, E. J., **15**:96(96), 97(96), 138(96)

Loehrke, R. I., **14**:140, 141, 146; **18**:66(121), 85(121); **20**:205(100), 210(100), 211(100), 212(100), 303(100)

Lof, G. O. G., **18**:19(45), 81(45)

Lofgran, L., **20**:259(250), 311(250)

Loftus, J. J., **10**:280(117, 118), 284

Logan, B. F., **30**:288(76), 310(76)

Logan, L. M., **10**:11, 32, 36, 37

Logvinenko, D. D., **25**:166(54), 245(54)

Logwinuk, A. K., **10**:201, 202, 203, 206, 209, 217; **14**:214, 218(148), 223(148), 226(148), 227, 246; **19**:133(138), 188(138)

Lohe, H., **13**:2(13), 43, 59

Lohe, P., **13**:211(18), 264; **18**:219(113), 220(115), 238(113, 115)

Lohneiss, W. H., **10**:253(49), 282

Lohrisch, W., **11**:218, 221, 256

Lohrisch, W. L., **8**:94(7), 158; **18**:87(7), 158(7)

Loitsyanskiy, L. G., **6**:511(22), 562; **21**:198(25), 237(25)

Loiziansky, L., **11**:218(90), 221(90), 236(90), 256

Lokai, V. I., **31**:160(48, 61), 324(48, 61)

Lokshin, V. A., **6**:558(96), 564; **7**:31(16), 33, 84; **8**:94(12), 159; **11**:241, 242, 260; **18**:88(12), 158(12); **30**:242(34), 251(34)

Lomax, H., **2**:430(31), 451; **25**:331, 414

Lombara, J. S., **26**:109(22), 113(22), 114(22), 115(22), 116(22), 118(22), 144(22), 149(22), 152(22), 153(22), 156(22), 157(22), 158(22), 159(22), 209(22)

Lombard, C. K., **24**:207, 274

Lombardi, C., **1**:361(13), 363(13), 378(13), 381(13, 40, 41), 413(13, 40, 41), 419(111), 421(13, 40, 41), 427(129), 434, 436(13),

Lombardi, C. (*continued*)
    440, 441, 441(13, 40, 40, 99), 442, 443;
    17:28(81), 59
Lombardi, G., **SUP**:195(382); **19**:309(93),
    312(93), 353(93)
Lommel, E., **6**:228, 365
Lommers, L., **29**:171(137), 209(137)
London, A. L. (Chapter Author), **5**:131(13), 244,
    247, 356, 508; **8**:94, 148, 157(18), 159;
    12:40(48), 44(48), 45, 49, 50, 51(61),
    52(48), 54(48), 61, 74, 75; **14**:97(187, 188,
    189), 104; **SUP**:2(6, 13), 3(13), 11(17),
    16(24), 29(13, 24), 31(6), 38(6), 44(6),
    46(54), 53(6), 57(54), 59(6), 89(13),
    144(13), 199(13), 200(13), 203(13),
    205(13), 224(6), 225(6), 270(13), 272(13),
    273(13), 286(13), 290(13), 291(13),
    293(13), 295(13), 350(13), 357(13), 366(6),
    382(13), 405(522, 523), 407(6), 409(6),
    415(6), 416(17), 418(54), **15**:10(30),
    23(48), 24(48), 27(30), 29(48), 32(48),
    55(30), 56(48); **17**:246, 261, 316;
    18:88(18), 146(18), 156(18), 158(18);
    19:248(1), 261(1), 268(1), 277(1), 278(1),
    279(1), 280(1), 281(1), 282(1), 289(1),
    291(1), 295(1), 297(1), 298(1), 300(1),
    306(1), 312(1), 349(1); **20**:133(3), 136(3),
    145(9), 155(3), 179(3, 9), 244(179),
    308(179); **26**:232(36), 237(36), 244(36),
    248(36), 256(36), 261(36), 263(36),
    279(36), 291(36), 305(36), 306(36),
    322(36), 323(36), 326(36); **31**:160(2, 4),
    322(2, 4)
London, H., **12**:233(64), 256(64), 280
London, J., **14**:254(17), 269(17), 273(17), 274,
    279
Long, E. L., **9**:3(84), 84, 109
Long, G., **10**:107(34), 164; **29**:130(2), 135(2),
    149(2), 166(2), 202(2)
Long, R. R., **5**:131, 160(52), 208, 216(10), 246,
    248; **30**:31(64), 90(64)
Longenbaugh, R. S., **23**:306(121, 122, 123),
    359(121, 122, 123)
Longeway, A., **28**:342(25), 416(25)
Longo, J., **1**:410, 434(93), 442
Longsderff, R. W., **7**:248(65), 275(65), 316
Longsworth, L. G., **12**:241(65), 280
Longwell, P. A., **SUP**:73(191), 91(191), 94(191),
    142(191), 161(191), 165(191), 166(191),
    167(191), 300(191)

Look, D. C., **31**:347(34), 426(34)
Loomba, R. P., **30**:279(57), 280(57), 309(57)
Lopata, S., **23**:298(91), 358(91)
Loper, D. E., **28**:252(41), 255(41), 288(41),
    326(41), 331(41)
Loper, J. L., **5**:471, 516
Lopez-Gomez, R., **30**:222(45), 223(45),
    224(45), 225(45), 251(45); **31**:468(59),
    473(59)
Lorass-Nagy, V., **8**:66, 90
Lord, E., **25**:292, 294, 317
Lord, H., **21**:246(18), 257(18), 275(18)
Lord, H. A., **28**:376(150), 378(150), 385(150),
    413(229), 421(150), 425(229)
Lorenz, G. C., **6**:41(84), 126; **7**:368, 379
Lorenz, J. J., **9**:205(23), 268
Lorenz, L., **4**:5, 62
Lorenzen, A., **21**:156, 181
Lorenzini, E., **18**:163(20), 236(20)
Loscutoff, W. V., **6**:474(305), 501
Lothe, J., **9**:202, 268; **10**:92, 164; **14**:282,
    287(12), 304(31), 306, 307(31, 48, 49, 64,
    67), 308(64, 67), 309(31, 48, 49, 64, 67),
    310(31), 314(31), 316(31), 317(31),
    319(31), 320(31), 341, 342, 343
Lotkin, M., **15**:325
Lotstedt, P., **24**:227, 230, 270
Louie, D. L. Y., **29**:234(77, 82), 275(77),
    325(82), 326(82), 339(77, 82)
Louis, J. F., **23**:298(90), 358(90)
Louis, L., **30**:205(69), 208(69), 216(69),
    221(69), 253(69)
Loureiro, C. O., **23**:370(17), 461(17)
Loutaty, R., **31**:453(25), 471(25)
Love, E. S., **6**:44, 126
Love, L., **22**:20(6, 7), 60(6, 7), 61(6), 69(6),
    70(6), 152(6, 7)
Love, M. D., **25**:355, 382, 399, 408, 414
Love, P. H., **23**:284(38), 314(139), 355(38),
    360(139)
Love, T. J., **8**:66, 90
Love, T. J., Jr., **3**:207, 250; **5**:28(40), 52;
    11:365(120, 121, 122), 366(121), 371(120,
    121, 122), 374(120), 437
Lovegrove, P. C., **1**:384(51), 385(51), 441
Lovell, B. J., **26**:182(139), 183(139), 184(139),
    215(139)
Lovenguth, R. F., **14**:133, 146
Lovering, T. S., **15**:325
Low, A. R., **4**:152, 223; **18**:167(50), 237(50)

Low, G. S., **23**:218(250), 219(250), 220(250), 277(250)
Low, J. W., **10**:71, 73(62), 76, 78, 82
Lowan, A. N., **15**:325
Lowder, J. E., **5**:293(38), 298(38), 322; **8**:245, 256, 282; **12**:143, 193
Lowdermilk, W. H., **2**:42(42), 107; **5**:449, 450, 515; **6**:538, 563; **19**:334(133), 335(133), 355(133)
Löwe, F., **6**:197(35), 364
Lowell, H. H., **11**:208, 261; **15**:325
Lowell, R. L., **9**:304(60, 61), 347
Lowenheim, F. A., **28**:341(12), 415(12)
Löwenhielm, G., **29**:151(72), 206(72), 234(90), 339(90)
Lowenstein, E. V., **11**:365(111), 437
Lowery, A. J., Jr., **5**:109, 128; **18**:255(23), 257(23), 293(23), 321(23)
Lowndes, D. H., **28**:76(46), 79(45, 67), 80(46), 140(45, 46), 141(67)
Lowndes, R. P., **10**:21, 23, 36
Lownertz, P., **31**:440(19), 471(19)
Lowry, S. A., **28**:277(110), 280(110), 334(110)
Lowry, W. E., **29**:69(24), 71(24), 124(24)
Loyalka, S. K., **26**:32(132), 99(132); **29**:244(96), 312(96), 320(96), 340(96)
Loytsyansky, L. G., **3**:34(1), 97
Lozonschi, Ch., **14**:231(167), 246
Lozonschi, Gh., **19**:147(164), 189(164)
Lu, J. W., **28**:282(117), 334(117)
Lu, P. C., **1**:298, 353; **SUP**:225(416); **15**:3(17), 55(17), 325
Lu, S. M., **29**:39(33), 40(33), 41(33), 44(33), 57(33)
Luangdilok, W., **29**:325(178), 326(178), 327(178), 344(178)
Lubanska, H., **28**:19(33), 71(33)
Lucas, A., **19**:102(42), 105(42), 107(42), 108(42), 180(42), 185(42); **25**:182(76), 183(76), 184(76, 79), 185(79), 186(76, 79), 187(79), 200(76, 79), 202(79), 224(79), 246(76, 79)
Lucas, G. E., **29**:194(207), 213(207)
Lucas, H. G., **9**:119, 177
Lucas, J. G., **7**:358, 379
Lucas, J. W., **10**:71, 73(63, 64, 65, 66, 67), 76, 82, 83
Lucas, K., **15**:232, 277
Lucas, P. G. J., **17**:102(82), 126(82), 127(82), 129, 130, 155, 156

Lucchesi, P. J., **25**:168(65), 169(65), 170(65), 175(65), 176(65), 177(65), 228(65), 232(65), 233(65), 245(65)
Lucero, D. A., **29**:221(25, 26), 233(26), 271(25), 303(26), 309(25), 335(25), 336(26)
Luchko, N. N., **21**:185(9, 10), 198(9, 10), 202(9, 10), 203(9, 10), 236(9, 10)
Lück, G., **26**:240(47), 253(47), 269(47), 326(47)
Ludford, G. S. S., **SUP**:69(144), 161(144); **26**:61(35), 94(35)
Ludtke, P. R., **21**:46(74), 47(74), 52(74)
Ludwieg, H., **2**:16, 105; **6**:515(30), 562
Ludwig, C. B., **5**:295(40, 43), 307(64), 308(64), 309(75), 322, 323; **12**:129, 137, 193
Ludwig, O., **3**:287(81), 301
Luffy, J. W., **23**:292(59), 356(59)
Lui, C. Y., **7**:171, 172, 173, 178, 212
Lui, D. D. S., **29**:307(168), 344(168)
Luijkx, J.-M., **28**:370(134), 421(134)
Luikov, A. V. (Chapter Author), **1**:123, 146(15), 173(19), 176(19), 177(20, 21), 182(25), 184; **6**:376, 377, 418(151), 464, 465(3, 4), 486, 493; **8**:3, 84; **9**:357(10), 373(41), 414, 415; **13**:125, 126(29), 133, 142, 185(29), 191, 201; **SUP**:12(18), 17(18), 55(18, 19), 56(18), 137(18, 19), 189(18, 19); **15**:289, 325; **19**:143(151), 188(151), 191(1, 2), 193(1, 12, 16), 195(12, 16), 197(16), 242(1, 2), 243(12, 16); **25**:11(38, 39, 40, 41, 43), 12(40), 31(39, 40), 32(40, 41), 112(40, 179), 141(38, 39, 40, 41, 43), 148(179), 257, 283, 285, 289, 317; **30**:168(134), 169(134), 195(134)
Lukin, V. N., **31**:287(195), 331(195)
Lukyanov, P. I., **14**:160(70), 161(70), 162(70), 163(70), 166(70), 167(70), 168(70), 169(70), 244
Lumley, J. L., **12**:77(2), 111; **21**:185(8, 11), 188(12, 13), 196(12), 199(12), 236(8, 11, 12, 13); **25**:322, 414; **30**:289(78), 310(78), 318(16), 329(16), 351(16), 362(16), 363(16), 364(16), 369(16), 427(16)
Lumley, O. L., **21**:263(34), 265(34), 276(34); **30**:5(86), 9(86), 91(86)
Lummer, M., **29**:155(83), 206(83)
Lummus, J. L., **2**:385(59), 395
Lumsdaine, E., **11**:318(23), 376(23), 384, 385(23, 136), 387, 388(23), 435, 438
Lunardini, V. J., **19**:1(1), 36(73), 83(1), 86(73)
Lundberg, D. D., **SUP**:191(377)

Lundberg, R. E., **3**:126(35), 127(35), 173;
    SUP:31(41, 42), 32(41, 42), 33(41, 42),
    106(42), 172(42), 284(452), 285(41, 42),
    289(41, 42), 292(41, 42), 301(41), 302(41),
    303(41), 306(41, 42), 307(41), 309(41),
    310(41, 42), 311(41), 312(41), 313(41),
    314(41, 42), 316(42); **17**:260, 316
Lundell, J. H., **19**:81(256), 94(256)
Lundgren, T. S., **1**:300(44), 353; **6**:415(79), 489;
    7:184(53), 191(74), 192(74), 213;
    SUP:42(51), 43(51), 71(162), 72(162),
    77(162), 87(162), 139(162), 145(162),
    163(162), 198(51), 199(51), 228(51),
    247(51), 248(51), 286(51), 287(51),
    288(51); **19**:268(48), 271(48), 272(60),
    274(60), 351(48), 352(60); **23**:207(164),
    274(164); **24**:113(104), 114(104), 188(104)
Lunev, V. V., **3**:230, 235, 251
Lunina, L. I., **1**:435(141), 444
Lunney, J. G., **28**:102(71), 141(71)
Lurie, H., **4**:43(76), 64; **9**:317(79), 347; **17**:22,
    59
Lushchikov, V. V., **14**:152(29, 30), 243
Lushnikov, A. A., **14**:318(87), 344
Luss, D., **SUP**:80(230), 120(230), 121(230),
    158(230), 159(230), 186(230); **15**:229(9),
    233(37), 234(37), 235(9, 37), 239(9), 276,
    277; **24**:102(13), 184(13)
Lustenader, E. L., **30**:239(35), 240(35), 251(35)
Luther, H. A., **30**:272(41), 309(41)
Lutwyche, M. I., **28**:103(59), 141(59)
Lux, I., **31**:334(10), 425(10)
Luyet, B. J., **22**:195(129, 130), 352(129, 130)
Lyall, E., **25**:166(57), 245(57)
Lyapin, M. B., **8**:148, 149, 160; **18**:146(80),
    147(80), 159(80)
Lyche, B. C., **2**:374, 395; **SUP**:109(275);
    15:99(100), 139(100)
Lyczkowski, R. W., **SUP**:28(38), 64(38),
    202(38), 208(38), 209(38), 214(38, 406),
    215(38), 216(38), 218(38), **19**:289(85),
    291(85), 353(85)
Lykoudis, P. S., **1**:274, 281(10), 287(20), 293,
    298(10, 32, 40), 330(60), 333(60), 334(69),
    335(69), 336, 337, 338, 339(60), 343(69),
    344, 345(40, 69), 346, 347(60), 348(78),
    351(40), 353, 354; **6**:6, 122; **14**:121, 146
Lykov, A. V., *see* Luikov, A. V.
Lykow, A. W., **24**:104(46), 186(46)
Lyman, F. A., **7**:307, 308, 319; **13**:164(51), 202

Lynch, D. R., **19**:22(45), 85(45)
Lynch, F. E., **9**:73(47), 76(47), 108
Lynch, T. D., **5**:65, 90, 125; **11**:105(96), 193
Lynes, L. L., **6**:116(183), 117(183), 118(183),
    131
Lyon, D. N., **4**:170, 225; **5**:405, 462, 463, 464,
    492, 493, 511, 514, 515; **11**:105(72, 120),
    111, 114, 192, 194; **16**:114(79), 155, 234;
    17:146, 157, 340
Lyon, J. B., **2**:377(33), 395; **13**:219(65), 265
Lyon, R. E., **11**:105(60), 110, 191
Lyon, R. N., **3**:123, 124, 131, 139(41), 171, 173,
    174; **6**:504, 519, 561; **30**:42(65), 90(65)
Lyons, **18**:328(20), 329(20), 350(20), 362(20)
Lyons, D. W., **13**:129, 202; **15**:183(49), 224(49);
    25:259, 281, 317
Lyons, R. E., **5**:74, 126
Lyons, W. E., **10**:16(22), 36
Lytle, D., **26**:169(121), 214(121)
Lytle, F. W., **5**:469, 516
Lyubimov, D. V., **20**:331(48), 332(50), 351(48, 50)
Lyubov, B. Ya., **19**:195(19), 243(19)
Lyusternik, L. A., **8**:13, 85
Lyutich, N. V., **14**:152(29, 30), 243
Lyzenga, G., **30**:356(106), 357(106), 431(106)

M

Ma, C.-F. (Chapter Author), **20**:244(201),
    252(201), 253(201), 254(201), 255(201),
    309(201); **23**:8(18), 12(18, 35, 49), 14(18,
    35, 49), 17(35), 25(49), 29(18), 30(18, 35,
    56), 32(18), 33(18, 35), 37(18), 39(18),
    44(35), 46(18, 35), 47(18), 49(18, 35),
    50(35), 53(18), 57(18), 58(18), 60(18),
    78(18), 86(18), 90(18), 127(18, 35),
    128(49), 129(56); **26**:105, 106(11),
    109(20), 113(20), 114(20), 115(20),
    116(20), 117(20), 118(20), 123(57, 58, 60,
    61), 125(58, 60, 61), 126(57), 127(57, 58,
    60, 61, 72), 128(57), 129(57, 58, 60, 61),
    130(57, 58, 60, 61), 131(57, 58, 61, 72),
    132(57), 133(60, 61), 149(20, 58), 151(20,
    58), 155(58, 61), 156(20), 182(141),
    183(141), 184(141), 185(141), 186(141),
    208(11), 209(20), 211(57, 58, 60, 61, 72),
    215(141)
Maa, H. R., **15**:230, 231(15), 233, 276
Maas, R., **21**:319, 320, 321, 325, 329, 331, 345

Maass, O., **2**:356; **25**:294, 319

Mabuchi, I., **7**:348, 353(43), 378; **9**:294(19), 346; **11**:203, 206, 254; **13**:2(15), 10(15), 11(15), 59; **23**:293(63), 340(269), 341(269), 356(63), 367(269)

MacAdams, J. N., **20**:14(18), 15(18), 80(18)

MacArthur, C. D., **23**:294(75), 357(75)

Macbeth, R. V., **4**:204, 226; **17**:2(5), 26(79), 27, 28, 30(91), 45, 56, 59, 60; **20**:250(213), 309(213); **29**:234(87), 249(87), 251(87), 259(113), 260(111, 113, 115), 262(113), 263(111, 113, 115), 264(111, 113), 266(113), 339(87), 341(111, 113, 115)

Macbeth, R. W., **1**:410, 442

Macbuchi, I., *see* Mabuchi, I.

MacCormack, R. W., **24**:234, 271

MacDavid, K. S., **23**:293(64), 307(64), 356(64)

Macdonald, H. M., **SUP**:322(476)

Macdonald, I. F., **23**:215(165), 274(165)

MacDonald, J. K. L., **1**:72(25), 75(25), 121; **19**:13(26), 19(26), 84(26)

Macdougall, G., **9**:190(7), 268

Macey, R. I., **1**:55, 120

MacFarlane, D. R., **22**:177(131), 196(132), 352(131, 132)

MacFarlane, R., **29**:76(35), 125(35)

MacGregor, R. K., **8**:195, 227; **9**:390(62), 396(74), 399(74), 416; **18**:18(37), 81(37)

Mach, L., **6**:195, 268, 363

Machac, I., **23**:193(33), 215(168), 218(167), 237(166), 238(166), 239(166), 240(166), 241(166), 242(166), 243(166), 246(166), 268(33), 274(166, 167, 168)

MacInnes, D. A., **12**:202(67), 224(67), 226(67), 229(67), 234(67), 235(67), 236(67), 241(67), 246(67), 253(67), 280

Mack, F. E., **9**:406(98), 417

Mack, L. M., **14**:284(7), 332, 333(7), 341

Mackay, D. B., **2**:434(34), 451

Mackenzie, J. E., **3**:287(84), 301

Mackewich, W. V., **31**:250(145), 251(145), 252(145), 328(145)

Mackey, C. O., **4**:67(4), 137

Mackie, J. W., **8**:13, 85

Maclaine-Cross, I. L., **SUP**:67(125), 280(125), 282(125)

Maclay, G. J., **5**:301(57), 323

Macleod, N., **21**:179, 180

Macosco, C. W., **13**:219(90), 266

Macosko, R. P., **4**:210, 211, 227; **9**:259(105), 271

Madan, S., **11**:220, 223, 228, 237, 248, 249, 258, 261

Madden, A. J., **4**:44, 64; **9**:304(57), 347; **11**:202(28), 205(28), 209(28), 213(28), 214(28), 253, 278(9), 279(9), 280(9), 314

Maddison, R. J., **29**:165(111), 207(111)

Maddock, B. J., **17**:147(172), 157

Maddock, J. L., **23**:222(120, 121), 226(120, 121), 272(120, 121)

Maddox, D. E., **26**:170(123), 214(123)

Madejski, J., **5**:414, 513; **6**:474(262), 499; **9**:3(85), 76(85), 109; **10**:152, 154, 155, 165; **SUP**:82(238); **28**:24(44), 27(44, 49), 28(44, 49), 36(44, 49), 53(44), 56(44, 49), 72(44, 49)

Mader, P. P., **4**:228

Madni, I. K., **16**:53, 57

Madonna, L. A., **14**:318(85), 321, 322(85), 323(85), 324, 344

Madsen, J., **7**:273(94), 317

Madsen, N., **16**:233

Madsen, R. A., **5**:484(272), 486, 487, 488, 489, 491, 492(272), 494(272), 517; **17**:117(107), 155

Madson, R. A., **11**:105(137), 195

Maecker, H., **1**:283(15), 284(15), 353; **4**:232(3), 233(3), 234(3), 243, 245, 248(3), 260, 262(25), 264, 293(3, 56), 313, 314

Maeda, M., **25**:258, 283, 285, 319; **28**:135(72, 73), 142(72, 73)

Maeder, P. F., **3**:16(31), 19(30, 33), 21(34), 23(30, 34), 24(30), 26, 30(50), 32; **7**:119, 161; **11**:231(164), 232(164), 259, 260

Maekawa, H., **24**:3(5), 36(5)

Maeno, N., **19**:37(83), 87(83); **24**:30(65), 38(65)

Maerefat, M., **21**:79(24), 134(24)

Maerker, J. M., **23**:226(169), 274(169)

Maevskü, E. M., **17**:32(104), 61; **21**:47(81), 53(81)

Maewal, A., **15**:325

Magallon, D., **29**:140(31), 156(87), 165(31), 178(31), 204(31), 206(87)

Magaritz, M., **30**:96(29), 190(29)

Magee, C. W., **28**:97(106), 143(106)

Magee, P. M., **6**:539, 542(57), 543(57), 563

Magirl, C. S., **28**:243(17, 18), 244(17, 18), 245(18), 246(18), 247(18), 248(18), 282(17, 18, 123), 283(18, 123), 284(18, 123), 285(18, 123), 286(18), 314(161), 329(17), 330(18), 334(123), 336(161)

Magnuson, R. A., **SUP**:198(388), 199(388), 209(388), 210(388), 211(388); **19**:268(59), 272(59), 273(59), 274(59), 352(59)

Magrakvalidze, T. Sh., **31**:304(207), 332(207)

Magrini, A., **26**:137(84, 90), 148(84, 90), 151(84, 90), 152(84), 154(84), 167(84), 212(84, 90)

Magueur, A., **23**:219(170), 225(170), 274(170)

Mahajan, R. L. (Chapter Author), **20**:316(12), 350(12); **28**:239(9), 329(9), 339, 344(63), 364(103), 367(103, 119), 368(119), 369(103), 373(103), 376(147, 151, 152), 379(151, 152), 380(151), 381(152), 383(152), 384(152), 390(176), 391(176, 178), 395(103), 398(178), 399(178), 400(178), 408(176, 224), 410(227), 411(103, 176, 228), 413(228, 230, 231), 418(63), 419(103), 420(119), 421(147), 422(151, 152), 423(176, 178), 425(224, 227, 228, 230, 231); **30**:362(125), 413(125), 432(125)

Mahalingam, M., **30**:173(147), 195(147)

Mahalingam, R., **15**:101(115), 102(115), 103(115), 104(115), 139(115)

Mahalingham, L. M., **20**:259(248), 262(248), 311(248)

Mahalingham, M., **20**:259(250), 311(250)

Mahaney, H. V., **19**:300(88), 301(88), 353(88)

Maheshri, J. C., **28**:190(47), 229(47)

Mahn, C., **4**:265(10), 268(10), 313

Mahony, J. J., **4**:156, 224; **8**:12, 85; **9**:302(49), 347; **11**:216(208), 262

Mai, K. L., **5**:131(12), 241, 247

Maidanik, V. N., **6**:539, 540(61), 541(61), 563

Maiman, T. H., **30**:319(17), 351(17), 427(17)

Mainster, M. A., **22**:266(133, 134, 135, 222), 268(133, 134, 135), 270(135), 271(135), 272(135), 273(134, 222), 274(134), 275(134), 276(134), 353(133, 134, 135), 357(222)

Maiorov, V. A., **31**:160(53), 324(53)

Maise, G., **7**:190, 214

Maisel, D. S., **3**:9, 10, 31; **11**:219, 222, 231, 232, 236, 257, 259

Maitland, G. C., **28**:364(99), 419(99)

Maj, S., **11**:437

Majda, A., **24**:221, 270

Majiros, P. G., **1**:369(19), 376(19), 380(19), 387(19), 390(19), 393, 440

Major, B. H., **15**:245(76a), 248, 279; **26**:47(99), 97(99)

Majumdar, A., **23**:139(18), 140(18), 170(18), 185(18); **30**:102(43), 142(43), 143(43), 145(43), 147(43), 190(43); **31**:408(72), 428(72)

Majumdar, A. K., **13**:110(56), 117

Majumdar, A. S., **19**:248(5), 281(83), 285(83), 319(5), 349(5), 353(83); **26**:132(80), 212(80)

Majumdar, S., **24**:199, 211, 213, 272, 273

Makarenko, G. I., **8**:34(178), 88

Makarevičius, M. M., **8**:114(42), 115(42), 159

Makarevičius, V., **18**:88(20), 107(36), 108(20), 109(36), 110(20), 113(36), 136(20), 141(20), 143(20), 147(20), 148(20), 150(20), 158(20, 36)

Makarevičius, V. J., **8**:95(20), 109(20), 113(20), 116(20), 138(20), 141(74), 142(74), 143(74), 145(74), 146(79), 151(20), 159, 160

Makarov, Y. N., **28**:390(174), 423(174)

Makel, D. B., **23**:330(238), 365(238)

Makhatkin, A. V., **14**:206(138), 246

Makhorin, K. E., **10**:208, 217; **14**:155(49, 53, 54), 243; **19**:164(204), 168(204), 171(204), 190(204)

Maki, H., **21**:130(127), 139(127)

Maki, M., **28**:386(157), 387(157), 422(157)

Makino, A., **26**:78(174), 100(174)

Makino, Y., **24**:6(25), 37(25)

Makkencherry, S., **17**:51, 64

Malan, D. H., **10**:243(33), 281

Malan, P. H., **17**:190(45), 315

Malashkin, I. I., **31**:265(163), 329(163)

Malcorps, H., **23**:292(60), 356(60)

Malenkov, I. G., **21**:328, 329, 331, 345

Malesinsky, W., **16**:66, 153

Malev, V. V., **15**:291(2), 321

Malhotra, A., **21**:22(38), 23(38), 51(38)

Malik, M. R., **30**:365(132), 432(132)

Malik, P. R., **20**:281(299), 313(299)

Malina, J. A., **6**:525(35), 526, 562

Malkin, A. Y., **13**:217(31), 264

Malkin, V. M., **6**:435(167), 437(167), 494

Malkmus, W., **5**:287, 295(39, 43), 322, 323; **12**:139, 193

Malkov, M. P., **5**:406(170), 513

Mal'kov, V. A., **20**:264(267), 312(267)

Malkovsky, V. I., **25**:51(80, 81), 143(80, 81)

Malkus, W., **30**:50(66, 67), 90(66, 67)

Malkus, W. V. R., **4**:152, 223; **14**:33, 102

Mall, B. K., **25**:260, 283, 317

Mallard, E., **17**:181, 314

Mallen, A. N., **31**:112(34, 35, 36, 37, 38), 119(34, 35, 36, 37, 38), 120(35, 36, 37, 38), 121(34), 128(34), 137(34, 38), 153(34, 35, 36, 37), 157(34, 35, 36, 37, 38)

Mallinson, G. D., **24**:195, 270

Mallock, A., **21**:143, 146, 181

Malloy, J. F., **9**:350(2), 352(2), 414

Malmuth, N. D., **8**:11, 12(35), 85

Maloletov, I. L., **25**:77(128), 146(128)

Malone, E. W., **23**:301(102), 302(102), 358(102)

Malone, M. F., **19**:19(35), 85(35)

Malozemov, V. V., **25**:4(31), 141(31)

Maltby, J. D., **31**:388(41), 427(41)

Malugin, Yu. S., **6**:539, 563

Malyukovich, S. A., **14**:236(181, 182), 247

Maman, N., **31**:407(70), 428(70)

Mamer, W. J., **31**:433(16), 471(16)

Mamontova, N. N., **17**:7(33), 57

Manabe, S., **8**:281, 283

Manaker, A. M., **15**:325

Manchester, F. D., **17**:69(31), 153

Mancuso, T., **23**:349(303), 368(303)

Mandell, D., **8**:264(41), 283

Mandell, W., **2**:294, 354

Mandeno, P., **25**:260, 279, 317

Mandhane, J. M., **20**:83, 131

Maneval, J. E., **23**:378(54), 463(54)

Manfredini, A., **29**:234(93), 340(93)

Manganaro, J. L., **9**:304(62), 347

Mangelsdorf, H. G., **12**:129(18), 133(18), 192

Mangler, W., **4**:331, 443; **26**:111(27), 209(27)

Manglik, R. M., **30**:198(8), 249(8)

Mangold, D., **24**:103(24), 185(24)

Mani, R., **15**:324

Mani, R. U. S., **7**:133, 161

Manjo, G., **22**:229(47), 348(47)

Manjunath, M., **23**:202(171), 208(171, 172), 211(171), 212(171), 274(171, 172)

Manke, C. W., **28**:376(145), 421(145)

Mankevich, V. N., **26**:122(50), 210(50)

Man'kovskij, D. N., **26**:314(86), 328(86)

Mann, D. B., **5**:389(114), 510; **9**:350(4), 352(4), 394(4), 414; **15**:50(83), 57(83)

Mann, J. B., **3**:274, 301

Mann, W. B., **2**:327, 355

Mann, W. R., **8**:36, 89

Mannes, R. L., **4**:162, 164, 191, 194, 224

Manneville, P., **25**:356, 417

Mannheimer, R. J., **13**:179(58), 202

Manning, F. S., **7**:89(5c, 5d), 160

Manning, M., **22**:421(34), 436(34)

Mannov, G., **17**:37(118), 61

Manohar, R., **SUP**:73(175, 180), 77(175), 89(175, 247), 93(175), 140(247), 141(247), 142(175, 247), 144(175), 145(175), 146(175, 247), 147(247), 288(180), 298(180)

Manohar, R. P., **26**:26(133), 99(133)

Manontova, N. N., **16**:141(114), 156

Manrique, L., **13**:219(57), 259, 265

Mansion, H. D., **18**:167(51), 237(51)

Manson, J. R., **28**:118(82), 119(82), 120(82), 142(82)

Manson, L., **5**:432, 514; **11**:96, 191

Mansour, I. A. S., **23**:230(232), 232(232), 277(232)

Mansour, N. N., **25**:341, 382, 414

Mantell, C., **24**:19(53), 38(53)

Mantzouranis, B. G., **1**:363(16), 377, 389, 390(16), 392, 427(128), 440, 441, 443

Manushin, E. A., **9**:95(55c), 101(55c), 108; **31**:160(49), 324(49)

Manwani, P., **26**:81(248), 104(248)

Maples, A. L., **28**:245(22), 330(22)

Marangozis, J., **21**:172, 181

Marayama, S., **30**:405(175), 434(175)

Marble, F. E., **2**:213, 269

Marchaterre, J. F., **1**:394(64), 398(64), 442

Marchello, J. M., **10**:189, 191, 217

Marchello, M., **15**:237, 278

Marchesi, R., **22**:421(33), 436(33)

Marchi, S. C., **28**:36(62), 53(62), 73(62)

Marchiano, S. L., **12**:233(4), 251(4), 270(4), 277

Marcillat, J., **8**:300, 301(49), 329, 350

Marcinauskas, K. F., **18**:142(77), 159(77)

Marco, S. M., **SUP**:62(70), 197(70), 205(70), 208(70), 223(70), 224(70), 234(70), 236(70); **19**:258(12), 350(12); **24**:19(49), 38(49)

Marcus, B. D., **7**:306, 319, 320b, 320e; **16**:234; **20**:24(50), 81(50)

Marcus, P. S., **21**:181

Maresca, M. W., **3**:106(6), 172

Margaritis, A., **23**:188(10), 253(10), 267(10)

Margenau, H., **3**:270(34), 300

Margolis, D. L., **22**:287(113), 288(113), 352(113)

Margolis, G., **11**:412(185), 439

Margulies, R. S., **1**:218(34), 264

Mariano, J., **8**:281, 283

Marinelli, V., **17**:2(10), 40(10), 56; **20**:250(218), 309(218)

Marinero, E. E., **28**:84(76), 142(76)

Marioka, M., **30**:323(25), 427(25)

Maris, H. J., **17**:68(6), 72, 153

Marity, A. H., **17**:31(101), 61

Marivoet, J., **23**:217(173), 274(173)

Markatos, N. C., **24**:302(25), 319(25)

Markels, M., **14**:129(65, 66), 130, 133, 134, 146

Markest, W., **1**:394(66), 398(66), 442

Markin, V. S., **15**:291(2), 321

Markov, S. B., **6**:390(18), 402, 486; **25**:81(129), 146(129)

Markovich, D. M., **26**:123(70), 124(70), 211(70)

Markovin, M. V., **6**:396(28a), 405, 487

Markovitz, H., **2**:358, 361, 363, 395; **15**:182(46), 224(46)

Markovsky, P. M., **25**:77(126, 127, 128), 83(126, 127), 84(167, 168, 169), 92(169), 96(178), 146(126, 127, 128), 148(167, 168, 169, 178)

Markowitz, A., **20**:244(184), 308(184)

Marks, C. H., **15**:250, 280

Marks, W. M., **19**:334(132), 335(132), 355(132)

Markson, A. H., **5**:453, 515

Markvart, M., **10**:169(4), 215

Markworth, A. J., **28**:24(45), 72(45)

Marle, C. M., **23**:374(39), 376(39, 48), 377(39), 378(39), 462(39, 48)

Marner, W. J., **15**:177(35), 198(88), 199(88), 200(88), 201(88), 202(89), 203(89), 204(89), 217(35), 219(88), 221(89), 224(35), 225(88, 89); **20**:187(41), 301(41); **23**:293(64), 307(64), 356(64)

Maron, D. Moalem, *see* Moalem-Maron, D.

Maroti, L. A., **5**:130(3), 177(3), 179(3), 180, 246

Maroudas, D., **30**:360(119), 432(119)

Marov, M. Y., **10**:53(34), 58(34), 64(34), 70(34), 71(34), 82

Marra, R. A., **13**:164(54), 202

Marrone, P. V., **2**:190, 269

Marrucci, G., **12**:103, 113; **15**:63(21), 64(21), 77(21), 135(21); **20**:100, 131; **24**:107(71), 187(71); **25**:258, 286, 317

Marschall, E., **9**:264(129), 265(142), 271, 272; **15**:232, 277; **16**:136, 155

Marsden, S. S., **23**:193(7), 267(7)

Marsh, B. D., **19**:82(272, 273), 95(272, 273)

Marshall, B. W., **29**:165(109, 114), 207(109), 208(114)

Marshall, J. H., **2**:354

Marshall, J. S., **29**:46(41), 48(41), 57(41)

Marshall, R. J., **23**:206(174), 218(174), 220(174), 274(174)

Marshall, W. R., **26**:53(179), 101(179); **28**:10(11), 71(11); **29**:285(154), 343(154); **30**:243(56), 252(56)

Marshall, W. R., Jr., **10**:174(39), 215

Marston, A. C., **10**:34, 37, 53(31), 68(31), 81

Marta, I. F., **25**:166(52, 53), 245(52, 53)

Martellucci, A., **6**:41(81), 65(81), 108(166), 126, 130

Martens, H. E., **10**:50(30), 51(30), 59(30), 68(30), 81

Martin, B., **13**:219(38, 60), 261(118), 264, 265, 266

Martin, B. W., **7**:367, 379; **8**:163, 226; **9**:3(9, 92), 9, 10(87), 11, 17, 18(87), 21(83, 87), 22, 23, 24, 25, 26, 27, 29, 35, 36, 38, 39, 46, 52, 55, 70, 99, 100, 106, 109; **SUP**:70(152), 87(152); **19**:334(143), 337(143), 339(143), 355(143); **21**:158, 178, 180, 181

Martin, D. G., **19**:334(143), 337(143), 339(143), 355(143)

Martin, H. (Chapter Author), **13**:1, 2(6, 7, 8, 9), 13(9), 22(9), 23(9), 24(9), 25, 26(9), 27(6, 9), 32(8, 9), 33, 34(9), 35(9), 36, 37(9), 46(9), 52(7, 9), 59; **19**:120(105, 106), 121(105, 106, 108, 109), 122(105, 106, 108, 109), 187(105, 106, 108, 109); **20**:255(235), 310(235); **23**:7(8), 126(8); **26**:106(7), 107(7), 129(7), 135(7), 136(7), 171(7), 208(7); **27**:192(40); **28**:54(85), 68(85), 74(85)

Martin, H. L., **23**:320(183), 362(183)

Martin, J. C., **14**:78, 104

Martin, J. H., **4**:43(75), 64; **9**:317(78), 347

Martin, M., **11**:220, 224, 237, 258; **30**:172(142), 195(142)

Martin, M. W., **1**:373, 441; **5**:451, 454, 515

Martin, W. R., **31**:408(72), 412(76), 428(72), 429(76)

Martinelli, L., **24**:217, 263, 272

Martinelli, R. C., **1**:242, 265, 357, 370, 371, 372, 382, 440, 441; **3**:102, 124, 131, 171, 172; **5**:451, 453, 454, 456, 457, 515; **11**:202, 204, 248, 253; **SUP**:414(531), **15**:195(72, 73), 224(72, 73); **31**:134(18), 156(18)

Martinez, F., **19**:62(144), 63(144), 64(144), 65(144), 66(144), 67(144), 89(144)

Martinez, O., **30**:204(68), 205(68), 208(68), 221(68), 253(68)

Martinez, O. M., **24**:102(15), 185(15)

Martinez-Sanchez, M., **23**:298(90), 358(90)

Martini, W. D., **8**:167, 170, 172, 212, 224, 226

Marto, P. J., **7**:256, 257, 317; **10**:138, 140, 142, 166; **20**:244(206), 273(284), 290(284), 295(284), 309(206), 312(284); **21**:125(120), 138(120)

Martyanova, T. S., **25**:4(29, 32), 141(29, 32)

Martynenko, O. G., **SUP**:2(9); **20**:217(122), 305(122); **21**:185(9), 198(9), 202(9), 203(9), 236(9)

Martynova, O. I., **31**:458(39), 472(39)

Martyushin, I. G., **10**:202, 218; **19**:117(97), 129(97), 130(97), 187(97)

Maruyama, Y., **29**:151(70), 165(116), 206(70), 208(116)

Maruyana, T., **25**:81(131), 146(131)

Marvin, M., **18**:44(86), 83(86)

Marwah, M., **28**:410(227), 425(227)

Marx, K. D., **29**:233(66, 67), 302(66), 303(161), 309(67), 338(66, 67), 343(161)

Marxman, G. A., **30**:64(68, 69), 65(68), 90(68, 69)

Mascarenhas, S., **14**:119, 120, 146

Mascarenhas, Y., **14**:119(70), 120, 146

Mascola, R. E., **1**:96(46), 121

Mashburn, D. N., **28**:76(46), 79(67), 80(46), 140(46), 141(67)

Mashelkar, R. A. (Chapter Author), **15**:69(45), 136(45), 143, 144(9), 157(19), 178(19), 179(39), 180(19), 181(19), 184(19), 189(67), 190(67), 191(67), 193(69), 194(69), 195(69), 214(114), 216(67, 69), 221(19), 223(9, 19), 224(39, 67, 69), 225(114); **18**:161, 162(14), 163(14), 209(98), 214(108, 121), 215(108), 224(98, 120), 226(121, 122, 123), 227(121), 229(121), 230(121), 233(108, 121), 234(98, 129), 236(14), 238(98, 108, 120, 121, 122, 123, 129); **19**:249(7), 349(7); **24**:102(4), 106(4), 184(4); **25**:254, 258, 260, 262, 269, 272, 278, 281, 286, 288, 315, 316, 317, 319

Masi, J. F., **2**:37(28, 30), 38(28), 106

Masi, M., **28**:375(141), 421(141)

Maskaev, V. K., **14**:151(21), 160(21), 161(21), 166(21), 176(161), 221(21), 225, 242, 246; **19**:100(21, 22), 103(21, 22), 116(22), 117(21), 118(21), 126(22), 127(22), 128(22), 165(22), 171(22), 185(21, 22)

Maslach, G. J., **7**:209, 218

Maslen, S. H., **7**:184, 197, 213, 215

Maslennikov, M. M., **6**:474(307), 501

Masleyeva, N. V., **8**:103(29), 104(29), 105(29), 132(29), 133(29), 159; **11**:228(226), 262; **18**:98(29), 99(29), 130(29), 131(29), 158(29)

Maslichenko, P. A., **11**:56(7, 8, 9, 10, 11), 58(7, 8, 9, 10, 11), 59(7, 8, 9, 10, 11), 60(7, 8, 9, 10, 11), 118, 189

Masliyah, J. H., **SUP**:67(127, 128), 368(128, 516), 369(516), 370(128, 516), 375(127, 128), 376(127, 128), 377(516), 378(128, 516), **24**:114(118), 188(118); **25**:293, 318

Maslov, A. M., **31**:160(27), 323(27)

Mason, A. D., Jr., **22**:243(169), 354(169)

Mason, B. J., **14**:318, 320(91), 321, 322, 344

Mason, D. M., **11**:219(100, 105), 222(100, 105), 236(100, 105), 256, 257

Mason, E. A., **3**:210(28, 32, 33), 263, 266, 267, 268, 269, 271(33), 272, 279, 280, 281, 281(90), 282(61), 291, 293(92, 93), 298(25, 96), 300, 301, 302

Mason, H. B., **15**:189(59), 224(59)

Mason, J. L., **3**:120(23), 173

Mason, J. N., **6**:474(238), 483(238), 498

Mason, J. P., **1**:232, 265; **15**:254, 280

Mason, P. J., **25**:342, 361, 366, 375, 376, 414, 415, 416

Mason, S. G., **20**:105, 131; **26**:29(235), 103(235)

Mason, W. E., **11**:202, 204, 248, 253

Mass, U., **29**:77(36), 125(36)

Massier, P. F., **2**:53(59), 54(59), 55(59), 56(59), 57(59), 60(59), 61(59), 62(59), 64(59), 65(59), 66(59), 68(59), 69(59), 70(59), 72(59), 73(59), 94(59), 107, 108

Masson, D. J., **3**:35(22), 98

Masson, H., **19**:125(128, 129), 188(128, 129)

Massons, J., **30**:396(162), 399(162), 434(162)

Mast, P. K., **29**:324(176), 325(176), 326(176), 327(176), 344(176)

Mastin, C. W., **19**:50(119), 57(119), 88(119); **26**:45(227), 103(227)

Mastrangelo, C. H., **28**:99(98), 143(98)

Masubuchi, M., **6**:474(267), 499

Masuda, H., **11**:203, 206, 220, 223, 226, 254, 258; **21**:120(109), 122(109), 138(109)

Masuda, K., **28**:135(26), 139(26)

Masumoto, H., **20**:287(312), 289(312), 314(312)

Masuoka, T., **14**:16(40), 19, 20(40), 71, 101; **24**:163(147), 190(147); **30**:121(81), 192(81)

Masuyama, T., **23**:193(175), 214(175), 216(175), 274(175)

Mata, R., **23**:306(121), 359(121)

Matekunas, F. A., **10**:156, 165

Mathers, W. G., **9**:304(57), 347

Matheson, C. C., **5**:481, 517

Matheson, G. L., **14**:232, 246

Mathieu, Ph., **19**:199(51), 200(51), 202(51), 244(51)

Mathiprakasam, B., **21**:168, 181

Mathur, A., **19**:102(38, 39, 40, 41, 43, 44), 103(41), 106(41, 43, 44), 107(41, 43, 44), 108(41, 44, 49), 109(43, 44), 110(41, 44, 64), 111(38, 39, 40, 41), 122(110, 111), 123(110, 111), 124(110), 126(111), 127(111), 128(111), 129(110), 130(110), 131(110), 136(40), 138(40), 139(40), 140(40), 141(40), 142(40), 163(221), 165(221), 166(111, 221), 167(221), 168(221), 169(221), 170(221), 171(221), 172(221), 174(221), 175(221), 176(221), 177(221), 180(40, 41, 43, 44, 110), 181(221), 185(38, 39, 40, 41, 43, 44, 49), 186(64), 187(110, 111), 190(221)

Mathur, P. C., **28**:23(41, 42), 64(41, 42, 83), 72(41, 42), 73(83)

Matin, S. A., **9**:253(82), 270

Matorin, A. S., **16**:127, 155

Matovic, M., **23**:307(124), 359(124)

Matricon, M., **15**:325

Matsuba, I., **28**:402(214), 424(214)

Matsuda, T., **24**:282(39), 284(39), 286(39), 317(39), 320(39)

Matsui, H., **24**:3(5), 36(5); **29**:96(64), 126(64)

Matsumoto, K., **28**:402(214), 424(214); **30**:323(25), 427(25)

Matsumoto, M., **8**:78(285), 91

Matsumoto, R., **2**:380(77), 396; **4**:26(44), 63; **15**:189(60), 224(60)

Matsumoto, S., **28**:96(68), 141(68)

Matsumura, H., **10**:151, 157(71), 165

Matsumura, M., **27**:ix(15), 187(36, 37, 38), 193(37, 38), 196(42)

Matsumura, S., **23**:58(94), 61(94), 77(94), 106(94), 131(94)

Matsunaga, A., **29**:160(94), 161(94), 207(94)

Matsuoka, H., **1**:231(52), 265; **11**:12(27), 48; **20**:5(3), 7(3), 9(3), 10(3), 11(3), 15(3), 79(3)

Matsuoka, K., **27**:124(27)

Matsuura, A., **19**:310(94), 312(94), 353(94); **30**:379(147), 433(147)

Matsuzaki, M., **29**:162(100), 207(100)

Matthaus, W., **18**:327(8), 347(8), 349(8), 350(8), 358(8), 359(8), 362(8)

Matthews, L. K., **23**:306(121, 122, 123), 359(121, 122, 123)

Matthews, R. K., **23**:280(14), 304(14), 315(14), 322(14), 325(14), 354(14)

Matthys, E. F., **28**:28(50, 51, 52), 36(50, 51, 52), 56(50, 51, 52), 59(52), 72(50, 51, 52)

Matting, F. W., **3**:51, 59, 98

Mattingly, G. E., **28**:19(30, 31), 71(30, 31)

Matulevicius, E. S., **25**:233(118), 234(118), 248(118)

Matveev, A. B., **21**:2(4), 3(4), 50(4)

Matveev, V. I., **31**:160(17), 323(17)

Matzen, E. J. P., **23**:217(66), 269(66)

Matzner, B., **1**:422(119), 443; **17**:39(129), 62

Maughan, J. R., **28**:370(132), 421(132)

Maulbetsch, J. S., **9**:259(108), 260(108), 271; **31**:433(8), 447(8), 471(8)

Maull, D. F., **31**:172(79), 325(79)

Maull, D. J., **6**:73(116), 127

Maurer, G. W., **1**:394(67), 398(67), 421(116), 442, 443

Maurer, J., **17**:133, 156

Mavriplis, D. J., **24**:217, 263, 272

Mawhinney, G. S., **18**:169(44), 170(44), 178(44), 220(44), 221(44), 233(44), 236(44)

Mawid, M., **26**:61(134), 99(134)

Maxwell, J. C., **2**:364, 395; **7**:163, 210; **18**:179(62), 181(62), 190(62), 237(62); **24**:109(88), 187(88)

May, D., **31**:402(59), 428(59)

May, M., **30**:256(4), 307(4)

May, S. C., **15**:236(44), 272(44), 278

Mayer, E., **15**:285, 287, 315, 326

Mayer, F. X., **25**:168(65), 169(65), 170(65), 175(65), 176(65), 177(65), 228(65), 232(65), 233(65), 245(65)

Mayer, H. L., **5**:320(94, 95), 324

Mayer, J. E., **14**:306, 342

Mayer, J. W., **28**:80(31), 103(87), 139(31), 142(87)

Mayer, M. G., **14**:306(52), 342

Mayhew, Y. R., **5**:166, 248

Mayinger, F. M., **8**:135, 137, 160; **16**:179, 238; **17**:31(100), 60; **18**:133(67), 134(67), 135(67), 159(67); **29**:5(3), 7(3), 15(3), 17(3), 18(3), 19(3), 24(3), 25(3), 26(3), 55(3)

Maykochin, A. S., **25**:4(15), 140(15)

Mayle, R. E., **23**:327(220), 344(295), 364(220), 368(295)

Mazo, R. M., **17**:67, 153

Mazumder, J., **28**:123(16), 135(20), 139(16, 20)

Mazur, A. I., **13**:43, 60; **31**:160(39), 323(39)

Mazur, P., **1**:182(26), 184; **8**:26(134), 29(134), 87; **12**:206(16), 264(16), 271(16), 278; **15**:11(31), 55(31); **22**:177(140), 187(139), 196(136, 137), 294(138), 295(136, 139), 312(140), 353(136, 137, 138, 139, 140, 141); **23**:397(60), 463(60); **31**:86(9), 102(9)

Mazur, Yu. S., **19**:164(210), 169(210), 170(210), 190(210)

Mazzola, M. S., **31**:461(40), 472(40)

McAdams, W., **9**:182(1), 255, 268

McAdams, W. H., **1**:10, 16(18), 49, 187(3), 190, 213, 263, 264; **2**:5, 105; **4**:155, 156, 223; **5**:66, 67, 99, 100, 104, 125, 127, 357, 358, 363, 365, 366, 367, 368, 375(74), 396(74), 423, 453, 508, 515; **11**:2, 47, 105(57, 53), 109(58), 110, 115(57), 191, 202, 205, 208(31), 212(31), 213, 214, 225, 253, 305, 306, 315; **SUP**:47(56), 48(56), 50(56), 108(56), 412(56); **19**:190(199); **20**:15(26), 80(26)

McAlister, A., **18**:171(74), 237(74)

McArdle, J. G., **4**:150(7), 222, 228

McAuliffe, C. D., **23**:193(176), 274(176)

McBirey, A. R., **10**:43, 81

McBride, B. J., **29**:62(11), 124(11)

McBride, G. B., **13**:67(7), 115

McCabe, S., **20**:286(310), 287(310), 313(310)

McCaffery, T. V., **22**:30(43), 154(43)

McCahan, S., **29**:185(180), 186(180), 187(183), 211(180, 183)

McCall, M. J., **17**:119(116), 156

McCardell, R. K., **10**:107(37), 164

McCarthy, H. E., **9**:137, 156, 161, 179

McCarthy, J. R., **5**:361, 363, 508; **6**:538, 563

McCarthy, M. J., **23**:378(54), 463(54)

McCartney, M. J., **18**:333(47), 336(47), 363(47)

McCarty, R. D., **17**:102(77, 78), 155

McCay, M. H., **28**:277(109, 110), 280(109, 110), 282(119), 309(109), 334(109, 110, 119)

McCay, T. D., **28**:277(109, 110), 280(109, 110), 282(119), 309(109), 334(109, 110, 119)

McClellan, R., **7**:200, 216

McClintock, F. A., **15**:26(51), 56(51); **16**:271, 364

McColl, W. F., **31**:402(60), 428(60)

McComas, S. T., **SUP**:41(48), 43(48), 73(48), 199(48), 213(48), 228(48), 248(48), 288(48); **19**:324(114), 334(139, 140), 336(139, 140), 338(139, 140), 354(114), 355(139, 140)

McConaghy, G. A., **15**:128(156), 141(156)

McConnell, D. G., **5**:479, 517

McConnell, J. C., **14**:267(44), 280

McCook, R. D., **4**:120(96), 140; **22**:30(43), 154(43)

McCord, G., **30**:382(150), 433(150)

McCord, T. B., **10**:7, 8(14), 9, 21, 22(14), 24(32, 37), 25(32, 37), 35, 36

McCormick, D. C., **23**:333(246), 365(246)

McCormick, J. L., **9**:207(29), 208, 210(29), 211, 214, 232(29), 234, 235, 236, 239, 240(70), 243, 265(33), 269, 270; **21**:80(32), 135(32)

McCormick, P. Y., **13**:120, 200

McCoy, J. J., **18**:190(82), 191(82), 199(82), 237(82)

McCoy, J. K., **28**:24(45), 72(45)

McCraw, D. A., **11**:391(145), 438

McCrea, W. H., **5**:40(73), 54

McCroskey, W. J., **7**:204, 217

McCuen, P. A., **3**:126(35), 127(35), 173; **SUP**:31(42), 32(42), 106(42), 153(332), 154(332), 169(332), 172(42), 177(332), 185(332), 284(452), 285(42), 289(42), 292(42), 306(42), 310(42), 314(42), 316(42); **17**:260(91), 316

McDonald, B. M., **24**:201, 275

McDonald, J. E., **14**:297, 298(27, 28), 302, 303, 342; **26**:12(135), 99(135)

McDonald, J. W., **SUP**:73(193), 92(193), 94(193), 165(193)

McDonald, R. A., **29**:188(186), 212(186)

McDonough, J. B., **5**:109, 128

McDonough, J. M., **25**:354, 355, 356, 415

McDougal, E. I., **12**:256(66), 280

McDougall, J. G., **7**:204(134), 217

McEligot, D. M., **6**:539, 542(57), 543, 563; **SUP**:18(32), 35(45), 71(45), 77(45),

McEligot, D. M. (*continued*)
  136(32), 145(45), 146(330), 147(330),
  148(330), 149(330), 150(330), 218(411),
  297(45), 319(45), 320(45), 321(45)
McElroy, C., **20**:264(266), 312(266)
McElroy, M. B., **14**:267(44, 47), 280
McElroy, W. D., **10**:118(50, 51, 52, 53), 121(50,
  51, 52, 53), 165
McEntire, J. A., **4**:206, 227; **7**:254, 255, 317
McFadden, G. B., **19**:79(205, 220), 80(205),
  92(205), 93(220)
McFadden, J. H., **31**:106(17), 138(17), 156(17)
McFadden, P. W., **5**:73, 126, 400, 480, 481(263),
  484(272), 486, 487, 488, 490, 491,
  492(272), 494, 495(263), 498, 499, 517;
  6:312(101), 366; **7**:50; **11**:67(32), 80, 82,
  105(137, 138, 139), 190, 195; **17**:119(113),
  156
McFarland, D., *see* McFarland, R. D.
McFarland, J. K., **30**:232(36), 234(36), 251(36)
McFarland, R. D., **18**:46(89, 90), 83(89, 90)
McFarlane, D. R., **22**:176(69), 350(69)
McFarlane, R., **29**:307(168), 344(168)
McGarthy, A. M., **28**:98(105), 143(105)
McGarvey, G. B., **31**:444(41), 472(41)
McGee, J. P., **9**:118(8), 177
McGee, T., **23**:288(48), 306(48), 356(48)
McGillis, W. R., **23**:12(74), 14(74), 53(74, 75),
  56(74), 57(75), 58(74), 61(74), 87(74),
  90(74), 106(74), 130(74, 75)
McGinley, M. B., **15**:287(97), 325
McGinty, D. J., **14**:306, 339(139), 343, 346
McGlone, B., **22**:20(5), 21(5), 23(5), 60(5),
  82(5), 152(5)
McGrary, D. M., **12**:87, 112
McGrath, J. J., **22**:292(142), 300(143), 321(150),
  353(142, 143, 150), 360(6), 435(6)
McGrew, J. L., **4**:191, 225
McGuinness, M. J., **30**:116(69), 192(69)
McGuire, J. H., **10**:279(113), 284
McGuirk, J. J., **13**:67(12, 13), 107, 115
McHale, S., **25**:261, 318
McIlhinney, A. E., **10**:203(101), 217
McInteer, B. B., **7**:279, 319
McIntire, L. V., **15**:208(96), 213(110, 111, 112),
  225(96, 110, 111, 112)
McIntosh, G. E., **15**:48(76), 57(76); **16**:233
McIntyre, J. T., **11**:218, 221, 236, 256
McKean, D. C., **5**:295(46), 323
McKee, H. R., **17**:4, 8, 57

McKenna, D. R., **28**:389(171), 390(171),
  391(171), 397(188), 398(171, 188),
  422(171), 423(188)
McKenzie, D. P., **19**:82(262), 94(262)
McKenzie, D. R., **18**:185(130, 131, 132),
  186(132, 137, 138), 191(131, 132, 137,
  138), 238(130, 131, 132), 239(137, 138);
  23:399(66, 67), 422(66, 67), 463(66, 67)
McKerslake, D., **11**:318(27), 376(27), 435
McKibbin, R., **20**:315(8), 316(8), 349(8)
McKibbins, S. W., **10**:172, 175, 177(21), 179,
  215
McKillop, A. A., **SUP**:80(231), 121(231),
  149(231), 150(231), **15**:101(111),
  139(111); **23**:327(217), 336(254, 255),
  364(217), 366(254, 255)
McKinley, R. M., **23**:274(177)
McKinney, B. G., **7**:271, 301, 315, 317
McKinney, G. W., **31**:402(61), 428(61)
McLaren, D. R., **23**:222(114), 223(114),
  225(114), 272(114)
McLaren, J., **14**:173, 182, 183(89), 184(89),
  244
McLaughlin, J., **31**:410(74), 428(74)
McLaughlin, M. H., **19**:98(7), 108(7), 137(7),
  160(7), 184(7)
McLeod, P. S., **28**:352(90), 419(90)
McMahon, H. O., **11**:365(113, 114), 371(113,
  114), 372, 375(114), 437
McMalion, M., **29**:71(27), 121(27), 125(27)
McMantus, W. R., **22**:229(144), 353(144)
McManus, H. N., **1**:376, 385(25), 388, 441
McMillan, H. K., **15**:202(89), 203(89), 204(89),
  221(89), 225(89); **20**:187(41), 301(41)
McMillan, O. J., **25**:341, 380, 415
McMillen, T. J., **18**:203(94, 95), 205(94),
  206(94, 95), 238(94, 95)
McMordie, P. K., **18**:36(70), 38(70), 82(70)
McMordie, R. K., **5**:451, 453, 515;
  SUP:77(214), 148(214)
McMullen, A. S., **31**:453(42), 472(42)
McMurray, D. C., **23**:101(105), 131(105);
  26:147(104), 163(104), 165(104),
  169(104), 186(104), 213(104)
McNabb, A., **14**:63(136), 103
McNally, W. A., **12**:103, 105, 106, 108, 113;
  15:63(19a), 64(19a), 114(19a), 135(19a)
McNaught, J. M., **16**:147, 156
McNelly, M. J., **16**:118, 155, 181(16), 238
McNulty, J. P., **28**:316(162), 336(162)

McPhedran, R. C., **23**:399(66, 67), 422(66, 67), 463(66, 67)

McPhedron, R. C., **18**:185(130, 131, 132), 186(132, 137, 138), 191(131, 132, 137, 138), 238(130, 131, 132), 239(137, 138)

McPherson, R. K., **4**:120(91), 140

McQuiston, F. C., **17**:317

McQuivey, R. S., **16**:6(4), 15(4), 16(4), 56

McReynolds, L. S., **5**:389, 510

McSweeney, T. I., **7**:240, 253, 266, 315

McUmber, L. M., **29**:220(42, 52), 221(42, 46, 47, 48, 49, 50, 51, 52, 53, 54), 255(52), 270(52), 271(141, 142), 278(52), 282(52), 290(52), 310(52), 315(42), 316(42), 323(42), 337(42, 46, 47, 48, 49, 50, 51, 52, 53, 54), 342(141, 142)

McVey, D. F., **24**:102(12), 184(12)

McWhirter, R. W. P., **4**:251(14, 15), 252(15), 313

McWhorster, D. B., **30**:101(37), 120(37), 121(37), 190(37)

Meachum, J. S., **28**:135(25), 139(25)

Meader, P. F., **2**:94(76, 77), 108

Mecham, W. J., **14**:232(168), 234(168), 246

Meckrios, A. A., **6**:550(88), 564

Medeiros, A. A., **5**:355(63), 360, 361, 362, 368, 370, 372, 373, 374, 445, 446, 447, 448, 449, 450, 451, 507; **7**:53(35), 85; **11**:170(180), 173(180), 174(180), 175(180), 177(180), 185(180), 186(180), 197

Medhekar, S., **29**:180(164, 165), 181(164), 199(164), 210(164, 165)

Medina, A. G., **24**:103(18), 185(18)

Mednick, R. L., **23**:207(13, 14), 235(13, 14), 267(13, 14)

Mednikova, N. M., **31**:160(35), 323(35)

Medvetskaya, N. V., **21**:7(15), 8(19, 20), 9(20), 11(20), 12(20, 24), 22(20), 29(20), 51(15, 19, 20, 24)

Meel, D. S., **8**:121, 160

Megaridis, C. M., **26**:78(136, 137), 99(136, 137); **28**:4(4), 6(4), 7(4), 8(4), 9(4), 10(4), 15(4), 36(57, 58), 37(57), 39(57), 40(57), 41(57), 42(57), 43(57, 58), 44(57), 45(57), 46(57), 47(57), 48(57), 54(57, 58), 70(4), 72(57, 58)

Megerlin, F., **20**:244(185), 308(185)

Meghreblian, R. V., **5**:28(36), 52

Mehl, J. B., **17**:76, 154

Mehra, V., **18**:245(5), 246(5), 320(5)

Mehra, V. K., **19**:81(248), 94(248)

Mehra, V. S., **5**:108, 127

Mehrabian, R., **28**:15(24), 18(24), 71(24), 231(3, 5, 6), 251(3, 6, 39), 266(5), 329(3, 5, 6), 330(39)

Mehrotra, S. P., **28**:306(147), 335(147)

Mehta, D., **23**:216(178), 217(178), 274(178)

Mehta, K. N., **20**:341(61), 352(61)

Mehta, N. C., **9**:156, 179

Mei, F. M., **18**:22(52), 81(52)

Meier, R., **13**:58, 60

Meigs, P., **4**:82(20), 138

Meijer, P. H. E., **17**:68(9), 153

Meijer, P. S., **24**:314(2), 318(2)

Meiklejohn, G. T., **15**:250, 280

Meilroy, W., **1**:333(62), 340, 354

Meisenburg, S. J., **9**:265(139), 272

Meissner, J., **13**:215, 216(28), 218(28), 257(113), 264, 266

Meixner, B. D., **24**:315(9), 319(9)

Meksyn, D., **4**:331, 335, 444; **8**:14, 15(76), 86

Meksyn, D. S., **21**:181

Melcher, J. R., **14**:123, 127(27), 145, 148

Melik-Akhnazarov, T. Kh., **19**:111(75), 186(75)

Melik-Pashaev, N. I., **6**:549, 551, 563

Melli, T. R., **30**:94(12), 189(12)

Melling, R., **11**:391(150), 411, 438

Mellink, J. H., **17**:94(67), 154

Melngailis, J., **28**:343(49), 417(49)

Melnichenko, V. V., **18**:169(46), 171(46), 176(46), 228(46), 229(46), 237(46)

Melnikov, Yu. A., **15**:326

Melosh, M. J., **19**:82(265), 94(265)

Melville, H. W., **3**:281(84), 301

Men', A. A., **11**:344(66b), 345(66b), 360(93d), 365(123), 374(123), 375(123), 392(154, 154a), 400(163), 401(163), 404(154, 170), 407(154), 436, 437, 438

Mena, B., **19**:317(107), 318(107), 354(107); **23**:193(12), 206(12), 226(12), 267(12); **25**:292, 308, 315

Menard, A. R., **17**:152(184), 158

Menard, W. A., **5**:293(35), 294(35), 309(80), 322, 324; **8**:241, 243(8, 9), 244, 282

Menard, W. E., **12**:138(29), 139, 140, 141, 146, 147, 193

Mendelson, E. S., **22**:20(7, 44), 21(44), 60(7), 152(7), 154(44)

Mendelssohn, K., **17**:82, 154

Mendenhall, W., **31**:349(35), 426(35)

Menderfield, E. L., **10**:172(22), 175(22), 177(22), 179(22), 215

Mendes, P. Souza, *see* Souza Mendes, P.

Mendoza, C. A., **30**:181(158), 196(158)

Mene, P. S., **25**:161(43), 162(43), 184(43), 200(43), 244(43)

Menegus, D. K., **30**:154(118), 194(118)

Meng, J. C. S., **6**:116(181b), 117(181b), 118, 131

Mengüç, M. P., **23**:136(8), 141(20), 184(8), 185(20); **27**:ix(12); **29**:311(171), 344(171); **31**:390(45), 427(45)

Menkes, H. R., **3**:35(17), 97

Mennicke, U., **26**:254(69, 70), 255(69, 70), 327(69, 70)

Mennig, G., **13**:219(72, 79), 248, 265

Mennig, J., **19**:35(70), 86(70)

Menold, E. R., **4**:38, 63; **6**:474(233), 480(233), 482, 497; **8**:170(33, 34), 173, 198, 201, 203, 204, 205, 206, 207, 208, 209, 210, 211(33), 216, 221, 226; **9**:311(69), 347

Menon, A. S., **19**:61(139), 62(139), 89(139)

Menon, S., **25**:394, 402, 417

Menzie, D. E., **23**:193(53), 204(53), 205(53), 206(53), 222(53), 226(53), 269(53)

Menzies, D. C., **29**:136(20), 143(20), 203(20)

Mercer, A. M., **SUP**:106(271), 172(348), 213(405)

Mercer, A. McD., **8**:70(268), 90

Mercer, J. W., **14**:6(26), 52(122), 53(122, 123), 54(122, 123), 78, 79(169), 83(169), 88, 100, 103, 104

Mercer, W. E., **6**:312(105), 366; **SUP**:190(376), 191(376), 194(376); **19**:308(92), 312(92), 353(92)

Mercier, C., **16**:72(24), 73(24), 74, 75, 80(24), 153, 154

Mercier, J. L., **21**:300, 345

Meredith, R. E., **12**:238(68), 280; **18**:162(15), 179(15), 181(66), 182(68), 183(15), 184(15), 185(66), 186(15), 187(15, 68), 188(15), 190(15), 236(15), 237(66, 68)

Meric, R. A., **20**:186(42), 187(42), 301(42)

Merilo, M., **17**:39(130, 131), 62; **29**:220(43), 221(43), 316(43), 337(43)

Merk, H. J., **3**:15, 32; **4**:9, 18, 33, 62, 63, 335, 446; **8**:19, 86, 118, 120, 132, 160; **11**:202, 203(38), 205(38), 209(38), 225(38), 253, 254; **15**:152(10), 223(10); **18**:113(46), 114(46), 159(46)

Merker, G. P., **24**:289(22), 319(22)

Merkle, C. L., **24**:193, 194, 272

Merkulov, A. P., **31**:160(7), 322(7)

Mermangen, W. H., **8**:273, 283

Merriam, R. L., **3**:163(44), 164(44), 174; **11**:350(68), 352(68), 436

Merril, J. A., **1**:410, 442

Merrill, E. W., **2**:385(101), 396; **12**:78(5), 88(5), 102(5), 106(5), 111; **15**:63(17), 64(17), 69(17), 92(87), 96(95), 97(95), 104(124), 110(17), 114(124), 135(17), 138(87, 95), 140(124); **18**:169(33), 170(33), 172(33), 208(97), 212(33), 236(33), 238(97)

Merring, R., **30**:222(45), 223(45), 224(45), 225(45), 251(45); **31**:468(59), 473(59)

Mersman, W. A., **15**:285, 287, 326

Mersmann, A., **10**:203, 217

Merte, H., Jr. (Chapter Author), **4**:161, 162, 169(39), 170(41), 171(41), 172(41), 173(41), 174(41), 206(88), 224, 226; **5**:90, 91, 126, 348, 375(87), 376, 377(89), 381, 395, 398, 400, 401, 407(87), 413, 414, 416, 419, 422, 423, 424(87), 425, 426(87), 427, 432, 433, 434, 435, 437, 441, 459, 460, 461, 462, 463, 464, 465, 467(236), 468, 507, 508, 509, 510, 513, 515, 516; **9**:181, 256(88), 270, 296, 346; **11**:60(27), 105(27, 67), 106, 107(93), 108(27), 112, 115, 118(27), 119(27, 93), 120(67), 121(27, 67, 93), 122(27, 67), 190, 192, 193; **16**:233; **18**:257(37, 46), 260(37, 46), 299(37), 302(37), 321(37), 322(46); **21**:56(5), 133(5); **23**:328(222), 364(222)

Mertl, J., **23**:202(135), 213(135), 218(136), 220(136), 273(135, 136)

Meryman, H. T., **4**:123(105), 141; **22**:176(69), 195(145), 350(69), 353(145)

Merzkirch, W., **30**:256(7), 264(26), 273(46), 274(46), 283(68, 69), 286(69), 287(74), 288(74), 289(80), 290(80), 291(85, 86, 87), 292(87, 89), 293(87), 294(7), 307(7), 308(26), 309(46), 310(68, 69, 74, 80), 311(85, 86, 87, 89)

Mesler, R., **29**:166(119), 208(119)

Mesler, R. B., **10**:109(40, 41), 164; **11**:17(54), 19, 49; **16**:101, 154; **17**:4, 57; **20**:23(44), 81(44)

Metais, B., **SUP**:411(524), 413(524)

Métais, O., **25**:345, 346, 361, 378, 392, 401, 406, 409, 415, 417

Metiu, H., **9**:215(38), 264(38), 269; **14**:293(25), 341

Metropolis, N., **5**:3, 50

Metz, B., **4**:120(92), 140

Metzger, D. E., **5**:199, 200, 201, 203, 249; **7**:348, 350, 354, 354(44), 355(45), 358, 361, 370, 375, 378, 379; **13**:43, 60; **20**:232(158), 306(158); **23**:322(202), 324(209), 333(247), 344(295), 363(202, 209), 365(247), 368(295); **26**:130(76), 149(76), 151(76), 161(76), 170(126, 127), 212(76), 214(126, 127)

Metzger, W., **17**:152(183), 158

Metzner, A. B. (Chapter Author), **2**:357, 357(65), 358, 361, 363, 364(65), 365(18), 371(74), 372(71, 74), 373(66), 375(124), 382(26, 65, 73), 383(26), 384, 385, 386, 387, 391(75, 76), 394, 395, 396, 397; **7**:103, 160; **12**:88, 101, 102(21), 112; **13**:207(5), 264; **15**:61(2), 63(20), 64(20), 66(2), 67(41), 71(19, 52), 77(20), 88(75), 92(86), 101(116), 102(116), 104(128), 105(128), 107(128), 111(123), 121(123), 132(2), 134(2), 135(20), 136(41, 51), 137(63, 75), 138(86), 139(116, 123), 140(128), 144(3), 149(3), 178(36), 182(47, 48), 190(68), 193(70), 196(84), 197(3), 217(84), 219(84), 220(68), 223(3), 224(36, 47, 48, 68, 70), 225(84); **19**:249(6), 264(27), 325(119), 338(153), 349(6), 350(27), 354(119), 356(153); **23**:188(180), 196(235), 198(180, 181), 206(47, 174, 179, 235), 216(47), 217(47), 218(174), 220(174), 226(48), 269(47, 48), 274(174, 179, 180), 275(181), 277(235); **24**:102(1), 184(1); **25**:253, 258, 286, 292, 296, 315, 318

Mewes, D., **13**:16(46, 47), 60; **30**:261(16), 308(16)

Mewis, J., **23**:192(182), 275(182)

Meyder, R., **SUP**:65(102), 355(102), 360(102)

Meyer, B. A., **18**:17(31), 25(54), 80(31), 81(54)

Meyer, G., **28**:370(127), 420(127)

Meyer, H., **14**:333, 334(121), 345

Meyer, J., **13**:169, 202

Meyer, J. E., **31**:121(19a), 139(19b), 156(19a, 19b)

Meyer, J. F., **23**:329(232), 365(232)

Meyer, J. L., **28**:318(182, 183, 196), 337(182, 183), 338(196)

Meyer, J. P., **15**:285(116), 287, 290, 326

Meyer, L., **29**:155(83), 206(83)

Meyer, R. C., **1**:335, 337, 340, 354

Meyer, R. T., **18**:28(63), 82(63)

Meyer, R. X., **1**:333(63), 354

Meyer, W. A., **12**:79, 111; **15**:105(137), 111(137), 140(137)

Meyers, P. S., **26**:147(104), 163(104), 165(104), 169(104), 186(104), 213(104)

Meyerson, B. S., **28**:352(89), 419(89)

Meynart, R., **30**:291(88), 311(88)

Meyvis, J., **17**:218(67), 231(73), 250(73), 258(73), 259(73), 260(73), 290(67, 73), 294(67, 73), 295(67, 73), 315, 316

Miao, H., **19**:310(101), 313(101), 353(101)

Miao, Y., **28**:370(131), 421(131)

Michael, Y. C., **30**:295(95), 311(95)

Michaeli, W., **28**:201(48), 230(48)

Michaelides, E. E., **25**:292, 296, 316

Michaelidis, M., **28**:394(184), 423(184)

Michaels, A. S., **10**:92, 164; **15**:287(206), 329

Michaels, J. N., **28**:343(51), 417(51)

Michaelson, A. A., **6**:196, 197, 363

Michailov, I. I., **17**:327(33, 34), 334(33, 34), 336(33, 34), 340

Michalik, E. R., **11**:376(131), 378(131), 382(131), 384, 385, 438

Michele, H., **23**:219(183), 220(183), 221(183), 275(183)

Michell, J. L., **26**:120(41), 210(41)

Michels, A., **7**:9

Michels, W. C., **2**:355

Michelsen, M. L., **SUP**:79(227), 118(227)

Michenko, N., **5**:409, 411, 413, 440, 513

Michiyoshi, I., **2**:380, 396

Michiyoski, I., **4**:7, 26(44), 33, 62, 63

Michlin, J., **23**:219(68), 222(68), 270(68)

Michniewicz, M., **23**:197(137), 199(137), 201(137), 202(137), 206(137), 218(138), 221(137), 273(137, 138); **24**:110(95), 111(95), 188(95)

Mickley, H. S., **3**:64, 71, 99; **10**:186, 187, 188, 189, 191, 200, 203, 204, 205, 206(81), 216, 217; **14**:151, 169, 207, 212, 214, 216(15), 217(15), 218(15, 145), 222(15), 223(145), 242, 246; **15**:87(73), 90(73), 92(73, 87), 93(73), 111(73), 137(73), 138(87), 236(39), 277; **16**:364; **19**:123(115, 116), 129(116), 130(116), 132(115), 134(115, 143), 173(143), 187(115, 116), 188(143), 330(126), 354(126)

Micol, J. R., **23**:320(159, 161), 361(159, 161)
Micuta, W. L., **17**:251, 316
Middis, J., **30**:210(37), 213(37), 233(37), 251(37)
Middlehoek, J., **28**:394(185), 395(185), 423(185)
Middleman, S., **13**:219(92), 225, 256, 266; **15**:70(48), 71(48), 77(48), 136(48), 144(6), 149(6), 179(6), 223(6); **23**:188(184), 193(43, 82), 198(43), 199(43), 200(43), 201(43), 220(82), 268(43), 270(82), 275(184); **24**:109(86), 110(94), 111(94), 112(94), 187(86), 188(94); **28**:402(204, 205, 206, 207), 404(204), 424(204, 205, 206, 207)
Mierswa, B., **28**:96(12), 138(12)
Migai, V. K., **31**:160(10, 26, 28), 165(26, 28), 196(99, 100), 197(26, 28), 200(100), 212(113), 224(116), 225(116), 283(189), 322(10), 323(26, 28), 326(99, 100), 327(113, 116), 331(189)
Migay, V. K., **SUP**:227(418)
Mighdoll, P., **5**:298(54), 323; **8**:252(33), 253(33), 258(33), 259(33), 278(49), 279(49), 283
Migiris, C. E., **28**:343(53), 417(53)
Mihail, A., **9**:3(97), 110
Mihe, J. P., **23**:338(263), 366(263)
Mihelčić, M., **28**:317(175), 337(175); **30**:348(151, 153), 383(151), 390(153), 394(151), 433(151, 153)
MiHero, F. J., **16**:53, 58
Mii, T., **19**:108(51), 186(51)
Mikami, H., **7**:181, 213
Mikesell, R. D., **1**:219, 220(41), 264; **15**:242(54), 278
Mikhail, N. R., **5**:406(153), 512
Mikhailov, A. I., **31**:165(72), 182(72), 186(72), 325(72)
Mikhailov, A. V., **31**:279(182), 283(189), 330(182), 331(189)
Mikhailov, I. G., **14**:259(25), 278(25), 279(25), 280
Mikhailov, M. D., **19**:192(10), 243(10); **24**:56(22), 99(22)
Mikhailov, M. S., **25**:131(204), 136(204), 150(204); **31**:160(23), 323(23)
Mikhailov, Yu. A., **19**:191(2), 242(2); **24**:42(15), 99(15)
Mikhailova, T. V., **25**:35(70a), 143(70a)
Mikhalchenko, R. S., **9**:406(99), 417
Mikhaylov, G. A., **8**:134(65), 135(65), 136(65),

160; **18**:132(65), 133(65), 134(65), 159(65)
Mikhaylov, Y. A., **1**:173(19), 176(19), 184; **8**:3, 84
Mikheyev, M. A., **1**:141(13), 184; **6**:532, 563; **8**:94(15), 114(15), 129(15), 130(15), 159; **11**:203, 205, 228, 253; **18**:88(15), 109(15, 41), 125(15), 127(15), 158(15, 41); **31**:236(123), 327(123)
Mikhlenov, I. P., **19**:110(72), 186(72)
Mikhlin, S. G., **8**:34(175), 57(175), 88, 258(38), 279(38), 283; **19**:195(22), 196(22), 237(22), 238(22), 243(22)
Mikic, B. B., **9**:205(23), 233(66), 242(66), 268, 270, 359(16), 414; **11**:14(48, 49), 15, 16, 49; **16**:95, 154; **20**:232(166), 243(174, 175, 176, 177), 259(246), 307(166, 174, 175, 176, 177), 311(246); **21**:86(50), 88(56), 125(114), 135(50), 136(56), 138(114); **22**:304(183), 305(183), 306(183), 307(183), 309(183), 311(183), 312(183), 355(183); **23**:298(89), 358(89)
Mikielewicz, J., **9**:3(85), 76(85), 109
Mikitenko, E. D., **17**:327(33, 34), 334(33, 34), 336(33, 34), 340
Mikk, I. R., **23**:305(112), 359(112)
Mikkelsen, C. D., **SUP**:95(254)
Miksis, M. J., **28**:41(69), 73(69)
Mikulecky, D. C., **22**:287(146), 353(146)
Mikulin, E. I., **31**:160(43), 324(43)
Mikus, T., **25**:168(66), 171(66), 173(66), 175(66), 176(66), 189(66), 196(66), 245(66)
Milanez, L. F., **19**:36(74), 86(74); **20**:203(92), 204(92), 205(92), 303(92)
Miles, D. N., **17**:30(93), 60
Miles, J. B., **SUP**:202(393), 203(393)
Miles, R. G., **9**:253(80), 270
Milford, C. M., **7**:358, 379
Milich, W., **5**:109, 128
Milionshchikov, M. D., **25**:76(123), 145(123)
Millard, R. C., Jr., **18**:340(60), 341(60), 342(60), 363(60)
Millazo, G., **12**:226(69), 229(69), 237(69), 242(69), 249(69), 250(69), 252(69), 255(69), 280
Miller, C., **23**:198(185), 275(185); **25**:299, 318
Miller, C. E., **12**:242(114), 282
Miller, C. G., **23**:315(147), 318(147), 320(147, 158, 159, 160), 360(147), 361(158, 159, 160)

Miller, C. O., **10**:201, 202, 203, 206, 209, 217; **14**:214, 218(148), 223(148), 226(148), 227, 246; **19**:133(138), 188(138)

Miller, D. C., **30**:396(160), 398(160), 434(160)

Miller, D. G., **12**:259(70), 261(70), 280

Miller, D. L., **7**:310, 319

Miller, D. W., **12**:18(30), 20(30), 21(30), 26(30), 74

Miller, E. N., **9**:121, 178

Miller, E. R., **30**:216(38), 251(38)

Miller, G. B., **2**:443(42), 444(42), 451

Miller, J. A., **SUP**:73(185), 191(377), 192(379), 213(185), 242(185); **28**:366(116, 118), 382(116), 393(116), 394(116), 420(116, 118)

Miller, J. C., **28**:135(69), 141(69)

Miller, J. H., **9**:262(121), 271; **14**:137, 138, 148

Miller, J. T., **7**:197, 215

Miller, L. N., **12**:22, 74

Miller, M., **18**:198(89), 237(89)

Miller, P. S., **7**:368, 379

Miller, R. N., **5**:347, 506

Miller, R. W., **10**:109(37), 164; **SUP**:71(158), 198(158), 199(158), 200(158), 210(158), 213(158), 225(416), 240(158), 241(158), 242(158, 427); **19**:268(49), 271(49), 272(49), 274(49), 351(49)

Miller, W. F., Jr., **31**:334(7), 346(7), 357(7), 365(7), 425(7)

Miller, W. S., **5**:330, 404, 405, 409, 411, 412, 413, 420, 421, 440, 445, 511

Millero, F. J., **18**:330(28, 29, 31), 333(28, 43, 44, 45, 46), 335(29), 336(43), 337(53), 348(28), 350(28), 359(28), 362(28, 29, 31), 363(43, 44, 45, 46, 53)

Millikan, C. B., **15**:88(76), 137(76); **19**:325(120), 354(120)

Millikan, R. A., **2**:355

Millikan, R. C., **8**:248(21, 22, 24, 26), 283

Milliken, F. P., **17**:100(94), 154

Mills, A. F., **9**:221, 258(98), 265(133), 269, 270, 272; **12**:136(24), 186(24), 193; **SUP**:73(193), 92(193), 94(193), 165(193); **15**:232(22), 277; **23**:285(40), 355(40); **26**:55(97), 78(114), 97(97), 98(114)

Mills, A. T., **21**:73(17), 134(17)

Mills, D. M., **28**:80(64), 141(64)

Mills, E. S., **4**:228

Mills, R., **30**:290(81), 310(81)

Millsaps, K., **3**:107, 173; **4**:24, 63; **6**:435(168),

437(168), 494; **9**:303(51, 52), 347; **11**:213, 214, 255; **SUP**:112(280), 113(280); **28**:389(162), 422(162)

Milly, P. C. D., **30**:96(27), 102(41), 189(27), 190(41)

Milman, O. O., **31**:160(50), 324(50)

Milne, E. A., **3**:176(4), 182(4), 189(4), 204, 210(4), 248; **11**:332(54), 436; **21**:246(24), 276(24)

Milne, P. A., **9**:29(6), 34, 106

Milne-Thompson, L. M., **5**:92, 117, 127; **20**:354, 359, 362, 388

Milne-Thomson, L. M., **8**:49(231), 89; **17**:11(52), 58; **23**:2(4), 126(4); **26**:119(38), 122(38), 188(38), 210(38)

Milovanov, Y. V., **9**:241(71), 242(71), 255(71), 270; **21**:92(66), 136(66)

Milton, G. W., **18**:199(150, 151, 152), 200(151), 239(150, 151, 152)

Min, J. H., **29**:26(22), 27(22), 28(22), 30(22), 56(22)

Min, K., **9**:134, 144, 179

Min, T. C., **SUP**:361(506, 507), 363(506, 507), 365(514, 515)

Minaeu, G. A., **25**:163(45), 244(45)

Minashin, M. E., **1**:435(140, 141), 444

Minashin, V. E., **SUP**:355(496), 356(496)

Minchenko, E. P., **16**:235

Minchenko, F. P., **16**:181(14), 226, 235, 238; **20**:14(20), 15(20), 16(20), 80(20)

Minden, C. S., **11**:2(2), 47

Miner, E. W., **25**:359, 415

Mingle, J. O., **11**:397(159), 401(159, 166), 403(166), 438

Minière, F., **23**:294(72), 357(72)

Miniovich, Y. M., **1**:178, 184

Minkowycz, W. J., **2**:434(32), 436(32), 451; **8**:35(182, 183), 88; **9**:258(94, 95), 265(94), 270; **14**:57(128, 129), 58(128, 129), 59(128, 129), 60(128, 129), 63, 75(128, 129), 103; **15**:124(149), 141(149), 229(8), 232(8, 21), 276, 277; **19**:79(203), 92(203); **21**:109(94), 137(94); **24**:194, 272; **31**:401(58), 428(58)

Minkwitz, G., **6**:225, 226(70), 227(70), 228(70), 365

Minnich, S. H., **15**:43(67), 57(67)

Mino, Y., **28**:345(64), 418(64)

Minto, R., **21**:287, 300, 345

Minton, P. F., **20**:76(68), 82(68)

Mintz, M. S., **12**:242(96), 247(96), 281
Mintz, Y., **14**:266(41), 280
Minyatov, A. V., **15**:326
Mirabel, P., **14**:327(105, 106), 328(105), 329(106), 334(105, 106), 339(105, 106), 345
Mirande, J., **6**:30(50), 31(51), 113(51), 124
Mirels, H., **2**:172(45), 176(45), 179(45), 269; **7**:206, 207, 217, 218; **8**:36(207), 42(207), 89
Mirenayat, **18**:36(71), 37(71), 82(71)
Mironov, A. I., **15**:328
Mironov, B. P., **6**:379(11), 380(11), 486; **25**:28(58), 142(58)
Mironov, V. F., **1**:144, 184
Mironov, V. N., **1**:435(141), 444
Mironov, Y. Y., **SUP**:336(488), 337(488)
Miropolsky, Z. L., **1**:244(85), 266, 396(76), 398(76), 412, 416(107, 108), 417(107, 108), 420(113, 114), 442, 443; **6**:551, 552, 564; **7**:30, 38(21), 39, 49, 50, 66, 67, 84, 85; **11**:170(173), 173, 176, 177(173), 185, 186(173), 196
Mirza, F. A., **28**:150(49), 230(49)
Mirzoyan, P. A., **21**:46(75), 47(75), 52(75)
Misenta, R., **17**:127, 156
Mishima, K., **17**:37, 62; **29**:271(139), 272(139), 342(139)
Mishra, I. M., **23**:213(186), 236(186), 238(186), 240(186), 275(186); **25**:258, 288, 289, 291, 318
Mishra, P., **23**:213(186), 214(236), 236(186, 236), 238(186), 240(186), 241(236), 275(186), 277(236); **25**:258, 288, 289, 291, 293, 297, 298, 300, 304, 305, 306, 317, 318
Mishra, S. P., **15**:179(40), 180(41), 224(40, 41)
Misra, B., **3**:108(14), 173; **5**:445, 446, 447, 448, 449, 514; **9**:251(79), 270
Misra, D. S., **28**:103(70), 141(70)
Mistra, B., **11**:170(185), 173(185), 175(185), 177(185), 197
Mital, U., **9**:80, 82, 83, 109
Mitalas, G. P., **6**:416(85), 490
Mitchell, D. E., **29**:146(42), 204(42)
Mitchell, J. E., **7**:88, 140, 160, 161; **8**:293(18), 349; **12**:233(71), 251(71), 270(71), 271(71), 280
Mitchell, J. W., **5**:131(9), 191, 192, 193, 199, 200, 201, 203, 246; **SUP**:18(26), 119(26); **18**:13(19d), 14(19d), 17(31), 25(54),

80(19d, 31), 81(54); **22**:59(45), 62(45), 63(45), 65(45), 66(45), 67(45), 68(45), 69(45), 90(45), 138(45), 154(45)
Mitchell, R., **9**:3(93), 109
Mitchell, S. J., **20**:138(8), 179(8)
Mitjaev, Ju. I., **1**:435(141), 444
Mitra, S. S., **11**:352(73), 436
Mitrofanos, V. V., **29**:98(65), 126(65)
Mitshell, J. W., **6**:474(314), 501
Mitskevich, A. I., **6**:474(287), 500
Mitson, A. E., **10**:200, 202, 217
Mitsoulis, E., **28**:150(49), 230(49)
Mitsuishi, N., **25**:258, 259, 270, 271, 275, 276, 277, 278, 319
Mitsumori, K., **21**:120(106), 122(106), 138(106)
Mitsumura, H., **5**:233(166), 234, 235(166), 251; **7**:129(38), 161; **21**:163, 182
Mittag, K., **17**:69(16), 70(16), 71(16), 74, 153
Mittal, R., **24**:227, 230, 231, 274
Mitteldorf, J., **28**:133(83), 142(83)
Mityaev, U. I., **1**:435(140), 444
Mix, T., **6**:82(138), 129
Mixon, F. O., Jr., **1**:232, 265; **14**:136, 146; **23**:230(187), 275(187); **25**:273, 318
Miyahara, S., **30**:329(33), 428(33)
Miyake, Y., **25**:384, 403, 413
Miyasaka, K., **26**:82(138), 84(246), 99(138), 104(246)
Miyasaka, Y., **23**:8(19), 12(19, 34), 14(19, 34), 16(34), 28(19, 34), 29(19, 34), 36(19, 34), 39(19, 34), 42(34), 46(34), 59(19, 34), 61(19, 34), 79(19), 80(19), 82(19, 95), 83(19), 84(19), 88(19), 93(19), 103(34), 106(19), 127(19, 34), 131(95); **26**:120(39), 121(39), 122(39), 123(39), 162(39, 117), 163(39, 117), 167(39), 169(39), 210(39), 214(117)
Miyatake, O., **11**:293, 315; **15**:166(25, 26, 27), 167(25, 26, 27, 28), 223(25, 26, 27), 224(28); **18**:73(143, 144), 86(143, 144); **20**:186(44, 45, 46), 187(43, 44, 45, 46), 191(43, 44, 45, 46), 301(43, 44, 45, 46); **28**:36(57, 58), 37(57), 39(57), 40(57), 41(57), 42(57), 43(57, 58), 44(57), 45(57), 46(57), 47(57), 48(57), 54(57, 58), 72(57, 58)
Miyauchi, T., **13**:164, 202
Miyazaki, H., **23**:5(6), 126(6); **26**:122(49), 123(49), 210(49)
Miyazaki, K., **29**:139(30), 204(30)

Miyazaki, Y., **20**:187(66, 67), 275(289), 276(289), 302(66, 67), 313(289)

Mizrahi, J., **21**:334, 344; **23**:207(15), 235(15), 267(15)

Mizuno, M., **23**:58(83), 60(83), 63(83), 77(83), 78(83), 85(83), 93(83), 109(113), 110(113), 112(113), 130(83), 132(113); **26**:181(134), 182(134), 201(134), 202(134), 214(134)

Mizushina, T. (Chapter Author), **7**:87, 108, 110, 122, 128, 160, 161; **11**:231, 232, 260; **12**:233(72), 237(72), 251(72), 270(72), 280; **15**:92(84), 93(84), 101(112), 102(112), 103(112), 105(131), 107(139), 108(139), 111(139), 113(139), 114(139), 124(139), 125(139), 128(158), 130(84, 139), 138(84), 139(112), 140(131, 139), 141(158); **21**:181; **25**:81(131), 146(131), 258, 260, 283, 284, 285, 318

Mizutani, H., **10**:64(50, 51), 67(51), 82

Mizutani, Y., **31**:334(12), 425(12)

Mladin, E. C., **26**:200(162), 216(162)

Mlodinski, B., **10**:177(52), 216

Mls, J., **23**:376(45), 462(45)

Mo, A., **28**:263(76), 306(146), 332(76), 335(146)

Moalem-Maron, D. (Chapter Author), **15**:227, 243, 244, 245, 246(78), 247(79, 80), 248, 250, 251(80), 252(80, 88, 90, 91), 253, 255(91), 256(91), 257(90, 91), 258(90, 91), 259(91), 260(91), 262(78), 278, 279, 280; **20**:111, 112, 130; **21**:56(6), 133(6), 141, 142, 179, 181, 183; **26**:2(203), 47(203), 102(203)

Moallemi, M. K., **19**:72(164, 165, 166), 73(165, 167), 90(164, 165, 166, 167); **24**:5(19), 6(24), 8(29, 30), 36(19), 37(24, 29, 30); **30**:335(66, 67), 337(66), 351(97, 98, 99), 352(97, 98, 99), 354(98, 99), 356(98, 99), 429(66, 67), 431(97, 98, 99)

Moan, M., **23**:219(170), 225(170), 226(199, 200), 228(199, 200), 274(170), 275(199, 200)

Mobbs, F. R., **21**:154, 155, 178

Močalov, V. A., **6**:312(100), 366

Mochida, A., **25**:396, 416; **27**:181(32), 186(32); **31**:334(12, 13), 391(48), 413(48), 418(48), 425(12, 13), 427(48)

Mock, W. C., Jr., **11**:232(169), 237(169), 260

Modest, M. F., **19**:80(244, 245), 94(244, 245); **30**:338(78), 339(78), 340(84), 342(84), 343(78), 430(78, 84); **31**:334(2), 425(2)

Moeck, E. O., **17**:39(129), 62

Moen, I. O., **29**:98(68, 69), 101(73), 102(73), 126(68, 69, 73)

Moen, R. H., **1**:357(3), 374(3), 375(3), 440

Moench, A. F., **14**:78, 80(166), 81(171), 82(171), 95, 104

Moeng, C.-H., **25**:344, 347, 362, 371, 374, 375, 376, 377, 401, 412, 413, 415, 416

Moffat, D. F., **20**:222(138, 139, 140), 223(139), 224(138, 139, 140), 225(140), 226(139), 256(138), 305(138), 306(139, 140)

Moffat, H., **28**:364(104, 105), 365(105), 369(104, 105), 370(104, 105), 371(104), 372(104), 373(104, 105), 374(104), 379(104), 389(105), 419(104, 105)

Moffat, R. J. (Chapter Author), **13**:81(44), 116; **16**:241, 242, 255, 265(13, 45), 270(44), 271(27), 280, 299(32), 302(9, 16, 28), 306(9), 310(23), 312(7, 11, 12, 18), 314(7), 320, 322(13), 328(46), 329, 334(44), 335(44), 339(44), 349(32, 49), 363, 364, 365; **18**:30(65), 32(65), 33(65), 82(65); **20**:196(80), 205(109, 110, 111, 112), 206(109), 207(109, 110, 111, 112), 208(110), 210(109), 211(109), 232(80, 153, 159), 233(110, 153, 159), 234(159), 235(159), 238(159), 242(153), 302(80), 304(109, 110, 111, 112), 306(153), 307(159); **23**:280(20, 21, 22), 287(22), 288(20), 289(20), 306(120), 309(21), 310(21), 322(22), 329(225), 330(20, 22, 225, 234), 333(250), 355(20, 21, 22), 359(120), 364(225), 365(234), 366(250); **30**:301(103), 311(103)

Moffatt, W. C., **1**:311, 312(49), 354

Moffette, R. T., **29**:68(18), 124(18)

Mogford, D. J., **29**:259(113), 260(113), 262(113), 263(113), 264(113), 266(113), 341(113)

Mohamed, S. A., **11**:298, 299, 315

Mohammed, M., **31**:459(43), 472(43)

Mohan, V., **23**:208(188, 189), 210(188), 275(188, 189)

Mohenski, T. A., **20**:248(211), 249(211), 309(211)

Mohtadi, M. F., **14**:21(55), 71(55), 72(55), 73(55), 101

Moin, P., **24**:192, 271; **25**:333, 336, 337, 342, 348, 351, 353, 354, 358, 360, 361, 362, 367, 369, 370, 374, 375, 378, 381, 382,

Moin, P. (*continued*)
    383, 385, 386, 392, 393, 394, 396, 397, 402, 403, 404, 408, 412, 413, 414, 415, 416, 417
Moissis, A., **5**:420, 422, 514
Moissis, R., **15**:244(70), 279; **20**:110, 111, 131
Mojtabi, A., **28**:370(127), 420(127)
Mojtehedi, W., **9**:227(59), 229(59), 269
Mokuya, K., **28**:402(214), 424(214)
Molchanov, A. M., **25**:35(70a), 143(70a)
Molerus, O., **19**:123(122), 125(122), 187(122); **23**:206(190), 275(190)
Molgaard, J., **20**:264(265), 312(265)
Mollard, J., **17**:322, 326(19), 339
Mollekopf, N., **26**:226(30), 227(30), 325(30)
Mollendorf, J. C. (Chapter Author), **9**:289, 290(12), 292(12), 311(73), 320(73), 334(73), 339(103), 346, 347, 348; **16**:24, 47, 54, 55, 57; **18**:325, 330(1), 331(1), 332(1), 350(1), 359(1), 361(1); **20**:203(88), 303(88)
Möller, F., **8**:281(50), 283
Moller, J. C., **23**:320(185), 362(185)
Molnar, D., **28**:11(15), 71(15)
Momenthy, A. M., **5**:381, 509
Monaghan, R., **3**:54, 60, 61, 98
Monberg, E. M., **30**:415(181), 435(181)
Monchick, L., **3**:266, 267, 268, 269, 279, 280, 281, 282(61), 298(25), 300, 301
Moncrief, J. A., **4**:120(89), 140; **22**:226(147), 353(147)
Monde, M., **17**:19(59), 20(59), 21(59), 24(59), 58, 59; **20**:252(239), 310(239); **23**:8(16, 17), 12(16, 17, 44, 45, 47, 51, 57, 64), 13(64), 14(16, 17, 44, 45, 51, 57), 15(47, 64), 22(16, 17, 44, 45, 47), 23(16, 47), 24(16), 26(16, 51), 27(16, 17, 47), 30(16, 57), 31(45, 57), 34(57), 35(64), 36(16, 17, 44), 37(16), 39(16, 17, 44), 40(64), 41(16), 55(47, 77, 78), 58(44), 59(16, 17, 45, 47, 51, 57, 64, 77, 78), 60(44), 61(16, 17, 45, 47, 51, 57, 64, 77, 78), 62(16, 44), 63(16, 17, 44, 77, 80), 64(16), 66(47, 80), 67(47), 69(77), 70(77, 78, 80), 71(78), 74(16, 44, 77, 78, 80), 75(16, 17, 44, 47, 77, 78, 80), 76(80), 77(16, 47, 77, 78), 78(16, 17, 45, 47, 78), 79(16, 47, 80), 85(45, 51, 80), 86(45), 87(16), 90(16), 93(45, 51), 95(78, 80), 96(80), 97(80), 99(64), 100(17, 64, 80, 104), 101(16, 64), 103(78), 106(16, 17, 44,
45, 47, 64), 107(16, 44, 57), 127(16, 17), 128(44, 45, 47, 51), 129(57, 64), 130(77, 78, 80), 131(104)
Mondin, H., **1**:384(49), 385(49), 388(49), 441
Monin, A. S., **6**:395(23), 398, 487, 488, 511(21), 513(21), 562; **8**:289(6), 291(6), 349; **14**:250(2), 263(2), 264(2), 265(2, 39), 266(2), 273(2), 279, 280; **16**:307, 364; **21**:200(27), 237(27); **25**:50(75), 143(75)
Monit, R., **15**:63(32), 64(32), 65(32), 136(32)
Monozaki, K., **31**:450(49), 473(49)
Monroe, S. G., **11**:318(13), 434
Mons, R. F., **16**:53, 57
Montagna, William, **4**:141
Montemango, C. D., **30**:96(30), 190(30)
Montgomery, R. S., **24**:8(35), 14(35), 37(35)
Monthie, A., **30**:232(39), 233(39), 251(39)
Monti, R., **12**:92(27), 103, 112
Montogmery, S. R., **SUP**:76(210), 77(219), 214(210), 217(210), 219(219), 220(219), 404(219)
Montoya, P. C., **30**:121(80), 192(80)
Montroll, E. W., **5**:48, 54; **8**:248, 282
Moody, F., **29**:251(105), 325(105), 340(105)
Moog, D. B., **19**:78(195), 92(195)
Moon, S. H., **28**:351(83), 418(83)
Moon, W. T., **11**:240(238), 263
Moon, Y., **24**:218, 272
Mooney, M., **2**:375(78), 396
Moore, B. C., **5**:478, 479, 516
Moore, B. J., **4**:331(25), 444
Moore, C. J., **15**:287, 289, 326
Moore, D. W., **20**:380, 381, 388; **26**:10(83), 96(83)
Moore, F. D., **11**:17(54), 19, 49; **17**:4, 57; **22**:226(46), 348(46)
Moore, F. E., **19**:70(160), 71(160), 90(160); **24**:3(10), 5(10), 36(10)
Moore, F. K. (Chapter Author), **2**:111(6), 266, 268; **4**:331, 334, 444; **6**:396(28), 487; **12**:1, 15, 16, 17(27), 18, 19, 24, 25(42), 26, 27, 31, 33(47), 34(47), 41, 42, 43, 48, 53, 55(57), 60, 61, 63(57), 65, 74, 75; **17**:164, 314
Moore, G. R., **10**:102, 164
Moore, J., **2**:231(65), 233(65), 269
Moore, J. P., **20**:264(266), 312(266)
Moore, R. A., **4**:283(48), 315
Moore, R. W., **5**:381, 509
Moore, T. W., **10**:179(60), 216

Moorhouse, W. E., **9**:3(94), 90, 109

Moo-Young, M., **25**:272, 273, 318

Moran, K. P., **20**:183(9), 244(195), 250(233), 251(233), 283(300), 299(9), 308(195), 310(233), 313(300)

Moran, W. R., **11**:315; **30**:280(62), 310(62)

Morbidelli, M., **26**:45(36), 94(36)

Morcos, S. M., **SUP**:412(528), 413(528)

Mordchelles-Regnier, G., **8**:184, 214, 227

Moreau, C., **28**:53(79), 73(79)

Moreau, R., **30**:333(57), 429(57)

Morehouse, K. A., **23**:320(174), 345(296), 362(174), 368(296)

Morel, T., **23**:343(290), 367(290)

Moreland, C., **18**:220(154), 221(154), 222(154), 233(154), 239(154)

Morel-Seytoux, H. J., **30**:94(9, 10), 102(10), 189(9, 10)

Moreno, G. F., **10**:261(73), 272(73), 283

Moreno, J., **23**:292(57), 356(57)

Moreno, L., **23**:218(250), 219(250), 220(250), 277(250)

Moresco, L. L., **16**:136, 155

Morese, **31**:459(44), 472(44)

Morette, R. A., **13**:219(70), 265

Moretti, P. M., **6**:404, 488; **16**:247(30), 280(30), 364

Morgan, A. N., **15**:63(22), 64(22), 135(22)

Morgan, G. W., **20**:364, 388

Morgan, I., **11**:105(100), 193

Morgan, V. T. (Chapter Author), **11**:199, 200(63), 211(63), 212, 255; **18**:123(55), 159(55)

Morgan, W. R., **2**:411, 412(10), 417(10), 450

Morgen, R. L., **11**:365(111), 437

Morgunov, N. G., **1**:435(141), 444

Mori, H., **11**:420(212), 440

Mori, S., **SUP**:12(20, 23), 13(20), 137(20), 188(23), 189(20, 23)

Mori, Y., **1**:298, 353; **5**:244, 251; **SUP**:413(530); **15**:215(115), 225(115); **19**:307(96), 313(96), 353(96); **21**:109(97, 99), 117(102, 103, 104, 105), 129(125), 130(126, 127), 137(97, 99, 102), 138(103, 104, 105), 139(125, 126, 127); **23**:293(66), 349(66, 305), 356(66), 368(305); **26**:44(88), 96(88)

Mori, Y. H., **21**:300, 345; **26**:49(228, 229), 51(228, 229), 103(228, 229)

Morie, F. H., **5**:390, 510

Morihara, H., **SUP**:73(194), 165(194, 343), 166(194), 169(194)

Morihira, M., **11**:231(242), 232(242), 263

Morinishi, Y., **25**:391, 393, 404, 406, 416

Morishita, T., **11**:231, 232, 237, 260

Morita, I., **23**:118(122), 132(122)

Moritz, A. R., **4**:119(82), 125(82), 140, 141; **22**:11(20), 17(20), 230(148), 231(148), 232(100), 314(148), 329(100), 351(100), 353(148)

Moritz, K., **7**:320c(148), 320d, 320f

Morizumi, S. J., **12**:143, 177, 178, 193

Morkovin, M. V., **13**:79(30), 116

Morley, E. W., **6**:197(31), 363

Morley, M. J., **9**:114, 116, 125, 127(1), 177

Morley, T. B., **26**:220(9), 237(9), 240(9), 246(9), 253(9), 274(9), 324(9)

Morooka, S., **23**:253(126), 254(126), 257(126), 261(126), 272(126)

Moros, A., **28**:318(192), 338(192)

Morosanu, C. E., **28**:339(1), 341(1), 343(1), 344(1), 350(1), 351(1, 72), 387(159), 388(159), 389(159), 415(1), 418(72), 422(159)

Morosova, N. A., **7**:9

Morozhenko, A. V., **10**:10, 36

Morozkin, V. I., **4**:206(84), 226

Morrin, E. H., **1**:357(1), 371(1), 440

Morris, B. W., **29**:234(87, 88, 89), 249(87), 251(87), 266(131), 339(87, 88, 89), 342(131)

Morris, D. J., **17**:23(73), 59

Morris, J. C., **4**:241, 242(7), 313

Morris, S., **25**:269, 318

Morris, W. D., **5**:242, 244(187), 251; **8**:6, 84; **9**:2, 3, 4, 73, 74, 98, 107, 108, 109, 110

Morrison, F. A., **14**:127, 146

Morrison, F. A., Jr., **26**:18(213), 28(139), 29(77, 78, 139, 202), 30(77, 139), 96(77, 78), 99(139), 102(202, 213); **30**:120(76), 121(76), 167(76), 192(76)

Morrison, J. F., **4**:120(90), 140

Morrison, R., **22**:406(22), 435(22)

Morrison, R. A., **20**:244(200), 309(200)

Morrisson, J. L., **2**:355

Morrow, J. D., **6**:29(44), 124

Morrow, W. B., **14**:92, 94(182), 104

Morse, A. P., **13**:75(23), 115

Morse, D. C., **12**:5(14), 7(14), 35(14), 73

Morse, F. H., **5**:348, 507

Morse, H. L., **9**:141, 179

Morse, P. M., **3**:156(43), 174; **8**:14(77), 33(77), 45(77), 86; **24**:40(7), 98(7)

Morse, R. D., **25**:161(39), 244(39)

Morse, T. F., **7**:176, 179, 212, 213

Mortenson, E. M., **9**:218(46), 269

Morton, B. R., **10**:255(51, 52, 53, 54), 282; **13**:66, 114; **16**:7, 9, 10, 11, 53, 56; **19**:77(177), 91(177); **20**:187(47), 301(47)

Morton, H., **28**:11(15), 71(15)

Morton, R. K., **15**:250, 280; **26**:49(80), 96(80)

Moser, M., **30**:172(142), 195(142)

Moser, R., **24**:192, 271

Moser, R. D., **25**:382, 392, 393, 403, 413, 416

Moses, C. A., **14**:333, 334(122), 345

Moses, G. A., **29**:149(52), 205(52), 233(69), 338(69)

Mosher, D. R., **1**:357(3), 374(3), 375(3), 440

Moskowitz, S. L., **9**:3(41), 91(41), 95(41), 97(41), 108, 110

Moskvicheva, V. N., **16**:235

Moslehy, F. A., **30**:261(19), 308(19)

Moss, R. A., **7**:258, 317

Mosteller, W. L., **7**:256, 257, 317

Mostinskij, I. L., **1**:416(107), 417(107), 420(114), 443

Mostinskiy, I. L., **9**:265(136), 272

Motakef, S., **30**:339(80), 358(111, 112), 360(80, 120, 121), 430(80), 431(111, 112), 432(120, 121)

Motamedi, M., **22**:277(219), 278(149), 353(149), 356(219), 421(35), 422(35), 423(35), 424(35), 436(35)

Motoki, Y., **26**:45(224), 103(224)

Mott, I. E. C., **31**:461(45), 472(45)

Mott, N. F., **2**:343, 354

Mottaghian, R., **SUP**:360(504), 361(504, 510), 363(504, 510, 512)

Motte, E. I., **5**:96, 97, 98, 127; **11**:134, 140, 142, 143(149), 144, 195

Motuleuich, V. P., **3**:35, 98

Moulton, R. W., **2**:386(58), 395; **7**:103(18), 108(18), 110(18), 160

Mountziaris, T. J., **28**:375(139, 140), 390(177), 421(139, 140), 423(177)

Mouri, A., **18**:59(112), 60(112), 61(112), 63(112), 85(112)

Moussa, N. A., **22**:321(150), 322(151), 323(151), 324(151), 325(151), 326(151), 353(150, 151)

Moussez, C., **9**:3(97), 110

Mow, K., **23**:215(165), 274(165)

Moyer, C. A., **22**:227(71), 350(71)

Moyls, A. L., **15**:63(29), 64(29), 136(29)

Moyne, C., **30**:172(141), 195(141); **31**:2(1, 13), 5(1, 13), 89(13), 94(1), 102(1, 13)

Mozhanov, E. V., **25**:92(170), 148(170)

Mozorov, M. G., **6**:69(117), 73(117), 127

Mrani, A., **23**:253(266), 254(266), 255(266), 256(266), 261(266), 278(266)

Mucciardi, A. N., **9**:236(68), 237(68), 270

Muchi, I., **28**:267(87), 333(87)

Muchlenov, I. P., **19**:114(94), 187(94)

Muchnik, G. F., **8**:24, 87; **SUP**:197(384), 209(384), 210(384); **15**:326, 330; **19**:197(34), 198(34), 244(34), 258(16), 259(16), 268(16), 350(16)

Muckenfuss, C., **3**:282(63), 301

Mudawar, I. A., **20**:183(13), 246(208), 247(208, 209), 248(208), 299(13), 309(208, 209); **23**:5(7), 13(7, 72), 15(7, 72), 19(7), 20(7), 32(7), 33(7), 39(7), 51(72), 52(72), 59(7, 109), 61(7, 109), 65(85), 88(7), 89(101), 90(7, 102), 91(7), 92(7), 93(7), 98(7, 102), 99(102), 100(85), 103(109), 105(109), 126(7), 130(72, 85), 131(101, 102, 109); **26**:135(82), 164(82), 188(153), 192(152, 153), 212(82), 215(152, 153)

Muddalli, Y. G., **31**:433(8), 447(8), 471(8)

Muehlbauer, J. C., **11**:411(184), 439; **19**:4(13), 5(13), 18(13), 84(13)

Mueller, A. C., **5**:56, 102, 104, 125; **11**:55(1), 105(1), 110, 189, 212(64), 213, 214, 241, 242, 255; **13**:18, 26(50), 60; **15**:34(59), 57(59); **26**:306(85), 328(85)

Mueller, H. A., **22**:350(86)

Mueller, H. F., **8**:11, 12(35), 85

Mueller, W. K., **5**:230(160), 251; **6**:474(236), 484(236), 497; **11**:281(15), 284(15), 290(23), 315; **20**:186(33, 53), 187(33, 53), 190(33), 300(33), 301(53)

Mueller, W. W., **20**:186(73), 187(73), 198(73), 200(73), 302(73)

Muffler, L. J. P., **14**:26(38), 78(38), 88(38), 100

Muhleberg, E., **28**:36(62), 53(62), 73(62)

Mujumdar, A. S., **11**:229, 231, 232, 237, 262; **19**:61(139), 62(139), 89(139); **23**:7(10), 126(10), 320(170), 337(262), 361(170), 366(262)

Mukerjee, T., **7**:367, 379

Mukhamedzyanov, R. A., **25**:132(206), 150(206)

Mukherjee, D. K., **30**:385(152), 386(152), 387(152), 396(152), 400(152), 401(152), 402(152, 168), 433(152), 434(168)

Mukhin, V. A., **6**:379(11), 380(11), 486; **25**:28(58), 142(58)

Mukhlenov, I. P., **10**:202(158), 203(102, 103, 107), 217, 218; **14**:158(62), 243

Mukoyed, N. I., **15**:326

Mukunda, H. S., **30**:65(74), 90(74)

Mulder, J., **SUP**:170(344)

Mulford, R. N., **5**:406(171), 513; **11**:105(83), 109, 112, 192

Mulholland, G. P., **15**:285, 286, 287, 289, 293, 299(124), 309, 310, 316, 323, 325, 326

Mull, W., **8**:165, 166, 176, 184, 226

Müller, G., **29**:163(102), 207(102); **30**:313(6), 320(23), 326(6), 332(47), 334(58), 346(47), 347(6), 349(58, 95), 350(95), 351(6, 96), 358(116), 360(116), 361(122), 365(131), 382(6, 149, 150), 383(6), 426(6), 427(23), 428(47), 429(58), 431(95, 96), 432(116, 122, 131), 433(149, 150)

Müller, K., **21**:319, 320, 321, 325, 329, 331, 345; **29**:178(158), 180(169), 181(158), 210(158), 211(169)

Müller, K. G., **5**:304(62), 309(62), 323; **9**:127(28), 178

Muller, R. S., **23**:281(29), 355(29); **28**:99(98), 143(98)

Muller, U., **19**:79(218, 219), 93(218, 219)

Müller-Schauenburg, W., **22**:18(28)

Müller-Steinhagen, H. M., **30**:210(37), 213(37), 233(37), 251(37); **31**:294(204), 332(204), 461(46), 472(46)

Mulligan, J. C., **10**:48, 81; **19**:74(170), 75(170), 90(170); **20**:348(69), 352(69); **24**:167(150), 170(150), 171(150), 190(150)

Mullin, J. B., **28**:342(34), 416(34)

Mullin, T., **21**:156, 178, 181

Mullin, T. E., **7**:66(49), 85

Mullins, W. W., **19**:79(204), 92(204)

Mullkin, T. W., **3**:246(100), 251

Mulpuru, S. R., **29**:79(41), 125(41)

Mumford, A. R., **5**:453, 515

Munakata, T., **SUP**:101(265), 113(265); **30**:393(158), 396(158), 399(158), 434(158)

Muncey, R. W., **10**:71, 72(56), 82

Mungan, N., **23**:222(192), 223(191), 275(191, 192)

Munro, A., **4**:120(91, 92), 140

Munro, W. D., **5**:230, 250; **21**:150, 151, 152, 182

Munroe, L. R., **4**:129(110), 135(110), 141

Muntz, E. P., **6**:41, 65(80), 126

Murakami, K., **15**:166(26, 27), 167(26, 27), 223(26, 27); **28**:135(26), 139(26), 285(124), 334(124)

Murakami, M., **20**:287(312), 289(312), 314(312); **31**:343(30), 426(30)

Murakami, S., **25**:396, 416

Murakami, Y., **2**:374, 396

Murakawa, K., **SUP**:190(374), 289(458, 459), 297(374), 301(459, 466), 319(466)

Muramatsu, K., **29**:325(180), 326(180), 344(180)

Muramoto, H., **7**:108(24), 110(24), 160

Murano, H., **20**:183(11), 299(11)

Murase, T., **10**:43, 81; **20**:277(293, 294), 278(295), 283(304), 313(293, 294, 295, 304)

Murata, K. K., **29**:234(82), 325(82), 326(82), 339(82)

Murcray, D. G., **10**:11, 36

Murcray, F. H., **10**:11, 36

Murgai, M. P., **8**:11, 13, 85; **10**:255(56), 282

Murgatroyd, D., **4**:121(97), 140

Murgatroyd, W., **17**:31(101), 61

Murnaghan, F. D., **SUP**:2(4), 197(4), 223(4), 234(4), 247(4), **19**:258(11), 350(11); **24**:40(6), 98(6)

Murphy, D. W., **4**:191(69), 225

Murphy, J. O., **15**:211(107), 212(107), 225(107)

Murphy, J. R., **18**:169(30), 170(30), 172(30), 209(30), 210(30), 236(30)

Murphy, N. F., **13**:219(94), 266

Murphy, R. W., **20**:244(189), 308(189)

Murray, B. C., **10**:71, 73(61), 82

Murray, D. O., **9**:409(105), 417

Murray, J. D., **8**:13, 85; **10**:170, 215; **14**:154(43), 243

Murray, W., **9**:261(118), 271

Murray, W. D., **19**:22(43), 45(43), 85(43)

Murthy, M. S., **23**:262(11), 267(11)

Murthy, V., **20**:209(117), 304(117)

Murti, P. S., **SUP**:271(448)

Murto, P. J., **9**:410(106), 417

Murty, S. S. R., **3**:210(42), 250

Murty Kanury, A., **10**:225, 229, 231, 243(39), 281, 282; **17**:175, 180, 184(22), 186, 190, 314, 315

Musienko, L. E., **17**:116(106), 155

Muskat, M., **20**:315(1), 349(1); **24**:104(43, 44), 112(44), 186(43, 44)

Muskhelishvili, N. I., **8**:34(176), 42, 53(176), 88, 89

Musonge, P., **21**:188(18), 237(18)

Musse, S., **9**:103, 106

Musters, J. J., **23**:193(76), 270(76)

Musters, S. M. P., **25**:181(74), 199(74), 246(74)

Muzichenko, L. B., **10**:177(51), 216

Muzio, L. J., **23**:293(64), 307(64), 356(64)

Myakichev, A. S., **21**:17(32), 51(32)

Myakochin, A. S., **25**:131(201, 205), 132(201), 133(205), 150(201, 205); **31**:228(118), 327(118)

Myers, G. E., **3**:93, 94, 100; **6**:474(314), 501; **13**:11(36), 60; **19**:81(250), 94(250); **22**:59(45), 62(45), 63(45), 65(45), 66(45), 67(45), 68(45), 69(45), 90(45), 138(45), 154(45)

Myers, J. A., **9**:258(100), 271

Myers, J. E., **1**:230, 265; **10**:172, 177(24), 215; **20**:15(25), 80(25); **21**:328, 345; **30**:168(130), 195(130)

Myers, P. S., **23**:101(105), 131(105)

Myrum, T. A., **19**:61(138), 89(138); **24**:6(20), 36(20)

Mysels, K. J., **12**:242(73, 93, 94, 95), 280, 281

N

Na, T. Y., **4**:33, 63; **9**:296, 346; **15**:153(13), 185(13), 223(13); **25**:262, 318

Nachman, A., **25**:262, 318

Nachtsheim, P. R., **4**:53, 64; **9**:324(94), 348; **15**:156(17), 223(17)

Nadebaum, P. R., **21**:175, 181

Nader, W. K., **24**:113(101), 188(101)

Nadhuker, M., **20**:259(250), 311(250)

Nafe, J. P., **4**:121(102), 123(102), 141

Nafziger, R., **26**:72(85), 73(85), 74(85, 183), 96(85), 101(183)

Nagae, O., **23**:55(77), 59(77), 61(77), 63(77), 69(77), 70(77), 74(77), 75(77), 77(77), 130(77)

Nagakubo, S., **24**:6(23), 37(23)

Nagamatsu, H. T., **4**:330, 444; **23**:320(164), 361(164)

Nagamine, K., **28**:342(26), 416(26)

Nagashima, A., **3**:257(13), 260, 300

Nagata, N., **21**:181

Nagata, S., **21**:91(62), 136(62)

Nagata, T., **27**:167(30)

Nagatome, S., **17**:21(64), 59

Nagatomo, H., **11**:105(99), 114, 193

Nagelhout, D., **28**:267(95), 333(95)

Nagendra, H. R., **11**:213, 214, 255

Nagle, M. E., **29**:8(8), 55(8); **30**:32(60), 90(60)

Nahas, N. C., **18**:179(58), 180(58), 237(58)

Nahme, R., **13**:219(44), 233, 258, 265

Nair, B. G., **29**:39(29), 56(29)

Naito, E., **SUP**:69(145, 146), 73(145), 160(145, 146), 192(146), 194(146); **19**:268(38, 39), 270(38), 351(38, 39)

Naitoh, M., **15**:270(109), 281

Naitou, T., **20**:186(48), 187(48), 301(48)

Najafi, B., **29**:147(48), 194(48), 205(48)

Nakabayashi, K., **21**:161, 181

Nakagawa, T., **21**:181; **26**:49(229), 51(229), 103(229)

Nakagawa, Y., **1**:288, 290(29), 353; **5**:205, 206, 249; **30**:50(70), 90(70)

Nakajima, T., **20**:75(65, 66), 82(65, 66), 244(205), 309(205)

Nakajima, Y., **7**:128(37), 161

Nakamachi, I., **27**:ix(6, 8, 11)

Nakamotto, M., **21**:29(47), 31(47), 52(47)

Nakamura, H., **SUP**:64(97, 98), 224(97), 237(97), 238(424), 260(97, 98), 262(98); **19**:310(94), 312(94), 353(94); **20**:186(48), 187(48), 301(48); **28**:376(142), 421(142)

Nakamura, M., **SUP**:413(530)

Nakamura, T., **27**:ix(8, 14)

Nakamura, Y., **31**:334(12, 13), 425(12, 13)

Nakanishi, E., **10**:109(39), 164

Nakanishi, K., **28**:318(188), 337(188)

Nakanishi, S., **23**:12(53), 14(53), 28(53), 29(53), 33(53), 36(53), 39(53), 58(53), 59(93), 60(53), 61(93), 77(93), 78(93), 84(53), 93(53), 106(53, 93), 108(53), 109(53, 93, 113), 110(53, 93, 113), 111(53), 112(53, 93, 113), 115(53), 118(53, 122), 129(53), 131(93), 132(113, 122); **26**:181(134), 182(134), 201(134), 202(134), 214(134)

Nakao, S., **20**:183(7), 299(7)

Nakata, H., **20**:183(7), 299(7)

Nakatogawa, T., **13**:11(39), 60; **20**:252(238), 255(238), 310(238)

Nakayama, A., **24**:120(128), 122(128, 129), 124(128), 125(128, 129, 132), 127(132), 129(134), 130(134), 132(134), 134(134),

135(128, 134), 136(128), 137(128), 145(128, 138), 147(128, 138), 148(128, 138), 151(138), 152(138), 155(145), 157(145), 158(145), 159(145), 160(146), 162(146), 163(146, 149), 165(149), 166(149), 189(128, 129, 132, 134, 138), 190(145, 146, 149); **25**:258, 262, 267, 270, 318, 319

Nakayama, P. I., **13**:80(31), 94(31), 116

Nakayama, W., **5**:245, 251; **16**:232, 233; **17**:140, 148, 157; **20**:75(66), 82(66), 221(135), 232(135, 152), 242(152), 244(205), 305(135), 306(152), 309(205); **21**:117(102, 103, 104, 105), 137(102), 138(103, 104, 105); **26**:123(57), 126(57), 127(57), 128(57), 129(57), 130(57), 131(57), 132(57), 211(57); **31**:160(64), 324(64)

Nakayama, Y., **20**:75(65), 82(65); **28**:135(72, 73), 142(72, 73)

Nakayanne, W., **5**:244, 251

Nakazawa, M., **23**:109(115), 110(115), 113(115), 115(115), 132(115)

Nakoryakov, V. E., **25**:4(10, 11), 33(10), 140(10, 11); **26**:109(24), 137(88), 149(24), 151(24), 201(24), 202(24), 203(24), 209(24), 212(88); **31**:160(9), 165(9), 182(9), 322(9)

Nallassamy, M., **25**:331, 416

Nam, S. W., **28**:343(53), 417(53)

Namba, S., **28**:343(50), 417(50)

Namkoong, D., Jr., **4**:210, 211, 212(96), 213(96), 227

Nanda, R. J., **4**:35, 63

Nandakumar, K., **SUP**:67(127, 128), 368(128), 370(128), 375(127, 128), 376(127, 128), 378(128); **20**:341(60, 61), 352(60, 61)

Nandapurkar, P. J., **28**:253(56), 256(56, 66), 309(156, 157), 331(56), 332(66), 336(156, 157)

Nandapurkar, S. S., **9**:261(115), 271

Nannagia, S., **15**:248(80a), 280

Nannei, E., **23**:9(24), 127(24)

Nansteel, M. W., **18**:63(115), 65(118), 69(118, 130), 70(118, 130), 85(115, 118, 130)

Nanzer, J. O., **26**:173(133, 136), 177(133), 178(133, 136), 181(136), 214(133), 215(136)

Napalkov, G. N., **10**:174(40), 175(40), 215

Napier, D. H., **15**:250, 280

Napolitano, L. G., **2**:390, 396

Naraghi, M. H. N., **30**:328(29), 338(29), 339(29, 81, 82, 83), 340(29), 341(29), 343(85), 344(29), 427(29), 430(81, 82, 83, 85)

Narang, B. S., **SUP**:74(198), 165(198), 166(198)

Narasimha, R., **4**:355, 446

Narasimhamurty, G. S. R., **21**:337, 339, 345; **26**:2(140), 99(140)

Narasimhan, T. N., **14**:16(39), 37, 51(39), 101, 102

Narasimha Rao, P. V., **SUP**:43(52)

Narayan, J., **28**:124(93), 143(93)

Narayan, K. A., **23**:208(91), 270(91)

Narayana, P., **20**:209(117), 304(117)

Narayan Rao, M. L., **SUP**:244(429)

Narazaki, K., **30**:121(81), 192(81)

Nardacci, J. L., **5**:233(169), 235(169), 251; **21**:151, 164, 166, 180, 181

Narita, H., **24**:30(65), 38(65)

Naritomi, M., **26**:45(224), 103(224)

Narusawa, U., **20**:328(38), 351(38); **21**:73(18), 74(18), 79(18), 134(18)

Narvaez, C., **23**:262(221b), 276(221b)

Naryanasamy, O. S., **11**:391(151), 438

Nash, C. A., **21**:103(84), 137(84)

Nash, D. B., **10**:21, 23, 24(32, 33, 34), 36

Nash, J. F., **6**:18, 28, 29(30), 123

Nassar, M. M., **23**:230(232), 232(232), 277(232)

Nassau, K., **30**:319(18), 367(134), 396(134), 397(134), 427(18), 432(134)

Nasser-Rafi, R., **28**:267(90), 333(90)

Natarajan, N. M., **SUP**:197(385), 202(392), 286(455); **19**:259(18), 262(18), 350(18)

Natarajan, R., **26**:71(141), 99(141)

Nathenson, M., **14**:98, 99, 105

Natusch, H. J., **14**:176(94), 202, 237(94, 132), 244, 245; **19**:148(175, 176), 189(175, 176)

Naudascher, E., **23**:222(193), 224(193), 225(193), 275(193)

Naumov, V. I., **25**:4(18, 19), 132(206), 140(18, 19), 150(206)

Naurits, L. P., **6**:390(19), 405(19), 486

Navins, R. G., **6**:474(332), 502

Navon, U., **9**:140(62), 179; **15**:229(9), 235(9), 239(9), 276

Nawate, Y., **16**:232, 233

Naxmi, F., **2**:354

Nayfeh, A. H., **8**:22(105), 86; **15**:326

Naylor, P., **29**:160(95), 207(95)

Naysmith, A., **6**:41(86, 87), 126

Nazarenko, V. S., **31**:160(10), 322(10)

Nazarov, N. I., **6**:418(138, 139), 492, 493; **8**:25, 87

Nazir, M., **26**:49(205), 102(205)

Neal, L. G., **7**:236(28), 237, 239, 249, 266(28), 304, 306, 314

Neale, G. H., **24**:113(101), 188(101)

Nece, R. E., **5**:175, 177, 178, 181, 248

Nechayev, V. M., **19**:193(15), 243(15)

Necmi, S., **21**:79(22), 92(22), 134(22)

Necodeme, P., **30**:338(74), 350(74), 403(173), 405(74, 173), 430(74), 434(173)

Nedderman, R. M., **8**:297(32), 349

Nee, V. W., **15**:189(61), 224(61)

Neeper, D. A., **17**:69(20), 70(20), 71(20), 153

Neff, J. J., **25**:190(89), 191(89), 192(89), 200(89), 204(89), 205(89), 222(89), 226(89), 227(89), 228(89), 246(89)

Negi, J. G., **8**:65, 90

Negus, K. J., **20**:269(275), 270(275), 312(275)

Neighbors, P. K., **19**:317(110), 324(110, 117), 339(110), 354(110, 117)

Neiland, V. Ya., **6**:396(32), 487

Neill, K. O', **19**:22(45), 85(45)

Neilson, D. G., **28**:275(103), 278(112, 113), 279(112, 113), 280(112, 113, 115), 281(113), 282(112), 286(131), 287(131), 288(131), 310(112, 159, 160), 311(159), 312(159, 160), 313(159), 314(159), 315(160), 333(103), 334(112, 113, 115), 335(131), 336(159, 160)

Neimark, A. V., **31**:160(32), 323(32)

Nein, M. E., **4**:150, 223; **5**:351(57), 378, 379, 380, 381, 383(57), 386(57), 387(57), 388(57), 507, 509, 510

Neindre, Le, **21**:2(9), 3(9), 51(9)

Nejat, Z., **17**:50(160, 162), 63

Nelson, D. A., **14**:252(5), 254(5), 279

Nelson, D. B., **1**:242(83), 265, 372, 382, 441; **5**:451, 453(218), 454, 456, 457, 515; **31**:134(18), 156(18)

Nelson, H. F., **5**:26, 51

Nelson, J. A., **12**:3(9), 14(9), 73

Nelson, K. E., **12**:156, 193

Nelson, L. A., **20**:182(3), 290(315), 291(315, 316, 317), 292(315), 299(3), 314(315, 316, 317)

Nelson, L. S., **29**:163(101), 164(107), 166(118, 120), 175(153), 190(199), 192(199), 207(101, 107), 208(118, 120), 210(153), 212(199), 268(134), 342(134)

Nelson, R. A., **11**:202, 204, 263; **26**:45(159, 160), 100(159, 160)

Nelson, R. G., **5**:36, 53

Nelson, W., **6**:98(160), 100(160), 101, 130; **10**:87(3), 107(32), 111, 163, 164, 165

Nemat-Nasser, S., **15**:321(83, 127), 324, 326

Nemchinov, I. V., **3**:225, 251; **11**:108(177), 439

Nens, S., **18**:214(106), 238(106)

Nereo, G. E., **28**:231(2, 3, 4), 251(2, 3, 4), 304(4), 329(2, 3, 4)

Nerushev, A. F., **14**:252(7), 255, 279

Nesis, E. I., **11**:11(24), 48

Nesselman, K., **26**:220(5), 324(5)

Nesterenko, A. V., **1**:131, 184

Nesterov, A. K., **15**:128(155), 141(155)

Nestler, D. E., **6**:41(89), 42, 69(89), 75(89), 82(89, 139), 94(89), 95(89), 126, 129

Nestor, O. H., **4**:295, 315

Neswald, R. G., **4**:115(69), 140

Nettleton, M. A., **29**:98(66), 126(66)

Neu, J. T., **11**:375(128), 437

Neu, R. F., **2**:74, 108

Neudeck, G. N., **28**:386(158), 388(158), 422(158)

Neudeck, G. W., **28**:365(106), 419(106)

Neugbuer, C. A., **20**:259(247), 311(247)

Neugebauer, F. J., **30**:239(35), 240(35), 251(35)

Neukirchen, B., **14**:176(99), 178, 202(99), 203, 244; **19**:146(162), 188(162)

Neuman, S. P., **14**:3(8), 100

Neumann, D., **23**:280(11), 299(11), 314(11), 315(11), 322(11), 325(11), 354(11)

Neumann, G., **16**:48, 57

Neumann, R. D., **23**:280(12), 298(12), 305(12), 354(12)

Neumann, R. J., **21**:3(7), 42(7), 43(7), 44(7), 51(7)

Neuringer, J. L., **1**:333(62), 340, 354

Neuroth, N., **11**:358(82, 83, 87), 360(93c), 375(127), 404(83), 436, 437

Neuskii, A. S., **3**:213, 250

Neverov, A. S., **25**:4(15), 130(198, 199, 200), 131(201, 205), 132(201), 133(198, 199, 200, 205), 140(15), 149(198, 199, 200), 150(201, 205); **31**:201(103, 104), 205(104), 279(181), 281(196), 305(209, 210), 326(103, 104), 330(181), 331(196), 332(209, 210)

Nevins, R. G., **4**:90(28), 91(28), 92, 95, 104(50), 106(50), 109(28), 110, 138, 139; **11**:105(87), 113, 118, 192

Nevins, R. G., Jr., **18**:257(45), 263(45), 294(45), 322(45); **23**:109(126), 110(126), 119(126), 132(126)

Nevstrueva, E. I., **1**:394(71), 398(71), 442

Nevzglyadov, W. G., **3**:35, 97

Newburgh, L. H., **4**:141

Newcomb, T. P., **15**:326

Newell, G. F., **3**:257, 300

Newell, M. E., **8**:195(45), 227; **11**:423(219), 440; **SUP**:64(95), 202(95), 203(95), 205(95), 224(95), 230(95), 234(95); **19**:279(65), 281(65), 282(65), 352(65)

Newell, W. C., **2**:327(98), 355

Newlander, R. A., **6**:94(147), 129

Newman, A. B., **13**:120, 201; **26**:19(142), 27(142), 99(142)

Newman, B. G., **11**:241, 242, 261

Newman, G. R., **21**:185(11), 236(11)

Newman, J., **7**:90(8), 160; **SUP**:75(202), 102(266), 106(202), 118(266), **28**:389(167), 394(167), 422(167)

Newman, J. S., **12**:226(76, 77), 228(76, 77), 229(77), 233(74, 76, 77, 78), 237(77, 78), 238(77, 79), 239(77), 241, 242(77), 248(77), 249(77), 250(77), 251(74, 75, 76, 77, 79, 99), 252(77), 254(77), 255(77), 261(76, 77), 269, 270(74, 76, 77, 79, 99), 271, 280, 281

Newman, W. A., **20**:186(26), 187(26), 300(26)

Newman, W. H., **29**:137(24), 203(24)

Newson, I., **15**:228(2), 268, 276

Newton, M., **22**:20(6), 60(6), 61(6), 69(6), 70(6), 152(6)

Ng, C. K., **4**:119(85), 140

Ng, K. H., **13**:81(41), 116

Ng, K. S., **15**:63(19), 64(19), 71(19, 52), 85(52, 72), 88(74), 90(77), 102(77), 104(77), 109(19), 110(140), 114(19), 118(19, 52, 72), 119(72), 131(74), 132(74), 135(19), 136(52), 137(72, 74), 138(77), 140(140); **19**:330(127), 355(127)

Ng, M. L., **25**:260, 279, 280, 281, 318

Ng, W. W. L., **17**:38(126), 62

Nguyen, A. T., **29**:32(25), 56(25)

Nguyen, H. D., **26**:16(145), 25(146), 31(143), 83(144), 99(143, 144, 145, 146)

Nguyen, Q. D., **23**:190(194), 275(194); **25**:256, 318

Nguyen, Q.-V., **26**:86(147), 99(147)

Nguyen-Tien, K., **23**:253(214), 254(195, 214), 255(214), 256(214), 258(195, 214), 259(214), 275(195), 276(214)

Ni, J., **28**:252(49, 52, 53), 257(49), 258(49), 259(49), 260(49), 261(49), 263(49, 52, 53, 81), 264(49), 267(49), 268(49, 52, 53, 81), 269(49), 285(49, 52, 53), 302(52, 53), 303(49, 81), 327(49, 52, 53), 331(49, 52, 53), 332(81)

Ni, J. R., **20**:336(54), 352(54)

Ni, R. H., **24**:272

Nicholas, D., **24**:3(3), 36(3)

Nicholl, C. I. H., **8**:299(45), 305, 350

Nicholls, C. M., **5**:327(17), 331, 333(17), 343(17), 505

Nicholls, J. A., **29**:169(130), 209(130)

Nichols, B. D., **28**:41(65), 73(65)

Nichols, L. D., **2**:445(43), 452; **5**:313(89), 324

Nichols, R. A., **8**:33, 88

Nichols, R. T., **29**:220(41), 221(26, 27, 29, 30, 31, 32, 33, 34, 35, 36, 37, 38, 39, 40, 41), 230(41), 233(26), 255(40), 260(116), 261(117), 263(116, 117), 264(116, 117, 126), 265(40, 41, 117), 269(40), 270(40, 41), 271(29, 30, 31), 276(41), 277(126), 278(40), 280(29, 31), 282(29, 31, 41), 290(40, 41), 293(29), 295(29), 300(40), 302(36, 37), 303(26), 306(41), 309(29, 38), 310(32, 33, 34, 35, 36, 37, 40), 311(32, 33, 34, 35, 36, 37), 315(30, 40, 172), 316(30, 40), 318(40), 319(40), 323(40, 172), 336(26, 27, 29, 30, 31, 32, 33, 34, 35, 36, 37, 38), 337(39, 40, 41), 341(116, 117, 126), 344(172)

Nicholson, J. H., **23**:320(188), 362(188)

Nicholson, M. K., **20**:84, 85, 88, 91, 93, 97, 99, 100, 101, 109, 110, 131, 132

Nicholson, S. B., **10**:40, 73(2), 76, 78, 80

Nicklin, D. J., **1**:360, 363(11), 440; **20**:100, 101, 105, 131

Niclause, M., **2**:353

Nicodeme, P., **30**:403(172), 404(172), 434(172)

Nicodemo, L., **15**:92(88), 138(88)

Nicol, A. A., **9**:261(117), 271

Nicolette, V. F., **24**:303(26), 319(26)

Nicolis, C., **19**:82(279), 95(279)

Nicoll, K. M., **6**:75(131), 76(131), 80, 128

Nicoll, W. B., **2**:43, 107; **7**:339(18, 19), 348, 359(18), 378; **11**:238, 263

Nield, D. A., **14**:19, 27, 29, 101; **20**:315(10), 316(10), 318(16), 321(16), 323(16),

Nield, D. A. (*continued*)
325(24), 347(67), 350(10, 16, 24), 352(67); **23**:217(196), 275(196), 369(4), 460(4); **24**:104(70), 113(103), 179(70), 187(70), 188(103); **30**:111(60), 122(60), 191(60)

Nielsen, H. J., **10**:255, 275(106), 282

Nielsen, J. N., **6**:109(171), 116, 117, 118, 130, 131

Nielsen, L. R., **18**:189(143), 239(143)

Nielson, D. W., **14**:43(102), 102

Niemi, R. O., **19**:334(135), 335(135), 338(135), 355(135)

Nienow, A. W., **25**:166(58), 199(58), 245(58)

Niethhammer, J. E., **20**:232(161), 237(161), 238(161), 239(161), 240(161), 307(161)

Nieuwstadt, F. T. M., **25**:375, 376, 416

Nievergeld, P., **17**:231(73), 250, 258(73), 259, 260, 290, 294(73), 295, 316

Nigam, K. D. P., **23**:237(157), 242(157), 254(231), 258(231), 274(157), 277(231)

Nigam, S. D., **8**:26, 87

Nigham, S. D., **5**:161, 248

Nigon, J. P., **5**:406(171), 513; **11**:105(83), 109, 112, 192

Nihoul, J. C. J., **13**:219(39, 47, 50), 264, 265

Nijaguna, B. T., **21**:168, 181

Nikai, I., **16**:95, 96, 101, 127, 154

Nikaidoh, H., **22**:290(152), 353(152)

Nikiforov, A. N., **25**:81(135, 143, 150), 146(135, 143), 147(150)

Nikitenko, N. I., **6**:435(165, 166), 436(165, 166), 494

Nikitin, I. K., **6**:400, 488

Nikitin, V. A., **25**:166(49), 245(49)

Nikitin, Yu. M., **25**:64(108), 145(108)

Nikitopoulos, C. P., **16**:53, 58

Nikneja, J., **21**:79(23), 92(23, 73, 74), 96(73, 74), 134(23), 136(73, 74)

Nikolaeva, A. D., **11**:358(90), 436

Nikolayev, V. S., **7**:206, 217

Nikuradse, J., **16**:302, 309, 364; **19**:324(115), 354(115)

Nillesen, J. C. M., **28**:103(101), 143(101)

Nilson, R. H., **29**:38(27), 40(27), 56(27); **30**:121(78, 79, 80), 134(91), 135(91), 192(78, 79, 80), 193(91)

Nilsson, E. K., **10**:275(106), 284

Nilsson, O., **12**:251(121), 282

Nimmo, B., **9**:256(87), 270

Ninić, N., **11**:8, 48

Nir, A., **18**:201(93), 202(93), 203(93), 204(93), 238(93)

Nirala, A. K., **30**:266(30, 31), 267(31, 32), 308(30, 31, 32)

Niro, A., **24**:8(39), 14(39), 37(39)

Nishhiyama, E., *see* Nishiyama, E.

Nishi, M., **17**:140, 157

Nishiguchi, A., **26**:45(150), 99(150)

Nishikawa, K. (Chapter Author), **1**:231(52), 265; **5**:67, 125, 409, 411, 413, 513; **6**:551, 555(82), 564; **9**:9(52), 10(52), 21(51, 52), 23(52), 36(51, 52), 37(52), 99(52), 108; **11**:12(27), 48, 65, 67, 68, 76(30), 77(30), 78(30), 80, 82(30), 83(30), 105(59, 99, 121), 110, 114, 126, 127, 131, 132, 134, 136(141), 190, 191, 193, 194, 195; **16**:136, 155, 166, 170, 171(3, 4), 173, 174, 187(3), 188, 189(3), 191(20), 193(3, 4, 20), 194(20), 203, 209, 214, 215, 216, 219(20), 220(3, 4, 20), 221, 222(3, 4), 226, 232, 233, 235, 236, 238; **17**:35, 61; **18**:256(29), 257(29), 262(29), 265(29), 267(29), 270(29), 271(29), 272(29), 274(61), 290(61), 292(29), 295(61), 296(61), 298(61), 304(61), 313(29), 321(29), 322(61); **20**:1, 3(1), 4(2), 5(3, 4, 5), 7(3, 9), 9(3), 10(3), 11(1, 3), 13(14), 14(9), 15(3, 9, 24), 16(35), 17(35), 18(35), 20(40), 22(40), 27(55), 33(55), 36(57), 42(57), 43(57), 55(59), 58(59, 60), 59(60), 64(61), 68(61), 70(62), 72(63), 76(70), 79(1, 2, 3, 4, 5, 9), 80(11, 14, 24), 81(35, 40, 55, 57), 82(59, 60, 61, 62, 63, 70); **21**:37(58), 38(63), 52(58, 63)

Nishikawa, S., **25**:358, 416

Nishimura, M., **11**:399, 408; **15**:328

Nishina, Y., **28**:135(26), 139(26)

Nishino, J., **23**:329(233), 365(233)

Nishino, S., **28**:342(27), 416(27)

Nishio, G., **26**:45(224), 103(224)

Nishio, S., **21**:128(123), 138(123); **29**:131(14), 203(14)

Nishioka, K., **14**:282, 341

Nishioka, M., **11**:220, 224, 240, 241, 243, 264

Nishiwaki, H., **13**:11(39), 60

Nishiwaki, N., **3**:81, 82(76), 84(76), 87, 99; **6**:546(65), 563; **7**:34, 84, 346, 353(41), 378; **11**:281(18), 315; **20**:252(238), 255(238), 310(238); **21**:4(10), 34(52), 35(52), 36(52), 37(59), 51(10), 52(52, 59)

Nishiyama, E., **16**:232; **20**:244(197), 308(197)

Nissan, A. H., **5**:230, 233, 235(161), 241, 251; **7**:308, 319; **21**:151, 164, 166, 180, 181

Nitsan, U., **19**:82(263), 94(263)

Niuman, F., **4**:23, 63

Nix, G. H., **11**:13(34), 48; **16**:214(24), 216(24), 233, 238

Nixon, J. A., **11**:318(13), 434

Noack, R., **14**:160(76), 161(76), 165, 165(76), 166, 167, 168(76), 169(76), 171(76), 172(76), 176(76), 244

Nobel, A. P. P., **10**:193(91), 199(91), 202(91), 204(91), 205(91), 206(91), 217

Nobel, P., **14**:151(13), 242

Noble, B., **8**:49(233), 50(233), 52, 90

Noble, C. E., Jr., **1**:284(16), 353

Noble, J. J., **11**:420(209), 439

Noda, I., **15**:326

Noda, K., **23**:193(175), 214(175), 216(175), 274(175)

Noë, A. R., **14**:210, 211(143), 226, 246; **19**:135(236), 190(236)

Noel, B. W., **23**:294(75), 357(75)

Noel, M. B., **31**:209(111), 327(111)

Noether, E., **24**:41(23), 99(23)

Noggle, T. S., **28**:80(64), 141(64)

Nogo, Y., **23**:293(66), 349(66), 356(66)

Nogotov, E. F., **14**:152(30), 243

Nogotov, I. L., **25**:81(156), 147(156)

Nohira, H., **7**:129(39), 161; **21**:165, 178

Nolan, P. F., **10**:229, 281

Noll, W., **2**:360, 396; **24**:107(73), 187(73)

Nomura, M., **11**:231, 232, 237, 260

Nonn, T., **23**:13(37, 46), 15(37, 46), 18(37), 19(37), 20(37), 22(37, 46), 26(37), 27(37, 46), 28(37), 34(37, 46), 38(46), 39(37, 46), 49(37, 46), 59(37, 46), 61(37, 46), 64(46), 65(37, 46), 77(37, 46), 79(46), 87(46), 90(46), 92(46), 100(37, 46), 101(37, 46), 128(37, 46)

Noordsij, P., **7**:130, 161

Nordberg, M., **24**:27(57), 38(57)

Norden, H. V., **6**:418(117), 491

Nordin, S., **31**:461(47), 472(47)

Nordlie, R. L., **5**:165, 166, 248

Norman, B. N., **11**:105(120), 114(120), 194

Norman, G. E., **5**:309(78), 310(82), 324

Norman, J. R., **9**:127(30), 178

Norman, R. F., **6**:474(314), 501

Normand, X., **25**:346, 364, 416

Normington, P. J. C., **30**:173(147), 195(147)

Noronha, R. I., **20**:187(69), 302(69)

Norris, R. H., **SUP**:154(334), 170(334)

Norris, W. T., **17**:147(172), 157

North, R. J., **6**:195(23), 363

Northby, J. A., **17**:100, 155

Northdurft, W., **2**:355

Northup, L. L., **8**:248, 282

Norton, B., **18**:25(56c), 82(56c)

Norton, M. G., **28**:103(87), 142(87)

Norton, R. C., **12**:3(8), 11(8), 14(8), 73

Norton, R. J. G., **23**:288(76), 295(76), 346(299, 300), 347(300), 357(76), 368(299, 300)

Nosach, V. G., **26**:252(63, 64), 253(63, 64), 261(63), 281(64), 316(64), 317(64), 318(63, 64), 327(63, 64)

Nosov, V. S., **9**:119(11), 178

Noto, K., **15**:189(60), 224(60)

Notter, R. H., **15**:122(145), 140(145); **19**:19(35), 85(35)

Nourbakash, H. P., **29**:171(137), 209(137), 233(63), 319(63), 338(63)

Nourgaliev, R. R., **29**:249(102), 340(102)

Novak, L., **31**:438(48), 449(48), 473(48)

Novakovic, M., **31**:343(28), 426(28)

Novichenok, L. N., **18**:163(22), 169(46), 171(46, 56), 176(46, 56), 177(56), 212(56), 228(46), 229(46), 236(22), 237(46, 56)

Novikov, I. I., **7**:10(5), 83

Novikov, P. A., **1**:153, 154(16), 156, 161, 184

Novikov, V. S., **19**:195(21), 240(21), 243(21)

Novotny, J. L., **2**:379(103), 396; **8**:12, 85, 257(37), 273(37), 283; **12**:141(36), 193; **13**:219(80), 265; **SUP**:176(353), 180(353); **19**:334(139, 140), 336(139, 140), 338(139, 140), 355(139, 140)

Novozhilov, I. F., **31**:160(10), 224(116), 225(116), 322(10), 327(116)

Nowikow, I. I., **20**:19(38), 81(38)

Noyes, R. C., **4**:167, 224; **5**:420, 421, 422, 424, 513; **11**:105(97), 193; **17**:22, 59

Noyes, R. N., **SUP**:18(30), 136(30)

Nozad, I., **18**:191(148, 149), 230(148), 239(148, 149); **23**:372(34), 386(34), 390(34), 391(34), 393(91), 398(34), 400(34), 404(34), 409(34), 422(34), 442(34), 452(91), 462(34), 464(91); **31**:82(14, 15), 92(14, 15), 102(14, 15)

Nozu, S., **21**:120(106, 107), 121(110), 122(106, 107, 111), 138(106, 107, 110, 111)

Nudelman, S., **11**:352(73), 436
Nugartnam, S., **31**:170(76), 325(76)
Nukiyama, S., **1**:187(2), 190, 263; **5**:67, 125;
    7:88, 160; **11**:2, 13, 47, 105(56), 109(56),
    110, 191; **17**:2, 56; **21**:85(46), 135(46);
    23:94(103), 131(103)
Numata, S., **17**:51(163), 64
Numerov, S. N., **24**:104(51), 186(51)
Nunamaker, R. R., **4**:150(7), 222, 228
Nunes, E. M., **30**:328(29), 338(29), 339(29),
    340(29), 341(29), 344(29), 427(29)
Nunge, R. J., **3**:167, 169, 170, 174;
    SUP:75(201), 106(201), 305(201),
    306(201), 309(201), 310(201)
Nunner, B., **22**:209(89), 210(89), 211(89),
    214(89), 215(89), 217(89), 218(89),
    219(89), 350(89)
Nunner, W., **31**:189(91), 190(91), 196(91),
    203(91), 207(91), 213(91), 214(91),
    224(91), 326(91)
Nunobe, M., **26**:142(100), 144(100), 157(100),
    158(100), 213(100)
Nunziato, J. W., **18**:163(19), 236(19)
Nurick, W. H., **6**:11(21), 30(21), 34(21), 123
Nusselt, W., **1**:433, 443; **4**:54, 64; **6**:504, 561;
    9:218, 244, 256, 258, 269; **11**:200(1), 202,
    204, 209, 212(1), 252; **13**:240, 266;
    SUP:2(2), 79(2), 99(2), 154(333),
    169(333), 170(333); **20**:133(1), 138(1),
    139(1), 141(1), 171(1), 179(1); **21**:55(2),
    65(2), 92(2), 104(2), 133(2); **26**:238(42),
    326(42); **30**:34(71), 90(71), 142(103),
    193(103)
Nussle, R. C., **4**:146(2), 150(8), 222, 223
Nutall, R. L., **2**:37(28), 38(28), 106
Nuttall, H., **SUP**:227(417), 346(491)
Nutter, G. D., **11**:423(225), 424(225), 440
Nuzhnyi, V. A., **11**:376(132a), 378(132a),
    381(132a), 438
Nye, J. F., **24**:7(27), 30(62), 31(27), 32(27),
    37(27), 38(62)
Nyholm, I. R., **3**:51(56), 59(56), 98
Nyren, R. H., **SUP**:120(295)

O

Oates, G., **1**:316(54), 349(54), 354
O'Bara, J. T., **9**:243(72), 270
Obata, M., **27**:ix(9), 174(9), 181(31);
    31:334(14), 425(14)

Oberai, M. M., **7**:197, 215
Oberbeck, A., **4**:4, 62; **17**:124, 156
Obermok, V. N., **25**:166(54), 245(54)
Oberste-Lehn, K., **30**:287(74), 288(74), 291(85),
    310(74), 311(85)
Obliven, A. N., **3**:76(71a), 99
Obot, N. T., **26**:132(80), 170(125), 212(80),
    214(125)
Oboukhov, A. M., **21**:267(36), 276(36);
    30:4(72), 74(72), 90(72)
O'Brien, J. E., **23**:320(173, 174, 175, 176),
    347(176), 362(173, 174, 175, 176)
O'Brien, P., **29**:263(124), 270(124), 271(124),
    272(124), 273(124), 341(124)
O'Brien, R. J., **6**:312(102), 366
O'Brien, R. W., **18**:185(133), 191(133, 144),
    230(144), 238(133), 239(144); **23**:369(2),
    407(2), 460(2)
O'Brien, T. P., **28**:102(71), 141(71)
O'Brien, V., **30**:290(81), 310(81)
O'Callaghan, M. G., **19**:79(209), 92(209)
O'Callaghan, P. W., **20**:268(274), 312(274)
Ocansey, P., **28**:267(92), 333(92)
Occelli, R., **30**:137(96), 193(96)
Ochi, T., **23**:12(53), 13(68), 14(53), 15(68),
    28(53), 29(53), 33(53), 36(53), 39(53),
    50(68), 58(53), 59(93), 60(53), 61(93),
    77(93), 78(93), 84(53), 93(53), 106(53, 68,
    93), 108(53), 109(53, 93), 110(53, 93),
    111(53), 112(53, 93), 115(53), 118(53,
    122), 129(53, 68), 131(93), 132(122)
Ochiai, J., **21**:84(42), 96(75), 97(75), 100(42),
    135(42), 136(75)
Ochiai, M., **29**:175(150), 210(150)
Ochoa-Tapia, J. A., **23**:398(63, 64), 415(64),
    463(63, 64); **31**:28(17, 18, 19, 20, 21),
    34(16), 80(11c, 16), 102(11c, 16), 103(17,
    18, 19, 20, 21)
Ockendon, J. R., **19**:2(5), 14(28, 29), 15(29),
    16(28, 29), 24(28), 25(28), 28(28), 31(29),
    32(29), 40(28, 29), 82(266), 83(5), 84(28,
    29), 95(266)
Ockrent, C., **9**:190(7), 268
Oda, K., **21**:105(90), 106(92), 137(90, 92)
Oda, O., **30**:320(24), 427(24)
Odeh, A. S., **23**:212(197), 275(197)
Oden, J. T., **31**:336(22), 339(22), 340(22),
    426(22)
O'Donnel, **18**:328(21), 329(21), 350(21),
    362(21)

Odum, T., **4**:109(55), 110(55), 139

Offergeld, E., **14**:192, 245

Ofi, O., **20**:186(49), 187(49), 301(49)

Ogale, V. A., **9**:3(93, 98), 36, 66, 71(98), 93(98), 95, 96(98), 97(98), 109, 110

O'Gallagher, J. J., **18**:20(48), 25(56a), 81(48), 82(56a)

Ogasawara, H., **15**:270(109), 281

Ogata, H., **16**:233; **17**:140, 148, 157, 323, 324, 326, 328(24), 335(24), 340

Ogawa, K., **25**:261, 269, 274, 275, 278, 318; **28**:345(64), 418(64)

Ogino, F., **7**:108(24), 110(24), 160

Ogiwara, S., **20**:277(294), 313(294)

Ogniewicz, Y., **29**:39(30), 56(30)

Ogston, A. G., **12**:242(80), 280

Oguchi, H., **7**:204, 204(129, 130), 216, 217

Oguma, M., **27**:ix(6)

Oguri, T., **27**:124(27)

Ogushi, T., **20**:287(312), 289(312), 314(312)

Oguz, H. A., **26**:11(148), 99(148)

Oguz, H. N., **26**:48(149, 191), 99(149), 101(191)

Oh, I., **28**:365(106), 386(158), 388(158), 419(106), 422(158)

Oh, M. D., **29**:136(21), 137(21), 139(21), 140(21), 147(21), 168(21), 203(21)

Ohara, M., **11**:412(185), 439

O'Hara, S., **11**:412(197, 200), 415(197), 439

Ohashi, H., **29**:151(69), 205(69)

Ohba, K., **26**:45(150), 99(150)

Öhman, G. A., **11**:209, 216(62), 255

Ohmi, M., **25**:81(130, 132, 134, 137, 138, 139, 157), 146(130, 132, 134, 137, 138, 139), 147(157)

Ohnishi, Y., **14**:97(192), 104

Ohno, H., **17**:26(78), 32, 35(78), 59

Ohring, G., **8**:281, 283

Ohta, H., **16**:166(3), 170(3, 4), 171(3, 4), 173(3, 4), 174(3, 4), 187(3), 188(3), 189(3), 193(3, 4), 203(3), 209(3), 214(3), 215(3), 220(3, 4), 221(3, 4), 222(3, 4), 232, 238; **20**:20(40), 22(40), 27(55), 33(55), 81(40, 55)

Ohta, M., **30**:320(24), 427(24)

Ohtaka, M., **27**:181(32), 186(32)

Oka, S., **8**:110(34), 159; **18**:104(34), 105(34), 158(34)

Oka, S. N., **11**:220, 224, 237, 258

Okada, K., **30**:334(62), 429(62)

Okada, M., **19**:62(148), 63(148), 64(148), 89(148)

Okada, S., **7**:132, 161

Okada, T., **28**:135(72, 73), 142(72, 73)

Okamoto, T., **28**:285(124), 334(124)

Okano, T., **30**:359(123), 361(123), 432(123)

Okano, Y., **30**:358(115), 432(115)

Okasha, O., **15**:48(77), 57(77)

Okazaki, **13**:164, 202

Oker, E., **16**:233; **23**:328(222), 364(222)

Okigami, N., **27**:ix(6)

Okkonen, T., **29**:249(101, 102), 340(101, 102)

Okoo-Kuak, W., **26**:239(45), 242(45), 251(45), 253(45), 326(45)

Okorounmu, O., **14**:334(125), 345

Oksanen, P., **24**:30(63), 38(63)

Oktay, S., **20**:183(9, 17), 184(17), 244(182, 194), 250(194), 299(9, 17), 308(182, 194)

Okubu, T., **28**:343(55), 417(55)

Okuma, Y., **23**:12(47), 15(47), 22(47), 23(47), 27(47), 55(47), 59(47), 61(47), 66(47), 67(47), 75(47), 77(47), 78(47), 79(47), 106(47), 128(47)

Okuno, A. F., **3**:71, 74, 99

Okutani, K., **20**:183(6), 299(6)

Olander, D. R., **5**:149, 247; **15**:290, 323; **28**:123(74), 142(74), 389(164), 422(164)

Olbricht, W. L., **23**:219(198), 275(198)

Ölcer, N. Y., **8**:67, 90; **15**:326

Oldfield, M. L. G., **23**:280(27), 320(187, 188), 321(190, 192, 194, 197, 198, 200), 346(190), 355(27), 362(187, 188, 190), 363(192, 194, 197, 198, 200)

Oldroyd, J. G., **13**:208(9), 209(10), 212(9), 264; **15**:62(8), 78(8), 135(8); **24**:109(87), 187(87)

O'Leary, B. T., **10**:5(8), 19, 21, 24(27, 30), 36, 57(45), 82

O'Leary, J. P., **9**:118(8, 9), 177; **30**:236(1), 249(1)

Oleinik, O. A., **6**:396(38), 487

Oleschuk, O. N., **3**:276(55), 301

Olfe, D. B., **3**:212(57), 250; **11**:318(6), 376(6), 388(6), 434

Olfie, D. B., **5**:254(7), 282(26), 305(26), 306(7), 310(26), 322

Olhoeft, J. E., **29**:71(31), 125(31)

Oliger, J., **24**:221, 272

Oliker, I., **15**:273, 281

Olin, H. L., **14**:217, 218(151), 222(151), 246; **19**:134(142), 188(142)

Olinger, S. J., **11**:105(123), 114, 194

Olive, K. W., **6**:406(61), 407(61), 409(61), 410(61), 489

Oliver, C. C., **3**:231, 234, 235, 251

Oliver, D. A., **5**:348, 506

Oliver, D. L. R., **26**:5(154), 8(153, 154, 157), 10(153), 15(54), 20(157), 22(152, 157), 23(155), 24(155), 25(155), 26(151, 155), 27(54, 151, 156), 28(54), 29(54), 30(54), 95(54), 99(151, 152, 153, 154), 100(155, 156, 157)

Oliver, D. R., **2**:372, 396; **13**:261(130), 267; **SUP**:412(527); **15**:64(38), 101(38), 102(38), 136(38), 194(71), 195(81), 196(85), 197(85), 224(71), 225(81, 85); **19**:317(105, 106), 318(105, 106), 354(105, 106)

Oliver, H., **15**:287, 327

Oliver, M. J., **23**:321(194), 363(194)

Oliver, M. S., **29**:221(25, 26), 233(26), 261(117), 263(117), 264(117), 265(117), 271(25), 303(26), 309(25), 315(172), 323(172), 335(25), 336(26), 341(117), 344(172)

Oliver, R. N., **2**:326, 327, 332, 355

Oliviera, A., **24**:227, 268

Olmer, F., **2**:355

Olmstead, W. E., **2**:445(44), 452

Olsen, A., **17**:37(118), 61

Olsen, D. A., **18**:53(101), 54(101), 55(105), 56(105), 84(101, 105)

Olsen, D. R., **20**:259(250), 311(250)

Olsen, R., **22**:290(152), 353(152)

Olsen, W., **5**:353, 507

Olsofka, F. A., **8**:257(37), 273(37), 283; **12**:141(36), 193

Olson, G. L., **28**:341(16), 416(16)

Olson, J. H., **9**:137, 156, 161, 179

Olson, R. L., **10**:172(25), 174(25), 175(25), 177(25), 215

Olsson, R. G., **20**:252(236), 310(236); **26**:141(94), 202(94), 212(94)

Olstad, R. A., **28**:123(74), 142(74)

Olstad, W. G., **8**:6, 84

Olunloyo, V. O. S., **SUP**:69(144), 161(144)

Oman, R. A., **7**:184(57), 192, 213

Omari, A., **23**:226(199, 200), 228(199, 200), 275(199, 200)

O'Melley, R., **8**:13, 85

Omelyuk, V. A., **9**:103, 104, 110

Omohundro, G. A., **8**:140(70), 160; **18**:139(71), 159(71)

Omori, T., **27**:ix(8, 10); **31**:334(16), 425(16)

O'Neal, C. O., Jr., **3**:276, 301

O'Neal, H. A., **4**:109(55), 110(55), 139

O'Neil, P. S., **20**:76(68, 69), 82(68, 69)

O'Neill, B. K., **23**:196(2), 206(2), 267(2)

O'Neill, K., **14**:11, 100

Onik, G., **22**:304(153), 353(153)

Onishi, S., **23**:296(78, 79, 80, 81), 297(79, 80), 298(81), 341(79), 343(79), 352(81), 357(78, 79, 80, 81)

Ono, A., **23**:284(36), 355(36)

Ono, N., **30**:333(52), 349(101), 351(101), 378(101), 407(52), 408(52), 428(52), 431(101)

Ono, Y., **31**:450(49), 473(49)

Onodera, S., **15**:286, 326

Onsager, **18**:328(17), 329(17), 350(17), 362(17)

Onsager, L., **12**:234(82), 235(82), 256(82), 259(82), 269(82), 280

Onufriev, A. T., **3**:243, 246, 251

Oolman, L. D., **22**:196(114), 352(114)

Oosthuizen, P. H., **11**:220, 223, 228, 237, 248, 249, 258, 261

Openhein, A. V., **22**:419(25), 435(25)

Oppenheim, A. K., **2**:407, 450; **5**:20, 51; **7**:187, 209, 214; **12**:129(14), 138(14), 169, 193

Oppenheim, U. P., **5**:308(74), 323

Oran, E. S., **29**:91(59), 126(59)

Ordin, P. M., **5**:381, 510

Orell, A., **15**:250(88, 89), 252(88), 253(89), 280; **20**:84, 85, 92, 97, 99, 131

Oreper, G. M., **19**:80(238), 93(238); **28**:317(172, 173), 337(172, 173); **30**:333(53), 334(53), 429(53)

Oreshkin, O. F., **25**:81(152), 147(152)

Organ, A. E., **28**:317(174), 337(174)

Oriani, R. A., **14**:305, 306, 342, 343

Oriolo, F., **29**:234(93), 340(93)

Orito, F., **30**:359(123), 361(123), 432(123)

Orlando, A. F., **16**:299, 311, 349, 352, 364

Orlichek, A., **10**:177(55), 216

Orlicki, D., **19**:79(201), 92(201)

Orloff, L., **10**:258(63), 260, 282; **30**:76(32), 77(32), 89(32)

Ormiston, S. J., **18**:73(141), 74(141), 86(141); **20**:186(50), 187(50), 301(50)

Ornatski, A. P., **11**:241, 242, 260

Ornatskii, A. P., **17**:32(104), 61

Orning, A. A., **25**:223(109), 247(109)

Ornstein, L. S., **2**:317, 355

Orochko, D. I., **14**:160(70), 161(70), 162(70), 163(70), 166(70), 167(70), 168(70), 169(70), 244

Orr, C., **18**:169(37), 170(37), 171(74), 172(37), 236(37), 237(74)

Orr, C., Jr., **2**:364, 396; **9**:171, 180

Orszag, S. A., **14**:51, 102; **25**:355, 361, 394, 402, 416, 418, 419

Ortabasi, U., **18**:20(49), 81(49)

Ortega, A. (Chapter Author), **20**:181, 196(80), 205(109, 110, 111, 112), 206(109), 207(109, 110, 111, 112), 208(110), 210(109), 211(109), 232(80, 159), 233(110, 159), 234(159), 235(159), 238(159), 302(80), 304(109, 110, 111, 112), 307(159)

Ortolano, D. J., **23**:293(62), 356(62)

Oruma, F. O., **19**:38(86, 87), 87(86, 87)

Osakabe, H., **23**:109(119), 110(119), 116(119), 132(119)

Osako, M., **10**:67(53), 82

Osberg, G. L., **10**:171(17), 189, 202, 203, 205, 215, 217; **14**:232(169), 246; **19**:123(121), 125(121), 187(121)

Osborn, J. E., **15**:321, 321(8)

Osborne, D. G., **19**:311(102), 313(102), 353(102)

Osborne, D. V., **17**:69(29), 153

Osborne, W. Z., **28**:124(78), 142(78)

Oseen, C. W., **21**:334, 345

Osgood, R. M., Jr., **28**:342(30, 37), 416(30, 37)

O'Shaughnessy, E. J., **22**:223(110), 228(110), 352(110)

Oshima, M., **30**:349(155), 390(155), 433(155)

Osipova, O. A., **19**:144(157), 188(157)

Osipova, V. A., **6**:416(87, 88), 490

Osipova, V. L., **31**:275(178), 281(178), 330(178)

Oslejsek, O., **20**:297(324), 314(324)

Ossin, A., **11**:203(46), 206(46), 254

Oster, G. F., **22**:287(154), 354(154)

Österdahl, C., **17**:29(86), 60

Ostergaard, K., **23**:253(201), 275(201)

Ostergren, G., **22**:223(110), 228(110), 352(110)

Osterle, J. F., **11**:290(25), 315; **SUP**:73(172), 160(172), 161(172), 163(172), 165(172), 190(172), 191(172), 299(172), 399(172); **18**:73(142), 86(142); **19**:268(52), 271(52), 273(52), 275(52), 351(52); **20**:186(24), 187(24), 300(24)

Ostermaier, B. J., **14**:325, 327(103), 329, 330, 344

Ostrach, S. (Chapter Author), **1**:41, 44, 50; **4**:1, 6, 14, 15, 62, 64, 152, 223; **5**:131, 134(26), 207, 246, 247, 250; **6**:396(28), 487; **8**:161, 162(4, 5), 163(7, 52), 164(52), 165, 166, 167, 169, 170(33, 34), 171, 172(36), 173, 192, 197, 201, 203210, 211(33), 212(35), 213(49), 214222, 224, 226, 227; **9**:20, 63, 110, 288, 296(28), 346; **11**:68, 190, 269, 272, 273(6), 314; **18**:11(10, 11), 58(107, 108), 59(107, 108), 72(137), 79(10, 11), 84(107, 108), 86(137); **19**:80(223, 235), 93(223, 235), 309(98, 99), 310(101), 313(98, 99, 101), 353(98, 99, 101); **20**:186(51), 187(51, 52), 301(51, 52); **24**:278(27), 280(27), 319(27); **28**:370(126, 131), 420(126), 421(131); **30**:51(50), 52(50), 89(50), 272(43), 277(43), 282(43), 283(43), 284(43), 285(43), 309(43), 364(129), 366(129), 432(129)

Ostrander, L. E., **6**:474(306), 501

Ostrogorsky, A., **30**:332(47), 346(47), 428(47)

Ostroumov, G. A., **6**:474(225, 246), 483(225, 246), 497, 498; **8**:164, 226

Ostrovskiy, J. N., **16**:95, 96, 97(55), 100, 101, 102, 114(71, 74), 116(71), 117(54), 125, 154, 155

Ostwald, W., **13**:217(30), 264; **24**:108(81, 82), 187(81, 82)

O'Sullivan, M. J., **14**:43, 44(105), 45(101), 46(105), 47, 48, 49(101), 50(101, 117, 118), 66, 72, 102, 103; **20**:315(7), 316(7), 349(7)

Oswatitsch, K., **2**:58, 108; **6**:3(5), 122; **14**:332, 345

Otaka, M., **27**:ix(19), 181(31); **31**:334(14), 388(39, 40), 391(48), 413(48), 418(48), 425(14), 427(39, 40, 48)

Otani, S., **20**:181(1), 299(1)

Otero, C. J., **23**:371(32), 462(32)

Othmer, D. F., **10**:169(7, 10), 215; **14**:151(8), 232, 242; **15**:228(1), 271, 276; **19**:99(11), 103(11), 184(11)

Otis, D. R., Jr., **20**:232(162, 164), 238(162, 164), 240(162), 241(162), 256(164), 307(162, 164); **24**:302(28, 29), 313(29), 319(28, 29)

O'Toole, J. L., **11**:300, 315

Ototake, N., **11**:226, 262

Otsuka, K., **20**:183(6), 299(6)

Ott, H. H., **13**:18, 60

Ottewill, R. H., **23**:262(32a), 268(32a)

Otto, E. W., **4**:146(2), 150(8), 222, 223; **9**:186(5), 268
Otto, S. W., **30**:356(106), 357(106), 431(106)
Ou, J. W., **6**:396, 397(40), 431, 488; **SUP**:80(228, 229), 83(239), 109(228), 110(228), 111(228), 129(239), 130(239), 131(239), 157(229), 176(229), 178(358), 184(229)
Ouazzani, M. T., **28**:370(127), 420(127)
Ouelhazi, N., **30**:172(143), 195(143)
Ouyang, F., **19**:179(234), 190(234)
Ovechkin, D. M., **17**:22(68), 59
Oven, M. J., **20**:256(244), 311(244); **23**:13(65), 15(65), 37(65), 109(65), 110(65), 112(65), 129(65)
Overbeek, J. T., **12**:242(83), 252(83), 255(83), 256(83), 259(83), 281
Overcamp, T. J., **12**:23(38), 74
Oversby, W. M., **19**:82(267), 95(267)
Owada, H., **19**:74(172), 76(172), 77(172), 91(172)
Owase, Y., **23**:8(19), 12(19), 14(19), 28(19), 29(19), 36(19), 39(19), 59(19), 61(19), 79(19), 80(19), 82(19), 83(19), 84(19), 88(19), 93(19), 106(19), 127(19)
Owen, B. B., **12**:228, 234(31), 235(31), 236(31), 237(31), 240(31), 256(31), 259(31), 269(31)
Owen, R. G., **23**:108(111), 109(111), 110(111), 132(111)
Owens, F. L., **16**:232
Owens, G. U., **5**:131, 208, 216(10), 246
Owens, W. L., Jr., **1**:239, 265
Oxenius, J., **8**:273, 283
Ozaki, R., **28**:28(53), 36(53), 56(53), 72(53)
Özgü, M. R., **SUP**:77(215), 139(215), 145(215)
Ozisik, M. H., **15**:289, 317, 326, 330
Özişik, M. N., **9**:265(146), 272; **11**:322(38), 326(48), 329(48), 388(138), 397(48, 160), 401(160), 408, 435, 438; **15**:26(52), 56(52); **17**:237; **19**:35(70), 38(86, 87), 74(170), 75(170), 78(187), 86(70), 87(86, 87), 90(170), 91(187), 281(75), 282(75), 284(75), 286(75), 294(75), 352(75); **23**:136(4), 184(4); **24**:56(21, 22), 99(21, 22); **28**:77(54, 75), 141(54), 142(75); **31**:343(27), 426(27)
Ozkaynak, T. F., **14**:226(162), 246; **19**:128(134), 164(215), 168(215), 171(215), 188(134), 190(215)

Ozoe, H., **SUP**:108(273), 128(306), 143(273), 148(306), 404(273, 306); **15**:208(98), 209(98), 210(98), 225(98); **18**:12(13, 18), 19(39), 59(112), 60(112), 61(112), 63(112), 79(13, 18), 81(39), 85(112); **20**:378, 382, 387; **24**:312(30), 313(30, 31), 320(30, 31); **30**:334(62, 63), 349(63), 396(165), 400(165), 429(62, 63), 434(165)

P

Paal, L. L., **6**:407(62), 408(62), 409(62), 489
Packard, R. E., **17**:89(60), 154
Paddleford, D. F., **29**:71(31), 125(31)
Padet, J., **23**:326(215), 364(215)
Padilla, A., Jr., **11**:91, 191; **30**:42(73), 90(73)
Padovan, J., **15**:320(133, 134, 135), 326
Paffenbarger, J., **26**:220(17), 255(17), 271(17), 304(17), 305(17), 315(17), 325(17)
Pagani, C. D., **7**:171(28, 29), 172(28, 29), 176(29), 178(29), 181, 212
Page, F., **6**:515(27), 562; **7**:52(28), 84
Page, F. M., **29**:130(3), 202(3)
Page, R. H., **6**:7(17, 18, 19), 15(17, 19), 123
Pahor, S., **3**:107, 173; **SUP**:79(223), 156(337, 338)
Pai, B. R., **7**:339(17, 23), 350, 355, 359, 378, 379
Pai, S. I., **1**:276(12), 300(43), 353; **3**:192, 249; **6**:488
Pai, V. K., **11**:105(124), 114, 194
Paik, S., **26**:25(146), 99(146)
Paine, M. A., **23**:447(90), 464(90)
Pais, M. R., **26**:123(65, 66), 125(65, 66), 148(107), 151(107), 154(107), 211(65, 66), 213(107)
Paivanas, J. A., **5**:343345, 506
Pak, M. I., **6**:416(87, 88), 490
Pakhomov, L. V., **14**:205, 245
Pakhomova, E. S., **15**:325
Palanikuman, P., **9**:80, 110
Palazzi, G., **17**:25(77), 59
Palchonok, G. I., **19**:148(170), 153(170), 154(170), 189(170)
Paleev, I. I., **11**:56(11), 58(11), 59(11), 60(11), 118, 189; **20**:19(39), 81(39)
Palen, J. W., **16**:117(84), 118, 123, 124, 125, 149, 150, 155, 156; **30**:202(58), 252(58); **31**:443(70), 474(70)

Palfrey, D. L., **28**:341(14), 415(14)

Palik, E. D., **23**:145(23), 185(23)

Palla, R. L., Jr., **22**:330(155), 342(155), 354(155)

Pallas, S. G., **17**:131(136), 156

Pallone, A., **2**:231, 233(65), 266, 269, 270; **4**:279, 314, 334, 445; **6**:41(77, 78), 65(77, 78), 87, 89(140), 90, 91(140), 126, 129

Palm, E., **14**:30, 31, 33, 101; **20**:330(47), 351(47)

Palma, G., **13**:219(51), 265

Palmateer, S. C., **28**:389(170), 422(170)

Palmer, D. C., **9**:364(32), 386(32), 415

Palmer, D. G., **15**:321

Palmer, L. D., **9**:73(47), 76(47), 108

Palmer, S. B., **28**:103(70), 141(70)

Palmer, T. D., **14**:14(34), 100

Palmer, W., **28**:84(76), 142(76)

Pan, F., **6**:61, 127

Pan, F. Y., **26**:12(158), 100(158)

Pan, R. B., **11**:203(45), 206(45), 254

Pan, Y., **26**:127(71), 137(92), 138(92), 139(92), 149(71), 151(71), 153(92), 154(71, 92), 155(92), 165(71, 92), 173(128, 129), 175(128, 129), 176(128), 177(128), 179(128, 129), 180(129), 181(128, 129), 211(71), 212(92), 214(128, 129)

Pan, Y. S., **7**:204, 206, 208, 217

Panaev, V. N., **6**:457(185), 495

Panaia, J. N., **SUP**:76(212), 171(212)

Panchal, C. B. (Chapter Author), **15**:232(17), 276; **30**:222(40, 41, 46, 47), 223(46, 47), 224(40, 41, 46, 47), 225(46), 226(40, 41), 251(40, 41, 46), 252(47); **31**:431, 436(50, 52), 438(50, 56), 439(50, 56), 443(51), 444(51), 447(50), 448(52), 454(50), 460(52), 473(50, 51, 52, 56)

Panchurin, N. A., **6**:405, 408, 410, 489; **25**:81(159), 147(159)

Panday, S., **30**:102(40), 185(40), 190(40)

Pandya, S. J., **10**:34, 37, 53(31), 68(31), 81

Panevin, V. I., **31**:265(159, 160), 329(159, 160)

Pang, Y., **28**:308(154), 336(154)

Pannett, R. F., **4**:114(67), 140

Pannu, J., **19**:58(132), 59(132), 61(132), 89(132); **24**:3(6), 36(6)

Panov, O. M., **19**:164(210), 169(210), 170(210), 190(210)

Pantaloni, J., **30**:137(96), 193(96)

Pantanzidou, M., **30**:181(155), 196(155)

Pantazelos, P. G., **7**:275(115), 318

Panteleev, A. A., **25**:127(194, 195), 129(195), 149(194, 195)

Pantolini, J., **26**:49(218), 50(218), 103(218)

Panton, R. L., **10**:231, 232(21), 233(21), 281; **28**:156(50), 230(50)

Pao, Y.-H., **25**:343, 416

Paoli, R., **10**:109(41), 164; **29**:166(119), 208(119)

Paolucci, S., **24**:300(33), 320(33)

Papanicolaou, G. C., **18**:200(153), 239(153)

Papell, S. S., **3**:84(85), 89, 100; **4**:164, 165, 202, 203, 204, 224, 226; **17**:37(121), 62; **23**:282(33), 355(33)

Pappas, C. C., **3**:55, 61, 71, 74, 98, 99

Pappell, S. S., **7**:338, 344, 352(39), 355, 358, 358(53), 359, 368(11), 370, 377, 378, 379

Parabrahmachary, S., **SUP**:217(410)

Paramonov, N. V., **31**:279(183), 281(196), 282(183, 188), 287(183, 188), 292(183, 188), 330(183), 331(188, 196)

Parczewski, K. I., **4**:108(53), 139

Parikh, P. G., **25**:81(145), 146(145)

Pariyskiy, Yu. M., **14**:97(191), 104

Park, C., **4**:280, 282, 314; **19**:81(256, 258), 94(256, 258)

Park, C. K., **29**:234(81), 339(81)

Park, E. L., **11**:105(70), 111(70), 192; **18**:255(22), 257(22), 262(22), 321(22)

Park, H. C., **23**:201(202), 202(202), 204(202), 205(202), 206(202), 216(202), 220(202), 275(202)

Park, H. S., **29**:180(159, 160), 210(159, 160)

Park, J. S., **24**:315(16, 32), 319(16, 32), 320(32); **31**:259(150, 151), 329(150, 151)

Park, K., **16**:46(20), 57

Park, K. A., **20**:203(99), 204(99), 210(99), 212(99), 214(99), 216(99), 217(99), 245(207), 246(207), 303(99), 309(207)

Park, M. G., **2**:385(72), 396

Park, W. H., **10**:171, 215; **19**:123(121), 125(121), 187(121)

Parker, F. R., **5**:48(77), 54

Parker, G. H., **7**:235, 284, 285, 314

Parker, H. W., **23**:229(215), 276(215)

Parker, J. A., **23**:193(203), 275(203)

Parker, J. C., **30**:98(33, 34), 190(33, 34)

Parker, J. D., **1**:429, 432, 443; **7**:66(49), 85; **9**:227(57), 269; **11**:203, 206, 263; **17**:317

Parker, S. P., **30**:255(2), 307(2)

Parker, W. J., **10**:269, 283

Parkinson, D. H., **5**:326(8), 505

Parks, C. J., **20**:294(319), 314(319)

Parks, J. C., **20**:183(12), 299(12)

Parks, P. E., **17**:75(46), 102(46), 154

Parl, S., **20**:152(17), 179(17)

Parlange, J. Y., **14**:306, 343

Parlar, M., **30**:114(66), 116(66), 191(66)

Parmentier, E. M., **15**:206(93, 94), 208(93, 94), 211(93, 94), 225(93, 94); **30**:148(108), 193(108)

Parmer, J. F., **5**:26, 51

Parmley, R. T., **9**:413(108), 417

Parnar, A. L., **14**:152(30), 243

Parnas, A. L., **6**:474(221), 479, 497; **11**:219, 223, 235, 238, 240, 258; **14**:151(1), 152(27, 29), 176(183), 236(183), 237(183), 242, 243, 247; **19**:102(36), 116(36), 117(36), 129(36), 132(36), 185(36)

Parrish, J. A., **22**:278(4), 347(4)

Parrot, J. E., **15**:48(80), 57(80)

Parrott, J. E., **18**:161(11), 236(11)

Partain, L. D., **28**:352(90), 419(90)

Parthasarathy, K., **7**:367, 379

Partington, J. R., **30**:268(33), 308(33)

Partridge, B. A., **25**:166(57), 245(57)

Partridge, E. P., **30**:243(74), 253(74)

Paruit, B. A., **SUP**:50(59), 51(59), 181(59)

Parygin, V. V., **17**:140, 327(28, 30), 328(28), 330(28, 30), 331(30), 332(28, 30), 335(28), 336(28, 30)

Pasamehmetoglu, K. O., **26**:45(159, 160), 100(159, 160)

Pasari, S. N., **23**:198(143), 206(143), 226(143), 228(143), 229(143), 273(143)

Pascal, F., **23**:212(211, 212, 213), 276(211, 212, 213)

Pascal, H., **23**:212(204, 205, 206, 207, 208, 209, 210, 211, 212, 213), 275(204, 205, 206), 276(207, 208, 209, 210, 211, 212, 213); **24**:111(125), 118(125), 119(125), 170(151), 172(151), 174(151), 189(125), 190(151)

Paschkis, V., **11**:105(89), 113, 193; **26**:238(43), 253(43), 269(43), 326(43)

Pasquino, A. D., **2**:365, 396

Patankar, S. V., **8**:35, 88; **13**:67(11, 15, 16, 17, 18), 69, 70(20), 71, 81(20), 84, 92, 102, 115; **SUP**:73(176), 89(248), 164(248), 211(176), 298(248); **16**:355, 365; **19**:19(37), 46(111), 50(111), 52(111), 85(37), 88(111), 268(54, 55), 271(54, 55), 351(54, 55); **20**:202(86, 87), 227(145), 229(145, 150, 151), 230(145), 231(145), 243(169), 303(86, 87), 306(145, 150, 151), 307(169); **24**:193, 194, 195, 196, 199, 201, 210, 211, 221, 247, 249, 271, 272, 273; **25**:331, 416; **28**:196(51), 230(51), 251(33), 255(64), 330(33), 332(64), 382(156), 422(156); **29**:25(20), 26(21), 56(20, 21); **30**:356(103), 431(103)

Patanker, S. V., **7**:339, 378

Patch, R. W., **5**:295(42), 323

Patel, B. M., **5**:90, 126

Patel, D. C., **19**:112(83), 113(84, 85), 180(83, 84, 85), 186(83, 84, 85)

Patel, K., **23**:193(87), 270(87)

Patel, P. D., **29**:169(128), 170(128, 135), 171(128, 135), 208(128), 209(135)

Patel, R. C., **19**:156(186), 179(186), 189(186)

Patel, R. D., **10**:189, 190, 191, 195, 217; **14**:151, 242

Patera, A. T., **20**:232(166), 243(174, 175, 176, 177), 307(166, 174, 175, 176, 177)

Patil, P. R., **20**:326(31), 328(36, 39), 350(31), 351(36, 39)

Patnaik, G., **26**:75(161), 100(161)

Patnaik, S., **28**:390(172), 391(172), 392(172), 422(172)

Patt, H. J., **4**:284(52), 315

Pattabhi Ramacharyulu, N.-C., **SUP**:251(436), 271(448)

Pattantyus, H. E., **15**:244(68), 279

Patten, T. D., **20**:186(36), 187(36), 300(36); **29**:39(29), 56(29)

Pattenden, R. F., **5**:241, 251

Patterson, C. V. S., **20**:217(125), 218(125), 219(125), 305(125)

Patterson, E. M., **14**:298(29), 342

Patterson, G. K., **15**:63(27), 64(27), 135(27)

Patterson, J. C., **24**:296(36), 299(36), 300(34, 35), 301(35), 303(40), 307(34, 36), 312(36), 320(34, 35, 36, 40)

Pattie, B. D., **1**:383(46), 441

Pattipati, R. R., **19**:99(19), 100(19), 108(19, 50), 185(19), 186(50)

Patton, A. J., **21**:72(14), 134(14)

Patton, R., **18**:25(56b), 82(56b)

Patureaux, T., **31**:453(25), 471(25)

Patwari, A. N., **23**:253(214), 254(195, 214), 255(214), 256(214), 258(195, 214),

259(214), 275(195), 276(214)
Patzek, T. W., **30**:185(162), 196(162)
Paul, F., **13**:217(32), 218(32), 264
Paul, P. J., **30**:65(74), 90(74)
Paulik, M. D., **12**:281
Paulon, J., **23**:280(9), 354(9)
Paulsen, K. D., **22**:290(156), 354(156)
Paulsen, M. P., **31**:106(20), 156(20)
Paulter, N., **28**:80(31), 139(31)
Paumerd, G., **31**:178(84), 196(84), 325(84)
Pavelescu, C., **28**:351(73), 418(73)
Pavlov, N. V., **30**:242(34), 251(34)
Pavlov, P. A., **10**:106(30), 107(30), 108, 164; **11**:8, 48
Pavlov, Y. M., **16**:216(27), 233, 234, 239; **17**:146(166), 157, 327(35), 335(28), 336(35), 337(35), 340; **18**:241(2), 249(2), 256(2), 310(2), 313(2), 320(2)
Pavlović, M., **26**:246(52), 253(52), 270(52), 274(52), 279(52), 314(52), 316(52), 326(52)
Pavlovskii, Iu. N., **6**:116, 131
Pavlovsky, G. I., **6**:418(101), 491
Pavlovsky, V. G., **31**:178(83), 213(83), 214(83), 325(83)
Pavshukov, A. V., **11**:318(20), 435
Pawelek, R. A., **SUP**:139(321)
Pawlowski, P. H., **7**:320e(165), 320g
Payne, A., **21**:158, 181
Payne, H., **7**:170(21), 211
Payne, L. E., **26**:4(162), 100(162)
Payne, L. W., **23**:229(215), 276(215)
Paynter, H. M., **22**:287(9, 157), 288(9, 157), 347(9), 354(157)
Paz, U., **12**:87, 101, 102, 112; **15**:105, 107(138), 111(138), 114(138), 140(138)
Pearce, C. A. R., **18**:187(69), 237(69)
Pearce, C. W., **28**:340(7), 415(7)
Pearce, D. L., **31**:456(53), 473(53)
Pearce, J. A., **22**:248(158), 277(219), 319(159, 206), 320(87), 321(159), 342(158), 350(87), 354(158, 159), 356(206, 217, 219), 421(32, 35, 36, 37), 422(35), 423(35), 424(35), 428(37), 436(32, 35, 36, 37)
Pearce, W. M., **6**:312(105), 366; **SUP**:190(376), 191(376), 194(376); **19**:308(92), 312(92), 353(92)
Pearcey, H. H., **6**:95(158), 130
Pearlman, H., **2**:352
Pearson, J. R. A., **13**:206(3), 233, 260(114),

261(3, 118), 264, 266; **24**:19(52), 38(52); **28**:149(68), 152(52), 230(52, 68); **30**:43(75), 90(75)
Pearson, R. A., **2**:411, 450
Pearson, R. G., **2**:123(11), 268
Pease, D. C., **10**:118(50, 51, 52), 121(50, 51, 52), 165
Peatman, J. B., **22**:411(23), 412(23), 435(23)
Peck, D. L., **14**:97(194), 105
Peck, H. L., **28**:365(110), 366(110), 367(110), 368(110), 420(110)
Peck, M. K., **18**:17(33), 80(33)
Peck, R., **15**:233(37), 234(37), 235(37), 277
Peck, R. E., **2**:316, 355; **SUP**:80(230), 120(230), 121(230), 158(230), 159(230), 186(230)
Peckover, R. S., **29**:158(88), 163(88), 206(88)
Pedersen, D. R., **7**:354(44), 378; **19**:79(202, 203), 92(202, 203); **21**:300, 344
Pedišius, A. A., **8**:146(78), 160; **18**:145(79), 159(79)
Pedroso, R. I., **19**:30(60), 31(62), 32(62), 86(60, 62)
Peebles, E. M., **3**:28(46), 32
Peebles, F. N., **SUP**:361(506), 363(506); **20**:6(7), 79(7)
Peebles, L. H., Jr., **15**:96(95), 97(95), 138(95)
Peek, C., **18**:20(48), 81(48)
Peercy, P. S., **28**:80(31, 100), 139(31), 143(100)
Pegg, D. E., **22**:177(45), 187(6), 188(161), 220(186), 290(163), 304(160, 162, 164, 186), 347(6), 348(45), 354(160, 161, 162, 163, 164), 355(186)
Pei, D. C. T., **13**:126, 153(31), 161, 162, 169, 191, 201; **15**:184(50), 185(50), 224(50); **23**:340(273, 274), 367(273, 274)
Pei, W., **30**:262(24), 308(24)
Peierls, R., **11**:354, 436
Peifer, W. A., **19**:80(233), 93(233)
Pelepeichenko, I. P., **6**:474(222), 480, 497
Pell, W. H., **26**:4(162), 100(162)
Pellew, A., **4**:152, 223
Pellow, A., **5**:204, 249
Penciner, J., **12**:226(27a), 278
Pendse, H., **15**:327
Penndorf, R. B., **23**:146(29), 185(29)
Pennell, K. D., **30**:111(58), 191(58)
Penner, S. S., **2**:114(9), 123(9), 268; **3**:212(57), 217(63), 250; **5**:254(2), 255(2), 266(2), 277(2), 279(2), 280(2), 283(2), 284(2), 291(2), 292, 293, 294(2), 307(66), 308(66),

Penner, S. S. (*continued*)
309(75), 313(2), 321, 323; **8**:233, 241, 282; **11**:318(6, 10, 16), 376(6, 10), 378, 381, 388(6, 10), 389, 434; **12**:129, 139, 162, 192; **15**:327

Pennes, H. H., **22**:8(29), 18(29), 19(46), 22(46), 24(46), 28(46), 29(46), 154(46), 245(165), 329(165), 354(165)

Penney, W. R., **11**:203(44), 206(44), 248(44), 254; **30**:230(42), 231(42), 251(42)

Pennington, R. H., **5**:121, 128, 434(203), 514

Pennucci, J., **23**:293(68), 357(68)

Penot, F., **20**:187(27), 300(27)

Penski, K., **4**:283, 314

Penycook, S. J., **28**:79(67), 141(67)

Pepov, K. M., **9**:103(9a), 106

Pera, L., **9**:297(36), 299(45), 300(45), 304(55, 67), 305(45), 310(67), 324, 326(36), 335(90), 339(102), 342(102), 343, 346, 347, 348; **11**:203, 206(58), 208, 209(58), 216(58), 248(58), 254, 278(10), 279(10), 314; **20**:203(94, 96), 303(94, 96); **30**:280(65), 310(65)

Peraldi, O., **29**:86(51), 88(51), 89(51), 125(51)

Perdikis, C. P., **24**:105(36), 186(36)

Perdon, A., **23**:418(75), 464(75)

Pereira, J. C. F., **24**:213, 271

Pereira, M., *see* Collares Pereira, M.

Pereira Duarte, S. I., **23**:370(11), 461(11); **24**:102(15), 185(15)

Perelman, T. L., **6**:376(2, 3, 4, 5), 377, 464, 465(3, 4, 192), 486, 495; **8**:36, 37, 38, 40, 42, 89; **25**:11(38, 41), 32(41), 141(38, 41)

Perepelitsa, B. V., **12**:103, 108(37), 113; **15**:91, 92(80), 107(80), 108(80), 138(80); **25**:81(162), 147(162)

Peretz, D., **9**:128(34, 35), 129(35), 178; **14**:231(167), 246; **19**:147(164), 189(164)

Pereverzev, D. A., **6**:418(131), 492

Perevozchikova, J. P., **14**:160(72), 161(72), 164(72), 166(72), 167(72), 168(72), 169(72), 170(72), 171(72), 172(72), 176(72), 206(72), 244

Perez, S. E., **29**:233(63), 319(63), 338(63)

Peri, J. S. J., **17**:68(9), 153

Peri, S. S., **23**:246(38b), 268(38b)

Peric, M., **24**:199, 247, 270, 271, 272

Pericleous, K. A., **24**:302(25), 319(25)

Perkins, A. S., **11**:105(125), 115, 194

Perkins, H. C., Jr., **3**:16, 26, 32; **6**:539, 542, 543(60), 563; **8**:104, 105(30), 159; **11**:219, 223, 228, 238, 257; **SUP**:2(12), 411(12), 414(12), 415(12), 416(12); **15**:5(20), 55(20), 121(143), 140(143); **18**:98(30), 99(30), 128(30), 158(30); **19**:304(89), 334(138), 336(138), 353(89), 355(138)

Perkins, K. R., **SUP**:218(411)

Perl, W., **22**:30(47, 48), 36(47, 48), 154(47, 48)

Perleson, A. S., **22**:287(154), 354(154)

Perlin, J., **18**:28(62), 82(62)

Perlmutter, M., **1**:15, 16(17), 49, 306(45), 308, 311, 354; **2**:422, 423(18), 424(16, 17, 18), 425(18), 432, 445(18), 450, 450(18); **3**:211(49), 250; **5**:28(42), 29, 30, 32, 33, 35, 40, 52, 53, 313(87), 324; **6**:425(159), 428, 430(159, 160), 433(160, 163), 434(163), 458, 459(159), 494; **7**:175, 212; **8**:54(240, 241), 90; **31**:334(17), 341(17), 425(17)

Perloff, D. S., **15**:327

Permyakov, V. A., **15**:273(116, 117), 281; **31**:160(42), 282(186), 283(189), 324(42), 331(186, 189)

Pernell, T. L., **30**:396(160), 398(160), 434(160)

Perng, C. Y., **24**:218, 272

Perona, J. P., **20**:186(29), 187(29), 300(29)

Perrakis, M., **31**:464(54), 473(54)

Perre, P., **30**:172(141, 142), 195(141, 142)

Perrins, W. T., **18**:186(137, 138), 191(137, 138), 239(137, 138); **23**:399(67), 422(67), 463(67)

Perron, G., **18**:337(53), 363(53)

Perroud, P., **1**:404(84), 411, 427, 434, 442, 443; **5**:412, 415, 416, 422, 513; **11**:105(68), 109(68), 111(68), 192

Perry, C., **19**:80(234), 93(234); **28**:246(24, 26), 317(26), 330(24, 26)

Perry, C. H., **10**:21, 23, 36

Perry, C. W., **1**:239, 265; **16**:125, 155

Perry, J. H., **1**:363(15), 440; **15**:64(40), 65(40), 136(40)

Perry, K. P., **13**:44(61), 45, 60

Perry, M. P., **14**:140, 145

Persh, I., **3**:53(47), 61(47), 98

Persh, J., **2**:16(18), 43(18), 105

Person, L. N., **31**:176(82), 325(82)

Pery, R., **6**:38(72), 125

Peshin, R. L., **9**:135, 179

Peshkov, V. P., **17**:97, 154

Peskin, R. L., **9**:123, 124, 125(25), 150, 170, 178, 179

Peskov, O. L., **17**:29, 39(85), 60

Pessers, T., **17**:296, 317

Petavel, J. E., **11**:202, 204, 209(6), 252

Petela, R., **15**:48(78), 57(78)

Peter, F. G., **1**:394(72), 398(72), 442

Peteriengo, G., **17**:28(81), 59

Peterka, V., **25**:4(28), 141(28)

Peterlin, A., **2**:363(13), 365(13), 394

Peterlongo, C., **1**:245(89), 266

Peterlongo, G., **1**:361(13), 363(13), 378(13), 381(13, 39, 40, 41), 411(13, 39, 40, 41), 413(13, 39, 40, 41), 421(13, 40, 41), 436(13), 440, 441

Peters, A. R., **19**:146(160), 149(160), 188(160)

Peters, D. L., **18**:156(84), 159(84), 328(23), 330(23), 362(23)

Peters, J. I., **5**:406(158), 512

Peters, K., **10**:177, 216

Peters, Th., **4**:237(4), 240(4), 313

Petersen, A. W., **2**:387, 396; **15**:105(130), 140(130)

Petersen, E. E., **2**:390(2, 100), 391(100), 394, 396; **25**:257, 262, 281, 282, 283, 284, 288, 289, 290, 315, 318

Petersen, H. L., **7**:181, 213

Peterson, A. C., **9**:208(32), 238(32), 239, 240, 269

Peterson, E. E., **15**:185(55), 224(55)

Peterson, E. G., **11**:232(170), 237, 260

Peterson, G. P. (Chapter Author), **20**:181, 260(251), 261(251), 262(251, 257), 263(257), 265(257), 266(269), 273(284, 285), 275(290, 291), 276(290), 277(291, 292), 278(291), 283(303), 284(303), 285(303), 290(284), 295(284), 311(251, 257), 312(269, 284, 285), 313(290, 291, 292, 303)

Peterson, K. E., **20**:274(287), 312(287)

Peterson, P. F., **29**:302(160), 303(162), 343(160, 162); **30**:173(146), 195(146)

Peterson, W. C., **16**:236

Petitt, E., **10**:40, 73(2), 76, 78, 80

Petrakis, D. N., **28**:309(158), 336(158)

Petrash, D. A., **4**:146(2), 150(8), 222, 223

Petree, D. K., **5**:233(172), 236, 251

Petrick, M., **1**:394(62), 398(62), 442; **9**:258(97), 270

Petrie, A. M., **11**:220, 224, 258

Petrie, C. J. S., **13**:260(116), 266; **15**:213(113), 225(113)

Petrie, D. J., **21**:287, 300, 344

Petrie, J. C., **14**:179(100), 188(100), 189, 191(100), 195, 245; **19**:113(89), 119(89), 148(89), 186(89)

Petro, W. G., **28**:342(24), 416(24)

Petropoulos, H. H., **15**:321

Petrov, N. E., **21**:34(55, 56), 35(55), 36(55, 56), 37(56), 52(55, 56)

Petrov, N. G., **9**:265(141), 272

Petrovichev, V. I., **17**:327(28), 328(28), 330(28), 332(28), 335(28), 336(28), 340; **21**:37(61), 39(61), 40(61), 41(61), 52(61)

Petrovsky, Yu. V., **31**:160(3), 245(134), 322(3), 328(134)

Petrusanskij, V. J., **13**:261(123), 266

Petschek, H. E., **1**:269(5), 352

Pettigrew, C. K., **10**:211, 218

Petty, C. A., **23**:193(42), 268(42); **25**:292, 308, 316

Pettyjohn, R. R., **9**:370(40), 372(40), 415

Petukhov, B. S. (Chapter Author), **1**:141(14), 184, 414(105), 443; **6**:378(9), 396, 397(9), 419, 466(9), 486, 504, 504(13, 14), 506, 521, 523(13), 526, 531, 533(14), 539, 540(61), 541(61), 562, 563, 564; **7**:48, 49, 50, 84; **11**:158(167), 196; **SUP**:2(8), 125(304), 133(310), 145(304), 193(8), 227(8), **17**:327(31, 32), 333, 334, 335, 336(31, 32), 340; **18**:255(15), 301(63), 307(63), 320(15), 322(63); **19**:191(4), 242(4); **21**:1(1), 4(1), 6(14), 7(15), 8(16, 17, 19, 20), 9(14, 20), 11(1, 16, 20), 12(16, 20), 13(1), 14(28, 29), 15(17, 30), 17(14, 33), 18(14), 19(14), 20(14, 33), 21(17), 22(14, 20, 33), 24(14), 28(14), 29(17, 20), 31(14), 36(28), 50(1), 51(14, 15, 16, 17, 19, 20, 28, 29, 30, 33); **25**:13(48), 33(48), 48(72, 74), 82(72), 122(72), 141(48), 143(72, 74); **31**:179(85), 195(98), 229(119), 256(149), 325(85), 326(98), 327(119), 329(149)

Petukhov, V. V., **6**:526, 562

Petuskey, W. T., **28**:351(82), 418(82)

Petzold, K., **13**:2(14), 10(14, 32), 11, 15, 17(32), 59, 60

Peube, J. L., **5**:177, 183(86), 185, 249; **20**:187(27), 300(27)

Pexton, A. F., **1**:409(90), 442

Peyret, R., **24**:194, 210, 213, 272

Pezzin, G., **13**:219(51), 265

Pfankuch, H., **14**:28, 101

Pfann, W. G., **30**:332(46), 428(46)

Pfeffer, R., **9**:118, 177; **23**:230(216), 276(216); **30**:236(43), 251(43)

Pfeil, D. R., **20**:205(103), 304(103)

Pfender, E. (Chapter Author), **4**:229, 248(13), 296(64), 300(64), 301(64), 303(64, 65), 304, 305(65), 313, 315

Pfister, G., **21**:156, 181

Pfotenhauer, J. M. (Chapter Author), **17**:65, 125, 129(131), 130(131), 131(132), 133(132), 135, 136, 137, 156

Pham, Q. T., **19**:27(56, 57), 85(56, 57)

Phan, R. T., **23**:196(78), 270(78)

Phan-Thien, N., **18**:199(151, 152), 200(151), 239(151, 152); **23**:219(115, 268), 272(115), 278(268)

Philbrook, W. O., **21**:300, 346

Philip, J. R., **15**:286, 327

Philippoff, W., **2**:376, 378, 395; **13**:219(68), 265; **15**:182(45), 224(45)

Phillip, J. R., **13**:125, 201

Phillips, E. C., **7**:236(29), 241, 243, 259, 261, 303, 314, 316

Phillips, H. M., **28**:102(37), 140(37)

Phillips, J. C., **11**:352(74), 355(74), 436

Phillips, R. E., **24**:249, 272

Phillips, T. J., **20**:217(130), 220(130), 305(130)

Phillips, W. F., **7**:178, 212; **21**:246(19), 273(19, 42, 43), 276(19, 42, 43)

Philpot, C. W., **10**:228(11), 234(13), 265(11), 281, 283

Philpot, J. St. L., **6**:162, 363

Phipps, C. R., Jr., **28**:124(77, 78), 142(77, 78)

Piacsek, S. A., **5**:216, 250

Piatt, J. D., **20**:232(160), 237(160), 307(160)

Piavanas, J. A., **15**:48(74), 57(74)

Piccone, M., **5**:82, 126, 412(148), 512; **11**:105(79, 80), 112(79, 80), 121(79), 122(79), 193; **16**:234; **20**:24(51), 81(51); **29**:91(59), 126(59)

Picha, K. G., **5**:177(83), 178, 249

Picot, J. J. C., **18**:163(24), 169(40, 44), 170(24, 40, 44), 177(24, 57), 178(40, 44), 220(24, 40, 44, 154), 221(24, 40, 44, 154), 222(154), 233(44, 154), 236(24, 40, 44), 237(57), 239(154)

Picraux, S. T., **28**:80(100), 143(100)

Picus, V. J., **7**:39, 66(50), 67(50), 86

Pielke, R. A., **25**:377, 419

Pierce, D. C., **17**:69(20), 70(20), 71(20), 153

Pierce, F. J., **23**:349(304), 368(304)

Piercy, N. A., **11**:226, 263

Piercy, N. A. V., **SUP**:62(74), 322(74), 324(74), 341(74)

Pierse, B. L., **11**:144(160), 146(160), 162(160), 164(160), 170, 177, 182, 196

Pierson, A. G., **11**:34(74), 49

Pierson, O. L., **8**:94(10), 159; **18**:88(10), 158(10)

Pierson, R., **14**:86(174), 87(174), 88, 104

Pierson, W. J., **16**:48, 57

Pieters, C., **10**:7, 8, 9, 36

Pietralla, M., **18**:161(10), 236(10)

Pigal'skaya, L. A., **11**:344(66, 66a), 345(66, 66a), 349, 412(201), 436, 439

Pigford, R. L., **2**:371, 396; **15**:101(108), 139(108), 195(74), 225(74)

Piggott, B. D. G., **23**:108(110), 109(110), 110(110), 132(110)

Pikashov, V. S., **14**:155(53, 54), 243

Pike, E. R., **30**:329(37), 332(37), 428(37)

Pike, J. G., **11**:105(140), 195

Pikus, I. F., **6**:418(126), 492

Pikus, V. U., **7**:38(21), 84

Pilch, M., **2**:391(90), 396

Pilch, M. M. (Chapter Author), **29**:62(7), 72(7), 124(7), 168(127), 169(127), 174(127), 208(127), 215, 216(5, 11), 217(11, 14), 218(18, 19, 23), 219(18, 19, 24), 220(41), 221(25, 27, 28, 29, 30, 31, 32, 33, 34, 35, 36, 37, 38, 39, 40, 41), 228(18), 230(41), 232(18, 23), 233(18, 19, 23, 60, 61, 67, 68), 234(23), 239(23), 241(23), 244(18), 248(23), 249(23, 100), 251(23, 100), 253(109), 255(40), 261(117), 263(117), 264(117, 126), 265(40, 41, 117), 268(134), 269(5, 24, 40), 270(19, 40, 41, 100), 271(25, 28, 29, 30, 31, 137), 273(145), 274(23), 275(23), 276(23, 41, 147, 148), 277(19, 24, 126), 278(40, 148), 279(18, 19, 23, 24), 280(28, 29, 31), 282(28, 29, 31, 41), 288(18), 290(40, 41), 293(29), 295(23, 29), 297(18), 299(159), 300(40), 301(23, 159), 302(36, 37, 159), 303(159), 304(159), 305(19, 159), 306(41, 159), 307(159), 308(159), 309(25, 29, 38, 67), 310(32, 33, 34, 35, 36, 37, 40), 311(23, 32, 33, 34, 35, 36, 37), 312(18, 24), 315(30, 40, 172), 316(30, 40), 318(40), 319(19, 23, 40), 322(19), 323(24, 40, 172), 324(175), 329(18, 147), 334(5), 335(11, 14, 18, 19,

23, 24, 25), 336(27, 28, 29, 30, 31, 32, 33, 34, 35, 36, 37, 38), 337(39, 40, 41), 338(60, 61, 67, 68), 340(100, 109), 341(117, 126), 342(134, 137), 343(145, 147, 148, 159), 344(172, 175)

Pilipenko, V. N., **15**:97(97), 138(97)

Pilitsis, S., **23**:219(115, 217), 272(115), 276(217)

Pillai, K. K., **23**:217(218), 276(218)

Pilling, N. B., **5**:65, 90, 125; **11**:105(96), 193

Pillow, A. F., **8**:165, 166, 226

Pilsworth, M. N., **2**:365, 396

Pimenta, M. M., **16**:301, 302(9, 33), 305, 306(9), 309, 310, 311, 349, 352, 363, 364

Pimputkar, S. M., **12**:41, 74

Pinchera, G. C., **11**:18(61), 49

Pinder, G. F., **14**:52(122, 123), 53(122, 123), 54(122, 123), 103; **30**:96(27), 98(36), 189(27), 190(36)

Pinder, K. L., **15**:269, 281; **23**:193(4), 196(4), 201(4), 202(4), 267(4); **24**:118(126), 189(126)

Pinheiro, J. de D. K. S., **31**:294(202), 295(202), 332(202)

Pinkus, O., **24**:1(1), 36(1)

Pinus, N. A., **14**:262(29), 280

Piomelli, U., **25**:337, 342, 351, 353, 354, 360, 361, 367, 369, 370, 375, 378, 382, 386, 392, 396, 403, 404, 412, 416

Pipkin, A. C., **18**:163(18), 236(18); **20**:364, 388

Piret, E. D., **20**:15(33), 81(33)

Piret, E. L., **4**:44, 64; **8**:128(61), 130(61), 160; **9**:304, 347; **11**:202(28), 205(28), 209(28), 213(28), 214(28), 218, 222, 253, 256, 278(9), 279(9), 280(9), 314; **18**:123(60), 124(60), 127(60), 159(60)

Pirkle, J. C., Jr., **SUP**:132(309); **25**:234(126, 127, 128), 248(126, 127, 128)

Pirri, A. N., **28**:133(83), 142(83)

Pirsko, A. R., **10**:224, 253(8), 281

Pirueva, L. V., **6**:418(118), 491

Pisani, A. P., **23**:281(29), 355(29)

Pisarev, V. Y., **16**:114(71, 74), 116(71), 155

Pisarevsky, V. M., **25**:81(151), 147(151)

Pishakov, V. S., **19**:164(204), 168(204), 171(204), 190(204)

Pišlar, V., **11**:20(70), 21(70, 71, 72), 31(71, 72, 73), 33(71), 34(72), 36(71), 37(71), 49

Piterskikh, G. P., **6**:504(12), 562

Pitter, R. L., **26**:8(167), 44(167), 100(167)

Pitteway, M. L. V., **14**:268(50), 270(50), 280

Pittman, J. F. T., **25**:288, 318

Pitts, C. C. (Chapter Author), **1**:185; **5**:107, 127; **7**:145(55), 161; **16**:128, 130, 155; **17**:2(15), 56

Pitts, D. R., **11**:105(126, 128), 115, 194; **23**:340(267), 366(267)

Piwonka, T. S., **28**:245(21), 266(21), 267(21), 330(21)

Placzek, G., **5**:310(84), 324

Planovski, A. N., **14**:160(70), 161(70), 162(70), 163, 166(70), 167(70), 168(70), 169(70), 244

Plapp, J. E., **4**:53, 64; **9**:323, 324(87), 348

Plass, G. N., **5**:37, 53, 287(31, 32), 288(31), 291(31), 301(55), 322, 323; **12**:136(23), 193

Platt, G. K., **5**:378, 379, 509

Platt, M., **26**:220(19), 221(19), 236(19), 243(19), 249(19), 250(19), 253(19), 269(19), 281(19), 300(19), 302(19), 325(19)

Platten, J. K., **28**:370(133, 134), 421(133, 134)

Pleshchenkov, G. A., **6**:416(95), 417(97), 490

Plesset, M. S., **1**:217, 218(31), 220, 221(31), 264; **4**:178, 225; **11**:13(38), 14, 15, 48; **15**:242(56), 243, 278; **16**:82, 83, 85, 87, 154; **26**:31(163), 50(163), 100(163)

Pletcher, R. H., **16**:53, 57; **24**:194, 268, 272; **25**:331, 417

Plitt, K. F., **10**:276(108), 284

Plumat, E., **11**:425(228), 426(228), 440

Plumb, O. A., **14**:16(42), 58, 69, 75(42), 76(42), 101; **15**:189(62), 224(62); **23**:371(27), 374(37), 435(27), 461(27), 462(37); **30**:95(23), 102(42), 142(101, 102), 143(42, 101), 145(101), 147(102), 167(126, 127), 169(136), 172(144), 189(23), 190(42), 193(101, 102), 194(126, 127), 195(136, 144)

Plumblee, H. E., **6**:73(123), 128

Plummer, P. L. M., **14**:306, 343

Plumpire, R. A., **17**:177, 314

Plyutinskii, V. I., **6**:447(179), 448(180), 474(285, 317), 494, 500, 501

Pnueli, D., **8**:33, 88, 213(49), 222, 227; **SUP**:67(122); **15**:238(46), 278

Podberezsky, A. I., **19**:100(28), 111(28), 115(28), 116(28), 117(28), 118(28), 126(131), 129(28), 130(28), 135(28), 136(28), 137(28, 131, 144), 139(28),

Podberezsky, A. I. (*continued*)
140(131, 144), 144(28), 145(131, 144, 158), 158(131), 159(28, 144), 161(144), 185(28), 188(131, 144, 158); **25**:205(104, 105), 247(104, 105)
Podporin, I. V., **31**:228(118), 327(118)
Podstrigach, Ja. S., **15**:327
Podstrigach, Ya. S., **11**:412(193), 439
Podzeev, V. G., **11**:56(11), 58(11), 59(11), 60(11), 118, 189
Poggi, D., **21**:287, 300, 345
Pogson, J. H., **9**:390(62), 396(74), 399(76), 416
Pogson, J. T., **20**:262(258), 311(258)
Pohl, H. A., **14**:112(74a), 147
Pohle, F. V., **1**:64, 121; **15**:287, 327
Pohlhausen, E., **3**:13, 21, 32
Pohlhausen, E. Z., **15**:14(38), 56(38)
Pohlhausen, K., **1**:53, 120; **3**:107, 173; **4**:23, 24, 63; **5**:177, 249; **6**:435(168), 437(168), 494; **9**:303(51, 52), 347; **11**:213, 214, 255; **SUP**: 112(280), 113(280), **28**:389(162), 422(162)
Poincaré, H., **8**:4, 84
Poirier, D. R., **28**:20(37), 21(37), 72(37), 245(22), 251(36, 37), 253(55, 56), 255(55), 256(55, 56, 66), 258(55), 262(72), 266(36, 37), 267(90, 91, 92, 93, 94, 95), 303(72, 138, 139), 309(36, 156, 157, 158), 330(22, 36, 37), 331(55, 56), 332(66, 72), 333(90, 91, 92, 93, 94, 95), 335(138, 139), 336(156, 157, 158)
Poiseuille, J., **SUP**:78(222)
Poisson, A., **18**:330(30, 31), 335(30), 362(30, 31)
Pokhvalov, Yu. Ye., **17**:37(122), 62
Pokryvailo, N. A., **6**:474(223), 480(223), 497; **15**:128(155), 130(155), 141(155)
Pokusaev, B. G., **26**:109(24), 137(88), 149(24), 151(24), 201(24), 202(24), 203(24), 209(24), 212(88)
Polak, M., **17**:152(182), 158
Polanyi, J. C., **28**:106(13), 139(13)
Polanyi, M., **2**:355
Polasek, F., **20**:279(297), 280(297, 298), 281(298), 282(298), 297(324), 313(297, 298), 314(324)
Polat, S., **23**:7(10), 126(10), 320(172), 362(172)
Poletavkin, P. G., **18**:256(25), 321(25)
Poletti, G., **24**:8(36), 14(36), 37(36)
Polezhaev, V. I., **6**:474(260), 483(260), 499; **8**:169, 226

Polgar, L. G., **5**:24, 25, 26, 51
Polhamus, G. D., **22**:248(2, 218), 250(2), 251(2), 252(2), 347(2), 356(218)
Poling, B. E., **23**:70(87), 130(87)
Polisevski, D., **24**:104(62), 186(62)
Poljak, G., **5**:20, 51
Poljak, M. P., **9**:71(117), 104(117), 111
Poll, A., **9**:127(29), 178
Pollack, G. L., **17**:70(22), 153
Pollard, D., **19**:82(277), 95(277)
Pollard, R., **28**:389(167), 394(167, 184), 422(167), 423(184)
Polley, G. T., **31**:453(42), 472(42)
Pollock, D., **23**:288(47), 356(47)
Pollock, D. W., **30**:168(128), 194(128)
Polnitskii, K. A., **11**:203, 206, 255
Polomik, E. E., **1**:428(130), 443; **5**:100, 101, 102, 127; **11**:170(178), 173, 175, 177(178), 197
Polonskaya, F. M., **1**:124, 184
Polonsky, V. S., **31**:265(163), 329(163)
Poltav'seva, L. L., **6**:474(310), 501
Poltz, H., **11**:363, 437
Polyaev, V. M., **31**:160(53), 324(53)
Polyak, G. L., **3**:212, 250; **11**:390(143), 397(143), 438
Polyak, M. P., **25**:127(196), 149(196)
Polyakov, A. F. (Chapter Author), **21**:1, 4(11), 6(14), 9(11, 14), 14(28), 17(14), 18(14), 19(14), 20(14), 22(14), 24(14), 28(14), 31(14), 32(50), 33(50), 36(28), 51(11, 14, 28), 52(50)
Polyakov, V. K., **31**:160(46), 324(46)
Polyakov, Y. A., **6**:416(91), 490; **8**:24, 87; **19**:197(34), 198(34), 244(34)
Polymeropoulos, C. E., **9**:317(81), 327(96), 328(97), 332, 333, 347, 348
Polymeropoulus, C. E., **6**:474(258), 480(258), 483(258), 499
Pomanenko, P. N., **3**:76(71a), 99
Pomarentsev, V. M., **10**:202(158), 218
Pomeau, Y., **25**:356, 417
Pomerantsev, V. M., **14**:158(62), 243
Pomerantsev, V. V., **3**:284(66), 288(66), 301
Pomerantz, M. L., **4**:206(87), 226; **5**:73, 126, 432, 462, 514; **11**:105(63), 106, 109, 111, 116, 191
Pomfret, M. J., **11**:203, 206, 262
Pompe, W., **28**:103(88), 142(88)
Pond, G. R., **11**:302(37), 315
Pong, L. C., **29**:233(69), 338(69)

Pong, L. T., **29**:189(193), 212(193)

Ponteduro, A. F., **30**:207(55), 208(55), 216(55), 218(55), 221(55), 223(55), 224(55), 252(55); **31**:456(68), 473(68)

Ponter, A. B., **9**:227(58, 59), 229(58, 59), 231, 269; **16**:70, 153, 235, 236

Poole, D. H., **29**:132(15), 175(15), 203(15)

Poots, G., **1**:116, 122, 291, 295, 296(30), 297, 353; **8**:70(268), 90, 166, 173, 176, 194, 226; **9**:253(80), 270; **SUP**:186(369), 213(405); **19**:31(65), 36(65), 37(65), 60(135), 86(65), 89(135)

Pop, I., **24**:122(129), 125(129), 189(129); **25**:291, 318

Pope, D., **9**:209(48), 210(34), 221(48), 233(34), 244(48), 264(130), 265(34, 48, 130), 269, 271; **21**:82(33), 83(38), 88(54), 91(33), 95(33), 135(33, 38, 54)

Pope, D. H., **5**:356(69), 508

Pope, G. A., **23**:206(102), 226(102), 228(102), 271(102)

Popov, D. N., **25**:4(20), 140(20)

Popov, V. I., **12**:103, 108, 113

Popov, V. N., **6**:504(14), 506, 521, 531, 533, 546, 549, 562, 563; **21**:2(3), 3(3), 13(26, 27), 14(27), 15(27), 22(40), 24(40, 41, 42, 43), 25(40), 26(40, 41, 42, 43), 27(41, 42, 43), 29(27), 34(3, 55, 56), 35(55), 36(55, 56), 37(3, 56), 38(3), 39(40, 41, 68, 69), 40(68, 69), 41(3, 69), 50(3), 51(26, 27, 40), 52(41, 42, 43, 55, 56, 68, 69); **25**:48(74), 143(74)

Popov, V. P., **6**:474(219, 223), 479, 480(223), 497

Popov, V. V., **21**:15(30), 51(30)

Popova, T. F., **25**:158(34, 35, 36, 37), 194(35, 36, 37), 201(35, 36, 37), 218(36), 229(37), 236(145), 244(34, 35, 36, 37), 249(145)

Popovich, A. T., **8**:293(21), 298(34), 303(21), 349

Poppendiak, H. F., **18**:169(30), 170(30), 172(30), 209(30), 210(30), 236(30)

Poppendiek, H. F., **3**:122, 173

Poreh, M. (Chapter Author), **12**:77, 77(4), 78(4), 80(4), 82, 86(13), 87, 101, 102, 111, 112; **13**:81(40), 116; **15**:63(26), 64(26), 105(138), 107(138), 111(138), 114(138), 135(26), 140(138); **24**:102(2), 184(2)

• Porro, A. R., **23**:326(214), 364(214)

Porta, E. W., **SUP**:75(201), 106(201), 305(201), 306(201), 309(201), 310(201)

Portat, M., **23**:280(9), 294(72, 73), 354(9), 357(72, 73)

Porter, J. E., **13**:207(6), 264; **14**:122, 124, 125(75, 76), 126, 130, 144, 145, 147; **SUP**:2(10), 3(10); **15**:144(4), 223(4); **25**:253, 318

Porter, J. W., **15**:267(102), 268(102), 269(102), 280

Porter, R. S., **13**:219(57), 259, 265

Porter, W. F., **21**:287, 293, 345

Portyanko, A. A., **30**:242(34), 251(34)

Poshkas, P., **31**:160(60), 324(60)

Poškas, P. S., **18**:135(70), 136(70), 142(70), 159(70)

Posnov, B. A., **1**:166, 184

Pospieszcyk, A., **28**:75(96), 112(79, 96), 142(79), 143(96)

Post, A. H., **5**:381, 509

Postma, A. K., **29**:71(28), 125(28)

Potemski, R., **28**:375(141), 421(141)

Pötke, W., **13**:10(31), 17, 41, 60

Potter, C., **12**:226(85), 252(85), 255(85), 281

Potter, C. J., **9**:209(48), 210(34), 221(48), 233(34), 244(48), 264(130), 265(34, 48, 130), 269, 271; **21**:82(33), 83(38), 88(54), 91(33), 95(33), 135(33, 38, 54)

Potter, J. L., **7**:197, 198, 215

Potter, R. E., **14**:97(86), 104

Potucek, F., **23**:230(218a), 232(218a), 276(218a)

Poulikakos, D. (Chapter Author), **18**:50(97, 99), 51(97, 99), 84(97, 99); **20**:327(32), 329(42, 45), 343(63), 344(63), 345(63), 346(63, 65), 347(65), 351(32, 42, 45), 352(63, 65); **24**:299(50), 300(50), 302(50), 307(50), 320(50); **28**:1, 24(43), 25(43), 26(43), 28(56), 29(43, 56), 31(56), 33(43), 35(43, 56), 36(43, 56, 57, 58), 37(43, 57), 38(43), 39(57), 40(57), 41(57), 42(57), 43(57, 58), 44(57), 45(57), 46(57), 47(57, 77), 48(57, 77, 78), 49(77), 50(77), 51(77), 52(78), 53(77, 78), 54(57, 58, 80, 81), 55(80), 56(56, 81), 57(81), 59(56, 81), 60(81), 61(81), 62(81), 63(81), 72(43, 56, 57, 58), 73(77, 78, 80, 81)

Poulter, R., **14**:124, 125(75), 147

Pound, G. M., **9**:202, 268; **10**:92(12), 164; **14**:282, 287(12), 304(31), 306, 307(31, 48, 49, 64, 66, 67), 308(64, 67), 309(31, 48, 49, 64, 67), 310(31), 314(31), 316(31), 317(31), 318(85), 319(31), 320(31), 321(85), 322(85), 323(85), 324(85), 341, 342, 343, 344

Pouring, A. A., **14**:331(110), 333(116, 117), 334(110), 345
Povarnitsyn, M. S., **SUP**:179(359)
Povsten, S. G., **18**:257(43), 259(43, 55), 260(43), 265(43, 55), 266(43, 55), 321(43), 322(55)
Powe, R. E., **11**:285(19, 20), 286(20), 287(20), 288(19, 20), 315
Powell, C. F., **14**:319, 321, 322, 323(90), 324, 344
Powell, R. L., **13**:219(92), 225, 256, 266; **15**:84(71), 137(37)
Powell, R. W., **24**:27(56), 38(56)
Powell, W. B., **2**:75, 76, 77, 95, 108
Powers, D. A., **29**:216(6), 268(134), 324(176), 325(176), 326(176), 327(176), 334(6), 342(134), 344(176)
Powers, J. H., **20**:244(181), 308(181)
Powers, S. E., **23**:370(17), 461(17); **30**:111(54, 57), 191(54, 57)
Pozvonkov, M. M., **17**:325(27), 340
Prabhakara, C., **8**:281, 283
Prabhu, A., **25**:383, 408
Pracht, W. E., **14**:97, 104
Prager, W., **13**:211(24), 264; **21**:247(27), 276(27)
Prakash, C., **18**:19(44), 48(44), 73(148), 81(44), 86(148); **20**:186(71, 72, 74), 187(71, 72, 74), 199(71, 72), 200(71), 201(74), 202(74), 211(71), 217(131), 219(131), 302(71, 72, 74), 305(131); **28**:252(46, 47, 48), 253(46, 47), 256(48), 258(48), 259(48), 260(48), 262(47), 264(47, 48), 266(46), 272(47), 326(46), 331(46, 47, 48)
Prakash, M. V. L., **16**:136, 156
Prakash, O., **25**:293, 297, 298, 300, 302, 303, 304, 305, 306, 307, 308, 309, 310, 311, 318
Prakash, S., **6**:435(171), 494; **26**:11(164), 20(164), 36(122), 66(122, 164, 165), 67(164, 165), 68(165), 98(122), 100(164, 165)
Pramuk, F. S., **5**:109, 128
Prandtl, L., **5**:219(140), 250; **6**:390, 487, 504, 561; **13**:80(33), 83, 91, 92, 116; **19**:18(31), 84(31)
Prantil, V. C., **28**:252(40), 330(40)
Prantz, J. F., **10**:172, 174, 177, 215
Prapas, D. E., **18**:25(56c), 82(56c)
Prasad, A., **24**:3(12, 13), 5(12, 13), 36(12, 13)
Prasad, B., **23**:214(236), 236(236), 241(236), 277(236)

Prasad, B. S. V., **26**:255(72, 73, 74, 75), 262(75), 304(72, 73), 305(72, 73), 327(72, 73), 328(74, 75)
Prasad, C., **10**:266(83), 283
Prasad, K. K., **11**:408(180), 409(181), 439
Prasad, V. (Chapter Author), **20**:217(128), 220(128), 305(128); **30**:313, 313(8), 323(26), 325(26), 327(27, 28), 328(29, 30), 329(36), 334(60, 61), 335(27, 28, 67, 68), 336(27, 28, 68), 337(27), 338(29), 339(29), 340(29), 341(29), 344(29), 345(26), 347(8), 350(26), 351(27, 28, 36, 68, 99), 352(27, 36, 99), 354(27, 99), 356(27, 99), 357(107, 108), 361(107, 108), 371(141), 372(141), 373(141), 374(141), 375(141), 376(141), 377(141), 378(30, 141), 379(141), 380(141), 381(141), 384(141), 385(152), 386(152), 387(152), 388(27), 389(27), 390(27), 391(60), 392(141), 396(152), 400(152), 401(152), 402(152), 406(28, 30, 141), 407(30, 107, 176), 408(30, 141, 176, 177), 409(30), 410(177), 411(177), 412(26, 28, 68, 176, 177), 413(26), 415(68, 182), 416(182), 417(26, 28, 68), 418(28), 420(28), 422(60, 107, 108, 184), 423(60), 426(8), 427(26, 27, 28, 29, 30), 428(36), 429(60, 61, 67, 68), 431(99, 107, 108), 433(141, 152), 434(176), 435(177, 182, 184)
Praslov, R. S., **20**:209(115), 304(115)
Prat, M., **23**:372(35, 36), 376(35), 431(35, 36, 84, 85), 436(35), 437(35), 441(35), 462(35, 36), 464(84, 85)
Prata, A. T., **20**:205(105), 212(105), 304(105)
Pratap, V. S., **13**:106(54), 117
Prater, P. G., **9**:405(92), 417
Prats, M., **14**:21, 101
Pratt, A. W., **6**:418(103), 491; **17**:180, 314
Prausnitz, J. M., **16**:61, 63, 153, 176, 184(8), 238; **23**:70(87), 130(87); **28**:7(5), 70(5)
Preckshot, G. W., **9**:261(116), 271; **10**:146, 165
Preibe, L. A., **22**:273(166, 167, 221), 279(167), 280(221), 315(166), 317(166), 318(166), 354(166, 167, 168), 356(221)
Preisendorfer, R. W., **3**:183, 249
Preobrazensky, S. S., **25**:4(30), 141(30)
Prepechenko, I., **18**:161(9), 236(9)
Prescott, P. J. (Chapter Author), **28**:231, 252(45), 253(45), 254(45), 255(45, 60, 61), 276(45),

290(134), 292(134), 298(134), 299(134), 300(134), 301(60, 136), 302(60, 61, 134, 136), 303(135, 136), 304(136), 306(136), 308(136), 311(45), 317(179), 319(45, 61, 200, 201), 320(61, 134, 135, 200), 321(201), 322(201), 323(200), 324(201), 325(134), 331(45, 60, 61), 335(134, 135, 136), 337(179), 338(200, 201); **30**:334(59), 429(59)

Prescott, R., **6**:202, 364

Presler, A. F., **6**:537, 563; **7**:52(31), 85

Press, W., **15**:48(79), 57(79)

Pressburg, B. S., **1**:367(18), 377, 383, 440

Presser, C., **28**:19(32), 71(32)

Pressley, R. J., **30**:319(19), 427(19)

Preston, J. S., **28**:102(94, 112), 143(94, 112)

Pretsch, J., **4**:331, 444

Preusser, P., **16**:66, 67, 68, 107, 108, 116, 117(17, 65), 119, 120, 123, 125, 127, 128, 129, 130, 153, 154

Price, E. W., **6**:211, 212, 365; **11**:318(7), 376(7), 434

Price, J. F., **5**:167, 168, 169, 248

Price, P. H., **3**:41, 42, 98; **SUP**:83(242), 85(242, 245); **25**:28(57), 142(57)

Price, T. T., **2**:76, 77, 108

Prichodko, I. M., **6**:418(120), 492

Pridantsev, A. I., **17**:327(28, 30), 328(28), 330(28, 30), 331(30), 332(28, 30), 335(28), 336(28, 30), 340

Pridantsev, A. S., **25**:92(172), 148(172)

Priebe, S. J., **14**:198(127, 128), 199(127, 128), 202, 245; **19**:148(174), 189(174)

Priedeman, D., **26**:191(150, 151), 195(150, 151), 215(150, 151)

Priemer, R., **17**:321(13), 339

Priestley, C. H. B., **11**:229(171), 260

Prigogine, I., **8**:26(133, 135), 27(138, 140), 32(137, 138), 87; **13**:130(40), 164(40), 202; **15**:11(32), 55(32), 229(14), 230(14), 276; **24**:40(3, 4), 46(3, 4), 98(3, 4); **31**:50(8), 102(8)

Prikhodtseva, T. V., **25**:92(172), 148(172)

Prim, P. C., **30**:331(42), 428(42)

Prinicero, X., **22**:203(20), 347(20)

Prins, J. A., **4**:9, 18, 62, 63; **9**:296, 346; **11**:202, 203(38), 205(38), 209(38), 225(38), 253, 254; **SUP**:170(344); **15**:152(10), 223(10)

Prisnyakov, V. F., **25**:4(24, 25), 140(24, 25)

Pritchard, A. M., **31**:439(55), 473(55)

Pritchett, J. W., **14**:78(165), 83(173), 84, 88(173), 90(180), 92(180), 93(180), 103, 104

Probert, S. D., **18**:25(56c), 68(128), 70(128), 82(56c), 85(128); **20**:268(274), 312(274)

Probstein, R. F., **1**:334, 335, 341(66), 354; **3**:61(62), 99, 211, 250; **4**:318, 443; **5**:28(39), 52; **6**:105(165), 130; **7**:164(3), 187, 187(3), 194(3), 195, 196(64), 197, 198(64), 199, 203(3), 204, 205, 206(64), 208, 211, 214, 215, 217; **14**:314, 343

Procter, W. S., **15**:195(79), 196(79), 225(79)

Proctor, M. P., **23**:319(155), 361(155)

Proctor, W. S., **SUP**:412(526)

Progelhof, R. C., **18**:228(125), 238(125)

Projahn, U., **19**:50(118), 51(118), 53(118), 57(118), 58(118, 129), 59(129), 88(118), 89(129); **24**:3(8), 36(8)

Prokhorov, A. M., **28**:103(88), 127(7), 138(7), 142(88)

Prokhorov, A. V., **14**:305, 343

Prokhorov, V. A., **31**:265(162), 329(162)

Prokofiev, V. E., **6**:418(129, 136), 492

Prokopets, S. I., **6**:418(105), 491

Pronko, V. G., **11**:168(168), 170(168, 169, 170), 196; **18**:264(57), 322(57); **31**:192(96), 326(96)

Proskuryakov, K. I., **6**:474(285), 500

Prosperetti, A., **26**:31(163), 50(163), 100(163)

Protopopov, V. S., **1**:414(105), 443; **6**:551, 552, 564; **7**:48(22), 49(22), 50, 84; **21**:10(22), 17(33), 20(33), 22(33), 34(53), 35(53), 37(60), 39(60, 65, 66, 70), 40(60), 41(66), 42(66), 51(22, 33), 52(53, 60, 65, 66, 70)

Protter, M. H., **15**:307(155), 327

Proudman, I., **5**:204, 249

Prudovsky, A. M., **25**:81(152), 147(152)

Prueger, G., **28**:402(208), 424(208)

Pruess, K., **30**:93(7), 120(77), 121(77), 140(97, 98), 141(97), 154(117), 181(157), 185(117, 162), 189(7), 192(77), 193(97, 98), 194(117), 196(157, 162)

Pruitt, B. A., Jr., **22**:222(189), 229(144), 243(169), 353(144), 354(169), 355(189)

Pruitt, G. T., **2**:385(87), 396; **15**:105(134), 107(134), 111(134), 114(134), 140(134)

Pruppacher, H. R., **21**:336, 337, 338, 339, 344; **26**:2(166), 8(124, 167), 12(124), 20(166), 44(124, 167), 53(27, 168), 94(27, 168), 98(124), 100(166, 167, 168)

Prusa, J. (Chapter Author), **19**:1, 12(22), 21(40),
    50(117), 51(120, 121), 52(120, 121),
    53(121, 124), 54(120, 123, 124), 55(120,
    124), 56(120, 124), 66(40, 123), 73(123),
    84(22), 85(40), 88(117, 120, 121, 123, 124)
Pruschek, R., **7**:320c(148), 320d(153, 154), 320f
Pruzan, D. A., **30**:173(148), 195(148)
Pryadko, N. A., **16**:127(90), 155
Pryputniewicz, R. J., **16**:15, 16, 57
Przekwas, A.-J., **24**:199, 275
Pshenichnikov, Yu. M., **25**:81(162), 147(162)
Ptacnik, J., **20**:280(298), 281(298), 282(298),
    313(298)
Pucci, P. F., **9**:102, 110
Puchkov, P. I., **31**:247(142), 328(142)
Pugh, E. R., **28**:133(83), 142(83)
Pugh, L. G. C. E., **4**:94(34), 138
Puiggali, J.-R., **31**:4(22), 98(22), 99(22), 103(22)
Puigjaner, L., **19**:102(42), 105(42), 107(42),
    108(42), 180(42), 185(42); **25**:182(76, 77),
    183(76), 184(76, 79), 185(79), 186(76, 79),
    187(79), 190(88), 191(88), 192(88),
    200(76, 79), 202(79), 204(88), 205(88),
    222(88), 224(79), 227(88), 228(88),
    246(76, 77, 79, 88)
Pujado, P. R., **19**:22(44), 85(44)
Pukhlyakov, V. P., **25**:81(163), 82(163), 147(163)
Pulling, D. J., **1**:424(121), 425, 426(121),
    428(121), 443; **7**:40(51), 42, 66(51), 67(51),
    86; **17**:37(120), 42(136, 138), 62;
    23:108(111), 109(111), 110(111), 132(111)
Pulsifer, A. H., **19**:108(54), 186(54)
Pun, W. M., **13**:12(43), 60, 113(59), 117,
    236(104), 237(104), 266
Puplett, E., **10**:34, 37, 53(31), 68(31), 81
Purcupile, J. C., **17**:30, 60
Purday, H. F. P., **SUP**:67(124), 170(124),
    197(124), 280(124), **19**:258(17), 350(17)
Purdy, K. R., **15**:195(80), 225(80)
Puri, O. P., **14**:298(29), 342
Puris, B. I., **25**:257, 283, 285, 289, 317
Purohit, A., **28**:135(20), 139(20)
Purohit, K. S., **18**:189(142), 239(142)
Purushothaman, A., **26**:2(140), 99(140)
Pusatcioglu, S. Y., **18**:169(41), 170(41), 236(41)
Pushkarev, L. I., **14**:205(137), 206(137),
    246(137)
Pushkina, O. I., **17**:51, 64
Putman, A. A., **10**:260, 283
Putman, J. L., **1**:397(77), 442

Putnam, G. L., **2**:386(57, 58), 395; **7**:88(2),
    91(2), 98(2), 103(18), 108(2, 18), 110(18),
    160; **12**:233(62), 251(62), 270(62), 280;
    15:128(151), 141(151)
Putterman, S. J., **17**:75, 154
Puzyrev, Y. M., **SUP**:124(302)
Puzyrewski, R., **9**:224(53), 269
Pyatt, K. D., Jr., **5**:320(91), 324
Pye, D. J., **23**:222(219), 223(219), 226(219),
    276(219)
Pyle, D. L., **17**:181, 185, 186, 314
Pytte, A., **4**:291, 315

Q

Qian, R., **22**:326(229), 327(229), 357(229)
Qiu, T. Q., **28**:77(80), 142(80)
Quan, G., **23**:193(175), 214(175), 216(175),
    274(175)
Quan, V., **9**:122, 127(21), 129(21), 178
Quarini, G. L., **25**:382, 414
Quarmby, A., **SUP**:72(164), 298(164), 306(469),
    310(469), 316(469), **19**:338(151, 152),
    340(151), 356(151, 152)
Quast, A., **7**:320c(169), 320g
Quataert, D., **7**:320c(170), 320g
Queiroz, M., **26**:200(160), 216(160)
Quelle, F. W., Jr., **11**:318(17), 434
Querry, M. R., **22**:426(39), 436(39)
Quick, K. S., **29**:218(19), 219(19), 233(19),
    270(19), 277(19), 279(19), 305(19),
    319(19), 322(19), 335(19)
Quinn, B. W., **23**:215(251), 278(251)
Quinn, C. P., **29**:101(72), 126(72)
Quintana, G. C., **26**:11(169), 100(169)
Quintard, M. (Chapter Author), **23**:369, 370(6),
    371(30, 31, 33), 373(6), 374(30, 43, 43a,
    43b, 44, 44a), 375(43, 43a, 43b, 44, 44a),
    376(43, 43a, 43b), 381(43, 43a, 43b, 44,
    44a), 382(43, 43a, 43b, 44, 44a), 383(43,
    43a, 43b, 44, 44a), 385(43, 43a, 43b, 44,
    44a), 398(65), 407(6), 409(43, 43a, 43b, 44,
    44a, 65), 411(31), 415(65), 428(43, 43a,
    43b), 430(31), 435(6, 88), 437(43, 43a, 43b,
    44, 44a), 441(43, 44), 461(6), 462(30, 31,
    33, 43, 43a, 43b, 44, 44a), 463(65),
    464(88); **30**:95(24), 110(53), 111(59),
    189(24), 191(53, 59); **31**:1(12), 4(22),
    18(25, 26), 19(28, 29, 30, 31, 32a), 20(28,

29, 30, 31, 32a), 22(28, 29, 30, 31, 32a), 23(28, 29, 30, 31, 32a, 32b), 34(27), 39(12, 24, 25, 26), 40(12), 43(27, 32b), 53(27, 32b), 54(27, 32b), 60(12), 68(28, 29, 30, 31, 32a), 69(32b), 72(23), 74(27), 76(27), 80(23), 82(27), 98(22), 99(22), 102(12), 103(22, 23, 24, 25, 26, 27, 28, 29, 30, 31, 32a, 32b)

Quintiere, J. G., 11:290(23), 315; 17:306, 317; 20:186(53), 187(53), 301(53)

R

Raaijmaker, M. J., 28:369(120), 373(120), 420(120)

Raasch, J., 28:11(14), 71(14)

Rabas, T. J., 30:200(44), 205(44), 222(45, 46, 47), 223(45, 46, 47), 224(45, 46, 47), 225(45, 46), 251(44, 45, 46), 252(47); 31:438(56), 439(56), 456(17), 468(57, 58, 59), 471(17), 473(56, 57, 58, 59)

Rabello, R. F., 14:119(70), 120(70), 146

Raben, I. A., 3:28, 32; 16:232; 20:14(22), 15(22, 22), 16(22), 17(22), 80(22)

Raber, J. L., 30:243(33), 251(33)

Rabinovich, G. D., 26:236(46), 239(46), 243(46), 247(56), 253(46), 269(46), 322(46), 323(46), 326(46), 327(56)

Rabinovich, L. B., 10:206, 217; 19:115(95), 144(155), 187(95), 188(155)

Rabl, A., 18:3(5), 4(5), 7(5), 9(5), 19(45), 20(47), 25(56a), 26(57), 79(5), 81(45, 47), 82(56a, 57)

Rabonovich, G. D., 6:474(275, 276, 283), 499, 500

Rachmile, I., 15:229(13), 237(13), 238(13), 276

Rackham, B., 13:119(1), 200

Rackley, R. L., 8:29, 87

Radcenco, V., 15:7(23), 27(23), 55(23)

Radchenko, M. I., 1:406(86), 442

Radd, F. J., 5:326(3), 505

Radebaugh, R., 17:102, 155

Radford, B. A., 1:376(27), 441

Radhadevi, P. V., 24:175(155), 177(155), 178(155), 190(155)

Radin, I., 15:63(27), 64(27), 135(27)

Radulovič, P. T., 25:262, 316; 30:302(105), 311(105)

Rae, W. J., 2:219, 231, 243, 269; 23:317(152),

318(152), 320(179, 184, 185), 361(152), 362(179, 184, 185)

Raelson, V. J., 1:315, 316(52), 354

Rafalski, P., 8:24, 33, 86, 88

Rafferty, D. A., 17:204(57), 219(57), 315

Ragent, Boris, 1:284(16), 353

Raghavan, V. R., 26:50(24, 25), 93(24), 94(25)

Raghaven, C., 18:58(107), 59(107), 84(107)

Raghaven, R., 14:13(33), 97(33), 100

Raghavendra, N. M., 15:250, 280

Ragheb, H. S., 18:257(47), 258(54), 262(47, 54), 267(47), 268(47), 269(47), 322(47, 54)

Raghuraman, J., 23:208(188, 189), 210(188), 275(188, 189)

Ragi, E. G., 20:76(67), 82(67)

Ragsdale, W. C., 2:52, 107

Rahm, L., 24:295(38), 296(37, 38), 320(37, 38)

Rahman, M. M., 26:201(166), 203(166, 168, 169, 170, 171, 172, 173, 174, 175), 216(166, 168, 169, 170, 171, 172, 173, 174), 217(175)

Rahman, S. H. (Chapter Author), 25:151

Rahn, H., 4:95(36), 138

Rai, G., 28:19(34), 71(34)

Rai, M. M., 24:218, 272

Rai, V. C., 15:269, 281

Rai-Choudhoury, P., 28:351(69), 418(69)

Raiff, R. J., 13:129, 202

Raina, G. K., 26:49(172), 50(170, 171), 100(170, 171, 172)

Raines, B., 2:312, 327, 355

Raisen, E., 31:458(60), 473(60)

Raiszadek, F., 26:45(63), 95(63)

Raithby, G. D. (Chapter Author), 4:304(66), 315; 11:266; 18:12(14), 13(19), 14(19), 15(20), 16(24, 25, 26), 17(25, 32), 65(119), 73(141), 74(141), 79(14), 80(19, 20, 24, 25, 26, 32), 85(119), 86(141); 20:186(50), 187(50), 301(50); 23:333(244), 365(244); 24:195, 201, 249, 272, 275; 26:79(86, 87), 80(86), 81(86), 96(86, 87)

Raizer, Yu. P., 8:246, 248, 282; 28:133(116), 144(116)

Rajagopalan, R., 15:327

Rajaram, H., 30:95(26), 189(26)

Rajpaul, V. K., 9:130, 175, 178

Raju, B. B., 15:327

Raju, G. J. V. J., 10:207(121), 217

Raju, K. S., 10:207, 217; 26:123(56), 126(56), 129(56), 133(56), 210(56)

Raju, M. S., **26**:75(44), 76(44), 85(173), 94(44), 100(173)

Rall, D. L., **23**:279(4), 285(43), 309(4), 354(4), 356(43)

Ramachandran, A., **11**:213(74), 214(74), 219, 223, 237, 241, 242, 255, 257, 261; **SUP**:201(390, 391), 228(391), 317(473)

Ramachandran, N., **19**:62(152), 67(152, 153), 90(152, 153); **28**:251(34, 35), 330(34, 35)

Ramachandran, P. A., **30**:339(79), 430(79)

Ramadhyani, S., **19**:19(37), 46(111), 50(111), 52(111), 58(128), 85(37), 88(111), 89(128); **20**:222(138, 139, 140), 223(139), 224(138, 139, 140), 225(140), 226(139), 256(138), 305(138), 306(139, 140); **26**:123(53, 54, 63, 64), 124(63, 64), 126(53, 54), 129(53, 54), 134(54), 135(54, 81, 83), 150(53, 54), 151(53, 54), 161(54), 163(81), 173(131), 177(131), 178(131), 188(146, 148, 154), 190(146), 191(148, 149), 192(154), 193(146), 210(53, 54), 211(63, 64), 212(81, 83), 214(131), 215(146, 148, 149, 154); **28**:251(33), 330(33)

Ramakrishna, B. B., **15**:320, 328

Ramakrishna, D., **15**:286, 301, 315, 327

Ramakrishna, J., **17**:300(106), 305, 306, 317

Ramakrishna, K., **20**:209(116), 304(116)

Ramamoorthy, M. V., **SUP**:70(149), 87(149)

Raman, J. R., **23**:208(39), 211(39), 218(39), 268(39)

Ramanjam, T. K., **26**:220(6), 324(6)

Raman Rao, M. V., **SUP**:249(433)

Ramaswami, D., **14**:152(37), 208(141), 243, 246

Rameprian, B. R., **25**:81(140, 141), 146(140, 141)

Ramesh, P. S., **23**:370(15), 461(15); **30**:101(39), 122(39, 86, 87), 123(86), 124(86), 125(86), 127(39, 87), 132(86), 190(39), 192(86, 87)

Ramey, H. J., **14**:13(33), 16(43), 92, 93, 94(181), 97(33), 99, 100, 101, 104, 105

Ramey, H. J., Jr., **23**:212(105), 271(105)

Ramgopal, A., **25**:258, 286, 318

Ramirez, W. F., **23**:370(12), 372(12), 411(12), 427(12), 438(12), 439(12), 442(12), 443(12), 461(12)

Ramos Berjano, F., **18**:73(140), 75(140), 86(140)

Rampf, H., **31**:247(139, 140), 328(139, 140)

Ramsey, J. C., **13**:211(23), 264; **18**:169(42), 170(42), 173(42), 214(42), 217(42), 220(42), 236(42)

Ramsey, J. W., **7**:341(27), 341(28), 348, 350, 370, 371, 371(28, 71, 72), 372, 373, 373(28), 374, 375(71), 378, 379; **19**:46(109, 110), 47(109, 113, 114), 48(114), 49(114), 55(113), 56(113), 57(113), 58(114, 127), 61(109), 65(109), 66(114), 88(109, 110, 113, 114), 89(127)

Ramzy, G. A., **29**:71(27), 121(27), 125(27)

Rana, R., **14**:48(111), 102

Randall, D. G., **6**:551, 564

Randall, D. S., **29**:299(157), 302(157), 343(157)

Randall, K. R., **18**:13(19d), 14(19d), 24(53), 80(19d), 81(53)

Randall, R., **18**:169(30), 170(30), 172(30), 209(30), 210(30), 236(30)

Randall, W. C., **4**:118(96), 120(96), 140

Randolph, A. L., **26**:78(174), 100(174)

Randolph, W. O., **5**:468, 516

Rangel, C., **23**:193(12), 206(12), 226(12), 267(12); **25**:292, 308, 315

Rangel, R. H., **26**:61(176), 70(177), 79(112, 113), 86(147), 88(175, 178), 98(112, 113), 99(147), 101(175, 176, 177, 178); **28**:36(60, 61, 62), 53(60, 62), 54(61), 69(60), 72(60, 61), 73(62)

Ranger, A. A., **29**:169(130), 209(130)

Ranken, W. A., **7**:240, 242, 247(48), 271, 315

Rannie, W. D., **3**:63(64b), 99

Ransom, V. H., **31**:106(21), 138(21), 156(21)

Rant, Z., **15**:7(25), 55(25)

Ranz, W. E., **7**:88, 145, 160; **10**:174(39), 215; **26**:53(179), 101(179); **28**:10(11), 71(11); **29**:285(154), 343(154)

Rao, A. R. K., **23**:198(145, 149), 206(145, 149), 226(145, 149), 229(145, 149), 273(145, 149)

Rao, B. K., **19**:264(28), 267(28), 311(103), 314(103), 319(103), 326(28, 122), 327(122), 328(122), 329(122), 330(103), 331(28), 332(28), 333(28, 103), 334(103), 337(103), 339(103), 340(103), 341(122), 343(103), 350(28), 353(103), 354(122)

Rao, C. V., **10**:207(121), 217; **26**:220(6), 324(6)

Rao, H. V., **26**:245(51), 253(51), 326(51)

Rao, K. S., **20**:281(299), 313(299)

Rao, M. A., **11**:203, 206, 254

Rao, M. H., **15**:250, 280

Rao, M. M., **24**:240, 274

Rao, M. S., **15**:327

Rao, P. K. M., **23**:58(100), 60(100), 90(100), 91(100), 106(100), 131(100)

Rao, P. S. C., **23**:370(14), 461(14)

Rao, S. P., **10**:200, 205, 217

Rao, V. D., **19**:37(75), 86(75)

Rao, V. V., **13**:10(29), 17, 18, 60; **26**:123(51), 126(51), 129(51), 210(51)

Rapp, G. M., **4**:104(47), 139

Rappaz, M., **28**:252(50), 262(69), 263(50, 77, 79, 80), 272(69), 327(50, 77), 331(50), 332(69, 77, 79, 80)

Raptis, A., **20**:336(55, 56), 352(55, 56)

Raptis, A. A., **24**:105(36), 186(36)

Rashevsky, N., **1**:55, 121

Rashidi, M., **25**:291, 318

Rask, R. B., **7**:348, 353(42), 359(42), 378

Rasmussen, R., **26**:53(168), 100(168)

Rasool, S. I., **8**:231(3), 282

Rassadkin, Y. P., **SUP**:81(236), 123(236), 124(236), 135(236, 314)

Rassolov, B. K., **31**:160(17), 323(17)

Rastegar, S., **22**:278(149), 353(149)

Rastogi, A., **10**:21, 23, 36

Rathkopf, J., **31**:408(72), 412(76), 428(72), 429(76)

Rating, W., **2**:298(119), 338(119), 355

Ratkowsky, D. A., **SUP**:66(112, 113), 346(112, 492), 347(492), 349(112), 350(113), 351(492)

Ratonyi, R., **7**:169, 183(12), 211

Ratten, G. M. J., **28**:369(120), 373(120), 420(120)

Ratulowski, J., **23**:193(76), 270(76)

Ratzel, A. C., **23**:320(166), 361(166)

Rau, G., **22**:209(89), 210(89), 211(89), 214(89), 215(89), 217(89), 218(89), 219(89), 220(10), 347(10), 350(89)

Rautenbach, R., **31**:462(61, 62, 63), 473(61, 62, 63)

Rauwendaal, C., **28**:149(53), 184(54), 230(53, 54)

Ravdel, A. A., **10**:177, 216

Ravese, T., **5**:453, 515

Ravetto, F. T., **19**:179(233), 190(233)

Ravin, V. S., **11**:412(192), 439

Ravine, T. L., **20**:203(93), 205(93), 303(93)

Ravi Prasad, A., **SUP**:249(432)

Raw, M. J., **19**:26(53), 85(53)

Rawdon, A. H., **5**:445, 514

Ray, J. A., **29**:101(75), 127(75)

Ray, J. R., **28**:80(56), 141(56)

Raychandhuri, B. C., **6**:474(245, 256), 483(245, 256), 498

Rayfield, G. W., **17**:90(61), 154

Rayleigh, J. W. S., **1**:237, 265; **5**:40, 53, 129, 246; **6**:197, 364; **18**:167(48), 179(63), 181(63), 183(63), 184(63), 237(48, 63); **21**:146, 148, 181

Rayleigh, L., **11**:13(36), 48; **24**:174(153), 190(153)

Rayleigh, R. S., **23**:439(89), 464(89)

Raymond, J. F., **19**:78(194), 92(194)

Rayner, S., **8**:49, 89

Raynor, S., **2**:445(44), 452

Razelos, P., **20**:148(10), 179(10)

Razumovskiy, V. G., **21**:47(81), 53(81)

Read, H. E., **2**:365(18), 394

Read, R. W., **8**:248(25), 283

Read, S. M., **18**:327(5), 362(5); **21**:103(81), 137(81); **30**:246(63), 252(63)

Rebenberg, J. M., **28**:306(148), 335(148)

Rebière, S., **1**:404(84), 411(84), 427(84), 442

Rebont, J., **7**:197, 215

Rebrov, A. K., **11**:203, 206, 209(42), 254

Rebuffet, R., **6**:29(49), 124

Rechard, O. W., **5**:3, 50

Recknagel, H., **17**:305(111), 317

Redberger, P. J., **SUP**:325(480)

Reddy, A. K. N., **12**:207(10), 226(10), 229(10), 237(10), 250(10), 252(10), 255(10), 256(10), 277; **17**:160(5), 163, 313, 314

Reddy, G. B., **20**:348(69), 352(69)

Reddy, K. C., **7**:192, 214

Reddy, P. J., **SUP**:271(448)

Redeker, E. R., **7**:260, 317, 368, 379

Redekopp, L. G., **6**:73, 75(130), 77(130), 80(130), 81(130), 82, 128

Redish, K. A., **10**:184(69), 185(75), 192(69), 216; **14**:151(18), 152(18, 33), 153(18), 154(33), 193(33), 217(18), 242, 243

Ree, F. H., **15**:79(70), 137(70)

Ree, T., **2**:361(40), 395; **15**:79(70), 137(70)

Reeber, M. D., **5**:406(161), 512; **20**:244(191), 262(259), 308(191), 311(259)

Reed, C. E., **15**:236(39), 277

Reed, J. C., **2**:363, 382(73), 396; **15**:67(41), 136(41); **19**:264(27), 350(27); **23**:198(181), 275(181)

Reed, R. E., **20**:285(308), 313(308)

Reed, S. G., **14**:306, 343

Reed, T. B., **11**:412(190), 439

Reed, T. M., **10**:203, 217
Reekie, J., **17**:95, 154
Rees, K. A., **SUP**:289(457)
Reese, W., **18**:161(5), 236(5), 238(105)
Reeves, B. L., **1**:298, 353; **4**:335, 445; **6**:14(25), 123
Reeves, E., **4**:90(27), 93(33), 94(27, 33), 95(33), 96(33), 138
Regalbuto, J. A., **9**:144, 179
Regan, J. D., **6**:29(39), 124
Reggio, M., **24**:218, 272
Regier, A., **5**:149, 153(44), 156, 167, 247; **21**:171, 182
Rego-Teixeira, A., **18**:25(55), 81(55)
Rehfuss, R. A., **15**:198(88), 199(88), 200(88), 201(88), 219(88), 225(88)
Rehm, T. R., **4**:195, 196, 225
Rehme, K., **SUP**:355(498, 499), 358(498, 499, 502), 359(499, 502), 360(499, 502)
Rehwinkel, H., **13**:219(84), 266, 267
Reichardt, H., **6**:504(7), 512, 514, 561; **13**:30, 35, 60; **25**:19(54), 24(54), 27(54), 75(54), 142(54)
Reichelt, W., **23**:216(220), 276(220)
Reichman, J., **10**:73(57, 58), 82
Reid, R. C., **5**:381, 509; **10**:109(39), 164; **16**:176, 184(8), 238; **23**:70(87), 130(87); **28**:7(5), 70(5); **29**:130(1), 138(1), 140(1), 158(1), 202(1)
Reid, R. E., **11**:412(185), 439
Reid, R. L., **21**:22(39), 23(39), 51(39)
Reid, W. H., **21**:188(19), 227(19), 237(19); **28**:370(136), 421(136)
Reid, W. P., **15**:287, 327
Reif, F., **17**:90(61), 154
Reiger, H., **24**:3(8), 36(8)
Reiher, H., **3**:8, 9, 31; **8**:94(6), 158, 165, 166, 176, 184, 226; **11**:218, 221, 231, 232, 236, 256; **18**:87(6), 157(6)
Reilly, I. G., **15**:162(22), 168(222), 223(22); **25**:286, 318
Reimann, J., **29**:253(106), 254(106), 340(106)
Reimann, M., **6**:317(88), 366; **16**:225(29), 227, 236, 239; **21**:327, 345
Reinder, M., **5**:177, 249
Reinecke, W. G., **29**:170(132), 209(132)
Reineke, H. H., **29**:5(3), 6(4, 5), 7(3), 15(3, 5), 17(3), 18(3), 19(3, 4), 20(5), 24(3), 25(3), 26(3), 31(23), 33(23), 55(3, 4, 5), 56(23)
Reiner, G., **11**:420(215), 423(215), 440

Reiner, M., **2**:358, 396; **15**:77(62), 137(62); **24**:108(84), 187(84)
Reiners, U., **23**:117(120), 132(120)
Reisen, L. I., **11**:158(167), 196
Reisman, J. I., **12**:70(63), 75
Reisman, O., **SUP**:136(318)
Reiss, E. E., **14**:140, 147
Reiss, F., **7**:320c(147), 320f
Reiss, H., **10**:92, 164; **14**:305(35), 306, 309(58, 59, 60, 61), 327, 327(36), 329(107), 330, 342, 343, 345
Reiss, L. P., **2**:388, 396; **7**:110(25), 160, 320e(168), 320g; **12**:233(86), 251(86), 270(86), 281; **15**:128(152), 141(152)
Reiss, M., **31**:461(46), 472(46)
Reistad, G. M., **15**:29(56), 56(56)
Reitano, R., **28**:76(55), 141(55)
Reiter, F. W., **3**:276(56), 301
Reitz, J. G., **4**:210(93), 214(93), 227
Reizes, J. A., **21**:161, 167, 181
Rektorys, K., **15**:41(63), 57(63)
Relf, E. F., **11**:217, 262
Rembrand, R., **20**:84, 85, 92, 97, 99, 131
Rempe, J. L., **29**:215(1), 248(1), 334(1)
Renfftlen, R. G., **31**:453(69), 454(64, 76), 473(64), 474(69, 76)
Renfro, D., **29**:69(22), 124(22)
Renkel, H., **5**:34(46), 34(48), 34(49), 36, 52
Renksizbulut, M., **26**:66(184), 72(85, 182, 184, 185), 73(85, 182), 74(85, 183), 78(180), 79(86, 87, 181), 80(86), 81(86), 96(85, 86, 87), 101(180, 181, 182, 183, 184, 185)
Rennie, D. W., **4**:120, 140
Reny, G. D., **9**:404(88), 417
Renz, U., **15**:232, 277; **21**:5(12), 22(12), 23(12), 51(12)
Renzoni, R., **1**:433(137), 443
Reocreux, M., **11**:44(76), 49
Reonigk, K. F., **28**:402(200), 405(200), 424(200)
Reppy, J. D., **17**:76, 77, 154
Resche, P. H., **11**:105(118), 114(118), 121(118), 194
Reshetov, V. A., **31**:160(46), 324(46)
Reshotko, E., **4**:330, 331, 444; **6**:111, 131
Resnick, W., **25**:194(95), 234(131), 247(95), 248(131)
Rettaliata, J. T., **14**:333, 334(123), 345
Rettig, R. B., **12**:208(61), 213(61), 237(61), 242(61), 251(61), 261(61), 265(61), 266(61), 268(61), 269(61), 271(61), 280

Reuter, H., **11**:318(29, 30), 376(29, 30), 378(29), 380(30), 435

Revankar, S. T., **29**:47(42), 57(42), 263(124), 270(124), 271(124, 140), 272(124), 273(124, 140), 341(124), 342(140)

Revelle, R., **17**:161(8), 313

Rex, J., **4**:165, 166, 224

Reynolds, **30**:78(49a)

Reynolds, A. B., **29**:273(145), 343(145)

Reynolds, J. M., **1**:409, 442

Reynolds, M. M., **5**:474, 516

Reynolds, O., **6**:504, 561; **24**:30(60), 38(60)

Reynolds, W. C., **1**:60, 63, 64, 67, 121; **2**:16(19), 25(19), 105; **3**:14, 23, 32, 81, 85, 86, 99, 106(7, 8), 126, 127, 172, 173; **5**:348, 506; **8**:291(13), 300(52), 301(52), 303(52), 304(52), 344(13, 52, 65), 349, 350; **SUP**:18(28), 31(41, 42), 32(41, 42), 33(41, 42), 35(44), 71(44), 77(44), 83(28, 241), 106(42), 145(44), 153(332), 154(332), 169(132), 172(42), 177(332), 185(332), 193(44), 194(44), 284(452), 285(41, 42), 287(44), 288(44), 289(41, 42), 292(41, 42), 297(44), 301(41), 302(41), 303(41), 306(41, 42), 307(41), 309(41), 310(41, 42), 311(41), 312(41), 313(41), 314(41, 42), 316(42), 319(44), 321(44); **15**:5(20), 55(20); **16**:245, 255, 256, 356, 364, 365; **17**:260(91), 316; **23**:284(37), 355(37); **25**:81(145), 146(145), 334, 340, 341, 349, 361, 363, 374, 380, 381, 382, 385, 386, 404, 408, 409, 410, 413, 414, 416; **26**:146(103), 213(103); **29**:62(10), 124(10); **31**:174(80), 325(80)

Reynolds, W. S., **25**:21(55), 142(55)

Rhea, L. G., **11**:105(87), 113, 118, 192; **18**:257(45), 263(45), 294(45), 322(45)

Rhee, K. T., **31**:392(49), 427(49)

Rhee, S. W., **28**:351(83), 418(83)

Rhie, C. M., **24**:193, 199, 201, 211, 249, 272

Rho, O., **14**:124, 147

Rhodes, E., **26**:39(111), 98(111)

Rhodes, P. C., **6**:474(208), 496; **8**:12, 85

Riazantsev, Iu. S., **26**:21(79), 96(79)

Ribando, R. J., **14**:19, 41(51, 94), 42(94), 43(98), 101, 102

Ribaud, G., **11**:226, 263; **30**:24(76), 91(76)

Ribbin, R. A., **21**:25(44), 52(44)

Rice, C. W., **11**:201(4), 202, 204(10), 209, 225, 252

Rice, M. H., **14**:90(180), 92(180), 93(180), 104

Rice, P., **16**:82(30), 117(30), 119, 125, 154, 237

Rice, P. A., **19**:58(132), 59(132), 61(132), 89(132); **24**:3(6), 36(6)

Rice, R., **26**:123(62), 125(62), 132(62), 211(62)

Rice, R. L., **14**:185, 245

Rice, W., **5**:177, 249

Rich, B. R., **9**:296(31), 346

Richard, C. C., **30**:222(30), 251(30)

Richard, J., **17**:71(33), 153

Richards, B. E., **23**:320(163), 361(163)

Richards, C. H., **4**:69(12), 75(12, 17), 81(18), 138

Richards, D. E., **20**:203(93), 205(93), 303(93)

Richards, J. I., **23**:376(47), 378(47), 436(47), 462(47)

Richards, L. A., **13**:121, 201

Richards, R. H., **5**:355, 375, 507

Richards, R. J., **5**:403, 511, 512

Richardson, B., **1**:394(63), 398(63), 442

Richardson, D. L., **4**:97, 98, 100, 130(112), 139, 141

Richardson, F. E., **21**:287, 293, 345

Richardson, G. T., **10**:270(102), 282

Richardson, J. A., **30**:243(61, 62), 252(61, 62)

Richardson, J. F., **10**:170, 172, 175, 177, 180, 200, 202, 215, 217; **19**:186(62); **20**:109, 113, 131; **23**:234(221), 242(139a, 221a), 244(221a), 261(221a), 262(139a, 221a), 273(139a), 276(221, 221a); **25**:166(55), 245(55), 261, 288, 318

Richardson, J. L., **18**:326(2), 328(2, 23), 329(2), 330(23), 350(2), 361(2), 362(23)

Richardson, P. D., **3**:36, 44, 45, 98; **4**:334, 445; **5**:144, 166, 178, 179(35), 180182, 247, 248; **6**:511(23), 562; **8**:24, 87; **11**:226, 259; **15**:12(36), 56(36); **30**:242(48), 252(48)

Richardson, S. M., **15**:101(110), 139(110); **19**:281(74), 284(74), 294(74), 352(74); **28**:152(52), 230(52)

Richart, D. S., **10**:211, 218

Richarz, F., **14**:319, 344

Richmond, J. C., **5**:26, 51; **11**:365(118), 437

Richter, A., **11**:217(252), 234(252), 264

Richter, H. J., **20**:90, 132

Richter, K., **25**:380, 417

Richter, R., **30**:239(35), 240(35), 251(35)

Richter, W., **22**:275(203), 356(203); **28**:369(121), 373(121), 420(121)

Ricou, F. B., **29**:266(130), 342(130)

Ricou, F. P., **13**:66, 114

Ricou, R., **28**:318(180, 181, 182, 183, 196), 337(180, 181, 182, 183), 338(196)

Ricque, R., **11**:44(76), 49

Riddel, F., **3**:34(4), 97

Riddell, F. R., **1**:284(17), 344(17), 346(17), 353; **2**:49, 107, 139, 196, 198, 199(24), 200(24), 201(24), 268; **4**:271, 314; **7**:196, 197, 214

Ridder, S. D., **28**:19(30, 31, 32), 71(30, 31, 32), 251(39), 330(39)

Riddiford, A. C., **12**:233(88), 248(87), 251(88), 270(88), 281; **21**:143, 182

Riddle, M. J., **23**:262(221b), 276(221b)

Rideal, E. K., **9**:184, 185, 198(4), 199(4), 202(4), 268; **13**:121, 201

Ridgway, K., **23**:217(222), 276(222)

Riede, P. M., **5**:345, 506

Riedel, E. P., **11**:318(18, 19), 435

Riedel, L., **18**:327(10), 362(10)

Rieger, H., **19**:50(118), 51(118), 53(118), 57(118), 58(118, 129), 59(129), 68(159), 69(159), 70(159), 88(118), 89(129), 90(159)

Riemann, J., *see* Reimann, J.

Riemersma, H., **5**:326(10), 505

Riemke, R. A., **31**:106(27), 138(27), 157(27)

Rienitz, J., **6**:135(7), 195(7), 363

Riester, J. B., **16**:10, 11, 14, 16, 19, 27, 29, 32, 57

Rietema, K., **10**:192(90), 217; **25**:181(74), 199(74), 246(74)

Rietschel, H., **8**:94(3), 158; **18**:87(3), 157(3)

Rifert, V. G., **31**:278(180), 330(180)

Rigbi, Z., **13**:219(78), 221(78), 265

Rigby, D. L., **23**:320(184), 362(184)

Rigby, G. R., **19**:108(52), 186(52)

Rigby, M. J., **23**:321(194), 363(194)

Rigdon, W. S., **20**:366, 371, 388

Rightly, M. J., **29**:190(198), 212(198)

Riha, P., **18**:210(99, 100), 238(99, 100)

Riherd, D. M., **17**:12(53), 58

Riley, D. S., **19**:31(65), 36(65), 37(65), 86(65)

Riley, N., **6**:474(202), 475, 496; **30**:333(50), 428(50)

Rimini, E., **28**:76(6), 138(6)

Rimlinger, M. J., **24**:218, 272

Rinaldo, P. M., **11**:105(58), 109(58), 110(58), 191; **20**:14(18), 15(18), 80(18)

Rinaloo, P. M., **5**:66, 67, 125

Rinderer, L., **5**:494, 517

Rineiskaya, G. V., **25**:81(155), 147(155)

Riney, T. D., **14**:90(180), 92(180), 93(180), 104

Rinfret, A. P., **5**:326(5), 505; **22**:195(170), 354(170)

Ringer, D. U., **26**:226(30), 227(30), 325(30)

Ringler, H., **4**:265(10), 268(10), 313

Ringwood, A. E., **19**:82(267, 269), 95(267, 269)

Rinkinen, W. J., **17**:306(113), 317

Ripkin, J. F., **2**:391(90), 396

Risaev, T. R., **31**:296(205), 332(205)

Rish, J. W., **23**:139(13), 184(13)

Ristorcelli, J. R., Jr., **28**:344(63), 367(119), 368(119), 418(63), 420(119); **30**:318(16), 329(16), 351(16), 362(16), 363(16), 364(16), 369(16), 427(16)

Ritchie, R. H., **28**:118(81, 82), 119(82), 120(82), 142(81, 82)

Ritook, Z., **22**:2(31), 18(31)

Ritter, G. L., **4**:170(53), 225; **5**:462, 463, 464, 515

Ritter, R. A., **2**:361, 396

Ritter, R. B., **30**:202(58), 252(58); **31**:443(70), 474(70)

Rittmann, J. C., **10**:231, 232(21), 233(21), 281

Ritz, H., **31**:245(135), 328(135)

Rivas, A., **6**:418(130), 492

Rivers, A. D., **28**:41(74), 73(74)

Rivers, W. J., **5**:480, 498, 499, 500, 517; **11**:105(139), 195

Rivir, R. B., **23**:319(154), 361(154)

Rivkin, S. L., **21**:2(5), 3(5), 38(5), 51(5)

Rivkind, V. Ya., **26**:8(186, 187), 101(186, 187)

Rivlin, R. S., **2**:360, 363, 396; **19**:268(30), 330(30), 350(30)

Rizk, M. K., **25**:394, 402, 417

Roache, P. J., **24**:226, 227, 272

Roald, B., **21**:171, 182

Robards, A. W., **22**:186(171), 354(171)

Robb, J., **9**:405(94), 417

Robb, W. M., **10**:115(48), 165

Robbers, J. A., **5**:92, 93, 94, 95, 127; **11**:134(147), 138(147), 139(147), 140(147), 141(147), 142(147), 195

Robbins, J. H., **5**:390, 510

Robbins, R. F., **5**:389(114), 510

Robert, F. W., **27**:27(22)

Roberts, A. F., **10**:230(19, 20), 281; **17**:175, 180(23), 184(23), 190, 191, 196, 198, 314, 315

Roberts, A. S., **19**:53(122), 88(122)

Roberts, D. N., **17**:34(112), 61

Roberts, G. J., **29**:234(88, 92), 321(92), 339(88), 340(92)

Roberts, J. J., **7**:275(108), 318

Roberts, J. K., **2**:310, 355

Roberts, L. G., **5**:159(51), 160, 161, 164, 165, 248

Roberts, O. P., **5**:343345, 506; **15**:48(74), 57(74)

Roberts, P. H., **14**:122, 126, 147; **17**:75(49), 82, 85, 86, 93, 94, 102, 154; **21**:149, 150, 182; **28**:252(41), 255(41), 288(41), 326(41), 331(41)

Roberts, T. R., **17**:323, 339

Robertson, A. F., **10**:230(19, 20), 235(25), 258, 280(117, 119), 281, 282, 284

Robertson, C. R., **15**:287(206), 329

Robertson, D., **11**:425(231), 440

Robertson, H. P., **21**:191(20), 237(20)

Robertson, J. M., **8**:305(57), 350; **26**:220(18), 227(18), 252(18), 325(18)

Robertson, M. W., **12**:50, 61, 75

Robie, R. A., **10**:44(17), 59, 60(16), 61, 63(17), 64(49), 69, 70, 81, 82

Robillard, L., **24**:303(44), 320(44)

Robinson, A. F., **9**:3(101), 93, 94, 110

Robinson, A. R., **5**:216, 250

Robinson, C. H., **29**:146(41), 204(41)

Robinson, J. L., **14**:66, 103

Robinson, J. M., **4**:228

Robinson, R. A., **12**:226(89), 229(89), 234(89), 235(89), 236(89), 237(89), 240(89), 256 (89), 261(89), 262(89), 263, 266(89), 281

Robinson, W., **11**:228, 259

Roblee, L. H. S., **9**:243, 270; **23**:217(223), 276(223)

Roblin, M. L., **30**:256(4), 307(4)

Robson, M. C., **22**:226(119), 247(172), 352(119), 354(172)

Rocca, J. J., **28**:343(44, 45, 46), 417(44, 45, 46)

Rocha, A., **18**:190(83, 84), 191(83, 84), 193(83), 194(83), 195(84, 86), 196(86), 197(86), 237(83, 84, 86)

Rocheleay, R. E., **28**:342(35), 416(35)

Rockett, J. A., **10**:251, 258(63), 282

Rodebush, W. H., **14**:307, 343

Rodgers, C. D., **5**:284(28), 322; **14**:252(6), 279

Rodi, W., **13**:75(23), 76, 78, 79, 115; **24**:199, 211, 247, 268, 273

Rodin, E. Y. (Chapter Author), **15**:283, 291, 302(210), 330

Rodney, W. S., **11**:374(125), 437

Rodriguez, F., **15**:63(31), 64(31), 94(92), 97(92), 136(31), 138(92)

Rodriguez, H. A., **1**:383(46), 441

Rodriguez, L., **15**:52(85), 58(85)

Rodriguez, M., **19**:74(171), 91(171)

Roemer, R. B., **22**:30(23, 49), 153(23, 28), 154(49), 290(173, 174), 354(173, 174)

Roenigk, K. F., **28**:402(201), 424(201)

Roesler, J., **24**:302(29), 313(29), 319(29)

Roetman, E. L., **14**:6(28), 100

Roffman, R., **12**:23(39), 74

Rogallo, R. S., **25**:333, 336, 341, 378, 380, 381, 414, 415, 417

Roge, J. A., **30**:171(140), 172(140), 195(140)

Rogers, A. C., **5**:390, 510

Rogers, A. N., **15**:236, 272(44), 278

Rogers, E. W. E., **6**:98, 130

Rogers, G. J., **15**:327

Rogers, J. H., **23**:222(117), 272(117)

Rogers, J. T., **17**:37, 39, 62

Rogers, J. V., **30**:227(21, 22), 228(21), 250(21, 22)

Rogers, M. H., **5**:149, 247

Rogers, M. M., **25**:382, 416

Rogovin, M., **29**:59(1), 72(1), 123(1)

Rohatgi, U. S., **31**:112(22, 38), 119(38), 120(38), 137(38), 156(22), 157(38)

Rohles, F. H., **4**:104(50), 106(50), 139

Rohonczy, G., **6**:532, 563

Rohr, J. J., **21**:185(4), 204(4), 206(4), 207(4), 236(4)

Rohsenow, W. M., **1**:195, 226, 228, 229(47), 231, 233, 234(60), 259, 263, 265; **3**:109, 173; **4**:159, 189, 207(91), 224, 225; **5**:62, 102, 125, 127, 342, 343(35), 344, 409, 411, 413, 414, 419(194), 420, 440, 441, 443, 445, 469, 506, 513, 514; **9**:211(37), 218, 220, 222, 249(75), 255(42), 259(108), 260(108), 264(131), 268, 269, 270, 271, 272; **10**:86, 134, 138, 140, 142(66, 67), 143, 145, 147, 149, 150, 151, 152(65), 157, 158, 161, 163, 165, 166; **11**:14(48, 49), 15(49), 16, 49, 144(161, 162, 164, 165, 166), 146(161, 162, 164, 165, 166), 147(161, 166), 148(161, 166), 160(164), 164, 167(166), 168(161, 166), 169(166), 170(161, 162, 164, 165, 166), 172, 173, 174(165), 177, 181, 184, 185(165), 196, 273(7), 274, 314; **SUP**:2(5), 139(5), 153(5),

Rohsenow, W. M. (*continued*)
190(5); **15**:17(41), 56(41); **16**:60, 95, 153, 154, 180, 181, 182, 238; **17**:5, 9, 57, 58, 140, 157; **20**:15(23), 80(23), 186(76), 187(76), 191(76), 194(76), 195(76), 202(76), 244(188), 302(76), 308(188); **21**:66(7, 8), 68(12), 73(20), 75(12, 20), 105(85), 133(7), 134(8, 12, 20), 137(85); **23**:9(21), 16(33), 22(42), 28(54), 58(108), 60(108), 102(108), 127(21, 33), 128(42), 129(54), 131(108); **26**:52(210), 102(210); **30**:277(53), 283(53), 309(53); **31**:137(4), 155(4)

Roidt, M., **SUP**:69(141), 161(141); **19**:268(43), 270(43), 351(43)

Roisen, L. I., **6**:526, 562

Roizen, L. I., **18**:257(44), 260(44), 267(44), 268(44), 312(44), 322(44); **31**:160(20), 245(134), 256(149), 323(20), 328(134), 329(149)

Roizman, D. Kh., **25**:81(159), 147(159)

Rojas, E., **23**:292(56, 57), 356(56, 57)

Rolin, M. N., **30**:296(99), 298(99), 311(99)

Roller, J. E., **6**:405, 406(60), 407(60), 408(60), 409(60), 410(60), 489

Roller, S. F., **29**:69(20), 124(20)

Rollin, A., **12**:85, 112

Rolling, R. E., **9**:362(24), 415

Rom, J., **6**:38, 41(88), 109(69), 125, 126

Roma, C., **12**:49, 52, 75

Romanenko, P. M., **3**:71, 74, 99

Romanenko, P. N., **13**:2(16), 59

Romanov, A. G., **9**:102, 103(9a), 106, 110

Romanov, V. A., **14**:97(191), 104

Romanov, V. I., **17**:327(35), 336, 337, 340

Romanova, N. A., **10**:202(155), 205(116), 206, 217, 218; **14**:176(163), 228(163), 237(163), 246; **19**:113(90), 187(90)

Romanova, N. N., **14**:268(53), 280

Romashin, A. G., **11**:360, 437

Romero, L. A., **30**:121(78, 79), 192(78, 79)

Romig, M. F. (Chapter Author), **1**:267; **3**:110(18), 173

Romig, M. R., **4**:28, 63, 230, 313

Ronchi, V., **6**:199, 364

Rooke, J. H., **18**:50(95), 51(95), 84(95); **30**:281(66), 310(66)

Roos, J. H., **5**:423, 512

Roos, J. J., **11**:105(65), 109(65), 111(65), 191; **16**:234

Roos, J. N., **6**:41(83), 68(83, 115), 69(115), 70(115), 73(83), 76(83), 78(83), 79(83), 82(83), 126, 127

Roper, C. H., **9**:137, 145, 156, 179

Rosa, R. J., **1**:299(41), 349, 353

Rosanov, V. I., *see* Rozanov, V. I.

Rose, H. E., **9**:135, 156, 179; **14**:4(12), 5, 100

Rose, J. W., **9**:210(35, 36), 231(63), 232(36), 240, 244(36), 264(63), 265(36, 132), 266, 269, 272; **21**:79(22, 23, 25, 26), 82(35, 36, 37), 83(39), 84(45), 88(51, 52, 53), 92(22, 23, 45, 63, 67, 68, 73, 74, 88), 93(67), 94(39, 53, 72), 95(35, 36, 37, 39, 67), 96(67, 73, 74, 77), 97(67, 77), 100(45), 105(87, 88), 109(95), 120(109), 122(109), 125(88), 134(22, 23, 25, 26), 135(35, 36, 37, 39, 45, 51, 52, 53), 136(63, 67, 68, 72, 73, 74, 77), 137(87, 88, 95), 138(109)

Rose, P. H., **3**:61, 99; **4**:281, 314, 334(44), 445

Rose, P. W., **29**:259(113), 260(112, 113, 115), 262(113), 263(113, 115), 264(112, 113), 266(113), 341(112, 113, 115)

Rose, R. K., **2**:74, 108

Rose, R. L., **3**:95(90), 100

Rose, R. P., **31**:121(19a), 139(19b), 156(19a, 19b)

Rose, W. J., **16**:144, 156

Rosen, D. I., **28**:133(83), 142(83)

Rosen, E. M., **SUP**:121(299); **15**:195(75), 225(75)

Rosen, R., **19**:198(44, 45), 244(44, 45)

Rosen, S. L., **15**:101(113), 104(113), 139(113)

Rosenberg, B., **20**:380, 388

Rosenberg, R. C., **22**:287(113), 288(113), 352(113)

Rosenberger, F., **28**:370(135), 421(135); **30**:365(131), 432(131)

Rosenbrock, H. H., **6**:474(279), 500

Rosenecker, C. N., **9**:119, 177

Rosenhead, L., **4**:331, 444; **6**:396(29), 487

Rosensweig, R. E., **25**:167(63, 64), 168(65, 66), 169(63, 64, 65), 170(63, 64, 65), 171(63, 66), 172(63), 173(66), 175(65, 66), 176(65, 66), 177(63, 65), 178(63, 64), 179(63), 180(63), 181(63), 187(80), 189(66), 193(94), 195(97), 196(66), 198(80), 203(97), 228(65), 232(65), 233(63, 65), 237(147), 245(63, 64, 65, 66), 246(80), 247(94, 97), 249(147)

Rosental, E. O., **10**:173, 177, 215

Rosenthal, D., **4**:296(63), 315

Rosenweig, M. L., **8**:70, 72, 91

Rosenzweig, M. L., **5**:147(38), 247

Rösgen, T., **30**:262(23), 308(23)

Roshko, A., **6**:6, 9(20), 29(20), 31(20), 33, 34, 48(56), 52(56), 69(118), 72(118), 77(56), 96, 109, 110, 112(172), 113(172), 122, 123, 124, 128, 130, 131; **8**:98, 99, 100, 159; **11**:216, 225, 258; **18**:92(25), 93(25), 94(25), 158(25); **20**:360, 387

Rosiczkowski, J., **23**:338(265), 366(265)

Rosler, R. S., **28**:402(190), 423(190)

Rosler, S., **30**:117(71), 118(71), 192(71)

Rosner, D. E., **2**:4, 46, 47, 48, 49, 50, 51, 86, 105, 111(4, 7), 164, 172, 183, 185(49), 247, 268, 269, 270; **8**:36(199), 89; **28**:28(54), 36(54), 56(54), 72(54)

Rosnovskiy, S. V., **21**:14(28), 32(50), 33(50), 36(28), 51(28), 52(50)

Ross, D. C., **22**:231(175), 239(176), 354(175, 176)

Ross, D. K., **10**:184(69), 192(69), 216

Ross, J., **3**:255(6), 256, 260(6, 8), 264, 265(20), 300; **14**:293(25), 341

Ross, J. W., **29**:216(11), 217(11), 221(25, 26), 233(26), 271(25), 303(26), 309(25), 315(172), 323(172), 335(11, 25), 336(26), 344(172)

Ross, O. K., **14**:152(33), 154(33), 193(33), 243

Ross, R. C., **16**:364

Ross, S. M., **27**:51(26)

Ross, T. K., **16**:70, 153

Rosse, Earl of, **10**:73(60), 82

Rosseland, S., **3**:190, 191(12), 209(12), 211(12), 249; **11**:332(53), 362, 435

Rossetti, S., **9**:118, 177; **30**:236(43), 251(43)

Rossie, A. N., **18**:69(129), 70(129), 85(129)

Rossie, J. P., **12**:3, 5, 6(11), 14(11), 15(11), 16, 18(28), 19, 59, 67, 68, 69(62), 73, 74, 75

Rossiter, J. E., **6**:73(124), 128

Rossler, W. V., **5**:131(9), 203, 246

Rosson, H. F., **9**:258(100), 259(107), 271

Rossow, V. J., **1**:269, 312(3), 322, 323, 324, 325(3, 58), 333(64), 352, 354

Rotem, A., **9**:296(35), 346

Roth, C., **11**:318(15), 434; **20**:325(23), 350(23)

Roth, J. A., **28**:341(16), 416(16)

Rothe, P. H., **15**:270(106), 271(106), 281

Rothenberg, J. E., **28**:106(51), 110(51), 116(52), 140(51), 141(52)

Rotherberg, J., **22**:229(64), 349(64)

Rothermel, R. C., **10**:267, 268, 269(99), 272(103), 283, 284

Rothfus, R. R., **SUP**:201(389), 202(389)

Rothgeb, T. M., **19**:53(122), 88(122)

Rothman, A. J., **3**:216(52), 301

Rothstein, W., **2**:58, 108

Rothwell, E., **10**:229(17), 281

Rott, N., **7**:197, 214

Rotta, J., **6**:398, 399, 488

Rotte, W., **7**:130, 161

Rottenkolber, H., **6**:202, 221, 364, 365

Roubeau, P., **5**:423, 511

Roudebush, W. H., **5**:381, 509

Rougeux, A., **23**:300(96), 358(96)

Rough, R. R., **11**:420(205), 439

Rougier, M., **10**:9, 36

Rouhani, S., **31**:112(9), 156(9)

Roulier, A., **22**:266(177), 355(177)

Rounthwaite, C., **1**:430, 443

Rouse, H., **9**:298, 347; **10**:250, 282; **16**:5(1), 9(1), 11(1), 31(1), 56; **31**:170(76), 325(76)

Rouse, P. E., Jr., **23**:224(224), 276(224)

Roux, J. A., **23**:139(13), 184(13)

Roux, R., **15**:322

Rowdon, A. H., **5**:102, 127

Rowe, P. N., **10**:176, 177, 178, 216; **14**:154(39), 243; **19**:125(127, 128, 129), 188(127, 128, 129); **25**:166(56, 57, 58), 199(58), 245(56, 57, 58)

Rowell, H. S., **28**:149(55), 230(55)

Rowell, L. B., **22**:360(2), 435(2)

Rowley, H. H., **2**:330, 355

Rowley, R. W., **2**:2(1), 105

Rowlinson, J. S., **7**:6, 6(2), 7(2), 10(1), 83

Roy, A. S. (Chapter Author), **12**:196, 200(91), 224(43), 227(91), 230(91a), 233(23, 90), 237(90), 238(90), 241(43), 246(43), 247(43), 251(90), 253(43), 256(23), 259(43), 269(91a), 278, 279, 281

Roy, B. V., **31**:294(201), 332(201), 456(37), 472(37)

Roy, D. N., **SUP**:71(161), 136(319), 145(329), 288(161), 298(161)

Roy, M. T., **1**:405(85), 409(85), 411(85), 442

Roy, P., **25**:384, 410

Roy, R. P., **25**:96(176), 148(176)

Roy, S., **9**:292(16), 346

Roy, S. K., **24**:3(15, 16), 5(15, 16), 36(15, 16)

Rozanov, V. I., **25**:114(182, 184), 148(182), 149(184)

Roze, N. V., **25**:5(35), 141(35)

Rozenbaum, R. B., **19**:101(33), 185(33); **25**:157(18), 243(18)

Rozenshtok, Y. L., **6**:418(104, 124), 474(199), 475, 491, 492, 496

Rubel, A., **26**:183(145), 215(145)

Rubesin, M. W., **2**:41, 106, 172, 173, 269; **3**:54, 98; **8**:22, 86; **16**:365

Rubin, H., **12**:88, 101, 102, 104, 105, 112; **14**:28, 101; **20**:324(21), 325(22, 23, 25, 26), 326(27, 28, 30), 350(21, 22, 23, 25, 26, 27, 28, 30)

Rubin, S. G., **SUP**:212(403)

Rubinshtein, L. J., **11**:411(182), 439

Rubinsky, B., **19**:78(193, 194, 195), 80(236), 92(193, 194, 195), 93(236); **22**:177(45), 187(178), 199(180), 200(181, 230), 202(182), 204(182), 205(182), 206(182), 207(182), 208(182), 209(182), 219(184), 220(179, 185, 186), 304(153, 164, 183, 186), 305(183), 306(183), 307(180, 183), 309(183), 311(183), 312(183), 348(45), 353(153), 354(164), 355(178, 179, 180, 181, 182, 183, 184, 185, 186, 187), 357(230); **25**:190(89), 191(89), 192(89), 200(89), 204(89), 205(89), 222(89), 226(89), 227(89), 228(89), 246(89); **28**:253(54), 331(54)

Rubinstein, J., **24**:114(105), 188(105)

Rubinstein, L. I., **19**:2(4), 4(4), 5(4), 9(4), 14(4), 15(4), 28(4), 83(4)

Rubtsov, N. A., **3**:36(36), 98; **6**:474(322), 502; **11**:408(178), 439

Ruby, W. A., **26**:130(76), 149(76), 151(76), 161(76), 212(76)

Ruccia, F. E., **5**:381, 475, 476, 509, 516

Ruch, M. A., **20**:255(242), 310(242); **23**:13(41), 15(41), 21(41), 26(41), 27(41), 30(41), 37(41), 39(41), 109(41), 110(41), 117(41), 128(41)

Ruckenstein, E., **9**:215(38), 264(38), 269; **10**:189, 202, 216, 218; **11**:15, 16, 49, 84(44), 87, 90(44), 191; **15**:104(125), 140(125), 185(54), 224(54), 244(71, 72), 279; **24**:148(141), 190(141); **25**:258, 283, 286, 318; **26**:23(188), 101(188)

Rudakov, R. N., **11**:297(28), 315

Rudd, M. J., **15**:91, 92(81), 138(81)

Rude, G. M., **29**:98(68), 126(68)

Ruder, J. M., **5**:390, 510

Rudnick, I., **17**:102, 155

Rudolf, C., **11**:2(2), 47

Rudraiah, N., **20**:327(33), 328(36), 329(41), 351(33, 36, 41); **24**:105(34), 106(40), 175(155), 177(155), 178(155), 185(34), 186(40), 190(155)

Rudy, T. M., **21**:120(108), 138(108)

Ruehle, R., **7**:320c, 320f

Ruelle, D., **25**:356, 417

Ruetsch, R. R., **18**:228(125), 238(125)

Rufer, C. E., **9**:261, 271

Ruffin, R. S., **22**:251(85), 350(85, 86)

Ruggeri, R. S., **5**:172(75, 76), 248

Ruhle, V. R., **7**:275(109), 318

Ruiz, X., **30**:396(162), 399(162), 434(162)

Rumble, E., **29**:147(48), 194(48), 205(48)

Rumford, F., **10**:203, 217

Rumore, F. C., **15**:47(73), 57(73)

Rumpel, W. F., **3**:284(67), 301

Rumpf, H., **28**:11(14), 71(14)

Rumynskii, A. N., **3**:230, 235, 236, 251

Runchal, A. K., **13**:12(43), 60, 81(38), 116, 236(104), 237(104), 266; **31**:171(77), 325(77)

Runcorn, S. K., **9**:3(102), 110

Rundle, P. C., **28**:365(111, 112), 366(112), 420(111, 112)

Runge, C. D., **26**:137(86), 212(86)

Runge, I., **18**:181(65), 185(65), 237(65)

Runge, J., **6**:136(10), 363

Runstadler, P. W., **25**:21(55), 142(55)

Runstadler, P. W., Jr., **8**:297(33), 305(58), 344(65), 349, 350

Rupe, J. H., **2**:75, 79, 86, 108

Rupinskas, V., **22**:10(12), 17(12)

Ruppel, H. M., **29**:184(176), 211(176)

Rusanov, K. V., **17**:140(155), 146(168), 157

Ruseckas, T., **8**:101, 159; **18**:96(28), 158(28)

Rushchinsky, V. M., **6**:474(311), 501

Rusinek, I., **23**:371(32), 462(32)

Rusinko, F., **2**:138(22), 268

Russel, L. M., **23**:331(240, 241, 242, 243), 332(243), 365(240, 241, 242, 243)

Russel, T. W. F., **28**:342(35), 416(35)

Russel, W. B., **23**:261(225), 276(225)

Russell, D. J., **9**:390(62), 416

Russell, F. B., **18**:16(26), 80(26)

Russell, H. W., **18**:179(60), 180(60), 237(60)

Russell, K. C., **9**:202(16, 18), 268; **14**:304(31), 307(31, 64, 67), 308(64, 67), 309(31, 64,

67), 310(31), 314(31), 316(31), 317(31), 319(31), 320(31), 332(115), 333(115, 118), 334(115, 118), 335(118), 342, 343, 345

Russell, R. D., **26**:304(82, 83), 328(82, 83)

Russo, A. J., **24**:102(12), 184(12)

Russo, R. E., **28**:80(113, 114), 81(114), 82(114), 83(114), 84(114), 85(114), 86(114), 87(113, 115), 88(114), 89(115), 90(115), 93(113), 94(113), 143(113, 114), 144(115)

Rust, W. M., **15**:327

Ruth, D. W., **18**:12(14), 79(14)

Rutherford, R., **3**:269(24), 300

Rutner, Ya. F., **8**:50, 90

Rutsky, J. N., **6**:474(218), 479, 497

Rutter, J. W., **15**:328

Ruttner, L. E., **20**:273(283), 291(316), 312(283), 314(316)

Ruzicka, J., **5**:464, 513; **11**:105(73, 74), 111, 121(73), 122(73), 192; **18**:257(41), 298(41), 299(41), 321(41)

Ryadno, A. A., **19**:195(27), 243(27); **25**:4(13), 31(13), 33(13), 64(13), 96(13), 127(13), 140(13)

Ryan, D., **23**:398(61), 407(61), 409(61), 421(61), 463(61); **31**:34(32c), 36(32c), 76(32c), 80(32c), 86(32c), 103(32c)

Ryan, J. M., **9**:401(82), 417; **21**:171, 172, 174, 182

Ryan, J. V., **10**:280(119), 284

Ryan, N. W., **2**:383, 396

Ryan, T. J., **22**:225(188), 355(188)

Ryan, W., **26**:2(12), 84(12), 87(12), 93(12)

Ryan, W. P., **3**:16, 32

Rybin, I. R., **18**:257(44), 260(44), 267(44), 268(44), 312(44), 322(44)

Rybski, W., **30**:279(58), 280(58), 309(58)

Ryckmans, Y., **30**:338(74), 350(74), 403(173), 405(74, 173), 430(74), 434(173)

Ryder, E. A., **23**:288(51), 306(51), 356(51)

Rykalin, N., **28**:127(84), 142(84)

Rynne, D. M., **28**:365(109), 420(109)

Ryskin, G. M., **26**:8(186, 187), 21(189), 22(189), 101(186, 187, 189)

Ryspaeu, N. S., **14**:203(133), 245

Ryspaev, N., **19**:149(178), 189(178)

Ryu, Z. M., **22**:421(29), 422(29), 423(29), 424(29), 425(29), 426(29), 427(29), 429(29), 430(29), 432(29), 436(29)

Ryvkin, V. B., **6**:376(5), 377(5), 465(192), 486, 495

Ryzhov, Y. A., **25**:4(12), 31(12), 35(70a), 140(12), 143(70a); **30**:83(41), 89(41); **31**:160(21), 182(21), 323(21)

S

Saad, M., **5**:348, 506

Saari, J. M., **10**:47, 57, 59, 60(48), 62, 72(54), 73(64, 65), 76, 78, 82, 83

Saari, S., **SUP**:213(403)

Saarlas, M., **7**:338, 377

Saaski, E. W., **14**:140, 147

Saavedra, J. J., **26**:203(176), 217(176)

Sabersky, R. H., **1**:206(14), 264; **6**:525(37), 526, 562; **7**:59, 60, 85; **15**:63(29), 64(29), 104(126), 121(126), 136(29), 140(126); **20**:24(49), 27(49), 81(49); **21**:168, 180; **31**:205(110), 209(110, 111), 211(110), 327(110, 111)

Sabhapathy, P., **19**:78(182), 91(182); **30**:351(100), 431(100)

Sabzevari, A., **8**:170, 172, 173, 210, 211, 212(35), 215, 216, 219, 221, 224, 227

Sachaev, A. I., **13**:261(123), 266

Sachs, E., **28**:402(208), 424(208)

Sackinger, P. A., **30**:350(92, 94), 366(94), 371(94), 379(94), 430(92), 431(94)

Saddy, M., **9**:258(94), 265(94), 270; **15**:232(21), 277

Sadegh, A. M., **19**:44(101), 78(101), 87(101)

Sadeghipour, M. S., **19**:74(170), 75(170), 90(170)

Sadek, S. E., **9**:265(143), 272

Sadhal, S. S., **26**:2(193), 3(193), 5(193), 6(192), 9(192), 11(148), 12(190, 194), 16(192), 17(192), 31(193), 32(52, 53), 34(52, 192), 35(53), 36(18, 52), 47(107), 48(107, 149, 191), 50(238, 239), 51(239), 59(52, 73, 192), 60(73), 63(73), 83(192), 93(18), 95(52, 53), 96(73), 97(107), 99(148, 149), 101(190, 191, 192, 193, 194), 104(238, 239)

Sadikov, I. N., **SUP**:6(15), 76(15, 208, 209), 178(208, 209), 217(15)

Sadovnikov, G. V., **6**:474(213), 478(213), 496

Sadowski, T. J., **23**:198(226), 199(226), 200(226), 204(226), 205(226), 206(226), 218(226), 220(226), 221(226), 223(226), 226(226), 276(226)

Saeki, T., **27**:196(42)

Sáez, A. E., **23**:371(32), 462(32); **30**:95(19), 189(19)

Safdari, Y. B., **11**:318(22), 385(22), 435

Saffman, P. G., **21**:197(24), 237(24)

Safran, S., **17**:131(135), 149(135), 156

Sagan, I. I., **16**:127(90), 155

Sagar, D. V., **26**:220(6), 324(6)

Sage, B. H., **3**:16, 32, 120(23), 173; **6**:515(27, 28), 562; **7**:52(28), 84; **10**:208, 217; **11**:219(100, 105), 220, 222(100, 105), 224, 236(100, 105), 237, 256, 257, 262; **19**:143(152), 188(152)

Sah, C. T., **15**:327

Saha, P., **31**:140(23), 142(23), 156(23)

Sahagian, Ch., **11**:412(191), 439

Sahara, K., **20**:183(6), 299(6)

Sahira, K., **30**:333(52), 349(101), 351(101), 378(101), 407(52), 408(52), 428(52), 431(101)

Said, E., **24**:36(18)

Saifullah, A. Z. A., **24**:3(5), 36(5)

Sailor, R. A., **2**:391(75), 396

Saito, A., **24**:3(5, 5, 14), 6(21, 22, 23, 26), 36(5, 5, 14, 21, 22), 37(23, 26)

Saito, H., **11**:220, 224, 263; **19**:81(249), 94(249)

Saito, M., **21**:86(49), 135(49); **26**:81(195), 101(195); **29**:131(12), 171(136), 203(12), 209(136)

Saito, T., **17**:42, 44(143), 63; **27**:ix(5)

Saitoh, T., **19**:19(36), 58(133), 59(133), 68(133), 70(133), 85(36), 89(133); **24**:3(7), 36(7)

Saiy, M., **13**:75, 86(24), 115

Sakaguchi, I., **7**:122(35), 161

Sakaguchi, T., **20**:110, 130

Sakai, K., **23**:349(305), 368(305)

Sakai, T., **26**:81(195), 101(195)

Sakakibara, M., **SUP**:12(20, 23), 13(20), 137(20), 188(23), 189(20, 23)

Sakane, J., **28**:318(195), 338(195)

Sakauchi, K., **28**:369(122), 420(122)

Sakhuja, R. K., **17**:50(158), 63; **20**:256(244), 311(244); **23**:13(65), 15(65), 37(65), 109(65), 110(65), 112(65), 129(65)

Sakisyan, L. A., **16**:116(81), 117(81), 125(81), 155

Sakly, M., **19**:82(260), 94(260)

Sakson, T. W., **11**:105(126), 115(126), 194

Sakurai, A., **16**:236; **17**:22, 24, 59; **18**:257(31, 32), 262(31, 32), 292(32), 295(32), 298(31), 300(31, 32), 304(31), 305(31), 321(31, 32); **23**:293(70, 71), 357(70, 71)

Sakurai, T., **24**:282(39), 284(39), 286(39), 317(39), 320(39)

Sakuta, K., **25**:51(87), 144(87)

Sala, R., **17**:28(82), 59

Salak, R., **24**:114(113), 188(113)

Salakhutdinov, D. Kh., **25**:61(97), 144(97)

Salami, E., **18**:337(50), 363(50)

Salariya, K. S., **17**:250, 316

Salcudean, M., **17**:37(125), 39(125), 62

Salcudean, M. E., **19**:80(240), 93(240); **30**:351(100), 431(100)

Salgers, E. L., **4**:114(67), 140

Salisbury, J. W., **10**:11, 32, 36, 37

Salisbury, R. E., **22**:222(189), 355(189)

Sallack, J. A., **10**:107(31), 164

Salmi, E. W., **7**:270(88), 317

Salmon, J., **30**:356(106), 357(106), 431(106)

Salnik, Z. A., **30**:390(154), 433(154)

Salomatov, V. V., **6**:418(145, 146, 150, 121, 133), 492, 493; **SUP**:124(302)

Salpeter, E. E., **5**:286(22), 303(22), 322

Saltanov, G. A., **31**:160(55), 324(55)

Saltiel, C., **30**:339(83), 430(83)

Saltsburg, H., **14**:305(35), 329, 330, 342

Saltzman, E. J., **6**:29(46), 124

Salzberg, C. D., **11**:374(126), 437

Salzberg, F., **5**:167, 170, 171, 172, 248

Samant, K. R., **20**:222(141), 225(141), 226(141), 306(141); **23**:328(223), 364(223)

Samarasekera, I. V., **28**:306(145), 335(145)

Sambasiva Rao, N. V., **SUP**:216(408)

Sameshima, T., **28**:96(85), 142(85)

Sammakia, B. G., **20**:232(155), 233(155), 242(155), 306(155), 316(12), 350(12); **28**:239(9), 329(9)

Sammanka, B., **30**:362(125), 413(125), 432(125)

Sammis, C. G., **14**:54(124), 103

Samoilovich, Yu. A., **24**:48(17), 99(17)

Samootsakorn, P., **17**:292(101), 295(101), 317

Samoška, P. S., **8**:114(40), 130(40), 145(76), 146(76), 147(79), 151(82), 152(82), 153(76, 82), 159, 160; **18**:109(40), 142(76), 151(76), 152(76), 153(76), 158(40), 159(76)

Samoto, N., **28**:343(50), 417(50)

Samoylovich, Yu. A., **19**:197(35), 198(35), 244(35)

Sample, A. K., **28**:277(107, 108), 280(107, 108),

303(108), 310(107, 108), 333(107), 334(108)

Sampson, D. H., **3**:183, 186(8), 188(8), 192, 249; **5**:254(8), 257, 259(10), 313(8), 316(10), 322; **8**:252, 254(31), 283

Sampson, P., **19**:74(174), 75(174), 91(174)

Sampson, R. L., **1**:34(31), 50

Sams, E. W., **5**:368, 508; **31**:219(115), 222(115), 261(115), 327(115)

Samson, T., **14**:160(78), 161(78), 166, 244

Samuel, A. E., **7**:346, 352(37), 378

Samuels, M. R., **SUP**:73(195), 77(195), 300(195), 314(195), 321(195)

Sanada, A., **17**:37(119), 61

Sanborn, H. S., **11**:218, 221, 255

Sanchez, T., **19**:317(107), 318(107), 354(107)

Sanchez-Martinez, R. A., **28**:351(77), 418(77)

Sandall, O. C., **SUP**:50(58), 51(58, 59), 126(58), 181(59), 420(58)

Sandall, O. G., **15**:122(148), 141(148)

Sandberg, R. O., **5**:102, 127; **7**:50

Sandbery, R. O., **11**:170(174), 173(174), 175(174), 176(174), 177(174), 185(174), 186(174), 196, 197

Sandborn, V. A., **7**:194(83), 214; **11**:226(144), 241, 242, 259, 261

Sander, A., **14**:318(86), 321, 322, 323(86), 324, 344

Sanders, B. R., **24**:213, 227, 270; **26**:45(63), 72(61, 62), 75(161), 95(61, 62, 63), 100(161)

Sanders, R. L., **29**:234(94), 340(94)

Sanderson, S., **31**:123(24), 134(24), 156(24)

Sandhu, K., **30**:209(19), 212(19), 213(19), 233(19), 250(19)

Sandiford, B. B., **23**:226(227), 276(227)

Sandman, W., **17**:72(35), 153

Sandrelli, G., **29**:234(93), 340(93)

Saner, J. A., **18**:220(118), 238(118)

Sanfeld, A., **12**:261(91b, c), 269(91b), 281

Sanford, C. E., **17**:143(161), 144(161), 150(161), 157

Sangani, A. S., **18**:191(135), 238(135); **23**:207(228), 277(228); **25**:293, 294, 318

Sangani, H. N., **28**:184(30), 193(31), 229(30, 31)

Sangen, E. (Chapter Author), **17**:159, 275(96), 301, 302(109), 317

Sani, R. L., **1**:427, 443; **17**:131, 156; **28**:390(176), 391(176), 408(176), 411(176), 423(176)

Saniei, N., **23**:282(32), 324(208), 330(208), 355(32), 363(208)

Sankara Rao, P. S., **SUP**:256(437), 258(437, 438)

Sano, Y., **23**:13(66), 15(66), 43(66), 109(66), 110(66), 117(66), 129(66)

Santangelo, J. C., **5**:102, 103(3), 104, 106, 114, 125; **11**:55(3), 56, 57(3), 105(3), 110, 189

Santangelo, J. G., **18**:258(48), 260(48), 322(48)

Santini, R., **26**:49(218), 50(218), 103(218); **30**:137(96), 193(96)

Santos, V. A., **SUP**:92(250), 96(250)

Sapareto, S. A., **22**:361(8), 435(8)

Sarafa, Z. N., **5**:178, 249

Saraie, J., **28**:342(27), 416(27)

Sarangi, S., **26**:220(8), 324(8)

Sarazin, J. R., **28**:280(114), 282(118), 303(114), 334(114, 118)

Sardesai, R. G., **16**:147(128), 148, 156

Sareen, S. S., **15**:287, 327

Sarhan, A., **21**:179

Saris, F. W., **28**:96(14), 139(14)

Sarkar, S., **10**:200, 218

Sarkicz, V. B., **19**:110(72), 114(94), 186(72), 187(94)

Sarkits, V. B., **10**:202(158), 203(102, 103, 107), 217, 218; **14**:158(62), 243

Sarma, G. N., **6**:396(36, 37), 474(201), 487, 496

Sarma, K. V. N., **SUP**:70(149), 87(149)

Sarma, P. K., **19**:37(75), 86(75); **20**:209(116, 117), 304(116, 117)

Sarnecki, A. J., **16**:363

Sarofim, A. F., **5**:473, 516; **9**:362(22), 363(22), 364(22), 365(22), 415; **11**:323(42), 325(42), 420(209), 435, 439; **17**:204(56), 315; **23**:139(12), 154(39), 168(39), 184(12), 185(39); **27**:36(25); **30**:338(76), 430(76)

Sarpkaya, T., **6**:411(74), 489

Sarti, G., **18**:163(17), 236(17)

Sartor, W. E., **30**:219(49), 252(49)

Sasagawa, Y., **11**:420(212), 440

Sasaguchi, K., **17**:51(163), 64

Sasaki, H., **30**:344(86), 430(86)

Sasaki, T., **20**:187(66, 67), 275(289), 276(289), 302(66, 67), 313(289)

Sasaki, Y., **15**:286, 326

Sasamori, T., **14**:254(17, 18), 255, 269(17, 18), 273(17, 18), 274, 279

Sasik, R., **28**:76(86), 142(86)
Sasscer, D. S., **30**:222(41, 46, 47), 223(46, 47), 224(41, 46, 47), 225(46), 226(41), 251(41, 46), 252(47); **31**:438(56), 439(56), 473(56)
Sastri, K. S., **20**:244(202), 250(202), 309(202)
Sastri, V. M. K., **14**:155, 156, 157, 230(166), 243, 246; **SUP**:216(407), 218(407), 219(407), 402(407); **19**:147(165), 149(165), 151(165), 153(165), 154(165), 174(225), 189(165), 190(225), 281(81, 82), 285(81, 82), 352(81), 353(82)
Sastrohartono, T., **28**:151(38, 58, 59), 154(57), 162(38), 163(38), 165(38), 166(59), 168(59), 169(59), 170(59), 171(59), 180(56), 183(56), 188(59), 189(59), 190(38, 57, 58), 193(38, 58), 198(57), 205(57), 206(57), 229(38), 230(56, 57, 58, 59)
Sastry, C. V. N., **20**:209(117), 304(117)
Sastry, U. A., **SUP**:43(52, 53), 63(80, 85, 86, 87, 88, 89), 279(85, 86), 328(485), 341(80, 89), 343(80), 349(88, 89), 352(87), 353(88)
Saterbak, R. T., **11**:220(126), 223(126), 237(126), 258
Sathiyamoorthy, D., **19**:178(231), 179(231), 190(231)
Sathyamurthy, P., **24**:247, 273
Satik, C., **30**:114(66), 116(66), 117(72), 118(72), 191(66), 192(72)
Satish, M. G., **23**:208(228a), 230(228a), 277(228a)
Sato, K., **3**:16(24), 32; **5**:445, 446, 447, 448, 449, 514; **11**:170(185), 173(185), 175(185), 177(185), 197; **29**:84(49), 93(49), 94(49), 95(49), 100(49), 125(49)
Sato, M., **28**:364(102), 373(137), 385(137), 419(102), 421(137)
Sato, N., **24**:312(30), 313(30, 31), 320(30, 31)
Sato, S., **10**:172, 215; **11**:203, 206, 209(39), 254; **17**:323, 324, 326, 328(24), 335(24), 340
Sato, T., **3**:14, 32; **10**:151, 157(71), 165; **23**:329(233), 365(233)
Satoh, K., **20**:183(6), 299(6)
Satterlee, H. M., **4**:153(24), 223; **5**:348, 390, 391(119), 392, 393, 394, 395, 506, 510
Sauer, E. T., **11**:105(57), 115, 191
Sauer, F. M., **7**:190, 193, 202(128), 214, 216
Sauerbrey, R., **28**:102(37), 140(37)
Saule, A. V., **4**:214(98a), 227
Saulnier, J. B., **23**:326(213), 364(213)

Saunders, A. H., **18**:167(51), 237(51)
Saunders, J. H., **28**:24(45), 72(45)
Saunders, O., **6**:538, 563
Saunders, O. A., **2**:52, 107; **4**:6, 15, 44, 59, 62, 63, 64, 152, 223; **5**:141, 144, 149, 150(31), 166(56), 178, 179(35), 180182, 247, 248; **8**:165, 172, 226; **9**:16, 44, 110, 296(32), 322(82), 346, 347; **11**:278(12), 279(12), 300, 305, 314, 315; **30**:280(60), 309(60)
Saunders, R. C., **2**:4(3), 86(3), 105
Saur, J., **22**:177(70), 178(70), 312(70), 350(70)
Savage, S. B., **16**:53, 57
Saver, F. M., **11**:225, 258
Savery, C. W., **SUP**:160(340), 168(340), 192(340); **19**:268(37), 270(37), 351(37)
Savic, P., **1**:218, 264
Savich, P., **11**:14(44), 48
Savill, A. M., **21**:185(2), 236(2)
Saville, D. A., **9**:304(65, 66), 347; **11**:203, 206, 254
Savin, V. K., **13**:11(41), 15, 17(44), 60; **25**:131(204), 136(204), 150(204); **31**:160(23), 323(23)
Savino, J. M., **8**:35, 88; **SUP**:205(395), 208(399, 400)
Savins, J. G., **2**:375(96), 385(95), 388, 396; **15**:63(30), 64(30), 136(30); **23**:192(229), 277(229); **24**:104(30), 185(30)
Savkar, S. D., **14**:125, 147; **SUP**:72(171), 75(206), 102(206), 178(206), **20**:187(54), 301(54)
Savr, L. E., **25**:81(147, 148), 147(147, 148)
Sawabe, A., **28**:343(47), 417(47)
Sawafa, S., **25**:51(87), 144(87)
Sawatzki, O., **11**:241, 243, 262
Sawde, J. A., **5**:423, 512
Sawdye, J. A., **11**:105(81), 112(81), 192
Sawin, H. H., **28**:351(78), 418(78)
Sawochka, S. G., **1**:428(130), 443; **5**:100, 101, 102, 127; **11**:170(178), 173(178), 175(178), 177(178), 197
Sawyer, D. H., **28**:351(76), 418(76)
Saxena, S. C. (Chapter Author), **3**:287(90), 302; **11**:230(176), 260; **14**:149; **19**:97, 98(3, 4, 8), 99(14, 18), 100(14, 18, 29), 102(14, 37, 38, 39, 40, 41, 43, 44), 103(14, 41), 106(41, 43, 44), 107(14, 41, 43, 44), 108(14, 18, 41, 44, 49, 58), 109(43, 44), 110(14, 18, 37, 41, 44, 64, 66), 111(3, 4, 14, 37, 38, 39, 40, 41), 112(3, 79, 80, 81, 83), 113(79, 80, 84,

85, 86, 87, 91), 114(14, 37), 116(37),
117(37, 80), 118(14), 119(79, 91), 120(79,
91), 121(107), 122(79, 81, 110, 111),
123(110, 111), 124(110), 126(111, 131),
127(111), 128(111), 129(81, 110),
130(110), 131(81, 110), 132(81), 133(81),
134(81), 135(81), 136(14, 29, 40), 137(29,
131, 144), 138(40), 139(29, 40), 140(29,
40, 131, 144, 147), 141(40), 142(40, 147,
149, 150), 143(147), 144(147), 145(3, 8,
29, 131, 144, 147), 146(159, 161), 147(159,
161), 149(159, 161), 150(159, 161, 166,
167), 151(159, 161, 166, 167), 152(159,
161), 153(159, 161), 154(3, 159, 161, 166),
155(190), 156(8, 185, 186, 187, 190),
157(29, 87, 188, 189, 190), 158(87, 131),
159(8, 29, 144, 185, 187, 188, 189),
160(14, 79, 187, 188, 189), 161(3, 4, 8, 14,
29, 144, 202), 162(14, 147), 163(221),
165(221), 166(111, 221), 167(221),
168(221), 169(221), 170(221), 171(221),
172(79, 221), 174(202, 221, 227),
175(221), 176(221), 177(221), 178(231,
232), 179(186, 231, 233), 180(14, 40, 41,
43, 44, 79, 80, 83, 84, 85, 110, 147),
181(161, 189, 221), 184(3, 4, 8), 185(14,
18, 29, 37, 38, 39, 40, 41, 43, 44, 49),
186(58, 64, 66, 79, 80, 81, 83, 84, 85, 86,
87), 187(91, 107, 110, 111), 188(131, 144,
147, 149, 150, 159, 161), 189(166, 167,
185, 186, 187, 188, 189, 190), 190(202,
221, 227, 231, 232, 233); 23:280(19),
306(19), 354(19); 25:151, 152(1, 2),
164(48), 187(81, 82), 188(81, 82), 191(90),
194(82), 195(81, 82), 201(81, 82), 202(82),
203(82), 204(1, 102), 205(1, 48, 81, 82, 90,
103, 104, 105), 206(81, 82), 214(2, 48, 90),
225(2, 112, 113), 226(115), 227(1), 228(81,
82), 229(112, 117), 230(1), 231(1),
232(82), 237(82), 238(1), 242(1, 2),
244(48), 246(81, 82, 90), 247(102, 103,
104, 105, 112, 113), 248(115, 117)
Saxsena, N., **28**:342(35), 416(35)
Sayama, H., **18**:12(13), 79(13)
Saydah, A. R., **6**:41(89), 42(89), 69(89), 75(89),
82(89), 94(89), 95(89), 126; **18**:41(84),
83(84)
Saykin, A. M., **21**:39(64), 40(64), 52(64)
Sazblewski, N., **3**:35(26), 98
Scadron, M. D., **11**:219, 222, 257

Scala, S. M., **2**:137, 146, 152, 196, 247, 251(25),
266, 268, 270; **4**:280, 314
Scanlan, J. A., **11**:285(19), 286, 287, 288(19,
21), 315
Scarfone, C., **28**:103(87), 142(87)
Scesa, S., **7**:343(32), 344, 351, 361, 379
Schaaf, S. A., **6**:38(73), 125; **7**:164(2), 165(2),
168, 187, 188, 190, 190(2), 202(128), 211,
216; **15**:285, 287, 327
Schach, W., **26**:109(14, 15), 208(14), 209(15)
Schack, A., **5**:294(50), 323; **12**:138, 193
Schad, O., **8**:135, 137, 160; **18**:133(67), 134(67),
135(67), 159(67)
Schaefer, H., **4**:331, 445
Schaefer, R., **30**:222(45, 46, 47), 223(45, 46, 47),
224(45, 46, 47), 225(45, 46), 251(45, 46),
252(47); **31**:438(56), 439(56), 468(57, 59),
473(56, 57, 59)
Schaefer, R. J., **23**:370(7), 427(7), 461(7)
Schaeffer, R. C., **14**:129, 132, 134, 135, 136,
145
Schaetzle, W. J., **11**:220, 223, 258
Schafer, D. M., **26**:135(81, 83), 163(81), 212(81,
83)
Schäfer, K., **2**:298, 338, 355; **3**:276(56), 301
Scháfer, N., **30**:320(23), 427(23)
Schafer, R. W., **22**:419(25), 435(25)
Schaffer, M. C., **22**:251(85), 350(85)
Schaffers, W. J., **1**:79, 121
Schamberg, R., **7**:170(17), 191, 192(75, 76),
211, 214
Schapink, F. W., **17**:127(127), 156
Schapker, R. L., **1**:59, 121
Schardin, H., **6**:135(2), 162, 166, 187(2), 189,
268, 278, 362
Scharge, R. W., **5**:347, 353, 506
Scharmann, A., **30**:396(161), 399(161),
434(161)
Scharrer, L., **14**:323, 324(98), 325, 344
Schatz, A., **29**:152(73), 178(158), 180(169),
181(158), 206(73), 210(158), 211(169)
Schatz, J. F., **11**:360, 437
Schatz, P. N., **5**:295(46), 323
Schatzmann, M., **16**:12, 57
Schauer, I. I., **3**:93(87), 94(87), 100
Schauer, J. J., **13**:11(36, 40), 60
Schechter, R. S., **2**:380, 396, 397; **8**:32, 87;
19:198(43), 244(43), 260(20), 262(20),
264(20), 265(20), 293(20), 350(20);
24:40(5), 46(5), 98(5)

Scheele, G. F., **SUP**:414(534); **15**:195(77), 196(82), 198(87), 199(87), 200(87), 202(87), 203(87), 204(87), 205(91), 220(87), 225(77, 82, 87, 91)

Scheer, B. T., **12**:242(92), 281

Scheibe, H.-J., **28**:103(88), 142(88)

Scheidegger, A. E., **7**:236, 242(34), 315; **13**:184, 203; **20**:315(3), 349(3); **24**:104(48), 112(48), 186(48); **30**:95(18), 189(18)

Scheiner, P., **23**:146(25), 185(25)

Scheiwe, M. W., **19**:79(197, 198), 92(197, 198); **22**:212(117), 293(224), 352(117), 357(224)

Schekman, A. I., **25**:252, 319

Schelkhin, K. I., **29**:89(55), 126(55)

Schelkhin, K. J., **29**:86(52), 126(52)

Schenk, J., **2**:374, 378, 396; **8**:70, 90; **9**:46(104), 47, 48, 110; **11**:203(38), 205(38), 209(38), 225(38), 253, 254; **13**:219(66), 265; **SUP**:120(294), 121(294), 170(344), 185(365), 186(365), 367, 368, 371), 249(435), 250(435), 251(435)

Schenkel, G., **13**:206(1, 2, 4), 210(4), 219(2), 248(109), 261(121), 264, 266

Schepers, H. J., **6**:71(121, 122), 76(121, 122), 79(122), 80(122), 128

Scherberg, M. G., **4**:46, 47, 64

Schets, G. A. C. M., **22**:279(209), 356(209), 426(38), 436(38)

Schetz, J. A., **4**:26, 27, 37, 63; **6**:474(235, 241), 480(235, 241), 482, 483(241), 498; **9**:311(68), 347; **13**:67(5), 114

Scheuing, R. A., **7**:184(57), 192, 213

Scheurer, G., **24**:247, 271

Schichtel, B. E., **28**:376(151), 379(151), 380(151), 422(151)

Schijf, J., **13**:248(108), 266

Schiller, L., **13**:229(101), 266; **SUP**:41(49), 69(49), 85(49), 145(49), 160(49); **19**:268(33), 270(33), 350(33)

Schimmel, W. P., **8**:257, 273, 283; **12**:141(36), 193; **15**:327

Schindler, M., **7**:297, 319, 320c(148), 320f

Schinkel, W. M. M., **18**:12(15), 13(15, 19c), 14 (19c), 18(38), 79(15, 18a), 80(19c), 81(38)

Schins, H., **7**:271(90), 274, 289, 317; **29**:140(31), 159(90), 165(31, 113), 173(145), 178(31, 158), 181(158), 204(31), 207(90), 208(113), 209(145), 210(158)

Schitsman, M. E., **21**:34(51), 52(51)

Schladow, S. G., **24**:303(40), 320(40)

Schleicher, K., **18**:332(32), 333(32), 362(32)

Schlichting, H., **1**:29(27), 35(27), 50, 53, 112, 114, 120; **2**:255(90), 270; **3**:5, 13(1), 14(1), 21(1), 23(1), 31, 34(2), 97, 218(64), 219(64), 223(64), 250; **4**:331, 445; **5**:78, 125, 131, 136(29), 150(15), 167, 168, 172(74), 247, 248; **7**:340, 378; **8**:97(23), 159; **9**:28, 110; **11**:13, 48; **13**:4, 11(19), 59, 79(27), 116, 219(42), 250(111), 265, 266; **SUP**:54(65), 69(65, 137), 160(65, 137), 164(65, 137), 192(137); **15**:11(35), 12(35), 56(35), 152(11), 174(11), 223(11); **16**:261, 309, 365; **18**:92(24), 158(24); **19**:268(41), 270(41), 351(41); **20**:367, 388; **21**:144, 182; **24**:115(121), 189(121), 306(41), 315(41), 320(41); **25**:283, 318; **28**:389(160), 422(160); **30**:143(104), 193(104), 273(47), 309(47); **31**:107(25), 138(25), 156(25)

Schlichting, J., **26**:190(147), 215(147)

Schlinger, W. G., **3**:120, 173; **6**:515(27, 28), 562; **7**:52(28), 84

Schlitt, K. R., **7**:320e(164), 320f

Schlitz, L. Z., **28**:316(162), 336(162)

Schloerb, O., **7**:320d(153), 320e

Schlosser, S., **28**:397(186), 423(186)

Schluderberg, D. C., **30**:236(50), 252(50)

Schlünder, E. U., **13**:2(1, 4, 5, 6, 8), 3, 6(1), 7, 8, 9(1), 10(1), 11(1, 5), 15, 16, 17(1), 18, 19(5), 25(1, 5), 27(6), 32(5, 8), 36(4), 37(4), 38(4), 39(4), 40(4), 41, 42(1), 46(8), 59, 60; **14**:151, 152, 242; **15**:232, 277; **23**:370(9), 427(9), 461(9); **31**:160(44), 324(44)

Schlüter, H., **13**:219(71), 265

Schmaderer, F., **28**:397(186), 423(186)

Schmadl, J., **16**:115(80), 117(80), 155

Schmall, R. A., **10**:209(134), 218; **14**:177(97), 204(97), 206(97), 207(97), 229(97), 244; **19**:120(102), 122(102), 129(102), 130(102), 131(102), 148(102), 187(102)

Schmid, J., **SUP**:355(497), 356(497); **15**:327

Schmidt, A., **10**:177(55), 216

Schmidt, A. F., **5**:469, 516

Schmidt, C., **17**:138(149), 149, 151, 152(149), 157

Schmidt, E., **1**:287, 353; **6**:189, 363; **7**:55, 85; **8**:130(63), 160, 182, 227; **9**:3(107), 11, 77, 87, 91, 93, 94, 110; **11**:218, 222, 236, 256; **12**:129, 133(20), 192; **14**:118(83, 84), 119, 124(83, 84), 147; **18**:127(62), 159(62); **20**:377, 388; **21**:55(3), 82(3), 133(3);

30:72(77), 91(77), 272(45), 273(45), 309(45)

Schmidt, E. H. W., **4**:5, 15, 62, 155, 223; **5**:207, 250

Schmidt, E. M., **6**:41(82), 65(82), 126

Schmidt, F. G., **22**:251(85), 350(85)

Schmidt, F. H., **22**:350(86)

Schmidt, F. W., **8**:195(45), 227; **11**:423(219), 440; **SUP**:57(66), 64(95), 73(188, 189), 88(252), 92(188, 189), 94(188, 189, 252), 95(252), 97(189), 112(66), 113(66), 114(252), 115(252), 116(66), 117(66), 140(327), 144(66, 327), 165(188, 189), 167(188, 189, 252), 168(189), 174(66, 252), 175(252), 202(95), 203(95, 252), 204(252), 205(95, 252), 206(252), 224(95), 230(95), 233(252), 234(95, 252), 235(252), 414(327); **15**:38(60), 57(60); **19**:279(65), 281(65), 282(65), 352(65); **20**:15(31), 80(31), 134(5), 150(5), 179(5); **24**:249, 272

Schmidt, G., **2**:297, 354

Schmidt, H., **25**:348, 376, 377, 401, 417

Schmidt, H. E., **11**:318(12), 434

Schmidt, J. F., **6**:404(58), 488

Schmidt, K. R., **6**:555(95), 564; **7**:31(14), 33, 84; **11**:170(175), 173, 176(175), 177(175), 185(175), 186(175), 197

Schmidt, P. F., **28**:341(13), 415(13)

Schmidt, P. S., **15**:52(86), 53(86), 58(86)

Schmidt, R., **10**:56(44), 82

Schmidt, R. C., **20**:227(145), 229(145), 230(145), 231(145), 306(145); **29**:25(20), 26(21), 56(20, 21), 324(175), 344(175)

Schmidt, R. J., **4**:152, 223; **8**:165, 172, 226; **10**:155, 156, 165

Schmidt, R. R., **19**:46(109), 47(109), 61(109), 65(109), 88(109)

Schmidt, W., **24**:234, 271

Schmidt, W. A., **30**:245(51), 252(51)

Schmidt, Z., **20**:84, 109, 116, 120, 131

Schmit, L. A., **1**:103, 122

Schmitt, L., **25**:380, 391, 406, 417

Schmitz, G., **4**:284(52), 315

Schmitz, R. A., **SUP**:139(323), 140(323), 142(323), 143(323), 145(323), 146(323)

Schnabel, P., **14**:119, 120(88), 124(88), 125, 147

Schnable, G. L., **28**:341(13), 415(13)

Schnautz, J. A., **11**:231, 232, 237, 259

Schneider, G. E., **19**:26(53), 62(150), 65(150),
85(53), 90(150); **24**:194, 247, 249, 272, 273

Schneider, K. J., **14**:71, 72(154), 73(154), 103

Schneider, L., **6**:23(37), 124

Schneider, M. C., **28**:262(71), 263(71), 332(71)

Schneider, P. J., **1**:58, 121; **3**:117, 173; **5**:108, 128, 337, 338, 506; **15**:322; **19**:192(9), 243(9); **22**:161(190), 355(190)

Schneider, R. E., **29**:233(64), 338(64)

Schneider, S. H., **12**:25(45), 74; **19**:82(275), 95(275)

Schneider, W. F., **9**:101, 110

Schneiderman, L. L., **17**:23(72), 59; **23**:82(98), 131(98)

Schneller, J., **26**:225(26), 227(26), 243(26), 253(26), 269(26), 325(26)

Schniewind, J., **15**:327

Schnurmann, R., **14**:125, 133, 147

Schober, T. E., **9**:3(41), 91(41), 95(41), 97(41), 108

Schödel, G., **6**:284(85), 286, 340(85, 92), 343(85), 344(85), 366; **11**:363(105), 437

Schoeck, P. A., **4**:293(57, 58, 59), 294(57, 58, 59), 295, 315

Schoene, T. W., **12**:5(14), 7(14), 35(14), 73

Schoenhals, R. J., **4**:47, 64; **6**:474(301), 501; **7**:45, 62, 63, 85; **26**:227(28), 252(28), 325(28)

Schoessow, G. J., **5**:90, 126

Scholander, P. F., **22**:60(50, 51, 52), 61(52), 70(50, 51, 52), 154(50, 51, 52)

Scholten, P. C., **12**:242(93, 94, 95), 281

Scholtz, M. T., **26**:110(25), 120(25), 153(25), 209(25)

Scholz, C., **10**:43(14), 62, 81

Scholz, F., **8**:145(77), 148, 149, 160; **18**:145(78), 146(78), 147(78), 159(78)

Scholz, H. W., **26**:220(21), 227(21), 325(21)

Scholz, M. T., **13**:6, 59

Scholz, R., **23**:117(120), 132(120)

Schonrng, B., **24**:247, 268

Schooley, J. F., **23**:288(46), 356(46)

Schornhorst, J. R., **5**:26, 51

Schorr, A. W., **9**:299(45), 300(45, 48), 301, 305(45), 347

Schorre, C. E., **5**:326(4), 505

Schowalter, W. R., **2**:390, 391, 394, 396; **SUP**:69(140), 161(140); **15**:144(8), 149(8), 208(96), 213(111), 223(8), 225(96, 111); **19**:268(42, 44), 269(42), 270(42, 44), 275(44), 277(42), 279(44), 351(42, 44);

Schowalter, W. R. (*continued*)
20:321(17), 350(17); 23:219(55), 226(230), 228(230), 269(55), 277(230); 24:107(74), 187(74); 25:253, 254, 255, 262, 318
Schrader, H., 13:2, 4, 5, 59
Schrage, R. W., 9:216(39), 217, 218, 269; 21:68(11), 93(11), 134(11)
Schramm, R., 29:31(23), 33(23), 56(23)
Schraub, F. A., 8:297(30, 33), 304(30), 344(65), 349; 31:174(80), 325(80)
Schraub, R. A., 16:248, 365
Schraul, F. A., 25:21(55), 142(55)
Schreiber, E., 10:43(14), 62, 81
Schretzmann, K., 7:274, 318, 320c(147), 320e(168), 320f, 320g
Schriven, L. E., 16:82, 83, 87, 90, 91, 92, 94, 119, 154
Schrock, S. L., 3:124(31), 173
Schrock, V. E., 1:402, 404(83), 426, 442; 5:115, 128; 11:105(119), 114, 194; 14:77, 103; 16:228, 239; 23:33(63), 129(63); 26:22(249), 53(249), 104(249); 29:47(42), 48(44), 50(46), 57(42, 44, 46); 30:151(110), 193(110)
Schrodt, J. E., 9:378(50), 381(50), 407(50), 415
Schroeder, H., 28:341(11), 415(11)
Schroeder, T., 14:54(125), 103
Schubauer, G. B., 11:232(169), 237(169), 260
Schubel, M. E., 26:220(20), 225(20), 245(20), 253(20), 325(20)
Schubert, G., 14:16(45), 17(45), 25(45), 26, 26(45), 101; 19:82(261, 264), 94(261, 264); 30:119(75), 123(88), 192(75, 88)
Schuderberg, D., 9:117(5), 129(5), 177
Schuh, H., 9:288, 298(6), 302(6), 346; 15:327
Schuk, H., 4:6, 62
Schüller, R. B., 16:205(28), 217(28), 239
Schultheiss, E., 30:396(161), 399(161), 434(161)
Schultz, D. L., 8:293(17), 349; 23:280(24), 309(24), 315(24), 317(24), 318(24), 319(24), 320(187, 188, 189), 321(24, 191, 195, 196), 322(24), 323(24), 325(24), 337(259), 346(191), 355(24), 362(187, 188, 189, 191), 363(195, 196), 366(259)
Schultz, R. R., 10:156, 165
Schultz-Grunow, F., 3:14, 32; 5:177, 182, 249, 251; 21:149, 155, 182
Schulz, A., 23:298(92, 93), 358(92, 93)
Schulz, G., 6:135(7), 195(7), 225, 226(70), 227(70), 228(70), 363, 365

Schumann, T. E. W., 20:171(19), 179(19)
Schumann, U., 21:185(7), 236(7); 25:337, 343, 347, 348, 367, 376, 377, 378, 379, 380, 397, 401, 405, 411, 412, 413, 416, 417
Schumpe, A., 23:253(214), 254(195, 214, 231), 255(214), 256(214), 258(195, 214, 231), 259(214), 275(195), 276(214), 277(231)
Schumway, R., 31:112(9), 156(9)
Schurig, O. R., 11:202, 204, 252
Schurig, W., 21:55(3), 82(3), 133(3)
Schuster, A., 3:203, 249
Schuster, J. R., 9:261(113), 271
Schütz, G., 7:101, 110, 160
Schwab, B., 11:218, 221, 236(92), 256
Schwabe, D., 30:396(161), 399(161), 434(161)
Schwalbe, W., 29:173(145), 181(171), 209(145), 211(171)
Schwartz, B. B., 15:43(64), 57(64)
Schwartz, F. L., 5:90, 126
Schwartz, J., 7:246, 274, 294, 316, 320a(58, 136), 320e
Schwartz, K. W., 5:237, 238, 251
Schwartz, L., 23:376(46), 378(46), 462(46)
Schwartz, S. H., 4:157, 158, 162, 164, 191, 194, 224, 228; 16:10(12), 11(12), 14(12), 16(12), 19(12), 27(12), 29(12), 32(12), 57; 23:13(50), 15(50), 25(50), 128(50)
Schwartz, W. H., 13:5(22), 11(22), 59
Schwarz, C. E., 21:279, 281, 287, 293, 301, 316, 344; 29:17(14), 56(14), 259(118), 261(118, 121), 263(118, 121), 264(118, 121), 265(118, 121), 282(118), 328(181), 341(118, 121), 344(181)
Schwarz, K. W., 17:90, 97, 99, 100(74), 101, 154; 21:149, 151, 153, 158, 179, 182
Schwarz, W. H., 11:240, 241, 242, 261; 15:84(71), 137(71)
Schwarzschild, K., 3:203, 249
Schweikle, J. D., 5:501, 517
Schweitzer, S., 14:76, 103
Schwerdtfeger, K., 28:318(184, 193, 194), 337(184), 338(193, 194)
Schwertz, F. A., 3:285, 301
Schwille, F., 30:180(154), 196(154)
Schwind, R. G., 5:389, 390, 510; 16:270(44), 334(44), 335(44), 339(44), 365; 18:30(65), 32(65), 33(65), 82(65)
Sciance, C. T., 5:414, 417, 418, 422, 438, 439, 513; 16:193, 220(21), 222, 223(21), 235, 236, 238; 20:16(37), 17(37), 81(37)

Sciulli, C. M., **14**:318(85), 321(85), 322(85), 323(85), 324(85), 344

Sclater, J. G., **14**:41(95), 42(95), 102

Scofield, M. P., **21**:300, 345

Scoppa, C. J., **14**:327(105, 106), 328(105), 329 (106), 334(105, 106), 339(105, 106), 345

Scorah, R. L., **5**:65, 66, 67, 104, 125; **11**:2, 47, 105(71), 109, 111, 191; **18**:257(33), 258(33), 259(33), 260(33), 264(33), 293(33), 306(33), 321(33); **20**:15(29), 80(29)

Scorer, R. S., **12**:36(47a), 74

Scott, **25**:295

Scott, C. I., **3**:71(68), 75, 99

Scott, C. J., **4**:274(37), 314; **6**:40(76), 41(76), 77(76), 126; **21**:168, 180; **23**:280(25), 355(25)

Scott, D., **18**:71(135), 72(135), 86(135); **31**:401(56), 402(56), 427(56)

Scott, D. S., **20**:113, 131

Scott, E., **29**:170(133), 185(133), 209(133)

Scott, E. J., **SUP**:121(299)

Scott, L. E., **5**:389(114), 510

Scott, R. A. M., **11**:412(187), 439

Scott, R. B., **5**:327(17), 331, 333(17), 335, 343, 505, 506; **9**:350(3), 352(3), 401(3), 414; **15**:44(68), 57(68)

Scott, S. L., **20**:115, 131, 132

Scriven, L. E., **1**:218, 264; **8**:2, 24(8), 25(126), 32(8, 126), 84, 87; **9**:234, 270; **11**:13(40), 15, 48; **SUP**:67(123); **15**:243, 278; **19**:198(36, 37), 244(36, 37); **28**:41(70, 73), 73(70, 73); **30**:94(12), 189(12)

Scrivener, C. T. J., **23**:321(192), 363(192)

Scurlock, R. G., **17**:146, 157

Sdobnov, G. N., **11**:144(155), 146(155), 151(155), 152(155), 156(155), 159(155), 170(155), 172(155), 173(155), 176(155), 177(155), 179(155), 180(155), 182(155), 183(155), 185(155), 195, 196

Seader, J. D., **5**:330, 404, 405, 409, 412, 413, 420, 421, 440, 441, 445, 511; **8**:26, 87

Sealey, C. J., **14**:155(55), 243

Sears, G. W., **28**:16(28), 71(28)

Sears, J. T., **25**:153(8), 243(8)

Sears, W. R., **11**:239, 261

Seasholtz, R. G., **23**:280(6), 305(6), 354(6)

Seaton, A. E., **30**:202(26), 250(26)

Seaton, R. E., **31**:441(29), 471(29)

Sebald, J. F., **12**:23(37), 74

Seban, R. A., **2**:25, 45(48), 94(75), 106, 107, 108; **3**:13, 31, 32, 80, 81(82), 84(81), 89, 91, 93(82), 94, 95, 99, 100, 122, 137, 138, 139, 173, 238(89), 251; **6**:42, 126, 504(9), 515(26), 525(9), 561, 562; **7**:339, 343, 344, 346, 351(32), 352, 354(14), 355, 361, 377, 378; **9**:221, 254, 269, 270, 379(53), 386(53), 389(53), 394(53), 397(53), 416; **11**:231, 232, 259; **13**:5(21), 11(21), 59; **SUP**:7(16), **15**:189(59), 224(59); **21**:73(17), 109(96), 110(96), 134(17), 137(96)

Sebits, D. R., **14**:140, 146, 147

Sechenov, G. P., **10**:206, 217; **19**:110(71), 111(74), 115(95), 144(155), 186(71, 74), 187(95), 188(155)

Seckendorff, R., **4**:118(75), 140

Sedahmed, G. H., **23**:230(232), 232(232), 277(232)

Sedahmed, H. G., **25**:259, 281, 286, 315

Sedgwick, T. O., **28**:351(66), 418(66)

Sedov, L. I., **15**:97(97), 138(97)

Sedova, T. K., **17**:328(36), 340

Seeley, L. E., **26**:8(196), 101(196)

Seely, J. H., **20**:221(136), 244(180), 290(313), 305(136), 308(180), 314(313)

Seeniraj, R. V., **19**:32(66), 33(68), 37(66), 42(66), 86(66, 68)

Seetharam, T. R., **21**:39(67), 40(67), 52(67)

Seetharama, K. N., **20**:209(116), 304(116)

Seetharamu, K. N., **16**:136, 156; **26**:50(24, 25), 52(197), 93(24), 94(25), 102(197)

Sega, S., **10**:269(94), 283

Segal, E., **28**:351(72, 73), 387(159), 388(159), 389(159), 418(72, 73), 422(159)

Segal, M. D., **25**:114(180, 181, 182), 121(192), 148(180, 181, 182), 149(192)

Seginer, A., **6**:41(88), 126

Segner, J., **28**:106(13), 139(13)

Segovia, G., **18**:326(2), 328(2), 329(2), 350(2), 361(2)

Sehgal, B. R., **29**:220(43), 221(43), 249(101, 102), 316(43), 337(43), 340(101, 102)

Seide, P., **15**:327

Seidl, A., **30**:349(95), 350(95), 382(150), 431(95), 433(150)

Seifert, A., **13**:219(81), 265

Seiler, L., **22**:70(36), 71(36), 72(36), 76(36), 79(36), 80(36), 81(36), 82(36), 83(36), 85(36), 90(36), 100(36), 108(36), 109(36), 154(36)

Seitz, F., **1**:214, 264
Sek, J., **23**:192(134), 197(134), 198(134), 206(134), 214(134), 272(134)
Sekerka, J., **28**:317(178), 337(178)
Sekerka, R. F., **15**:328
Sekerka, R. K., **19**:79(204), 92(204)
Sekhon, K. S., **20**:182(3), 290(315), 291(315, 316, 317), 292(315), 299(3), 314(315, 316, 317)
Seki, N., **19**:38(84), 43(97), 77(178), 79(213), 87(84, 97), 91(178), 92(213), 99(13), 185(13); **30**:151(111), 152(111), 194(111)
Sekulić, D. P. (Chapter Author), **26**:219, 220(13), 222(24), 231(34), 234(34), 236(34), 237(41), 241(41), 246(34, 54, 55), 247(13, 41), 248(41), 251(34), 253(13, 34), 261(34), 269(34), 270(34), 271(34), 276(13), 278(13), 281(34, 80), 282(34), 283(34), 287(34), 291(34), 292(41), 294(41), 295(41), 296(41), 297(80, 81), 298(41), 299(41), 300(80), 314(34), 316(34), 320(34), 322(34), 323(34), 325(13, 24), 326(34, 41), 327(54, 55), 328(80, 81)
Selamet, A., **21**:240(10, 11, 12), 246(12), 263(11), 275(10, 11, 12); **23**:146(27, 28), 185(27, 28); **30**:65(15), 72(15), 88(15)
Selby, R. J., **23**:193(107), 271(107)
Selcukogh, Y. A., **9**:407(101), 408(101), 417
Self, S., **29**:130(10), 203(10)
Selig, K. P., **17**:152(183), 158
Selikhovkin, S. V., **6**:409, 410, 489
Selimos, B., **20**:329(45), 351(45)
Selin, G., **9**:255, 257, 272
Seliverstov, B. N., **6**:418(119), 491
Sellars, J. R., **8**:22(102), 86; **15**:100(101), 139(101)
Sellens, R. W., **26**:88(198), 89(198, 199), 102(198, 199)
Sellers, J., **3**:82(78), 99
Sellers, J. P., Jr., **2**:75, 108
Sellers, J. R., **SUP**:18(25), 21(25), 79(25), 101(25), 102(25), 119(25), 120(25), 125(25), 137(25), 170(25), 171(25)
Sellers, W. D., **12**:7, 11, 24(15), 25(15), 73
Sellers, W. H., **2**:434(35), 451
Sellin, R. H. J., **15**:96(96), 97(96), 138(96); **25**:286, 317
Sello, H., **15**:327
Sellschop, W., **21**:55(3), 82(3), 133(3)

Semenov, E. G., **11**:423(221, 222, 223), 440
Semenov, K. N., **14**:121, 130, 144
Semenov, N. I., **1**:398(79), 399(79), 442; **2**:353
Semeria, R., **11**:18(65, 66), 31(65, 66), 49
Semiat, R., **20**:84, 85, 92, 95, 97, 99, 108, 131
Semichev, S. A., **25**:92(170, 171), 148(170, 171)
Semilet, Z. V., **SUP**:240(425)
Semjonow, V., **13**:216(29), 217(32), 218, 264
Semura, J. S., **17**:142(159), 143(161, 162), 144(159, 161), 145(162), 147(159), 149(162), 150(161, 162), 151(178, 179), 157
Sen, P. K., **15**:289, 328
Senda, J., **28**:36(63), 39(63), 73(63)
Senfteleben, H., **11**:202, 205, 213, 214, 253; **14**:107, 119(86a, 87), 120, 124(86a, 87, 88), 125, 147
Senftleben, H., **1**:287, 353; **4**:156, 223
Sengelin, M., **18**:171(119), 223(119), 225(119), 238(119)
Sengers, J. V., **7**:9; **21**:2(9), 3(9), 51(9)
Sengupta, S., **24**:3(12, 13, 15, 16), 5(12, 13, 15, 16), 36(12, 13, 15, 16, 18)
Senkin, V. I., **6**:474(288), 500
Sephton, H. H., **30**:246(52), 252(52)
Seppa, I., **6**:418(117), 491
Sepri, P., **30**:290(82), 310(82)
Serafimidis, R., **7**:133(44), 161
Serazetdinov, A. Z., **5**:132, 133, 161, 162, 163, 164, 247
Serebryakov, B. R., **14**:234(176), 235(176), 247
Sereeteriin, B., **14**:209(142), 210(142), 246
Sereeterin, B., **10**:209, 218
Sergazin, Z. F., **9**:265(135), 272
Sergeev, G. T., **1**:124, 184
Sergeev, O. A., **11**:358(90), 360(93d), 404(170), 436, 437, 438
Sergeev, V. L., **25**:31(59), 32(59), 142(59)
Sergeeva, L. A., **6**:296, 474(213), 478(213); **25**:31(59), 32(59), 142(59)
Sergeyev, N. D., **17**:29(85), 39(85), 60
Sergienko, A. A., **2**:68, 108
Sergius, L. M., **10**:224(8), 253(8), 281
Sergovsky, P. S., **1**:178, 184
Series, R. W., **30**:333(55), 390(156), 429(55), 434(156)
Sernas, S., **11**:290(24), 315
Sernas, V., **18**:36(72), 73(145), 83(72), 86(145); **20**:186(22, 75), 187(22, 75), 190(22), 191(22), 192(22), 197(22), 201(75),

300(22), 302(75); **28**:150(14, 15), 151(59),
154(14, 57), 166(59), 168(59), 169(59),
170(59), 171(7, 59), 173(7), 174(7), 175(7),
184(14, 15), 185(7, 14, 15), 186(15),
187(15), 188(59), 189(59), 190(57), 195(5),
197(5), 198(5, 57), 200(5, 34, 35), 202(5),
205(57), 206(57), 207(5), 208(22), 211(22),
228(5, 7, 14, 15, 22), 229(34, 35), 230(57,
59)

Serov, E. P., **6**:474, 501; **20**:152(16), 179(16);
25:4(26), 140(26)

Serrudo, G., **18**:25(55), 81(55)

Serth, R. W., **15**:92(89), 138(89); **25**:262, 318

Servouze, Y., **23**:294(74), 357(74)

Sesonske, A., **3**:124(31), 173

Seth, A. K., **14**:139, 147

Seth, B., **26**:89(200), 102(200)

Settari, A., **26**:234(37, 39), 254(37, 39), 269(37),
326(37, 39)

Settarova, Z. S., **11**:358(90), 365(123), 374(123),
375(123), 436, 437

Sevastyanov, V. V., **21**:46(77, 79), 52(77, 79)

Severin, P. I. W., **28**:365(110), 366(110),
367(110), 368(110), 420(110)

Severn, R. T., **1**:119, 122

Severs, E. T., **15**:70(50), 136(50)

Sevigny, E., **5**:172, 248

Sevruk, I. G., **9**:299(41), 347

Sevryugin, A. L., **17**:327(35), 336(35), 337(35),
340

Seward, T. P., III, **28**:270(100), 333(100)

Seyer, F. A., **12**:85, 112; **15**:63(20), 64(20),
77(20), 91(79), 92(79), 135(20), 138(79),
193(70), 224(70); **23**:193(256), 220(256),
221(256), 278(256); **25**:292, 300, 308, 319

Seyfert, P., **17**:108(96, 98), 109(98), 113(98),
114(98), 119(116), 120(117), 121, 122,
155, 156

Seymour, P. J., **23**:317(152), 318(152), 320(185),
345(298), 361(152), 362(185), 368(298)

Sforza, P. M., **16**:53, 57

Shabott, A. L., **5**:301(57), 323

Shabrin, A. N., **6**:402, 488

Shack, A., **20**:148(12), 179(12)

Shackleford, W. L., **11**:358(89), 359(89), 436

Shade, K. W., **SUP**:218(411)

Shaffer, E. C., **SUP**:177(355), 179(360),
180(360), 185(360), 363)

Shaffer, L. H., **12**:242(96), 247(96), 281

Shafrin, E. G., **9**:229, 230, 269

Shah, B., **19**:108(57), 186(57)

Shah, H. J., **20**:268(272), 312(272)

Shah, J. M., **26**:220(15), 234(15), 254(15),
269(15), 305(15), 325(15)

Shah, M. J., **2**:390, 391, 394, 396; **15**:185(55),
224(55); **25**:257, 262, 281, 282, 283, 284,
288, 289, 290, 315, 318

Shah, M. M., **17**:32, 61

Shah, P. J., **19**:113(86), 186(86)

Shah, R. K. (Chapter Author), **SUP**:2(13), 3(13),
11(17), 16(24), 29(13, 24), 46(54), 57(54),
66(115, 116), 89(13), 98(259), 103(270),
104(270), 105(270), 107(270), 125(270),
126(270), 127(270), 128(270), 144(13),
168(259), 171(270), 172(270), 173(270),
174(270), 180(270), 181(270), 182(270),
199(13), 200(13), 203(13), 205(13,
116), 207(116), 211(270), 225(116),
226(116), 229(116), 230(116), 232(116),
254(116), 256(116), 258(116), 259(116),
260(116), 270(13), 272(13), 273(13),
286(13), 290(13), 291(13), 293(13),
295(13), 301(259), 350(13), 357(13),
382(13), 397(116), 398(259), 399(259),
405(522), 416(17), 418(54); **17**:246, 261,
316; **19**:248(1, 2, 3), 261(1), 268(1), 277(1,
2), 278(1, 64), 279(1), 280(1), 281(1),
282(1), 289(1), 291(1, 2), 292(2), 294(86),
295(1, 2), 297(1, 2), 298(1), 299(2), 300(1),
306(1, 2), 312(1, 2), 322(3), 338(149),
349(1, 2, 3), 352(64), 353(86), 356(149);
20:136(6), 138(6), 148(6), 179(6),
296(322), 314(322); **26**:219, 219(2, 3, 4),
255(76, 77), 274(79), 291(79), 304(4),
305(3), 306(85), 324(2, 3, 4), 328(76, 77,
79, 85); **31**:160(65), 325(65)

Shah, S., **18**:73(148), 86(148); **20**:186(74),
187(74), 201(74), 202(74), 302(74)

Shah, S. J., **22**:229(1), 347(1)

Shah, V. L., **SUP**:73(181), 88(181), 89(181),
164(181), 298(181), **19**:268(53), 271(53),
281(80), 285(80), 286(80), 287(80),
351(53), 352(80); **22**:84(58), 155(58)

Shah, Y. T., **13**:260(114), 266; **15**:287(163), 328;
30:93(5), 168(5), 189(5)

Shahani, H., **28**:262(70), 267(70), 289(70),
290(70), 291(70), 292(70), 294(70),
295(70), 296(70), 332(70)

Shahbenderian, A. P., **21**:180

Shai, I., **10**:142, 143, 145, 147, 165

Shair, F. H., 7:144(53), 161
Shakerin, S., 18:66(121), 85(121); 20:205(100), 210(100), 211(100), 212(100), 303(100)
Shakher, C., 30:266(30, 31), 267(31, 32), 308(30, 31, 32)
Shakhin, V. M., 25:81(136, 153), 146(136), 147(153)
Shakhova, N. A., 10:173, 215
Shakir, S., 16:72(24), 73(24), 74(23), 80(24), 125, 153, 155
Shalaev, S. P., 10:53(34), 58(34), 64(34), 70(34), 71(34), 82
Shallcross, D. C., 23:293(65), 356(65)
Shaller, D. V., 17:298, 317
Shampine, L. F., 11:397, 438
Shamsi, M. R. R. I., 28:306(147), 335(147)
Shamsundar, N., 19:26(51), 79(212), 85(51), 92(212); 24:3(4), 36(4); 28:77(89), 142(89); 31:343(26), 426(26)
Shang, H. M., 24:247, 269
Shangareev, K. R., 25:51(76), 61(95, 96), 143(76), 144(95, 96)
Shankar, P. N., 26:32(201), 102(201)
Shanny, R., 9:140(62), 179
Shao, B. Y., 30:175(151), 196(151)
Shapira, M., 23:409(71), 463(71)
Shapiro, A. B., 17:175, 314; 30:338(72), 339(72), 430(72)
Shapiro, A. H., 2:13(16), 17(16), 18(16), 105; SUP:6(14), 69(14), 86(14), 87(14), 398(14)
Shapiro, J., 24:19(52), 38(52)
Shapiro, M., 23:409(71), 463(71)
Shapiro, Z. M., 29:68(18), 124(18)
Shaposhnikov, V. A., 17:327(33, 34), 334(33, 34), 336(33, 34), 340
Shapovalov, V. V., SUP:112(282), 113(282), 119(292), 136(315)
Sharan, A., 23:74(91), 75(91), 79(91), 94(91), 131(91)
Sharapov, B. N., 1:435(140), 444
Sharif, A. O., 25:288, 318
Sharikov, J. V., 10:177(47), 216
Sharma, C. L., 25:51(82, 83), 143(82, 83)
Sharma, D., SUP:73(177), 209(177), 211(177), 212(177), 398(177), 19:268(57), 269(57), 271(57), 272(57), 273(57), 274(57), 276(57), 351(57)
Sharma, G. K., 11:247, 248, 249, 261; 21:5(13), 37(60), 39(60, 67, 70), 40(60, 67), 51(13), 52(60, 67, 70)

Sharma, K. K., 25:290, 319
Sharma, K. N., 20:329(43), 351(43)
Sharma, M. K., 23:193(233), 215(233), 237(233), 239(233), 277(233)
Sharma, O. P., 11:318(10, 16), 376(10), 378, 381, 388(10), 389, 434; 25:259, 269, 319
Sharma, R. C., 20:329(43), 351(43)
Sharma, V. P., 4:35, 63
Sharon, A., 29:170(134), 185(177), 209(134), 211(177), 277(149), 278(149), 343(149)
Sharpe, L., 26:29(202), 102(202)
Shastry, U. A., 8:44, 89
Shaukatullah, H., 20:185(108), 203(96), 205(108), 303(96), 304(108)
Shaumeyer, J. N., 17:131(137), 132(137), 133(137, 140, 141), 134(137), 135(137), 156
Shaver, C. R., 2:385(101), 396
Shaver, R. G., 15:63(17), 64(17), 69(17), 110(17), 135(17)
Shavit, G., 7:331(3), 332, 338, 346, 353, 353(3), 354, 377; SUP:42(50), 43(50), 209(50), 210(50), 211(50); 19:258(15), 259(15), 268(15), 272(15), 274(15), 350(15)
Shaw, D. A., 15:122(146), 128(146), 140(146)
Shaw, D. J., 12:252(97), 255(97), 256(97), 281
Shaw, D. T., 28:135(118, 119), 144(118, 119)
Shaw, D. W., 28:351(85, 86, 88), 419(85, 86, 88)
Shaw, E. A. G., 12:22(35), 74
Shaw, H. R., 14:97(185), 104
Shaw, M. T., 15:79(69), 137(69); 23:218(69), 270(69)
Shaw, P. V., 7:108, 110, 112, 160
Shaw, R. P., 19:44(100), 87(100)
Shaw, W. A., 25:295, 316
Shay, R. J., 20:283(301), 293(327), 294(301), 313(301)
Shchitnikov, V. K., 1:141, 184; 11:219, 223, 237, 241, 242, 257
Shchokin, V. K., 15:328
Shchukin, A. V., 31:160(48, 61), 324(48, 61)
Shchukin, V. K., 25:92(173), 148(173); 31:160(8, 25, 34), 181(8, 25), 322(8), 323(25, 34)
Shea, J. J., 1:78, 110, 121; 8:4, 84
Shea, M. J., 28:113(90), 118(90), 119(90), 120(90), 142(90)
Sheaffer, R. L., 31:349(35), 426(35)
Sheard, A. G., 23:321(200), 363(200)
Shedd, G. M., 28:343(49), 417(49)

Sheeman, T. V., **8**:149, 160; **18**:147(81), 159(81)

Sheer, R. E., **23**:320(169), 361(169)

Sheer, T. J., **12**:3(7), 14(7), 51(7), 59(7), 68(7), 69(7), 73

Sheffield, R. E., **23**:196(235), 206(234, 235), 226(234), 277(234, 235)

Shefizadeh, F., **17**:185(39), 314

Shefsiek, P. K., **7**:248(67), 316

Shehu Diso, I., **26**:49(218), 50(218), 103(218)

Sheikholeslami, R., **30**:216(53), 252(53)

Sheindlin, B. S., **14**:176(119), 188(119), 191(119), 237(119), 245; **19**:160(193), 189(193)

Shekarriz, A., **30**:102(42), 143(42), 190(42)

Shekriladze, I. G., **9**:258(99), 271; **14**:133, 148

Sheldon, D. B., **7**:182(51), 213

Sheldon, M., **29**:84(47), 125(47)

Sheldon, R. P., **18**:214(107), 238(107)

Shelkov, A. N., **25**:131(203), 132(203), 150(203)

Shelton, J., **17**:175, 314

Shelton, J. T., **9**:244(73), 261(73), 270

Shelyag, V. R., **SUP**:12(21), 188(21)

Shelyakov, O. P., **25**:166(54), 245(54)

Shemer, L., **20**:105, 110, 112, 130, 132

Shen, Chzha-Yuan, **1**:420(113), 443

Shen, S. F., **7**:170(18), 211; **9**:294(20), 346; **28**:167(39), 201(39), 204(39), 229(39)

Shen, T., **10**:269, 283

Shen, T. B. C., **26**:250(59, 60), 253(60), 261(59), 281(60), 316(60), 320(60), 327(59, 60)

Shen, Y. C., **26**:109(17), 209(17)

Sheng, P., **28**:41(75), 73(75)

Shenkman, S., **4**:205(80), 206(80), 226

Shenoy, A. V. (Chapter Author), **15**:143, 157(19, 20), 158(20), 160(20), 161(21), 162(20), 163(20, 21), 164(21), 165(21), 172(32), 173(32), 178(19), 180(19), 181(19), 184(19), 185(52), 187(56), 188(56), 190(67), 191(67), 193(69), 194(69), 195(69), 216(67, 69), 217(32), 218(32), 221(19), 223(19, 20, 21), 224(32, 52, 56, 67, 69); **19**:249(7), 349(7); **24**:101, 102(4, 6, 7), 106(4), 112(99), 119(127), 120(128), 122(128), 123(127), 124(128), 125(128), 129(99), 130(99), 132(99), 135(128), 136(128), 137(128), 142(127), 143(127), 145(128, 138), 147(99, 128, 138), 148(99, 128, 138, 142, 143, 144), 149(99), 151(138), 152(127, 138), 154(127), 155(145), 157(145), 158(145), 159(145), 184(4, 6, 7), 188(99), 189(127, 128, 138), 190(142, 143, 144, 145); **25**:254, 258, 262, 267, 270, 281, 283, 286, 287, 288, 318, 319

Shenoy, S. U., **21**:5(13), 51(13)

Shepard, C. E., **4**:283(46, 47), 314

Shepard, J. E., *see* Shepherd, J. E.

Shepard, M. L., **22**:188(44, 102), 201(191), 348(44), 351(102), 355(191)

Shepherd, J. E., **23**:305(115), 320(166), 359(115), 361(166); **29**:85(50), 90(58), 102(78), 104(78), 125(50), 126(58), 127(78), 185(180), 186(180), 187(183), 211(180, 183), 299(155, 158), 343(155, 158)

Shepherd, W. H., **28**:365(107), 420(107)

Shepp, L. A., **30**:288(76), 310(76)

Sheppard, T. S., **5**:390, 510

Sher, N. C., **1**:394(68), 398(68), 442

Sherberg, M. G., **6**:38, 125

Shercliff, J. A., **1**:299(42), 353

Sheridan, R. A., **2**:156(29), 269

Sheriff, H. S., **26**:199(159), 216(159)

Sheriff, N., **31**:247(141), 248(143), 250(143), 261(152), 328(141, 143), 329(152)

Sherley, J. E., **4**:161, 167, 224; **5**:464, 466, 512, 516

Sherlock, R. A., **17**:69(21), 70(21), 71(21), 153

Sherman, A., **1**:316(53), 354; **28**:344(60), 418(60)

Sherman, B. A., **23**:13(50), 15(50), 25(50), 128(50)

Sherman, F. S., **7**:182, 200, 213, 216; **18**:36(75), 38(77), 39(79), 83(75, 77, 79)

Sherman, H., **30**:204(23), 250(23)

Sherman, M., **8**:213, 227

Sherman, M. P., **29**:61(2), 62(9), 72(2), 102(77), 108(81), 122(87), 123(2), 124(9), 127(77, 81, 87)

Sherman, S., **15**:327

Sherony, D. F., **SUP**:24(36), 64(36), 254(36), 255(36), 256(36)

Sherrard, C. P., **25**:288, 318

Sherry, R. R., **29**:233(64), 338(64)

Shertzer, C. R., **2**:361, 396; **15**:182(48), 224(48)

Sherwood, A. E., **14**:19(49), 22, 101

Sherwood, L. T., **23**:280(7), 354(7)

Sherwood, T. K., **3**:9, 10, 31; **11**:219, 222, 231, 232, 236, 257, 259; **13**:120, 121, 175, 200, 201; **15**:128(150), 141(150), 236(39), 264(96), 267(96), 277, 280; **16**:176, 184(8), 238; **21**:171, 172, 174, 182; **28**:7(5), 70(5)

Shestov, E. D., **6**:474(310), 501
Shetz, A., **9**:54, 110
Sheu, J.-P., **SUP**:142(324), 143(324), 146(324)
Sheu, J. P., **30**:119(74), 192(74)
Shevich, Yu. A., **31**:160(43), 324(43)
Shevyakov, A. A., **6**:474, 501; **25**:4(29, 32), 141(29, 32)
Shewen, E. C., **23**:333(244), 365(244)
Sheynkman, A. G., **9**:256(92), 270
Shibamaru, N., **28**:135(72), 142(72)
Shibanov, B. V., **11**:376(132a), 378(132a), 381(132a), 438
Shibata, H., **20**:183(7), 299(7)
Shibata, Y., **21**:91(60), 136(60)
Shibayama, S., **23**:13(30), 15(30), 16(30), 17(30), 23(30), 24(30), 25(30), 27(30), 127(30)
Shih, C. J., **14**:332, 345
Shih, F. S., **SUP**:65(108), 199(108), 247(108), 262(108), 263(108), 277(108); **19**:262(24), 350(24)
Shih, J. S., **SUP**:202(393), 203(393)
Shih, P., **28**:376(146), 421(146)
Shih, T., **31**:401(58), 428(58)
Shih, T. I.-P., **24**:218, 272
Shih, Y. P., **15**:101(109), 139(109); **25**:259, 290, 317
Shiiba, Y., **28**:36(58), 43(58), 54(58), 72(58)
Shiina, Y., **19**:78(183), 91(183)
Shil'dkret, V. M., **17**:327(31, 32), 333(31), 334(31, 32), 335(32), 336(31, 32), 340
Shimada, R., **21**:125(116), 138(116)
Shimamoto, S., **17**:140(158), 157
Shimanski, J. N., **10**:177, 216
Shimazaki, T. T., **3**:238(89), 251; **6**:504(9), 515(26), 525(9), 561, 562; **SUP**:7(16)
Shimiza, M., **17**:21, 22, 59
Shimizu, M., **14**:267(45), 280; **23**:55(76), 58(76), 60(76), 62(76), 63(76), 67(76), 69(76), 70(76), 71(76), 74(76), 75(76), 77(76), 95(76), 130(76)
Shimizu, R., **28**:343(50), 417(50)
Shimizu, S., **21**:85(47), 91(61), 135(47), 136(61); **25**:51(87), 144(87)
Shimizu, S. S., **23**:330(237), 365(237); **26**:132(78), 212(78)
Shimizu, Y., **21**:124(113), 138(113)
Shimomura, R., **11**:105(99, 121), 114(99), 193, 194
Shimonis, V. M., **31**:159(1), 160(1, 56), 161(1), 177(1), 182(1), 247(1), 262(1), 263(1), 322(1), 324(56)
Shimoyama, T., **23**:12(62), 14(62), 32(62), 58(62), 60(62), 89(62), 109(62), 110(62), 120(62), 129(62)
Shimshoni, M., **SUP**:69(139), 85(139)
Shimura, F., **30**:313(2), 315(2), 330(2), 378(2), 382(2), 426(2)
Shin, C. T., **24**:192, 244, 270
Shine, A. J., **6**:312(102), 366; **11**:248, 261
Shingu, P. H., **28**:28(53), 36(53), 56(53), 72(53)
Shinke, T., **SUP**:12(23), 188(23), 189(23)
Shinnar, R., **31**:462(65), 473(65)
Shinohara, K., **14**:99, 105
Shiotsu, M., **16**:236; **17**:22(70), 24(70), 59; **18**:257(31, 32), 262(31, 32), 292(32), 295(32), 298(31), 300(31, 32), 304(31), 305(31), 321(31, 32)
Shipley, G. H., **15**:264(96), 267(96), 280
Shirai, T., **10**:169(9), 172(28), 215
Shiralkar, B. S., **7**:37, 40, 84
Shiralkar, G. S., **18**:58(109), 60(109), 84(109)
Shirazi, M. A., **16**:6, 10(13), 11(13), 15, 16, 56, 57
Shiren, N. S., **17**:68(10), 153
Shires, J. W., Jr., **20**:244(187), 308(187)
Shiroki, K., **30**:396(159), 397(159), 434(159)
Shirshov, A. N., **25**:75(122), 145(122)
Shirtliff, C. G., **3**:75(71), 99
Shitsman, M. E., **6**:551, 552, 555, 557(77), 558(77), 564; **7**:30, 31(13), 32, 33, 36, 37, 39, 49, 50, 66, 66(50), 67, 67(50), 84
Shitzer, A., **22**:1(33), 14(33), 18(32, 33), 19(54), 28(53), 30(25, 54), 153(25), 154(53), 155(54), 158(192, 193), 355(187, 192, 193)
Shivakamar, P. N., **SUP**:341(489), 342(489), 349(489, 494), 350(489, 494)
Shklover, G. G., **31**:160(50), 324(50)
Shlanchyauskas, A. A., **11**:203(43), 206(43), 254
Shlapkova, Ya. P., **14**:176(158, 159), 221, 225, 236(158, 159), 246
Shlosinger, A. P., **7**:238, 254, 266, 267, 277, 277(44), 315
Shlyakntina, K. I., **6**:418(152), 493
Shmeter, S. M., **14**:262(29), 280
Shock, R. A. W. (Chapter Author), **16**:1, 66, 68, 69, 70, 71, 72(23), 77(23), 104(23), 107, 108, 142(116, 117), 143(117), 144(23), 145, 146, 147(128), 148(128), 153, 155, 156

Shoenhals, R. J., **11**:105(64), 111, 191

Shoham, O., **20**:84, 85, 90, 108, 115, 119, 121, 122, 123, 128, 130, 132

Shohet, J. L., **1**:311, 354

Sholokhov, A. A., **8**:49, 89; **SUP**:355(496), 356(496)

Shonnard, D. R., **23**:407(69), 422(69), 435(69), 442(69), 463(69)

Shonung, B., **24**:199, 211, 273

Shorenstein, M. L., **7**:204, 205, 217

Shorin, G. N., **6**:390(15), 405(15), 486

Shorin, S. N., **3**:212, 215, 218(60), 242, 243, 244(60), 245(60), 250, 251

Shorley, G., **24**:19(50), 38(50)

Short, B. E., **7**:52(29), 84

Short, B. I., **3**:54, 60, 61, 98

Short, W. Leyh, **1**:376(28), 441

Short, W. W., **3**:16(25), 32

Shoukri, M., **16**:232

Shoulberg, R. H., **3**:55, 98; **13**:211(15), 264; **18**:169(45), 171(45), 176(45), 236(45)

Shouman, A. R., **19**:79(203), 92(203)

Shpanskii, S. P., **17**:328(36), 340

Shpil'rain, E. E., **26**:221(23), 246(23), 253(23), 325(23)

Shrago, Z. K., **18**:340(58), 341(58), 342(58), 363(58)

Shreiber, I. R., **26**:137(88), 212(88)

Shreshta, G. M., **8**:11(33), 85

Shrier, A. L., **20**:109, 113, 131

Shrivastava, S., **19**:110(64), 121(107), 186(64), 187(107); **25**:187(81, 82), 188(81, 82), 194(82), 195(81, 82), 201(81, 82), 202(82), 203(82), 205(81, 82), 206(81, 82), 228(81, 82), 232(82), 237(82), 246(81, 82)

Shshitnikov, V. K., **6**:474(213, 214), 478(213, 214), 496

Shtrikman, S., **18**:237(88)

Shu, J.-Z., **30**:264(27), 265(27), 273(27), 299(100), 300(100), 308(27), 311(100)

Shuen, J.-S., **24**:193, 199, 273, 275

Shuler, K. E., **5**:48, 54

Shulman, Z. P., **15**:128(160), 130(160), 141(160); **18**:169(46), 171(46, 56), 176(46, 56), 177(56), 212(56), 228(46), 229(46), 237(46, 56); **25**:257, 283, 285, 289, 317

Shumakov, N. V., **6**:376(6, 7), 414(6), 415(6), 463(6, 7), 464(6, 7), 467(6), 486; **25**:4(14), 11(37), 32(14, 37), 140(14), 141(37)

Shuman, H. L., **16**:98, 154

Shumkov, S. H., **25**:158(27, 28, 29, 30, 31, 32, 38), 159(27, 32), 160(29), 161(28, 29, 38), 163(29), 165(28, 29, 38), 168(32), 200(27, 32), 201(32), 202(27), 243(27), 244(28, 29, 30, 31, 32, 38)

Shumway, R. W., **SUP**:35(45), 71(45), 77(45), 145(45), 297(45), 319(45), 320(45), 321(45, 475)

Shur, G. N., **14**:262(29), 280

Shure, R. I., **4**:214(98a), 227

Shuy, E. B., **25**:81(160), 147(160)

Shvab, V. A., **8**:94(14), 121, 159

Shvarev, K. M., **28**:78(91), 85(91), 142(91)

Shvartsman, V. L., **25**:203(101), 247(101)

Shved, G. M. (Chapter Author), **14**:249, 252(4, 7), 253(11), 254(4), 255, 257(24), 260(4), 261(4), 263, 264(4), 265, 266, 267(38, 48), 268(38), 269(56), 271, 275(11), 276(48, 56), 277, 278(56), 279, 280

Shvets, I. T., **25**:4(33), 141(33)

Shvets, M. E., **1**:83, 121

Shwab, V. A., **18**:88(14), 116(14), 158(14)

Shyiskya, K. F., **25**:28(57a), 142(57a)

Shyu, R. J., **20**:232(157), 306(157)

Shyy, W. (Chapter Author), **24**:191, 192, 193, 194, 196, 199, 200, 201, 202, 206, 207, 210, 211, 213, 214, 215, 217, 218, 219, 221, 222, 225, 226, 227, 230, 231, 234, 235, 238, 239, 240, 242, 244, 247, 249, 255, 258, 259, 262, 263, 264, 265, 266, 267, 269, 273, 274, 275; **28**:308(154), 336(154)

Siao, T. T., **31**:170(76), 325(76)

Sibbitt, W. L., **2**:364(51), 395; **18**:169(27), 170(27), 173(27), 218(27), 223(27), 224(27), 236(27)

Sibitt, N. L., **18**:179(61), 180(61), 237(61)

Sibold, D., **28**:111(92), 142(92)

Sibulkin, M., **2**:6(8), 9, 23, 24(8), 105; **7**:146, 161; **26**:109(16), 209(16)

Sickinger, A., **28**:36(62), 53(62), 73(62)

Sickmann, J., **5**:172, 173, 174, 248

Šicman, M. E., **1**:412, 416(107), 417(107), 443

Sidahmed, G. H., **15**:128(153), 141(153)

Sidall, R. G., **8**:295(26), 349

Sideman, S. (Chapter Author), **SUP**:80(230), 120(230), 121(230), 158(230), 159(230), 186(230); **15**:227, 228(4, 5), 242, 245, 246(75, 78), 247(79, 80), 248, 249(77), 250(88, 89), 251(80), 252(80, 88, 90, 91),

Sideman, S. (*continued*)
253, 255(91), 256(91), 257(90, 91), 258(91), 260(91), 262(78), 271(4), 276, 279, 280; **21**:56(6), 133(6); **26**:2(203), 47(203), 49(204), 102(203, 204)

Sidljar, M. M., **15**:323

Sidlyar, M. M., **6**:418(110, 111), 491

Sidorov, E. A., **1**:23, 50; **3**:230, 251; **6**:440, 441, 442(174, 176), 494

Siebers, D. L., **16**:270, 334, 335, 339(44), 365; **18**:30(65), 32(65), 33(65), 82(65)

Sieder, E. N., **6**:531, 532, 562; **15**:102(117), 139(117); **25**:295, 319

Siegal, R., **3**:126(34), 132(34), 173; **12**:96(29), 113

Siegel, A., **8**:291(11), 344(11), 349

Siegel, B. L., **5**:449, 450, 515

Siegel, R. (Chapter Author), **1**:234, 239, 265, 306(45), 307, 309, 309(47), 310, 311, 354; **2**:422, 423(18), 424(16, 17, 18, 20), 425(18, 20), 432, 445(18), 450, 451; **4**:33, 42(60), 58, 63, 64, 144, 147(5), 149(5), 157, 160(37), 162, 164, 167, 169, 175, 176, 179, 180, 181, 182, 183(60), 184, 185, 186, 190, 191, 209, 222, 224, 225, 227; **5**:462, 467, 468, 515, 516; **6**:420, 421(155), 422(154), 424(154), 425(159), 428, 430(154, 159, 160), 431, 433(160, 163), 434(163), 442, 444(177), 445(177), 446(177), 447, 458, 459(159), 469(177), 474(226), 480, 493, 494, 495, 497, 524(31), 525(33), 562; **8**:33, 34, 35, 49, 53, 54(240, 241, 242, 244), 55, 88, 89, 90; **11**:11(25), 12(25), 48, 105(129), 115, 194, 319, 321(33), 329(33), 335(33), 338(33), 339(33), 435; **SUP**:6(14), 18(29), 66(119), 69(14), 75(204), 86(14), 87(14), 125(29), 129(29, 307), 135(29), 136(29), 145(29), 179(29), 180(204), 188(29), 192(378), 205(119, 395), 208(399, 400), 217(204), 266(119), 398(14), **15**:100(102), 120(142), 121(142), 139(102), 140(142); **16**:97, 98, 154; **17**:7, 57; **19**:39(88, 89), 87(88, 89), 278(62), 279(62), 352(62); **23**:135(1), 145(1), 148(1), 149(1), 150(1), 151(1), 164(1), 184(1); **27**:23(21); **30**:338(77), 339(77), 430(77); **31**:334(3), 371(38), 425(3), 427(38)

Siegel, R. S., **9**:169, 180, 256(89), 270, 312(74), 347

Siegel, R. T., **12**:124, 168(9), 192

Siegell, J. H., **25**:153(10), 168(66), 171(66), 173(66), 175(66), 176(66, 70), 181(75), 182(75), 189(66, 83, 84, 85), 193(93), 196(66), 234(83, 84, 85, 124, 125, 126, 127, 128, 129, 130), 236(124), 243(10), 245(66), 246(70, 75, 83, 84, 85), 247(93), 248(124, 125, 126, 127, 128, 129, 130)

Siegman, A. E., **28**:102(28), 139(28)

Sielcken, M. O., **17**:201(53), 231(74), 260, 275(96), 293, 296, 301, 302(74), 303(74), 315, 316, 317

Siemes, W., **21**:319, 320, 321, 325, 329, 331, 345

Sienicki, J. J., **29**:151(72), 206(72), 215(1), 216(3, 4), 220(42, 43), 221(42, 43), 233(65), 248(1), 249(99), 251(104), 258(3, 4), 260(3, 4), 262(3, 4), 264(3, 4), 271(143), 273(143), 280(104), 315(42), 316(42, 43), 320(104), 323(42), 334(1, 3, 4), 337(42, 43), 338(65), 340(99, 104), 342(143)

Sieniutycz, S., **24**:42(16), 99(16)

Sigalla, A., **3**:93, 100

Sigeman, S., **6**:474(207), 496

Sigillito, V. G., **SUP**:132(309)

Sigmon, T. W., **28**:96(63), 98(105), 141(63), 143(105)

Signh, S. N., **5**:161, 248

Sigurgeirsson, T., **29**:130(9), 184(9), 203(9)

Sih, G. C., **8**:53, 90

Sikarskie, D. L., **19**:16(30), 84(30)

Sikka, S., **SUP**:124(303)

Silberman, E., **23**:5(6), 126(6); **26**:122(49), 123(49), 210(49)

Silberstein, L., **6**:205, 364

Sill, K. H., **23**:298(92), 358(92)

Silliman, W. J., **28**:41(73), 73(73)

Silveira-Neto, A., **25**:346, 392, 406, 417

Silver, A. H., **7**:344, 352, 378

Silver, L., **16**:147, 148, 149, 156

Silver, R. S., **9**:258, 265(102), 271

Silver, S., **2**:7(14), 9(14), 27(14), 62(14), 105

Silverman, W. B., **11**:420(205), 439

Silverstein, M., **20**:244(199), 309(199)

Silveston, P. L., **11**:300, 315; **30**:28(78), 91(78)

Silvestri, M. (Chapter Author), **1**:245, 266, 355, 361(13), 363(13), 378(13), 379(33, 34, 35, 36), 381(13), 385(34), 389(34, 36), 390(34, 36, 57), 393(36), 401(34), 402, 410(96), 411(13), 413(13), 421(13, 82), 427(129),

430(82), 433(137), 436(13), 440, 441, 442, 443; **17**:28(81), 59; **24**:8(39), 14(39), 37(39)

Simbirsky, D. F., **6**:474(222), 480, 497

Simchenko, L. E., **14**:152(26), 242

Simeonides, G., **23**:326(212), 364(212)

Simka, H., **28**:375(141), 421(141)

Simmers, D. A., **21**:169, 182

Simmons, F. S., **5**:301(56), 323

Simmons, G., **10**:43(15), 44(15), 45(15), 54(36), 60(15), 62(15), 64(15), 65(15, 52), 67(52), 74(36), 81, 82; **11**:360, 437

Simmons, G. M., **5**:28(41), 52; **8**:70, 90

Simmons, L. F. G., **11**:241, 242, 261

Simmons, S. G., **23**:296(79), 297(79), 341(79), 343(79), 347(302), 348(302), 349(302), 357(79), 368(302)

Simmons, W. E., **1**:394(70), 398(70), 442

Simms, D. L., **10**:243(35), 281, 282

Simms, S. D., **11**:318(15), 434

Simon, F. F., **4**:202, 226; **11**:59, 190

Simon, G., **28**:111(29, 30), 139(29, 30)

Simon, H. A., **4**:31, 63; **6**:312(98), 366; **7**:56, 57, 85

Simon, I., **10**:48, 51, 52, 81

Simon, N. J., **21**:149, 153, 161, 179

Simon, R., **3**:183, 186(7), 188(7), 192(7), 248

Simon, T., **16**:265, 320, 365

Simon, T. W., **20**:222(141), 225(141), 226(141), 244(190), 249(190), 250(190), 306(141), 308(190); **23**:46(67), 47(67), 48(67), 64(84), 77(84), 100(84), 121(67), 129(67), 130(84), 327(216), 328(216, 223), 364(216, 223)

Simonds, R. R., **SUP**:413(529)

Simoneau, R. J., **4**:202(77), 203(77), 226; **11**:17(59), 18(59), 49, 59, 84(42), 85(42), 86, 190; **17**:37(121), 62; **23**:320(174), 345(296, 297), 350(306), 351(297, 306), 362(174), 368(296, 297, 306)

Simonich, J. C., **16**:328(46), 329, 365; **23**:330(234), 365(234)

Simonov, M. N., **29**:96(61), 126(61)

Simons, R. E., **20**:183(10), 244(180), 262(260), 283(300), 299(10), 308(180), 311(260), 313(300)

Simpkins, P. G., **2**:270; **29**:145(37), 171(37), 204(37)

Simpson, C. J. C. M., **8**:248(23), 283

Simpson, C. J. S. M., **14**:256(23), 279

Simpson, H. C., **11**:220, 224, 258; **26**:49(205), 102(205)

Simpson, R. L., **16**:279, 349, 355, 356, 365

Sims, G. E., **21**:328, 345

Sims, K., **31**:410(74), 428(74)

Singer, R. J., **28**:22(39), 72(39)

Singer, R. M., **3**:110, 173; **4**:228; **9**:261, 271; **10**:159(83, 84), 166

Singh, B., **19**:108(52), 186(52); **25**:258, 288, 289, 290, 318, 319

Singh, B. N., **16**:225(32), 227(32), 236, 239

Singh, B. P. (Chapter Author), **23**:133, 136(9), 156(9), 161(9), 170(49), 172(49), 179(52), 184(9), 186(49, 52)

Singh, B. S., **13**:164, 202

Singh, D., **23**:213(186), 214(236), 236(186, 236), 238(186), 240(186), 241(236), 275(186), 277(236)

Singh, G., **20**:100, 102, 132

Singh, I., **15**:329

Singh, R. K., **28**:124(93), 143(93)

Singh, R. N., **8**:65, 90

Singh, R. P., **24**:103(18), 185(18)

Singh, S., **1**:410(94), 442

Singh, S. N., **3**:107, 173; **5**:183, 249; **SUP**:112(281), 113(281), 117(281)

Singhal, A. K., **13**:81(42), 116

Singham, J. R., **SUP**:64(99), 379(99), 380(99), 381(99), 382(99)

Singleton, E. B., **5**:279(24), 322

Sinha, D. B., **14**:135, 145

Sinha, D. N., **17**:142, 143(161), 144(161), 145, 147, 149, 150(161), 151, 157

Sinha, S. N., **5**:159(51), 165, 248

Sinitcin, A. T., **21**:46(76, 77, 78, 79), 52(76, 77, 78, 79)

Sinitsyn, E. N., **10**:103, 104, 105(20), 106, 164

Sinitsyn, I. T., **17**:319(4), 327(4), 339

Sinke, W., **28**:96(14), 139(14)

Sinnitsin, I. T., **21**:32(48), 52(48)

Sinton, W. M., **10**:76, 78, 82

Sipavičius, Č. J., **18**:139(75), 159(75)

Sipe, J. E., **28**:102(94, 112), 143(94, 112)

Siple, P. A., **4**:93(31), 94(31), 96(31), 138

Sirieix, M., **6**:17, 27, 29(40), 30(50), 31(51), 34, 51(96), 65(96), 66(96), 113(51), 123, 124, 127

Sirignano, W. A., **10**:267, 283; **26**:2(206), 11(164), 20(164), 36(122), 45(220, 221), 53(206), 55(123, 206), 56(206), 66(115,

Sirignano, W. A. (*continued*)
   122, 164, 165), 67(4, 115, 164, 165, 209,
   230, 231, 232, 233, 234), 68(165), 69(4,
   206, 209), 71(4), 72(4), 75(44, 161), 76(44,
   45), 78(115, 136, 137, 232, 233), 79(4),
   84(206, 208), 85(173), 86(45), 87(45),
   88(178, 206, 207), 89(7, 8, 9, 10, 56, 200),
   92(4), 93(7, 8, 9, 10), 94(44, 45), 95(56),
   98(115, 122, 123), 99(136, 137), 100(161,
   164, 165, 173), 101(178), 102(200, 206,
   207, 208, 209), 103(220, 221, 230, 231,
   232, 233, 234); 28:3(3), 70(3)
Sirivat, A., 21:218(35), 219(35), 220(35),
   237(35)
Sirota, A. M., 18:340(58), 341(58), 342(58),
   363(58)
Sirtl, E., 28:351(74, 75, 76), 418(74, 75, 76)
Siskovic, N., 13:219(95), 266; 23:193(237),
   218(237), 220(237), 277(237)
Sislian, J. P., 14:336, 345
Sissenwine, N., 4:82(20), 138
Sitar, N., 30:111(55), 154(55), 155(55), 156(55),
   181(155), 191(55), 196(155)
Sitenok, N. A., 15:324
Sitharamayya, S., 26:123(56), 126(56), 129(56),
   133(56), 210(56)
Sivaji, K., 26:2(140), 99(140)
Sivasegaram, S., 7:357, 358, 379
Sivasubramanian, M. S., 30:262(20), 308(20)
Sivertson, E. P., 23:296(82), 352(82), 357(82)
Siviour, J. B., 11:105(88), 113, 121, 192
Siwari, I., 11:420(206), 439
Siwatibau, S., 17:283, 317
Sjorgreen, B., 24:227, 230, 270
Skalak, R., 22:30(14), 153(14)
Skan, S. W., 4:336, 446; 26:111(26), 120(26),
   209(26)
Skartsis, L., 25:252, 292, 293, 297, 298, 308, 319
Skartvedt, G., 5:390, 510
Skelland, A. H. P., 15:61(3), 62(3), 63(3), 67(3),
   70(3), 71(3), 76(3), 90(3), 130(159), 134(3),
   141(159), 144(5), 223(5); 18:162(13),
   236(13); 19:259(19), 283(19), 350(19);
   23:188(238), 189(238), 190(238),
   197(238), 198(238), 277(238); 24:108(79),
   159(79), 187(79); 25:259, 290, 319;
   28:153(60), 230(60)
Skelton, W. T. W., 29:167(123), 208(123)
Skéma, R. K., 23:59(99), 61(99), 82(99), 87(99),
   91(99), 101(99), 131(99)

Skinner, L. A., 16:87, 154
Skipper, R. G. S., 11:278, 279, 305, 314
Skiruin, S. C., 3:81(84), 99
Skirvin, S. C., 7:344, 352(35), 371(38), 378
Skadzień, J., 26:251(61, 62, 62a), 253(61),
   281(61, 62), 327(61, 62, 62a)
Sklyar, F. R., 6:418(143), 493
Skocypec, R. D., 31:334(4), 335(4), 388(4),
   389(4), 390(4), 392(4), 395(4), 425(4)
Skoglund, V. J., 6:474(281), 500
Skornyakova, L. G., 30:242(34), 251(34)
Skramstad, H. K., 11:232(169), 237(169), 260
Skrinska, A., 31:160(16), 323(16)
Skripov, V. P., 7:57, 58, 60, 85; 10:103, 104(22),
   105, 106(30), 107(29), 108, 164; 11:8, 48;
   29:138(28), 204(28)
Skryabina, L. P., 8:50, 90
Slack, E. G., 3:55, 98
Slack, G. A., 28:99(95), 143(95)
Slack, M., 31:245(137), 328(137)
Slagh, T. D., 20:182(4), 299(4); 23:32(60),
   129(60)
Slančiauskas, A. A., 8:95(20), 109(20), 110(20),
   115(43), 116(20), 138(20), 142, 144(75),
   145(75), 146(78), 147(79), 151(20), 159,
   160; 18:88(20), 108(20), 109(42), 110(20),
   136(20), 141(20), 143(20), 145(79),
   147(20), 148(20), 150(20), 158(20, 42),
   159(79); 23:59(99), 61(99), 82(99), 87(99),
   91(99), 101(99), 131(99); 31:159(1),
   160(1), 161(1), 177(1), 182(1), 247(1),
   262(1), 263(1), 322(1)
Slaney, M., 30:287(73), 310(73)
Slattery, J. C., 13:129(39), 131, 133, 134,
   136(39), 137, 138(39), 153(39), 158, 175,
   176(39), 202; 14:6, 11(30), 100;
   SUP:70(148), 87(148), 160(148); 15:62(5),
   69(44), 135(5), 136(44); 19:251(8),
   268(36), 270(36), 349(8), 351(36);
   23:188(239), 277(239), 374(40), 462(40);
   24:40(8), 98(8), 114(110), 188(110);
   25:253, 262, 270, 272, 273, 315, 319;
   31:27(33), 103(33)
Slattery, J. L., 24:108(76), 187(76)
Slattery, R. E., 6:36(63), 125
Slayzak, S. J., 26:173(137, 138), 182(137, 138),
   215(137, 138)
Sleep, B. E., 30:121(82), 192(82)
Slegel, D. L., 16:10(13), 11(13), 15(13), 16(13),
   57

Slegers, L., **9**:254, 270; **21**:109(96), 110(96), 137(96)

Sleicher, C. A., Jr., **3**:114, 115(20), 126, 147, 150, 173; **6**:515(29), 525(32), 562; **7**:52(27), 84; **8**:22(103), 86; **9**:174, 180; **11**:200(61), 240(61), 241(61), 242(61), 255; **15**:121(141), 122(145), 140(141, 145)

Slesarev, V. A., **25**:127(195), 129(195), 149(195)

Sleytr, U. B., **22**:186(171), 354(171)

Slezak, S. E., **29**:89(56), 126(56)

Slezkin, N. A., **SUP**:71(159a), 87(159a), 190(373), 298(159a)

Slichter, W. P., **30**:331(42, 43), 428(42, 43)

Sliepcevich, C. H., **16**:193, 220(21), 222(21), 223(21), 238

Sliepcevich, C. M., **5**:414, 417, 418, 422, 438, 439, 513; **10**:109, 164, 260, 282; **11**:105(69, 70), 111(69, 70), 116(69), 192; **16**:235, 236; **29**:130(7), 203(7)

Sliff, H. T., **4**:228

Slinko, M. G., **19**:110(70), 111(70), 186(70)

Slitenko, A. F., **25**:127(197), 149(197)

Slobodchuk, V. I., **25**:51(79), 143(79)

Slobodyanyuk, L. I., **9**:103, 104, 110

Sluyter, W. M., **11**:18(64), 31(64), 49; **16**:88(39), 100, 138(108), 154, 156

Slyshenkov, V. A., **25**:81(151), 147(151)

Smagorinsky, J., **25**:332, 340, 370, 417

Small, J., **3**:16, 32; **11**:218, 221, 236, 256

Small, M. B., **28**:341(15), 415(15)

Small, S., **9**:147, 179

Small, W., **16**:117(84), 118, 123, 125, 155

Smalley, J. L., **23**:299(94), 358(94)

Smart, D. R., **18**:17(32), 80(32)

Smeets, G., **6**:194, 199(21), 363

Smelt, R., **14**:317, 344

Smeltzer, W. M., **20**:264(265), 312(265)

Smetana, P., **28**:316(168), 336(168)

Smiley, E. F., **3**:275, 301

Smirnov, A. A., **6**:376(8), 463(8), 464, 466, 474(215, 216, 217), 478, 479, 480, 486, 496

Smirnov, N. N., **15**:287(165), 328

Smirnov, O. K., **21**:44(72), 52(72); **25**:73(116, 117, 118, 119), 145(116, 117, 118, 119)

Smirnov, V. A., **13**:15, 17(45), 60; **26**:123(52), 126(52), 210(52)

Smirnov, V. P., **SUP**:336(488), 337(488)

Smit, J. A. M., **15**:285(189), 287, 290, 324, 329

Smith, A. M. O., **4**:331, 334, 444, 445; **13**:79(29), 116

Smith, A. R., **5**:227(142), 250

Smith, B., **14**:125, 147

Smith, B. B., **5**:10(10), 11(12), 50

Smith, C. E., **SUP**:135(313)

Smith, D., **6**:549, 563; **23**:296(79), 297(79), 341(79), 343(79), 357(79)

Smith, D. R., **11**:365(111), 437

Smith, E. B., **28**:364(99), 419(99)

Smith, E. C., **12**:3(10), 14(10), 73

Smith, E. G., **23**:329(225), 330(225), 364(225)

Smith, F., **3**:60, 99

Smith, F. G., **7**:4, 83

Smith, F. H., **6**:197, 364

Smith, F. T., **19**:31(65), 36(65), 37(65), 86(65)

Smith, F. W., **23**:222(192), 227(240), 275(192), 277(240)

Smith, G. M., **1**:382(42), 441

Smith, G. P., **21**:154, 182

Smith, I., **12**:256(98), 281

Smith, J. L., **17**:146, 147, 157

Smith, J. L., Jr., **15**:40(61), 44(69), 47(73), 57(61, 69, 73)

Smith, J. M., **2**:52, 107; **5**:233(172), 236, 251; **6**:549, 551, 563, 564; **7**:49, 84; **10**:172(22), 175(22), 177(22), 179(22), 215; **SUP**:97(257), 98(257), 168(257); **19**:175(229), 177(229), 190(229); **23**:371(24), 461(24)

Smith, J. W., **8**:298(35), 300(54), 301(54), 349, 350; **15**:91(82), 138(82); **26**:8(67, 196), 95(67), 101(196); **31**:191(94), 211(94), 326(94)

Smith, K. A., **12**:78, 88, 102, 106, 107, 108, 111; **15**:87(73), 90(73), 92(73, 87), 93(73), 96(95), 97(95), 104(124), 111(73), 114(124), 137(73), 138(95), 140(124); **18**:208(97), 238(97); **19**:330(126), 354(126)

Smith, K. L., **19**:36(71), 86(71)

Smith, K. R., **17**:300(106, 108), 305(108), 306, 317

Smith, L. H., **22**:353(141)

Smith, L. M., **28**:277(110), 280(110), 334(110)

Smith, M., **28**:410(226), 425(226)

Smith, M. C., **11**:229(172), 231, 232, 260; **14**:97(186), 104

Smith, M. E., **6**:38, 125; **9**:406(98), 417

Smith, M. K., **23**:341(277), 343(277), 367(277)

Smith, N., **11**:226(151), 259

Smith, N. D., **18**:337(48), 363(48)

Smith, P. J., **23**:139(17), 184(17)

Smith, P. N., **29**:244(96), 266(131), 312(96), 320(96), 340(96), 342(131)

Smith, R., **15**:136(42)

Smith, R. A., **11**:240(238), 263

Smith, R. C., **26**:52(210), 102(210); **29**:234(94), 340(94)

Smith, R. K., **10**:256, 282

Smith, R. L., **14**:96, 104

Smith, R. N., **15**:128(154), 141(154); **19**:26(52), 27(52), 37(82), 85(52), 87(82); **21**:182

Smith, R. V., **5**:403(139), 404, 405, 406, 407(139), 408(139), 428, 440, 445, 469, 511, 516; **11**:105(104), 175, 177(177), 185, 187(177), 193, 197; **17**:138(146), 147, 148, 156, 157

Smith, R. W., **24**:222, 225, 269

Smith, S. A., **31**:433(16), 471(16)

Smith, V. C., **12**:129, 133(21), 192

Smith, V. G., **15**:328

Smith, W., **6**:211, 365; **9**:127(29), 178, 220(47), 269

Smithells, C. J., **30**:346(88), 430(88)

Smolak, G. R., **4**:228

Smolim, V. N., **1**:435(141), 444

Smolin, V. N., **17**:328, 340

Smolsky, B. M., **1**:130, 184; **6**:474(214, 223), 478(214), 480(223), 496, 497; **25**:31(59), 32(59), 142(59)

Smorchkov, V. N., **11**:423(220), 440

Smorodinsky, E. L., **13**:219(74), 265

Smuda, P. A., **9**:390(61), 391(61), 416

Smulders, P. T., **17**:164(18), 314

Smylie, D. E., **14**:122, 126(90), 144, 147

Smyrl, W. H., **12**:237(99), 251(99), 270(99), 281

Snail, K., **18**:20(48), 81(48)

Sneddon, I. N., **8**:78(280), 91

Snedeker, R. S., **26**:183(144), 215(144)

Snedon, I. N., **22**:258(194), 355(194)

Snider, D. M., **11**:376(133), 385(133), 387(133), 388(133), 438

Snittina, O. F., **21**:21(37), 51(37)

Snow, D. F., **21**:125(114), 138(114)

Snow, S. D., **29**:215(1), 248(1), 334(1)

Snyder, H. A., **5**:233, 236, 238, 251; **21**:153, 157, 182

Snyder, N. W., **5**:409, 411, 420, 422, 513; **11**:219, 222, 257

Snyder, W. T., **3**:106, 172; **SUP**:216(409), 325(481)

Snytin, S. Yu., **23**:9(25), 116(25), 127(25)

So, D. S., **29**:259(122), 261(122), 262(122), 263(122), 264(122), 266(122), 341(122)

So, R. M. E., **23**:320(169), 361(169)

Sobel, N., **20**:186(73), 187(73), 198(73), 200(73), 302(73)

Sobey, I. J., **25**:324, 417

Sobolev, V. V., **3**:176(3), 178(3), 186(3), 204, 248; **5**:13, 51; **11**:325(45), 435

Sobolik, K. B., **23**:301(105), 358(105)

Socolow, R. H., **15**:2(8), 54(8)

Soddy, F., **2**:325, 355

Soden, R. R., **30**:319(18), 427(18)

Soderman, E., **29**:234(90), 339(90)

Soderstrom, G. F., **4**:67(6), 116(73), 117(73), 137, 140; **22**:20(30), 21(30), 82(30), 153(30)

Soehngen, E., **9**:322(83), 333, 348

Soehngen, E. E., **4**:51, 64; **6**:211(57), 312(94), 364, 366; **25**:276, 319

Soehngren, E., **11**:219, 222, 257

Soengen, E., **8**:120(50), 121(50), 160; **18**:114(49), 115(49), 159(49)

Softley, E. J., **6**:41, 65(80), 126

Sogin, H. H., **3**:14, 16(33), 19(33), 26, 32; **11**:231(164), 232(164), 259, 260

Soh, W. H., **24**:306(42), 320(42)

Sohal, M. S. (Chapter Author), **16**:88(39), 100, 101, 154; **17**:319

Sohrab, S. H., **21**:263(32), 276(32)

Sokolov, M., **15**:211(104), 225(104)

Sokolov, V. S., **6**:418(118), 491

Sokolova, I. N., **1**:110, 122

Sokovishin, Y. A., **20**:217(122), 305(122)

Sola, A., **10**:107(37), 164

Solberg, H. L., **SUP**:291(462)

Solbrig, C. W., **SUP**:24(36), 28(38), 64(36, 38), 202(38), 208(38), 209(38), 214(38), 215(38), 216(38), 218(38), 254(36), 255(36), 256(36); **19**:289(85), 291(85), 353(85)

Soldaini, G., **1**:245(89), 266, 361(13, 14), 363(13), 378(13), 381(13, 39, 40, 41), 410(96), 411(13, 39, 40, 41), 413(13, 39, 40, 41), 419(111), 421(13, 40, 41), 427(129), 436(13), 440, 441, 442, 443

Soliman, M., **6**:460, 461, 467, 474(200, 203), 477, 495, 496; **9**:261, 271

Soliman, M. M., **7**:320b, 320e

Solnechnyi, E. M., **25**:4(30), 141(30)

Solntsev, V. P., **25**:35(70a), 143(70a)
Solomon, A. D., **19**:2(6), 83(6)
Solomon, J. M., **7**:206, 217
Solomonov, S. D., **SUP**:197(384), 209(384),
   210(384); **19**:258(16), 259(16), 268(16),
   350(16)
Soloukhin, R. I., **29**:98(65), 126(65); **30**:293(90),
   296(99), 298(99), 311(90, 99)
Soloviev, B. S., **25**:81(156), 147(156)
Solovyev, V. A., **14**:259(25), 278(25), 279(25),
   280
Solvason, K. R., **18**:67(123), 72(123), 85(123)
Someah, K., **31**:453(66), 473(66)
Somers, E. V., **9**:3(68), 109, 304(56), 347;
   SUP:414(533)
Somers, R. R., II, **15**:38(60), 57(60)
Somerscales, E. F. C. (Chapter Author),
   18:44(88), 83(88); **30**:28(79), 91(79), 197,
   198(7, 8), 207(55), 208(55), 216(55),
   218(55), 221(55), 223(55), 224(55),
   232(14, 54), 234(14, 54), 249(7, 8),
   250(14), 252(54, 55); **31**:443(67), 456(68),
   473(67, 68)
Someswara Rao, S., **SUP**:249(433), 251(436),
   271(448)
Sommer, S. C., **3**:54, 60, 61, 98
Sommerfeld, A., **6**:136, 170(9), 363; **15**:285, 328
Sommeria, G., **25**:371, 417
Sommerscales, E., **11**:298, 299, 315
Somon, F. F., **11**:17(59), 18(59), 49
Son, J. E., **23**:198(146), 206(146), 226(146),
   229(146), 273(146)
Son, J. S., **7**:108, 110, 114, 160
Sondergeld, C. H., **14**:88, 104
Sondergeld, C. K., **30**:122(84), 124(84), 125(84),
   126(84), 127(84), 128(84, 90), 129(84, 90),
   131(84), 192(84), 193(90)
Sone, M., **23**:109(113), 110(113), 112(113),
   132(113)
Song, F., **30**:262(24), 308(24)
Song, S., **20**:270(276), 312(276)
Song, S. H., **4**:90(26), 92(26), 95(26, 36),
   120(86), 138, 140
Song, W. J., **22**:19(55, 56), 69(56), 71(55, 56),
   96(55, 56), 131(55), 132(55), 133(55),
   134(55), 135(55), 136(55), 137(55),
   151(55, 56), 155(55, 56)
Song, Y.-Z., **30**:274(48), 275(48), 276(48),
   309(48)
Sonin, A. A., **15**:270(107), 281

Sonin, A. S., **11**:423(221, 222, 223), 440
Sonju, O. K., **29**:66(15), 68(15), 124(15)
Sonnemann, G., **5**:369, 508; **7**:53, 85
Sonnenmeier, J. R., **23**:317(152), 318(152),
   361(152)
Sonntag, R. E., **15**:5(21), 6(21), 55(21)
Sonolikar, R. L., **25**:153(11), 161(43), 162(43),
   184(43), 198(11), 200(43), 201(11),
   232(11), 236(11), 243(11), 244(43)
Soo, S. L., **5**:175, 176, 178, 248; **9**:121, 135,
   144, 145, 147, 156, 178, 179
Soong, T. T., **25**:252, 292, 319
Sorbie, K. S., **23**:188(241), 228(241), 277(241);
   24:104(68), 187(68)
Sorensen, J. P., **15**:287(173), 328
Sorenson, E. S., **19**:111(76), 121(76), 122(76),
   186(76)
Sorey, M., **14**:49, 102
Sorey, M. L., **14**:3(8), 17(46), 37, 49(114), 100,
   101, 102
Sorlie, T., **26**:220(11), 236(11), 237(11), 240(11),
   242(11), 243(11), 244(11), 245(11),
   248(11), 253(11), 269(11), 274(11),
   279(11), 281(11), 291(11), 292(11),
   296(11), 297(11), 300(11), 301(11),
   306(11), 309(11), 310(11), 312(11),
   313(11), 314(11), 322(11), 325(11)
Sorokin, A. G., **25**:28(57a), 142(57a)
Sorokin, D. N., **11**:18(62, 63), 19(62, 63), 31(62,
   63), 49; **17**:22(68), 59
Sorokin, Yu. L., **17**:51, 64
Sostman, H. E., **22**:364(9), 435(9)
Soundalgekar, V. M., **15**:180(42, 43), 224(42, 43)
Souquet, J., **2**:324(28), 353
South, V., **15**:232, 277
Southwell, C. J., **15**:195(73), 225(73)
Southwell, R. V., **4**:152, 223; **5**:204, 249
Souvaliotis, A., **25**:305, 319
Souza Mendes, P., **20**:205(105), 212(105),
   304(105)
Sovershenny, V. D., **8**:19, 86
Sovran, G., **13**:79(30), 116
Sözen, M., **23**:370(16), 461(16); **30**:94(13a),
   189(13a)
Spalding, D. B. (Chapter Author), **2**:111(2),
   123(12), 268; **3**:35(15), 36, 50, 56, 57, 58,
   59, 60, 67, 68, 82(77), 97, 98, 99; **4**:331,
   444; **7**:132(43), 161, 339, 377, 378; **8**:35,
   36, 58, 288(4), 348; **9**:299(43), 347;
   12:80(12), 112, 237(40), 251(40), 279;

Spalding, D. B. (*continued*)

13:12(43), 60, 62, 65, 66, 67(9, 10, 11, 12, 15, 16, 17, 18), 68(19), 69, 70(20), 71, 72, 73, 75, 76, 80(32), 81, 83, 84, 91, 92, 96, 97, 102, 105, 106(53, 54), 107, 110, 113(59), 114, 115, 116, 117, 236(104), 237(104), 266; **SUP**:73(176), 89(248), 164(248), 211(176, 402), 298(248); **16**:355, 365; **17**:187, 192, 195, 196, 203, 204, 315; **19**:268(54, 55, 56), 271(54, 55, 56), 351(54, 55, 56); **24**:193, 194, 195, 211, 264, 271, 272, 274; **25**:323, 417; **26**:53(211), 102(211); **28**:308(153), 319(153), 336(153); **29**:266(130), 342(130); **30**:52(80), 55(80), 58(80), 60(80), 65(80), 71(80), 91(80); **31**:160(44), 324(44)

Spalding, E. C., **15**:328

Spangenberg, W. G., **7**:200(122), 216

Spangler, J. G., **12**:79(8a), 80, 112

Sparks, D. L., **24**:104(29), 185(29)

Sparnaay, M. J., **12**:252(100), 255(100), 281

Sparrow, E. M. (Chapter Author), **1**:2, 12, 13(6), 33(29), 40, 41, 48, 49, 50, 292, 294(31), 353; **2**:379, 396, 399, 407(7), 415(12), 417(12), 418(12), 420(13), 421(14, 15), 424(21), 425(23), 427(13, 14, 28, 29, 30), 430(15, 30, 21), 434(32), 436(32), 437(39), 439(39), 441(40), 443(40, 42), 444(42), 445(45, 46), 448(46), 449(46), 450, 451, 452; **3**:108(11, 12), 126, 132, 173, 211(48), 250; **4**:7, 12, 15, 17, 21, 23, 25, 29, 35, 42(18), 44, 62, 63, 64, 209, 227, 274, 314; **5**:25, 27, 28(31), 29, 40(68, 71), 44, 51, 52, 53, 54, 82, 96, 99, 126, 127, 132(16), 134(16), 144, 145, 148, 149, 230, 247, 250, 254(3), 255(3), 306(3), 312(3), 313(3), 316(3), 321; **6**:415, 431, 435(172), 442, 444(177), 445(177), 446(177), 447, 467, 469(177), 474, 476, 480(228), 481, 489, 494, 495, 496, 497, 524, 525(33, 35), 526, 562; **7**:169, 184, 185, 186, 191, 192, 211, 213, 214; **8**:22, 33, 34, 74, 86, 88, 250(30), 252(30), 256(30), 264(30), 283; **9**:169, 180, 248, 249, 250(77), 251(76, 77), 256(89), 258(94, 95), 261, 264(129), 265(94, 140), 270, 271, 272, 286(3), 287(5), 294(18), 295(23), 300(47), 301, 303(54), 325(5), 345, 346, 347, 348, 361(21), 363(21), 364(21), 365(21), 377(21), 379(21), 396(21), 398(21), 415; **11**:67, 73, 74, 80,

82(33), 83, 127, 129, 131, 134, 143, 144(143), 190, 195, 213, 214, 247, 255, 261, 294, 302(38), 315, 326(47), 329(47), 403(47), 435; **12**:96, 113; **13**:6, 59, 219(80), 265; **14**:5, 69(150), 71(150), 100, 103; **SUP**:18(29), 42(50, 51), 43(50, 51), 65(103, 104, 105), 66(114, 119), 71(162), 72(162, 163, 166), 75(204), 77(162), 87(162), 125(29), 129(29, 307), 135(29), 136(29), 139(162), 145(29, 162), 163(162), 176(353), 179(29), 180(204, 353), 188(29), 190(372), 192(372, 378), 194(372), 195(372, 382), 198(51, 388), 199(51, 166, 388), 205(119), 209(50, 388), 210(50, 166, 388), 211(50, 388), 213(166), 217(204), 227(105), 228(51), 229(105), 232(105, 423), 236(423), 237(423), 240(166), 241(166), 242(166), 247(51), 248(51), 262(114), 264(423), 266(119, 423), 267(114), 268(114), 269(447), 286(51), 287(51, 163), 288(51, 163), 298(163), 299(163), 325(482), 336(163), 355(103), 356(103), 357(103), 363(104), 364(104); **15**:100(102), 120(142), 121(142), 124(149), 139(102), 140(142), 141(149), 185(51), 224(51), 229(8), 232, 276, 277, 325, 328; **16**:136, 156; **17**:242(79), 316; **18**:15(100), 19(44), 48(44), 52(100), 53(100), 73(147, 148, 149), 74(149), 81(44), 84(100), 86(147, 148, 149), 156(83), 159(83); **19**:19(37), 26(51), 46(109, 110, 111), 47(109, 112, 114), 48(114), 49(114), 50(111), 52(111), 58(114, 127), 61(109, 137, 138), 62(137, 140, 141), 65(109), 66(114), 71(162), 72(162), 79(212), 85(37, 51), 88(109, 110, 111, 112, 114), 89(127, 137, 138, 140, 141), 90(162), 92(212), 258(15), 259(15), 268(15, 48, 50, 59), 271(48, 50), 272(15, 50, 59, 60), 273(59), 274(15, 50, 59, 60), 278(62), 279(62), 309(93), 312(93), 334(139, 140, 142, 147), 336(139, 140, 142), 337(147), 338(139, 140), 340(142), 350(15), 351(48, 50), 352(59, 60, 62), 353(93), 355(139, 140, 142, 147); **20**:186(56, 57, 71, 72, 74), 187(55, 56, 57, 70, 71, 72, 74), 191(55, 78), 198(56), 199(71, 72), 200(71), 201(74), 202(74), 205(102, 103, 104, 105, 106), 207(102), 211(71), 212(102, 104, 105, 106),

217(131), 219(131), 227(146), 229(146), 230(146), 232(161, 162, 163, 164), 237(161), 238(161, 162, 163, 164), 239(161), 240(161, 162, 163), 241(162, 163), 256(164), 301(55, 56, 57), 302(70, 71, 72, 74, 78), 304(102, 103, 104, 105, 106), 305(131), 306(146), 307(161, 162, 163, 164), 377, 383, 388; **21**:67(9), 109(94), 134(9), 137(94), 150, 151, 152, 182; **23**:136(3), 184(3); **24**:3(4), 6(20), 36(4, 20), 112(100), 146(139), 188(100), 189(139), 194, 272; **26**:120(43), 165(43), 182(139), 183(139), 184(139), 210(43), 215(139); **28**:77(89), 142(89), 251(33), 330(33), 369(123), 420(123); **30**:34(54, 81), 38(54), 39(54), 90(54), 91(81), 272(42), 280(63), 309(42), 310(63); **31**:343(26), 344(31), 426(26, 31)

Specchia, V., **30**:168(131), 195(131)

Spector, W. S., **4**:118(76), 140

Spee, B. M., **6**:73(127), 128

Spektor, É. L., **SUP**:81(236), 123(236), 124(236), 135(236, 314)

Spence, C. M., **18**:327(4), 362(4)

Spence, D. A., **3**:35(27), 60, 98

Spence, S. P., **30**:243(56), 252(56)

Spencer, B. W., **8**:292(14), 349; **29**:151(71, 72), 206(71, 72), 216(3, 4), 218(19), 219(19), 220(42, 43, 45, 52), 221(42, 43, 45, 46, 47, 48, 49, 50, 51, 52, 53, 54), 233(19, 65), 249(99), 251(104), 255(52), 258(3, 4), 260(3, 4), 262(3, 4), 264(3, 4), 270(19, 52), 271(141, 142, 143), 273(143), 277(19), 278(52), 279(19), 280(104), 282(52), 290(52), 305(19), 310(52), 315(42), 316(42, 43, 45), 319(19), 320(104), 322(19), 323(42), 334(3, 4), 335(19), 337(42, 43, 45, 46, 47, 48, 49, 50, 51, 52, 53, 54), 338(65), 340(99, 104), 342(141, 142, 143)

Spencer, D. L., **9**:253, 265(147), 270, 272

Spencer, J. D., **9**:119(9a), 177

Spencer, R. C., **12**:5(13), 73

Spencer, R. S., **13**:209, 210, 264; **18**:216(112), 238(112)

Spera, F. J., **19**:82(270), 95(270)

Sperling, J., **6**:176, 188(17), 363

Speziale, C. G., **25**:337, 362, 375, 382, 405, 411, 416, 418; **30**:329(32), 427(32)

Spicochi, K. C., **28**:124(78), 142(78)

Spiegel, E. A., **21**:246(23), 276(23)

Spielman, L., **25**:252, 319

Spielman, M., **3**:95(90), 100

Spiers, D. M., **7**:358, 379

Spiga, M., **18**:163(20), 236(20)

Spindler, R. J., **11**:374(125), 376(132), 378(132), 388(132, 137), 389(132), 390(132), 437, 438

Spindler, W., **23**:344(293), 368(293)

Spitzer, K.-H., **28**:318(184, 193, 194), 337(184), 338(193, 194)

Spitzer, L., **1**:271(6), 352; **4**:264(26), 314

Spitzer, W. G., **11**:357(80), 436

Spivak, G. V., **2**:355

Spizzichino, A., **9**:363(28), 415

Sposito, G., **23**:378(52), 463(52)

Spotz, E. L., **3**:268(22), 300

Spradley, L., **24**:312(43), 320(43)

Sprague, R. W., **18**:328(23), 330(23), 362(23)

Sprenger, E., **17**:305(111), 317

Spriggs, T. W., **15**:62(9), 78(9), 135(9)

Springer, G. S. (Chapter Author), **7**:163, 169, 174, 180, 181(47), 182(36, 51), 183(12), 211, 212, 213; **9**:357(11), 358(11), 414; **14**:281; **18**:189(140), 239(140); **21**:56(4), 72(14), 73(18), 74(18), 79(18), 133(4), 134(14, 18), 273(44), 276(44)

Springer, W., **4**:104(50), 106(50), 139, 259(23), 262(23), 263(23), 313

Springett, B. E., **5**:237, 238, 251; **21**:153, 158, 182

Spurlock, J. M., **15**:195(80), 225(80)

Squarer, D., **29**:47(42), 57(42)

Squire, H. B., **2**:45(47), 107; **4**:9, 62; **30**:14(82), 46(82), 91(82)

Squire, L. C., **16**:365

Squire, W., **1**:84, 121

Squires, K., **25**:362, 416, 418

Squyers, A. L., **16**:364

Sraporough, C. E., **19**:113(93), 187(93)

Sreenivasan, K., **15**:287, 328

Srenivasan, K., **11**:219, 223, 237, 257

Sridhar, T., **23**:206(92), 219(92), 223(92), 270(92)

Sridharan, K., **SUP**:68(135), 97(135), 168(135)

Srimani, P. K., **20**:327(33), 351(33)

Srinivas, B. K., **23**:193(40), 199(243), 204(243), 205(243), 206(243), 216(243), 217(243), 237(243), 239(242), 240(242), 241(243), 243(242, 243), 244(242), 251(242),

Srinivas, B. K. (continued)
268(40), 277(242, 243)
Srinivasan, J., 14:279(63), 280; 19:80(246),
94(246)
Srivastava, K. K., 23:280(19), 306(19), 354(19)
Srivastava, R. C., 2:390, 395
Stachiewicz, J. W., 9:7, 106
Stacy, E., 8:128(61), 130(61), 160; 18:123(60),
124(60), 127(60), 159(60)
Stacy, M., 11:218(96), 222(96), 256
Stafford, E. M., 15:328
Stalder, J. R., 2:287, 355; 7:189, 190, 200, 201,
214
Stallybrass, M. P., 8:33, 88
Stambler, I., 4:114(66), 139
Stamm, H. G., 6:224, 365
Stamps, D. W., 29:61(6), 66(6), 68(6), 72(6),
84(6), 85(6), 123(6), 218(18, 19), 219(18,
19), 228(18), 232(18), 233(18, 19),
244(18), 268(134), 270(19), 277(19),
279(18, 19), 288(18), 297(18), 304(166),
305(19), 307(166), 312(18), 319(19),
322(19), 329(18), 335(18, 19), 342(134),
344(166)
Standart, G. L., 15:232(16), 276
Stanek, M. J., 23:320(183), 362(183)
Stanek, V., 28:250(31), 330(31)
Stangle, G. C., 24:102(11), 184(11)
Stanislav, J. F., 20:93, 100, 132
Staniszewski, B. E., 1:220, 264; 16:98, 154
Stankevics, J. O., 4:281, 314
Stanley, E. M., 18:327(6), 328(12, 13), 362(6,
12, 13)
Stanmore, B. R., 29:131(13), 203(13)
Stanners, C. D., 28:106(13), 139(13)
Stappenbeck, A., 9:110
Starner, K. E., 7:244, 245, 316; 23:315(145),
360(145); 30:199(57), 252(57)
Starovskij, A. A., 1:416(107), 417(107), 443
Starr, J. B., SUP:42(51), 43(51), 198(51),
199(51), 228(51), 247(51), 248(51),
286(51), 287(51), 288(51), 19:272(60),
274(60), 352(60)
Stasiek, J. A., 25:392, 394, 410, 418
Stasilyavichus, Yu., 31:160(16), 323(16)
Stasiulevičius, J. K., 8:114(40), 130(40),
145(76), 146(76), 151(82), 152(82),
153(76, 82), 159, 160; 18:109(40), 142(76),
151(76), 152(76), 153(76), 158(40),
159(76)

Stassen, E. G., 22:279(209), 356(209), 426(38),
436(38)
States, M. W., 2:355
Staub, F. W., 17:30(90), 60; 19:100(26),
145(183), 149(179), 160(179, 183),
161(26), 185(26), 189(179, 183);
31:142(46), 158(46)
Stauffer, D., 14:298, 339(137), 342, 346
Stàvall, W. B., 1:358(7), 440
Stearns, S. D., 22:419(26), 436(26)
Steckler, K., 23:343(285), 367(285)
Steed, R. C., 17:119(114), 156
Steegmaier, K., 6:265(80), 365
Steehle, B., 29:155(83), 206(83)
Steele, H. S., 28:124(78), 142(78)
Steen, H. W., 30:243(56), 252(56)
Steere, R. C., 11:360(97), 437
Stefan, J., 3:287(89), 302; 19:5(14), 41(14),
84(14)
Stefanescu, D. M., 28:263(78), 332(78)
Stefanović, M., 11:4(5), 8, 20(70), 21(70, 71,
72), 31(71, 72, 73), 33(71), 34(72), 36(71),
37(71), 47, 48, 49
Stefany, N. E., 6:474(254), 483(254), 484, 498
Stegelman, A. F., 31:453(69), 474(69)
Steger, J. L., 24:218, 269, 274
Stegun, I. A., 9:374(42), 415; 11:337(59),
378(59), 436; 20:357, 360, 361, 364, 367,
370, 386
Steidler, F. E., 13:219(67), 248(67), 265
Steigelmann, W. H., 9:118(6), 177
Steiger, M. H., 2:266, 270
Steigmann, G. A., 10:19(28), 36
Stein, A., 11:318(15), 434
Stein, B., 7:320b(143), 320e
Stein, G. D., 14:333, 334(122), 345
Stein, H. N., 16:66, 153
Stein, N., 3:284(67), 301
Stein, R. P. (Chapter Author), 3:101, 120(22),
122, 126(36), 138(40), 156(42), 173, 174;
8:70, 90
Steinberg, D. S., 20:221(134), 305(134)
Steinberg, V., 20:323(18, 19), 328(37, 40),
350(18, 19), 351(37, 40)
Steinbrenner, U., 29:5(3), 6(5), 7(3), 15(3, 5),
17(3), 18(3), 19(3), 20(5), 24(3), 25(3),
26(3), 55(3, 5)
Steiner, G., 7:320c(146), 320f
Steiner, J. M., 15:211(107), 212(107), 225(107)
Steiner, L., 26:2(212), 102(212)

Steingrimsson, V., 3:284(67), 301
Steinhagen, H. P., 19:81(250), 94(250)
Steinle, H. F., 4:167, 225
Steinle, H. I., 11:105(130), 115, 194
Stejskal, J., 23:230(218a), 232(218a), 276(218a)
Stempel, F. C., 23:342(282), 367(282)
Stenuit, R., 4:109, 139
Stepanchuk, A. V., 25:166(60, 61, 62), 167(60, 61, 62), 245(60, 61, 62)
Stepanov, B. M., 11:423(221, 222, 223), 440
Stepanova, G. I., 14:253(11), 275(11), 279
Stephan, K., 13:219(69), 243(69), 265;
    SUP:139(322), 190(322), 191(322),
    194(381); 16:66, 67, 68, 95, 99, 107, 108,
    114(73), 116, 117(14, 17, 65, 73), 118, 119,
    120, 123, 124, 125, 126, 153, 154, 155,
    181(17), 238; 18:255(17, 24), 256(24),
    257(17, 24), 261(17), 320(17), 321(24)
Stephen, K., 15:287, 328
Stephens, D. R., 10:43, 44(12), 45(12), 81
Stephens, R. W. B., 2:354
Stephenson, D. G., 6:416(85), 490
Stepka, F. S., 23:331(240), 365(240)
Stepleman, R. S., 25:189(84), 234(84), 246(84)
Sterling, C. W., 8:22(103), 86
Sterman, L. S., 1:233, 265; 6:526, 562; 16:110,
    111, 114(68), 128, 131, 155
Stermole, F. J., 6:474(282), 500; 19:22(44),
    85(44)
Stern, E. J., 30:350(93), 430(93)
Stern, F., 9:265(144), 272; 11:352(75), 436
Stern, O., 2:353
Stern, S. C., 25:252, 319
Sternlicht, B., 24:8(33), 14(33), 37(33)
Sternling, C. V., 16:107, 154
Sterret, J. R., 6:94(146), 95(149), 129
Stetekee, J. A., 1:276(13), 353
Steube, R. S., 28:282(118), 285(127), 334(118, 127)
Stevens, G. F., 17:29, 30(91), 60
Stevens, G. T., 4:214(98a), 227
Stevens, J., 26:127(71), 137(91, 92), 138(92),
    139(92), 140(93), 141(91), 142(101),
    143(91, 101), 144(101), 146(91, 93),
    149(71), 150(111, 112), 151(71, 111, 112),
    153(92, 93), 154(71, 92, 93, 111), 155(92,
    111), 158(91, 101), 159(111), 160(111),
    162(111), 165(71, 92), 182(112), 183(112),
    184(112), 185(112), 198(91), 211(71),
    212(91, 92, 93), 213(101, 111, 112)
Stevens, J. W., 18:257(38, 42), 261(38, 42),
    264(42), 304(38), 305(38), 321(38, 42),
    322(60)
Stevens, R. B., SUP:76(212), 171(212)
Stevenson, A. C., SUP:324(477)
Stevenson, J. F., 15:287(172), 289, 328
Stevenson, R., 16:233, 234; 17:321(16, 17), 339
Stevenson, T. N., 16:365
Stevenson, W. H., 11:404(173), 407(173),
    423(173, 219a), 439, 440
Stever, H. G., 14:284(8), 341
Stevie, F. A., 28:97(106), 143(106)
Steward, F. B., 11:376(130), 437
Steward, F. R., 17:190, 205, 212, 315
Steward, W. E., 19:36(71), 86(71)
Steward, W. G., 5:355, 375, 403, 507, 511;
    17:149, 150, 152(176), 157
Stewart, F. R., 10:243(36), 269(96), 282, 283
Stewart, G. N., 18:188(72), 237(72)
Stewart, J. C., 5:320(91), 324
Stewart, J. E., 2:374(21), 395
Stewart, L. D., 30:154(119), 194(119)
Stewart, M. B., 26:18(213), 102(213)
Stewart, M. J., 28:249(27, 28), 330(27, 28)
Stewart, W. E., 2:358(7), 368(7), 371(7);
    3:192(17), 249; 11:269(6a), 314; 12:197(9),
    198(9), 206(9), 207(9), 208(9), 213(9),
    219(9), 232(9), 244(9), 277; 13:7(27), 60,
    126(33), 128(33), 131(33), 169(33), 201,
    208(8), 209(8), 214(8), 264; 14:194(124),
    206(139), 245, 246; SUP:54(62), 67(121);
    15:13(37), 56(37), 98(98), 100(98),
    138(98), 287(173), 328; 16:147(126), 156,
    364; 18:216(110), 238(110); 21:240(3),
    275(3); 23:189(21), 195(21), 197(21),
    267(21); 25:261, 263, 264, 267, 319;
    28:267(88), 333(88), 353(93), 355(93),
    356(93), 358(93), 419(93); 29:251(108),
    293(108), 296(108), 340(108);
    30:364(130), 432(130); 31:9(2), 102(2)
Stewartson, K., 4:331, 335, 444, 445; 5:177, 249;
    6:13, 123, 396, 487; 7:206, 217; 9:296(33),
    343, 346
Stickler, D. J., 31:461(45), 472(45)
Stickler, L. A., 29:215(1), 248(1), 334(1)
Stiefel, E., 23:418(73), 464(73)
Stiffler, A. K., 24:8(37, 38), 14(37, 38), 37(37, 38)
Still, R., 16:46(20), 57
Stillinger, F. H., 14:309, 343

Stimpson, L. D., **10**:73(67), 76, 83
Stinchcombe, R. A., **17**:37(123), 50(123), 62
Stine, H. A., **4**:283(45, 47), 314, 315;
    7:200(123), 216
Stitt, F., **2**:354
St. Jeronimo, M. A. Da, **19**:186(62)
Stock, L., **28**:369(121), 373(121), 420(121)
Stoddard, F. J., **2**:266, 270; **23**:320(178),
    362(178)
Stoddart, D. E., **9**:29(6), 34, 106
Stodola, A., **15**:2(2), 54(2)
Stoecker, W. F., **15**:3(19), 55(19)
Stokes, A. G., **11**:365(112), 437
Stokes, G. G., **5**:159, 248
Stokes, H. A., **5**:356, 363, 364, 508
Stokes, R. H., **12**:226(89), 229(89), 234(89),
    235(89), 236(89), 237(89), 240(89),
    256(89), 261(89), 262(89), 263, 266(89),
    281; **18**:328(19), 329(19), 350(19), 362(19)
Stoll, A. M. (Chapter Author), **4**:65, 69, 71(13),
    75(12, 17), 81(18), 82(19), 83, 117(72),
    118(72), 119(81), 121(81, 100), 123(81,
    100), 124(81), 125(81, 108), 127(81),
    129(109, 110), 130(111), 135(110), 138,
    140, 141; **11**:318(28), 376(28), 435;
    22:1(35), 11(34), 18(34, 35), 158(196),
    230(196), 231(199, 200), 240(197, 198,
    199, 200, 213, 214), 241(214), 242(196,
    214), 314(195), 330(213), 331(213),
    332(213), 355(195, 196, 197, 198, 199),
    356(200, 213, 214)
Stollery, J. L., **7**:331, 333(5), 335, 342(29), 350,
    352(29), 353, 359, 361, 377, 378
Stolwijk, J. A. J., **11**:318(26), 376(26), 435;
    22:30(57), 134(57), 155(57)
Stolz, G., **6**:416(80), 489; **11**:105(89), 113, 193
Stolzenbach, K. D., **12**:12, 74; **13**:108, 109,
    117
Stone, A. A., **9**:3(111), 75(111), 110
Stone, C. R., **16**:93, 94, 117(49), 154
Stone, F. S., **24**:104(47), 186(47)
Stone, H. L., **24**:247, 274
Stoner, J. T., **5**:469, 516
Stops, D. W., **3**:276(50), 301
Storm, F. K., **22**:290(201), 356(201)
Storz, R., **28**:117(27), 139(27)
Stosiċ, N., **25**:367, 412
Stotler, H. H., **19**:126(130), 188(130)
Stout, E., **4**:51(93), 64
Stow, F. S., Jr., **15**:92(93), 97(93), 138(93)

Stoy, R. L., **13**:67(6), 115
St. Pierre, C., **15**:208(99), 225(99)
Stracham, C., **2**:354, 355
Strack, S. L., **6**:41(84, 85), 126
Straight, D. M., **5**:447, 514
Stralen, S. J. D. van, **30**:173(145), 195(145)
Strand, J., **SUP**:79(223), 156(337, 338)
Strand, T., **26**:109(18), 209(18)
Strandberg, M. W. P., **3**:287, 301
Strang, G., **24**:241, 274
Strange, R. R., **23**:280(15), 298(84, 85, 86, 87,
    88), 306(84, 85, 86, 87, 120), 354(15),
    357(84, 85, 86), 358(87, 88), 359(120)
Stranski, I. N., **14**:287, 298, 341
Strasser, A., **30**:76(20), 88(20)
Stratford, B. S., **3**:35(28), 98
Stratton, J. A., **1**:273(9), 274(9), 275(9), 279(9),
    353; **14**:109, 147; **21**:126(121), 138(121),
    247(26), 276(26)
Straub, J., **24**:289(22), 319(22)
Straught, D. M., **11**:170(183), 173(183),
    174(183), 177(183), 185(183), 197
Straus, J. M., **14**:16(45), 17(45), 25(45), 26(45),
    31, 33(77), 49(77), 72, 101; **30**:119(75),
    123(88), 192(75, 88)
Strauss, A. M., **19**:199(49), 244(49); **24**:41(11),
    98(11)
Streckert, J. H., **7**:304, 319
Street, R. E., **7**:204, 217
Street, R. L., **24**:218, 272, 303(40), 320(40)
Strehlow, R. A., **3**:284, 301
Streid, D. D., **SUP**:154(334), 170(334)
Strenge, P. H., **16**:64(8), 153
Strenge, P. S., **1**:209, 213, 264; **10**:119, 131(56),
    132(56), 165; **11**:9(17), 48
Strickland, R. D., **12**:252(101), 256(101), 281
Strickler, R. F., **8**:281(51), 283
Stringfellow, G. B., **28**:351(81), 373(138),
    418(81), 421(138)
Strite, M. K., **5**:34(46), 34(47), 34(49), 36, 52
Stritzker, B., **28**:75(96), 112(79, 96), 142(79),
    143(96)
Strnad, J., **3**:107, 173
Strobridge, T. R., **15**:45(71), 57(71)
Stroeve, P., **23**:398(63), 463(63); **31**:34(16),
    80(16), 102(16)
Stroh, N., **22**:296(39), 348(39)
Strom, B. D., **23**:53(75), 57(75), 130(75)
Strong, H. M., **9**:376(44), 415
Strong, P. F., **11**:318(14), 434

Struble, C. L., **23**:13(39), 15(39), 20(39), 32(39), 128(39)

Struminskii, V. V., **6**:396, 397(27), 487

Struthers, J. O., **30**:331(43), 428(43)

Stuart, D. S., **29**:234(94), 340(94)

Stuart, J. T., **4**:152, 223; **5**:149, 156, 182, 247; **9**:323, 348; **30**:50(83), 91(83)

Stuart, R. V., **23**:328(223), 364(223)

Stuart, S. T., **21**:164, 182

Stuart, W. B., **15**:272(114), 281

Stuben, K., **24**:251, 274

Stubos, A. K., **30**:117(72), 118(72), 136(94, 95), 192(72), 193(94, 95)

Stuckles, A. D., **18**:161(11), 236(11)

Studt, F. E., **14**:14(35), 100

Stulc, P., **20**:278(296), 279(297), 280(297), 296(323), 313(296, 297), 314(323)

Stull, R. B., **25**:376, 377, 411, 413

Sturas, J. I., **4**:211, 227

Sturat, J. T., **21**:153, 179

Sturges, W., **18**:333(37), 362(37)

Stutzman, R. J., **20**:186(59), 187(59), 191(59), 195(59), 301(59); **30**:272(39), 308(39)

Stylianou, S. A., **21**:79(25), 88(53), 94(53, 72), 134(25), 135(53), 136(72)

Styrikovich, M. A., **1**:244(85), 266, 396(76), 398(76), 399, 412, 416(106, 107, 108), 417(107, 108), 420(113), 429(80), 442, 443; **31**:265(163), 329(163)

Su, C. C., **23**:333(248), 365(248); **26**:170(126), 214(126)

Su, C. L., **7**:169, 178, 211

Su, C. C., **26**:123(66), 125(66), 148(107), 151(107), 154(107), 162(116), 163(116), 170(116), 211(66), 213(107, 116)

Su, K. C., **23**:142(21), 185(21)

Su, M. D., **25**:380, 412

Suarez, E., **23**:340(267), 366(267)

Subbotin, V. I., **9**:241(71), 242(71), 255(71), 270; **11**:18(62, 63), 19(62, 63), 31(62, 63), 49; **17**:22, 29(85), 39(85), 59, 60; **21**:92(66), 136(66); **31**:236(124), 317(215), 327(124), 332(215)

Subrahmanyam, S. S., **SUP**:267(445)

Subramanian, C., **23**:340(268), 367(268)

Subramanian, C. S., **23**:327(221), 364(221)

Subramanian, K. N., **21**:287, 293, 345

Subramanian, K. R., **31**:342(25), 426(25)

Subramanian, R. S., **13**:164, 202

Subramanian, V. S., **3**:14, 26, 32; **11**:231, 232, 259

Subramaniyan, V., **5**:155, 157, 248

Sucio, S. N., **9**:3(112), 91(112), 97, 110

Sucker, D., **20**:377, 386

Suckow, W. H., **15**:62(4), 134(4); **19**:281(71), 284(71), 294(71), 352(71)

Suematsu, H., **17**:50(161), 63

Suetin, O. N., **6**:418(136), 492

Sugawara, A., **11**:360, 437

Sugawara, K., **28**:366(115), 369(115), 389(165, 168), 420(115), 422(165, 168)

Sugawara, M., **19**:38(84), 43(97), 79(213), 87(84, 97), 92(213)

Sugawara, S., **3**:14, 32; **4**:7, 9, 24, 26, 33, 62, 63; **9**:207(31), 210(31), 242, 269

Sugden, R. P., **5**:451, 515

Sugimoto, J., **29**:151(70), 165(116), 206(70), 208(116)

Sugino, E., **SUP**:71(157), 287(157), 288(157), 289(157), 297(157)

Sugiyama, H., **19**:74(172), 76(172), 77(172), 91(172)

Sugiyama, K., **20**:387; **21**:76(21), 77(21), 134(21); **27**:viii(4)

Sugiyama, S., **11**:399(161), 438; **15**:328

Suh, C. S., **5**:396(130A), 511

Suh, I.-S., **23**:254(124), 261(124), 272(124)

Suita, M., **25**:81(144), 146(144)

Suitor, J. W., **31**:433(16), 471(16)

Sukanek, P. C., **13**:219(40, 41, 48, 55), 221(55), 259, 264, 265; **30**:262(20), 308(20)

Sukhatme, S. P., **3**:109, 173; **9**:218, 220, 269; **11**:247, 248, 249, 261

Sukomel, A. S., **9**:120, 129(12), 178; **19**:144(157), 188(157); **31**:160(18), 275(178), 281(178), 323(18), 330(178)

Sulainistras, A., **29**:98(69), 126(69)

Sulilatu, W., **17**:218(67), 231(73), 250(73), 258(73), 259(73), 260(73), 290(67, 73), 291(99), 294(67, 73), 295(67, 73), 301, 302(99, 110), 315, 316, 317

Sullivan, F. W., **8**:140(71), 144(71), 160; **18**:139(72), 140(72), 146(72), 147(72), 159(72); **25**:292, 293, 315

Sullivan, G. A., **1**:376(29), 441

Sullivan, J. A., **5**:159(51), 160, 161, 164, 165, 248

Sullivan, J. S., **29**:299(157), 302(157), 343(157)

Sullivan, P. F., **26**:188(146), 190(146), 191(149), 193(146), 215(146, 149)

Sullivan, R. R., **25**:294, 319

Sultan, M., **16**:232
Sulzmann, K. G. P., **5**:309(75), 323
Sumarsano, M., **31**:388(39), 427(39)
Sumarsono, M., **27**:ix(19), 181(31)
Sumi, K., **10**:259, 282
Summerfield, M., **2**:372, 395; **6**:531, 562, 563
Summers, R. M., **29**:234(94), 340(94)
Summitt, R. L., **29**:234(85), 269(85), 339(85)
Sun, C.-S., **24**:193, 202, 214, 255, 263, 264, 269, 274
Sun, H., **26**:123(57, 60, 61), 125(60, 61), 126(57), 127(57, 60, 61), 128(57), 129(57, 60, 61, 75), 130(57, 60, 61, 75), 131(57, 61), 132(57), 133(60, 61), 155(61), 182(142), 183(142), 184(142), 185(142), 186(142), 211(57, 60, 61, 75), 215(142)
Sun, K., **11**:57, 58(22), 190
Sun, K. H., **17**:7, 57; **20**:250(231), 310(231)
Sun, Y. Z., **23**:307(127), 359(127)
Sun, Z. S., **15**:208(97), 209(97), 210(97), 225(97)
Sunada, D. K., **30**:101(37), 120(37), 121(37), 190(37)
Sundaram, C. V., **19**:178(231), 179(231), 190(231)
Sundaram, R., **2**:426(25), 450(25), 451
Sundaram, T. R., **3**:35(24), 36(24), 98
Sundararajan, T., **23**:202(112), 208(108, 109, 112), 210(112), 211(112), 212(108, 111, 112), 214(108), 237(111), 238(111), 239(111), 240(111), 244(110), 246(111), 247(110, 111), 248(111), 249(111), 250(110, 111), 251(111), 271(108, 109, 110, 111, 112); **26**:32(216), 36(214, 215, 216, 217), 37(216), 39(216, 217), 40(216), 42(215), 43(215), 59(73), 60(73), 63(73), 96(73), 102(214, 215, 216, 217)
Sundaresan, S., **30**:112(64, 65), 191(64, 65)
Sundarraj, S., **28**:263(73, 74, 75), 332(73, 74, 75)
Sunderland, J. E., **11**:220(126), 223(126), 237(126), 258, 411(184), 439; **13**:129(37), 202; **19**:4(13), 5(13), 12(20), 13(20), 18(13), 79(207), 84(13, 20), 92(207)
Sundheim, B. R., **12**:224(102), 247(102), 259(103, 104), 269(103, 104), 281
Sundquist, B. E., **14**:305, 306, 342, 343
Sundstrom, A., **24**:221, 272
Sundstrom, D. W., **18**:189(141), 228(127), 238(127), 239(141)

Sunkoori, N. R., **10**:173, 177, 179, 215
Suo, M., **17**:34, 61
Suo-Antilla, F. A., **29**:9(11), 55(11)
Suo-Antilla, A. J., **21**:301, 345
Suplin, V. Z., **25**:203(101), 247(101)
Suprum, V. M., **14**:158(60), 243
Suprun, N. M., **19**:100(23), 102(23), 185(23)
Suraiya, T., **15**:128(157), 131(157), 132(157), 141(157)
Surinov, Yu. A., **3**:196(20), 249
Surkov, G. A., **8**:75, 76, 77(277, 278), 91; **14**:152(26), 242
Surnov, A. V., **1**:399(80), 429(80), 442
Survila, V. J., **18**:135(70), 136(70), 142(70), 159(70)
Suryanarayana, C., **28**:8(9), 26(48), 70(9), 72(48)
Suryanarayana, N. V., **11**:105(67), 115, 120(67), 121(67), 122(67), 192
Susott, R. A., **10**:229(12), 281
Suter, A., **21**:287, 290, 300, 345
Sutera, S. P., **2**:94(77), 108; **3**:30, 32
Sutey, A. M., **7**:97, 98(7), 160
Sutherland, J. P., **10**:203(101), 217; **14**:232, 246
Sutherland, W. A., **3**:106(9), 172; **31**:250(146, 147), 251(146), 328(146), 329(147)
Sutterby, J. L., **24**:108(77), 187(77)
Sutterby, J. S., **15**:62(6), 135(6); **19**:251(9), 350(9)
Sutton, F. M., **14**:22, 23(58), 101
Sutton, G. W., **1**:77, 78, 121, 269(4), 316(53), 352, 354; **2**:202, 205(54), 206(54), 269
Sutugin, A. G., **14**:318(87), 344
Suwanprateep, T., **26**:203(176), 217(176)
Suzaki, K., **11**:213(75), 215(75), 255
Suzuki, K., **23**:12(61), 13(30), 14(61), 15(30), 16(30), 17(30), 23(30), 24(30), 25(30), 27(30), 32(61), 35(61), 39(61), 49(61), 53(61), 54(61), 58(61), 60(61), 89(61), 92(61), 106(61), 127(30), 129(61), 329(233), 365(233)
Suzuki, M., **26**:105(5), 123(5), 126(5), 208(5); **28**:364(102), 373(137), 385(137), 419(102), 421(137)
Suzuki, T., **28**:386(157), 387(157), 422(157)
Svalov, G. G., **20**:24(54), 27(54), 81(54)
Svehla, R., **28**:364(101), 419(101)
Svehla, R. A., **2**:37, 39(31), 106; **3**:263, 264, 265, 267, 269, 280, 300
Svenčianas, P. P., **8**:114(42), 115(42), 159
Svensson, H., **6**:162, 363

Svensson, U., **13**:94, 97, 116, 117
Sverdrup, H. U., **18**:339(54), 363(54)
Svetlova, L., **16**:235
Swalley, F. E., **5**:378(93), 509
Swan, D. W., **14**:120, 144
Swan, G. W., **15**:328
Swanberg, C. A., **14**:3(5), 100
Swank, L. R., **7**:348, 358(54), 361(54), 379
Swanson, J. L., **16**:234
Swanson, T. D., **26**:201(164), 203(173, 174), 216(164, 173, 174)
Swartz, J. A., **10**:275(106), 284
Swean, T. F., Jr., **25**:359, 415
Sweet, D. W., **29**:221(27), 233(68), 234(92), 321(92), 336(27), 338(68), 340(92)
Swensen, H. S., **7**:50
Swenson, D. V., **29**:194(203, 205), 212(203), 213(205)
Swenson, H. S., **6**:549(67), 551, 553, 563; **11**:170(176), 173, 176(176), 177(176), 185(176), 186(176), 197
Swift, D. L., **14**:158, 244
Swigart, R. J., **8**:13, 85
Swigert, P., **15**:156(17), 223(17)
Swindell, W., **30**:289(77), 310(77)
Swinney, H. L., **21**:147, 153, 154, 155, 178, 179, 180, 182
Swire, H. W., **11**:203, 206(56), 210(56), 220(56), 223(56), 235, 238(56), 248(56), 254
Swisher, S. D., **12**:24(44), 74
Swisher, S. E., **23**:349(304), 368(304)
Syamala Rao, C., **SUP**:310(472)
Sydoriak, S. G., **17**:146, 157, 321, 323, 326, 337, 339
Sykes, J. A., **15**:237, 278
Sykes, J. F., **30**:121(82), 181(156), 192(82), 196(156)
Sykes, R. J., **25**:375, 418
Sylvester, D. W., **12**:82, 112
Sylvester, N. D., **15**:101(113), 104(113), 139(113); **20**:84, 85, 92, 95, 132
Symons, J. G., **18**:17(33, 34), 80(33, 34)
Synge, J. L., **3**:183, 248
Synge, R. L. M., **12**:256(66), 280
Syrnikov, Yu., **14**:259(25), 278(25), 279(25), 280
Syromyatnikov, N. I., **9**:119(11), 178; **10**:208, 218; **19**:149(168), 150(168), 151(168), 153(168), 189(168)
Syutkin, S. V., **25**:153(9), 162(9, 44), 163(44), 164(9, 44, 47), 165(9, 44, 47), 166(49, 50, 51), 197(47), 201(44), 226(9), 243(9), 244(44, 47), 245(49, 50, 51)
Syverud, A. N., **29**:188(186), 212(186)
Szabo, M. T., **23**:227(244), 277(244)
Sze, N. D., **14**:267(47), 280
Szechenyi, E., **23**:280(9), 354(9)
Szego, J., **15**:38(60), 57(60)
Szekely, J., **14**:155(56), 157, 243; **19**:44(99), 80(238), 87(99), 93(238), 164(205), 171(205), 173(205), 190(205); **21**:280, 287, 299, 300, 312, 345; **28**:1(1), 10(12), 11(12, 13), 12(13), 13(13), 14(13), 15(13), 17(12, 13), 18(12, 13), 19(12), 23(12), 36(59), 39(59), 43(59), 46(59), 64(82), 69(59), 70(1), 71(12, 13), 72(59), 73(82), 238(8), 240(11), 241(11, 14), 250(30, 31, 32), 251(38), 275(38), 276(38), 317(172, 173), 318(182, 188, 190, 191, 196), 319(197, 198, 199), 329(8, 11, 14), 330(30, 31, 32, 38), 337(172, 173, 182, 188, 190), 338(191, 196, 197, 198, 199), 376(153, 154), 385(153, 154), 422(153, 154); **30**:333(53, 56), 334(53), 429(53, 56)
Szembek-Stoeger, M., **23**:230(264), 231(264), 278(264)
Szetcla, E. J., **5**:355(62), 507
Szetela, E. J., **6**:549(71), 550(71), 563
Szewczyk, A. A., **4**:16, 51, 52, 53, 63, 64; **9**:324(88), 348; **25**:282, 319
Szwarcbaum, G., **6**:474(207), 496
Szydlowski, W., **28**:190(61), 230(61)

T

Tabaczynski, R. J., **21**:246(20, 21), 263(20, 21), 276(20, 21)
Tabakoff, W., **26**:170(124), 214(124)
Tabbey, J., **9**:3(40), 91(40), 95(40), 97(40), 107
Tabor, D., **28**:127(97), 143(97)
Taborek, J. T., **15**:21(47), 56(47); **16**:149(131), 150(131, 133), 156; **26**:219(1), 324(1); **30**:202(58), 252(58); **31**:160(44), 324(44), 443(70), 450(71), 474(70, 71)
Tacconi, E. A., **17**:28(81), 59
Tachibana, F., **5**:74, 91, 126, 233(166), 234, 235(166), 251; **7**:129(38), 161; **11**:68(38), 69, 70, 83, 105(38), 112, 121(38), 122(38), 123, 190; **21**:163, 182
Tachibana, M., **30**:396(166), 400(166), 434(166)

Tacke, K.-H., **28**:318(184, 194), 337(184), 338(194)

Tadaki, J., **28**:369(122), 420(122)

Tadios, E. L., **29**:218(18, 19, 22), 219(18, 19), 228(18), 232(18), 233(18, 19), 234(22), 244(18, 22), 270(19), 275(22), 277(19), 279(18, 19), 280(22), 281(22), 288(18), 295(22), 296(22), 297(18, 22), 301(22), 305(19), 306(22), 307(22), 308(22), 310(22), 311(22), 312(18), 319(19, 22), 321(22), 322(19), 329(18), 335(18, 19, 22)

Tadmor, Z., **13**:261(120, 127), 266, 267; **28**:146(62), 149(62, 63), 150(62), 153(62), 184(62), 223(62), 224(63), 230(62, 63)

Tadrist, L., **26**:49(218), 50(218), 103(218); **30**:137(96), 193(96)

Tadros, Th. F., **23**:262(32a), 268(32a)

Taga, M., **23**:13(68), 15(68), 50(68), 106(68), 129(68)

Taghavi-Tafreshi, K., **19**:80(230, 231), 93(230, 231)

Tagle, J. A., **28**:75(96), 112(96), 143(96)

Tahir, A. E., **17**:37(125), 39(125), 62

Tahir, M. I., **25**:293, 316

Tai, Y. C., **28**:99(98), 143(98)

Taikeff, E. A., **2**:38(36), 106

Tait, R. W. F., **18**:167(47), 237(47)

Tait, S., **28**:282(116), 334(116)

Taitel, Y. (Chapter Author), **9**:265, 272; **11**:363(107), 437; **SUP**:112(283), 114(283); **15**:232, 240, 242(52), 277, 278; **20**:83, 84, 85, 89, 90, 93, 95, 99, 108, 111, 112, 115, 119, 120, 121, 122, 123, 128, 130, 132; **26**:49(204), 102(204)

Tajima, O., **11**:220, 224, 237, 263

Takagaki, T., **28**:402(216), 425(216)

Takagi, M., **25**:358, 416

Takahashi, K., **11**:420(206), 439; **25**:258, 283, 285, 319; **29**:162(100), 207(100)

Takahashi, M., **29**:162(100), 207(100); **31**:450(49), 473(49)

Takahashi, P. K., **31**:436(52), 448(52), 460(52), 473(52)

Takahashi, R., **28**:366(115), 369(115), 389(168), 420(115), 422(168)

Takahashi, R. K., **23**:327(217, 218), 364(217, 218)

Takahasi, K., **28**:342(26), 416(26)

Takakura, N., **28**:343(50), 417(50)

Takamatsu, T., **26**:14(219), 103(219)

Takami, H., **6**:2, 122

Takano, K., **21**:128(123), 138(123)

Takano, T., **29**:151(69), 205(69)

Takao, K., **7**:193, 214

Takara, E., **29**:153(78), 206(78)

Takashima, M., **14**:126, 147

Takashima, Y., **7**:181(50), 213

Takata, A. N., **22**:275(203), 330(202), 331(202), 332(202), 356(202, 203)

Takata, K., **9**:224(54), 269; **10**:100, 101, 104, 164; **11**:8, 48

Takegawa, T., **17**:7(36), 57

Takenchi, M., **20**:209(113), 304(113)

Takens, F., **25**:356, 417

Taketani, T., **21**:129(125), 130(126, 127), 139(125, 126, 127)

Takeuchi, D. I., **26**:170(127), 214(127)

Takeuchi, M., **11**:213(75), 215(75), 255

Takeuti, Y., **15**:328

Takeyama, T., **21**:85(47), 91(61), 125(115), 135(47), 136(61), 138(115); **23**:12(58, 59), 14(58, 59), 31(58, 59), 34(58, 59), 42(58, 59), 58(58, 59, 94), 60(58, 59), 61(94), 77(94), 87(58, 59), 91(58, 59), 92(59), 103(58, 59), 106(94), 109(58, 59), 110(58, 59), 117(58, 59), 129(58, 59), 131(94)

Takhar, H. S., **9**:295(22), 346

Takigawa, T., **21**:174, 180

Takoudis, C. G., **28**:365(106), 386(158), 388(158), 402(203), 419(106), 422(158), 424(203)

Takserman, U., **23**:219(68), 222(68), 270(68)

Takserman-Krozer, R., **13**:219(73, 78), 221(78), 265

Takuda, H., **23**:9(20), 109(20, 115, 117, 118), 110(20, 115, 117, 118), 113(20, 115, 117), 115(20, 115, 117), 116(118), 119(20), 127(20), 132(115, 117, 118)

Tal, R., **26**:45(220, 221), 103(220, 221)

Talbot, L., **1**:347, 354; **2**:355

Taliaferro, S., **25**:262, 318

Talintsev, S. N., **25**:4(22), 140(22)

Talley, D. G., **26**:69(222), 103(222)

Talmor, E., **6**:390(16), 405, 486; **7**:348, 355, 379

Talty, R. D., **1**:426, 443

Talwalkar, A., **19**:108(57), 186(57)

Tam, C. K. W., **23**:207(245), 277(245)

Tam, H. S., **22**:30(32), 134(32), 153(32)

Tamamushi, R., **12**:250(106), 251(106), 281

Tamarin, A. I., **14**:155, 160(79), 161(79),

167(79), 169(79), 176(177, 178), 232(79), 233(79, 173), 236(181), 237(79, 177, 178), 243, 244, 246, 247, 335, 335(173); **19**:148(170), 153(170), 154(170), 189(170)

Tambovtsev, Yu. I., **25**:166(59), 245(59)

Tamir, A., **9**:265(145), 272; **SUP**:112(283), 114(283); **15**:229(10, 12, 13), 230, 232, 235(12), 237, 238(13), 240, 242, 276, 277, 278

Tamm, H., **29**:69(21), 71(33), 84(48), 110(21), 124(21), 125(33, 48), 307(168, 169), 344(168, 169)

Tamonis, M. M., **8**:114(42), 115(42), 159

Tamura, M., **31**:334(13), 425(13)

Tan, C. W., **SUP**:118(289)

Tan, H. M., **19**:334(145), 337(145), 355(145)

Tan, M. J., **29**:168(126), 172(126), 208(126)

Tan, Z., **31**:395(51), 427(51)

Tanaguchi, H., **23**:161(46), 186(46)

Tanaka, H., **6**:546, 563; **7**:34, 84; **15**:166(25, 26, 27), 167(25, 26, 27, 28), 223(25, 26, 27), 224(28); **18**:73(144), 86(144); **20**:186(46), 187(46), 191(46), 301(46); **21**:4(10), 34(52), 35(52), 36(52), 51(10), 52(52), 73(19), 74(19), 79(19), 83(40, 41, 41), 84(43, 44), 88(57), 89(41, 41, 57), 90(41, 41), 92(43, 44, 69, 70, 71), 96(43, 44, 69, 70, 71), 134(19), 135(40, 41, 41, 43, 44), 136(57, 69, 70, 71); **25**:51(84), 143(84)

Tanaka, K., **20**:55(59), 58(59), 70(62), 82(59, 62)

Tanaka, M., **8**:232(4), 243(4), 282; **26**:45(223, 224), 103(223, 224)

Tanaka, O., **20**:181(1), 299(1)

Tanaka, S., **20**:283(304), 313(304); **21**:125(116), 138(116); **23**:340(269), 341(269), 367(269)

Tanaka, T., **11**:203, 206, 254; **25**:51(87), 144(87)

Tanaka, Y., **23**:109(118), 110(118), 116(118), 132(118)

Tananaiko, I. M., **20**:42(58), 43(58), 49(58), 82(58)

Tanasawa, I. (Chapter Author), **21**:55, 80(28), 82(28), 84(42), 85(48), 86(49), 91(60, 62), 96(75), 97(75), 100(42), 128(123), 134(28), 135(42, 48, 49), 136(60, 62, 75), 138(123); **29**:131(14), 164(106), 203(14), 207(106); **30**:393(158), 396(158), 399(158), 434(158)

Tanasawa, Y., **23**:94(103), 131(103)

Tanazawa, Y., **7**:88(5b), 160

Tanes, M. Y., **16**:214(23), 215(23), 232, 238

Tang, H. H., **6**:17, 18, 25(27), 27(27), 123

Tang, H. T., **24**:102(9, 10), 184(9, 10)

Tang, J., **29**:175(154), 181(154), 210(154)

Tanger, G. E., **16**:214(24), 216(24), 233, 238

Tangerman, F. M., **30**:356(104), 431(104)

Tangroth, B., **22**:347(12)

Tani, I., **1**:103, 122; **4**:334, 445; **6**:69(119), 109, 128, 130

Tanigawa, Y., **15**:328

Taniguchi, H. (Chapter Author), **27**:viii(1, 2, 3, 4, 5, 6, 7), ix(8, 9, 10, 11, 13, 14, 15, 17, 18, 19), 46(6, 7), 146(28, 29), 147(28), 167(30), 174(9), 181(31, 32), 186(32), 187(36, 37, 38), 193(37, 38, 39, 40, 41); **31**:334(11, 14, 16), 335(18), 365(11), 388(39, 40), 391(48), 413(48), 418(48), 425(11, 14, 16, 18), 427(39, 40, 48)

Taniguchi, K., **27**:viii(4)

Taniguchi, N., **30**:349(155), 390(155), 433(155)

Tanimoto, A., **SUP**:12(20, 23), 13(20), 137(20), 188(23), 189(20, 23)

Tanimoto, S., **8**:299(40), 305, 349

Tanneberger, H., **7**:8(3), 83

Tannehill, J. C., **24**:194, 268

Tanner, D. W., **9**:209, 210(34), 221(48), 233(34), 244(48), 264(130), 265, 269, 271; **21**:82(33), 83(38), 88(54), 91(33), 95(33), 135(33, 38, 54)

Tanner, J. M., **4**:94(34), 138

Tanner, L. H., **6**:201(46), 364

Tanner, R. I., **2**:361, 396; **15**:77(65), 87(65), 89(65), 137(65), 211(104), 225(104); **23**:188(247), 189(247), 190(246), 208(89a), 219(268), 245(89a), 246(89a), 270(89a), 277(246, 247), 278(268); **24**:107(75), 187(75); **25**:256, 319

Tantam, D. H., **9**:405(94), 417

Tanzawa, T., **30**:295(94), 311(94)

Tanzer, H., **20**:286(310), 287(310), 313(310)

Tao, L. C., **19**:31(63), 86(63)

Tao, L. N., **6**:474(237, 263), 484(237), 497, 499; **8**:42, 43, 44(216, 217, 218), 89; **10**:256, 282; **SUP**:63(79, 82, 83, 84), 75(205), 82(79), 102(205), 158(79), 224(79), 225(79), 247(79), 249(79), 251(205), 262(82), 277(82, 83), 278(84), **19**:12(23, 24, 25), 13(24, 25), 19(25), 84(23, 24, 25), 202(57), 245(57); **20**:187(58), 301(58)

Tarapore, E. D., **28**:318(189), 337(189)

Tarasov, A. I., **25**:127(197), 149(197)

Tarasova, N. V., **6**:551, 554, 563, 564
Tarbell, W. W., **29**:216(5, 6, 7, 8, 11), 217(11, 14), 221(25, 26, 27), 233(26, 67), 249(100), 251(100), 260(116), 263(116), 264(116), 268(134), 269(5), 270(100), 271(25, 137), 276(147, 148), 278(148), 303(26), 309(25, 67), 315(172), 323(172), 329(147), 334(5, 6, 7, 8), 335(11, 14, 25), 336(26, 27), 338(67), 340(100), 341(116), 342(134, 137), 343(147, 148), 344(172)
Tarbuck, K. J., **23**:217(222), 276(222)
Targ, S. M., **SUP**:71(159), 87(159), 298(159)
Tarifa, S. C., **10**:261, 272, 283
Tarleev, P. N., **25**:4(11), 140(11)
Tarleyev, P. N., **31**:160(9), 165(9), 182(9), 322(9)
Tarshis, L. A., **11**:318(5), 412(5, 197, 200), 413(5), 415(197), 417(5), 434, 439
Tarunin, E. L., **8**:168, 173, 226
Tasdemiroglu, E., **18**:73(140), 75(140), 86(140)
Taslim, M. E., **20**:328(38), 351(38); **23**:330(235), 365(235)
Tassart, M., **23**:206(24), 218(24), 237(24), 239(24), 240(24), 242(24), 244(24), 267(24)
Tassopoulos, M., **30**:136(95), 193(95)
Tatchell, D. G., **9**:99(43a), 100(43a), 108; **SUP**:73(177), 209(177), 211(177), 212(177), 398(177); **19**:268(57), 269(57), 271(57), 272(57), 273(57), 274(57), 276(57), 351(57)
Tate, G. E., **15**:102(117), 139(117)
Tate, G. N., **6**:531, 532, 562
Tate, N. A., **25**:319
Tatom, J. W., **2**:434(38), 451; **5**:381, 390, 509, 510; **6**:474(252), 483(252), 484, 498
Tattersall, R. B., **29**:165(111), 207(111)
Tauber, R. N., **28**:98(107), 143(107)
Taunton, J. W., **14**:28, 101; **20**:323(20), 350(20)
Taussig, R. T., **18**:41(82), 83(82)
Taussky, O., **5**:10, 11(11), 50
Tavoularis, S., **21**:204(31, 32), 206(31, 33), 207(31, 33), 208(31, 32), 209(31), 222(32), 223(32), 237(31, 32, 33)
Tay, A. O., **SUP**:77(213), 185(213)
Taylor, A. B., **19**:6(16), 9(16), 82(266), 84(16), 95(266)
Taylor, G., **5**:110, 121, 128
Taylor, G. (F. R. S.), Sir, **20**:100, 101, 130
Taylor, G. I., **5**:130, 204, 228, 246, 249; **8**:286(1), 348; **13**:66(4), 114; **16**:7, 9(5),

10(5), 11(5), 53(5), 56; **20**:105, 132; **21**:143, 146, 148, 149, 151, 161, 182; **26**:14(225), 15(225), 18(225), 103(225); **29**:169(131), 209(131); **30**:6(84), 91(84)
Taylor, G. Y., **6**:504, 561
Taylor, H. S., **2**:355
Taylor, J. F., **3**:8(4), 9(4), 31; **7**:119(32), 161; **11**:219(97), 222(97), 229(97), 231(97), 232(97), 236(97), 256
Taylor, J. H., **5**:130(2), 141, 149, 150, 155, 159(2), 240, 241, 242, 246, 251
Taylor, J. J., **26**:137(86), 212(86)
Taylor, J. K., **6**:293, 366
Taylor, M. F., **5**:363, 508; **6**:538, 539, 563; **19**:324(113), 354(113)
Taylor, M. H., **3**:269(24), 300
Taylor, M. J., **22**:187(204), 295(204), 356(204)
Taylor, R. B., **4**:214(98a), 227
Taylor, R. J., **31**:447(72), 474(72)
Taylor, R. L., **2**:127(16), 131(16), 134(16), 268; **28**:41(66, 68), 73(66, 68), 342(18), 348(18), 350(18), 416(18)
Taylor, R. P., **23**:284(38), 314(139), 355(38), 360(139)
Taylor, T. D., **24**:194, 210, 213, 272; **26**:5(226), 6(226), 19(226), 21(226), 34(5), 93(5), 103(226)
Taylor, T. H. M., **1**:357(1), 371(1), 440
Taylor, W. J., **2**:327(57), 330(57), 354
Tchen, C. M., **9**:135, 179
Teagan, W. P., **7**:174(36), 182(36), 212
Teal, G. K., **30**:313(9), 314(10), 426(9, 10)
Teare, J. D., **2**:127(15, 16), 131(15, 16, 17), 134(15, 16, 17), 268
Tebenekhin, E. F., **31**:458(39), 472(39)
Teckchandani, L., **14**:17(47), 47, 49(47), 50(47), 101
Teeter, B. C., **15**:287(173), 328
Teilsch, H., **7**:8(3), 83
Teissèdre, C., **25**:345, 418
Teixeira, J. C., **21**:300, 345
Telgenkamp, J. A. H., **13**:248(108), 266
Telionis, D. P., **23**:280(18), 326(18), 337(18), 354(18)
Tell, E. N., **22**:322(151), 323(151), 324(151), 325(151), 326(151), 353(151)
Tellup, D. M., **5**:390, 391(120), 510
Temkin, A. G., **6**:412, 416(86, 90), 489, 490
Temperton, C., **13**:112(58), 117
Temple, E. B., **6**:202, 364

Teng, J. T., **15**:128(154), 141(154)

Tennekes, H., **21**:263(34), 265(34), 276(34); **30**:5(86), 7(85), 9(86), 49(85), 91(85, 86), 289(78), 310(78)

Teodovair, P., **23**:217(173), 274(173)

Teoman, Y., **19**:99(15, 17), 100(15, 17, 31, 32), 101(17), 102(32), 105(17, 32), 107(17, 32), 108(17, 32), 110(65), 164(213), 165(213), 169(213), 170(213), 171(213), 180(17, 32), 185(15, 17, 31, 32), 186(65), 190(213)

Tepper, K. A., **15**:47(73), 57(73)

Teptin, G. M., **14**:80, 262(30), 273(30)

Teraoka, K., **17**:7(37), 57

Terashima, K., **30**:344(86), 430(86)

Terbot, J. W., **20**:76(69), 82(69)

Ter Haar, L. W., **11**:203(38), 205(38), 209(38), 225(38), 253, 254

Ter-Mkrtchyan, A. A., **6**:376(1), 486; **25**:10(36), 141(36)

Ternovskaya, A. N., **14**:173(90), 176(90), 244; **19**:120(103), 187(103)

Terranova, B. E., **25**:235(142), 249(142)

Terrell, J. P., **23**:296(80, 81, 82), 297(80), 298(81), 352(81, 82), 357(80, 81, 82)

Terrill, R. M., **8**:11(33), 85

Test, F. L., **23**:333(246), 365(246)

Tester, J., **11**:412(185), 439

Tetervin, N., **2**:6(12), 105

Teuscher, K. L., **26**:188(154), 192(154), 215(154)

Tewari, P. K., **30**:126(89), 192(89)

Tewari, S. S., **20**:243(172), 307(172)

Tewfik, O. E., **3**:75, 99; **7**:202, 216

Thakur, S. S., **24**:226, 227, 231, 238, 239, 240, 242, 244, 274

Thal-Larsen, H., **6**:474(305), 501

Thames, F. C., **19**:50(119), 57(119), 88(119); **26**:45(227), 103(227)

Thauer, R., **4**:121(101), 140

Thelen, E., **21**:102(80), 103(80), 136(80)

Theoclitus, G., **20**:150(15), 179(15)

Theodorsen, T., **5**:149, 153(44), 156, 167, 247; **21**:171, 182

Theofanous, T. C., **17**:2(12), 40(12), 56

Theofanous, T. G. (Chapter Author), **15**:243, 278; **20**:250(220), 310(220); **29**:20(15), 56(15), 62(7), 72(7), 124(7), 129, 131(11, 12), 147(48), 150(60, 62, 63), 151(64), 153(78, 79, 80, 81), 157(79, 108), 164(108), 169(128), 170(128, 135), 171(128, 135, 136, 137), 174(108), 180(164, 165, 166, 167, 168), 181(108, 164), 182(166, 167, 168), 191(200), 193(202), 194(48, 62, 206, 207), 195(60, 167), 199(164), 200(108, 168), 201(168), 203(11, 12), 205(48, 60, 62, 63, 64), 206(78, 79, 80, 81), 207(108), 208(128), 209(135, 136, 137), 210(164, 165, 166), 211(167, 168), 212(200, 202), 213(206, 207), 218(23), 232(23), 233(23), 234(23, 72), 239(23), 241(23, 72), 248(23), 249(23), 251(23), 274(23), 275(23, 72), 276(23), 279(23), 295(23), 301(23), 311(23, 72), 319(23), 335(23), 338(72)

Theofilis, V., **25**:364, 412

Thevoz, P., **28**:263(77), 327(77), 332(77)

Thibault, J., **23**:305(116), 359(116)

Thibault, P. A., **29**:98(68, 69), 126(68, 69)

Thielbahr, W. H., **16**:280, 365

Thinakaran, R., **23**:216(79), 270(79)

Thinnes, G. L., **29**:215(1), 248(1), 334(1)

Thirriot, C., **18**:171(119), 223(119), 225(119), 238(119)

Thirsk, H. K., **12**:242(22), 278

Thiruvenkatachar, V. R., **8**:78(283), 91; **15**:320, 328

Thiry, F., **6**:73, 80(129), 128

Thodes, G., **25**:192(91), 246(91)

Thoenes, D., **25**:176(71), 246(71)

Thom, A., **11**:216, 262

Thoma, H., **8**:94(5), 158; **18**:87(5), 157(5)

Thomann, H., **6**:82(137), 128; **16**:319, 365

Thomas, A. G., **3**:51(56), 59(56), 98

Thomas, D., **11**:105(13), 111, 194

Thomas, D. B., **8**:299(39), 349; **30**:26(87), 91(87)

Thomas, D. G., **2**:364, 373, 382(106), 384, 387, 392(105), 396, 397; **9**:132, 134, 179; **21**:124(112), 138(112)

Thomas, E., **15**:287, 327

Thomas, G. M., **5**:309(80), 324

Thomas, G. O., **29**:98(66), 126(66)

Thomas, G. P., **29**:98(69), 126(69)

Thomas, G. R., **29**:216(4), 258(4), 260(4), 262(4), 264(4), 334(4)

Thomas, J. R., **15**:287, 328

Thomas, K. D., **15**:267(100), 269, 280

Thomas, L. B., **2**:295, 326, 327, 337, 355; **7**:168, 211

Thomas, L. K., **14**:86(174), 87(174), 88, 104

Thomas, M., **3**:217(63), 250

Thomas, M. A., **9**:232(65), 270; **SUP**:412(525)

Thomas, P. D., **1**:12, 49; **5**:348, 507; **24**:207, 274

Thomas, P. H., **10**:221, 243, 259, 281, 282

Thomas, R. A., **5**:177, 249

Thomas, R. B., Jr., **1**:350(81), 354

Thomas, R. N., **8**:273(44), 283

Thomas, R. W., **21**:161, 182

Thomas, S., **23**:193(248), 277(248); **26**:201(164, 165), 202(165), 203(169), 216(164, 165, 169)

Thomas, T. H., **3**:186, 249

Thomasson, R. K., **6**:474(291), 500

Thome, J. R. (Chapter Author), **16**:59, 72, 73, 74, 80, 89, 91, 92, 93, 95, 97, 98, 99, 100, 102, 103, 105, 107, 109, 110(43, 67), 117(40, 83), 120, 121, 122, 123, 124, 125, 126, 127, 150(132), 153, 154, 155, 156

Thomke, G. J., **6**:9(20), 29(20), 31(20), 33, 34, 48(56), 52(56), 77(56), 96, 123, 131

Thompson, A., **5**:287(29), 322

Thompson, B., **17**:26(79), 27, 28, 45, 59

Thompson, C. F., **24**:247, 274

Thompson, D. R., **19**:81(251), 94(251)

Thompson, G. E. K., **14**:14(35), 100

Thompson, H. D., **24**:218, 270

Thompson, J. C., **20**:269(275), 270(275), 312(275)

Thompson, J. F., **5**:351(57), 380, 383(57), 386(57), 387(57), 388(57, 113), 507, 510; **19**:50(119), 57(119), 88(119); **24**:213, 274; **26**:45(227), 103(227)

Thompson, J. L., **23**:222(192), 275(192)

Thompson, K. F., **22**:239(205), 356(205)

Thompson, L. B., **29**:299(157), 302(157), 343(157)

Thompson, L. R., **28**:343(44, 45, 46), 417(44, 45, 46)

Thompson, M. C., **24**:218, 275

Thompson, M. E., **28**:240(11), 241(11, 14), 329(11, 14)

Thompson, M. O., **28**:76(55), 80(31, 100), 96(42), 139(31), 140(42), 141(55), 143(100)

Thompson, R. H., **4**:120(88), 140

Thompson, R. T., **29**:299(157), 302(157), 343(157)

Thompson, S. L., **29**:234(94), 340(94)

Thompson, W. D., **4**:255(19), 313

Thompson, W. G., **18**:257(39), 261(39), 295(39), 298(39), 321(39)

Thompson, W. G., Jr., **23**:9(27), 127(27)

Thompson, W. P., **2**:164(33), 269; **23**:279(1), 312(1), 315(1), 354(1)

Thompson, W. R., **5**:359, 360, 508

Thomsen, E. G., **1**:357(1), 371(1), 440

Thomsen, S., **22**:319(206), 356(206)

Thomson, D. J., **25**:361, 375, 415

Thomson, J., **24**:30(59), 38(59)

Thomson, W., **9**:192, 268; **14**:290, 341

Thomson, W. J., **14**:202, 245

Thonglimp, V., **19**:108(59), 186(59)

Thorgerson, E. J., **17**:32(105), 61

Thornton, G. K., **17**:146, 157

Thornton, J. D., **1**:384(50), 441; **14**:124, 125, 133, 147; **26**:14(22, 23), 93(22, 23)

Thornton, P. R., **5**:134(26), 247; **8**:163, 226; **9**:20, 63, 110

Thorp, A. G., II, **1**:408(88), 411(88), 442

Thorpe, G. R., **23**:427(82), 464(82)

Thors, P., **30**:222(45), 223(45), 224(45), 225(45), 251(45); **31**:468(59), 473(59)

Threlfall, D. C., **30**:28(88), 91(88)

Thring, M. W., **17**:187, 192(42), 315

Thring, R. H., **19**:164(208), 166(208), 170(208), 172(208), 175(208), 190(208)

Throne, J. L., **15**:60(1), 134(1); **18**:228(125), 238(125)

Throne, W. L., **17**:39(128), 62

Thun, R. E., **20**:244(198), 308(198)

Thurnay, K., **29**:155(83), 206(83)

Thwaites, B., **4**:331, 334, 444, 445

Thyagaraja, A., **29**:150(58, 59), 205(58, 59)

Thynell, S. T., **31**:343(27), 426(27)

Tichacek, L. J., **16**:107, 154

Tichonov, A. I., **21**:17(32), 51(32)

Tidwell, V. C., **30**:96(31), 190(31)

Tiedt, W., **SUP**:289(456), 322(456), 324(456), 325(456), 326(456, 484), 328(456)

Tien, C., **2**:376, 390, 397; **SUP**:139(321); **15**:157(18), 162(18, 22), 165(18), 168(22, 29), 175(34), 176(34), 180(44), 208(97), 209(97), 210(97), 223(18, 22), 224(29, 34, 44), 225(97, 99, 100), 327, 328; **19**:44(102), 87(102), 260(21), 268(40), 269(40), 270(40), 272(40), 275(40), 276(40), 279(40), 281(40, 70, 78), 284(70), 285(40, 78), 293(70), 300(40, 78), 350(21), 351(40), 352(70, 78); **25**:234(119, 120, 121, 122, 123), 248(119, 120, 121, 122, 123), 259, 264, 265, 266, 268, 286, 315, 318; **30**:94(13), 169(13), 189(13)

Tien, C. L. (Chapter Author), **1**:230, 231, 265; **2**:388(119), 397; **5**:132(17), 132(19), 135, 141, 149, 150(28), 167, 168, 206(6), 247, 248, 254, 254(6), 284, 293(38), 298(51), 299, 308(51), 311(52), 312(6), 314(9), 316(9), 322, 323, 479, 517; **7**:273, 317; **8**:230, 231(1), 245, 256, 282; **9**:121, 122, 127(21), 129(17, 21), 167, 172(17), 178, 180, 350, 351(8), 357(13), 358(13), 361(13, 20), 362(24), 363(25, 27), 364(25, 31, 32, 33), 375(43), 377(46), 379(46, 52), 382(20), 384(13, 20), 385(31, 56), 386(31, 32), 387(33), 388(59), 390, 391(59, 60, 68), 393(64, 65, 68), 394(56, 69, 72), 395(64), 397(56), 398(70, 72), 399(33), 414, 415, 416; **11**:12, 48, 203(46), 206(46), 254, 278, 280, 314, 397(157), 438; **12**:117(1), 119, 120, 129, 134(1), 138(17), 143, 147(17), 177, 192, 193; **17**:51, 52, 64, 204, 218, 219, 236(68), 315; **18**:36(73), 38(78), 39(78), 58(109), 60(109), 83(73, 78), 84(109); **20**:317(15), 350(15); **21**:263(33), 276(33), 328, 345; **23**:136(7), 138(11), 139(18), 140(18), 153(11, 32, 33, 35), 154(37, 38, 40, 42), 155(43), 161(43, 44), 166(43), 167(43), 168(38, 43, 48), 170(18), 184(7, 11), 185(18, 32, 33, 35, 37, 38, 40, 42), 186(43, 44, 48); **24**:106(41), 186(41); **27**:ix(16); **28**:77(80), 142(80); **29**:39(30), 56(30); **30**:102(43), 142(43), 143(43), 145(43), 147(43, 107), 173(146), 190(43), 193(107), 195(146)

Tien, L. C., **19**:20(38), 85(38)

Tien, R. H., **19**:79(206), 92(206)

Tien, Y. C., **26**:123(60), 125(60), 127(60), 129(60), 130(60), 133(60), 211(60)

Tier, W., **11**:420(208), 439

Tierney, J. K., **20**:187(68), 302(68)

Tierney, J. W., **23**:217(223), 276(223)

Tieszen, S. R., **29**:102(77), 107(80), 127(77, 80)

Tifford, A. N., **2**:168, 269; **5**:167, 248; **8**:36, 88

Tikhonov, A. M., **31**:160(22), 236(22), 323(22)

Tiller, W. A., **15**:328; **30**:329(39), 428(39)

Tillman, W., **2**:16, 105

Tills, J. L., **29**:234(73, 76), 276(147), 282(76), 288(76), 313(76), 329(147), 338(73), 339(76), 343(147)

Tilton, L. O., **15**:101(115), 102(115), 103(115), 104(115), 139(115)

Tilton, L. W., **6**:293, 366

Timmerhaus, D., **26**:227(28), 252(28), 325(28)

Timmerhaus, K. D., **5**:326(2), 327(16), 451, 505, 515; **9**:360(19), 414

Timmeriaos, K. D., **5**:451, 515

Timmers, J. F., **23**:329(231), 365(231)

Timmons, D. H., **11**:397(159), 401(159, 166), 403(166), 438

Timofeeva, F. A., **6**:474(211), 477, 496

Timoshenko, S. P., **SUP**:62(72, 75), 67(72), 197(72), 264(72), 272(72)

Timson, W. J., **5**:423, 512; **11**:105(81), 112(81), 192

Tinaut, D., **18**:73(140), 75(140), 86(140)

Tindall, J. W., **5**:24, 51

Ting, L., **7**:197(89, 103), 215

Ting, R. Y., **15**:92(94), 97(94), 138(94)

Tingwaldt, C., **12**:129, 192

Tinney, E. R., **10**:269, 283

Tippets, F. E., **1**:240(80, 81), 241, 242, 243(82), 244, 245(80, 82), 265

Tips, J. H., **22**:266(134, 135, 222), 268(134, 135), 270(135), 271(135), 272(135), 273(134, 222), 274(134), 275(134), 276(134), 353(134, 135), 357(222)

Tironi, G., **7**:172, 175, 212

Tirrell, M., **23**:193(8), 267(8)

Tirski, G. A., **5**:191, 249; **8**:19, 86

Tirunarayanan, M. A., **11**:213(74), 214(74), 255; **SUP**:201(390, 391), 228(391)

Tishchenko, A. T., **14**:176(91), 244

Titomanlio, G., **18**:234(128), 238(128)

Tittle, C. W., **15**:285, 298, 321, 328

Titulaer, C., **10**:19(28), 36

Tiu, C., **19**:262(25, 26), 263(25), 264(25, 26), 265(25), 326(25), 330(25), 350(25, 26); **23**:196(148), 197(148), 198(142, 144, 145, 147, 148, 149), 201(148), 203(144, 249), 204(148), 206(142, 144, 145, 147, 148, 149), 215(251), 216(95), 218(250), 219(250), 220(250), 222(95), 223(142), 226(95, 142, 144, 145, 147, 148, 149), 228(142), 229(95, 142, 144, 145, 147, 149), 271(95), 273(142, 144, 145, 147, 148, 149), 277(249, 250), 278(251); **24**:105(38), 186(38)

Tiwari, S. N. (Chapter Author), **5**:298(54), 323; **8**:229, 248(29), 252(33), 253(33, 35), 256(35), 258(33), 259(33), 260, 263(39), 267(32), 276(48), 283; **12**:119, 120, 143(4), 192; **31**:391(46, 47), 427(46, 47)

Tjerkstra, H. H., **17**:128, 156
Tjoa, R., **14**:325(102), 344
Tjoan, L. T., **11**:203(38), 205(38), 209(38), 225(38), 253, 254
Tjoe, T., **31**:431(73), 474(73)
Tkachenko, G. M., **9**:71(117), 103, 104, 110, 111
Tkachuk, A. Y., **31**:238(125), 327(125)
Tleimat, B. W., **30**:240(59), 252(59)
To, W. M., **18**:39(79), 83(79)
Tobias, C. W., **4**:16(25), 63; **7**:89(6), 96(6), 98(6, 11), 100, 130(40), 160, 161; **12**:233(3, 28, 78, 105), 237(78), 238(68, 79), 251(3, 28, 79, 105, 117, 118), 269(78), 270(3, 28, 79, 105), 277, 278, 280, 281, 282; **18**:162(15), 179(15), 181(66), 182(68), 183(15), 184(15), 185(66), 186(15), 187(15, 68, 70), 188(15), 190(15), 236(15), 237(66, 69, 70); **21**:170, 171, 172, 174, 179
Tobilevich, N. Y., **16**:127, 155; **18**:255(21), 257(21), 263(21), 321(21)
Tobin, R. D., **9**:394(75), 395(75), 416
Tochitani, Y., **26**:49(228, 229), 51(228, 229), 103(228, 229)
Točigin, A. A., **1**:398(79), 399(79), 442
Toda, S., **19**:74(172), 76(172), 77(172), 91(172); **23**:11(28, 29), 13(28), 15(28), 20(28), 35(28), 36(28), 39(28), 45(28), 127(28, 29)
Todd, D. B., **28**:190(64), 230(64)
Todd, J., **5**:10, 11(11), 40(74), 50, 54
Todd, M. R., **28**:250(32), 330(32)
Todes, D. M., **19**:101(33), 102(45), 185(33, 45)
Todes, O. M., **14**:247; **25**:157(18), 223(110), 243(18), 247(110)
Todisco, A., **6**:41(77, 78), 65(77, 78), 126
Todsen, M. T., **14**:30(71), 101
Togashi, S., **21**:88(58), 89(58), 136(58)
Toh, K., **30**:396(165), 400(165), 434(165)
Tohda, Y., **24**:163(147), 190(147)
Tohuda, N., **6**:396(30), 474(30), 476(30), 487
Tokarenko, A. V., **15**:325; **19**:202(59), 245(59)
Tokarenko, V. F., **25**:73(115), 145(115)
Tokarev, V. N., **28**:102(99), 143(99)
Token, K., **20**:286(309), 313(309)
Tokihiro, Y., **24**:3(14), 36(14)
Tokuda, N., **15**:242(59a), 278; **19**:12(21), 32(21), 84(21)
Tokuda, S., **SUP**:413(530); **30**:28(46), 29(46), 89(46)
Tokura, I., **19**:81(249), 94(249)
Tokuyama, K., **17**:115(105), 116(105), 155
Tolansky, S., **6**:135(6), 195, 363
Tolchinskij, A. R., **26**:314(86), 328(86)
Tollmien, W., **13**:30, 35, 60
Tolman, R. C., **9**:196(11), 268
Tolubinski, E. V., **8**:35, 88
Tolubinskiy, V. I., **16**:114(71), 116(71), 155; **18**:257(43), 259(43), 260(43), 265(43), 266(43), 321(43)
Tolubinskiy, V. J., **16**:95, 96, 97, 100, 101, 102, 114(74), 117(54), 125, 154, 155
Tom, H. W. K., **28**:117(27), 139(27)
Tomboulides, A. G., **25**:394, 418
Tomita, Y., **2**:390, 391, 397; **12**:80, 112; **25**:272, 273, 319
Tomlinson, J. N., **18**:220(118), 238(118)
Tomotika, S., **20**:360, 388
Tompkins, F. C., **2**:353
Tomzig, E., **30**:382(149), 433(149)
Tonaka, N., **12**:250(106), 251(106), 281
Tong, A. Y., **26**:67(230, 231, 232, 233, 234), 78(232, 233), 89(10), 93(10), 103(230, 231, 232, 233, 234)
Tong, L. S., **1**:408(88), 411(88), 442; **5**:102, 109, 127, 128, 404, 511; **7**:50; **11**:170(174), 173(174), 175(174), 176(174), 177(174), 185(174), 186(174), 196, 197; **17**:2(4), 30(95), 40, 46, 47, 56, 60, 62; **20**:250(212), 309(212); **29**:148(51), 166(51), 205(51)
Tong, S. S., **24**:199, 200, 201, 206, 210, 213, 274
Tong, T. W., **23**:138(11), 153(11, 33), 184(11), 185(33)
Tong, W. T., **31**:334(4), 335(4), 388(4), 389(4), 390(4), 392(4), 395(4), 425(4)
Toni, T., **25**:51(87), 144(87)
Tonini, R. D., **23**:235(252), 236(253), 252(253), 278(252, 253); **24**:105(37), 186(37)
Toole, L. E., **5**:378(93), 509
Toomey, R. D., **10**:199, 204, 205, 207, 217; **14**:211, 216, 217, 218(144), 227, 246; **19**:128(135), 134(135), 188(135)
Toong, T., **2**:213, 217(61), 218(61), 269
Toor, H. L., **2**:376, 378, 397; **10**:189, 191, 217; **13**:209(11), 219(64), 264, 265; **SUP**:81(234), 109(279)
Toor, J. S., **5**:26, 51; **11**:318(32), 376(32), 428(238), 430(238), 435, 440
Topakoglu, H. C., **SUP**:341(490), 342(490), 343(490), 344(490)

Topmac, A. J., **28**:344(61), 345(61), 402(61), 418(61)

Topper, L., **SUP**:109(278)

Toral, H., **16**:72, 77, 104, 144, 145, 153

Toreas, N. E., **15**:270(107), 281

Torii, K., **13**:11(39), 60; **20**:252(238), 255(238), 310(238)

Torikoshi, K., **19**:99(13), 185(13)

Tornberg, N. E., **10**:21, 23, 36

Torner, R. V., **13**:261(122), 266

Torok, D., **20**:217(127, 132), 220(127, 132), 305(127, 132)

Torok, R. C., **29**:299(157), 302(157), 343(157)

Toronyi, R. M., **14**:77(160), 78, 82, 83(160), 103

Torrance, K. E., **10**:251, 252(46), 253(46), 258, 282; **14**:19, 41(51, 94), 42(94), 43(98), 101, 102; **15**:206(94), 208(94), 210(94), 225(94); **19**:78(185), 82(261), 91(185), 94(261); **28**:160(27), 229(27); **29**:38(28), 40(28), 56(28); **30**:101(39), 116(68), 117(68), 119(74), 122(39, 68, 86, 87), 123(86), 124(68, 85, 86), 125(86), 126(68, 85), 127(39, 68, 85, 87), 129(85), 130(85), 131(85), 132(86), 173(148), 190(39), 192(68, 74, 85, 86, 87), 195(148)

Torrence, K. E. E., **23**:370(15), 461(15)

Torres, F. E., **31**:40(34), 103(34)

Torres, F. J., **23**:331(241, 242), 365(241, 242)

Tory, E. M., **23**:161(47), 186(47)

Torza, S., **26**:29(235), 103(235)

Torzecki, J., **23**:218(138), 273(138)

Tosioki, U., **6**:474(302), 501

Toskubaev, I. N., **14**:195(125), 203(133, 134), 206(133, 134), 245; **19**:146(163), 149(177, 178), 188(163), 189(177, 178)

Totten, H. C., **5**:413, 513

Touart, C. N., **14**:253(8), 279

Touba, R. F., **7**:50

Tough, J. T., **17**:97, 98, 99, 100, 101, 102, 154

Touloukian, Y. S., **2**:37(28), 38(28), 106; **11**:213, 214, 230(176, 177), 255, 260; **24**:27(56), 38(56)

Toupin, R., **13**:134, 159(42), 202; **31**:15(35), 103(35)

Touryan, K., **7**:190, 214

Tousignant, L., **23**:22(43), 128(43)

Townsend, A. A., **3**:35(29), 98; **7**:24, 83; **8**:291, 299(39), 349; **14**:259(26), 260(26), 264, 280; **21**:154, 182, 183; **30**:26(87), 91(87)

Townsend, P., **19**:268(32), 317(32), 350(32)

Townsend, R. W., **30**:244(60), 252(60)

Tozeren, A., **24**:114(113), 188(113)

Tozzi, A., **17**:28(82), 59

Traber, D. G., **10**:202, 203(102, 103, 107), 217, 218; **14**:158(62), 243; **19**:110(72), 114(94), 186(72), 187(94)

Trabold, T. A., **26**:170(125), 214(125)

Trachtenberg, I., **28**:344(61), 345(61), 399(189), 400(189), 401(189), 402(61, 189, 209, 210, 211), 403(211), 404(211), 405(189), 418(61), 423(189), 424(209, 210, 211)

Trafton, L. M., **8**:231(2), 282

Träger, F., **28**:118(40), 119(40), 120(40), 140(40)

Tranter, C. J., **15**:285, 328, 329

Trapaga, G. M., **28**:10(12), 11(12), 17(12), 18(12), 19(12), 23(12), 36(59), 39(59), 43(59), 46(59), 64(82), 69(59), 71(12), 72(59), 73(82)

Trapp, J. A., **31**:106(27), 138(27), 157(27)

Trass, O., **13**:6, 10(29), 17, 18, 59, 60; **26**:110(25), 120(25), 123(51), 126(51), 129(51), 153(25), 209(25), 210(51)

Tratz, H., **SUP**:21(34), 76(34), 102(34), 107(34), 119(34), 126(34), 128(34)

Traugott, S. C., **3**:210(43, 46), 250; **14**:264(34), 265(34), 267, 280; **21**:246(13), 275(13)

Trautz, M., **3**:287(81), 301

Travis, L. P., **12**:138(27), 147(27), 193

Treadwell, K. M., **6**:474(265), 499

Tree, D. R., **23**:343(290), 367(290)

Trefethan, L., **9**:205(22), 233(22), 268

Trefethen, L., **7**:219, 313; **10**:99, 164; **14**:124, 147

Tremere, D. A., **15**:327

Trenberth, R., **29**:260(111), 263(111), 264(111), 341(111)

Trense, R. V., **14**:217, 246

Trepanier, J., **24**:218, 272

Trcpaud, P., **6**:38(72), 125

Treshchev, G. G., **1**:375, 390(24), 441

Trevino, C., **19**:74(171), 91(171)

Trevisan, O. V. (Chapter Author), **20**:315, 329(46), 330(46), 337(57), 339(57), 340(57, 58), 341(58, 59), 348(57, 58), 351(46), 352(57, 58, 59); **24**:106(42), 186(42)

Treybal, R. E., **21**:336, 345

Trezek, G. J., **9**:144(67), 147(67), 156(67), 179; **22**:10(36), 18(36), 360(7), 435(7)

Tribus, M., **1**:34(30), 50, 234(63, 64), 238(63, 64), 239(63, 64), 259(63, 64), 265; **3**:114, 115(20), 126, 132, 147, 150, 174; **4**:25, 63; **5**:83, 126, 419(193), 421, 512, 514; **6**:525(32), 562; **7**:81, 86, 330, 377; **8**:22(102), 86; **9**:174(77), 180; **SUP**:18(25), 21(25), 79(25), 101(25), 102(25), 119(25), 120(25), 125(25), 137(25), 170(25), 171(25), 177(354); **15**:2(9), 8(9), 27(54), 28(54), 55(9), 56(54), 100(101), 121(141), 139(101), 140(141); **16**:139(112), 156, 225, 239; **17**:11(51), 23(51), 58, 148(175), 157; **20**:250(229), 310(229); **23**:82(97), 131(97)

Tricomi, F. G., **8**:34(177), 88

Trier, W., **11**:391(144), 397(144), 438

Trilikauskas, V. V., **25**:114(180), 148(180)

Trilling, C. A., **14**:212, 214, 217, 218(145), 223(145), 246; **19**:123(115), 132(115), 134(115), 187(115)

Trilling, L., **6**:95(156), 130

Trilling, T. A., **10**:200, 203, 204, 217

Trimmer, L. L., **23**:280(14), 304(14), 315(14), 322(14), 325(14), 354(14)

Tripathi, A., **25**:293, 294, 295, 297, 298, 310, 319

Tripathi, G., **25**:259, 286, 316

Tripathi, P. K., **25**:257, 272, 274, 275, 317

Tritton, D. J., **4**:52, 64; **11**:217, 262; **30**:32(89), 91(89)

Trivedi, V. K., **8**:42, 89

Trohan, A. M., **8**:298(36), 349

Troitsky, V. S., **10**:41, 71, 81; **11**:360(92), 436

Trombetta, M. L., **SUP**:66(117), 325(117), 328(117), 329(486), 330(486), 331(486), 332(486), 333(486)

Tropea, C., **25**:390, 411

Troshin, Y. K., **29**:91(60), 126(60)

Trottenberg, U., **24**:251, 274

Trotter, D. P., **9**:144, 179

Troup, D. H., **30**:243(61, 62), 252(61, 62)

Trout, A. M., **3**:84(85), 89, 100; **7**:344, 352(39), 355, 358, 378

Troy, J. S., **21**:246(22), 257(22), 276(22)

Troy, M., **1**:245(86), 266

Troyan, E. N., **26**:109(24), 149(24), 151(24), 201(24), 202(24), 203(24), 209(24)

Truang, H. V., **23**:369(1), 460(1)

Trüb, J., **7**:116(28), 118(28), 120(28), 160

Trucco, A., **6**:41(81), 65(81), 126

Truckenbrodt, E., **5**:167, 168, 248

Truelove, J. S., **23**:139(15), 184(15); **31**:371(37), 426(37)

Truesdell, A. H., **14**:26(38), 78(38), 79, 88(38), 100, 104; **30**:140(99), 193(99)

Truesdell, C., **2**:363(116), 397; **12**:269(107), 281; **13**:134, 159(42), 202; **24**:107(73), 187(73); **31**:15(35), 50(36), 103(35, 36)

Truong, H. V., **15**:329

Trupp, A. C., **10**:180, 216

Trusela, R. A., **4**:210, 227

Trushin, V. A., **25**:127(194, 195), 129(195), 149(194, 195)

Tsai, D. H., **2**:37(30), 106

Tsai, F. P., **29**:51(48), 57(48)

Tsai, H. L., **28**:253(58), 304(58, 140, 141, 143, 144), 331(58), 335(140, 141, 143, 144)

Tsai, H. M., **25**:361, 382, 383, 394, 412, 413, 418, 419

Tsai, J. Y., **28**:343(39), 417(39)

Tsai, S., **5**:149, 152, 248

Tsai, S. W., **18**:189(140), 239(140)

Tsai, T. P., **30**:118(73), 120(73), 135(73), 192(73)

Tsang, C. F., **14**:51(120), 54(125), 102, 103; **24**:103(23, 24), 185(23, 24)

Tsang, C.-H., **26**:123(67), 124(67), 211(67); **28**:394(183), 423(183)

Tsantker, K. L., **25**:166(54), 245(54)

Tsao, C.-Y., **28**:20(36), 21(36), 22(36), 71(36)

Tsao, G. T., **18**:179(75), 180(75), 237(75)

Tsao, J. Y., **28**:80(100), 143(100), 342(32), 343(38), 416(32, 38)

Tsao, S., **5**:231, 232(155), 250; **21**:149, 150, 160, 183

Tsao, Y. H., **20**:271(278), 312(278)

Tsapatsis, M., **28**:343(58), 417(58)

Tsarev, I. V., **14**:205, 245

Tsay, S.-Y., **25**:259, 319

Tsederberg, N. V., **8**:266, 283

Tseitlin, L. M., **9**:96, 103(112b), 110

Tseng, A. A., **23**:12(71), 14(71), 51(71), 129(71)

Tseng, H. H., **26**:170(126), 214(126)

Tseng, T. P., **14**:311, 343

Tsenoglou, C., **28**:193(31), 229(31)

Tsetserin, V. A., **6**:408, 489

Tsien, H. S., **4**:331, 444; **8**:4, 84

Tsiganok, A. A., **11**:18(62, 63), 19(62, 63), 31(62, 63), 49

Tsirelman, N. M. (Chapter Author), **19**:191, 202(60, 61, 62, 63, 64, 65, 66), 207(66),

245(60, 61, 62, 63, 64, 65, 66)

Tsitovich, N. A., **1**:170, 184

Tsoi, P. V., **6**:425, 435(158), 493; **8**:33, 62, 88, 90; **19**:192(7), 195(7), 243(7)

Tsoi, V. I., **31**:281(185), 331(185)

Tsongas, G. A., **12**:7(17), 73

Tsou, F. K., **4**:7(8), 54(99), 62, 64; **7**:346, 361(61, 62, 63), 364, 365, 366, 367(61, 62, 63), 379; **15**:329; **23**:12(49), 14(49), 25(49), 128(49)

Tsou, F. T., **23**:320(162), 361(162)

Tsou, J. D., **15**:101(109), 139(109)

Tsubouchi, T., **11**:203, 206, 209(39), 220, 223, 226, 254, 258

Tsuchida, A., **3**:81(76), 82(76), 84(76), 87(76), 99; **7**:346, 353(41), 378

Tsuchiya, Y., **10**:259, 282

Tsuei, H. S., **15**:22(29), 168(29), 208(97, 100), 209(97), 210(97, 101), 224(29), 225(97, 100, 101)

Tsuei, Y. M., **24**:215, 271

Tsuge, A., **21**:4(10), 51(10)

Tsuji, I. J., **5**:167, 168, 248

Tsukada, T., **30**:358(115), 432(115)

Tsukamoto, O., **17**:152(181), 157

Tsuruga, H., **11**:231(241), 232(241), 263; **16**:234

Tsuruta, T., **21**:83(40), 88(57, 58), 89(57, 58), 90(59), 135(40), 136(57, 58, 59); **24**:163(147), 190(147); **30**:121(81), 192(81)

Tsurutani, K., **28**:36(63), 39(63), 73(63)

Tsutsumi, K., **20**:183(7), 299(7)

Tsvetkov, F. F., **9**:120, 129(12), 178; **SUP**:133(310); **31**:160(18), 323(18)

Tsyurik, V. N., **6**:474(311), 501

Tu, C., **24**:122(131), 129(131, 135), 130(135), 139(131), 141(131), 144(135), 145(135), 146(135), 147(135), 148(135), 149(135), 174(131), 189(131, 135)

Tu, C. W., **28**:351(70), 418(70)

Tu, P. S., **3**:106, 172

Tu, S. W., **25**:81(140, 141), 146(140, 141)

Tuchscher, J. S., **5**:293(36), 301(36), 322; **8**:254(10), 256(10), 257(10), 266(10), 282; **31**:389(42), 427(42)

Tucker, B. E., **18**:333(39), 342(39), 362(39)

Tucker, M., **2**:6(10), 105; **3**:35(32), 60, 98

Tucker, P. K., **24**:226, 227, 271

Tucker, T. C., **SUP**:361(506), 363(506)

Tuckerman, D. B., **20**:273(282), 312(282)

Tudhope, J. S., **18**:327(4), 328(11), 362(4, 11)

Tuesdell, C., *see* Truesdell, C.

Tufeu, R., **21**:2(9), 3(9), 51(9)

Tung, T. T., **15**:71(52), 85(52), 88, 90(77), 102(77), 104(77), 118(52), 131(74), 132(74), 136(52), 137(74), 138(77); **19**:330(127), 355(127)

Tung, V. X., **29**:43(38, 39), 45(38, 39), 57(38, 39); **30**:112(63), 134(92), 135(92), 151(112), 191(63), 193(92), 194(112)

Tunnicliffe, J., **28**:342(34), 416(34)

Tuoc, T. K., **25**:260, 279, 317

Tuomisto, H., **29**:20(15), 56(15)

Turchin, I. A., **6**:474(247), 480(247), 483(247), 498

Turcotte, D. L., **3**:63(64d), 99; **14**:41(94), 42(94), 43, 88, 102, 104; **15**:206(94), 208(94), 210(94), 225(94); **19**:73(168), 82(261, 266, 271), 90(168), 94(261), 95(266, 271); **24**:7(28), 37(28); **30**:119(74), 122(84), 124(84), 125(84), 126(84), 127(84), 128(84, 90), 129(84, 90), 131(84), 192(74, 84), 193(90)

Turian, R. M., **2**:379, 397; **13**:219(46, 52, 53, 54), 253, 265; **15**:76(58, 59), 137(58, 59), 327

Turilin, S. I., **8**:114(39), 148, 149, 159; **18**:109(39), 146(39), 147(39), 158(39)

Turkdogan, E. T., **26**:141(94), 202(94), 212(94)

Turkel, E., **24**:234, 271

Turkmenov, Kh. I., **31**:271(173), 330(173)

Turland, B. D., **29**:189(190), 194(208), 212(190), 213(208)

Turley, W. D., **23**:294(75), 357(75)

Turnbull, D., **10**:111, 165

Turnbull, R. J., **14**:123(95, 96, 97, 98, 99, 100, 101, 102, 103), 124, 125, 126(95, 96, 97, 98, 99, 100, 101, 102, 103), 147, 148

Turner, B. L., **20**:217(126, 133), 219(126, 133), 305(126, 133)

Turner, C. W., **31**:444(41), 472(41)

Turner, D. T., **2**:363(13), 365(13), 394

Turner, J. S., **13**:66(4), 96, 99, 114, 116; **14**:28(62), 101; **16**:7(5), 9(5), 10(5), 11(5), 53(5), 56; **20**:316(14), 347(14), 350(14); **28**:240(10), 244(10), 316(10), 329(10)

Turner, K. A., **28**:402(203), 424(203)

Turner, L. A., **2**:354

Turner, M. M., **15**:232(35), 277

Turner, R. C., **7**:240, 274(51), 315, 316

Turner, T. P., **28**:124(78), 142(78)
Turton, J. S., **5**:191(103), 249; **6**:474(273), 499; **SUP**:177(356); **16**:225(31), 226, 236, 239
Turton, R., **21**:334, 345
Tusima, K., **24**:30(64), 38(64)
Tuthill, E. J., **25**:180(72), 246(72)
Tuthill, W. E., **5**:515(230)
Tuttle, F., **13**:123, 201
Tutu, N. K., **29**:233(62, 70), 234(81), 259(118), 261(118, 119, 120, 121), 262(119), 263(118, 119, 120, 121), 264(118, 119, 120, 121), 265(118, 119, 121), 269(135), 274(146), 277(62), 282(118, 152), 328(181), 338(62, 70), 339(81), 341(118, 119, 120, 121), 342(135), 343(146, 152), 344(181)
Tuzla, K., **19**:164(219), 169(219), 171(219), 190(219)
Twyman, F., **6**:197(32), 364
Tyagi, V. P., **8**:44(220, 222, 223), 89; **SUP**:54(63, 64), 58(64), 63(81), 64(63, 64, 90, 91), 82(63), 158(63), 224(63), 225(63, 415), 249(63), 277(64), 278(90, 91)
Tyan-Tsi, E., **6**:474(320), 501
Tychesen, W., **13**:248(107), 266
Tye, R. P., **11**:352(72), 353(72), 436
Tyler, A. P., **29**:155(84), 156(84), 165(84), 206(84)
Tyomkin, A. G., **19**:192(5), 243(5)
Tyson, H., **2**:354
Tyson, T. J., **6**:107, 132
Tyurin, Y. A., **20**:262(261), 270(261), 311(261)
Tyvand, P. A., **20**:326(29), 350(29); **24**:31(67), 32(67, 68), 33(68), 34(68), 38(67)
Tzederberg, N. V., **7**:9
Tzivanidis, G., **20**:336(55, 56), 352(55, 56)
Tzou, D. Y., **28**:77(75), 142(75)

U

Uanezmoreno, A. A., **18**:156(83), 159(83)
Uberoi, M. S., **6**:390, 487; **30**:289(79), 310(79)
Uberreiter, K., **18**:214(106), 238(106)
Uchida, H., **23**:11(28), 13(28), 15(28), 20(28), 35(28), 36(28), 39(28), 45(28), 127(28)
Uchida, S., **20**:27(55), 33(55), 81(55)
Uchida, Y., **19**:307(96), 313(96), 353(96); **23**:349(305), 368(305)
Uchima, B., **21**:122(111), 138(111)

Udagawa, Y., **26**:105(5), 123(5), 126(5), 208(5)
Udaykumar, H. S., **24**:227, 230, 231, 274
Udell, K. S., **23**:281(29), 355(29); **29**:39(31), 56(31); **30**:98(32), 111(55), 114(32), 117(32), 154(55, 116, 118, 119), 155(55, 116, 121), 156(55, 116), 157(116), 163(124), 165(124), 166(124, 125), 169(139), 170(139), 175(150), 184(161), 190(32), 191(55), 194(116, 118, 119, 121, 124, 125), 195(139, 150), 196(161)
Ueda, H., **11**:231(168), 232(168), 260
Ueda, T., **17**:21, 42, 59, 62
Uehara, H., **11**:213(75), 215(75), 255; **21**:105(86), 137(86); **23**:12(64), 13(64), 15(64), 35(64), 40(64), 59(64), 61(64), 99(64), 100(64), 101(64), 106(64), 129(64)
Uehara, O. A., *see* Uyehara, O. A.
Ueno, S., **5**:13, 51
Ueno, T., **14**:155(47), 157(47), 243; **19**:164(207), 173(207), 174(207), 175(207), 176(207), 190(207)
Uglov, A., **28**:127(84), 142(84)
Uglum, J. R., **5**:417, 473, 516
Uhl, U. W., **16**:144(122), 156
Uhland, E., **13**:229(100), 261(100), 266
Uhlenbeck, G. E., **7**:170(16), 171, 172, 211, 212; **29**:120(86), 127(86)
Uhlherr, P. H. T., **23**:190(194), 215(251), 262(5, 6), 267(5, 6), 275(194), 278(251); **25**:256, 318
Uhlig, H. H., **12**:250(108), 281
Uhlmann, D. R., **28**:270(100), 333(100)
Ulam, S., **5**:3, 50
Ulanov, G. M., **25**:4(30, 32), 141(30, 32)
Ulbrecht, J. J., **15**:157(20), 158(20), 160(20), 162(20), 163(20), 179(39), 185(20), 223(20), 224(39); **23**:201(127), 208(127, 128), 210(127), 214(128), 230(127, 128, 130, 130a, 132), 231(128), 237(131), 238(131), 239(131), 240(131), 241(131), 243(131), 246(131), 252(131), 262(129), 272(127, 128, 129, 130, 130a, 131, 132); **25**:260, 261, 269, 271, 272, 273, 278, 279, 281, 283, 286, 287, 315, 317, 319
Ulbrecht, U., **15**:69(45), 136(45)
Ulinskas, P., **31**:160(52), 324(52)
Ulinskas, R., **31**:263(157), 284(157), 329(157)
Ulinskas, R. V., **8**:115(43), 159; **18**:109(42), 139(75), 142(77), 158(42), 159(75, 77)
Ulitko, A. F., **8**:35, 88

Ullah, N., **1**:86, 89(45), 95(45), 121
Ullrich, R., **SUP**:66(118), 355(118), 360(118)
Ulrichs, J., **10**:73(59), 82
Ulrichson, D. L., **SUP**:139(323), 140(323), 142(323), 143(323), 145(323), 146(323)
Ulsamer, J., **11**:218, 221, 225, 256
Umaya, M., **17**:38, 62
Umemiya, N., **11**:231(168), 232(168), 260
Umur, A., **9**:204(19), 206(19), 207, 210, 241, 268; **21**:80(31), 134(31)
Unger, H., **29**:163(102), 172(143), 173(143, 144, 145), 181(171), 207(102), 209(143, 144, 145), 211(171)
Ungurian, M., **29**:84(48), 125(48), 307(169), 344(169)
Unnikrishnan, A., **23**:262(41), 263(41), 264(41), 268(41)
Unnikrishnan Nair, V. R., **23**:262(41), 263(41), 264(41), 268(41)
Unno, W., **21**:246(23), 276(23)
Unny, T. E., **18**:15(20), 16(24, 25), 17(25), 80(20, 24, 25)
Unsal, E., **23**:193(254), 226(254), 278(254)
Unsöld, A., **3**:218, 250; **4**:245, 313; **5**:303(60), 323
Unterberg, W., **4**:146, 222, 228
Upadhya, G., **28**:263(78), 332(78)
Upadhyay, S. N. (Chapter Author), **19**:98(8), 113(86), 126(131), 137(131, 144), 140(131, 144, 147), 142(147), 143(147), 144(147), 145(8, 131, 144, 147), 156(8, 187), 158(131), 159(8, 144, 187), 160(187), 161(8, 144), 162(147), 179(233), 180(147), 184(8), 186(86), 188(131, 144, 147), 189(187), 190(233); **23**:192(152), 212(153), 213(153), 215(153), 230(153), 231(153), 236(153, 154), 238(153), 240(153), 242(153), 252(153, 154), 273(152, 153, 154); **25**:205(103, 104, 105), 226(115), 247(103, 104, 105), 248(115), 251, 257, 260, 272, 274, 275, 276, 277, 278, 279, 283, 284, 285, 289, 290, 291, 298, 306, 310, 316, 317, 318, 319
Upatnieks, J., **6**:200, 364
Urakawa, K., **20**:15(24), 80(24)
Uranata, I., **25**:81(138), 146(138)
Urbanovich, L. I., **SUP**:294(464, 465), 296(465)
Urbassek, H. M., **28**:111(92), 142(92)
Urbonas, P. A., **25**:114(180, 182), 148(180, 182)
Urdaneta, A., **15**:52(86), 53(86), 58(86)

Urushiyama, S., **7**:145, 161
Usagi, R., **20**:195(79), 302(79), 382, 387; **26**:131(77), 159(77), 212(77)
Usanov, V. V., **6**:390(15, 19), 405(15, 19), 486, 487; **31**:192(96), 326(96)
Useller, J. W., **4**:150(6a), 222
Ushakov, P. A., **31**:236(124), 317(215), 327(124), 332(215)
Ushida, K., **23**:253(126), 254(126), 257(126), 261(126), 272(126)
Ushijima, K., **15**:286, 329; **25**:394, 418
Ushijima, T., **31**:334(14), 425(14)
Usiskin, C. M., **1**:2, 12, 13(6), 49, 234, 239, 265; **2**:424(20), 425(20), 451; **3**:211(48), 250; **4**:160(37), 167, 169, 191, 224; **5**:27, 28(31), 29, 51, 52, 462, 467, 515
Uskov, I. B., **9**:71(117), 96, 103(112b), 104(117), 110, 111
Usmanov, A. G., **16**:116(81), 117(81), 125, 155
Ustimenko, B. P., **21**:166, 183; **31**:160(24), 323(24)
Ustinov, V. A., **26**:114(35), 202(35), 209(35)
Usui, H., **15**:92(84, 85), 93(84, 85), 107(139), 108(139), 111(139), 113(139), 114(139), 124(139), 125(139), 130(84, 139), 138(84, 85), 140(139); **25**:258, 260, 283, 284, 285, 318
Usui, S., **28**:96(85), 142(85)
Usui, T., **25**:81(134), 146(134)
Usyugin, I. P., **16**:114(78), 155
Utaka, Y., **21**:84(42), 85(48), 100(42), 135(42, 48); **24**:3(5, 14), 6(21, 22, 26), 36(5, 14, 21, 22), 37(26)
Utech, H. P., **10**:276(107, 109), 284
Utkin, O. N., **31**:242(130), 328(130)
Utsunomiya, K., **21**:109(99), 137(99)
Uwasawa, K., **28**:96(68), 141(68)
Uyehara, O. A., **23**:101(105), 131(105); **26**:147(104), 163(104), 165(104), 169(104), 186(104), 213(104)

V

Vachon, R. I., **11**:13(34), 48; **16**:214(24), 216, 227, 233, 238; **18**:179(76), 180(76), 228(76), 237(76)
Vader, D. T., **23**:8(12), 10(12), 11(12), 13(12, 36), 15(12, 36), 18(12, 36), 29(12), 33(12), 43(12), 44(12), 46(12), 126(12), 128(36);

Vader, D. T. (*continued*)
   26:122(47), 162(47, 118), 163(47, 118),
   165(47), 167(47, 118), 169(47), 187(47),
   189(47), 210(47), 214(118)
Vadivel, R., **19**:164(211), 168(211), 170(211),
   171(211), 190(211); **23**:280(19), 306(19),
   354(19)
Vadovic, C. S. J., **12**:235(109), 282
Vafai, K., **20**:317(15), 350(15); **23**:370(16),
   461(16); **24**:106(41), 186(41); **30**:93(4a),
   94(13, 13a), 169(13), 189(4a, 13, 13a)
Vagt, J. D., **8**:295(25), 349
Vaidyanathan, G., **20**:326(31), 350(31)
Vaihinger, K., **20**:23(45), 81(45)
Vajravelu, K., **20**:244(202), 250(202), 309(202)
Vakhrushev, I. A., **14**:176(93), 178, 244
Valensi, J., **7**:197, 215
Valent, V., **16**:125, 155
Valette, M., **29**:151(66), 157(66), 205(66)
Valisi, M., **29**:234(93), 340(93)
Valkenberg, W. G. J. N., **28**:402(193), 423(193)
Valuyeva, E. P., **21**:13(27), 14(27), 15(27),
   22(40), 24(40, 41, 42, 43), 25(40), 26(40,
   41, 42, 43), 27(41, 42, 43), 29(27), 39(40,
   41), 51(27, 40), 52(41, 42, 43)
Valvano, J. W. (Chapter Author), **22**:187(207,
   208), 277(219), 342(208), 356(207, 208,
   219), 359, 381(19, 20), 384(19, 20),
   394(19, 20), 435(19, 20)
Van Atta, C. W., **8**:291, 349; **21**:185(4), 204(4),
   206(4), 207(4), 236(4); **30**:290(83), 310(83)
Van Beek, H. J. M., **17**:127(127), 156
Van Buren, P. D., **19**:49(115), 62(143), 63(143),
   88(115), 89(143)
van Bylevelt, J. S., **11**:202(7), 204(7), 209(7),
   218(7), 221(7), 252
Vance, R. W., **5**:326(7), 331, 333(7), 355, 356,
   505; **9**:405(95), 416
Van Cleave, A. B., **2**:356
Van Cleve, J., **17**:143(160), 151(160), 157
van Dam, J., **13**:219(75), 234(75), 249(75), 265
van Dam, L., **22**:60(50), 70(50), 154(50)
Vandantseveeiy, B., **10**:207(123), 217
Van de Hulst, H. C., **5**:254(11), 322; **9**:363(29),
   365(29), 415; **23**:141(19), 145(19),
   147(19), 150(19), 164(19), 185(19);
   30:257(8), 307(8)
Vandekamp, L., **19**:53(124), 54(124), 55(124),
   56(124), 88(124)
Vandekoppel, R., **5**:390, 510

van den Berg, G. J., **17**:127(127), 156
VandenBerghe, T. M., **23**:333(249, 251),
   334(251), 335(251), 365(249), 366(251)
van den Bogaert, N., **30**:313(7), 338(7), 350(7),
   358(7), 403(174), 404(7), 406(7, 174),
   426(7), 434(174)
Vandenbulcke, L., **28**:394(182), 423(182)
Vanden Eykel, E. E., **6**:82(139), 129
Van der Borght, R., **15**:211(107), 212(107),
   225(107)
van der Brekel, C. H. J., **28**:402(192, 193, 198),
   423(192, 193), 424(198)
van der Does de Bye, J. A. W., **SUP**:186(365)
van der Eerden, J. P., **30**:335(64), 429(64)
Van der Graaf, F., **23**:280(17), 354(17)
van der Heeden, D. J., **17**:291, 302(99), 317
van der Hegge Zijnen, B. G., **3**:11, 26, 31; **7**:118,
   161; **11**:203, 205, 219, 222, 225, 229, 231,
   232, 237, 253, 259; **21**:43(71), 52(71)
van der Hegge Zijnen, B. S., **8**:127(57), 160;
   18:123(56), 159(56)
Van der Held, E. F. M., **3**:216, 250; **8**:176, 227;
   11:332(56), 436; **18**:13(19a), 14(19a),
   80(19a), 167(54), 237(54)
Vanderhoof, D. W., **19**:149(182), 189(182)
Van de Riet, E., **28**:103(101), 143(101)
van der Mast, V. C., **30**:246(63), 252(63)
van der Meer, H., **28**:356(96), 394(96), 419(96)
van der Molen, S. B., **17**:47, 63
Vanderslice, J. T., **3**:270(28, 33), 271, 300
van der Sluijs, J. C. A., **17**:68, 70(15), 153
Vander Stichele, S., **23**:326(212), 364(212)
van der Velden, H. A., **1**:79, 121
van der Velden, J. H., **17**:297, 317
Van der Vorst, H. A., **23**:418(76), 436(76),
   464(76)
van der Walt, N. T., **12**:3(7), 14(7), 51(7), 59(7),
   68, 69(7), 73
Vanderwater, R., **1**:410(94), 442
van de Ven, J., **28**:369(120), 373(120), 420(120)
van Dilla, M., **4**:93, 94, 96(31), 138
van Donselaar, R., **13**:219(75), 234(75), 249(75),
   265
Van Doormal, J. P., **24**:201, 275
Van Dorth, A. C., **23**:293(69), 357(69)
van Driel, H. M., **28**:102(94, 112), 143(94, 112)
Van Driest, E. R., **2**:41, 42, 45, 107; **3**:35, 60, 97,
   99; **4**:331, 444; **7**:53, 85, 133, 161; **16**:344,
   365; **21**:183; **25**:358, 359, 418
Van Drunen, F. G., **18**:167(54), 237(54)

Van Dyke, K. S., **2**:356

Van Dyke, M., **2**:257(92), 260(92), 270; **4**:341, 446; **7**:187, 197, 214, 215; **8**:4, 7, 12, 14, 84, 85, 214, 227; **9**:303(53), 347; **14**:69(147), 103; **SUP**:69(142), 161(142), 166(142)

Van Dyke, M. D., **19**:31(64), 86(64); **26**:6(236), 103(236)

Vanecek, V., **10**:169(4), 215

Van Es, H. E., **15**:78(66), 137(66)

Van-Fan, **6**:551(84), 564

VanFossen, G. J., **23**:345(296), 350(306), 351(306), 368(296, 306)

van Gemert, M. J. C., **22**:248(220), 277(220), 279(209, 210, 220), 280(220), 282(210), 284(210), 285(210), 286(210), 287(210), 356(209, 210, 220), 426(38), 436(38)

Van Gundy, D. A., **5**:471, 473, 516

van Hal, R. P. M., **28**:96(14), 139(14)

Van Heerden, C., **14**:151, 242

Van Heerden, G., **10**:193, 199, 202, 204, 205, 206, 217

Van Heiningen, A. R. P., **23**:320(170, 172), 337(262), 361(170), 362(172), 366(262)

Van Hise, V., **6**:35(58), 125

Vaniman, J. L., **5**:378, 379, 468, 509, 516

Van Itterbeck, A., **17**:127(127), 156

Vanka, S. P., **24**:196, 226, 227, 244, 247, 249, 262, 263, 271, 274, 275

Vanko, V. I., **8**:65, 90

Van Krevelen, D. W., **10**:193(91), 199(91), 202(91), 204(91), 205(91), 206(91), 217; **14**:151(13), 242

van Laar, J., **2**:374, 378, 396; **13**:219(66), 265

Van Laethen, R., **11**:425(228), 426(228), 440

Van Leer, B., **24**:234, 275

van Leeuwen, J., **13**:248, 266

Van Lierde, P., **23**:326(212), 364(212)

Van Loan, C. F., **23**:418(74), 464(74)

van Meel, D. A., **11**:219, 223, 237, 257

Vannerus, T., **5**:198, 249

Van Ouwerkerk, H. J., **11**:17(55), 49; **16**:87, 93, 101, 154

van Poollen, H. K., **23**:212(255), 278(255)

Van Ravenstein, Th. N. M., **19**:281(73), 284(73), 294(73), 295(73), 308(73), 312(73), 352(73)

van Schravendijk, B. J., **28**:402(215), 424(215)

Van Sciver, S. W., **17**:72(37), 73(37, 38, 43), 74(37, 43), 106(94), 107(94), 108(37), 110, 112, 113, 114(100), 117, 118, 120(43), 122, 123, 124, 154, 155, 156

Van Shaw, P., **12**:233(110), 251(110), 270(110), 282; **15**:128(152), 141(152)

van Stralen, S. J. D., **11**:14(47), 15, 16, 18(64), 31(64), 49; **15**:244, 279; **16**:75, 76, 82(28), 83, 84, 85, 86, 87, 88, 91, 92, 95(28), 100, 101, 104, 107(28), 119, 127(28), 132, 138, 139, 154

Van Tassell, W., **4**:279, 314

van Thinh, N., **8**:300, 301(55), 344(55), 350

VanVoorhis, C. C., **6**:268, 274, 365

van Wijk, W. R., **1**:239, 265; **16**:82, 86, 92(45), 95, 107, 127, 128, 154

Van Wyck, W. R., **2**:317, 355

Van Wylen, G. J., **5**:375(86), 381, 508, 509; **9**:199(13), 200, 201, 268; **15**:5(21), 6(21), 55(21)

van Zoonen, D., **9**:131, 179

Varadi, T., **19**:111(73), 118(73), 186(73)

Varagaftik, N. B., **3**:276(55), 301

Varanasi, P., **5**:293(37), 307(66), 308(66), 322, 323

Vardelle, M., **28**:53(79), 73(79)

Vardi, J., **4**:26, 58, 63

Varfolomeev, N. M., **31**:240(126), 328(126)

Varga, R. S., **24**:195, 275

Varker, C. J., **30**:318(13), 427(13)

Varma, A., **28**:402(199), 424(199)

Varma, R. K., **8**:11, 13, 85; **10**:255(56), 282

Varshavski, G. A., **8**:49, 89

Varygin, N. N., **10**:202, 218; **19**:117(97), 129(97), 130(97), 187(97)

Vas, I. E., **6**:95(155), 130

Vasalos, I. A., **23**:139(12), 154(39), 168(39), 184(12), 185(39)

Vasanova, L. K., **10**:177, 216

Vasantha, S. S., **4**:355, 446

Vasetskaya, N. G., **12**:80(11), 112; **15**:97(97), 138(97)

Vasilesco, V., **3**:256, 300

Vasileva, A. B., **8**:13, 85

Vasilevski, K. K., **6**:474(331), 502; **8**:35, 88

Vasiliev, A. A., **6**:418(142), 493; **29**:101(74), 126(74)

Vasiliev, G. M., **14**:176(183), 236(183), 237(183), 247

Vasiliev, L. L., **31**:160(31, 40, 53), 323(31), 324(40, 53)

Vasiliev, O. F., **6**:402, 488; **25**:81(142), 146(142)

Vasiliev, S. K., **14**:203(133, 134), 206(134), 245; **19**:146(163), 149(178), 188(163), 189(178)

Vasseur, P., **24**:303(44), 320(44)

Vassicek, F., **20**:297(324), 314(324)

Vassilatos, G., **14**:232(169), 246

Vassoille, R., **22**:187(211), 356(211)

Väth, L., **29**:155(83), 206(83)

Vattierra, M. L., **10**:243(38), 282

Vauclin, M., **19**:78(190), 91(190)

Vaughn, R. D., **2**:371(74), 372(74), 379, 396, 397; **15**:101(116), 102(116), 139(116)

Vazquez, J. E., **28**:102(61), 141(61)

Vedamurthy, V. N., **19**:164(211), 168(211), 170(211), 171(211), 174(225), 190(211, 225)

Veduta, A. P., **11**:423(220, 224), 440

Veeraburus, M., **21**:300, 346

Veghte, J. H., **4**:103, 139

Veinik, A. I., **1**:55, 212; **19**:195(25), 243(25)

Veistinen, M., **28**:15(20, 21), 22(21), 71(20, 21)

Vekshin, V. S., **6**:418(123), 492

Velev, K., **25**:158(31), 244(31)

Velkoff, H. R., **9**:262(121), 271; **14**:121, 137, 138, 148

Vella, S., **1**:392(60), 442

Vel'tishchev, N. A., **6**:551, 564

Veltkamp, G. W., **28**:402(198), 424(198)

Veluswami, M. A., **15**:324

Vemuri, S. B., **20**:232(163), 238(163), 240(163), 241(163), 307(163)

Venart, J. E. S., **26**:234(37), 254(37), 269(37), 326(37)

Vendrik, A. J. H., **4**:118(79), 140

Vengerov, I. R., **15**:329

Venkatapathy, E., **24**:215, 270

Venkata Rao, C., **SUP**:172(349), 310(472); **26**:239(44), 241(44), 253(44), 274(44), 326(44)

Venkateswaran, S., **24**:194, 272

Venkateswarlu, D., **21**:337, 339, 345

Venkatesworly, P., **20**:281(299), 313(299)

Verdaasdonk, R., **22**:279(209), 356(209), 426(38), 436(38)

Verdier, J., **14**:120, 121(16, 17, 18), 128, 129(16, 17, 18), 130(16, 17, 18), 144

Veres, D. R., **18**:295(62), 298(62), 300(62), 322(62)

Verevochkin, G. E., **13**:15(45), 17(45), 60; **26**:123(52), 126(52), 210(52)

Vergnes, F., **23**:338(263), 366(263)

Verhaart, P., **17**:164(16), 167(16), 241(77), 245, 248(16), 250, 276, 277, 286, 296, 314, 316

Verhoeven, D. D., **30**:296(96), 300(96), 311(96)

Verhoff, F. H., **SUP**:115(288), 118(288), 135(288)

Verhoven, J. P., **12**:242(111), 282

Verkin, B. I., **17**:140, 157

Verloop, J., **25**:181(73), 199(73), 246(73)

Verma, A. K., **30**:148(109), 149(109), 193(109)

Verma, R. L., **25**:262, 319

Verma, R. S., **19**:112(81), 121(107), 122(81), 129(81), 131(81), 132(81), 133(81), 134(81), 135(81), 186(81), 187(107)

Vermeer, N. J., **17**:231(74), 260, 293, 296, 302(74), 303(74), 316

Vermeulen, J. R., **SUP**:197(383), 198(383); **19**:258(13), 259(13), 268(13), 350(13)

Vernon, H. M., **4**:67, 137

Vernotte, M., **11**:219(110), 222(110), 237(110), 257

Veronis, G., **14**:31, 32(75), 101; **24**:286(45), 320(45); **30**:50(67, 90), 90(67), 91(90)

Verschoor, J. D., **9**:369(39), 415

Versluijs, C. W., **15**:285(189), 287, 290, 324, 329

Vertes, A., **28**:124(102, 103), 143(102, 103)

Verwoerd, M., **17**:218(67), 290(67), 294(67), 295(67), 315

Veslocki, T. A., **23**:123(128), 132(128)

Vest, C. M., **15**:211(105), 225(105); **30**:264(28), 268(28), 273(28), 302(28), 308(28)

Veter, A. M., **11**:144(156), 145(156), 146(156), 147(156), 170(156), 184(156), 196

Vetlutskii, V. N., **3**:243, 251

Vetrov, V. I., **21**:47(81), 53(81)

Vetter, K. J., **12**:226(112), 237(112), 249(112), 251(112), 255(112), 282

Vichrev, Y. V., **21**:32(48), 52(48)

Victor, M., **15**:228(3), 242(3), 271(3), 276

Victor, S. A., **22**:84(58), 155(58)

vida Brekel, C. H. J., **28**:365(110), 366(110), 367(110), 368(110), 420(110)

Vidal, R. J., **7**:205, 217

Vidil, R., **31**:463(74), 474(74)

Vidin, Y. V., **6**:418(108, 116, 141), 474(326), 491, 493, 502

Vidyanidhi, V., **SUP**:264(440), 267(440)

Vierkandt, S., **20**:84, 119, 121, 122, 123, 128, 132

Vieweg, F., **12**:24(41), 74

Vigier, G., **22**:187(211), 356(211)

Vihrev, Y. V., **21**:21(35), 51(35)

Vijayaraghavan, M. R., **14**:230(166), 246; **19**:147(165), 149(165), 151(165), 153(165), 154(165), 189(165)

Vikhrev, Y. V., **6**:551, 555(74), 558(96), 563, 564; **7**:31(16), 33, 84

Vilemas, J. V., **16**:110(68), 111(68), 114(68), 128(68), 131(68), 155; **31**:159(1), 160(1), 161(1), 177(1), 182(1), 247(1), 262(1), 263(1), 322(1)

Vilemas, Yu., **31**:160(56, 60), 324(56, 60)

Vilensky, V. D., **6**:378, 384, 434, 438, 486; **SUP**:336(488), 337(488); **21**:7(15), 51(15); **25**:17(51, 52), 32(61), 142(51, 52, 61)

Villadsen, J., **SUP**:79(227), 118(227)

Villaneuve, J., **1**:384(49), 385(49), 388(49), 441

Villani, S., **1**:379(33, 35), 394(73), 399(73), 433(137), 441, 442, 443

Vincent, R. K., **10**:11, 36

Vincenti, W. G., **3**:210(44), 250; **5**:255(13), 322; **7**:170(22), 171(22), 211; **8**:14, 86, 248, 249, 256(36), 273, 282, 283; **11**:217, 228, 259; **14**:279(63), 280; **21**:246(14), 275(14)

Vinen, W. F., **17**:75(50), 87(50), 90(50), 91(50), 92(50), 93(50), 94(50), 97, 102(50), 105, 111, 154

Vines, R. G., **3**:276(51), 301; **10**:265, 283

Vingerhoet, P., **20**:244(185), 308(185)

Vinnichenko, N. K., **14**:262(29), 280

Vinogradov, A. P., **10**:53, 64, 70, 81

Vinogradov, G. V., **13**:217(31), 264

Vinogradov, O. S., **31**:247(142), 328(142)

Vinokur, Y. G., **1**:399(80), 429(80), 442; **11**:170(169), 196; **20**:24(54), 27(54), 81(54)

Vintsevich, N. A., **25**:4(32), 141(32)

Vinzelberg, H., **16**:114(78), 155, 233; **17**:6(30), 57, 140(152), 142(152), 146(152), 157

Virk, P. S., **12**:78(5), 79, 80, 83, 84, 88(5), 102(5), 106(5), 111, 112; **15**:87(73), 90(73), 92(73, 87), 93(73), 104(124), 111(73), 114(124), 128(157), 131(157), 137(73), 138(87), 140(124), 141(157); **19**:330(126), 354(126)

Virkler, T. L., **14**:327(106), 329(106), 334(106), 339(106), 345

Virto, L., **23**:193(139), 273(139)

Virzi, A., **30**:382(148), 433(148)

Visbisky, R. F., **12**:23(37), 74

Vishich, M., Jr., **5**:172, 248

Vishnepolsky, I. M., **25**:4(32), 141(32)

Vishnev, I. P., **20**:24(54), 27(54), 81(54)

Visik, M. I., **8**:13, 85

Viskanta, R. (Chapter Author), **1**:2, 12, 13(15), 14(15), 15, 16(15, 16), 17, 18(16), 19(16), 21, 22(16), 23, 26, 33, 49, 50; **3**:107, 173, 175, 198(22), 203(24), 205(34), 211(24), 219(34), 220, 222(10), 223(70), 224(70, 71), 225(70, 71), 226, 229(77), 230, 231, 232(77), 233(77), 238, 239(90, 91, 94), 240(90, 91), 241(90, 91), 242(90, 91), 249, 251; **5**:26, 27, 28(33), 51, 52, 254(5), 306(5), 312(5), 321; **6**:474(325, 329, 330), 502; **8**:34(167, 168), 88; **9**:379(51), 415; **11**:317, 318(11, 32), 326(46), 330(46), 331(46), 332(46), 350(68), 351(71), 352, 361, 363(106), 365(99, 110, 124), 368(124), 369, 370, 372, 376(32, 133), 385(133), 387(133), 388(133), 390(140), 392(152, 153), 395(152, 153), 397, 398, 400, 401(152, 165), 402, 403, 404(153, 173), 405(46), 406, 407(173), 408, 412(197, 199, 200), 415(197, 199), 416, 417(46), 418, 419, 423(173, 219a), 428(237, 238, 239a), 430(238, 239a, 241a, 242, 243), 431(242), 434, 435, 436, 437, 438, 439, 440, 441; **SUP**:302(467, 468), 310(467); **18**:1(2), 19(43), 48(43), 79(2), 81(43); **19**:3(8, 9), 42(96), 43(98), 45(105, 106), 46(106, 107, 108), 47(106), 49(9, 115), 54(108), 55(106, 108), 56(106, 108), 57(106, 108, 125), 58(131), 59(131), 60(131, 134), 62(142, 143, 145, 146), 63(142, 143, 145, 146), 65(145), 66(145), 67(146, 154, 155), 68(155, 156, 158), 69(158), 72(164, 165, 166), 73(165, 167), 78(184), 79(210, 221), 83(8, 9), 87(96, 98), 88(105, 106, 107, 108, 115), 89(125, 131, 134, 142, 143, 145, 146), 90(154, 155, 156, 158, 164, 165, 166, 167), 91(184), 92(210), 93(221); **23**:1, 1(3), 7(3), 8(12), 10(12), 11(12), 13(12, 36, 69), 15(12, 36, 69), 18(12, 36), 29(12), 33(12), 43(12), 44(12), 46(12), 51(69), 53(73), 118(121), 120(121), 123(3, 128), 126(3, 12), 128(36), 129(69), 130(73), 132(121, 128), 136(8), 141(20), 184(8), 185(20); **24**:3(9), 5(19), 6(24), 8(29, 30), 36(9, 19), 37(24, 29, 30); **26**:105(3), 106(9), 120(40), 121(46), 122(40, 47), 144(102), 145(46), 146(46), 147(40, 46,

Viskanta, R. (*continued*)
   102), 155(113), 162(40, 47, 118), 163(47,
   118), 164(40, 102, 113), 165(46, 47, 102,
   113), 166(113), 167(40, 47, 118), 168(40,
   102), 169(47), 170(40, 113), 173(137, 138),
   182(137, 138), 187(47), 189(47), 204(178,
   180), 208(3, 9), 210(40, 46, 47), 213(102,
   113), 214(118), 215(137, 138), 217(178,
   180); **27**:ix(12); **28**:241(12), 242(12),
   249(29), 252(12), 253(12), 255(12),
   262(12), 264(12, 82), 266(12), 269(12, 82),
   275(12), 276(12, 106), 326(12), 329(12),
   330(29), 332(82), 333(106); **29**:171(137),
   209(137), 344(171); **31**:390(45), 427(45)
Visser, P. (Chapter Author), **17**:159, 222(69),
   223(69), 225(69), 228, 236, 241(77), 250,
   260, 275(96), 315, 316
Vitagliano, **18**:328(20), 329(20), 350(20),
   362(20)
Vitkin, D. E., **30**:291(87), 292(87, 89), 293(87),
   296(99), 298(99), 311(87, 89, 99)
Vitkus, G., **10**:73, 82
Vitt, O. K., **14**:151(21), 160(21), 161(21),
   166(21), 221(21), 242; **19**:100(22),
   103(22), 116(22), 126(22), 127(22),
   128(22), 165(22), 171(22), 185(22)
Vives, C., **19**:80(234), 93(234); **28**:246(24, 25,
   26), 317(26), 318(180, 181, 182, 183, 186,
   187, 196), 330(24, 25, 26), 337(180, 181,
   182, 183, 186, 187), 338(196)
Viviand, H., **5**:177, 182, 183(88), 184, 190, 191,
   249; **6**:38(72), 125
Vizel', Y. M., **9**:265(136), 272
Vlachopoulos, J., **13**:219(82), 243(82), 265;
   19:281(72), 284(72), 286(72), 294(72),
   295(72), 296(72), 297(72), 302(72),
   352(72); **28**:150(2, 49), 227(2), 230(49)
Vliet, G. C., **1**:245, 246(90), 249(90), 254(90),
   256(91), 257(90), 258(90), 259(91),
   260(91), 261(91), 262(91), 266; **5**:389, 390,
   510; **9**:384(55), 416; **17**:8, 10, 24(42), 58
Vodicka, V., **8**:65, 90; **15**:285, 287, 289, 316,
   320, 329
Vogel, G. J., **19**:99(18), 100(18), 108(18),
   110(18), 185(18); **25**:225(112), 229(112),
   247(112)
Vogel, M., **10**:269, 283
Vogelpohl, A., **16**:144(121), 156
Vogenitz, F. W., **6**:109(174), 110(174), 112(174),
   113(174), 131

Vogtländer, P. H., **7**:116, 118, 160
Vohr, J. H., **1**:358, 367(9), 440
Voishel, V., **3**:35, 60, 97
Voke, P. R., **25**:333, 340, 343, 357, 360, 361,
   382, 383, 394, 396, 397, 398, 399, 402,
   404, 406, 410, 412, 413, 418, 419
Volik, B. G., **6**:474(296), 500
Volk, W., **19**:126(130), 188(130)
Volkert, H., **25**:376, 401, 417
Völkl, J., **30**:358(116, 117), 360(116, 117),
   382(149), 432(116, 117), 433(149)
Völkl, V., **30**:361(122), 432(122)
Volkov, P. M., **6**:526, 539, 562
Volkov, V. N., **6**:418(137), 492
Voller, V. R., **19**:26(54), 27(54), 85(54);
   28:252(46, 47), 253(46, 47), 262(47, 69),
   263(73, 74, 75), 264(47), 266(46), 272(47,
   69), 308(155), 326(46), 331(46, 47),
   332(69, 73, 74, 75), 336(155)
Vollmer, M., **28**:118(40), 119(40), 120(40),
   140(40)
Volmer, M., **1**:208, 264; **9**:202, 268; **10**:92, 101,
   111, 164; **11**:6, 48; **14**:287, 298, 304, 321,
   322(95), 323, 324(95), 325, 341, 344
Volodin, Yu. G., **25**:61(99, 100), 144(99, 100)
Volotskaya, V. G., **17**:116, 155
Volyanyak, G. A., **25**:92(172), 148(172)
von Allmen, M., **28**:127(104), 143(104)
Von Brachel, H., **10**:210, 218
Von Elbe, G., **10**:247, 282; **29**:83(45), 125(45)
Von Glahn, U. H., **11**:170(182), 174, 177(182),
   185, 197
Von Herzen, R. P., **14**:41(95), 42(95), 102
von Kármán, T., **3**:14, 21, 32, 60, 99; **4**:8, 62,
   331, 444; **5**:130, 134(24), 149, 153, 154,
   246, 247, 248; **19**:323(112), 354(112)
von Neida, A. R., **30**:358(109, 110), 431(109,
   110)
von Obermayer, A., **3**:287(82), 301
VonPeschke, J., **11**:420(207), 439
von Smoluchowski, M., **2**:292, 293, 355
von Strahlen, S. J. D., **1**:239, 265
von Ubisch, L., **2**:356
Von Wieselberger, C., **11**:217, 262
Vorob'ov, V. S., **5**:309(78), 310(82), 324
Voronin, G. I., **31**:159(1), 160(1, 14), 161(1),
   177(1, 14), 182(1), 236(14), 247(1), 262(1),
   263(1), 322(1), 323(14)
Voropaev, A. R., **6**:418(113), 491
Vorotilin, V. P., **26**:23(237), 104(237)

Vortmeyer, D., **20**:329(41), 351(41); **23**:136(6), 166(6), 184(6), 217(36), 268(36), 370(7, 8), 427(7, 8), 461(7, 8)

Vos, A. S., **1**:239, 265; **16**:82(28), 86(28), 95(28), 107(28), 127(28), 154

Vos, B., **12**:262(46), 279

Vos, J. J., **4**:118(79), 140; **22**:251(212), 253(212), 356(212)

Voss, G. D., **17**:183, 314

Vossoughi, S., **23**:193(256), 220(256), 221(256), 278(256); **25**:292, 300, 308, 319

Votta, F., Jr., **9**:265(144), 272

Vrebalovich, T., **7**:200, 216

Vreedenberg, H. A., **10**:201, 209(131, 135), 218; **14**:172, 174(87, 88), 175(87, 88), 177(87, 88), 185, 187, 188(87, 88), 189(87, 88), 215(87, 88), 216, 218(149), 219(152), 220, 223(149, 152), 226(152), 227, 234(152), 244, 246; **19**:113(88), 119(88), 133(139, 140), 134(139), 186(88), 188(139, 140)

Vreman, A. W., **25**:364, 419

Vreman, B., **25**:364, 412

Vrentas, J. S., **SUP**:73(186), 92(186), 93(186), 96(186, 255), 99(186), 300(186); **19**:19(35), 22(46), 28(46), 85(35, 46)

Vu, T. C., **24**:199, 200, 207, 210, 211, 274

Vuillard, G., **28**:394(182), 423(182)

Vujanovič, B. D. (Chapter Author), **19**:199(47, 48, 49), 244(47, 48, 49); **24**:39, 41(9, 10, 11, 13), 48(18), 98(9, 10, 11, 13), 99(18)

Vuong, S. T., **26**:50(238, 239), 51(239), 104(238, 239)

W

Wachman, H. Y., **7**:168, 169, 211

Wacholder, E., **15**:324

Wachtell, G. P., **6**:244(73), 365; **9**:118(6), 177

Wackerly, D. D., **31**:349(35), 426(35)

Wada, E., **SUP**:103(269)

Wada, J., **28**:96(68), 141(68)

Wadsworth, D. C., **23**:5(7), 13(7, 38, 72), 15(7, 38, 72), 19(7, 38), 20(7), 32(7), 33(7), 34(38), 39(7), 51(72), 52(72), 59(7, 38, 109), 61(7, 38, 109), 88(7), 89(38), 90(7), 91(7), 92(7), 93(7), 98(7), 103(109), 105(109), 126(7), 128(38), 130(72), 131(109); **26**:135(82), 164(82), 188(153), 192(152, 153), 212(82), 215(152, 153)

Wageman, W. E., **7**:280, 319

Waggener, J. P., **9**:118(6), 177

Wagner, C., **7**:98, 160; **12**:233(113), 251(113), 270(113), 282

Wagner, H. G., **29**:87(53), 126(53)

Wagner, M. H., **13**:260(117), 266

Wagner, R. S., **12**:242(114), 282

Wagner, W., **17**:178, 202(28), 314

Wahi, M. K., **9**:130, 179

Wahiduzzaman, S., **23**:343(290), 367(290)

Wahl, G., **28**:389(169), 397(169, 186), 422(169), 423(186)

Wahlroos, J., **28**:20(36), 21(36), 22(36), 71(36)

Wainwright, T. E., **5**:48(76), 54

Wairegi, T., **26**:12(75), 96(75)

Waje, S. J., **23**:217(173), 274(173)

Wakao, N., **23**:371(24), 461(24); **24**:102(14), 184(14)

Wakehima, H., **14**:314, 317(79), 344

Wakeshima, H., **9**:224(54), 269; **10**:100, 101, 104, 164; **11**:8, 48

Wakimoto, T., **26**:142(100), 144(100), 157(100), 158(100), 213(100)

Wakiyama, Y., **17**:50(161), 63

Walch, M., **31**:461(40), 472(40)

Wald, A., **9**:159, 180

Walden, R. W., **17**:133, 156; **21**:154, 155, 183

Waldman, G. D., **29**:170(132), 209(132)

Waldmann, H., **7**:320c, 320f

Waldmann, L., **3**:301, 301(79)

Waldram, S. P., **23**:196(80), 270(80); **25**:199(100), 247(100)

Waldron, H. F., **7**:206, 218

Waldvogel, J. (Chapter Author), **28**:1, 54(80), 55(80), 73(80)

Walin, G., **24**:295(46), 296(37, 46), 302(46), 320(37, 46)

Walitza, E., **22**:296(39), 348(39)

Walker, D., **30**:356(106), 357(106), 431(106)

Walker, G., **10**:19(28), 36

Walker, G. H., **14**:298(29), 342

Walker, J. L., **11**:318(5), 412(5), 413(5), 417(5), 434

Walker, J. S., **30**:333(51, 54), 334(51), 351(54), 371(142), 390(54), 428(51), 429(54), 433(142)

Walker, K., **14**:16(41), 19, 20(41), 75(41), 101

Walker, K. L., **14**:34(86), 67, 102

Walker, L. F., **9**:3(113), 66(113), 110

Walker, P. L., **2**:138(22), 268

Walker, P. T., **15**:228(2), 268, 276

Walker, R. E., **3**:274, 285, 286, 287(78, 80), 288, 293(78, 91), 299(99), 301, 302

Walker, S. M., **15**:63(22), 64(22), 135(22)

Walker, W. S., **5**:149, 156, 182, 247

Walko, R. L., **25**:377, 419

Wall, F. T., **3**:287(88), 302

Wall, T. F., **31**:371(37), 426(37)

Wallace, D. J., **18**:220(154), 221(154), 222(154), 233(154), 239(154)

Wallace, J., **22**:421(31), 436(31)

Waller, P. R., **9**:129(37), 178

Wallick, G. C., **2**:375(96), 396

Wallis, G. B., **1**:359, 361, 390(59), 392, 440, 441; **9**:258, 265(102), 271; **15**:228(7), 270 (106), 271(106), 276, 281; **17**:44, 49, 51, 63, 64; **20**:90, 100, 107, 108, 132; **21**:331, 346

Wallis, J., **11**:9(16), 10, 48

Wallis, J. D., **1**:202(12), 210, 214, 264; **10**:122, 123, 125, 126, 127, 128, 130, 131, 132, 133, 152, 165; **20**:15(30), 80(30)

Walls, H. A., **20**:24(52), 81(52)

Walowit, J. A., **21**:149, 150, 160, 183

Walowitt, J., **5**:229(155), 230(155), 231, 232(155), 250

Walraven, Th., **3**:192(16), 249

Walrond, K. W., **23**:193(30), 206(30), 223(30), 226(30), 268(30)

Walsh, T. E., **30**:258(12), 259(12), 271(12), 279(56), 286(71), 294(12), 295(12), 296(56), 297(56), 300(71), 301(71), 302(12), 307(12), 309(56), 310(71)

Walshaw, C. D., **14**:252(6), 279

Walters, D. V., **31**:347(32, 33), 364(33), 367(32), 426(32, 33)

Walters, H., **6**:538, 563

Walters, H. H., **5**:402, 403, 440, 445, 446, 447, 448, 449, 508, 511; **11**:170(180), 173, 174, 175(181), 177(181), 197

Walters, K., **15**:70(49), 76, 77(49), 136(49); **19**:268(32), 317(32), 350(32); **23**:206(122), 223(122), 272(122); **24**:108(80), 187(80); **25**:292, 308, 317; **28**:152(10), 228(10)

Walters, S., **9**:104, 110

Walther, A., **11**:402(167), 438; **12**:124, 192

Walton, J. S., **10**:172, 174, 175, 177, 181, 182, 199, 203, 206, 215, 216; **14**:151(10), 242; **19**:144(153), 188(153)

Walz, A., **13**:79(28), 116

Wampler, F. C., **23**:193(257), 278(257)

Wamsler, F., **11**:202, 204, 209, 252

Wamsley, W. W., **10**:173, 215

Wan, S. F., **7**:169, 180, 211

Wanchoo, R. K., **26**:49(172), 100(172)

Wanford, D. W. B., **2**:353

Wang, A. B., **25**:390, 411

Wang, C., **12**:84(14a), 112; **23**:229(258), 278(258); **24**:122(131), 129(131, 135), 130(135), 139(131), 141(131), 144(135), 145(135), 146(135), 147(135), 148(135), 149(135), 174(131), 189(131, 135)

Wang, C. A., **28**:389(170), 390(172), 391(172), 392(172), 422(170, 172)

Wang, C. C., **30**:83(62), 90(62)

Wang, C. H., **26**:77(240), 104(240); **29**:47(42), 57(42)

Wang, C. Y. (Chapter Author), **28**:252(51), 264(51, 83, 84, 85, 86), 265(83), 266(84), 268(98), 327(51, 84), 331(51), 332(83, 84), 333(85, 86, 98); **30**:93, 94(16), 95(25), 103(16), 104(16), 105(16), 108(16, 44, 45, 46, 46a, 47, 48, 51), 117(71), 118(71), 122(44, 45), 127(44, 51), 128(44), 129(44), 130(44), 131(44), 132(44, 45), 133(45), 137(46), 138(46), 139(46, 51), 140(46a), 143(105), 144(105, 106), 145(106), 146(106), 147(106), 148(106), 149(106), 150(106), 151(105), 154(48), 156(48), 157(48), 158(48), 160(48), 161(48), 162(46a, 48), 177(47), 178(47), 179(47), 180(47), 181(47), 182(159), 189(16, 25), 190(44, 45, 46), 191(46a, 47, 48, 51), 192(71), 193(105, 106), 196(159)

Wang, D. I., **9**:376(45), 415; **15**:48(74), 57(74)

Wang, D. I. J., **5**:343345, 506

Wang, D. P., **18**:333(44), 363(44)

Wang, G.-X., **28**:28(50, 51, 52), 36(50, 51, 52), 56(50, 51, 52), 59(52), 72(50, 51, 52); **30**:422(184), 435(184)

Wang, H. E., **2**:94(76), 108; **3**:16(32, 34), 21(34), 23(34), 32, 256, 257, 300

Wang, J. Y., **11**:426, 428(235), 440

Wang, K. C., **3**:210(43), 250

Wang, K. K., **28**:167(39), 201(39), 204(39), 229(39)

Wang, K. Y., **18**:34(69a), 35(69a), 82(69a); **23**:153(32), 185(32)

Wang, L. S., **5**:254(6), 298(52), 299, 306(6), 311(52), 312(6), 322, 323; **8**:247, 273, 282; **9**:377(46), 379(46, 52), 390, 415; **11**:397(157), 438; **12**:177, 178, 180, 193

Wang, P., **19**:80(239), 93(239)
Wang, Q.-J., **18**:12(17), 79(17)
Wang, R. C., **19**:179(235), 190(235)
Wang, S., **5**:91, 126; **11**:105(94), 113(94), 193
Wang, S. K., **29**:151(71), 206(71), 215(1), 248(1), 334(1)
Wang, S. S., **28**:218(65), 230(65)
Wang, S. Y., **30**:96(28), 190(28)
Wang, T., **23**:327(216), 328(216), 364(216)
Wang, T. G., **24**:3(17), 5(17), 36(17)
Wang, T. P., **18**:333(45), 363(45)
Wang, T. Y., **17**:140, 141, 157; **25**:260, 270, 283, 319
Wang, W., **25**:234(132), 235(132), 248(132)
Wang, W. C., **21**:124(113), 138(113)
Wang, X. A., **28**:376(151), 379(151), 380(151), 408(224), 411(228), 413(228), 422(151), 425(224, 228)
Wang, X. S., **26**:109(21), 113(21, 28), 114(33), 209(21, 28, 33)
Wang, Y., **25**:234(134, 135, 136), 235(134, 135), 248(134, 135), 249(136); **27**:187(38), 193(38); **28**:190(66), 230(66)
Wang, Y. L., **SUP**:73(191), 91(191), 94(191), 142(191), 161(191), 165(191), 166(191), 167(191), 300(191)
Wang, Z., **23**:324(210), 363(210)
Wang-Chang, C. S., **7**:170(16), 171, 172, 211, 212
Wankat, P. C., **20**:321(17), 350(17)
Wanniarachchi, A. S., **26**:219(2), 324(2)
Wansbrough, R. W., **4**:206(82), 226; **5**:131(14), 247
Wansink, D. H. N., **17**:97, 154
Warburg, E., **2**:354
Warburton, F. L., **25**:294, 315
Ward, C. A., **10**:121, 165
Ward, J., **18**:68(128), 70(128), 85(128)
Ward, J. A., **31**:123(2, 24), 134(2, 24), 155(2), 156(24)
Ward, J. C., **14**:4(14), 5(14), 100; **24**:112(98), 188(98)
Ward, S., **29**:98(68), 126(68)
Ward, W. D., **5**:378(93), 509
Warhaft, J., **21**:185(8), 236(8)
Warhaft, Z., **21**:218(35), 219(35), 220(35), 237(35)
Warkentin, P. A., **17**:126(123), 156
Warming, R. F., **5**:28(35), 29, 52; **8**:263(40), 283; **11**:344(61), 436; **24**:226, 275

Warnatz, J., **29**:77(36), 125(36)
Warner, M. E., **12**:18(31), 23(31), 74
Warner, W. H., **20**:364, 388
Warn-Varnas, A., **24**:290(15), 319(15)
Warren, N., **10**:43, 62, 81
Warren, W. A., **5**:381, 510
Warren, W. R., **4**:280(41), 314
Warren, W. R., Jr., **5**:309(79), 324
Warrington, R. O., **26**:221(22), 325(22)
Warrior, M., **25**:234(121), 248(121)
Warshavsky, M., **15**:63(26), 64(26), 135(26)
Warshawsky, I., **11**:219, 222, 257
Warsi, Z. U. A., **15**:329
Wartenstein, **2**:356
Wartique, J. M., **13**:219(50), 265
Wasan, D. T., **2**:388, 397; **14**:151(12), 242; **15**:321; **19**:144(154), 188(154); **26**:13(11), 93(11)
Washington, K. E., **29**:108(82), 127(82), 218(22), 233(67, 68), 234(22, 73, 75, 76), 243(75), 244(22), 275(22), 280(22, 75), 281(22), 282(76), 288(76), 295(22), 296(22), 297(22), 301(22), 306(22), 307(22), 308(22), 309(67), 310(22), 311(22), 313(76), 319(22), 321(22), 335(22), 338(67, 68, 73), 339(75, 76)
Washington, L., **19**:334(132), 335(132), 355(132)
Washington, W. M., **12**:24(43), 74
Wasilik, R. M., **17**:69(20), 70(20), 71(20), 153
Waslo, S., **23**:193(81), 208(81), 270(81)
Wassel, A. T., **11**:344(64, 64a), 348, 436; **12**:183(62), 193
Wasserman, B., **15**:329
Wasserman, M. L., **15**:69(44), 136(44); **25**:270, 272, 273, 319
Wasserman, P. D., **28**:410(225), 425(225)
Wassilew, P., **16**:114(77), 155
Watada, H., **26**:19(241), 104(241)
Watanabe, H., **15**:328; **26**:45(224), 103(224); **28**:351(71), 418(71); **30**:392(157), 393 (157), 396(163), 399(163), 434(157, 163)
Watanabe, K., **4**:167, 171, 224
Watanabe, M., **30**:396(164), 399(164), 434(164)
Watanabe, Y., **23**:293(63), 356(63); **25**:358, 416
Watari, T., **20**:183(11), 299(11)
Waterland, L. R., **15**:287(206), 329
Waterman, T. E., **11**:354(76), 436
Waters, E. D., **7**:320a, 320e; **17**:39, 62
Waters, P. L., **14**:180, 245

Watkins, W. B., **24**:294(47), 320(47)

Watkinson, A. P., **30**:200(66), 201(2, 65, 66), 202(66), 203(2, 66), 204(66, 68), 205(68, 69), 208(2, 66, 68, 69), 214(64), 215(64), 216(53, 64, 69), 220(64), 221(68, 69), 227(66), 249(2), 252(53, 64, 65, 66, 67), 253(68, 69); **31**:456(75), 474(75)

Watson, A., **7**:31, 50, 52(12), 54(12), 66(12), 84; **21**:46(73), 52(73)

Watson, E. J., **4**:331, 444; **26**:114(29), 141(29), 142(29), 202(29), 203(29), 209(29)

Watson, K., **10**:46, 48, 50, 51, 52, 73(19), 81

Watson, K. M., **5**:255(14), 259(14), 264(14), 303(14), 304(14), 305(14), 310(14), 319(14), 320(14), 322

Watson, K. P., **14**:119, 128, 130(108), 135, 148

Watson, R. G. H., **9**:244(73), 261(73), 270

Watson, V. R., **4**:283(45, 46, 37), 284(51), 314, 315

Watts, J., **23**:333(245), 365(245); **31**:174(81), 325(81)

Watts, M. J., **21**:20(34), 22(34), 25(34), 26(34), 51(34)

Watts, P., **29**:163(104), 176(104, 155), 191(155), 207(104), 210(155)

Watts, R. G., **20**:371, 388

Watts, R. L., **31**:433(16), 471(16)

Watwe, A. A., **23**:329(226, 227), 364(226, 227)

Waymouth, J. F., **24**:275

Wayner, P. C., **11**:105(132, 133), 111, 194; **20**:183(12), 294(319, 320), 299(12), 314(319, 320)

Wayner, P. C., Jr., **13**:129, 202

Weakliem, H. A., **28**:342(25), 416(25)

Weast, R. C., **22**:373(17), 381(17), 394(17), 435(17); **23**:142(22), 185(22)

Weatherford, W. A., **10**:243(38), 282

Weaver, J. A., **19**:78(184), 91(184); **22**:240(213, 214), 241(214), 242(214), 330(213), 331(213), 332(213), 356(213, 214)

Webb, B. W. (Chapter Author), **19**:68(156), 72(164), 79(210), 90(156, 164), 92(210); **24**:5(19), 6(24), 36(19), 37(24); **26**:105, 123(55), 125(55), 127(71), 129(55), 137(91, 92), 138(92), 139(92), 140(93), 141(91), 142(101), 143(91, 101), 144(101), 146(91, 93), 148(55), 149(71), 150(111, 112), 151(55, 71, 111, 112), 153(92, 93), 154(71, 92, 93, 111), 155(92, 111), 158(91, 101), 159(111), 160(111), 162(111),

165(71, 92), 169(121), 173(128, 129), 175(128, 129), 176(128), 177(128), 179(128, 129), 180(129), 181(128, 129), 182(112), 183(112), 184(112), 185(112), 191(150, 151), 195(150, 151), 198(91), 200(160), 210(55), 211(71), 212(91, 92, 93), 213(101, 111, 112), 214(121, 128, 129), 215(150, 151), 216(160)

Webb, D. R., **16**:147, 156

Webb, P., **4**:116(74), 140

Webb, R. L., **15**:21(46), 56(46); **18**:156(85), 159(85); **20**:244(203), 309(203); **21**:120(108), 125(118, 119), 138(108, 118, 119); **30**:198(70, 71, 73), 221(72), 223(27, 72), 234(71), 250(27), 253(70, 71, 72, 73); **31**:160(62, 70), 294(62), 304(208), 324(62), 325(70), 332(208), 453(77), 454(76), 456(77, 78), 474(76, 77, 78)

Webb, W. H., **6**:109(174, 176), 110, 112, 113, 131

Webb, W. L., **14**:269(60), 280

Webb, W. W., **6**:474(261), 483(261), 499

Webber, J. H., **21**:66(8), 134(8)

Weber, A., **14**:287, 298, 341

Weber, G. J., **4**:114(67), 140

Weber, H., **15**:285, 329

Weber, H. E., **4**:283(50), 315

Weber, J., **17**:72(36), 153, 154

Weber, J. E., **14**:20, 30(72), 31(72), 33(72), 66, 101, 103; **20**:330(47), 351(47)

Weber, K. H., **14**:118, 148

Weber, M. E., **16**:133(98), 135, 136, 137, 138, 139, 140, 141, 155, 156; **19**:61(139), 62(139), 89(139), 103(47), 185(47); **20**:100, 101, 132, 380, 388; **21**:334, 344; **25**:252, 316; **26**:2(55), 20(55), 23(55), 95(55); **28**:7(7), 70(7)

Weber, N., **11**:285, 288, 315

Weber, S., **2**:294, 356

Weber, W. J., Jr., **23**:370(17), 461(17); **30**:111(54, 57), 191(54, 57)

Webster, C. A. G., **11**:241, 242, 261

Webster, J. G., **22**:360(4), 367(4), 370(4), 373(4), 384(4), 435(4)

Wechsler, A. E., **10**:42, 43, 45, 48, 50, 51, 52, 54(36), 58(22), 59(22), 64, 68, 74(36), 81, 82

Weckermann, F. J., **1**:361(13), 363(13), 378(13), 381(13, 40, 41), 411(13, 40, 41), 413(13, 40, 41), 421(13, 40, 41), 436(13), 440, 441

Wedemeyer, E. H., **24**:294(48), 320(48)

Weder, E., **11**:203, 206, 254

Wedlake, E. T., **23**:320(171), 361(171)

Weekman, V. W., Jr., **30**:168(130), 195(130)

Wegener, P. P., **14**:284(7), 317, 331(110), 332, 333(7), 334(110, 126), 335, 341, 344, 345

Wegner, R. E., **14**:78(168), 104

Wehrmann, O. H., **11**:200(61), 240(61), 241(61), 242(61), 255

Wei, C., **28**:364(103), 367(103), 369(103), 373(103), 395(103), 411(103), 419(103)

Wei, C.-C., **10**:146, 165

Wei, D. Y., **28**:308(154), 336(154)

Wei, L. H., **29**:47(42), 57(42)

Weidman, P. D., **28**:268(96, 97), 333(96, 97)

Weidnnann, M. L., **2**:328, 356

Weihs, D., **8**:29, 87

Weikle, D. H., **23**:342(280), 367(280)

Weil, J., **5**:131, 246; **11**:105(107), 113(107), 193

Weil, L., **5**:412, 415, 416, 422, 512, 513; **11**:105(68, 84, 85, 86), 109(68, 84), 111(68), 112, 192; **14**:120, 121(14, 15, 18), 127, 128, 129(18), 130(14, 15, 18), 144

Weiland, W. F., **19**:334(133), 335(133), 355(133)

Weinbaum, S., **6**:45, 46(93), 47, 48, 49, 50, 52(93), 53, 55(98), 82(98), 126, 127; **8**:169, 173, 202(31), 226; **19**:28(58), 29(58), 30(58, 61), 33(58, 61), 44(101), 78(101, 191, 192), 86(58, 61), 87(101), 91(191, 192); **22**:14(26, 37), 18(26, 37), 19(24, 34, 55, 56, 59, 60, 61, 62, 63, 70), 24(41, 60), 26(60, 61), 33(61, 63), 45(34, 61), 69(56), 71(19, 34, 55, 56, 59, 60, 61, 62, 63, 70), 96(34, 55, 56, 59, 60, 61, 62, 63, 70), 97(59), 99(59), 101(59), 103(34, 60), 104(59, 60), 105(59, 60), 106(60), 107(41, 59, 60), 108(34, 60), 109(34, 60), 110(34, 60), 111(34), 113(34), 114(34), 116(34), 117(34), 118(34), 119(34), 120(34), 121(34), 122(24, 41, 60), 123(34, 41, 59, 60, 61), 124(34, 60, 61), 125(34, 59, 60, 61), 126(34, 60, 61), 127(70), 128(61), 129(34, 60), 130(34, 60, 61), 131(24, 34, 55, 60), 132(24, 34, 41, 55, 60), 133(34, 55), 134(34, 55, 60), 135(55), 136(55), 137(55, 62, 63), 138(70), 139(60, 62), 140(61, 62), 141(41, 70), 142(61, 70), 144(63, 70), 145(70), 146(63, 70), 147(63), 150(63), 151(19, 24, 55, 56), 152(19, 41), 153(19, 24), 154(34, 41), 155(55, 56, 59, 60, 61, 62, 63, 70)

Weinberg, E. B., **15**:195(73), 224(73)

Weinberg, F., **28**:249(27, 28), 330(27, 28)

Weinberg, F. J., **6**:135(3), 162, 166, 195, 362; **17**:181(33), 314; **30**:301(104), 311(104)

Weinberg, S., **15**:235(38), 236(38), 277

Weinberger, H. F., **15**:307(155), 327

Weiner, A., **20**:380, 381, 388

Weiner, J. H., **22**:307(18), 308(18), 312(18), 347(18)

Weiner, K. H., **28**:98(105), 143(105)

Weiner, M. M., **5**:24, 51, 301(58), 323; **12**:139, 141(35), 147(32), 177(51), 186(35), 193

Weiner, R. K., **22**:222(63), 349(63)

Weiner, S. C., **19**:113(93), 145(184), 187(93), 189(184)

Weingaertner, E., **10**:204, 217

Weinstein, H. G., **24**:104(27), 185(27)

Weintraub, M., **10**:181(64), 216

Weise, A., **6**:265(79), 365

Weise, N. O., **19**:82(262), 94(262)

Weisman, J., **20**:83, 132; **30**:353(102), 354(102), 431(102)

Weiss, A., **1**:245(86), 266, 405(85), 409(85), 411(85), 421(116), 442, 443

Weiss, H.-J., **28**:103(88), 142(88)

Weiss, P. A., **19**:79(199), 92(199)

Weiss, R. F., **6**:45, 48, 50, 53, 57, 58, 61, 62, 64(101), 65(101), 98(160), 100(160), 101, 107, 126, 127, 130

Weiss, R. O., **8**:24, 87

Weiss, S., **5**:381, 510

Weissman, S., **3**:270(32, 33), 271(33), 272, 297, 298(96), 300, 302

Welander, P., **7**:170(15), 171, 211; **14**:44(107), 102

Welch, A. J., **22**:248(218, 220), 273(166, 167, 221), 277(40, 215, 216, 219, 220), 278(149, 215), 279(167, 210, 220), 280(220, 221), 282(210), 284(210, 216), 285(210), 286(210), 287(210), 315(166), 317(166), 318(166), 342(216), 348(31, 40), 353(149), 354(166, 167, 168), 356(210, 215, 216, 217, 218, 219, 220, 221), 386(21), 387(21), 404(21), 408(21), 421(35), 422(35), 423(35), 424(35), 435(21), 436(35)

Welch, A. U., **20**:186(39), 187(39), 300(39)

Welch, C. P., **6**:551, 564; **7**:49, 84

Welch, J. A., Jr., **5**:255(14), 259(14), 264(14), 303(14), 304(14), 305(14), 310(14), 319(14), 320(14), 322

Welch, J. E., **24**:194, 270

Welch, J. F., **9**:204, 206(20), 231(20), 268; **21**:80(30), 134(30)

Welch, W. E., **8**:163, 226

Welder, B. Q., **30**:243(74), 253(74)

Welker, J. R., **10**:260, 282

Welleck, R. M., **26**:19(242), 104(242)

Wells, C. S., Jr., **2**:385(120), 397; **12**:79(8a), 80, 88, 101, 102, 104, 105, 112; **15**:105(135), 107(135), 111(135), 114(135), 140(135)

Wells, E. N., **10**:19(25), 24(25), 36

Wells, M. B., **5**:37, 53

Wells, S. C., **25**:262, 319

Welsh, W. E., Jr., **2**:51, 52, 68, 75, 78(52), 80(52), 83(52), 84(52), 87, 107, 108, 153, 156, 269; **5**:177, 249; **9**:20, 33, 35(50), 40(50), 108

Weltmann, R. N., **7**:200(121), 216

Welty, J. R., **SUP**:135(313); **15**:101(114), 103(114), 139(114); **18**:246(8), 255(8), 257(8), 261(8), 265(59), 266(59), 270(8), 271(8), 272(8), 320(8), 322(59); **19**:164(214, 217, 218), 168(222), 169(214), 170(214), 171(217), 172(217), 190(214, 217, 218, 222)

Wen, C. Y., **9**:121, 178; **10**:189, 217; **14**:154(45), 243; **19**:99(19), 100(19), 108(19, 50), 185(19), 186(50, 60); **23**:193(259, 260), 215(265), 229(259, 260), 236(259, 265), 238(265), 239(265), 240(265), 250(259), 278(259, 260, 265)

Wendel, M. M., **SUP**:73(190), 164(190), 165(190)

Wender, L., **10**:200, 201, 205, 209, 217; **14**:187, 217, 218(117), 225(117), 226(117), 227, 229, 245; **19**:134(141), 135(141), 180(141), 188(141)

Wendt, J. F., **23**:326(212), 364(212)

Wendt, R. L., **SUP**:72(165, 170), 266(165, 170), 267(170, 444)

Wengle, H., **25**:360, 389, 402, 406, 419

Wenner, K., **8**:130(63), 160; **11**:218, 222, 236, 256; **18**:127(62), 159(62); **20**:377, 388

Wenzel, H., **9**:221(50), 269

Wenzel, L. A., **5**:413, 513

Wepfer, W., **15**:52(85), 58(85)

Wepfer, W. J., **15**:2(12), 55(12)

Werle, H., **19**:80(227), 93(227); **21**:281, 283, 287, 293, 301, 346

Werlein, P. P., **2**:316(108), 355

Wernekinck, U., **30**:273(46), 274(46), 283(69), 286(69), 309(46), 310(69)

Werner, H., **25**:360, 389, 402, 406, 419

Werner, R. A., **10**:56(44), 82

Werner, R. W., **7**:274, 276, 294(118), 295, 318, 319

Werther, J., **19**:123(122, 123, 124), 125(122, 123, 124), 187(122, 123, 124)

Weske, J. R., **11**:241, 242, 260

Wesley, D. A., **23**:284(35), 355(35)

Wesselink, A. J., **10**:40, 46, 71, 72, 75, 78, 81

West, D., **9**:209(48), 210(34), 221(48), 233(34), 244(48), 264(130), 265(34, 48, 130), 269, 271; **21**:82(33), 83(38), 88(54), 91(33), 95(33), 135(33, 38, 54)

West, E. A., **10**:48, 51, 52, 56, 57, 81

West, J., **2**:294, 354

West, J. T., **31**:402(61), 428(61)

West, L. A., **12**:3(7), 14(7), 51(7), 59(7), 68(7), 69(7), 73

West, R., **28**:267(89), 333(89)

West, T. J., **23**:222(117), 272(117)

Westenberg, A. A. (Chapter Author), **3**:253, 254(4), 274, 276(41, 53), 282(41, 53), 285, 287(78, 80, 83), 288, 293(78, 91), 299(99), 300, 301, 302

Wester, M. J., **29**:108(81), 127(81)

Westerberg, K. W., **25**:260, 268, 319

Westergren, L., **31**:461(47), 472(47)

Westhaver, J. M., **12**:237(115), 282

Westkaemper, J. C., **23**:285(42), 356(42)

Westman, A. E. R., **13**:121, 201

Weston, J. A., **26**:71(64), 95(64)

Westre, W. J., **12**:3(8), 11(8), 14(8), 73

Westwater, J. W., **1**:209, 213, 214, 216, 234(64), 238(64), 239(64), 259(64), 264, 265; **4**:159, 173, 224, 225; **5**:57, 66, 67(13), 67(20, 21), 68(13), 69(13), 71(13), 73, 75, 78, 81, 82, 88, 90, 102, 103(3), 104, 106, 109, 112, 113, 114, 125, 126, 128, 403, 404, 405, 406, 419, 428, 432, 436, 438, 439, 464, 495, 511, 512, 514; **7**:81, 86; **9**:204, 206(20), 207(29), 208, 210(29), 211, 214, 231(20), 232(29), 234, 235, 236, 238(32), 239, 240, 265(33), 268, 269; **10**:119, 131(56), 132(56), 165; **11**:9(17), 48, 55(3, 4), 56(21), 57(3, 12), 58(21), 60(4), 91, 95(25), 96(12), 98(25), 99, 100(25), 104, 105(3, 4, 12, 21, 117, 125), 108, 109, 110, 111, 112, 114, 115, 116, 118, 120(4, 25),

121(12, 25, 117), 122(12, 25, 117), 189, 190, 194; **16**:64(8), 90, 91, 133, 134, 138, 153, 154, 155, 156; **17**:3(16), 5, 8, 11(51), 16, 19, 23(51), 56, 57, 58, 147, 148(175), 157; **18**:247(11), 255(23), 257(23), 258(48), 259(56), 260(48), 293(23), 320(11), 321(23), 322(48, 56); **19**:44(103), 45(103, 104), 87(103), 88(104); **20**:15(32), 80(32), 250(229), 310(229); **21**:80(30, 32), 103(83, 84), 134(30), 135(32), 137(83, 84); **23**:82(97), 131(97)

Wetherald, R. T., **8**:281(52), 283

Weyburne, D. W., **28**:389(170), 422(170)

Weyl, E. J., **6**:135(5), 362

Weyl, F. J., **30**:262(25), 308(25)

Whalen, R. J., **4**:349, 446

Whaley, F. R., **2**:356

Whalley, E., **18**:334(57), 340(57), 341(57), 342(57), 363(57)

Whalley, P. B., **17**:42, 43, 44(140, 141), 62, 63; **29**:266(133), 342(133); **31**:123(28), 134(28), 157(28)

Wheatley, J. C., **17**:69(27), 72(27), 126(123), 153, 156

Whedon, G. D., **4**:120(88), 140

Wheeler, J. A., **15**:69(43), 136(43); **19**:261(22), 262(22), 264(22), 265(22), 267(22), 268(22, 31), 317(31), 330(31), 350(22, 31); **24**:104(27), 185(27)

Whellock, J. G., **14**:180(102), 245

Whipple, F. J. W., **5**:40(73), 54

Whitaker, S. (Chapter Author), **13**:119, 128(34, 36), 130(36), 135(34), 137, 138(43), 140(43), 144, 146(43, 49), 159, 161(43), 164, 170, 172(44), 173(44), 176(59), 202, 203; **14**:6(27), 8(22, 29), 12(22, 29), 100; **SUP**:73(190), 164(190), 165(190); **18**:191(148, 149), 230(148), 239(148, 149); **23**:369, 370(5, 6, 18, 22), 371(26, 27, 30, 31), 372(34), 373(6), 374(30, 37, 41, 42, 43, 43a, 43b, 44, 44a), 375(43, 43a, 43b, 44, 44a), 376(43, 43a, 43b), 377(49), 378(49, 54, 55), 381(43, 43a, 43b, 44, 44a, 58), 382(43, 43a, 43b, 44, 44a), 383(43, 43a, 43b, 44, 44a), 385(43, 43a, 43b, 44, 44a), 386(34), 390(34), 391(34), 394(18), 398(34, 61, 62, 63, 64), 400(34), 404(34), 407(6, 61, 69), 409(34, 43, 43a, 43b, 44, 44a, 61, 62, 70), 411(31), 415(64), 421(61), 422(34, 69), 423(26, 78), 426(79, 80),
427(18, 81, 82), 428(43, 43a, 43b), 430(31), 435(6, 27, 69, 88), 437(43, 43a, 43b, 44, 44a), 441(43, 44), 442(34, 69), 447(90), 451(18, 80), 452(22), 458(18), 461(5, 6, 18, 22, 26, 27), 462(30, 31, 34, 37, 41, 42, 43, 43a, 43b, 44, 44a), 463(49, 54, 55, 58, 61, 62, 63, 64, 69, 70), 464(78, 79, 80, 81, 82, 88, 90); **24**:103(17), 114(107, 108, 109, 119), 185(17), 188(107, 108, 109), 189(119); **25**:298, 319; **28**:11(18), 71(18); **30**:93(4, 4a), 94(4), 95(21, 23, 24), 98(21), 110(52, 53), 111(59), 169(4, 138), 188(4), 189(4a, 21, 23, 24), 191(52, 53, 59), 195(138); **31**:1, 1(12), 4(22), 7(40), 9(40, 45, 47, 48), 10(45), 13(11b), 17(44), 18(25, 26), 19(28, 29, 30, 31, 32a), 20(28, 29, 30, 31, 32a), 22(28, 29, 30, 31, 32a), 23(28, 29, 30, 31, 32a, 32b, 37), 26(50), 28(17, 18, 19, 20, 21), 30(5), 32(6), 33(6), 34(6, 16, 27, 32c, 40), 36(32c, 47), 39(12, 24, 25, 26, 41, 42, 51), 40(12, 41, 42, 46, 51), 43(27, 32b, 49), 46(38), 52(41, 42), 53(27, 32b), 54(27, 32b, 38, 39, 43, 49), 56(38), 60(12), 68(28, 29, 30, 31, 32a), 69(32b), 72(6, 40, 41, 42, 43), 74(27, 41, 42), 76(27, 32c, 40, 41, 42, 43), 80(11c, 16, 32c), 82(6, 14, 15, 27), 86(32c, 40, 41, 42, 43), 92(14, 15), 98(22), 99(22), 102(5, 6, 11b, 11c, 12, 14, 15, 16), 103(17, 18, 19, 20, 21, 22, 24, 25, 26, 27, 28, 29, 30, 31, 32a, 32b, 32c), 104(37, 38, 39, 40, 41, 42, 43, 44, 45, 46, 47, 48, 49, 50, 51)

White, A. J., **23**:315(148, 149), 360(148, 149)

White, C. W., **28**:80(64), 141(64)

White, D., **15**:69(46), 136(46); **19**:68(158), 69(158), 90(158)

White, D. A., **23**:212(261), 278(261)

White, D. E., **14**:2(1), 15(1), 26(38), 77(161), 78, 79, 88, 100, 103, 104

White, D. R., **8**:248(22), 283

White, E. P., **23**:108(110), 109(110), 110(110), 132(110)

White, F. M., **26**:118(37), 122(37), 129(37), 210(37); **28**:389(161), 422(161)

White, J. E., **10**:2(4, 5, 6), 12, 13, 23(6), 31(4, 6), 34(4, 6), 35, 54(39), 70(39), 75(39, 70, 71), 76(70), 77(39), 78(39, 70, 71), 80(39, 70), 82, 83; **19**:81(247), 94(247)

White, J. L., **2**:358, 361, 375(124), 378(121), 391(75, 121), 396, 397; **15**:77(63), 137(63),

White, J. L. (*continued*)
178(36, 37), 224(36, 37); **25**:262, 319;
28:151(67), 190(61, 66), 230(61, 66, 67)
White, J. W., **15**:183(49), 224(49); **25**:259, 281,
317
White, R. A., **6**:29(47), 124
White, R. M., **23**:281(29), 355(29)
White, S. M., **30**:147(107), 193(107)
White, S. P., **20**:347(67), 352(67)
White, T. J., **22**:266(133, 134, 135, 222),
268(133, 134, 135), 270(135), 271(135),
272(135), 273(134, 222), 274(134),
275(134), 276(134), 353(133, 134, 135),
357(222)
White, T. R., **19**:122(111), 123(111), 126(111),
127(111), 128(111), 166(111), 187(111)
White, Z., **24**:104(61), 186(61)
Whitehead, A. B., **19**:123(117, 118, 119, 120),
125(117, 118, 119, 120), 187(117, 118,
119, 120)
Whitehead, J. C., **19**:108(55), 186(55)
Whitehead, L. G., **5**:168(73), 248
Whitehouse, R. H., **4**:94(34), 138
Whitelaw, J. H., **7**:339, 348, 350, 352, 352(36),
355, 356, 357, 358, 359, 378, 379; **8**:298,
342(64), 349, 350; **18**:343(61), 344(61),
345(61), 346(61), 348(61), 349(61),
363(61)
Whitelaw, R., **9**:117(5), 129(5), 177
Whitelaw, R. L., **30**:236(50), 252(50)
Whiteley, A. H., **10**:118(50, 51, 52, 53), 121(50,
51, 52, 53), 165
Whiteman, I. R., **2**:374, 397; **SUP**:109(276)
Whiting, G. H., **7**:223(9), 235(21), 273(9), 274,
314
Whiting, R. L., **14**:92, 93, 94(181), 104
Whitlock, M., **2**:363(69), 396
Whitlock, P. A., **31**:334(8), 381(8), 425(8)
Whitmore, R. A., **23**:342(278), 367(278)
Whitmore, R. L., **22**:50(64), 99(64), 155(64)
Whitsitt, N. F., **15**:105(134), 107(134), 111(134),
114(134), 140(134)
Whitten, D. G., **16**:278, 349(49), 355, 357, 365
Whitty, G. F., **10**:270(102), 283
Wianachchi, S., **28**:135(119), 144(119)
Wibulswas, P., **SUP**:76(210), 77(219, 220),
214(210, 220), 215(220), 216(220),
217(210), 218(220), 219(219, 220),
220(219, 220), 221(220), 240(220),
242(220), 243(220), 244(220), 245(220),

402(220), 404(219); **19**:289(84), 291(84),
300(84), 307(84), 312(84), 353(84)
Wicke, E., **10**:193, 202, 203, 205, 206, 217;
14:218(153, 155, 156), 222(155, 156),
223(153), 246
Wickenhauser, G., **16**:171(5), 172(5), 173(5),
193(5), 215(5), 220(5), 221(5), 234, 238;
20:23(47), 81(47)
Wickett, A. J., **29**:194(205), 213(205)
Wickey, R. O., **1**:357(3), 374(3), 375(3), 440
Wicks, M., **1**:376(26), 387(26), 390(26), 392,
441
Wicks, M., III, **5**:451, 515
Widmer, A. E., **28**:402(191), 404(191), 423(191)
Widtsoe, J. A., **13**:121, 201
Widua, J., **31**:462(63), 473(63)
Wiederecht, D. A., **5**:369, 508; **7**:53, 85
Wiegand, J. H., **15**:330
Wiegandt, H. F., **15**:267, 268, 269, 270(103),
280, 281
Wieghardt, K., **7**:336, 337(10), 338, 343, 344,
370, 377
Wiehe, I. A., **13**:219(93), 266
Wieland, W. F., **6**:539, 563
Wieme, R. J., **12**:256(116), 282
Wien, W., **SUP**:285(453)
Wienecke, R., **4**:265(10), 267, 268(10, 31), 313,
314
Wiener, O., **6**:162, 363
Wierzba, I., **29**:68(17), 124(17)
Wiesinger, H., **22**:251(85), 350(85)
Wiggins, E. J., **25**:294, 319
Wiggs, M. M., **6**:79(134), 128
Wightmann, J. C., **20**:243(178), 308(178)
Wiginton, C. I., **19**:268(51), 271(51), 272(51),
274(51), 278(51), 351(51)
Wiginton, C. L., **SUP**:72(165, 168, 169, 170),
199(168), 210(168), 213(168), 266(165,
169, 170), 267(170)
Wigner, E. P., **2**:355
Wijeysundera, N. E., **18**:21(50), 22(50), 23(50),
26(50), 81(50)
Wikhammer, G. A., **17**:39(129), 62
Wilcox, S. J., **21**:73(20), 75(20), 134(20)
Wilcox, W. R., **2**:388, 397; **9**:304(58), 347;
11:412(186, 188), 439; **30**:402(169),
434(169)
Wild, N. E., **3**:284, 301
Wildey, R. L., **10**:46, 71(61), 73(62), 82
Wildin, M. W., **23**:301(103), 358(103)

Wilhelm, D. J., **9**:218, 219(45), 269

Wilhelm, R. H., **7**:89(5c, 5d); 160; **19**:177(230), 190(230)

Wilke, C. H., **21**:170, 171, 172, 174, 179

Wilke, C. R., **2**:38, 106, 388(119), 397; **3**:269, 299(98), 300, 302; **4**:16, 63; **7**:89, 96, 98, 98(11), 130(40), 160, 161; **12**:233(105), 251(105, 117, 118), 270(105), 281, 282; **15**:267(102), 268(102), 269, 280

Wilke, G. R., **15**:236, 278

Wilke, J. O., **30**:272(41), 309(41)

Wilke, T. E., **28**:402(203), 424(203)

Wilkes, J. O., **8**:168, 173, 193, 194, 226; **20**:100, 101, 105, 131

Wilkes, N. S., **25**:329, 384, 385, 398, 409, 419

Wilkie, D., **9**:3(114), 110; **31**:246(153), 261(153), 329(153)

Wilkins, D. G., **18**:337(50), 363(50); **21**:88(55), 103(81), 136(55), 137(81)

Wilkins, J E., **8**:33(158), 88

Wilkins, J. E., Jr., **2**:434(37), 451

Wilkins, M. D., **30**:111(58), 191(58)

Wilkinson, G. T., **9**:127(30), 178

Wilkinson, R. G., **18**:16(26), 80(26)

Wilkinson, S. P., **30**:365(132), 432(132)

Wilkinson, W., **2**:357(127), 358, 397

Wilkinson, W. L., **13**:219(76), 243(76), 265; **23**:193(23), 201(23), 202(23), 212(23), 213(23), 236(23), 238(23), 240(23), 242(23), 246(23), 267(23); **24**:108(78), 187(78)

Wilks, J., **5**:487(269), 517; **17**:68(13), 69(28), 70(13), 72(13), 75, 102, 153, 154

Wilkson, D. B., **7**:206, 218

Wille, R., **30**:32(35), 89(35)

Willhite, G. P., **23**:193(58), 226(262), 227(262), 269(58), 278(262)

William, D. W., **10**:270, 283

Williams, A. R., **4**:291, 315

Williams, B., **1**:389, 441

Williams, C. C., **11**:376(129), 437

Williams, C. G., **4**:120(90), 140

Williams, D., **5**:279(24), 298(45), 322, 323; **12**:129, 138(12), 192; **24**:19(50), 38(50)

Williams, D. C. (Chapter Author), **29**:62(8), 124(8), 146(43), 204(43), 215, 216(10), 217(13), 218(22), 221(33), 233(59), 234(22, 73, 75, 76, 77, 78, 79, 80), 243(75), 244(22), 266(132), 275(22, 77), 280(22, 75), 281(22), 282(76), 288(76), 291(80),

293(80), 294(80), 295(22, 80), 296(22), 297(22), 301(22), 306(22), 307(22), 308(22), 310(22, 33), 311(22, 33), 313(76), 319(22), 321(22), 328(78), 329(78), 335(10, 13, 22), 336(33), 338(59, 73), 339(75, 76, 77, 78, 79, 80), 342(132)

Williams, D. F., **14**:166, 173, 182, 183(89), 184(89), 244

Williams, D. L., **14**:41, 42(95), 102

Williams, E. J., **14**:320, 344

Williams, F., **17**:182(37), 184(37), 186, 192, 193, 314; **23**:333(245), 365(245); **31**:174(81), 325(81)

Williams, F. A., **10**:243(37), 269, 282, 283; **26**:81(15), 82(15), 88(243), 93(15), 104(243)

Williams, G. C., **10**:269(93), 283

Williams, J. J., **7**:350, 358, 379

Williams, J. R., **10**:184(69, 70), 192(69), 216; **14**:152(33), 154(33), 193(33), 243

Williams, M. C., **15**:63(24), 64(24), 135(24); **16**:134(99), 155, 233

Williams, M. J., **8**:127(58), 128, 130(58), 160; **11**:208, 216, 219, 223, 235, 237, 238, 248, 257; **18**:123(57), 124(57), 127(57), 159(57)

Williams, R. C., **22**:251(85), 350(85, 86)

Williams, R. H., **17**:160(6), 313

Williams, R. J., **22**:177(70), 178(70), 312(70), 350(70); **28**:106(13), 139(13)

Williams, W. A., Jr., **12**:5(28), 6(28), 16(28), 18(28), 19, 74

Williams, W. J., **10**:11, 36

Williams-Leir, G., **10**:279(115), 284

Williamson, J. W., **SUP**:70(151), 160(151); **19**:268(35), 270(35), 271(35), 351(35)

Williamson, K. D., **14**:232, 246

Williamson, K. D., Jr., **5**:356, 508

Williamson, N. V., **20**:182(4), 252(241), 257(241), 258(241), 299(4), 310(241); **23**:32(60), 129(60)

Willie, R., **29**:7(7), 55(7)

Willing, H., **9**:199(14), 200(14), 201(14), 203(14), 268

Willis, A. J., **31**:449(79), 474(79)

Willis, D. R., **7**:169, 171, 175, 177, 178, 179, 198, 211, 212, 213, 216

Willis, G. E., **25**:371, 401, 411

Willis, N. C., Jr., **26**:231(32), 236(40), 237(40), 247(32, 40), 248(32, 40), 249(40), 250(40), 253(32, 40), 261(32, 40), 268(32), 269(32,

Willis, N. C. (*continued*)
  40), 281(32, 40), 282(32, 40), 291(32, 40),
  296(40), 297(32, 40), 299(40), 300(32),
  303(40), 304(40), 316(32), 320(40),
  322(40), 325(32), 326(40)
Willmarth, W. W., 3:21(37), 32; 8:293(16), 349
Willmott, A. J., 20:134(5), 150(5), 179(5)
Willoughby, P. G., 28:11(16), 71(16)
Wills, J. A. B., 8:296, 314, 349
Willson, E. D., 9:202(16), 268
Willson, E. J., 14:333(118), 334(118), 335(118),
  345
Wilmoth, R. G., 31:406(67), 411(67, 75),
  428(67), 429(75)
Wilmshurst, R., 21:83(39), 94(39), 95(39),
  135(39)
Wilski, H., 13:211(22), 264
Wilson, A. H., 2:123(13), 268; 9:364, 415
Wilson, C. T. R., 9:199(12), 268; 14:318, 323,
  323(88), 324, 344
Wilson, D. G., 19:2(6), 83(6)
Wilson, D. J., 7:340(24), 348, 368(24), 378
Wilson, E. J., 14:332(115), 333(115), 334(115),
  345
Wilson, H. A., 3:274(42), 301; 28:16(26), 71(26)
Wilson, J. S., 4:120(89), 140
Wilson, K. H., 5:309(81), 310(83), 324
Wilson, L. O., 30:332(44, 45, 48), 428(44, 45,
  48)
Wilson, M. N., 17:323, 339
Wilson, M. P., Jr., 11:412(191), 439
Wilson, M. R., 7:197, 215
Wilson, N. W., 19:68(157), 90(157)
Wilson, P. W., 22:266(134, 135, 222), 268(134,
  135), 270(135), 271(135), 272(135),
  273(134, 222), 274(134), 275(134),
  276(134), 353(134, 135), 357(222)
Wilson, R. E., 3:35(14), 53, 60, 97
Wilson, R. G., 28:97(106), 143(106)
Wilson, S. D. R., SUP:69(143), 161(143)
Wilson, W., 18:333(33, 38), 334(33, 38),
  335(33), 362(33, 38)
Wilson, W. H., 10:44(17), 59(17), 60(17),
  63(17), 64(17), 69(17), 70(17), 81
Wilson, W. R. D., 24:8(34), 14(34), 37(34)
Winbrow, W. R., 6:29(43), 124
Winckler, J., 6:244(74), 268(78), 274(78), 365
Winding, C. C., 8:135, 160; 11:219, 222,
  236(98), 256; 18:133(69), 134(69), 159(69)
Wineman, S., 25:252, 292, 293, 316

Winer, M., 14:132(110), 148
Winer, W. O., 5:381, 382, 383, 384, 386,
  387(95), 509
Wing, G. M., 3:184(10), 249
Wingerath, K., 28:317(175), 337(175);
  30:348(151, 153), 383(151), 390(153),
  394(151), 433(151, 153)
Winkel, P., 17:97, 154
Winkelmann, A., 3:287(87), 302
Winkler, E. M., 2:16(18), 43(18), 105, 164, 269;
  3:53(47), 60, 61(47), 98, 99
Winkler, F., 11:203, 206, 254
Winnikow, S., 21:335, 337, 346; 26:10(244),
  104(244)
Winning, M. D., 2:382(128), 397
Winny, H. F., 11:226, 263; SUP:62(74), 322(74),
  324(74), 341(74)
Winovich, W., 7:200(123), 216
Winovieh, W., 19:81(256), 94(256)
Winslow, C. A., 22:20(27), 21(27), 28(27),
  82(27), 153(27)
Winslow, C.-E. A., 4:67(5, 9), 111(63), 118(78),
  137, 138, 139, 140
Winstead, T. W., 5:389, 510
Winston, R., 18:19(45), 20(46, 48), 25(56a),
  81(45, 46, 48), 82(56a)
Winter, D. F., 10:47, 57, 59, 60(48), 72, 73(48),
  82
Winter, E. R. F. (Chapter Author), 7:219; 9:5,
  10(63), 26, 28, 29, 30, 31, 35, 36, 46, 49,
  50(60, 64), 51(64), 52, 53(64), 54(64),
  55(64), 56(64), 57(64), 62(60, 64), 66,
  67(64), 69(64), 70(64), 71(64), 103(64),
  109; 10:156, 165; 19:79(222), 93(222)
Winter, H. H. (Chapter Author), 13:205, 219(77,
  85, 87, 88, 91), 221(77), 234(85), 237(85),
  248(91), 249, 256, 261(126, 131), 265, 266,
  267; 15:76(60), 137(60)
Winter, K. G., 23:337(260), 366(260)
Wintrich, H., 30:291(87), 292(87, 89), 293(87),
  311(87, 89)
Wirth, P. E. (Chapter Author), 15:283, 291,
  302(210), 330
Wirtz, R. A., 20:186(59), 187(59), 191(59),
  195(59), 227(147, 148), 228(147, 148),
  229(147, 148), 230(147, 148), 232(165),
  233(165), 236(165), 237(147, 148, 165),
  238(165), 243(173), 256(165), 301(59),
  306(147, 148), 307(165, 173); 30:272(39),
  308(39)

Wise, H., 3:287, 297, 298, 301
Wise, J. D., 14:298(29), 342
Wismer, K. L., 1:201(10), 264; 10:96, 97(14, 16), 98(14), 99(14), 164
Wissler, E. H., 2:380, 396, 397; 15:69(43), 136(43); 19:261(22), 262(22), 264(22), 265(22), 267(22), 268(22, 31), 317(31), 330(31), 350(22, 31); 22:9(38), 18(38), 30(65), 71(66, 68), 134(65), 137(66, 67), 140(67), 150(67), 152(67), 155(65, 66, 67, 68), 258(223), 260(223), 265(223), 266(223), 267(223), 268(223), 273(221), 280(221), 290(152), 353(152), 354(168), 356(221), 357(223); 23:218(263), 219(263), 221(263), 278(263)
Withers, J. G., Jr., 14:228, 246; 19:122(112), 129(112), 131(112), 149(180), 150(169), 153(169), 154(169), 187(112), 189(169, 180); 20:76(67), 82(67); 30:219(75, 76), 222(10), 223(10), 250(10), 253(75, 76)
Witherspoon, P. A., 14:3, 6(8), 16(39), 37, 51(39, 120), 97(192), 100, 101, 102, 104; 20:348(68), 352(68); 24:103(23), 185(23); 30:93(7), 154(117), 181(157), 185(117), 189(7), 194(117), 196(157)
Withington, J. P., 24:193, 275
Withrow, S. P., 28:79(67), 141(67)
Witkowski, S., 4:265(10), 267(28), 268(10), 313, 314
Witt, A. F., 30:339(80), 360(80), 402(170), 430(80), 434(170)
Witt, P. J., 29:152(76), 206(76)
Witt, R. J., 29:215(1), 248(1), 334(1)
Witte, A. B., 2:51, 52, 68, 75, 76(68), 78(52), 80(52), 81(52), 83(52), 84(52, 73), 87, 88(68), 89(68), 90(68), 91(68), 92, 93(68), 107, 108; 4:352, 446
Witte, L. C., 10:107(33), 164; 11:139, 140, 146(151), 195; 18:257(38, 42), 261(38, 42), 264(42), 304(38), 305(38), 321(38, 42), 322(60); 20:379, 388; 23:13(39), 15(39), 20(39), 32(39), 128(39); 29:150(61), 205(61)
Witteborn, F. C., 3:275(46), 276(46), 301
Wittig, S., 23:298(92, 93), 358(92, 93)
Witting, H., 5:228(150), 250
Wittke, D. D., 15:243(59), 244, 245, 249(59), 278
Wittliff, C. E., 7:197, 215
Witzell, O. W., 2:364(51), 395

Witzwell, O. W., 18:179(61), 180(61), 237(61)
Wodley, F. A., 10:229(14), 281
Wohl, K., 3:286, 288(72), 301
Woinowsky-Krieger, S., SUP:62(75)
Wolbers, S., 31:405(65), 428(65)
Wolf, C. C., 2:385(76), 391(76), 396
Wolf, C. J., 25:282, 319
Wolf, D. A., 20:274(288), 275(288), 312(288)
Wolf, D. H. (Chapter Author), 23:1, 53(73), 130(73); 26:106(9), 121(46), 144(102), 145(46), 146(46), 147(46, 102), 155(113), 164(102, 113), 165(46, 102, 113), 166(113), 168(102), 170(113), 208(9), 210(46), 213(102, 113)
Wolf, E., 6:135(1), 169, 179(1), 195, 219(1), 220(1), 292(1), 362; 11:320(34), 401, 435; 23:147(30), 185(30); 30:295(92), 311(92)
Wolf, F., 8:36, 89
Wolf, G., 5:412, 413, 513
Wolf, H., 2:16(20), 25, 43, 106; 5:361, 363, 508; 6:416(81), 490, 538, 563
Wolf, J., 26:220(14), 234(14), 254(14, 67, 68), 269(14), 325(14), 327(67, 68)
Wolf, L., SUP:360(504), 361(504, 510), 363(504, 510)
Wolf, S., 12:51(61), 75; 28:98(107), 143(107)
Wolff, F., 28:249(29), 330(29)
Wolff, H. G., 4:121(103), 141
Wolff, H. S., 4:94(34), 138
Wolfram, S., 23:434(87), 464(87)
Wolfshtein, M., 7:339(21), 378; 13:12, 60, 81(37), 116, 236(104), 237(104), 266
Wollersheim, D. E., 15:156(16), 157(16), 170(30), 171(30, 31), 172(31), 218(31), 223(16), 224(30, 31); 25:280, 281, 286, 316, 317
Wollhöver, K., 19:79(197, 198), 92(197, 198); 22:293(224), 357(224)
Wolter, H., 6:135(4), 162, 175, 178, 187(4), 201(4), 362, 363
Womac, D. J., 26:123(53, 54), 126(53, 54), 129(53, 54), 134(54), 135(54), 150(53, 54), 151(53, 54), 161(54), 173(131), 177(131), 178(131), 210(53, 54), 214(131)
Wones, D., 10:43(15), 44(15), 45(15), 60(15), 62(15), 64(15), 65(15), 81
Wong, C. C., 29:111(83), 112(83), 113(84), 115(85), 116(83, 84), 127(83, 84, 85), 303(164), 304(164), 305(164), 343(164)

Wong, P. T. Y., **4**:199, 225; **5**:113, 114, 115, 116(74), 117(74), 128, 403, 407, 419(143), 421, 425, 436, 439, 511; **16**:139(110), 156; **18**:249(14), 320(14)

Wong, T. C., **13**:6, 59

Wong, V. M., **10**:179(60), 216

Woo, K. N., **18**:25(56), 82(56)

Woo, S. E., **26**:53(245), 104(245)

Wood, A. D., **5**:309(81), 324

Wood, B. D., **10**:260, 282

Wood, C. C., **5**:378, 379, 509

Wood, D. G., **23**:293(65), 356(65)

Wood, H. L., **10**:50(30), 51(30), 59(30), 68(30), 81

Wood, J., **31**:413(77), 429(77)

Wood, R. D., **6**:551, 564

Wood, R. F., **19**:37(80), 86(80); **28**:76(46, 108, 109, 110), 80(46), 99(111), 140(46), 143(108, 109, 110, 111)

Wood, R. T., **8**:122(55), 160; **11**:229(167, 248), 231, 232, 237, 260, 263; **26**:166(119), 214(119)

Wood, R. W., **2**:356; **17**:29(84), 30, 60

Wood, W. W., **5**:48(77), 54

Woodburne, R. T., **22**:250(225), 357(225)

Woodcock, A. H., **4**:87, 138

Woodgen, A., **6**:95(157), 130

Wooding, R. A., **14**:19, 22, 30, 33, 60, 101, 103; **24**:104(61), 186(61); **30**:94(9), 189(9)

Woodroffe, I. A., **28**:133(23), 139(23)

Woodruff, D. W., **21**:103(83), 137(83)

Woodruff, L. W., **7**:368, 379; **23**:304(110), 359(110)

Woods, E. G., **24**:104(27), 185(27)

Woods, M., **22**:17(3), 158(24), 246(24), 342(24), 348(24)

Woods, M. W., **14**:333, 334(120), 345

Woods, W. K., **5**:99, 100, 127, 453, 515

Woodward, R. M., **5**:49, 54

Woodward, S. H., **23**:320(185), 345(298), 362(185), 368(298)

Woolley, H. W., **2**:37(28), 38(28), 106

Wooton, R. O., **1**:394(72), 398(72), 442

Worsøe-Schmidt, P. M., **6**:539, 542, 543(60), 563; **SUP**:75(200), 106(200), 126(200), 172(200), 173(200), 182(200), 302(200), 303(200), 306(200), 307(200), 309(200), 312(200), 414(535)

Worster, M. G., **28**:282(120), 334(120)

Worthington, W. H., **9**:3(115), 75(115), 110

Woschni, G., **23**:344(293), 368(293)

Wössner, G., **7**:297, 319, 320c

Wouters, P. J., **30**:338(74), 348(140), 350(74), 371(140), 382(140), 405(74), 430(74), 433(140)

Wozniak, G., **30**:262(23), 308(23)

Wozniak, K., **30**:262(23), 308(23)

Wragg, A. A., **7**:133, 161; **30**:279(57), 280(57), 309(57)

Wranglen, G., **12**:233(19), 251(19, 119, 120, 121), 270(19), 278, 282

Wray, J. H., **11**:375(128), 437

Wray, J. L., **22**:249(226), 251(226), 252(226), 357(226)

Wray, K., **2**:127(15), 131(15), 134(15), 268

Wray, R. C., Jr., **22**:226(119), 352(119)

Wray, W. O., **18**:46(89), 83(89)

Wright, C. A., **11**:202(7), 204(7), 209(7), 218(7), 221(7), 252

Wright, C. C., **5**:356(65), 445, 446, 447, 448, 449, 507, 508

Wright, D. E., **14**:5(15), 100

Wright, E. G., **28**:365(109), 420(109)

Wright, F. C., **6**:538, 563

Wright, J. A., **24**:218, 219, 226, 227, 238, 239, 244, 274, 275

Wright, J. S., **18**:337(49, 51), 363(49, 51)

Wright, L. T., **4**:67(4), 137

Wright, R. D., **11**:105(134, 135), 113, 195; **16**:114(76), 134, 137, 155

Wright, R. E., **23**:301(106), 358(106)

Wright, R. W., **29**:136(23), 137(23), 203(23)

Wright, S. J., **14**:156, 157, 180, 243, 245; **16**:53, 57; **19**:164(206), 170(206), 171(206), 190(206)

Wronski, S., **23**:230(264), 231(264), 278(264)

Wu, B. J., **14**:334(126), 335(126), 345

Wu, B. J. C., **14**:336, 346

Wu, C., **5**:135, 247

Wu, C.-H., **13**:67, 115

Wu, C. K., **29**:77(39), 125(39)

Wu, C. S., **1**:334, 335, 337, 338, 341, 344(67), 348, 354

Wu, F., **16**:53, 57

Wu, K., **23**:8(13), 12(13), 14(13), 25(13), 58(13), 60(13), 65(13), 78(13), 106(13), 126(13)

Wu, Q., **29**:263(124, 125), 266(125), 270(124), 271(124, 125, 138, 140), 272(124), 273(124, 125, 140), 278(125), 280(125), 341(124, 125), 342(138, 140)

Wu, R. L., **23**:340(270, 271), 367(270, 271)
Wu, R. S., **SUP**:156(336), 178(336, 358)
Wu, S. H., **30**:102(40), 185(40), 190(40)
Wu, Y. C., **11**:84(43), 93(43), 95(43), 97(43), 105(43), 112(43), 117(43), 190; **22**:330(228), 331(228), 332(228), 357(227, 228)
Wuest, W., **4**:331, 445
Wukalowitsch, M. P., **20**:19(38), 81(38)
Wulff, W. (Chapter Author), **22**:26(69), 30(69), 35(69), 37(69), 46(69), 48(69), 49(69), 55(69), 58(69), 155(69); **31**:105, 107(40), 108(31), 112(22, 34, 35, 36, 37, 38, 42), 113(40), 119(34, 35, 36, 37, 38, 42), 120(32, 35, 36, 37, 38), 121(34, 42), 128(33, 34), 129(41), 130(39), 133(41), 134(41), 136(41), 137(33, 34, 38, 42), 138(29, 30, 33, 40, 42), 139(29, 42), 141(42), 143(40), 144(40, 41), 145(40, 41), 149(41), 150(41), 151(41), 152(41), 153(34, 35, 36, 37), 156(22), 157(29, 30, 31, 32, 33, 34, 35, 36, 37, 38, 39, 40, 41, 42)
Wunder, R., **19**:128(136), 188(136)
Wunderlich, A., **25**:292, 316
Wunschmann, J., **14**:151, 152, 242
Wupperman, G., **3**:287(85), 301
Wurster, R. D., **4**:120(96), 140
Würtz, I., **17**:42, 43(142), 44(142), 45(142), 63
Wyatt, P. W., **7**:308, 309, 319
Wyatt, T., **7**:276, 319
Wyborny, H. P., **6**:71, 76(121, 122), 79, 80, 128
Wygnanski, I., **13**:86(47), 116
Wylie, R. G., **11**:220, 224, 262
Wyman, C. E., **28**:190(47), 229(47)
Wyndham, C. H., **4**:120(95), 140
Wyngaard, J. C., **25**:333, 344, 362, 371, 374, 375, 401, 415, 419
Wyss, C. R., **22**:360(2), 435(2)
Wyss, E., **15**:254, 280

X

Xavier, A. M., **19**:115(96), 116(96), 138(146), 140(146), 160(196), 187(96), 188(146), 189(196)
Xia, Q., **24**:304(49), 320(49)
Xiao, Q., **30**:338(75), 339(75), 356(105), 394(105), 395(105), 430(75), 431(105)

Xie, C., **19**:164(220), 168(220), 171(220), 190(220), 311(104), 314(104), 315(104), 316(104), 319(104), 321(104), 322(104), 354(104)
Xiong, T. Y., **26**:84(246), 104(246)
Xu, D., **28**:245(19), 253(59), 255(19, 59), 305(19, 59), 330(19), 331(59)
Xu, X. (Chapter Author), **28**:75, 80(113, 114), 81(114), 82(114), 83(114), 84(114), 85(114), 86(114), 87(113, 115), 88(114), 89(115), 90(115), 93(113), 94(113), 143(113, 114), 144(115)
Xu, Y., **22**:326(229), 327(229), 357(229)

Y

Yabe, A., **21**:129(125), 130(126, 127), 139(125, 126, 127)
Yacoub, N., **21**:142, 183
Yadigaroglu, G., **21**:287, 290, 300, 345; **25**:96(176), 148(176); **31**:138(44), 157(44)
Yadrevskaya, N. L., **30**:296(99), 298(99), 311(99)
Yaeger, D., **25**:13(47), 141(47)
Yaeger, J. C., *see* Jaeger, J. C.
Yaffee, M. L., **9**:98, 111
Yagi, S., **1**:382(45), 441; **14**:159, 244
Yaglom, A. M., **6**:395(23), 487, 511(21), 513(21), 562; **14**:250(2), 263(2), 264(2), 265(2), 266(2), 273(2), 279; **16**:307, 364; **21**:200(28), 201(30), 237(28, 30); **25**:50(75), 143(75)
Yaglou, C. P., **4**:104(46, 48), 139; **11**:241, 242, 260
Yahata, H., **21**:154, 183
Yahikazawa, K., **15**:101(112), 102(112), 103(112), 139(112)
Yajima, F., **30**:359(123), 361(123), 432(123)
Yakabou, I. T., **5**:309(78), 324
Yakaytis, F. L., **21**:46(77, 79), 52(77, 79)
Yakhot, A., **25**:361, 419
Yakhot, V., **25**:355, 361, 394, 419
Yakimovich, K. A., **26**:221(23), 246(23), 253(23), 325(23)
Yakovlev, A. T., **21**:166, 183
Yakovlev, S. I., **25**:4(22), 140(22)
Yakovlev, V. V., **6**:525(36), 526, 531, 562
Yakovleva, R. V., **6**:474, 501
Yakura, J. K., **6**:33, 35(53), 44, 51(53), 124

Yakush, E. V., **25**:4(12), 31(12), 140(12); **31**:160(21), 182(21), 323(21)

Yamada, A., **17**:35(113), 61; **28**:342(26), 416(26)

Yamada, H., **1**:101, 121

Yamada, I., **SUP**:64(97, 98), 224(97), 237(97), 260(97, 98), 262(98); **19**:310(94), 312(94), 353(94)

Yamada, Y., **21**:161, 181; **23**:154(37, 42), 185(37, 42)

Yamagata, K., **1**:231, 265; **5**:409, 411, 413, 513; **6**:551, 555(82), 564; **9**:9(52), 10(52), 21(51, 52), 23(52), 36(51, 52), 37(52), 99(52), 108; **11**:12(27), 48; **20**:3(1), 5(3), 7(3), 9(3), 10(3), 11(1, 3), 15(3), 79(1, 3), 80(11)

Yamaguchi, K., **18**:55(104), 67(125), 68(125), 84(104), 85(125)

Yamaguchi, M., **26**:14(219), 103(219)

Yamaguchi, Y., **30**:359(123), 361(123), 432(123)

Yamaka, J., **5**:167, 172(71), 248

Yamamoto, G., **8**:232, 243, 282

Yamamoto, H., **26**:105(5), 123(5), 126(5), 208(5)

Yamamoto, T., **28**:36(58), 43(58), 54(58), 72(58)

Yamanaka, A., **25**:258, 259, 270, 271, 275, 276, 277, 278, 319

Yamanaka, G., **16**:232; **20**:244(197), 308(197)

Yamanishi, T., **21**:129(125), 139(125)

Yamano, N., **29**:151(70), 165(116), 206(70), 208(116)

Yamasaki, K., **20**:55(59), 58(59), 82(59)

Yamauchi, A., **21**:125(115), 138(115)

Yamazaki, R., **19**:186(63)

Yan, H., **29**:62(7), 72(7), 124(7), 153(80), 206(80), 218(23), 232(23), 233(23), 234(23, 72), 239(23), 241(23, 72), 248(23), 249(23), 251(23), 274(23), 275(23, 72), 276(23), 279(23), 295(23), 301(23), 311(23, 72), 319(23), 335(23), 338(72)

Yan, M. M., **19**:33(67), 37(67), 42(67), 86(67)

Yan, X., **23**:330(236), 365(236)

Yanezmoreno, A. A., **20**:232(164), 238(164), 256(164), 307(164)

Yang, C. C., **16**:149(131), 150(131), 156

Yang, H. Q., **24**:199, 211, 275, 304(49), 320(49)

Yang, H. T., **23**:212(197), 275(197)

Yang, J. W., **9**:262(124), 271; **26**:34(247), 104(247); **29**:172(140, 142), 209(140, 142)

Yang, K., **31**:401(58), 428(58)

Yang, K. T., **1**:70, 106, 109, 121, 122; **4**:24, 36, 38, 44, 46, 63, 64, 334, 445; **6**:396(35), 474(229, 233, 248), 480(233, 248), 482,

487, 497, 498; **8**:12, 85; **9**:253(81), 270, 286(4), 295(21), 311(69), 312(4), 346, 347; **14**:66, 103; **15**:189(61), 224(61), 330; **18**:69(132), 70(132), 86(132); **19**:19(34), 51(34), 85(34); **24**:199, 211, 275, 303(26), 304(49), 319(26), 320(49)

Yang, L., **28**:376(147, 152), 379(152), 381(152), 383(152), 384(152), 421(147), 422(152)

Yang, Q. S., **23**:153(33), 185(33)

Yang, S. C., **24**:218, 275

Yang, S. S., **8**:293(16), 349

Yang, T. L., **28**:282(117), 334(117)

Yang, V., **24**:193, 275

Yang, W., **20**:328(34, 35), 351(34, 35); **23**:161(46), 186(46)

Yang, W.-J. (Chapter Author), **5**:348, 349, 350, 351(52), 353(53), 354, 507; **6**:396, 397(40), 431, 474(30, 31, 266, 268, 269, 270, 271, 292), 476(30, 31), 487, 488, 499, 500; **15**:242(59a), 278; **22**:18(39); **27**:ix(6, 7, 11, 15, 17, 18), 46(6, 7), 146(28), 147(28), 187(36, 37, 38), 193(37, 38, 39, 40, 41); **31**:334(11), 365(11), 425(11)

Yang, W. T., **14**:285(10), 304, 341

Yang, Y., **29**:151(69), 205(69)

Yang, Y. S., **23**:161(45), 186(45)

Yang, Z., **22**:326(229), 327(229), 357(229); **25**:360, 383, 419

Yankov, G. G., **21**:2(3), 3(3), 34(3), 37(3), 38(3), 39(68, 69), 40(68, 69), 41(3, 69), 50(3), 52(68, 69)

Yanovitskii, E. G., **10**:10, 36

Yanovskiy, L. S., **21**:17(32), 48(83), 51(32), 53(83); **31**:228(118), 327(118)

Yanowitz, H., **6**:474(253), 483(253), 498

Yao, L. S. (Chapter Author), **19**:1, 11(18), 22(42), 23(18, 42), 50(18, 116, 117), 51(18, 120, 121), 52(120, 121), 53(121), 54(18, 120, 123), 55(120), 56(120), 60(136), 61(136), 66(123), 67(136), 73(123), 74(136), 77(136), 79(214, 215), 80(136), 84(18), 85(42), 88(116, 117, 120, 121, 123), 89(136), 92(214, 215); **20**:187(60), 301(60)

Yao, M., **28**:36(63), 39(63), 73(63)

Yao, S. C., **20**:248(210), 309(210); **26**:22(249), 53(249), 69(222), 81(248), 103(222), 104(248, 249)

Yap, C., **23**:336(257), 366(257)

Yap, L. T., **26**:77(250), 104(250)

Yarishev, I. A., **6**:418(122), 492
Yarkho, S. A., **6**:376(1), 486; **11**:133(146),
 144(146, 154, 155), 146(146, 154, 155),
 151(155), 152(155), 156(155), 159(155),
 170(154, 155), 172(155), 173(155),
 176(155), 177(155), 179(155), 180(155),
 182(155), 183(155), 185(155), 195, 196;
 25:4(3), 10(36), 12(3), 22(3), 33(3), 43(3),
 52(3), 64(3), 92(3), 99(3), 106(3), 108(3),
 120(3), 130(3), 139(3), 141(36); **31**:159(1),
 160(1, 12, 29, 57), 161(1), 177(1), 182(1),
 183(12), 187(12, 87), 188(88, 89), 191(93,
 95), 198(93, 95), 200(95), 201(105, 106),
 205(105, 106), 207(106), 208(95), 209(93),
 217(105), 220(88), 221(105, 106), 222(105,
 106), 224(105, 106), 225(117), 232(105),
 243(12), 247(1), 253(12), 254(148), 262(1),
 263(1), 265(158, 161, 167), 266(29, 57),
 270(29, 57, 158), 271(57), 279(57),
 287(57), 292(57), 294(57), 301(57),
 305(57), 322(1, 12), 323(29), 324(57),
 326(87, 88, 89, 93, 95, 105), 327(106, 117,
 122), 329(148, 158, 161), 330(167)
Yarmchuck, E. J., **17**:89(60), 154
Yaronis, E. P., **11**:203(43), 206(43), 254
Yaroshenko, Y. G., **6**:418(143), 493
Yaschchenko, A. I., **23**:187(43a), 188(43a),
 193(43a), 268(43a)
Yaskin, L. A., **21**:16(31), 21(36, 37), 22(36),
 51(31, 36, 37)
Yastrebenetsky, A. R., **31**:160(15), 323(15)
Yasuda, Y., **24**:163(147), 190(147)
Yasuhara, M., **7**:206, 217
Yasuhira, N., **23**:9(20), 109(20), 110(20),
 113(20), 115(20), 119(20), 127(20)
Yasukochi, K., **17**:115(105), 116(105), 117, 118,
 119, 155
Yasunaka, M., **17**:4(22), 5(22), 22(22), 23(22), 57
Yasuzawa, K., **27**:ix(8)
Yatabe, J. M., **16**:91, 154
Yates, B., **13**:261(118), 266
Yates, H. W., **11**:426(234), 440
Yau, J., **2**:390, 397; **19**:268(40), 269(40),
 270(40), 272(40), 275(40), 276(40),
 279(40), 281(40, 78), 285(40, 78), 300(40,
 78), 351(40), 352(78)
Yavas, O., **28**:96(12), 138(12)
Yeager, E. B., **2**:4(5), 86(5), 105
Yeckel, A., **28**:402(204, 205, 206, 207),
 404(204), 424(204, 205, 206, 207)

Yedamurthy, V. N., **14**:155, 156, 157, 243
Yee, H. C., **24**:241, 275
Yeh, A. I., **28**:218(65), 230(65)
Yeh, F. C., **23**:306(117), 359(117)
Yeh, H. C., **15**:286, 287, 289, 330; **30**:239(77),
 253(77)
Yeh, K. C., **14**:268(52), 280
Yeh, L. T., **19**:37(76), 39(76), 42(76), 86(76)
Yeh, S. L., **26**:244(49), 245(50), 253(49),
 291(49), 296(49), 323(49), 326(49, 50)
Yeh, T. T., **8**:291, 349; **30**:290(83), 310(83)
Yeller, G., **28**:402(210), 424(210)
Yellot, J. I., **14**:333, 334(124), 345
Yeman, A. J., **20**:259(247), 311(247)
Yen, C. T., **30**:329(39), 428(39)
Yen, H. H., **11**:105(126), 115(126), 194
Yen, J. T., **SUP**:18(27), 62(27), 264(27), 266(27)
Yen, Y. C., **14**:72, 103; **19**:44(102), 87(102)
Yener, Y., **15**:330; **25**:51(88), 144(88)
Yeo, K. S., **18**:21(50), 22(50), 23(50), 26(50),
 81(50)
Yerbury, M., **10**:27, 37, 57(47), 82
Yerkess, A., **29**:140(31), 165(31), 178(31),
 204(31)
Yermakov, S. V., **17**:37(122), 62
Yeum, K., **28**:245(22), 330(22)
Yeung, E. S., **28**:135(19), 139(19)
Yeung, W.-S., **26**:2(251), 104(251)
Yewell, R., **24**:299(50), 300(50), 302(50),
 307(50), 320(50)
Yhland, C. H., **3**:219(67), 250
Yih, C. S., **5**:230(156), 250; **8**:19(80), 86; **9**:298,
 302(38), 347; **10**:250(45), 282; **14**:60, 103;
 SUP:170(345), 177(345); **16**:53, 57;
 24:286(51), 320(51)
Yih, I., **SUP**:120(296)
Yilmaz, S., **17**:5, 8, 57
Yilmazer, U., **28**:193(31), 229(31)
Yim, J., **23**:193(260), 229(260), 278(260)
Yimer, B., **20**:287(311), 288(311), 313(311)
Yin, S., **11**:288(21), 315
Ying, Q. Y., **28**:135(119), 144(119)
Ying, S. J., **10**:257, 282
Yoga, K., **28**:376(142), 421(142)
Yokogawa, M., **30**:323(25), 427(25)
Yokoya, S., **17**:4(22), 5(22), 7(37), 22(22),
 23(22), 39(132), 40(132), 57, 62;
 18:258(51), 261(51), 313(51), 322(51);
 23:75(92), 79(92), 97(92), 131(92)
Yokoyam, S., **21**:181

Yokoyama, S., **7**:128(37), 161
Yoler, Y. A., **1**:314, 316(51), 354
Yonehara, N., **26**:109(19), 113(19), 114(19, 34), 118(19, 34), 157(19), 171(19), 173(19), 174(19), 175(19), 177(19), 209(19, 34)
Yonge, T., **29**:91(59), 126(59)
Yoo, H., **28**:276(106), 277(106), 333(106)
Yoo, J., **22**:200(230), 357(230)
Yoo, K. J., **29**:259(127), 266(127), 269(127), 342(127)
Yoo, S. S., **12**:103, 105, 106, 108, 113; **15**:62(15), 63(15), 64(15), 65(15), 66(15), 70(15), 88(15), 89(15), 104(118), 105(15), 107(15), 111(15), 116(15), 117(15), 121(15), 124(15), 135(15), 139(118); **19**:327(124), 329(125), 340(124), 354(124, 125)
Yoon, C., **29**:180(159, 160), 210(159, 160)
Yoon, C. Y., **14**:193, 245
Yoon, D. Y., **28**:351(84), 419(84)
Yoon, G., **22**:356(217)
York, G. W., **28**:124(78), 142(78)
Yortsos, Y. C., **30**:108(50), 114(66), 116(66), 117(72), 118(72), 191(50, 66), 192(72)
Yos, J. M., **3**:270(28), 300; **4**:241(7), 242(7), 313
Yoshida, H., **20**:287(312), 289(312), 314(312)
Yoshida, K., **10**:195, 196, 217; **14**:151, 155(47), 157, 242, 243; **19**:108(51), 164(207), 173(207), 174(207), 175(207), 176(207), 186(51), 190(207); **20**:277(293, 294), 313(293, 294)
Yoshida, M., **28**:351(71), 418(71)
Yoshida, S., **17**:35(113), 61; **20**:76(70), 82(70)
Yoshida, T., **15**:92(84), 93(84), 130(84), 138(84)
Yoshihara, H., **7**:197(96), 215
Yoshikata, K., **7**:88(5a), 160
Yoshinaka, A., **28**:402(214), 424(214)
Yoshioka, K., **21**:38(62), 52(62)
Yoshioka, S., **28**:96(68), 141(68)
Yoshizawa, A., **25**:333, 338, 341, 342, 344, 392, 419
Yoshizawa, S., **7**:132(42), 161
Yosinobu, H., **20**:360, 388
Yoskioka, K., **7**:66(48), 85
Youn, H. K., **23**:376(47), 378(47), 436(47), 462(47)
Younan, G. W., **19**:77(178), 91(178)
Young, A. D., **19**:123(117, 118, 120), 125(117, 118, 120), 187(117, 118, 120)
Young, A. T., **10**:76(73), 78(73), 83

Young, C. B. M., **3**:35(33), 60, 98
Young, G. B. W., **2**:41, 106; **4**:331, 444
Young, J. F., **28**:102(94, 112), 143(94, 112)
Young, J. P., **2**:378(39), 395
Young, L., **28**:341(14), 415(14)
Young, L. A., **28**:133(23), 139(23)
Young, M. F., **23**:305(113), 359(113); **29**:151(65), 189(193), 205(65), 212(193)
Young, M. J., **23**:217(66), 269(66)
Young, P. A., **28**:106(13), 139(13)
Young, R. A., **3**:287, 301
Young, R. K., **20**:15(34), 81(34)
Young, R. L., **5**:477, 479, 516
Young, R. O., **12**:3(11), 14(11), 15(11), 59(11), 73
Yovanovich, M. M., **9**:359(16), 414; **20**:258(245), 259(246, 249), 261(252), 262(256), 266(268), 268(273), 269(275), 270(245, 275, 276), 311(245, 246, 249, 252, 256), 312(268, 273, 275, 276); **24**:17(43, 44, 45), 37(43, 44, 45)
Yu, C. P., **1**:287(20), 353; **14**:121, 146; **25**:252, 292, 319
Yu, J., **23**:12(49), 14(49), 25(49), 128(49)
Yu, Y. H., **19**:186(60); **23**:215(265), 236(265), 238(265), 239(265), 240(265), 278(265)
Yuan, S. W., **3**:79, 80, 99; **7**:279, 319
Yuan, T., **30**:385(152), 386(152), 387(152), 396(152), 400(152), 401(152), 402(152), 433(152)
Yuan, X.-D., **18**:55(105), 56(105), 84(105)
Yuan, Z. G., **30**:166(125), 194(125)
Yudaev, B. N., **13**:11(41), 60; **25**:131(204), 136(204), 150(204); **31**:160(23), 323(23)
Yudin, V. F., **31**:160(37), 323(37)
Yudovich, V. I., **6**:474(259), 483(259), 499
Yue, P. L., **16**:133(98), 135, 136, 137, 138, 139, 140, 141, 155, 156
Yuen, D. A., **19**:82(264), 94(264)
Yuen, M. C., **26**:53(252, 253), 66(184), 72(184, 185), 101(184, 185), 104(252, 253)
Yuen, W. W., **29**:150(60), 153(78, 79, 80, 81), 157(79, 108), 164(108), 174(108), 180(166, 167, 168), 181(108), 182(166, 167, 168), 195(60, 167), 200(108, 168), 201(168), 205(60), 206(78, 79, 80, 81), 207(108), 210(166), 211(167, 168)
Yueng, K. C., **14**:38(92), 39(92), 40(92), 102
Yuge, T., **20**:366, 370, 377, 383, 388
Yuill, C. H., **10**:242, 243, 281

Yuki, T., **25**:259, 319

Yun, K. S., **3**:272, 282(61), 298(96), 300, 301, 302

Yung, A. D., **6**:398(42), 411(42), 488

Yuregir, K. R., **19**:99(15, 17), 100(15, 17, 32), 101(17), 102(32), 105(17, 32), 107(17, 32), 108(17, 32), 110(65), 164(213), 165(213), 169(213), 170(213), 171(213), 180(17, 32), 185(15, 17, 32), 186(65), 190(213)

Yurich, S. P., **19**:149(182), 189(182)

Yurlova, E. V., **SUP**:179(359)

Yushin, A. Ya., **31**:195(98), 326(98)

Yuzhang, C., **23**:295(77), 346(299), 357(77), 368(299)

Z

Zaalouk, M. G., **16**:236

Zabetakis, M. G., **29**:69(26), 85(26), 124(26), 299(156), 343(156)

Zabrodsky, S. S. (Chapter Author), **10**:169(3), 176, 202, 206, 214, 216, 217; **14**:149, 151(1, 5), 152(29), 154(38), 155(54), 158(5), 160, 162, 176(178, 180, 181, 183, 184), 187(116), 188(110), 189(116), 233(173), 235(173, 178), 236(5, 180, 181, 182, 183, 184), 242, 243, 245, 246, 247; **SUP**:17(33); **19**:98(1, 3), 100(1, 29), 101(34), 102(36), 111(1, 3), 112(1, 3), 116(1, 36), 117(1, 36), 129(1, 36), 130(1), 132(36), 136(29), 137(29), 139(1, 29), 140(1, 29), 145(3, 29), 148(170), 153(170), 154(3, 170), 157(29), 159(29), 161(1, 3, 29), 178(232), 184(1, 3), 185(29, 34, 36), 189(170), 190(232); **25**:164(48), 166(59), 205(48), 214(48, 106), 244(48), 245(59), 247(106)

Zagainov, V. A., **14**:318, 344

Zahavi, E., **10**:179(61), 216

Zahn, M., **25**:187(80), 198(80), 246(80)

Zaichik, L. I., **31**:160(47), 324(47)

Zaidenman, I. A., **15**:326, 330

Zaidi, A., **23**:253(266, 267), 254(266, 267), 255(266, 267), 256(266, 267), 260(267), 261(266, 267), 278(266, 267)

Zaikovski, A. V., **14**:106(71, 72), 160(71, 72), 161(71, 72), 163(71, 72), 164(71, 72), 166(71, 72), 167(71, 72), 168(71, 72), 169(71, 72), 170(71, 72), 171(71, 72), 172(71, 72), 176(71, 72, 106), 180(106), 188(106), 191(106), 237(106), 244, 245; **19**:159(191), 189(191)

Zakharov, V. L., **1**:135, 184

Zakhem, R., **28**:268(96, 97), 333(96, 97)

Zaki, W. N., **23**:242(221a), 244(221a), 261(221a), 262(221a), 276(221a)

Zakin, J. L., **15**:63(26, 27), 64(26, 27), 135(26, 27)

Zakirov, I. P., **31**:281(185), 331(185)

Zakirov, S. G., **31**:159(1), 160(1), 161(1), 177(1), 182(1), 247(1), 262(1), 263(1), 271(173), 284(190, 191), 287(195), 322(1), 330(173), 331(190, 191, 195)

Zakkay, V., **6**:41(79), 65(79), 126; **7**:197(89, 102, 103), 215, 367, 379

Zaks, M. B., **19**:245(67)

Zaleski, T., **26**:234(38), 254(38), 326(38)

Zaloudek, F. R., **5**:151(45a), 247

Zalutskii, E. V., **6**:390(17), 486

Zamodits, H. J., **28**:149(68), 230(68)

Zamora, M., **23**:292(56, 57), 356(56, 57)

Zamuraev, V. P., **3**:230, 231, 232(83), 233(83), 251

Zandbergen, P. J., **25**:364, 419

Zaneveld, L., **22**:275(203), 356(203)

Zang, T. A., **25**:362, 382, 405, 411, 416, 418, 419

Zanker, A., **21**:334, 346

Zankl, G., **4**:265(10), 268(10), 313

Zanoli, A., **19**:62(151), 66(151), 90(151)

Zanotti, F., **23**:370(19, 20, 21), 390(21), 440(21), 443(21), 461(19, 20, 21)

Zapas, L. J., **15**:62(10), 78(10), 135(10)

Zapp, G. M., **3**:32; **8**:122(53), 160; **11**:219, 222, 231, 232, 236, 256; **18**:116(51), 117(51), 159(51)

Zara, E. A., **4**:164, 165, 224, 228

Zarafonitis, C., **29**:183(173), 185(173), 211(173)

Zarani, H., **28**:343(45), 417(45)

Zarič, Z. (Chapter Author), **6**:403, 488; **8**:285, 293(22), 299(42), 306(59), 307(22, 42, 60, 61), 308(22), 337(62, 63), 349; **11**:12(29), 48

Zarling, J. P., **SUP**:68(134), 273(134), 274(134), 361(509), 362(511), 363(509, 511), 364(509), 365(509, 514, 515)

Zaror, C. A., **17**:181, 185, 186, 314

Zarraga, M. N., **30**:32(89), 91(89)

Zarubin, V. S., **19**:245(68)

Zaslavsky, D., **24**:104(53), 186(53)

Zavargo, Z., **31**:343(28), 426(28)

Zavattarelli, R., **1**:245(89), 266, 381(39, 40, 41), 392(60), 410(96), 411(39, 40, 41), 413(39, 40, 41), 421(40, 41), 427(129), 441, 442, 443

Zavorka, I., **25**:4(28), 141(28)

Zaworski, R. J., **18**:343(62), 345(62), 363(62)

Zebib, A., **14**:25, 31, 66, 101, 103; **20**:226(143), 243(168), 306(143), 307(168), 331(49), 351(49); **30**:371(143, 144), 382(143), 394(144), 433(143, 144)

Zecchin, R., **9**:261(110), 271; **15**:48(75), 57(75)

Zedan, M., **24**:247, 273

Zeedijk, H., **17**:300, 317

Zeev, P., **30**:96(29), 190(29)

Zeh, D. W., **9**:296(34), 304(59), 343(34), 346, 347

Zehnder, L., **6**:195, 363

Zeiberg, S. L., **6**:474(236), 484(236), 497

Zeibig, H., **13**:219(58), 265

Zeiger, H. J., **28**:343(39), 417(39)

Zeldin, B., **SUP**:57(66), 73(188, 189), 92(188, 189), 94(188, 189), 97(189), 112(66), 113(66), 116(66), 117(66), 140(327), 144(66, 327, 328), 165(188, 189), 167(188, 189), 168(189), 174(66), 414(327)

Zeldovich, J., **11**:7, 48; **14**:287, 298, 304, 314, 341

Zeldovich, Y.(I.) B., **8**:246, 248, 282; **9**:298(37), 347; **28**:133(116), 144(116); **29**:96(61), 126(61), 140(36), 204(36)

Zeldovitch, A. G., **5**:406(170), 513

Zeller, H. W., **25**:252, 319

Zellnik, H. E., **2**:16(21), 43, 106

Zeman, O., **21**:188(14), 236(14); **25**:362, 418

Zemanski, M. W., **31**:105(43), 157(43)

Zemansky, N. W., **12**:202(122), 282

Zenchenkov, N. G., **14**:176(93), 178(93), 244

Zener, C., **2**:341, 343, 356; **19**:18(32), 84(32)

Zeng, X., **28**:255(65), 276(65, 105), 332(65), 333(105)

Zenkevich, B. A., **17**:29(85), 32(102), 39(85), 60, 61

Zenner, G. H., **5**:326(13), 327, 505

Zenz, F. A., **10**:169(7), 215; **14**:151(8), 232, 242; **19**:99(11), 103(11), 184(11)

Zerbe, J. E., **8**:135(68), 160; **18**:133(68), 134(68), 159(68)

Zerkle, R. S., **7**:236(27), 247, 277(27), 314

Zerlaut, G. A., **5**:26, 51

Zernike, F., **6**:201, 364

Zettlemoyer, A. C., **10**:92, 164

Zeugin, L., **16**:101, 154

Zeyen, R., **29**:178(158), 181(158), 210(158)

Zhang, G. J., **29**:263(124), 270(124), 271(124, 140), 272(124), 273(124, 140, 144), 341(124), 342(140), 343(144)

Zhang, G.-P., **19**:78(191, 192), 91(191, 192)

Zhang, G. T., **19**:179(234), 190(234), **25**:234(134, 135, 136), 235(134, 135), 248(134, 135), 249(136)

Zhang, H. (Chapter Author), **19**:164(216, 220), 168(220), 170(216), 171(216, 220), 174(216), 190(216, 220); **30**:313, 313(8), 327(27, 28), 328(29), 329(36), 334(60, 61), 335(27, 28, 66, 67, 68), 336(27, 28, 68), 337(27, 66), 338(29), 339(29), 340(29), 341(29), 344(29), 347(8), 351(27, 28, 36, 68, 97, 98, 99), 352(27, 36, 97, 98, 99), 354(27, 98, 99), 356(27, 98, 99, 104), 357(107, 108), 361(107, 108), 388(27), 389(27), 390(27), 391(60), 406(28), 407(107), 412(28, 68), 415(68, 182), 416(182), 417(28, 68), 418(28), 420(28), 422(60, 107, 108, 184), 423(60), 426(8), 427(27, 28, 29), 428(36), 429(60, 61, 66, 67, 68), 431(97, 98, 99, 104, 107, 108), 435(182, 184)

Zhang, J. P., **28**:76(6), 138(6)

Zhang, T., **30**:329(36), 351(36), 352(36), 428(36)

Zhang, X. (Chapter Author), **24**:129(135), 130(135), 144(135), 145(135), 146(135), 147(135), 148(135), 149(135), 189(135); **28**:75, 96(117), 97(117), 99(117), 100(117), 101(117), 102(117), 144(117)

Zhang, Y.-Z., **27**:ix(15), 187(37), 193(37)

Zhang, Z., **20**:202(87), 303(87), 342(62), 343(62), 352(62)

Zhang, Z. F., **19**:102(43, 44), 106(43, 44), 107(43, 44), 108(44), 109(43, 44), 110(44), 180(43, 44), 185(43, 44)

Zhao, B., **28**:218(65), 230(65)

Zhao, H. Y., **26**:109(20), 113(20), 114(20), 115(20), 116(20), 117(20), 118(20), 149(20), 151(20), 156(20), 209(20)

Zhao, Q., **31**:461(46), 472(46)

Zhao, T. S., **30**:182(159), 196(159)

Zhao, X.-W., **30**:274(48), 275(48), 276(48), 309(48)

Zhao, Z., **28**:36(57, 58), 37(57), 39(57), 40(57), 41(57), 42(57), 43(57, 58), 44(57), 45(57), 46(57), 47(57, 77), 48(57, 77, 78), 49(77), 50(77), 51(77), 52(78), 53(77, 78), 54(57, 58, 81), 56(81), 57(81), 59(81), 60(81), 61 (81), 62(81), 63(81), 72(57, 58), 73(77, 78, 81)

Zhdanovich, N. V., **25**:257, 283, 285, 289, 317

Zheleznyak, A. S., **26**:23(33), 94(33)

Zheltov, A. I., **19**:156(187), 159(187), 160(187), 189(187)

Zhemkov, L. I., **6**:416(92), 418(123), 490, 492

Zheng, J. P., **28**:135(118, 119), 144(118, 119)

Zheng, R., **23**:219(268), 278(268)

Zheng, Z. X., **19**:186(63)

Zho, D. L., **29**:77(40), 125(40)

Zhomakin, A. I., **28**:390(174), 423(174)

Zhong, T., **19**:311(104), 314(104), 315(104), 316(104), 319(104), 321(104), 322(104), 354(104)

Zhorzholiani, A. G., **14**:133, 148

Zhou, M., **28**:41(75), 73(75)

Zhou, Z., **31**:138(44), 157(44)

Zhu, J., **23**:208(228a), 211(269), 219(269), 230(228a), 277(228a), 278(269)

Zhu, M., **22**:19(70), 71(70), 96(70), 127(70), 138(70), 141(70), 142(70), 144(70), 145(70), 146(70), 155(70)

Zhuikov, V. V., **31**:160(48, 61), 324(48, 61)

Zhuk, V. I., **6**:416(25), 418(152), 487, 493; **8**:36(206), 89

Zhukhovitskii, E. M., **1**:295, 298, 353; **8**:164, 168, 173(26), 212, 226, 227; **20**:331(48), 332(50), 351(48, 50)

Zhukhovushii, E. M., **11**:297(28), 315

Zhukov, A. V., **31**:236(124), 327(124)

Zhukov, S. A., **18**:322(64)

Zhukov, V. M., **17**:327(31, 32), 333(31), 334(31, 32), 335(32), 336(31, 32), 340; **18**:264(58), 322(58)

Zhukov, V. P., **25**:4(30), 141(30)

Zhuravlenka, V. Y., **15**:326

Zick, A. A., **23**:207(270), 278(270), 399(68), 463(68)

Ziegenhagen, A. J., **SUP**:310(471)

Ziegler, E. N., **10**:191, 203(109), 208, 217; **14**:151, 152(34), 158(14, 34, 59, 65), 192, 193, 196, 201(34), 204, 209, 234, 242, 243, 244, 246

Ziegler, N., **19**:128(137), 188(137)

Ziegler, W. T., **11**:360(96), 437

Ziemer, R. W., **1**:350, 354

Zienkiewicz, O. C., **24**:227, 268; **28**:41(66, 68), 73(66, 68)

Ziering, S., **7**:171, 172, 212

Zierman, C. A., **5**:477, 478, 479, 516; **9**:402(85), 406(100), 417

Zijl, W., **15**:243, 244, 278, 279

Zilitinkevich, S. S., **14**:262(28), 266(28), 268(28), 273(28), 276(28), 280

Zimm, B. H., **23**:193(98), 223(98), 271(98)

Zimmels, Y., **23**:207(271), 278(271); **25**:194(95), 234(131), 247(95), 248(131)

Zimmerman, W., **17**:76, 154

Zimmermann, P., **7**:320c(149, 152), 320d, 320e(160, 161, 166), 320f, 320g

Zinemeister, G. E., **18**:189(142), 239(142)

Zino, I. E., **15**:330

Zinov'eva, K. N., **17**:69(30), 153

Zinsmeister, G. E., **SUP**:76(212), 171(212); **15**:324, 329; **23**:369(1), 460(1)

Zinukov, I. A., **25**:4(18), 140(18)

Zinyavichyus, F., **31**:263(157), 284(157), 329(157)

Zisman, W. A., **9**:190, 227(8), 229, 230, 231(62), 268, 269

Žiugžda, J., **18**:91(23), 92(23), 94(26), 107(38), 116(26), 117(26), 118(38), 119(23, 38), 120(23, 38), 121(38), 125(23, 38), 126(23, 38), 127(26, 23, 38), 128(26), 129(26), 130(26), 131(26), 158(23, 26, 38)

Žiugžda, V., **8**:101, 112(35), 113(38), 115(35), 123(38), 124(38), 125(38), 130(38), 147(79), 159, 160; **18**:96(28), 158(28)

Zmeikov, V. M., **21**:166, 183

Zmola, P. C., **1**:223, 264

Zoglin, P., **19**:30(59), 86(59)

Zolotogorov, M. S., **6**:418(114, 115), 491

Zommer, N., **15**:330

Zorbil, V., **20**:278(296), 279(297), 280(297), 296(323), 313(296, 297), 314(323)

Zou, Y. F., **30**:357(107, 108), 361(107, 108), 407(107), 422(107, 108, 183, 184), 431(107, 108), 435(183, 184)

Zozulya, N. V., **31**:245(133), 278(179), 282(186), 328(133), 330(179), 331(186)

Zrunchev, I. A., **25**:158(24, 33, 34, 35, 36, 37), 159(24), 194(35, 36, 37), 201(35, 36, 37), 218(36), 229(37), 236(144, 145, 146),

Zrunchev, I. A. (*continued*)
243(24), 244(33, 34, 35, 36, 37), 249(144, 145, 146)
Zuber, N., **1**:216, 217, 218(32), 221(28, 32), 224(28), 225(28), 230, 234, 238(63, 64), 239, 259, 264, 265; **4**:167, 178, 181, 224, 225; **5**:83, 107, 110, 126, 403, 405, 407, 409, 411, 413, 419(193), 420, 421, 422, 425, 426, 436, 439, 440, 467, 511, 513, 514; **7**:81, 86; **9**:227, 269; **10**:132, 133, 165; **11**:11, 13(39), 14(45), 15, 16, 48, 49; **14**:129(112), 148; **15**:242(53a, 55), 244(69), 278, 279; **16**:130, 132, 133, 139(111, 112), 155, 156, 181(12), 225, 238, 239; **17**:3, 11(51), 23, 56, 58, 148, 157; **18**:249(6), 320(6); **20**:80(12), 96, 107, 108, 132, 250(228, 229), 310(228, 229); **23**:82(97), 131(97); **29**:218(20), 219(20), 220(20), 228(20), 249(20), 256(20), 288(20), 312(20), 335(20); **31**:109(45), 110(45), 140(23), 142(23, 46), 145(47), 156(23), 158(45, 46, 47)
Zuev, I., **28**:127(84), 142(84)
Zueva, G. K., **6**:418(118), 491
Žukauskas, A. A. (Chapter Author), **8**:93, 95, 109(20), 112(35, 36), 113(20, 37, 38), 114(37, 42), 115(35, 42), 116(20), 123(38), 124(38), 125(38), 128, 129(37, 38), 130(37, 38), 138(20), 141(74), 142(74), 143(74), 144(75), 145(74, 75), 146(78), 147(79), 151(20), 159, 160; **11**:203, 206, 215, 220, 224, 226, 236, 241, 242, 254, 262; **18**:87, 88(20, 21), 91(23), 92(23), 94(26), 104(35), 105(35), 106(35), 107(35, 37, 38), 108(20), 109(35, 37, 42), 110(20), 113(35), 116(26), 117(26), 118(38), 120(23, 38), 121(38), 124(37), 125(23, 37, 38), 126(23, 37, 38), 127(23, 26, 37, 38), 128(26, 35), 129(26, 35), 130(26), 131(26, 35), 135(35, 70), 136(20, 35, 70), 139(75), 140(35), 141(20), 142(70, 77), 143(20, 35), 144(35), 145(35, 79), 147(20), 148(20), 150(20, 35), 156(35), 158(20, 21, 23, 26, 35, 37, 38, 42), 159(70, 75, 77, 79); **25**:252, 292, 296, 300, 319; **31**:159(1), 160(1, 36, 44, 52, 67, 68, 69), 161(1), 177(1), 182(1), 247(1), 262(1), 263(1), 294(36), 322(1), 323(36), 324(44, 52), 325(67, 68, 69)
Zukoski, E. E., **6**:92, 94(148), 108(184), 129; **20**:100, 102, 103, 104, 105, 132
Zumbrunnen, D. A., **23**:13(69), 15(69), 51(69), 118(121), 120(121), 129(69), 132(121); **26**:120(40), 122(40), 147(40), 162(40), 164(40), 167(40), 168(40), 170(40), 198(157, 158), 199(157, 159), 200(157, 162), 204(178, 179, 180), 210(40), 216(157, 158, 159, 162), 217(178, 179, 180)
Zunduggiyn, T. S., **10**:207(123), 217
Zuzovski, M., **18**:186(136), 191(136), 239(136)
Zvereva, T. V., **25**:4(22), 140(22)
Zwanzig, R. W., **2**:349, 356
Zweifach, B. W., **22**:225(231), 357(231)
Zwick, E. B., **4**:228
Zwick, S. A., **1**:217, 218(31), 220, 221(31), 264; **4**:178, 225; **11**:13(38), 14, 15, 48; **15**:242(56), 243, 278; **16**:82, 83, 85, 87, 154
Zycina-Molozhen, L. N., **3**:35(30), 98
Zysina-Molozhen, L. M., **9**:71(117), 104, 111; **25**:127(196), 149(196)
Zyszkowski, W., **8**:24, 33, 87, 88; **29**:163(103), 207(103)

SUBJECT INDEX

Bold text indicates volume number. The supplementary volume
published between volumes 14 and 15 is indicated by **SUP**.

A

ABAQUS (software), **30**:351
Abbe's theory, **6**:186
Ablation, **8**:32; **19**:19, 81; **22**:280, 319; **28**:109–
110; *see also* Phase change
direct containment heating, **29**:249–253
Ablation plume, **28**:105–107
Ablative material, **16**:300
Absolute temperature factor, **15**:8–10
Absolute transference number, **12**:246
Absorbing emitting media, **5**:27
Absorbing medium, **1**:2
Absorptance, **2**:401f
apparent, **2**:425ff
influence of cryo deposits on, **5**:477
Absorption, **5**:18
by solid walls, **27**:61–85
gas absorption
Monte Carlo method, **27**:62–66, 71–74,
80–85
radiative heat transfer, **27**:93–95
scattering and, **27**:92–99
simulation, **27**:50–56
Absorption bands, in molecular gas band
radiation, **12**:132–133
Absorption characteristics of cavities, *see*
Cavities
Absorption coefficient, **1**:3, 9ff, 14f; **3**:180f,
197; **8**:232; **9**:369; **12**:180; *see also* Gas
absorption coefficient
in Elsasser band model, **8**:237
influence of spectral, temperature and pressure
dependence on, **5**:33
volumetric absorption coefficient, **3**:181
Absorption of light, coefficient in tissue,
22:279
Absorption probability, **27**:51–52
Absorption spectra, for semitransparent solids,
11:358–359

Absorption suppression, **31**:365
Absorptive power, **27**:11
Absorptive properties, spectral variation in,
31:388
Absorptivity, **27**:1, 15
ABSORP variable, **27**:53, 204
ABSPR variable, **27**:53, 55, 204
Acceleration
combined with roughness and transpiration,
effects on boundary layer, **16**:306
effect on turbulent boundary layer, **16**:280–
297
effects of on turbulence, **2**:67f, 72ff
Acceleration pressure drop
in impinging flow, **13**:31
slug flow modeling, **20**:114
Accommodation coefficient(s), **2**:273, 277ff,
340, 342, 344, 345f, 348; **7**:165
devices employed in measuring, **2**:311ff
effects on
internal energy, **2**:337ff
rotational energy, **2**:338
vibrational energy, **2**:338
experimental determination of, **2**:304ff
factors modifying, **2**:328f
influence of pressure on, **2**:335f
measurements, **2**:304ff, 307ff, 311ff, 317ff,
319ff, 322ff, 324ff, 326ff
of
air, **2**:327
argon, **2**:327
carbon monoxide, **2**:327
helium, **2**:326f, 330
hydrogen, **2**:327
nitrogen, **2**:327
oxygen, **2**:327
triatomic gases, **2**:328
surface temperature effects, **2**:329ff
theoretical calculations of, **2**:339ff
Accumulation term, **31**:46, 61

Accuracy improvement in transient heat transfer problems, 1:96ff
Accurate film profiles, slug flow calculations, 20:99–100
Acetone-butanol, A_0, 16:119
Acetone-cyclohexanol, film boiling, 16:133, 138
Acetone-ethanol, A_0, 16:119
Acetone-ethanol-water, boiling, heat transfer, 16:116–117
Acetone-methanol, slope of saturation curve, 16:67
Acetone-methanol-water, boiling, 16:116
Acetone-water
　　A_0, 16:119
　　heat transfer coefficient, 16:107–108
　　prediction, 16:127
　　slope of saturation curve, 16:67
Acoustic Froude number, 24:312–313
Acoustic-gravity waves
　　damping coefficients for, 14:271–272
　　damping of, 14:268–279
　　damping rates for, 14:272–279
　　equations for, 14:269–271
Acoustic methods, transition boiling studies, 18:265–267
Acoustic Reynolds number, 24:312–316
Acoustic waves
　　caused by laser heating, 22:321
　　effects, 4:47ff
　　propagating, 24:312–316
Action integral, 24:40, 43, 44, 46, 71, 78, 79, 85, 95
Action of freezing, 25:235
Activated complex, 2:127ff
Activated molecules, stability of, 10:93–94
Activation energy
　　asymptotics, 26:61
　　in shear flow, 13:216–218
Active cavities, nonuniform superheat of, 10:127–128
Active length, 25:289
Adaptive grid, 24:213
Additives, heat transfer enhancement with, 30:235–237, 245–246
Adhesive force, 21:63–64
Adiabatic flame temperature, 29:62, 63
Adiabatic heat-transfer coefficient, forced convection, three-dimensional arrays, 20:232–238

Adiabatic isochoric complete combustion pressure, 29:62
Adiabatic stratification, 9:278
Adiabatic surface, 19:30
Adiabatic wall temperature, 7:326
ADI technique, 26:22, 23, 27
Adjoint heat conduction problem, 1:68
Admittance matrix, 31:129
Admixture, magnetic and nonmagnetic particles, 25:167, 177, 186, 191, 193, 194, 233, 235, 239
Aerodynamic heating, 1:269, 331ff, 349f
　　entropy production, 21:252–254
Aerosol filters, 25:292
Aerospace industry, heat exchangers, 26:220–221
Aerowindow, 18:41
Affinity chromatography, 25:235
　　ligand, 25:236
Age–density distribution function, 10:189
Agglomeration of particles, 25:165, 234
Aggregates, formation of, 14:284
Aggregation of particles, 25:202
Agitated tank, 25:260, 261, 273
　　vessel, 25:279
Agitation of bed particles, 25:192
AGW, see Acoustic-gravity waves
Air
　　control, woodburning cookstove and, 17:286–289
　　indoor quality, stoves and, 17:299–307
　　stoichiometric and excess, open fires and, 17:200–202
Air boundary layer
　　above lake, 13:79–82
　　boundary conditions for, 13:80–81
　　geometry and physics of, 13:79–80
Air cooling, thermal control of electronic components, 20:182–183
Aircraft engines, combustion chambers, 27:173–181
Air experiments, 2:52ff
　　nozzle configurations, 2:52f
Air jet cooling, 26:205
Airplane trajectory method of producing reduced gravity, 4:150
Air suction, influence on thermal characteristics of heat exchangers, 31:278
Air–water layer
　　geometry and physics of, 13:86–87

in THIRBLE classification, **13**:85–88

Alaska, magma energy in, **14**:97

Albedo
bond, **10**:13
geometric, **10**:13–14
of moon, **10**:7
normal, **10**:14
scattering, **27**:17–18, 95, 152–154, 190
for single scattering, **3**:201, 214

Algorithm, *see* Numerical method

Alloys, **28**:26
macrosegregation, **28**:243, 251, 266, 327
double diffusive convection, **28**:240–241
electromagnetic stirring, **28**:320–325, 327
inhibition, **28**:308–310, 316
semitransparent analog alloys, **28**:272
shrinkage and buoyancy, **28**:306, 308
tin–lead alloys, **28**:291–292, 295–304, 317
solidification, **28**:231–238, 326, 328
from bottom wall, **28**:277–282
continuum model, **28**:270–288, 311
freckle formation, **28**:243–244
mathematical models, **28**:249–269
metal alloys, **28**:288–308, 316, 327
multidirectional solidification, **28**:282–285
physical phenomena, **28**:238–249
process control strategies, **28**:235, 237, 238, 308–326, 327–328
semitransparent analog alloys, **28**:270–288
from sidewall of rectangular cavity, **28**:270–277
unidirectional solidification, **28**:277–282
zero gravity, **28**:309

α-mode failure, **29**:147, 149–151, 193–195

Altered free-volume state model, **18**:226–227

Alternating direction implicit technique, *see* ADI technique

Aluminized Mylar, **9**:382

Aluminum
laser vaporization, **28**:129–131
molten, interaction with water, **29**:188, 190–192

Aluminum–aluminum foils, thermal contact conductance, **20**:264

Aluminum antimonide, properties, **30**:321

Aluminum arsenide, properties, **30**:321

Aluminum–copper alloys, **28**:233, 236, 239, 304, 317

Aluminum industry, steam explosions, **29**:135, 188

Aluminum nitride, properties, **30**:321

Aluminum oxide, **9**:120

Aluminum phosphide, properties, **30**:321

Ambient media
characteristics, **16**:2–3
density stratification, determination, **16**:17–19
flow conditions, **16**:3
flowing
differential modeling of jets in, **16**:10–13
entrainment functions for jets in, **16**:12
properties, **16**:17–20
quiescent
differential modeling of jets in, **16**:9–10
entrainment functions for discharges into, **16**:11
stratification modeling, **16**:17–20
stratifications, effects on jets, **16**:42–51
stratified quiescent
jet behavior in, **16**:43–46
nondimensional differential form of governing equations, **16**:23
unstratified
flowing jets in, comparative calculations for, **16**:36–42
nondimensional differential form of governing equations, **16**:22
unstratified quiescent
modeling of jets in, **16**:25–36
nondimensional differential form of governing equations for, **16**:22

Amdahl's law, **31**:408

American Petroleum Institute (API), **31**:465–467

Amioca, conversion during extrusion, **28**:219–220, 221, 222

Ammonia, synthesis, **25**:152, 236

Ammonium chloride analog alloy, **28**:270, 286, 311–316

Amplification rates, **9**:326

Analog alloys, **28**:240

Analog to digital converter, **22**:365, 411–412, 415–416

Analogy, **25**:267, 275, 312
spinup and heatup, **24**:286

Analogy method, exact or approximate solutions by, **SUP**:62–63

Analysis
Acrivos, **25**:257, 261
Dienemann, **25**:283
Karman–Pohlhousen, **25**:283

Analytical solution methods, **22**:261, 271

Analytic iteration, **8**:56
Angle
 interstitial velocity, **25**:184
 phase, **25**:197
 polar, **25**:169
 repose, **25**:166, 181
Angle factors, **2**:400, 405f, 420f
 diffuse, **2**:411ff
Angle of illumination, **10**:28–30
Angular velocity, critical, induced rotational
 flow, **21**:148–149
Anisotropic mass diffusion, natural convection,
 20:325–327
Anisotropic permeability, alloy solidification,
 28:276
Anisotropic scattering, **19**:38–39; **27**:146–154;
 31:364–373
Anisotropic turbulence, inhomogeneous medium
 and, **14**:262–263
Annular channels
 heat transfer enhancement in tube bundles in,
 31:232–259
 with an inner grooved tube, **31**:241–245
 with one-sided protrusion groove turbulators,
 31:253–259
 transverse finning in, **31**:245–249
Annular-dispersed flows, **1**:355f
 applications of, **1**:435ff
Annular flow, **1**:355f, 358
 steady-state slugs, **20**:85–87
Annular grooves, **31**:160
 heat transfer enhancement with vapor
 condensation on, **31**:278–287
 salt deposition on, **31**:299–304
Annular sector ducts, **SUP**:269–272
Annular space, *see* Liquid metal heat transfer
Annulus, **6**:312
 film boiling of water in, **5**:100, 101
Anomalous-skin-effect, **9**:364, 385
Aperiodic regenerator operation, unidirectional
 regenerators, **20**:154–158
Apollo 11, **10**:3–6, 20, 23–26, 29–31, 33–34, 49,
 53–55, 59, 67–69, 74, 76, 78
Apollo 12, **10**:3, 5, 21–27, 29–34, 49, 54–55, 57,
 59, 67, 70, 72, 74, 76
Apollo 13, **10**:2, 41, 73
Apollo 14, **10**:3, 6, 23–34, 54–55
Apollo 15, **10**:2–3, 23–25, 32, 41, 74–75
Apollo 16, **10**:25, 41
Apollo 17, **10**:41

Apparent absorptance, **2**:425ff
Apparent emittance, **2**:425ff
Apparent friction factors, *see also* Friction
 factor–Reynolds number product
 for circular tube, **SUP**:88, 95, 97–98
 comparison of, **SUP**:397–399
 for concentric annular ducts, **SUP**:299–301
 definition of, **SUP**:40
 for parallel plates, **SUP**:167–168
 for rectangular ducts, **SUP**:211–212
 for triangular ducts, **SUP**:241
Apparent heat capacity, **22**:296
Apparent mass velocity in annular-dispersed
 flows, **1**:365
Apparent thermal conductivity, **5**:343
Apparent viscosity, **24**:108
Applications, magnetofluidized bed, **25**:232
Applied mathematicians, overview for,
 SUP:385–420
Approximate methods, *see* direct methods
Aqueous polyacrylamide, pH vs. shear velocity
 for, **15**:83
Aqueous solutions, transfer coefficients for,
 12:239
Arbitrary axial wall temperature distribution
 in circular ducts, **SUP**:118–120
 in concentric annular ducts, **SUP**:316–318
 in parallel plates, **SUP**:177–179
Arbitrary heat flux distribution
 in circular ducts, **SUP**:136
 in concentric annular ducts, **SUP**:316–318
 in parallel plates, **SUP**:185
Arbitrary shape, **24**:124, 132, 148
Arbitrary triangular ducts, **SUP**:237–240; *see
 also* Triangular ducts
Archimedes number, **24**:4; **25**:157, 158, 165,
 204, 225, 228
 bulk bed voidage and, **19**:110
 gas–solid system, **19**:106–107
 maximum Nusselt number and, **19**:102, 116
Arcs
 cathode axis parallel to a plane anode, **4**:295ff
 cylinder geometry with annular anode,
 4:320ff
 free-burning with plane anodes, **4**:293ff
Arctic ocean, temperature, salinity, and density,
 16:48–51
Arctube, **24**:265, 266, 267
Area goodness factor, **SUP**:393–395
Area integrals, estimate of, **31**:89

Areas of future research, **SUP**:419–420
Argonne National Laboratory (ANL) studies
α-mode failure, **29**:147
ANL/CWTI studies, **29**:147, 220, 221, 222, 225
ANL/CWTI-5, **29**:226, 316, 317, 323
ANL/CWTI-6, **29**:316, 317
ANL/CWTI-12, **29**:226, 316, 317, 323
ANL/CWTI-13, **29**:271
ANL/IET studies, **29**:220, 221, 222, 228, 310
ANL/IET-1RR, **29**:229
ANL/IET-3, **29**:229
ANL/IET-6, **29**:229
ANL/U studies, **29**:221, 222, 229–230
direct containment heating, **29**:216, 219, 220–222, 225
ARN, *see* Array of round nozzles
Aroclor solution
Raschig rings vs. flow rates for, **15**:269
steam condensation in countercurrent flow with, **15**:269
steam condensation in pipe with, **15**:263–264
ARO, *see* Array of round orifices
Arrangement
ordered particle, **25**:229
random particle, **25**:229
Array correction function, for impinging flow, **13**:21–22
Array of round nozzles, **13**:46–47
integral mean transfer coefficient for, **13**:18–22
optional spatial arrangements for, **13**:51
transfer coefficients vs. outlet flow conditions in, **13**:27–41
Array of round orifices
integral mean transfer coefficient for, **13**:22
Sherwood numbers and, **13**:37
Array of slot nozzles
high-performance, **13**:52
integral mean transfer coefficient for, **13**:22–26
optional spatial arrangements in, **13**:51
variation of transfer coefficients for, **13**:34–36
Arrays aligned, **25**:292
of cylinders, **25**:252, 296
model, **25**:293
periodic, **25**:292, 297
of rods, **25**:292
of spheres, **25**:252
square, **25**:292–294, 306

staggered, **25**:308
staggered square, **25**:293
triangular, **25**:293, 294, 308
Arrhenius function, **22**:293, 313, 319
comparison of alternative models, **22**:328
data for injury model coefficients, **22**:320, 324, 330
Arrhenius type reaction, **2**:214
Arteriovenous anastomoses, **22**:226
Artificial cavities, thermal cooling, boiling curve, **20**:247–248
Artificial neural networks (ANN), chemical vapor deposition modeling, **28**:408–413
Ascending center descending wall, **19**:125
Ascending wall descending center, **19**:125
ASN, *see* Array of slot nozzles
ASN transfer coefficients, **13**:36
Aspect ratio, **25**:152, 289
flat-plate collectors, **18**:12–15
influence of, on free convection inside rectangular cavities, **8**:184
Asymmetric-balanced regenerator, **20**:137–138
unidirectional operation, **20**:157–158
Asymmetric-unbalanced regenerator, **20**:137–138
unidirectional operation, **20**:157–158
Asymptotic, **25**:267, 272, 288
behavior, **25**:257
solution, **25**:262
values, **25**:286
Asymptotic expansion, **8**:13; **19**:28–33
of local similarity equation, **4**:335ff
Asymptotic methods, **8**:14, 37
Asymptotic suction layer, **16**:253
Asymptotic velocity profiles, for minimum drag asymptotic case, **15**:93
Atlantic ocean, tropical, temperature, salinity, and density, **16**:48–50
Atmosphere
heat transfer rate in, **12**:9
as reservoir, **12**:9
waste heat cooling and, **12**:2
Atmospheric absorption, blackbody curves for, **12**:8; *see also* Blackbody radiation
Atmospheric reduced pressure chemical vapor deposition (APCVD), **28**:344
Atomic and Molecular Spectra, **5**:267
electronic spectra, **5**:274
rotational spectra, **5**:267

Atomic and Molecular Spectra (*continued*)
  vibrational rotational spectra, **5**:269
Atomic bomb explosion, thermal burn from, **22**:237
Atomization, spray deposition, **28**:10, 11, 18–21
Atoms conservation equation, *see* Conservation of atoms equation
Attagel solutions, **15**:64–66, 89; **19**:249, 326
Attenuation
  inhomogeneous, **31**:376–379
  radiation intensity, **27**:148–149
Attenuation constant, gas–particle mixture, **27**:17
Attenuation distance, **27**:93, 95
Attraction, interparticle, **25**:165, 173
Augmentation of heat transfer, **9**:114
Augmentation technique, in heat transfer, **15**:21
Autocorrelation function, **8**:344, 345
Autoignition, **29**:307
Autoignition limit, **29**:67
Autoignition temperature, **29**:84–86
Auxiliary variables, slug flow, **20**:95–97
Availability, in heat transfer, **15**:7–8
Available work, vs. energy, **15**:8
Average, average of the, **31**:22–23
Average heat transfer
  free-surface jet, **26**:161–162, 170
  submerged jet, **26**:133–135
Average mass fraction, gradient of, **31**:33
Average of the gradient, **31**:13
Average temperature, **23**:374; **31**:50
  generalized volume average, **23**:376
  intrinsic phase average, **23**:375
  superficial phase average, **23**:374
  weighting function, **23**:377
Average velocity, fluid flow in a restricted channel, **18**:99
Averaging volume, variations in, **31**:16–17
Axial averaged velocity component, **31**:168
Axial conduction, effects of, **3**:106f
Axial dispersion, *see also* Mixing; Péclet number
  liquid–solid fluidized beds, **23**:250
  packed beds, **23**:229
  three-phase fluidized beds, **23**:257
Axial distance, dimensionless, **SUP**:41, 50
Axial flow effect, induced rotational flow, **21**:156–159
Axial heat diffusion, **3**:116ff
Axial momentum equation, liquid metal droplet, **28**:4
Axial velocity, mean, **SUP**:38

Axial wall heat flux, *see* Thermal boundary conditions
Axis of symmetry, liquid metal droplet, **28**:6
Axisymmetric flow
  in MHD, **1**:314ff
  potential flow convection, **20**:365–366
    isothermal spheroids of revolution, **20**:372–374
    planar flow, **20**:369
Axisymmetric jet, **26**:107
  array, **26**:171
  flow structure, **26**:136–144
  free-surface liquid jet, **26**:109–119
  heat transfer, **26**:124–126, 131–135, 137–162, 189–193, 204
  hydraulic jump, **26**:114, 201–203
  jet inclination, **26**:182–186
  motion of impingement surface, **26**:203–204
  stagnation region, Falkner–Skan flow, **26**:111, 120, 122
  submerged jet, **26**:123–135
  surface liquid jet, **26**:107–108

B

Back melting, **19**:43
Backward-facing step, **24**:222; **31**:167–169
Baffles effect, induced rotational flow, **21**:174–176
Baffle-tray column condensers, **15**:262–263
Baish models
  composite model, **22**:149
  experimental observations, **22**:151
  superposition model, **22**:71, 137–139
Baker plot, **5**:452, 453
Baker's boiling curves, immersion cooling techniques, **20**:245
Balanced-symmetric regenerator
  three-dimensional effectiveness plot, **20**:175
  unidirectional operation, **20**:173–174
Balance equation
  "boxed" regenerator, **20**:144–146
  heat transfer in incompressible liquid media, **18**:162–163
Balmer series, **5**:275
Band absorptance, **5**:292
  comparison of correlations, **5**:299
Band absorptance correlation, **8**:256
Band absorptance models, **8**:230

Band emissivity, **5**:292
  for box model, **5**:293
  exponential wide band model, **5**:294
Band location
  effective transmittance and, **12**:184
  in nonisothermal gas radiation, **12**:182
Band models, **5**:287
Band radiation, **5**:287
  from nonhomogeneous gases, **5**:300
Bandwise approach, **31**:394–395, 397–400, 423
Bank of cylinders, **25**:254, 297
  tubes, **25**:252
Banks
  closely spaced, heat transfer, **18**:145–147
  in crossflow, heat transfer calculations,
    18:154–156
  hydraulic drag
    calculation methods, **18**:148–150
    mean drag, **18**:150–153
    proposals for calculation of, **18**:153–154
  in-line, fluid flow pattern changes, **18**:103
  staggered, **18**:100
Barrel chemical vapor deposition reactors,
  28:346, 375–386
Barycentric velocity, **12**:204
Basalt
  lunar, **10**:44–45, 59–60, 62
  thermal conductivity of, **10**:57
Batchelor scale, **30**:10, 16, 63, 74
Bath temperature, in gas-liquid heat exchanger,
  15:34–35
Baule's theory, **2**:339ff
Bechtel Corporation, **15**:272
Bed
  aspect ratio, *see closest ratio*
  behavior, **25**:157, 194, 195, 205, 228, 238
  diameter, **25**:152
  effective permeability, **25**:157
  expansion, **25**:152, 155, 166, 190, 193, 201,
    236
  fluidized, **25**:252, 254, 273
    fibrous, **25**:292–294
    packed, **25**:252, 270, 274
    particulate, **25**:292, 304
    three-phase fluidized, **25**:254
  fluidlike properties, **25**:168
  geometry, **25**:152, 167
  height, **25**:152, 167, 174
  instability, **25**:168
  nonuniformity, **25**:168

  porosity, **25**:168
  premagnetized, **25**:201
  pressure drop, *see* Pressure drop
  stabilization, *see* Stabilization
  uniformity, **25**:163
  viscosity, **25**:161
  voidage, *see* Voidage
Bed expansion
  liquid fluidized beds, **23**:241
  three-phase systems, **23**:253
Bed fluidization, *see* Fluidization, bed
Beds, gas-fluidized
  high-temperature heat transfer process models,
    19:173–178, 181
  hydrodynamics, influence of internals,
    19:125–126
Bed voidage
  as fluidization commences, **19**:144
  at minimum fluidization, **19**:105–108
  for orthorhombic arrangement of particles,
    19:142
  on radiation, **19**:171–172
  surface orientation and, **19**:112–113
  turbulent flow and, **19**:143
BEER program, **27**:53–55, 203
Beer's law, **8**:239; **12**:132–133; **27**:14, 17, 34
  light absorption, **22**:249, 278
Behavior, bed, **25**:157, 194, 195, 205, 228, 238
  mass transfer, **25**:190
Bellows heat pipe, indirect cooling with, **20**:275–
  277
Bénard cells, cryogenic
  heat transfer in rotating helium I, **17**:133–137
  observations on heat transport in, **17**:130–137
  steady convection, **17**:130–131
  time-dependent convection, **17**:131–133
Bénard convection, **24**:174
Bénard-type convection, heat- and mass-transfer,
  natural convection onset, **20**:318–319
Benzene
  compound droplet, vaporization, **26**:78
  forced convection boiling of, **5**:96, 97, 99
  heat flow rates, effect of heater shape and
    material, **16**:191, 192
Benzene-diphenyl, boiling, effect of subcooling,
  16:110–111
Benzene-toluene, A_0, **16**:119
Benzoic acid, **25**:267, 272
Bernard countercurrent studies, **22**:19, 59
Bernoulli equation, **SUP**:69

Bernoulli equation (*continued*)
  Taylor bubble translational velocity, **20**:101–102
Bessel functions, **25**:100, 108
  unidirectional regenerator operation, **20**:152–153
Bethe–Teller relation, **8**:248
Bidirectional reflectance, **10**:19–21
Biharmonic equation, **8**:42
BILLEAU apparatus, **29**:155
Binary boundary layers, **8**:19
Binary droplet, vaporization, **26**:78–79
Binary mixture
  of gases, **3**:66, 290ff, 296ff
  pool boiling, **30**:173–174
Binary scaling law, **2**:246
Bingham plastic fluids, **2**:359, 373, 381f, 384; **23**:190
  equation, **2**:362, 383
  fluidized beds, **23**:239, 241
  packed beds, **23**:202, 214
  parameters, **2**:373
  transient laminar thermal convection from flat vertical plate to, **15**:177–178
Bingham plastics, **24**:107, 109
Biocide, **31**:438, 448
Bioclimate, **4**:67
Bioclimatological data, **4**:84f
Bioclimatology, **4**:67
Bioengineering, **22**:1, 2
Biofouling, **31**:435–439
  mitigation of, **31**:447–449
Bioheat equation, simplified, *see* Model; Weinbaum–Jiji models
Bioheat model, *see* Model
Bioheat transfer, *see* Heat, transfer
Biological liquids, thermal conductivity, **18**:209–211
Biotechnology, **22**:1, 14–16
Biotechnology, heat transfer in, **4**:65ff
  artificial environments, **4**:103ff
    indoors, **4**:103ff
    space vehicles, **4**:111ff
    underwater vessels, **4**:109ff
  natural environments, **4**:66ff
    extraterrestrial, **4**:97ff
    terraqueous, **4**:90ff, 98f
    terrestrial, **4**:66ff, 98f
  skin, its role in heat transfer, **4**:115ff
    characteristics, **4**:115ff

functions, **4**:115ff
    injury, **4**:123ff
    protection, **4**:128ff
    thermal sensation, **4**:121ff
Biot number, **1**:175; **13**:207, 238–239, 262; **SUP**:55–56; **19**:66, 152; **24**:44, 79
  Bi, **25**:99, 112, 113, 114, 138
  calculation of, **13**:223
  Nusselt number and, **13**:243
  in shear flow, **13**:222–225
  wall boundary condition and, **13**:224
Biot's method, **1**:79ff
Biot's variational method, for Stefan problems, **19**:39–40
Biot's variational principle, **8**:23; **19**:197–198
Birefringence, loss as indicator of thermal damage, **22**:319
Birnbrier data, vertical plates and channels, **20**:196
Black body, **5**:6, 259
Blackbody absorption, in atmosphere, **12**:8
Blackbody cavity, in thermal radiation, **12**:129
Blackbody intensity, for gas temperature, **12**:130
Blackbody radiation, **10**:11, 15–16; **12**:11; **27**:7–9, 10, 11, 12, 63; **28**:89
  in a plasma, **4**:238ff
  wavenumber ratios and, **12**:125
Blackbody radiosity
  defined, **12**:124
  temperature and, **12**:133
Blade, **25**:128
Blake–Kozeny equation, **28**:266
Blake–Plummer equation, **25**:151
Blasius correlation, hydrodynamics, slug flow, **20**:89
Blasius equation, **2**:21, 142, 203f
Blasius method, in film-boiling heat transfer, **11**:65
Blasius–Polhausen similarity transformation, in film-boiling heat transfer, **11**:126
Blasius's law, **31**:226
Blasius solution, **8**:67
Blasius-type expression, for drag-reducing fluid flow, **15**:193
Blister, due to burn, **22**:277
Blockage effect, **25**:275
Blockage ratio, **8**:103
  fluid flow in a restricted channel, **18**:97–98
Blood
  flow

in burned tissue, **22**:228
convective cooling of tissue, **22**:267, 276
effect on temperature during laser irradiation, **22**:266
freezing, **22**:12
heat conductivity, **18**:210–211
perfusion, **22**:8, 9
Blowdown
 airborne debris interactions, **29**:274–275
 reactor coolant system, **29**:235, 237, 238, 248, 256, 257, 310
Blowing, *see also* Transpiration
 definition, **16**:272
Blowing parameter, **3**:73f, 79f; **4**:322, 326, 330f
Blowout process, severe slugging, **20**:120–121
Blowout solids, **25**:233
Blowthrough, **29**:248–249, 253–256
Blunt bodies, **3**:5; **8**:13
Blunt body flows, **2**:234ff, 246ff
Blunt body (hypersonic), *see* Locally similar solution
Blunt needle of finite length, potential flow convection, **20**:372–373
Board–Hall model, steam explosions, **29**:141, 142, 145, 146, 185
Bodoia velocity profile, **SUP**:163, 191
Body-fitted coordinates, **19**:57
Body forces, **8**:164; **9**:276
Boe's criterion, severe slugging, **20**:121–122
 classical vs. quasi-equilibrium operations, **20**:128
Boiler furnaces, **27**:158–167
Boiler tubes, steam generation in, **14**:150
Boiling, **1**:185ff; **3**:109f; **7**:74; **16**:60; **22**:252, 280, 319; *see also* Film boiling; Film-boiling heat transfer; Nucleate pool boiling
 bubbles in, **16**:231–232
 convection and two-phase flow, **30**:122–141
 convective, **16**:142–150
  heat transfer coefficient, **16**:144
  heat transfer region, **16**:145
  research recommendations, **16**:150
 of cryogenics, **5**:400
 effects of gravity on, **5**:457
 film, **7**:78; **16**:133–142
  on flat plates and tubes, **16**:133–138
  at Leidenfrost point, **16**:138–142
  minimum heat flux, **16**:139–141
  mixture effect, **16**:133–138
  of pure fluids, **16**:139

 pure fluid vs. mixture, **16**:135–136
 on thin wires, **16**:138–139
 transition from stable nucleate boiling to, **16**:132
forced convection, **5**:403; **11**:43–45
inception, **16**:64–80
 effect of composition, **16**:68–70
 effect of contact angle, **16**:68–70
 effect of prepressurization, **16**:72–75
 effect of temperature, **16**:68–70
 experimental studies, **16**:70–75
intermediate and forced flow, **17**:8–9
liquid superheat in, **11**:9–17
of multicomponent liquid mixtures, **16**:59–156
of multicomponent liquids, **30**:173–174
nature of, **11**:1–2
nucleate, **7**:76; **11**:2
pool boiling, **5**:401
in porous media, **30**:122–141, 148–152
pseudo, **7**:87
Saturated, **1**:186ff
 Burnout, **1**:190
 Film boiling, **1**:190
 Nucleate, **1**:186ff
single-component
 peak nucleate heat flux, prediction, **16**:130–132
 vs. multicomponent, **16**:60
subcooled, **1**:191f
 of liquid mixtures, **16**:110–113
Boiling boundary layer, two-phase, **11**:32–38; *see also* Thermal boundary layer
Boiling curve
 saturated liquid, **23**:6
 transition boiling, **23**:111
Boiling heat transfer, **10**:110; **11**:2–3; **14**:127–139; *see also* Heat transfer
Boiling hysteresis, electric field and, **14**:132
Boiling incipient temperature, thermal boundary layer and, **11**:11
Boiling in reduced gravity, *see* Reduced gravity
Boiling liquid superheat, **11**:1–46; *see also* Liquid superheat
Boiling nucleation, **10**:85–162; *see also* Nucleation
 engineering significance of, **10**:107–111
 heat flow densities in, **10**:147
 Madejski analysis in, **10**:152–156
 model of, **10**:134
Boiling number, **5**:447

Boiling process, as heat transfer process, **11**:30–31

Boiling region diagram, narrow space nucleate boiling, **20**:47–49

Boiling site, density
experimental studies, **16**:75–80
variation with composition, **16**:70

Boiling situations with liquid splashing, CHF and, **17**:19–21
special, CHF in, **17**:7–9

Boiling superheat, **10**:148–162

Boiling surface, in nucleation, **10**:133–134

Boiling temperature, incipient, **11**:3–4

Boiling terminology, **1**:197ff

Boiling water reactors (BWR), direct containment heating, **29**:219, 324–328

Boltzmann constant, **2**:279, 288, 345; **3**:183; **9**:362; **15**:48; **22**:360, 384, 408, 421
thermistor, **22**:362–363

Boltzmann distribution, **4**:235, 245; **8**:246

Boltzmann equation, **3**:183; **5**:39, 40

Boltzmann number, **3**:218, 244
flame, **21**:262

Boltzmann principle, in liquid superheat, **11**:6

Boltzmann probability distribution, **11**:10

Boltzmann's intergradifferential equation, **3**:186

Boltzmann's neutron transfer equation, **3**:186

Boltzmann transformation, **19**:5
in heat conduction, **30**:120

Bond albedo, **10**:13

Bond graph, **22**:287

Bond number, **9**:186; **21**:117

Boric oxide, in Czochralski process, **30**:322, 323

Boron arsenide, properties, **30**:321

Boron nitride, properties, **30**:321

Boron phosphide, properties, **30**:321

Borosilicate glass, **9**:367

Bose–Einstein distribution, **11**:321

Bose–Einstein statistics, **12**:120

Bottling effects, **14**:69

Bottom heating, natural convection and, **30**:122–134

Bouguer number, **3**:215

Boundary
bubbling-stabilized transition, **25**:174
of macroscopic system, **31**:28
phase, **25**:170

Boundary collocation technique, **SUP**:65

Boundary condition effects

on combined forced convection and radiation, **1**:33ff

on combined free convection and radiation, **1**:38ff

perscribed surface heat flux, **1**:81, 84f

pulselike inputs, **1**:85ff, 93ff

step in surface temperature, **1**:59f, 63, 83f

Boundary conditions, **24**:44, 46, 67, 76, 82; **31**:54–56, 59–60, 106; *see also* Thermal boundary conditions
asymmetric, **24**:84
chemical vapor deposition, **28**:358, 382
dimensionless groups, **19**:254–257
dimensionless peripheral coordinates, **19**:254
dimensionless velocities, **19**:255
for energy closure problem, **31**:65–68
in energy/mass transport equations, **13**:133–137
error, **19**:22
first-kind, **24**:60, 62, 65, 74, 76, 78, 88
for fluids flowing through rectangular ducts, **19**:253–254
fourth-kind, **25**:4
general, **24**:44
governing, **31**:23–28
H, **19**:295
*H*1, *see H*1 boundary condition
*H*2, **19**:253, 254
heat- and mass-transfer
horizontal direction, **20**:331–332
natural convection, **20**:322–323
in heat conduction problems
in MHD problems, **1**:279
on porous bodies, **1**:173f
inlet, **25**:398, 405
laser evaporation, **28**:125
liquid metal droplet, **28**:5
mechanical, time-dependent, **24**:305–307
mismatch in, **31**:82
nonlinear radiation, **19**:32
outlet, **25**:398, 405
periodic, **25**:324, 405
potential flow convection, planar motion, **20**:356
realistic, **24**:307–312
second-kind, **24**:53; **25**:3
in shear flow, **13**:222–227
splat quenching, **28**:29, 47, 57
for temperature problem, **SUP**:16–36
thermal

continuously changing, **24**:303–304
  step change, **24**:280–294
third-kind, **24**:67, 70; **25**:3, 10, 11
T, see T boundary condition
for velocity problem, **SUP**:8
wall, **13**:224; **25**:364–370, 398
  global, **25**:367–370
  local, **25**:364–366
Boundary element method, **19**:44
Boundary integral method, **19**:44
Boundary layer, **8**:9–12, 14, 29; **25**:258, 261–
  263, 267, 270, 278–280, 286, 312; *see also*
  Turbulent boundary layers
  adverse-pressure gradient
    predictions, examples, **16**:358–359
    with strong blowing, predictions, examples,
      16:358–359
    with sudden nonequilibrium conditions,
      predictions, examples, **16**:358–360
  asymptotic, **16**:290
  calculations, **16**:295
  chemically reacting, **2**:110ff
  with chemical reactions, **8**:36, 40
  in compressible flow, **8**:35
  for convection inside rectangular cavities,
    8:181, 184
  damping function, **16**:344
  definition of, **SUP**:9
  in dissociating gases, **8**:36
  with dissociation, **8**:42
  droplet vaporization, **26**:66–71
  elastic, **25**:258
  equations, **9**:282
    continuity, **24**:116
    momentum, **24**:116, 117
    thermal, **24**:118
  equilibrium, **16**:242, 248–249, 249–254
    acceleration parameter, significance,
      16:252
    adverse-pressure gradient, **16**:242
      predictions, examples, **16**:357–358
    blowing parameter, **16**:250
    flows with axial pressure gradients, **16**:251–
      254
    predictions, examples, **16**:355–357
    pressure-gradient parameter, **16**:297
      definition, **16**:251
    transpired flow parameters, **16**:250
  friction factor, in acceleration, **16**:288–291
  with gas phase reactions, **2**:192ff

with heat transfer, **8**:35
heat transfer, and full-coverage film cooling,
  16:312–319
with injection, **8**:11, 12, 40
inlet conditions, overshoot, **16**:288, 291
inner region, **16**:343, 346–347
laminar, **16**:247, 249; **25**:270, 281, 282, 288
in laminar flow, **8**:35
laminarization, **16**:345
laminar–turbulent transition in fluid flow,
  18:91–92
local equilibrium, **16**:273, 278
mass diffusion, **16**:272
mass transfer, **16**:272
mixing length, **16**:344
mixing-length model, **16**:248, 342
model, for solution of momentum and energy
  equations, **16**:341–355
momentum, **25**:265, 283, 288
  development, **16**:251
  equilibrium, **16**:253–254
  momentum transfer, effects of acceleration,
    16:288–290
  under strong acceleration, **16**:247–248
momentum equation, **16**:342
at nozzle inlet, **2**:59, 63f
outer region, **16**:343
  model, **16**:347–348
Prandtl mixing length, **16**:343
predictions, examples, **16**:355–361
with pressure gradient, **8**:12, 35
pumping mechanism, **24**:283
with radiation, **8**:19
with rough surfaces, predictions, examples,
  16:360–361
separation in fluid flow past a single tube,
  18:90
shape parameters, **2**:99ff
stability, in fluid flow past a single tube, **18**:92
thermal, **25**:265, 282, 283, 288
  asymptotic, **16**:253
  equilibrium, **16**:253–254
thickness, **9**:287; **24**:287; **25**:13
  and heat transfer coefficient, **16**:298–299
transpired, **16**:242, 272; *see also* Boundary
  layer, laminar; Boundary layer, turbulent
turbulent, **25**:259, 291
  accelerated, **16**:242
  bursts, from wall region, **16**:248
  effect of acceleration, **16**:280–297

Boundary Layer (*continued*)
    with convex surface curvature, **16**:323–325
    effect of deceleration, **16**:297–301
    effect of free convection, **16**:331–335
    effect of roughness, **16**:300–311
    effect of surface curvature, **16**:319–331
    effect of transpiration, **16**:272–282
    factors affecting, **16**:272–341
    flow visualization, **16**:329–331, 332–333
    friction factor, and roughness, **16**:306, 307
    heat transfer
        effect of convex curvature, **16**:319–327
        effect of roughness, **16**:300
        effects of mixed convection, **16**:336–340
    hydrodynamics, effect of roughness, **16**:300
    inner region, similarity, **16**:251
    laminarization, **16**:248
    mean temperature profile, effect of roughness, **16**:307–310
    mean velocity profile
        effect of acceleration, **16**:290–297
        effect of roughness, **16**:307–310
    momentum transfer, effect of roughness, **16**:306–307
    negative-pressure, *see* Boundary layer, turbulent, accelerated
    outer region, similarity, **16**:251–252
    relaminarization, **16**:281
    retransition, **16**:248
    temperature profile
        effect of acceleration, **16**:290–297
        effect of deceleration, **16**:299–301
        effect of mixed convection, **16**:340–341
        effect of transpiration, **16**:278–282
    turbulence properties, effect of roughness, **16**:310–311
    velocity profile
        effect of deceleration, **16**:299–301
        effect of mixed convection, **16**:340–341
        effect of transpiration, **16**:278–282
    zero-pressure gradient, **16**:242
    in turbulent flow, **8**:35
    turbulent in rocket nozzle, **2**:2ff
    turbulent shear stress, **16**:342
    with variable wall temperatures, **8**:35
    vertical sidewall, **24**:284
    viscoelastic, **25**:278
    viscous-dominated region, **16**:343
    viscous sublayer

    model, **16**:344
    thickness, **16**:344–346
    in water at lake surface, **13**:82–85
Boundary layer analysis, **31**:138
Boundary layer approximations
    in forced convection flows, **15**:178–179
    refinements in, **14**:69
Boundary layer effects in MHD two-dimensional channel flow, **1**:311ff
    compressible flow, **1**:312ff
    incompressible flow, **1**:311f
Boundary layer equations, **8**:15, 117
    dedimensionalization of, **11**:269
    transformation into incompressible form, **2**:170ff
Boundary layer flow
    critical region, **7**:17
    heat- and mass-transfer, horizontal direction, **20**:332–336
    non-Newtonian fluid mechanics in, **2**:389f
    non-Newtonian heat transfer in, **2**:389ff
    in solar buildings, **18**:68–72
    in thermally active fault zones, **14**:66–67
    in thermosyphons, **9**:8, 39
    two-dimensional buoyancy-induced, **14**:69
Boundary layer separation, **8**:100, 101
Boundary layer stripping mechanism, cyclic thermal, **16**:109, 120–121
Boundary layer theory, **8**:162
Boundary layer thickness, **2**:101f, 195
    energy, **2**:14, 15
    momentum, **2**:13f, 15
    temperature, **2**:11
    velocity, **2**:11
Boundary layer type idealizations
    in hydrodynamically developing flow, **SUP**:9, 68–73
    in thermally developing flow, **SUP**:15
Boundary–value problems, **19**:195
Bound–bound radiation in a plasma, **4**:234f
Bound–free processes, **5**:302
Bounds, thermal conductivity of suspensions, **18**:197–200
Boussinesq approximation, **5**:217; **8**:175; **9**:277, 279; **24**:117
Boussinesq fluid, **14**:26
Boussinesq fluid assumption, **24**:280
Boussinesq relation, **21**:7
Boussinesq transformation, potential flow convection

axisymmetric potential flow, **20**:365–366
  planar motion, **20**:354
Box model, **8**:259
Brackett Series, **5**:275
Brayton cycle, **12**:7
Brayton cycle heat engine, counterflow heat
  exchanger for, **15**:26–27, 33
Breccia, lunar, **10**:59, 61
Breeder reactors
  loss-of-flow accidents, **29**:1–2
  postaccident heat removal, **29**:9
  transient-over-power accidents, **29**:1, 2
Bremsstrahlung, **5**:263
  in a plasma, **4**:237ff
Bridgman model, **18**:216–217
Brinkman–Darcy–Forchheimer equation,
  24:156
Brinkman equation, **24**:113, 114
Brinkman–extended Darcy model, **24**:113
Brinkman numbers, **SUP**:53–55; see also
  Viscous dissipation
  in steady shear flow, **13**:234
Bromley's Theory of film boiling, **5**:57
  limits of applicability, **5**:67
Bronze, horizontal tubes v. staggered tube
  bundles, **19**:157
Brownian motion, thermal conductivity of
  suspensions and, **18**:206–207
Brun number, **SUP**:55–56; **25**:11
Brunt equation, **12**:11
Brunt–Väisälä frequency, **5**:216; **24**:281, 287
Bubble
  in boiling, **16**:231–232
  classical gas, **25**:152
  coalescence, **19**:99; **25**:163
    peak heat flux and, **16**:132
  erupting, **25**:163
  flow pattern
    in-line tube bank, **19**:155
    shallow and tall beds, **19**:123–125
    staggered tube bank, **19**:156
  formation, **25**:168
    latent heat in, **16**:66–67
    work of, **16**:66–68, 107
  in immiscible liquid, **26**:47–50
  origins, **16**:64
  velocity, **19**:98
Bubble and packet contact times, **19**:127
Bubble behavior
  narrow space nucleate boiling, **20**:36–40

coalesced bubble region, **20**:36, 38–39,
  42–49
  isolated bubble region, **20**:36, 40–41
  nucleate boiling, **20**:2
  agitation model, **20**:3–4
  liquid and vapor profiles, **20**:33–34
  liquid film thickness, **20**:33–34
  surface configuration, **20**:23–27
Bubble columns
  condensation in, **15**:242–262
  multibubble systems in, **15**:248–260
  multitrain systems in, **15**:255–260
  single bubbles in motion in, **15**:244–254
  single stagnant bubbles in, **15**:242–244
Bubble condensation, **15**:242–262
  diffusion-controlled, **15**:243
  inertia-controlled, **15**:243
  noncondensables and, **15**:247
  pure and impure vapor factors in, **15**:251
Bubble density, velocity decrease due to,
  15:257
Bubble departure, **16**:95–105
  diameter, **16**:95–101
    variation with bubble inertia force, **16**:97–
    99
  frequency, **16**:96, 102–105
  inertia controlled, heat transfer coefficient,
    prediction, **16**:121
  surface tension controlled, **16**:99–100
    heat transfer coefficient, prediction, **16**:122
Bubble dynamics in boiling, **1**:215ff
Bubble evaporation mechanism, **16**:109
Bubble flow
  countercurrent, **15**:259
  volumetric transfer coefficient in, **15**:258–259
Bubble formation, **9**:295
  on horizontal cylinders, **5**:57
Bubble growth, **4**:174ff; **16**:80–94
  criteria for, **10**:128–129
  from cylindrical cavity, experimental data,
    16:92
  effective superheat, theoretical model, **16**:89
  from electrically heated wires, experimental
    data, **16**:92
  experimental studies, **16**:90–94
  on a heated surface, **1**:219ff
  on heated wall, radius of dry area underneath,
    16:100
  high-speed movies of, **11**:22–24
  high-speed photography and, **10**:146–147

Bubble growth (*continued*)
  liquid microlayer in, **11**:17
  in liquid superheat, **11**:11–17
  local surface heat flow and temperature in, **10**:140
  in nucleation studies, **10**:100–105
  rise in vapor temperature with, **16**:93–94
  stages, **16**:229–230
  at superheated wall, **16**:87–89
  surface tension controlled, **16**:103
  temperature field in, **11**:15–16, 18–30
  theoretical models, **16**:82–90
    Scriven, **16**:82–83
    Skinner and Bankoff, **16**:87
    Van Ouwerkerk, **16**:87
    Van Stralen, **16**:83–87, 87–89
  in a uniformly superheated liquid, **1**:216ff
  variation, with vapor-liquid composition, **16**:92–93
  waiting time, before departure, **16**:102–104
  wall surface temperature in, **10**:142
Bubble-induced entrainment
  efficiency, **21**:314
  initially stratified liquid layers, **21**:299–311
    background, **21**:299–300
    bubble Reynolds number, **21**:309
    bubble rise velocity, **21**:302–303
    correlation of data, **21**:307–310
    driving force, **21**:308
    efficiency, **21**:308–309
    experiment, **21**:301–302
    force balance, **21**:308
    heavy-liquid density effect, **21**:303–304
    heavy-liquid viscosity effect, **21**:305–307
    interfacial tension effect, **21**:306
    light-liquid density effect, **21**:304, 306
    light-liquid viscosity effect, **21**:307
    phenomenon observations, **21**:302–303
  interlayer heat transfer, **21**:314–317
  onset, stratified liquids, **21**:313
  rate, stratified liquids, **21**:313–314
Bubble nucleation, *see also* Boiling, inception
  model, at conical cavity, **16**:64–65
Bubble Nusselt number, **1**:228ff
Bubble patterns in boiling, **1**:247ff
Bubble population density, nucleate boiling, **20**:5–8, 13–15
Bubble radius, volume-equivalent, **21**:281
Bubble Reynolds number, **1**:227ff
Bubble rise velocity, **21**:302–303
Bubble train, **15**:248–254

collapse history in, **15**:252–253
multitrain systems in, **15**:255–260
Bubble velocities
  in condensing bubble train, **15**:250
  dispersed in liquid slug, **20**:107–108
  surface tensions parameters, **20**:104–105
Bubble volume
  entrainment onset, **21**:313
  maximum entrainable, **21**:314
  minimum, entrainment onset, **21**:295, 297
Bubbling, **25**:166, 168, 173, 194, 204
  fluidized bed, **25**:174, 175, 176, 190, 204, 222
Bubnov–Galerkin method, **19**:195
Buckingham potential, **3**:263, 291, 295
Buckley–Leverett case, **30**:121
Buffer layer, **3**:44
Building codes, fire research and, **10**:277–279
Bulk cavitation, **9**:191
Bulk compressibility, **7**:7
  isentropic, **7**:7
  isothermal, **7**:7
Bulk density, spectral directional reflectance and, **10**:26–27
Bulk density driven flow, in solar buildings, **18**:67–68
Bulk mean temperature, **SUP**:45
Bulk temperature, **3**:137, 168; **7**:185
Bundle
  elastic, **25**:258
  elliptic cylinder, **25**:293
  equilateral triangular, **25**:308
  in-line, **25**:306
  in-line square, **25**:308
  rod, **25**:256, 292, 295, 297, 298, 299, 305, 306, 308–312
  staggered triangular, **25**:308
  tube, **25**:296–307, 311
    staggered, **25**:300, 306
  turbulent, **25**:259, 291
  viscoelastic, **25**:278
Buoyancy, **9**:275
  combined mechanisms of, **9**:303
  coupling, **9**:324, 327
  effects, **16**:2, 243; **24**:280–293
  induced flow, **24**:215, 267
  mass-diffusion-induced, **9**:286
  normal component of, **9**:296
Buoyancy-driven flow, thermal microscales and, **30**:13–19
Buoyancy effects, **7**:17, 20, 69

forced convection, augmented heat transfer, **20**:226

heat- and mass-transfer
  boundary-layer profiles, **20**:333–335
  convection suppression, **20**:339
  enclosed porous layers, **20**:337–339
  horizontal direction, **20**:333–334
  horizontal line source, **20**:346–347
  multilayer structures, **20**:347
  vertical direction, **20**:318–319
  wall inclination, **20**:336

Buoyancy forces, **25**:133
  natural convection and, **30**:384–385, 386
  nucleate boiling, **20**:2–3
  in phase-change processes, **19**:50
  tangential, **9**:296

Buoyancy-induced flow and heat transfer
  asymptotic limiting relations, **20**:193, 195
  vertical plates and channels, **20**:185–187

Buoyancy-induced secondary flow, **15**:214

Buoyancy parameter, **28**:239

Buoyancy ratio, heat- and mass-transfer, **20**:320–321

Buoyant jet, **9**:282, 297

Burger equation, **25**:355

Burgers variables, **21**:231

Burn, **22**:11, 14
  atomic bomb explosion, **22**:237
  blister, **22**:277
  degree, **22**:226, 331
  electrical, **22**:243
  flame source, **22**:241
  laser irradiation source, **22**:248
  pain intensity due to, **22**:231
  thermal, **22**:222, 229, 328

Burning process, time-dependent, on open fires, **17**:228–236

Burning velocity, **29**:75–78

Burnout, **5**:99; **9**:79
  CHF in helium and, **17**:327, 328–329, 330, 331, 333, 334, 336
  heat flux of methane, **5**:424
  in saturated boiling, **1**:190

Butane, in water-ice slurries, **15**:268

BWR, *see* Boiling water reactors

C

Cadmium telluride
  properties, **30**:321

transition temperatures to transparency, **30**:340

Calcium carbonate, **25**:152
  precipitation fouling, **30**:206–207

Calculation method, quasisteady, **25**:58

Calculation procedures, slug flow, **20**:97–100

Calculus of variations, **8**:22

Calibration
  heat flux measurements, **23**:342–343
  of hot wire probes, **8**:312, 313

Calorimeters, **9**:401
  null-point, **23**:311–313
  response to constant heat transfer coefficient, **23**:309–310
  slug, **23**:307–311

Calorimetry, **9**:400; **22**:7

Candle flames
  axisymmetric, speckle photography, **30**:296–299
  combustion in, **10**:221–222

Capacitance parameter C, in molten polymer flow problems, **13**:207

Capacity rate imbalance, entropy generation due to, **15**:31

Capillarity, **9**:186

Capillary action, in porous media, **13**:121, 124

Capillary boundary layer, **30**:116

Capillary effect, on condensation heat transfer, **30**:145

Capillary flow, **SUP**:53

Capillary flow method, **3**:255f, 260

Capillary forces, multiphase flow, **30**:93, 94, 96–98

Capillary-porous bodies, heat and mass transfer in, **1**:123ff

Capillary pressure, **30**:96–97; **31**:41

Capillary tubes
  steady-state heat transfer in helium II and, **17**:104–107
  wettability, **21**:61–62

Capillary tube viscometer, **15**:70–72, 113

Capture particulate, **25**:189

Carbon dioxide
  gaseous radiation in, **8**:273
    radiation-conduction interaction in, **8**:266–273
    radiative transfer in, **8**:256, 259, 260
    thermodynamic nonequilibrium in, **8**:273
    variation with temperature of line half widths in, **8**:232
  physical properties, **21**:2–3
  total emissivity of, **12**:160

Carbon–iron alloy, solidification, **28**:288
Carbonization, of tissue during burn, **22**:280, 319
Carbon monoxide
    collisional relaxation times in, **8**:248
    comparison of thermodynamic equilibrium
        and nonequilibrium radiation results for,
        8:274, 275, 276
    conduction–radiation interaction for thermo-
        dynamic nonequilibrium in, **8**:277, 278
    infrared bands in, **8**:231
    radiation–conduction interaction in, **8**:266,
        267, 272
    radiative transfer in, **8**:259
    woodburning stoves and, **17**:299, 301–309
Carbon tetrachloride
    in forced convection film boiling, **5**:96, 98
    soil contamination by, **30**:177–181
Carbon tetrachloride-R-113, film boiling,
    16:133–134
Carbopol
    elastic behavior, **19**:330
    heat transfer studies in rectangular duct,
        19:321–322
    turbulent channel flow, **19**:326
    viscoelasticity, **19**:249–250
Carbopol (carboxypolymethylene), **25**:264, 280,
    281
Carbopol solution, **15**:64–65, 88–89
    Nusselt number for, **15**:102
    as viscous fluid, **15**:102
Carbose D (sodium carboxymethyl cellulose),
    25:264
Carboxymethylcellulose (CMC), **19**:329
Carboxy methyl cellulose (CMC), **25**:267,
    276–278, 283, 287–290, 308
Carboxymethylcellulose solutions, **15**:64, 167
    falling spheres in, **15**:73
    in laminar thermal convection studies,
        15:163–165
Carboxypolymethylene, *see* Carbopol
Cardioid ducts, **SUP**:277–278
Carman–Kozeny expression, **25**:196, 201, 223
Carnot efficiency, defined, **15**:45
Carreau model, **28**:153, 154
Carreau model B viscoelastic fluid model, stress
    tensor in, **15**:78, 80–86
Cartesian, **24**:205, 206
Cascade, infinite, **SUP**:161
Casting process
    alloy solidification, **28**:244–245, 306

natural convection and, **19**:60–61
Catalysis heterogeneous, **25**:236
Catalyst
    alumina-silica, **9**:114
    ferrochrome, **25**:158
    impregnated nickel, **25**:178
    zeolite, **25**:178
Catalytic reaction, *see* Locally similar solution
Catalytic surface reactions, **2**:135ff, 146ff
Catalytic wall, **2**:46; **8**:8
Cauchy-Green strain tensors, **15**:78
Cauterization, **22**:5, 11
Cavities
    absorption characteristics of, **2**:425ff
    active, **10**:127–128
    circular arc, **2**:421
    circular cylinder, **2**:419f, 424, 425, 427f
    conical, **2**:427ff
    emission characteristics of, **2**:425ff
    nucleation criteria for, **10**:129
    parallel-walled, **2**:424, 425, 428
    reciprocity theorem for, **2**:426f
    rectangular-groove, **2**:427f
    sidewall heated, **24**:296–302
    spherical, **2**:421, 427ff
    stability of in nucleation, **10**:134–148
    vee-groove, **2**:427ff
Cavity circular foil gage, **23**:305–306
Cavity dispersal phenomena, direct containment
    heating, **29**:256, 275
Cavity model, **25**:198
Cavity receivers
    central receiver systems, **18**:35–41
    flow pattern, **18**:35
    parabolic dish systems, **18**:41–46
Cavity sizes, **9**:213
CCM experiments, **29**:147
CDF, *see* Cumulative distribution function
Cell, **22**:293
Cell models
    free surface, **23**:207, 209–210
    packed beds, **23**:207, 209
    zero vorticity, **23**:207, 209, 211
Cell Péclet Number, **24**:195, 226, 228
Cellular convection, **8**:165
Cellulose, fire and, **10**:225–226
Center-of-mass velocities, **31**:110
Center-of-volume velocities, **31**:110
Central difference scheme, **24**:226, 237
Central receiver systems

aerowindow, **18**:41
cavity receivers, **18**:35–41
convective losses from, **18**:36–41
external receivers, **18**:30–34
schematic diagram of external and cavity
receivers, **18**:4
Centrifugal bed, **25**:153
Centrifugal forces, **8**:164
Ceramic, stove design and, **17**:297
CFV, *see* Curvilinear finite volume
discretization
Chain branching, effect on thermal conductivity
of polymeric liquids, **18**:218
Chains of particles, **25**:183
straight, **25**:202
vertical, **25**:185
CHAMP/VE model, steam explosions, **29**:151
Change, equations of, **12**:197–203
Change of phase, **8**:75
Channel, **25**:202
complex geometry, **25**:114
diameter, **25**:9
equivalent, perimeter, **25**:9, 10
flat, **25**:127
flow, **25**:4, 8, 22
geometry, transient heat transfer in helium II
enthalpy considerations, **17**:120–122
q, Δt conditions for burnout, **17**:119–120
recovery conditions, **17**:122–123
heating, one-sided, **25**:121
interstitial, **25**:184
length, CHF of helium and, **17**:322–323
steady-state heat transfer in helium II and
critical heat flux, **17**:110–113
heat transfer with mass flow, **17**:114–115
recovery from film boiling, **17**:113–114
temperature profiles and effective thermal
conductivity, **17**:107–110
straight interstitial, **25**:194, 229
tortuous interstitial, **25**:194
Channel emittance, vs. wall layer transmittance,
12:185
Channel flow, **8**:11, 32; **24**:155
chemical reactions, **6**:418
condensation on surface of cold water in,
15:238–239
critical region, **7**:19
defined, **13**:206
developed laminar flow in, **8**:43
in elliptic ducts, **8**:70

forced convection in, **8**:42
with MHD, **8**:56
one-dimensional model, **6**:374
in rectangular ducts, **8**:70
with slug flow, **8**:54
thermal conductivity variable, **6**:533
viscosity variable, **6**:528, 533
Channel flow studies
forced convection
flush heat sources, **20**:221–226
two-dimensional protruding elements,
20:227–232
Channeling in magnetically stabilized fluidized
bed (MSFB), **25**:175
Channels
alloy solidification, **28**:244–245, 310,
312–313, 316
boiling heat transfer enhancements in,
31:264–274; *see also* Annular channels
film boiling and, **11**:144–187; *see also* Forced
flow film boiling
restricted
average velocity, **18**:99
blockage ratio, **18**:97–98
flow in, **18**:97–99
mean velocity in minimum free cross
section, **18**:99
reference velocity, **18**:99
single-tube heat transfer in, **18**:129–131
Channel spacing, forced convection,
20:230–231
Chapman–Enskog kinetic theory, **3**:261
Chapman–Jouguet (CJ) point, **29**:144
Chapman–Rubesin constant, **4**:324
Chapman's formula, **2**:297
Characteristic dissociation temperature,
2:134
Characteristic length, **24**:111, 168
Characteristics
flowability, **25**:232
heat transfer, **25**:238
hydrodynamic, **25**:152
thermal bed, **25**:152
Characteristic velocity, **24**:168
Charcoal, as fuel, **17**:180–181, 228–231
Charge, conservation of, **12**:247–248
Charge transport, **12**:244–251
boundary conditions in, **12**:248–251
mass transport and, **12**:244–247
Charge uniformity, **9**:134

Charny models
  vascular heat transfer, **22**:71, 127, 151–152
  whole body heat transfer, **22**:134
Chato model
  countercurrent vessel pair
    superposition model, **22**:88–90, 105, 106,
      109, 111, 122
    variable blood flow, **22**:91–92
  isolated vessel
    governing equations, **22**:85–86
    solution, effectiveness, **22**:86–87
  surface effects
    solution with metabolism, **22**:194–96
    superposition model, **22**:92–93
  thermal equilibration length, **22**:86, 139
Chebyshev integration, unidirectional
    regenerator operation, **20**:154
Chemical diagnostics, **2**:110, 164
Chemical kinetics, **2**:122ff
Chemically frozen, **2**:46
Chemically frozen flow, **8**:8
Chemically reacting boundary layers, **8**:40; *see
    also* Boundary layer
Chemical potential
  defined, **12**:261
  as driving force for diffusion, **12**:263–268
Chemical reaction
  effects at stagnation point in MHD, **1**:346f
  heterogeneous, **2**:135ff
  homogeneous, **2**:123ff
Chemical reaction fouling, heat transfer
    enhancement, **30**:201, 208
Chemical–thermal explosions, **29**:187–192
Chemical vapor deposition (CVD), **28**:339–344,
    413–414
  equations
    multicomponent mixture, **28**:353–356
    simplified, **28**:356–362
  models, artificial neural network models,
    28:408–413
  reactors, **28**:344–345
    barrel reactor, **28**:346, 375–386
    horizontal reactor, **28**:346, 365–375
    hot-wall LPCVD reactor, **28**:399–408
    impinging-jet reactor, **28**:345, 346, 394–
      395
    pancake reactor, **28**:346, 386–389
    planar-stagnation-flow reactor, **28**:397–399
    rotating disk reactor, **28**:389–394
  transport phenomena, **28**:346–353, 362–365
Chen and Holmes continuum model

comparison with other models, **22**:45, 46,
    52–59, 68, 85, 139
conduction, **22**:47–48
  final form, **22**:57
  metabolism, **22**:47–48
governing equations
  blood flow, effective conductivity
    estimation, **22**:56
  formulation, **22**:56
  integral approximation, **22**:48
  perfusion term
    Fourier representation, **22**:52
    three subproblems, **22**:53–55
  thermal equilibration length, **22**:49–51, 74, 87,
    96, 104, 106, 139
Chenoweth-Martin correlation, **1**:374
CHF, *see* Critical heat flux
Chicago tap water, chemical analysis of, **15**:80–
    82
Chilton–Colbourn *j* factor relation, **21**:171
Chimney, stove design and, **17**:299, 300, 304
Chip-in-cavity electronic packages, forced con-
    vection, **20**:221
Chlorination, **31**:447, 449
Choice of velocity variables, **24**:205
Choroid, **22**:265, 273
Chromate, **31**:449
Churn turbulent region, bubble velocity,
    20:108
CHYMES model, steam explosions, **29**:150,
    151, 152, 155, 156
Circuit card channels, forced convection,
    20:242–243
Circular-arc cavities, *see* Cavities
Circular cavities, convection, **29**:17–20
Circular cylinders
  laminar heat transfer from, **11**:277–280
  longitudinal flow over, **SUP**:354–365
    central, corner, and wall channels,
      SUP:358–359
    concentric rod bundles, **SUP**:360–363
    exponential axial wall heat flux for,
      SUP:364
    fluid flow over, **SUP**:355–363
    fully developed flow in, **SUP**:355–365
    heat transfer in, **SUP**:363–365
    Nusselt numbers for, **SUP**:357
    thermally developing and hydrodynami-
      cally developed flow in, **SUP**:365
    triangular and square array of rod bundles,
      SUP:355–360

potential flow convection, **20**:359–360
  isothermal cylinder, **20**:359–360
  uniformly heated cylinder, **20**:360–361
Circular cylindrical cavities, *see* Cavities
Circular ducts, **SUP**:78–152
  arbitrary axial wall temperature distribution in, **SUP**:118–120
  arbitrary heat flux distribution in, **SUP**:136
  axially matched solutions for, **SUP**:85
  with central regular polygonal cores, **SUP**:350–352
  conjugated problems for, **SUP**:136–138
  constant heat flux in, **SUP**:82–83, 124–135, 144–149
  constant wall temperature in, **SUP**:79–80, 99–118, 138–144
  creeping flow for, **SUP**:91, 95–96, 98–99
  with diametrically opposite flat sides, **SUP**:273
  energy content of fluid in Graetz problem for, **SUP**:113–115
  exponential axial wall heat flux in, **SUP**:83–85, 135–136
  finite difference solutions for, **SUP**:88–89
  fluid axial heat conduction in, **SUP**:79, 81, 83, 111–118, 132–135
  friction factors for, **SUP**:78, 88, 95, 97–98
  fully developed flow in, **SUP**:78–85
  Graetz problem for, **SUP**:99–105
  hydrodynamically developing flow in, **SUP**:85–99
  infinite series solution functions for thermal entrance problem in, **SUP**:101, 126
  integral solutions for, **SUP**:85–87
  Lévêque solution for, **SUP**:105–108
  linearized solutions for, **SUP**:87
  longitudinal thin V-shaped fins within, **SUP**:374–375
  longitudinal thin fins within, **SUP**:367–370
  longitudinal triangular fins within, **SUP**:375–378
  Nusselt numbers for, **SUP**:79–85, 103–104, 108, 116, 122, 123, 127–128, 134–135, 140–141, 143, 147–151
  pressure drop for, **SUP**:97–98
  radiant flux boundary condition, **SUP**:81–82, 123–124
  simultaneously developing flow in, **SUP**:138–152
  solutions of Navier–Stokes equations for, **SUP**:89–97

  specified heat flux distribution in, **SUP**:82–83, 124–136
  specified wall temperature distribution in, **SUP**:99–120
  thermal entrance lengths for, **SUP**:105, 115, 126, 133, 143, 146
  thermally developing and hydrodynamically developed flow in, **SUP**:99–138
  with a twisted tape, *see* Twisted tapes, within circular ducts
  velocity overshoots in, **SUP**:91–92
  viscous dissipation, effect of, **SUP**:80, 82, 109–111, 129–132, 142, 146, 148–149
  wall thermal resistance in, **SUP**:80–81, 120–123, 149–152
Circular foil gage, **23**:300–306
  analytical models, **23**:301–302
  cavity, **23**:305–306
  convection correction, **23**:303–304
  design chart, **23**:302
  exponential time constant, **23**:301
  schematic, **23**:301
  temperature distribution, **23**:302–303
  thermopile, **23**:304
  water-cooled, **23**:305
Circular pipe flow, non-Newtonian fluids in, **15**:59–132
Circular sector ducts, **SUP**:264–267
Circular segment ducts, **SUP**:267–269
Circular tube passages, *see* Passages
Circulating bed, **25**:189
Circulating fluidized bed boiler (CFBB), **27**:187–193
C–J deflagration speed, **29**:87
Clapeyron equation, **1**:212, 217; **16**:66
Classical formulations, **19**:24
Clauser shape factor, **16**:252, 297
Clausius–Clapeyron effect, **31**:88, 98
Clausius–Clapeyron equation, **9**:198; **10**:95–96, 159; **16**:64, 231
Clausius–Clapeyron Kelvin equation, **31**:11, 48, 50–52
Clausius–Clapeyron relation, **11**:10, 14; **21**:70
CLCH model, **29**:218, 241, 242, 243, 301, 311, 320
Clearance, narrow space nucleate boiling, **20**:37–39
  coalesced bubble region, **20**:41–49
  isolated bubble region, **20**:40–41
  pressure and heat flux, **20**:47–79
Clinopyroxine, in lunar fines and rocks, **10**:7

Clo, **4**:91
Close-contact melting
in concentric annular regions, **19**:71
definition, **19**:70
in descent of heated horizontal cylinder
through solid PCM, **19**:72–73
for hot sphere melting its way through solid,
19:73
inside horizontal tubes, **19**:71
inside sphere, **19**:70–71
Closed-cycle gas turbine plant, designing,
31:182–185
Closure
coupled problem in, **31**:75–76, 77–80
for energy, **31**:60–68
gas phase diffusion problem of, **31**:68–72
for mass, **31**:56–60
quasi-steady problem of, **31**:69
thermal energy problem of, **31**:72–74
Closure level, **31**:2
Closure scale, **31**:1
Closure variables, **31**:76–77
Clothing, protection against burn injury, **22**:234,
240
Cluster of particles, **25**:231
CMC (carboxymethylcellulose), **19**:329; see also
Carboxymethylcellulose solutions
CM experiments, **29**:165
Coagulation, tissue, **22**:277, 319
Coal, **25**:152
combustion, **25**:152
fluidized-bed combustion, bed, **19**:97; see also
Fluidization
atmospheric, **19**:97
large particle systems, **19**:98, 100
pressurized, **19**:97
radiation in, **19**:98–99
small particle systems, **19**:98
resources in U.S., **19**:97
Coalesced bubble region
heat flux, emission frequency and, **20**:49, 52
low liquid level nucleate boiling, **20**:56–57
narrow space nucleate boiling, **20**:36, 38–39,
42–49
emission frequency, **20**:43–45
one-dimensional model, **20**:49–54
unsteady heat conduction model, **20**:49, 53
Coaxial cylinder method, thermal conductivity of
structured liquids, **18**:172–173
Coaxial thermocouple, **23**:313–314

Cocurrent-countercurrent parallel flow heat exchanger, **26**:228, 229
Cocurrent cross-flow heat exchanger, **26**:230–
231
Cocurrent parallel flow heat exchanger, **26**:228–
229, 245
temperature effectiveness, **26**:300–302
Cocurrent pipe contactor, condensation in,
15:263–264
Coefficient
convective transfer, **25**:312
demagnetizing, **25**:198
diffusion, **25**:312
drag, **25**:252, 312
friction, **25**:48, 84
gas
convective component heat transfer, **25**:222
radial dispersion, **25**:175, 176
heat transfer, **25**:3, 99, 103, 164, 177, 190,
191, 206, 230, 252, 286, 310, 312, 313;
see also Heat transfer coefficient
conventional reduced, **25**:103
hydraulic resistance, **25**:3
local, **25**:9
mass transfer, **25**:177, 252, 286, 289, 298
particle
convective component heat transfer, **25**:222
discharge, **25**:171
resistance, **25**:16, 76
thermal activity, insulation, **25**:112
volumetric expansion, **25**:23, 25
total heat transfer, **25**:222
Coefficient of nonstationary heat transfer, **21**:45
Coefficient of thermal expansion, **5**:142
Coefficient of turbulent viscosity, **21**:23
Coherence function, heat flux and velocity,
23:348–349
Coherent structure, **25**:324, 369–370
Cohesion, work of, **9**:189
Cohesive force, **21**:64
Colburn correlation, **30**:20
Colburn equation, **19**:161
Colburn factor, **SUP**:53
Colburn form of Reynolds Analogy, see Rey-
nolds analogy
Cold space, refrigerator power and, **15**:47
Coles' skin-friction coefficient, see Skin friction
coefficient
Collisional relaxation times
in carbon monoxide, **8**:248

in diatomic gases, **8**:248
Collision-based methods, **31**:364–366, 373, 422
 performance of, **31**:348–364
 of ray tracing, **31**:344–346
Collision broadening, **5**:278
Collision integral, **3**:262f, 267f
Collocation, method of, **1**:98ff
COLSYS, **26**:304–305
Combined air–water layer, in THIRBLE classi-
 fication, **13**:85–88
Combined buoyancy mechanisms, **9**:303
Combined entry length problem, **SUP**:16
Combined entry length solutions, *see*
 Simultaneously developing flow
Combustible gases, reaction of injected, *see*
 Reactions, of injected combustible gases
Combustion
 completeness, **29**:71–75, 111
 efficiency, **2**:51
 fast flames and detonation, **29**:86–102
 flaming
  model of, **17**:193–194
  of wood, **17**:188
 heat flux measurements, **23**:352–353
 hydrogen, *see* Hydrogen combustion
 of injected premixed fuel, **2**:202ff
 interface conditions and, **17**:195–196
 in reduced gravity, *see* Reduced gravity
 solid-phase
  in fuelbed, **17**:192–193
  of single piece of wood, **17**:190–192
  some example solutions for, **17**:196–200
 tissue, **22**:321
 volumetric, direct containment heating,
  29:307–308
Combustion chamber heat flux, **2**:79ff
Combustion chambers, jet engines, **27**:173–181
Combustion gas, **27**:181–187
Combustion parameters, **29**:62–65
 autoignition temperature, **29**:84–86
 burning velocity, **29**:75–78
 diffusion flame, **29**:62
 flammability limits, **29**:65–71
 ignition sensitivity, **29**:82–84
 premixed combustion, **29**:62
 turbulent burning rates, **29**:78–82
Combustion sheet
 off the wall, **2**:211f
 on the wall, **2**:212f
Combustion spray system, **26**:54, 59

slurry fuel droplet, **26**:81–82
Combustor, pressured fluidized bed, **25**:233
Common mode rejection ratio, **22**:406–407
Compact heat exchangers
 for cryogenics, **5**:356
 influence of nonuniform passages, **SUP**:405–
  410
 influence of superimposed free convection,
  SUP:410–414
 influence of temperature-dependent fluid
  properties, **SUP**:414–416
Comparisons of boundary conditions, **SUP**:389–
 392
Comparisons of solutions, **SUP**:392–405
Compatibility, chemical, **9**:40
Complete analytical solutions, **SUP**:419
Complete similitude, **31**:150
Complex variables, **8**:42
Complex variables method, exact solutions by,
 SUP:63
Complex velocity gradient, dimensionless,
 19:255
Components
 in equations of change, **12**:200–201
 independent number of, **12**:200
Composite grid, **24**:217
Composite media, **8**:62
Composites, **25**:292
Composite structure, **9**:352
Compound drop, transfer processes, **26**:47–52
Compound parabolic collectors
 calculation of natural convection losses,
  18:20–21
 commercial design, **18**:22–24
 geometries and orientation for natural con-
  vection losses, **18**:9
 natural convection in, **18**:20–25
 overall heat loss coefficient, **18**:25
Compressibility, **31**:10
 bulk, **7**:7
 gas, **25**:50
Compressibility effect, **24**:201, 262
Compressible flow, **25**:362–364, 396
 applications, **25**:364, 396
 Favre filtering, **25**:362
Compressible fluid, defined, **13**:209
Computational
 domain, **25**:324–325, 371, 375–396
 grid, **25**:324–330, 372, 375–396
Computational bands, **31**:394

Computation fluid dynamics, **31**:411
Computer generated analytic transformations, **19**:230–233
Computer graphics, **31**:410
Computer programs, **27**:203
  BEER, **27**:53–55
  printing instructions, **27**:203
  RADIAN1, **27**:203
  RADIAN, **27**:28–29, 46, 107–133, 158–167
  RADIANW, **27**:130–146
  RAT1, **27**:64–75, 80, 83, 88, 122
  RAT2, **27**:76–84, 88, 109, 122
  TFM, **27**:27–29
  variables list, **27**:204–208
  ZM, **27**:39–42
Computers
  parallel, **31**:400–405
  strategies for high performance, **31**:400–421
Computer simulations, *see* Simulations
Concave surfaces, impinging flow in, **13**:43
Concave velocity profile, *see* Velocity overshoots
Concentration boundary layer, **6**:182
Concentration diffusivity, *see* Transport properties of dilute gases
Concentration effects, on thermal conductivity of polymeric solutions, **18**:223–225
Concentration gradient, **22**:212
Concentration-integral method, **2**:138ff
Concentric annular ducts, **SUP**:284–321
  arbitrary axial wall temperature or heat flux distribution for, **SUP**:316–318
  constant heat flux on both walls for, **SUP**:294–295, 315
  constant temperature on both walls for, **SUP**:292–293, 314–315
  constant temperature on one wall, constant heat flux on the other wall in, **SUP**:296, 316
  exponential axial wall heat flux in, **SUP**:296–297
  finite difference solutions for, **SUP**:298–299
  fluid axial heat conduction in, **SUP**:309, 314–315
  fluid flow, **SUP**:284–289
  friction factors for, **SUP**:285–287, 299–301
  fully developed flow in, **SUP**:284–297
  fundamental solutions of four kinds for, **SUP**:289–291
    fundamental solutions of the first kind for, **SUP**:301–305
    fundamental solutions of the fourth kind for, **SUP**:312
    fundamental solutions of the second kind for, **SUP**:305–309
    fundamental solutions of the third kind for, **SUP**:309–311
  heat transfer, **SUP**:289–297
  hydrodynamically developing flow for, **SUP**:297–301
  hydrodynamic entrance lengths for, **SUP**:288–289
  linearized solutions for, **SUP**:297–298
  Nusselt numbers, **SUP**:286–287, 290–293, 295–296, 303–316, 320–321
  simultaneously developing flow in, **SUP**:319–321
  solutions of Navier–Stokes equations for, **SUP**:300
  specified wall temperature and heat flux distribution in, **SUP**:292–296, 312–318
  thermal entrance lengths for, **SUP**:318–319
  thermally developing and hydrodynamically developed flow in, **SUP**:301–319
  wall thermal resistance in, **SUP**:318–319
Concentric cylinder method, **3**:273
Concentric cylinders, convective heat transfer between, **11**:281–285
Concentric rod bundles, in cylindrical tube, **SUP**:360–363
Concentric rotating cylinders, **5**:130
Concentric spheres, heat transfer between, **11**:285–288
Condensation, **9**:81, 87; **16**:147; *see also* Liquid metal heat transfer
  adhesive force, **21**:63–64
  balance of interfacial forces, **21**:62–63
  bulk, **9**:223
  cohesive force, **21**:64
  contact angle, **21**:62–63
  correlations for, **9**:238
  in cryogenic systems, **5**:351
  direct contact, *see* Direct contact condensation
  dropwise, **9**:182, 204
  film, **9**:182, 244
  in forced convection, **9**:257
  gas bubble in immiscible liquid, **26**:47–48
  importance of, **31**:5

Condensation (*continued*)
  laminar film, **9**:79, 244
  in liquid bulk, **9**:226
  of mixtures, **9**:264
  moving liquid droplet, **26**:32–43
    compound drop, **26**:47–48
    droplet shape deformation, **26**:44
    electric field, **26**:17, 83–84
    gaseous environment, **26**:6, 31–32
    spherically symmetric, **26**:31–32
    spray of drops, **26**:44–47
    surfactants, **26**:43–44
  nuclei for, **9**:197
  in reduced gravity, *see* Reduced gravity
  rotating, **9**:261
  turbulent film, **9**:262
  types, **21**:57–59
  wettability, **21**:61–65
Condensation coefficient, **9**:219; **15**:228–229
  dependence on system pressure, **21**:73–75
  experimental determination, **21**:73–80
  metal vapor, **21**:73–75
  polyatomic fluid, **21**:79–80
Condensation curves, heat transfer coefficient,
    21:85–86
Condensation heat transfer, **21**:55–132; **31**:275–
    294; *see also* Dropwise condensation; Film
    condensation
  electric field enhancement of, **21**:125–131
    EHD instability of vapor–liquid interfaces,
      21:126–131
    principle, **21**:125–126
  noncondensable gas effect, **21**:107–108
  surface tension force enhancement of, **21**:115–
    125
    fin geometry for vertical condenser tube,
      21:117–119
    horizontal low-fin tube, **21**:120–123
    principle, **21**:115–117
    wire wound around tube, **21**:124–125
  theory, **21**:55
  transport at vapor–liquid interfaces, **21**:65–60
    condensation coefficient determination,
      21:73–80
    interphase mass transfer theory, **21**:67–72
    modification of Schrage's theory, **21**:72–73
    Nusselt equation, **21**:66–67
Condensation parameter, **26**:32
Condensation velocity, **26**:32
Condensers, **9**:79, **31**:319–320

Conditional distribution function, **5**:9
Conductance, semiconductor melting, **28**:81–87
Conductance parameter, **2**:434
Conduction, **5**:40; **8**:19, 33, 36, 166; **9**:35, 43,
    44; **25**:269, 281
  advantage of Monte Carlo method in treatment
    of, **5**:45
  application of Monte Carlo methods to, **5**:40
  free-molecule, **2**:273, 275f
  nonlinear problems in, **8**:40
  transient heat transfer in helium I and, **17**:149
  transient regime, **9**:311
Conduction and radiation, *see* Radiation heat
    transfer
Conduction-controlled freezing, **19**:58
Conduction equation, **1**:56ff; **5**:41
  one-dimensional, **23**:291
Conduction freezing, metal drops, **28**:17–18
Conduction heat transfer, **5**:333
  effect of variable properties on, **5**:333
Conduction of Heat in Solids (Carslaw and
    Jaeger), **15**:285
Conduction–radiation interaction
  for carbon dioxide, **8**:268, 269
  for carbon monoxide, **8**:270, 277, 278
  comparison of CO, CO_2, H_2O, and CH_4
    results, **8**:271, 272
  comparisons for large path length limit,
    8:268
  experimental measurements of, **8**:273
  thermodynamic nonequilibrium effects on,
    8:276–278
  for water vapor, **8**:270
Conduction regime
  limits of, **8**:179
  thermally thick, **31**:144–145
  thermally thin, **31**:144
Conduction term, **31**:61
Conductive effects, in convective heat transfer,
    15:12–13
Conductive energy transfer, **11**:320
Conductive greases, thermal contact conduc-
    tance, **20**:263–264
Conductive–radiative coupled problems, **5**:34
Conductivity
  effective thermal, **25**:190
  gas thermal, **25**:177
  solid thermal, **25**:226
Cone-and-plate geometry, normal stress mea-
    surements in, **15**:76–77

Cone flows, **9**:296
Cones, **25**:252, 291
  potential flow convection, **20**:367–368
  supersonic
    flow about, **2**:167f, 171, 173, 174ff,
      231ff
    initial relaxation near leading edge of,
      2:219ff
Conflagration, defined, **10**:223
Confluence analysis, **9**:159
Confocal elliptical ducts, **SUP**:341–346
  fluid flow in, **SUP**:341–343
  fundamental solutions of the fifth kind,
    SUP:343–344
  heat transfer in, **SUP**:343–346
Conformal mapping, exact and approximate
  solutions by, **SUP**:63–64, 75
Conical cavities, **5**:24; *see also* Cavities
  augmented nucleate boiling, **20**:64–67
  directional emissivity of, **5**:24
Conical rocket nozzle, **5**:36
  radiation exchange in, **5**:36
Conjugate, **25**:2, 3, 7
  unsteady, **25**:4, 31, 98
Conjugated problems, **6**:368, 373, 411, 457;
  SUP:11–13, 59–60, 419–420
  for circular tube, **SUP**:136–138
  for parallel plates, **SUP**:188–189
Conjugate gradient method, **23**:418
Conjugate heat transfer, forced convection,
  20:222–223
Conjugate problems, **8**:37, 40
  moving liquid droplet, **26**:22–26, 30–31
Conservation equations, **8**:27, 43; **31**:106, 111–
  113; *see also* Radiation–convection heat
  transfer
  Czochralski process, **30**:327–328
  derivation of from porous media by volumetric
    averaging, **14**:6–14
  ESPROSE.m field model, **29**:196–198
  liquid metal droplet in flight, **28**:3–5
  multiphase flow, **30**:100–101, 105–107
Conservation laws, **13**:62–63; **14**:6–14; **24**:206
Conservation of atoms equation, **2**:197, 227, 237
Conservation of energy, applied to film boiling in
  cylinder, **5**:62
Conservation of energy equation, **4**:4, 277, 285,
  308, 319
Conservation of mass equation, **4**:3, 276, 285,
  319

Conservation of momentum equations, **4**:3f,
  276f, 285, 319
Conservation of reactants equation, **2**:214
Conservation of species equation, **2**:112, 116,
  140, 147, 171, 193
  nonsimilar solutions of, **2**:164ff
  self-similar solutions of, **2**:144f
Consistency of the fluid, **24**:110
Constant-film-thickness, slug flow calculations,
  20:99
Constant-flux expressions, heat- and mass-transfer, **20**:341
Constant heat flux, **8**:119, 124
Constant heat flux case, in laminar thermal
  convection studies, **15**:163–165
Constant heat flux method, **23**:326–331
  color interpretation, **23**:330
  composite heated wall, **23**:327–328
  liquid crystals, **23**:328–331
Constant heat flux mode, **12**:100
  entrance region in, **12**:95–98
  of heat transfer, **12**:78
Constant heat flux problem, thermal flows and,
  12:91
"Constant q" insulation design, **15**:41–42
Constants and exponents, nucleate boiling,
  20:9–11
Constant shear stress approximation, **12**:81
Constant surface temperature, **8**:119
Constant surface temperature method, **23**:331,
  333–342
  convection loss, **23**:334–335
  differential thermocouple gage, **23**:336
  heated ribbon gage, **23**:336–337
  heated thin-film heat flux gage,
    23:338–340
  hot wire anemometer bridges, **23**:340–341
  segmented plate heaters, **23**:333
  skin friction gage, **23**:337–338
  time-averaged measurements, **23**:333
  unsteady calibration, **23**:341–342
Constant temperature mode, entrance region in,
  12:98–100
Constant temperature problem, thermal flows
  and, **12**:91
Constant wall temperature, in thermal convection
  in vertical tubes, **15**:202–205
Constant wall temperature mode, of heat transfer,
  12:78, 100
Constitutive equations, **24**:107–109

Constitutive laws, ESPROSE.m field model, **29**:198–201

Contact angle, **9**:229; **21**:62–63

Contact melting, **24**:1–38
  asperities, **24**:15–18
  correlation, **24**:11–13
  embedded object, **24**:6–8
  flat surface, **24**:6, 8–29, 33
  heat transfer in the solid parts, **24**:18–21
  ice, **24**:7, 29–34
  inside capsules, **24**:2–6
  pressure melting, **24**:29–34
  rolling contact, **24**:21–26
  scale analysis, **24**:11–13

Contact resistance, in heat conduction, **15**:285

CONTAIN code
  debris dispersal, **29**:275
  deflagration, **29**:306, 307
  direct containment heating, **29**:218, 232, 234, 241–245, 275, 283, 301, 311, 320–321, 328–329
  dome carryover, **29**:280
  HECTR model, **29**:108, 117
  ice condenser containment, **29**:328–329

CONTAIN/CORDE code, direct containment heating, **29**:232, 234, 243

Containment
  direct containment heating, **29**:215–330
  failure, **29**:216
  ice condenser containment, **29**:328–329
  problems of, in thermosyphons, **9**:40
  studies, **29**:216–330

Contamination effects, on transition boiling, **18**:295–296

Continuity equations, **2**:112, 171; **5**:61; **12**:197–200; **SUP**:9–10; **19**:25, 251; **24**:196
  chemical vapor deposition, **28**:353, 359, 361
  ESPROSE.m field model, **29**:196
  liquid metal droplet, **28**:3

Continuous Czochralski system (CCz)
  buoyant-induced flow oscillation and, **30**:378
  convection in, **30**:406–408
  simulation studies, **30**:348–349

Continuum model, alloy solidification, **28**:270–288, 311

Continuum physics, equations in, **13**:126

Continuum radiation, **5**:301

Contour integral method, **2**:414ff, 444

Contraction rate, ratio of to system volumetric elasticity, **31**:117

Contravarient velocity, **24**:205

Controlled variation scheme (CVS), **24**:244

Control volume formulation, **24**:195

Convection, *see also* Boiling, convective; Convective heat transfer; Forced convection; Natural convection
  alloy solidification, **28**:235, 237, 246–249
    ammonium chloride solution, **28**:311–316
    double-diffusive convection, **28**:240–241, 244, 277
    models, **28**:264–269
    thermosolutal convection, **28**:239–243
  analogies of nucleate boiling, **1**:226ff
  buoyancy-driven, **24**:280–303
  in circular cavities, **29**:17–20
  combined natural and forced, **11**:244–250
  convection–diffusion equation, **24**:195
  convection treatment, **24**:226
  in Czochralski process, **30**:315–318, 362, 406–423
  forced, **SUP**:6–7; **25**:259–261, 268, 270, 280–283, 286–288, 290
  free, **25**:259–262, 279–281, 286, 290
    effect on boundary layer heat transfer, **16**:331–335
  free thermal, **25**:262, 264
  gas, **25**:205, 234
  heat transfer to pan and, **17**:220–223
  laminar, **11**:265–310
    natural, **25**:254, 259
    thermal, **25**:286
  laminar forced, **25**:257–260
  microscales
    forced convection, **30**:19–24
    natural convection, **30**:25–33
  mixed, **24**:306
    effect on boundary layer heat transfer, **16**:335–341
  mixed free and forced, **25**:260
  natural, **11**:200–212; **24**:127–143, 214; **25**:257–263, 267, 268, 272, 289
  natural or free, **SUP**:6
  natural, transient heat transfer in helium I and, **17**:149–150
  nonbuoyancy-driven, **24**:312–316
  nonsteady, **1**:110ff
  in nuclear reactors, **29**:53
    dryout heat flux, **29**:34–52
    molten pool heat transfer, **29**:3–33
    transient reactors, **29**:27–31

Convection (*continued*)
  oscillatory, **24**:174–178
  particle, **25**:205, 231
  in porous media
    forced convection in, **30**:136–141
    natural convection, **30**:122–136
  and radiation, *see* Radiation heat transfer
  in rectangular cavities, **29**:4–17
  in spherical cavities, **29**:20–27
  steady-state forced, **24**:121–126
  steady-state mixed, **24**:144–155
  steady-state natural, **24**:127–143
  in stratified media, **9**:290
  superimposed free, **SUP**:410–414
  in superposed layers, **29**:31–33
  term, **9**:279
  thermoacoustic, **24**:312–316
  thermosolutal, **30**:94
  transient, **24**:167–174
  in trickle bed reactors, **30**:168
  turbulent-free, **11**:265–310; **25**:258, 267
  in unsaturated porous media, **30**:167–168
  velocity, **9**:277, 325
Convection boiling, temperature fluctuation in,
    11:43–45
Convection heat transfer, **SUP**:6
  in structured liquids, **18**:167
Convection-limited containment heating (CLCH)
    model, **29**:218, 241, 242, 243, 301, 311,
    320
Convection models, **9**:57, 58
Convection phase-change models, **19**:3
Convection problems
  approximate solutions to, **11**:265–310
  categories for, **SUP**:10, 15
  conjugated, **SUP**:11–13, 59–60
  conventional, **SUP**:10–11, 13–16, 57–59
  thermally developed flow, **SUP**:13–15
  thermally developing flow, **SUP**:15–16
Convective boundary layer, **25**:377
Convective cooling
  modeling, **28**:13–14
  single liquid metal droplet, **28**:3–11
Convective droplet, vaporization, **26**:58–59,
    70–71
Convective heat flux, **21**:260
Convective heat transfer, **5**:355
  aspect ratio effect in, **11**:208
  between asymmetrically heated vertical plates,
    11:293–294

combined natural and forced convection in,
    11:244–250
  between concentric cylinders, **11**:281–285
  between concentric spheres, **11**:285–288
  conductive vs. viscous effects in, **15**:12–13
  from cylinders in air, **11**:233–234
  dielectric liquids and, **14**:121
  entropy generation profiles and maps in,
    15:13–16
  equations for, **11**:200
  forced convection in, **11**:212–243
  free-stream turbulence from cylinders in
    crossflow, **11**:231–233
  in geothermal systems, **14**:3–15
  governing equations in, **14**:3–15
  initial and boundary conditions in, **14**:14–15
  integral methods for, **14**:67–68
  laminar flow of cryogenics in ducts, **5**:356
  local entropy generation in, **15**:11–25
  natural convection and, **11**:200–212
  from nonisothermal vertical plate, **11**:294–295
  from smooth circular cylinders, **11**:199–251
  between symmetrically heated vertical plates,
    11:290–292, 308–310
  temperature loading factor in, **11**:238
  thermal conductivity in, **11**:286
  turbulent flow of cryogenics in ducts, **5**:358
  between vertical flat plates, **11**:290–292
  between vertically eccentric spheres, **11**:288–
    290
  in wind-tunnel measurements, **11**:236–239
  yawed cylinder in, **11**:239–243
Convective heat transfer coefficient, **27**:48
Convective heat transfer rate, **27**:98
Convective inertia term, **24**:115
Convective–radiative coupled problems, **5**:34
Convective transfer coefficient, **25**:312
  heat, **25**:256, 262, 279
  mass, **25**:256, 262, 268, 279
  transport, **25**:252, 286
Convective transport
  and constitutive equation for forces acting on
    liquid phase, **13**:184–192
  Darcy's law in, **13**:175–184
  in gas phase, **13**:169–175
  in liquid phase, **13**:175–192
Convective transport term, **31**:44, 47
Conventional fluidized bed, **25**:225
Convergence, **8**:59, 60; **24**:246
Converging branch, flows for, **31**:131–132

Converging–diverging flows, **6**:403; **25**:92, 300, 304, 305
  laminar thermal, **25**:286
  turbulent free, **25**:258, 267
Conversion efficiency, **25**:152
Cooking
  appliances, costs of, **17**:173
  energy requirements for, **17**:167
Coolant, primary reactor, **9**:117
Cool-down process, **24**:313–315
Cooling, **19**:5; *see also* Subcooling;
    Supercooling
  air jet, **26**:205
  approximation of to space, **14**:252
  convection, **22**:276, 280
  of cryogens, **5**:468
  of electrical machine rotors, **9**:3
  gas, **25**:55, 96
  of gas turbine blades, **9**:3
  of internal combustion engines, **9**:3, 75
  liquid jets, **26**:105, 170, 205
  of nuclear reactors, **9**:3
  "once-through", **12**:12–13, 20–21, 24
  postburn, **22**:235
  rate, **22**:163–167, 183, 195, 203–207, 300
    average, **22**:196, 214
    definition, **22**:196
    dimensionless, **22**:207, 215
    location for optimum measurement, **22**:218–222
    representative, **22**:209, 214–219
  spray deposition
    convective cooling, **28**:3–11
    in flight, **28**:11–18
    splat cooling, **28**:23–54
  of transformers, **9**:3
  turbulent heat transfer under, **21**:34–37
  undercooling, **22**:176
Cooling systems
  economic optimization of, **12**:15–20
  power generation cost and, **12**:15–16
Cooling towers
  climatic impact of, **12**:28
  dry, *see* Dry cooling towers
  dry vs. wet, **12**:3–4
  fans for, **12**:13–14
  horizon pollution by, **12**:29
  humidity and air flow in, **12**:13
  plumes and drift in, **12**:23
  relative cost of cooling with, **12**:18

Coolocations and moments method, **19**:195
Coordinate transformations, **19**:16–24
  disadvantages, **19**:22–23
  moving-grid method, **19**:22
Coordination number, **25**:190
Copper, laser vaporization, **28**:129–131
Copper–aluminum alloys, **28**:233, 236, 239, 304
Copper foils, enhancement of thermal contact conductance, **20**:264
Copper–nickel alloys, **28**:233, 236
Copper surface
  augmented nucleate boiling, **20**:70–72
  wettability characteristics, **20**:76–77
CORDE models, **29**:232, 234, 243, 266, 321
"Core catchers", **19**:80
Core region in annular-dispersed flows, **1**:389
Core–tube densities, of lunar fines, **10**:55–56
Coriolis effects, **5**:209; **9**:80
Coriolis force, **8**:164; **9**:37, 84; **SUP**:7
  influence on liquid film, **9**:80
Corner zones, presence of narrow, **31**:181
Cornu's spiral, **6**:172
Corotating disks, **5**:175
  analysis of, **5**:183
  Nusselt numbers for, **5**:184
Corotating twin-screw extruder, **28**:187, 189–190, 191
Correlating equations
  heat transfer generalization, **20**:11–13
  nucleate boiling formulas, **20**:19–20
  potential flow convection, **20**:382–383
Correlation of contact mating, **24**:11–13
Corresponding states principle, **7**:5
Corrosion fouling, **31**:443–444
Corrugated ducts, **SUP**:275–276
Cosine heat flux, **3**:135f
Couette flow, **5**:228; **8**:29; **13**:219–222, 250, 257–258; **16**:278
  analysis of stability of, **5**:229
  converging and diverging, **13**:261
  critical Taylor number in, **5**:230
  with heat transfer by radiation, **3**:219ff
  induced rotational flow, **21**:144–147, 150–151
  plane and circular, **13**:254–259
  prediction of heat transfer in, **5**:228
  stability of, **5**:130
  stability studies, **21**:150–151
Couette flow in MHD, **1**:317ff
  hypersonic couette flow, **1**:322
  incompressible couette flow, **1**:317ff

Couette flow in MHD (*continued*)
    drag and Reynolds' analogy, **1**:322
    heat transfer, **1**:321f
    induced magnetic field, **1**:319ff
Couette rheometer, **13**:259
Couette systems, **13**:253–259
Couette viscometer, **15**:76
Countercurrent-cocurrent parallel flow heat exchanger, **26**:228, 229
Countercurrent cross-flow heat exchanger, **26**:230
Countercurrent heat transfer
    early studies
        by Bazett, **22**:60, 70
        by Scholander, **22**:60–62, 65, 70
    Keller and Seiler model, *see also* Model
        comparison with experiment, **22**:82
        comparison with other models, **22**:84, 85, 90–91, 100
        effective conductivity
            blood flow dependence, **22**:78–79
            definition, **22**:77
            limiting value, **22**:77–78
        governing equations, **22**:72–75, 108, 109
        parameter estimation, **22**:83
        physical basis, **22**:70–71
        solution, **22**:75–76
        temperature profiles, **22**:80
    Mitchell and Myers model, *see also* Model
        comparison with experiment, **22**:69–70
        governing equations, **22**:62–64, 90, 138
        parameter estimation, **22**:68–69
        solutions, **22**:65–67
Countercurrent parallel flow heat exchanger, **26**:228, 229, 242, 245
    temperature effectiveness, **26**:300–302
Counterflow heat exchanger
    for Brayton cycle heat engine, **15**:26
    capacity rate imbalance in, **15**:31
    entropy generation rate in, **15**:28
    intermediate cooling of, **15**:50
    large end-to-end temperature ratio for, **15**:48–51
Counterrotating twin-screw extruder, **28**:187, 190–191
Coupled closure problem, **31**:68–85
Coupling
    at closure level, **31**:92
    estimates of effects of at closure level, **31**:92–94

estimates of effects of at macroscopic level, **31**:94–95
    at macroscopic level, **31**:92
Coupling permeability tensors for two-phase flow, **31**:1
Covariant velocity, **24**:205
Coward and Jones apparatus, **29**:66–67, 105
CPO variable, **27**:161, 204
CP variable, **27**:161, 204
Crank–Nicolson finite-difference method, **19**:21
Creeping flow, **SUP**:91, 95–96, 98–99, 167, 213
Creeping flow solution, moving liquid droplets, **26**:4
Cribs, burning rate for, **17**:200
Crises, *see* Critical heat flux
Criterion, stability, **25**:195
Critical diameter in film boiling, **5**:68
Critical flow
    concepts, **18**:91
    heat transfer in, **18**:142–145
    mean drag of banks, **18**:151–153
Critical flow regime, **8**:145
Critical heat flux, **23**:53, 55–107
    in countercurrent internal flow under gravity, **17**:48–49
        flooding-controlled CHF, **17**:49–50
        in transitional intermediate region, **17**:50–52
    definition of, **17**:2
    direct cooling with heat pipes, **20**:290
    droplet splashing from free-surface jet, **23**:106–107
    in external flow and other conditions
        effects of subcooling, **17**:23–24
        heaters cooled by a jet of saturated liquid, **17**:19–22
        saturated pool boiling of liquid metal and nonmetal fluid at very low temperatures, **17**:22–23
    in flow boiling helium I, experimental data, **17**:321–337
    general image of phenomenon, **17**:52–54
    jet suction effects, **23**:106
    mechanism in forced internal flow, **17**:40–41
        annular flow regime, **17**:41–46
        subcooled and low-quality regime, **17**:46–47
        very high-pressure regime, **17**:47–48
    models of, general trends, **17**:9–11
    multiple-jet impingement, **23**:99–101

phenomenology in forced internal flow
   fluid-to-fluid modeling, **17**:30–31
   in vertical tubes, **17**:24–30
prime cause in annular regime, **17**:41–42
research needs, **23**:122
single-jet impingement
   I- and HP-regimes, **23**:67, 69–70
   L- and V-regimes, **23**:55, 57
single-jet impingement, **23**:55–99
   boiling curves, **23**:80–82
   CHF correlations, **23**:65–67, 74–76, 95–98
   CHF model, **23**:62–64
   critical Weber number, **23**:69
   data correlation, **23**:93–99
   differences between circular and planar jets,
      23:93
   differences between free-surface and sub-
      merged jets, **23**:93
   FC-72, **23**:64–65
   fluid properties, **23**:77–78
   free-surface, circular jets, **23**:55, 57, 62–79
   free-surface, planar jets, **23**:79–85
   Helmholtz critical wavelength, **23**:97
   investigations, **23**:58–61
   jet velocity, **23**:62–76, 79–90
   liquid film thickness, **23**:97–99
   liquid–vapor structure, **23**:98
   liquid viscosity-controlled regime, **23**:95
   mechanical energy balance, **23**:93–94
   nozzle/heater dimensions, **23**:78, 91
   nozzle-to-surface spacing, **23**:79, 91–93
   pool boiling, **23**:80–83
   pool height and jet velocity, **23**:85–86
   stagnation point, **23**:95
   subcooling, **23**:77, 84–85, 90–91
   submerged, confined, and plunging jets,
      23:85–92
   surface orientation, **23**:79
   surface tension-controlled regime, **23**:95
   transition regions, **23**:83–84
surface conditions effects, **23**:103–106
theoretical analysis in annular flow regime,
   17:42–46
thermal cooling of electronic components
   heat transfer characteristics, **20**:250–251
   high flux surfaces, **20**:249–250
   macrogeometry, **20**:247–248
   narrow spaces, **20**:248
trends in studies of, **17**:2–3
view of forced internal flow, **17**:31–32

generalized correlation in vertical tubes,
   17:32–37
generalized correlation of hydrodynamic
   nature, **17**:32
under outskirt conditions, **17**:37–40
wall jets, **23**:101–103
Critical heat flux in annular-dispersed flows,
   1:402ff
   correlations of, **1**:407ff
   in more complicated geometries, **1**:420ff
   peculiarities of, **1**:413ff
      with constant power, **1**:413ff
      with uneven power distribution, **1**:419ff
Critical heat flux in boiling, **1**:232ff
   external flow, **1**:256ff
   internal flow, **1**:241f
Critical isochor, **7**:6
Critical isotherm, **7**:4
Critical point, **8**:19; **9**:93
   forced convection near, **5**:355
Critical radius, **9**:196
   of cavities in nucleate boiling, **1**:209ff
Critical Rayleigh number, **8**:189; **14**:20–24, 29;
   24:177–178; *see also* Rayleigh number
   for rotating fluids heated from below, **5**:205
Critical region
   boiling near, **7**:74
   boundary layer flow, **7**:17
   buoyancy effects near, **7**:17, 20, 69
   channel flow, **7**:19
   compressibility near, **7**:7
   defined, **7**:2
   energy equation, **7**:15
   equation of motion, **7**:15
   forced and free convection combined near,
      7:66
   forced convection near, **7**:25
   free convection near, **7**:55
   heat capacity near, **7**:6
   heat transfer near, **7**:1–86
   molecular structure near, **7**:8
   physical properties near, **7**:3–15
   thermodynamic properties near, **7**:3
   transport properties near, **7**:8
Critical state, **9**:90; **22**:321
Critical thermal load, **4**:124; **22**:229, 314
Critical tube diameter, detonation, **29**:97–100
Critical wavelength, **5**:72
Critical wave number, **24**:176
Crocco integral, **3**:56

Cross-cocurrent flow heat exchanger, 26:230–231, 282, 284, 286
  temperature effectiveness, 26:302–304
Cross-countercurrent flow heat exchanger, 26:230, 231, 285–286
  temperature effectiveness, 26:302–304
Cross-coupling, 31:148
Cross flow, 8:93, 94, 99
Cross flow effects, 3:63f
Cross-flow heat exchanger, 26:229–231
  boundary conditions, 26:264
  design, 26:315–319
  energy equations, 26:261–263, 266–267
  exit temperature, 26:290
  literature review, 26:247–252
  mathematical model, solution, 26:281–290
  temperature effectiveness, 26:302–304
Cross flow in banks of tubes, heat transfer calculation, 18:154–156
Cross-flow-magnetically fluidized bed, 25:234
Cryobiology, 22:4, 12
Cryogenic fluids, boiling of, 5:403
Cryogenic heat transfer, 5:325
Cryogenic liquids, discharge from containers, 5:378
Cryogenic mechanical supports
  design of, 15:44–45
  for superconducting magnets, 15:43–45
Cryogenic rockets, 5:327
Cryogenics, 5:325
  application of, in thermosyphons, 9:3
  application to rocket propulsion, 5:327
  heat exchangers, 26:220, 222, 252, 305
  origin of word, 5:325
  shipping of, 5:326
  storage tank design, 5:378
  thermodynamic properties of, 5:327
  transport properties of, 5:328
Cryogenics industry, 5:326
Cryogenics systems
  frost formation in, 5:469
  interfacial phenomena in, 5:347
  interfacial temperature in, 5:352
  mass transfer in, 5:348
Cryogenic vessels, stratification in, 5:388
Cryogen refrigerator, 9:410
Cryogens
  heat flow rates, 16:194
  nucleate pool boiling, data available, 16:169, 170

Cryoinjury, 22:291
Cryopreservation, 22:12, 176, 187, 212, 290, 296
Cryoprotective agent, 22:187, 213, 291
  addition and removal from tissue, 22:304
Cryopumping, 5:326; 9:182
Cryosurgery, 22:11, 12, 296
  probe, 22:300
Crystal formation, heat transfer process during, 11:411
Crystal growth, temperature distribution in, 11:412
Crystallization fouling, 31:439–441
  mitigation of, 31:449–450
Crystals
  semitransparent, 11:413–419
  solidification and melting of, 11:416–419
Cubical element array, forced convection
  hydrodynamic data, 20:233–234
  Nusselt number, 20:233, 235, 237
Cubic packing, 9:373
CULDESAC model, steam explosions, 29:180, 186
Cumulative distribution function, 5:8; 31:390–400, 423
  for scattering direction, 31:369–370
Curtis–Godson approximation, 5:300
Curtis–Godson narrow-band scaling, in nonisothermal gas radiation, 12:176–177
Curved surfaces, 19:143, 145
Curvilinear coordinates, 24:196, 199
Curvilinear finite volume discretization (CFV), 30:352, 354–355
Cushion, 25:130
  closed, volume, 25:137
Cusped ducts, SUP:276–277
Cyclic equilibrium, regenerator theory, 20:135–136
Cyclic steady state (equilibrium), "boxed" regenerator, 20:147–148
Cylinder band absorptance, in isothermal gas radiation, 12:166–167
Cylinders, 9:303; 25:252, 254, 257–262, 270, 285, 286, 291, 292, 295, 298–280, 313; see also Turbulence, free-stream
  combined natural and forced convection in, 11:244–250
  convective heat transfer from, 11:199–251
  crossflow forced convection from, 11:234–235

half, **25**:128
heat transfer between, **11**:281–285
heat transfer to, **1**:344f
horizontal, **25**:259, 261, 279–283, 286
density anomaly and melting, **19**:69–70
inward solidification, **19**:60
melting of ice inside, **19**:69–70
melting of saturated ice, **19**:69
natural convection in melting, **19**:45–46
increase in free-stream turbulence on heat
transfer from, **11**:231–232
isothermally heated, melt volume v. time,
19:52–53
laminar heat transfer from, **11**:277–280
natural convection
effects on phase-change processes, **19**:49–
57
fins, effects on phase-change processes,
19:57–62
one-phase problems with convective and
radiative heating, **19**:32–33
rotating, **25**:186, 259, 261, 281
solidification time, **19**:31
tubes, *see* Tubes
vertical
effects of natural convection in melting,
19:46–47
inward melting, **19**:61
Cylindrical cavities, convection, **29**:17–20
Cylindrical channel, **3**:242ff
Cylindrical coordinate system, radiative heat
transfer, **27**:99–102
Cylindrical jets, condensation on, **15**:233
Czochralski crystal growth, **28**:247, 316
Czochralski process
as batch process, **30**:314
continuous growth system, **30**:348–349, 378,
406–408
as continuous process, **30**:378, 406–408
convection, **30**:315–318
continuous Czochralski growth system,
30:406–408
in Czochralski melt with partition, **30**:408–
412
flows driven by buoyancy, **30**:315, 384–390
flows driven by rotation, **30**:364–365, 367–
369, 385–390
flows driven by surface tension, **30**:316,
364, 365–367, 369, 384–390
global heat transfer, **30**:403–406

growth experiments, **30**:402–403
high-pressure liquid-encapsulated, **30**:412–
423
magnetic field and, **30**:318, 390–391, 422–
423
natural convection, **30**:263–264, 265–269,
270–283, 315, 385–386
simulation experiments, **30**:395–402
three-dimensional effects, **30**:391–395
Czochralski melt with partition, **30**:408–
412
elemental semiconductors, **30**:313–319
Group III–V compounds, **30**:319–325
heat transfer in, **30**:338–344
high-pressure liquid-encapsulated process,
30:323–325, 351, 361, 412–423, 424
liquid-encapsulated process, **30**:322, 323, 335,
337, 338, 348–349, 424
melt flow, **30**:362–370
models, **30**:326–327
magnetohydrodynamic, **30**:333–334, 351
mathematical, **30**:327–346
multizone adaptive curvilinear finite
volume scheme, **30**:352–354
thermomechanical, **30**:357–362
transport simulations, **30**:347–352
oxide crystals, **30**:319–320
steps in, **30**:314
stress calculations, **30**:359

D

DAK variable, **27**:161, 204
d'Alembert paradox, **8**:96; **18**:90
Damköhler number, **2**:143f, 147, 154, 156, 158,
185, 194f, 197, 208, 219, 220, 228, 237,
247, 249f
Damköhler number, **26**:60–61
Damping coefficients, of gravity waves, **14**:271–
272
Darcian velocity, **24**:128, 133, 138, 144
Darcy flow model, heat- and mass-transfer,
20:325–327
Darcy–Forchheimer equation, **24**:111
Darcy friction factor, **SUP**:40
Darcy number, **24**:120, 158
Darcy Pair, **25**:151
Darcy scale, **31**:1, 2

Darcy's constant
  heat- and mass-transfer
    natural convection, **20**:317
    vertical direction, **20**:320–321
Darcy's law, **14**:4–6, 56; **23**:212; **28**:264, 317;
  31:1, 31
  breakdown, **24**:111, 112
  in convective transport, **13**:175–184
  elastic Boger fluids, **24**:118
  flow regime, **24**:120, 121
  Hershel–Bulkley fluids, **24**:118
  non-Darcy effects, **24**:111–118
  power-law fluids, **24**:110
DBAs, *see* Design-basis accidents
DCH-2 study, **29**:316
DCH-3 study, **29**:316
DCH-4 study, **29**:316
DCHCVIM model, direct containment heating,
  29:233
DCH, *see* Direct containment heating
DCL, *see* Direct contact condensation
Deactivation superheat, of liquid mixtures,
  experimental data, **16**:72–74
Deborah number, **15**:77, 86–87, 89, 192;
  19:256–257; **25**:258, 286
  critical value, **23**:217–218
  defined, **15**:193
  definition, **23**:217, 220
Debris
  cavity dispersal phenomena, **29**:256–275
  chemical and thermal interactions with gas,
    29:281–297
  oxidation chemistry, **29**:282–284
  transport phenomena, **29**:275–281
  water, interaction with, **29**:312–324
Debye length, **4**:254
Debye temperature, **9**:364
n-Decane, vaporization, **26**:67, 78–79
Decay of melting process, **19**:66
Deceleration, effects on turbulent boundary layer,
  16:297–301
Defects, solidified alloys, **28**:237, 245
Deficiency thickness, **2**:11ff
Deflagration
  direct containment heating, **29**:303–306
  hydrogen combustion, **29**:108–114
Defocusing distance, speckle photography,
  30:260
Deformation rate tensor, **2**:358
Deformation tensor, **24**:108

DEF, *see* Method of discrete exchange
  factor
Degraded heat transfer, supercritical pressures,
  21:47–48
Degraded heat transfer regime, **31**:229
Degrading polymer solutions, turbulent heat
  transfer in, **15**:115–120
Degrees of freedom, **8**:231
Dehydration, **22**:280, 294
Deissler–Taylor analysis, **19**:324
Del, **27**:19
DELTAT variable, **27**:161, 204
Denaturation of tissue, **22**:280
Dendrites, **19**:58, 60
  alloy solidification, **28**:232–233, 234, 247–
    248, 266, 326–327
Dense plasmas, **4**:242ff
Density, **25**:314
  bed, **25**:161
  critical magnetic flux, **25**:101
  fluctuation, **25**:160
  fluidizing gas, **25**:167
  heat source, distribution, **25**:6
  longitudinal magnetic flux, **25**:162
  magnetic
    field energy, **25**:165, 197
    flux, **25**:158, 160, 162, 163, 164, 175, 179,
      239
  mass force, distribution, **25**:9
  mean fluid, **25**:9
  pure and saline water, **18**:330–332
Density anomaly, water, **19**:68
Density-based methods, **24**:193, 218, 240
Density change
  heat-transfer characteristics and, **19**:54
  Stefan number definition and, **19**:11
  two-phase melting and, **19**:65–66
Density correlation factor, defined, **30**:104
Density difference, between jet and ambient,
  determination, **16**:17
Density gradient, alloy solidification, **28**:238–
  239
Density measurements by radiation technique,
  1:394ff
Density ratio, **19**:11, 99
Departure from nucleate boiling, **1**:198, 232f
Dependence-included discrete-ordinates method,
  23:173–179
  bed properties, **23**:176–177
  single particle properties, **23**:174–175

Deposition
  process of, **31**:441–442
  rate of, **31**:295, 442
  and removal, **31**:440
Depressurization
  scaling of global system, **31**:150
  system, **31**:146
Depth of penetration, **8**:180
Dermis, **22**:224
Desalination, **31**:434
Desiccation of tissue, **22**:319
Design, stove optimization and, **17**:296–299
Design-basis accidents (DBAs), **29**:327
Designer, overview for, **SUP**:385–420
Design standards, **31**:464–467
Detonation, **29**:90–102
  critical energy for initiation, **29**:94–97
  critical tube diameter, **29**:97–100
  qualitative assessment of possibility of local
    detonations, **29**:122
  steam explosions, **29**:137, 140–147
Detonation limits, **29**:100–102
Detonation velocity, **29**:62, 63
Developing flows, single-screw extruder,
  28:161–164; *see also* Hydrodynamically
  developing flow; Simultaneously developing flow; Thermally developing flow; *and*
  appropriate ducts
  comparisons of f_{app} Re and $Nu_{x,bc}$ for,
    SUP:397–405
  in heat exchangers with multi-geometry
    passages, **SUP**:409–410
Deviatoric stress tensor, **2**:358
Devienne's method, **2**:304, 319ff, 324
Devonshire's theory, **2**:346ff
DHEAT2 model, direct containment heating,
  29:233
Diameter effect in thermosyphon, **9**:30
Diameter equivalent, **25**:312
  column or tube, **25**:312
  cylinder, **25**:284, 295, 312
  sphere, **25**:312
Diaphragms
  annular, **31**:160
  height and pitch of, **31**:192–193, 219–225
  height and shape of, **31**:213–219
  shape of and hydraulic resistance law, **31**:205
Diathermanous material, energy transport in,
  11:318
Diathermy, **22**:5, 10

Diatomic gases
  collisional relaxation times in, **8**:248
  dissociation-recombination of, *see* Dissocia-
    tion-recombination, of diatomic gases
  local thermodynamic nonequilibrium, **8**:249
  radiation results for thermodynamic none-
    quilibrium, **8**:275
Diatomic molecules, **8**:230
Diebond material thickness, thermal contact
  resistance, **20**:260–261
Dielectric fluids, direct cooling with heat pipes,
  20:289–290
Dielectric liquids, nonuniform electric fields and,
  14:121
Dielectric solids
  optical dispersion in, **11**:355–358
  phonons and, **11**:353–354
Dies, flow in, **28**:201–212, 226
Difference, phase, **25**:197
Differential equations
  hydrodynamics, slug flow, **20**:91–92
  for velocity and temperature problems,
    SUP:5–36
Differential thermocouple gage, **23**:336
Diffuse emission, **2**:403, 405, 426f
Diffuse reflection, **2**:402f, 426f
Diffusion, **25**:267
  chemical potential and, **12**:263–268
  Fick's first law and, **12**:204
  ordinary, *see* Ordinary diffusion
  in porous media, **13**:120
Diffusion approximation, **5**:30
Diffusion cloud chamber experiments, **14**:325–
  330
Diffusion coefficient, **2**:48, 114; **6**:164
  pseudo-binary, **31**:8
Diffusion coefficients, *see* Transport properties
  of dilute gases
Diffusion-controlled bubble condensation, **15**:243
Diffusion equation, **8**:27; **9**:304, 305
Diffusion flame burn (DFB), **29**:301
Diffusion flames
  development of, **2**:210f
  forced flame, **30**:60–72
  natural flame, **30**:65–78
  in nuclear reactor accidents, **29**:62
Diffusion flux, **12**:198
Diffusion in ionized gases, **1**:284
Diffusion model of drying, **31**:2
Diffusion model of mass transfer, **31**:57

Diffusion process, dilute solution, **31**:39
Diffusion theory, of drying, **31**:85–99
Diffusion velocity, **2**:113ff; **31**:7
Diffusive flux, multiphase mixture, **30**:105
Diffusive force, **31**:9
Diffusive transport, **31**:32
Diffusivity(-ies), **31**:4
  effective, **12**:237, 242
    in gas phase diffusion equation, **13**:168
  measurement of, **12**:220–221
  NaCl solution at atmospheric pressure for
      various salinities, **18**:329
  Stefan–Maxwell equations and, **12**:213–219
  in thermodynamic approach, **12**:270–271
  water, **18**:328–330
Digital
  filtering
    causal, **22**:416
    finite impulse response, **22**:417–418
    infinite impulse response, **22**:417–420
    median, **22**:416, 430–433
    nonlinear, **22**:416
    signal processing, **22**:365, 416–420
    60 Hz notch, **22**:417–420
Dilatant fluids, **2**:359, 363; **24**:107
Dilute solution constraint, **31**:69
Dilute solution diffusion process, **31**:39, 85
Dilute solutions, transfer equation and, **12**:234
Dimensional analysis applied to pool boiling,
    1:233f
Dimensionless axial distance, **SUP**:41, 50
Dimensionless bubble collapse history
  of bubble trains, **15**:253
  in multibubble train system, **15**:255–256
Dimensionless bubble propagation velocity,
    20:103–104
Dimensionless distance, **25**:314
Dimensionless groups
  from boundary conditions, **SUP**:47
  comments about confusing, **SUP**:418
  for fluid flow, **SUP**:39–44
  and generalized solutions, **SUP**:37–60
  for heat transfer, **SUP**:46–57
  three-fluid heat exchanger, **26**:234–237
Dimensionless heat flux, **SUP**:49
Dimensionless inlet temperature, three-fluid heat
    exchanger, **26**:268
Dimensionless number II_N, **30**:13–19
  buoyancy-driven diffusion flames, **30**:66, 67
  two-phase film, **30**:34–37

Dimensionless parameters
  "boxed" regenerator, **20**:146
  in shear flow, **13**:233–235, 252–253, 262
Dimensionless temperature, definition of,
    SUP:14
Dimensionless temperature correlation, **SUP**:58–
    59
Dimensionless temperature profile, **5**:370
Diode behavior, **9**:82
Dipole moment, **3**:266; **8**:231
Direct-chill (DC) casting, alloy solidification,
    28:306, 308
Direct condensation
  on cylindrical jets, **15**:233
  on fan spray sheet, **15**:234–235
  on unbounded liquid interfaces, **15**:233–242
Direct contact condensation, **15**:227–273
  at agitated liquid surface, **15**:239
  applications of, **15**:270–273
  on bubble columns, **15**:242–262
  on cocurrent pipe contactor, **15**:263
  on cold water in channel flow, **15**:238–239
  defined, **15**:228
  in emergency core cooling systems, **15**:270
  on falling supported beam, **15**:237–238
  in geothermal energy recovery processes,
      15:272–273
  one- and two-component systems in, **15**:260–
      262
  in presence of noncondensables, **15**:239–242
  schematic diagram of, **15**:273
  in spray and baffle-tray column condensers,
      15:262–264
  on surface of drops, **15**:235–336
  transfer unit approach in, **15**:265
  tray columns and packed bed contactors in,
      15:264–270
  in various contacting devices, **15**:262–270
  in water desalination schemes, **15**:271
Direct-contact packed-bed condenser, **15**:265
Direct-contact transfer
  fluid mechanics, **26**:3–19
    drop deformation, **26**:12–13, 18–19
    electric field effects, **26**:13–19
    inertialess translation, **26**:4
    Reynolds number translations, **26**:5–11,
        16–18
    surface-viscous effects, **26**:13–16
    surfactants, effect on drop motion, **26**:11–
        12

weakly inertial translation, **26**:5
heat and mass transfer, **26**:19–89
Direct containment heating (DCH), **29**:215–219,
320–330
debris
cavity dispersal, **29**:256–275
discharge phenomena, **29**:248–256
interactions with gas, **29**:235, 281–297
oxidation chemistry, **29**:282–284
transport phenomena, **29**:275–281
experiments, **29**:219–231
heat transfer to structures, **29**:308–312
hydrogen combustion, **29**:297–308
modeling, **29**:232–234
CONTAIN code, **29**:232, 234, 241–245
convection-limited containment heating
(CLCH) model, **29**:241, 242, 243, 301,
311, 320
MAAP code, **29**:232, 234, 247–248, 320
MELCOR code, **29**:232, 234, 245–247, 320
single-cell equilibrium (SCE) model,
29:232, 235–238
two-cell equilibrium (TCE) model, **29**:238–
240, 301, 311
nuclear safety, **29**:215–216, 235–236
in pressure-suppression containments
boiling-water reactors, **29**:324–327
ice-condenser containments, **29**:322, 328–
329
reaction time scales, **29**:286–288, 309
water and, **29**:312–324
Direct cooling techniques
heat pipes, **20**:289–295
thermal control of electronic components,
20:182–183
Direct exchange area, **27**:36–38
Directional absorptivity, **27**:15
Directional emissivity, **27**:14
Directional emittance, lunar, **10**:15–16
Directional reflectance apparatus, for lunar
materials, **10**:16–17
Directional reflectance coefficients, for lunar
materials, **10**:31
Directional reflectance equation, **10**:31
Directional spectral emittance, **10**:15–16
Directional spectral specular reflectivity, **23**:148
Direction of magnetization, **25**:169
Direct methods, **24**:46
partial integration method, **24**:40, 47
Rayleigh–Ritz method, **24**:40, 49, 79

Direct simulation, **25**:324–331
computational requirements, **25**:324–330
Direct simulation Monte Carlo approach to
radiative transfer (DSMC-R/T), **31**:334
DIRHET model, direct containment heating,
29:233
Dirichlet condition, **31**:80, 82
Discharge, circular, characteristics, **16**:3–6
Discharge coefficient, **28**:20
Discharge lamp, **24**:265
Discharge phenomena, reactor pressure vessel,
29:248–256
Discrete coordinate method, **3**:206f
Discrete-ordinate equations, solution, **23**:140–
141
Discrete-ordinates approximation, **23**:139–141
Discrete thermal sources
forced convection, **20**:221–226
liquid cooling, **20**:213–217
multiple flush-mounted sources, **20**:220
natural convection, in enclosures, **20**:217–220
vertical surfaces, **20**:202–209
cubical elements, **20**:205–206
heat-transfer coefficient variations, **20**:205–
206, 208–209
Nusselt numbers, **20**:204–205, 207–208
surface temperature variation, **20**:203–204
Disequilibrium, statement of thermodynamic,
31:137
Disks, **25**:291
Dispersed bubble velocity, slug flow, **20**:95–97
Dispersed film regime, **31**:265, 266–269
Dispersed flow, **1**:355f
Dispersion, **8**:327, 328
axial liquid, **25**:193
gas, **25**:167
in magnetically stabilized fluidized bed,
25:176
radial in magnetically stabilized bed,
25:175, 177
Dispersion relation, **24**:290
Dispersive transport, **31**:32
Displacement thickness, **2**:13; **4**:329, 354
Disposal of heat-releasing waste, **24**:1–38
Disruptive burning, **26**:81
Dissipation effects, **7**:8, 21
Dissipationless flow, **3**:231ff
Dissociated boundary layers, **8**:8, 9, 14, 19, 42
Dissociated gases, methods for determining
concentration diffusivity, **3**:297f

Dissociated gases, methods for determining
    concentration diffusivity (*continued*)
    thermal conductivity, **3**:276
    viscosity, **3**:257f
Dissociation-recombination, **2**:223ff
    of diatomic gases, **2**:132f
    kinetics, **2**:195f
Dissociation temperature, *see* Characteristic
    dissociation temperature
Dissociative relaxation, **2**:227ff, 231ff, 243ff
Dissolution, **25**:260, 272, 279
Distributed memory, **31**:403–404
Distribution, temperature, **25**:190
Distributor, **25**:162
    perforated plate, **25**:187, 189, 194
    porous plate, **25**:166, 178
    screen, **25**:177
Disturbance
    amplification characteristics of, **9**:326
    amplification of, **9**:328
    amplitude distributions of, **9**:326
    frequency of, **9**:325
    hydrodynamic, **25**:169
    symmetric, **9**:335
    temperature, **9**:323
    two-dimensional velocity, **9**:323
Disturbing factors, **3**:45, 46
Divergence, **24**:202, 204
    radiative heat flux equation, **27**:20, 45
Diverging branch, flows for, **31**:131–132
Diver's wet suit, insulating properties of, **4**:94ff
DLW variable, **27**:161, 204
DND, *see* Departure from nucleate boiling
Dodge–Metzner equation, **15**:89; **19**:326–328,
    340
Dolomito, **25**:152
Domain decomposition, **31**:405–406, 424
Dopant, Czochral ski process, **30**:330
Doping, ultrashallow p^+-junction formation,
    28:96–101
Doppler broadening, **5**:279
Doppler profile, **5**:284
Doppler shift velometer, **9**:141
Dorfman's method, **8**:74
Dorn effect, **12**:258–259
Dorodnitsyn integral method, **1**:109f; **2**:231f
Double correlations, of temperature and velocity
    components, **8**:305
Double-diffusive convection, **28**:240–241, 244,
    277

Double eigenvalue, **3**:167
Double integral method, **9**:312
Doubly connected ducts, **SUP**:31, 62, 284–340,
    341–353
    thermal boundary conditions for, **SUP**:31–36
Downward-facing surface
    heat transfer models, **20**:28–36
    liquid and vapor periods, **20**:29–31
Draft equation, for dry cooling towers, **12**:30–35
Drag, **8**:99
    bed, **25**:152
    convective heat transfer in, **8**:33
    effect of roughness on, **8**:112
    fluid, **25**:152
    hydraulic, *see* Hydraulic drag of banks
    influence of free stream turbulence on, **8**:103
    influence of heating on, **8**:101
    influence of Reynolds number on, **8**:102, 103
    influence of surface roughness on, **8**:103
    single tubes, **8**:100; **18**:95–97
        drag coefficient variation with Reynolds
            number, **18**:96–97
        heating effect, **18**:96
        pressure drag, **18**:96
        surface roughness and, **18**:97
        turbulence and, **18**:97
    in tube banks, **8**:106, 111
    tubes in a bank, **18**:106
Drag coefficient, **2**:381ff, 384, 387; **3**:4ff, 19, 37,
    39f, 93f; **21**:333–334; **25**:252, 286, 294,
    312; *see also* Coefficient
    in annular-dispersed flows, **1**:363
    force, **25**:312
    liquid droplets, **21**:334–337
    liquid metal droplet, **28**:7–8, 10
    local, **3**:46
    oscillation and instability effect, **21**:339
    packed beds, **23**:207
    reducing, **25**:267, 281, 286
    solid spheres, **21**:334
    versus Reynolds number, **21**:338–339
Drag flow, **13**:221–222
Drag force
    compound drop, vaporization, **26**:50
    moving liquid droplet
        electric field, **26**:17
        gaseous environment, **26**:7–9
        inertialess system, **26**:4
        intermediate Reynolds number, **26**:9, 10
        low Reynolds number, **26**:34

multicomponent drop, **26**:79
Drag in MHD couette flow, **1**:322
Drag ratio, **3**:37, 46f, 66ff
Drag-reducing fluids, **15**:192–195
Drag-reducing polymers, **19**:345
  mass transfer, **23**:222
  packed beds, **23**:232
Drag reduction
  defined, **12**:77
  heat transfer in flows with, **12**:77–110
  models of flow with, **12**:79–85
  Stanton number and, **12**:97
  velocity profiles in, **12**:84–85
Drag reduction asymptote, maximum, **12**:80
Draining, vessel, **25**:130
Drift-flux model, **31**:110–111
Drift-flux parameters, two-phase flow and,
    31:109–111
Drift velocity
  Taylor bubbles, **20**:100–101
    dimensionless, **20**:103–104
    horizontal slug flow, **20**:101–102
    inclined case, **20**:102–103
    surface tension parameter variation,
      20:103–104
    void-fraction-weighted area-averaged, **31**:110
Driving force, nucleate boiling, **20**:2–3
Driving potential, **2**:44ff
  without chemical reaction, **2**:44f
  with chemical reaction, **2**:45ff
Drop and transient regime, **31**:293
Drop bulk temperature, **26**:39, 40
Drop condensation, **31**:276
Drop deformation
  n-heptane, **26**:80
  moving liquid droplet, **26**:12–13
    condensation, **26**:44
    electric field, **26**:18–19
    evaporation, **26**:79–81
    low Reynolds number translation, **26**:6
    weakly inertial translation, **26**:5
Drop geometry, **26**:50
Droplet array, moving liquid droplet, **26**:84
Droplet capture model, steam explosions, **29**:140
Droplet evaporation time, **1**:137
Droplet formation, **9**:80
Droplet fragmentation
  hydrodynamic, **29**:168–175
  melt fragmentation, **29**:270–274
  thermal, **29**:175–176

Droplet group, moving liquid droplet, **26**:84
Droplet lifetime, **26**:56–57
Droplet oscillation, **26**:44–49
Droplet removal, **9**:234
Droplets, *see also* Moving liquid droplet
  binary, vaporization, **26**:78–79
  compound, **26**:47–52
  convective, vaporization, **26**:58–59, 70–71
  direct-contact transfer, **26**:1–3
    fluid mechanics, **26**:3–19
    heat and mass transport, **26**:19–89
  electrohydrodynamics, **26**:14–19
  formation of in homogeneous nucleation,
      14:284–285
  interactions, **26**:84–89
  lifetime, **26**:56–57
  molten metal, evaporation, **26**:83
  motion, **26**:11–19, 43–44
  multicomponent, vaporization, **26**:66, 77–79
  oscillation, **26**:44–49
  slurry fuel, vaporization, **26**:81–82
  spherical, inertialess terminal velocity, **26**:4
  spray, **26**:44–47, 53–57, 81–82
    deformation, **26**:79–80
    interaction, **26**:84
    transport, **26**:87–89
  spray deposition
    convective cooling, **28**:3–11
    multiple liquid metal droplets, **28**:54–69
    single liquid metal droplets, **28**:3–18, 23–54
    splat cooling, **28**:23–54
    sprays, **28**:18–22
    transport, **26**:43–44
Drop motion
  electric field effect, **26**:13–19
  surface-viscous effect, **26**:13
  surfactants, **26**:11–12, 43–44
Drops in suspension, thermal conductivity
  deformed, **18**:205–206
  spherical, **18**:203–204
Drop surfaces, direct contact condensation on,
    15:235–236
Drop tower method of producing reduced
    gravity, **4**:146ff
Drop transport, surfactants, **26**:43–44
Drop vaporization, **10**:101–102
Dropwise condensation, **21**:59–61, 80–104;
    31:275
  drop departure, **21**:82
  drop growth, **21**:82

Dropwise condensation (*continued*)
   drop size distribution, **21**:96–97
      timewise change, **21**:99–100
   heat transfer coefficient measurement,
         21:82–98
      condensation curves, **21**:85–86
      drop size effect, **21**:83–84
      high-pressure steam, **21**:90
      low-pressure steam, **21**:83
      material thermal properties effect, **21**:86–90
      metal vapors, **21**:92
      organic vapors, **21**:91–92
      small vapor-to-surface temperature differ-
         ences, **21**:91
      steam at near-atmospheric pressure,
         21:82–83
      surface thermal conductivity, **21**:88–89
      thermal resistance, **21**:87–88
   heat transfer theory, **21**:92–100
      Le Fevre–Rose theory, **21**:92–96
      Tanaka's theory, **21**:96–100
   high-performance condensing surfaces,
         21:101
   initial droplet formation, **21**:80
   maintaining for long periods, **21**:100–104
   microscopic mechanisms, **21**:80–82
   polymer coating, **21**:103–104
   surface heat transfer coefficient, **21**:94
   surface subcooling, **21**:93
   things not known about mechanism, **21**:81
"Dry body", **1**:123
Dry cooling
   advantages of, **12**:20–22
   competitive position of, **12**:29
   defined, **12**:13–14
   heat release level in, **12**:28
   hydrosphere involvement in, **12**:3
   for nuclear power plants, **12**:3
   resources, impacts, and regulation in, **12**:20–29
   results and implications for, **12**:18–20
   solar input in, **12**:26
   status and needs of, **12**:29
Dry cooling towers, **12**:1–71
   ambient temperature fluctuations in, **12**:67
   aspect ratio for, **12**:38, 42
   cost optimizing process in, **12**:64–66
   draft equation for, **12**:30–35
   draft-height correction for, **12**:34
   draft maintenance for, **12**:36–38
   fan equations for, **12**:34

   fans for, **12**:14, 33, 41–44
   flow of air in, **12**:14
   flow uniformity in, **12**:37
   free plume in, **12**:36
   heat exchanger arrangements and sizes for,
         12:39–52
   heat exchanger solutions for, **12**:56–61
   heights and fan powers in, **12**:33
   initial temperature difference for, **12**:66–70
   internal aerodynamics of, **12**:30–38
   low-tower draft maintenance in, **12**:37
   mechanical draft for, **12**:36–37, 60
   minimization principles for, **12**:41–45
   minimum tower size or fan power for,
         12:41–44
   module length and number for, **12**:59–60
   natural vs. mechanical draft for, **12**:61–64
   one-dimensional analysis of, **12**:31–35
   parameter changes for, **12**:61–62
   performance parameters for, **12**:53
   performance variations and losses in, **12**:59,
         66–70
   plumes and drift for, **12**:23, 37–38
   qualitative description of, **12**:30–31
   representative surfaces for, **12**:45–50
   Rugeley and tower losses for, **12**:59
   temperature inversions in, **12**:68
   tower calculations for, **12**:52–66
   tower size and shape for, **12**:35–38
   variation of parameters and similarity rules for,
         12:61–64
   vs. wet cooling towers, **12**:4, 18
   wind effects on, **12**:68–70
Drying, **25**:190
   of moist porous solids, **1**:128ff, 139ff
   of porous materials, **30**:168–172
   volume averaging, **30**:169
Drying process, **13**:119–200
   beginning and end of, **13**:196–197
   and convective transport in liquid phase,
         13:175–192
   diffusion-controlled, **31**:54, 85–99
   diffusion theory of, **13**:124–125, 194–198
   effective thermal conductivity in, **13**:140,
         158–164
   energy transport in, **13**:153–165
   enthalpy of vaporization in, **13**:157
   heat capacity in, **13**:155
   liquid phase in, **13**:184–192
   and mass transport in gas phase, **13**:165–175

moisture distribution in, **13**:125
problem and solution in, **13**:192–194
temperature gradients in, **13**:191–192
thermodynamic relations in, **13**:164–165
Dryout, **9**:87, 88
Dryout heat flux, **29**:34–35, 52–53
  forced flow and, **29**:50
  liquid subcooling, **29**:49
  mechanistic model, **29**:42–47
  multidimensional flows, **29**:50–52
  one-dimensional models, **29**:35–42
  overlying liquid layer depth, **29**:48–49
  system pressure, **29**:49–50
  two-phase flow, **30**:118
  unconsolidated porous media, **29**:47–48
Dryout phenomena, narrow space nucleate boil-
    ing, **20**:39–40
Ducts, *see* Liquid metal heat transfer; *and*
    appropriate ducts in question
Dufour energy flux, **28**:353
Duhamel integral, **21**:33
Duhamel's principle, **8**:78
Duhamel's superposition theorem, **SUP**:119,
    316–317
Duhamel's theorem, **3**:131f, 144, 148
DXG variable, **27**:161, 204
DYG variable, **27**:161, 204

                          E

early phases, **16**:243–249
  uncertainty analysis, **16**:270–271
Earth
  age of, **15**:285–286
  atmosphere of, **14**:271, 274–275
  core formation, **19**:82
  formation, **19**:1
  ice cap interaction, **19**:82
EBAL model, direct containment heating,
    29:233
Eccentric annular ducts, **SUP**:322–340
  fluid flow, **SUP**:322–328
  friction factors for, **SUP**:326–327
  fully developed flow in, **SUP**:322–333
  fundamental solutions, **SUP**:328–332, 336–
    337
  heat transfer, **SUP**:328–333
  hydrodynamically developing flow in,
    SUP:333–336

Nusselt numbers for, **SUP**:329–333, 338–340
  simultaneously developing flow in, **SUP**:337–
    340
  thermal entrance lengths for, **SUP**:339
  thermally developing and hydrodynamically
    developed flow in, **SUP**:336–337
Eccentric annulus, **3**:106
Eccentricity effect, induced rotational flow,
    21:157–160
ECC, *see* Emergency core cooling systems
Eckert numbers, **1**:281; **3**:2; **SUP**:53–55; **15**:12
Eckert reference temperature, **4**:326
Eckert–Schneider condition, **4**:322
Eddington approximation, **21**:247, 249
Eddy diffusivities, **3**:123ff, 171; **6**:510, 560;
    9:172, 174, 176
  of momentum, heat, and transfer, **15**:130–132
Eddy-diffusivity model, forced convection,
    20:224
Eddy viscosity, **9**:163; **21**:13
Edge-cooled heat-pipe, **20**:286–287
Effect
  core, **25**:200
  dipole, **25**:181
  end, **25**:311
  extensional, **25**:292, 308
  hysteresis, **25**:150, 167, 193
  stabilizing, **25**:158, 160
Effective band widths, **5**:294
  for carbon monoxide, **5**:295
Effective emission coefficient, **3**:186
Effective entrainment interval, **29**:327
Effective heat capacity, **24**:168
Effective length, **25**:298, 300
Effectiveness coefficient, *see* Liquid metal heat
    transfer
Effectiveness concept, "boxed" regenerator,
    20:145–146
Effectiveness–NTU approach, **26**:222, 256, 267
Effectiveness of heat exchanger, **9**:120
Effective packet density, **19**:175
Effective sticking coefficient, **28**:127
Effective temperature, **4**:104ff
Effective thermal conductivity, **3**:282f
  in drying process, **13**:140, 158–164
Effective transmittance, band location and,
    12:184
Effective viscosity, **24**:118, 175
Effects of surface roughness on boiling,
    1:230f

Efficiency
  of convective heat transfer, in closed stoves,
    17:256–259
  conversion, 25:152
  filtration, 25:232
  of heat transfer to pan, 17:223–226
  mass transfer, 25:190
  particulate capture, 25:233, 234
  vapor capture, 25:233
  of woodburning cookstoves
    critique of measurements of, 17:263–275
    fuel economy and, 17:275–284
Effusion-type gauge, 3:258
EHD, see Electrohydrodynamic heat transfer
Eigenfunction expansions, 8:66
Eigenfunctions, 24:57, 62, 65, 68, 97
Eigenvalues, 24:57, 60, 70; 31:149
Eikonal equation, 6:136
Einstein coefficients, 5:264; 8:232; 14:257
Einstein photoelectric law, 12:120; see also
    Nernst–Einstein equation
Einstein relation, 4:239
Ejection, 25:22, 23
  mean frequency, 25:21
Ekman layers, 28:310
  development of, 13:93–95
Ekman number, 5:218
Elasticity, 25:260
  number, 25:278, 312
  system, 31:150
Elastic sublayer, 12:79
Electrical charge relaxation time, 14:110
Electrical conductivity, 4:231
Electrically conducting fluids, heat transfer in,
    1:268ff
Electrical potential, distribution of in solution,
    12:196
Electric analogy, 8:111
Electric charges
  mass transport and, 12:196, 244–251
  redistribution, 21:126–127
Electric conductivity, 1:284f
Electric field effects
  moving liquid droplet, 26:13–19, 26–31, 83–
    84
  at stagnation point, 1:347
Electric field(s)
  boiling hysteresis and, 14:132
  condensation heat transfer enhancement,
    21:125–131

dielectric liquids and, 14:121
  in film boiling vs. nucleate boiling, 14:131
Electric power generation, cost of, 12:15–17; see
    also Power parks
Electric Power Research Institute, direct con-
    tainment heating studies, 29:216
Electrochemical diffusion, 12:242–244; see also
    Diffusion
Electrochemical method in transport phenomena,
    7:87–162
  application to mass transfer measurements,
    7:94–136
    artificially waved liquid layer, 7:133
    concentric rotating cylinder annulus, 7:128
    cross flow, 7:116
      mass transfer coefficient
        local, 7:120
        space-averaged, 7:118
    falling liquid film, 7:132
    in forced convection, 7:103–136
      local mass transfer fluctuation intensity,
        7:125
      transfer coefficient, local, 7:125
      transfer factor, space-averaged, 7:124
    in free convection, 7:98–103
      cylinder, 7:101
      plate
        horizontal, 7:100
        vertical, 7:98
      sphere, 7:103
    jet flow, 7:135
    packed beds, 7:134
    rotating and vibrating bodies, 7:130
    tube flow, 7:103
      fluctuation of mass transfer rates, 7:112
      laminar, 7:112
      mass transfer coefficient, dynamic
        response of, 7:116
      turbulent
        entry region, 7:110
        fully developed, 7:103
  application to shear stress measurements,
    7:136–144
    in boundary layer, 7:140, 142
    in well-developed flow, 7:138, 140
  application to velocity measurements,
    7:144–158
    fluctuating velocity
      agitated vessel, 7:155
      tube flow, 7:153

time-smoothed velocity boundary layer, 7:151
    tube flow, 7:147
Electrochemical potentials, in thermodynamic approach, 12:268–270
Electrochemical systems
    ion transport in, 12:196
    multicomponent solutions in, 12:196
Electrochemical transport, 12:195–271
    charge transport and, 12:244–251
    dilute solutions in, 12:234–237
    electrokinetic phenomena in, 12:251–260
    electrostatic potential and, 12:225–256
    equations of change in, 12:197–203
    general transport equation in, 12:231–244
    migration in, 12:221–231
    mobility in, 12:222
    ordinary diffusion in, 12:203–221
    practical vs. thermodynamic model in, 12:270–271
    supporting electrolytes in, 12:241
    terminology of, 12:242–244
    thermodynamic approach in, 12:261–172
Electrocrystallization, 12:242–244
Electrocution, 22:246
Electrodecantation, 12:242–244
Electrodeposition, 28:340
Electrodiffusion, 12:242–244; see also Diffusion
Electrodyalysis, 12:242–244
Electrofinishing, 31:461
Electrohydrodynamically coupled condensation heat transfer, 14:137–139
Electrohydrodynamically coupled convection heat transfer, 14:116–127, 142
Electrohydrodynamically coupled film boiling, photographs of, 14:135
Electrohydrodynamically coupled forced convection heat transfer, 14:124
Electrohydrodynamically coupled free convection, 14:126
Electrohydrodynamically enhanced heat transfer, in liquids, 14:107–143
Electrohydrodynamic coupling mechanisms, 14:108–112
Electrohydrodynamic devices and concepts, 14:139–141
Electrohydrodynamic effects, classes of, 14:110
Electrohydrodynamic forces, as controllable laboratory analogy, 14:122
Electrohydrodynamic heat pipe, 14:139–140

Electrohydrodynamic heat transfer
    boiling water and, 14:129–130
    categorization of papers on, 14:115
    coupled with convective heat transfer, 14:116–127
    experimental observations in, 14:141–142
Electrohydrodynamic heat transfer experiments, liquids used in, 14:117
Electrohydrodynamic heat transfer research parameters, 14:113–116
Electrohydrodynamic instability, vapor–liquid interface, condensation enhancement, 21:126–131
Electrohydrodynamics, droplet, 26:14–19
Electrokinetic phenomena, 12:251–260
    electroosmosis as, 12:252
    electroosmotic counterpressure in, 12:257–258
    electrophoresis in, 12:255–256
    electrosorption and, 12:259
    migration or sedimentation potential in, 12:258–259
    streaming potential as, 12:256–257
    terminology of, 12:260
    zeta potential in, 12:253–255
Electrolysis, as synthetic boiling, 14:136
Electrolytes, system design and control for, 12:196
Electromagnet, 25:162, 179, 180, 181, 194
    field, 25:7
Electromagnetic effects, 31:10
Electromagnetic fields, heat transfer enhancement by, 30:243–244
Electromagnetic rheocasting, 28:318
Electromagnetics, lost work, entropy production, 21:270–272
Electromagnetic spectrum, 27:6–7
Electromagnetic waves, 27:3–4
Electromembrane process, 12:242–244
Electromigration, 12:242–244; see also Diffusion
Electron
    number density of, 9:364
    relaxation time of, 9:364
Electron-beam assisted chemical vapor deposition, 28:343
Electronic ablation, 28:110
Electron microscopy, cryopreparation, 22:186
Electroosmosis, 12:252
Electroosmotic counterpressure, 12:257–258

Electrophoresis, **12**:242–244
Electroplating, **28**:340–341
Electrostatic potential
  measurement of, **12**:225–226
  reference electrode in, **12**:225
Electrostatic probe, **9**:144
Electrostriction, **1**:273f
Electrosurgery, **22**:248
Electrotransport, **12**:242–244; *see also* Diffusion
Elements
  effects of different, **31**:359–362
  types of in Monte Carlo methods, **31**:337–341
Elenbaas–Heller differential equation, **4**:263
Ellipsoid, potential flow convection, **20**:373–374
Elliptical channel, **8**:44
Elliptical cylinder, potential flow convection, **20**:361–362
Elliptical ducts, **SUP**:247–252
  with central circular cores, **SUP**:349–350
  confocal, **SUP**:341–346
  friction factors for, **SUP**:248, 250
  fully developed flow in, **SUP**:247–251
    fluid flow, **SUP**:247–249
    heat transfer, **SUP**:249–251
  Nusselt numbers for, **SUP**:248–250
  thermally developing flow in, **SUP**:251–252
Elliptical equations, **8**:13
Ellis fluid, as three-parameter model, **15**:168
Ellis model, **2**:362; **28**:153
Elongated bubbles, *see* Taylor bubbles
Elongation flow, **13**:260–262
Elsasser model, **5**:288; **8**:237
Elutrication, **25**:153
Embedding methods, **19**:15–16
Embryos
  equilibrium size distribution of, **14**:293–298
  formation of in homogeneous nucleation, **14**:284
  free energy of formation of, **14**:306
Emergency core cooling systems, direct contact condensation in, **15**:270
Emergent superheat, **10**:122
  determination of, **10**:126
Emission, **27**:4
  from gas volume, **27**:56–59
  from solid walls, **27**:59–61
Emission characteristics of cavities, *see* Cavities
Emission coefficients, **3**:180, 197
  mass emission coefficient, **3**:180
  volumetric emission coefficient, **3**:180

Emission frequency, narrow space nucleate boiling
  clearance, **20**:43–45
  coalesced bubble region, **20**:49, 51
  equivalent heat source, **20**:54–55
Emission path method, **31**:347
Emissive power
  inhomogeneous, **31**:376
  thermal radiation, **27**:10
Emissive power distribution, **5**:29
Emissivity, **27**:11, 12, 14; **28**:91
  total normal, **9**:386
Emittance, **2**:401
  apparent, **2**:425ff
  in semitransparent solids, **11**:373–374
Emitting-absorbing gas, **2**:400
Emitting medium, **1**:2
  two-flux approximation, **23**:138–139
Empirical convective factor, **26**:45
Emulsion packet, **10**:186–188, 196
Enclosed apparatus, thermal conductivity of structured liquids, **18**:173–174
Enclosed porous layers, heat- and mass-transfer, **20**:336–341
Enclosures, natural convection, **20**:217–220
End-to-end temperature ratio, for counterflow heat exchangers, **15**:48–50
Energetic transfer of heating a counter-current ambient effect (ETHICCA effect), **29**:153
Energy, **31**:27–28
  backscatter, **25**:350, 360–361
  calculation of heat balances and, **17**:294–296
  cascade, **25**:337
  closure for, **31**:60–68
  conservation of, **13**:62; **14**:11–14
  costs delivered to cooking pan, **17**:172
  equation, **25**:5, 18, 26, 45, 99, 323, 338, 345
  interphase transfer laws for, **31**:109
  Joulean dissipation by diffusion, **21**:272
  kinetic, **25**:165
  magnetic, **25**:155
    kinetic, **25**:197, 203
    magnetic, **25**:204
  volume-averaged equation of, **31**:9–10
  volume-averaged transport equations for, **31**:41–48
Energy and momentum transfer, constitutive relations for, **31**:108–109

Energy balance equations, **12**:197–200; **27**:19–20

Energy balance in annular dispersed flows, **1**:366

Energy balances, open fires
  flames and, **17**:203–216
  fuelbed, **17**:202–203

Energy conservation
  Czochralski process, **30**:328
  multiphase flow, **30**:101, 107

Energy consumption, per capita gross national product and, **17**:160

Energy content, **SUP**:56–57, 113–115, 174–175

Energy efficiency, **31**:433

Energy equations, **2**:8f, 19f, 28f, 112, 116, 140, 147, 171, 193f, 197, 214, 236, 377; **3**:5, 178; **5**:133; **8**:8, 21, 27, 43, 67, 71; **11**:324–325; **SUP**:13–15, **19**:250, 251, 257; **24**:45, 90
  chemical vapor deposition, **28**:353, 360, 362
  convective cooling, **28**:13
  derivatives of, **9**:326
  ESPROSE.m field model, **29**:197–198
  laser vaporization, **28**:128, 129
  Orr–Sommerfeld, **9**:326
  for radiatively participating gas, **8**:246
  spray deposition
    liquid metal droplet, **28**:4, 5
    recalescence, **28**:16
    solidification, **28**:17
    splat cooling, **28**:46
  three-fluid heat exchanger, **26**:258–264
  total thermal, **13**:154–158
  for transient conditions, **8**:25

Energy expenditures, allowed for heat transfer enhancement, **31**:162–165

Energy flux continuity equation, **26**:34

Energy flux vector, **3**:188

Energy levels of hydrogen atom, **5**:275

Energy/mass transport equations, **13**:126–153; see also Energy equations; Mass/energy transport equations

Energy requirements, for cooking food
  absorbed by chemical process of cooking, **17**:276, 277–278
  compensating for heat losses from pan, **17**:276–277, 278–279
  in production of water vapor, **17**:276
  to raise temperature of food ingredients, **17**:275–276
  to raise temperature of medium, **17**:275, 277

Energy storage, **24**:1–38
  latent, **22**:253, 277
  sensible heat units for, **15**:34–38
  specific, **22**:253

Energy thickness, see Boundary layer

Energy transfer, **11**:319–324; see also Heat transfer
  basic equations for, **11**:324–352
  conduction as, **11**:321–322
  physical processes and idealizations in, **11**:319–321
  radiation as, **11**:322–324
  radiative flux in, **11**:329–352

Energy transport, see also Mass/energy transport equations
  in drying process, **13**:153–165
  effective thermal conductivity in, **13**:158–164
  thermodynamic relations in, **13**:164–165
  total thermal energy equation in, **13**:154–158

Energy transport equation, **31**:4–5, 89–95

Engineering
  molecular gas radiation in, **12**:116
  second-law analysis in, **15**:2–3

Enhanced diffusion, **31**:87, 95
  effective, **31**:80
    estimate of tensor of, **31**:88

Enhanced heat conduction, **31**:92

Enhancement effect, **31**:36

Enhancement efficiency, versus Reynolds number, **31**:198–201

Enhancement factor, **18**:207–208

Enthalpy, **2**:112; **28**:64, 76–79
  calculation, **18**:360
  exit gas, furnace, **27**:157
  gas–wall system, **27**:48
  mean-mass, **25**:8, 20
  mean-mass gas, **25**:20
  specific, **25**:6
  total specific, **3**:39
  of vaporization per unit mass, **13**:157
  water
    changes at various temperatures and pressures, **18**:344
    changes for various temperatures and salinities, **18**:356
    pure and saline, **18**:343–344

Enthalpy changes, "boxed" regenerator, **20**:147–148

Enthalpy distribution, profile for, **31**:141–142

Enthalpy flux, **2**:19, 102

Enthalpy jump condition, 31:27–28
Enthalpy method, for Stefan problems, 19:24–28
Enthalpy of evaporation, reduced, 7:6
Enthalpy state, 2:46f
Entrainment
  in annular-dispersed flows, 1:389ff, 392f
  bubble-induced, *see* Bubble-induced entrain-
    ment
  immiscible liquid layers, 21:286–299
    background, 21:287
    criterion formulation, 21:289–293
    experiment, 21:293
    experimental results, 21:295–298
    penetration criterion, 21:290
    physical phenomena, 21:288–289
    threshold criterion, 21:295–296
    wake column, 21:291
    wake volume, 21:291
  particle, 25:233
Entrainment concept, 16:7
  in modeling schemes for jets, 16:7–13
Entrainment functions
  for buoyant jets, in flowing ambients, 16:12
  for use with quiescent ambients, 16:11
Entrainment models
  environmental, 16:46–51
  for jets, in unstratified quiescent ambients,
    16:25–36
  limitations, 16:50–52
Entrance effects, 6:378, 396, 410, 431, 525, 540
Entrance lengths, 8:32, 33; *see also* Hydrody-
    namic entrance length; Thermal entrance
    length
Entrance orifice, influence of, on thermosyphon,
    9:24
Entrance region of a tube, 3:41f
Entrance section, 25:53
  thermal, hydrodynamic, 25:125
Entrapment, of cells in ice, 22:293
Entropy, 8:23
  calculation, 18:360–361
  minimum, 19:198
  minimum production of, 8:23
  production, 25:9
  production of, 8:27
  water
    changes at various temperatures and pres-
      sures, 18:346
    changes at various temperatures and sali-
      nities, 18:357

pure and saline water, 18:345
Entropy balance, 21:248
Entropy flow, 21:241
Entropy generation
  capacity rate imbalance and, 15:31
  in continuous one-dimensional thermal insu-
    lation system, 15:40
  in heat transfer, 15:4–10, 21–25
  rate per unit length, 15:17
  reduction of in thermodynamic system, 15:10
Entropy generation analysis, in heat transfer and
    thermal design, 15:3–4
Entropy generation minimization
  as fixed-identity insulation system, 15:39–43
  in heat exchanger design, 15:25–38
  heat transfer augmentation and, 15:24
Entropy generation number, 15:16–21
  for general duct, 15:19
Entropy generation profiles and maps, 15:13–16
  laminar pipe flow and, 15:14
Entropy generation rate
  local, 15:118
  zero pressure drop and, 15:28
Entropy generation surface, for laminar boundary
    flow, 15:15
Entropy generation units
  number of, 15:29–32
  optimum duct geometry and, 15:32–33
Entropy increase cost, 15:52
Entropy production, 21:239–240
  flame quenching, 21:261–263
  fundamental difference of power, 21:248
  heat transfer, 21:259–260
  microscales, 21:263–268
  local, 21:247–249
  low electromagnetic work, 21:270–272
  number, 21:254–255
  Nusselt number, 21:260
  qualitative radiative, 21:254–259
    entropy production number, 21:254–255
    isotropic scattering, 21:255–256
    wall effect, 21:256–259
  radiation-affected turbulence, 21:268–270
  radiative stress, 21:245–247
  stagnant gas, 21:249–254
    aerodynamic heating, 21:252–254
    gas emissive power, 21:250–251
    heated plasma, 21:252–253
    radiative balance, 21:249
    radiative boundary conditions, 21:250–251

thermal balance, 21:249
thermodynamic foundations, 21:240–245
  control volume, 21:242–244
  heat flow, 21:241–242
  laws, 21:240–243
  mechanical energy balance, 21:242
  wall, 21:259–260
Environmental regulations, 31:469
Environmental temperature, 4:107
Environment Protection Agency, U.S., 12:3
Enzian injector, 2:77ff
Eötvös number, 26:36
Epiderms, 22:224
EPRI experiments, 29:216
Equation
  continuity, 25:5, 323, 338, 362
  energy, 25:5, 18, 26, 45, 99, 323, 338, 345
    turbulent, 25:50
  filtered, 25:338, 339
  Fourier, 25:6
  governing, 25:322–323, 338, 362
  heat balance, 25:103
  heat conduction, 25:7, 12
  motion, 25:5, 9
  Navier–Stokes, 25:5, 323, 338, 339, 362
  primitive, 25:323
  state, 25:10
Equations of change, 12:197–203
  ionic components and, 12:200–203
  stoichiometric relations and, 12:201
Equations of state, 2:112
  for gases, use of Monte Carlo method, 5:48
  van der Waals, 7:3
Equations of transfer, 3:183ff, 199ff, 203ff; 5:13;
  11:326
  formal solution of, 11:327–329
  for local thermodynamic nonequilibrium,
    8:248
  from spherical shell, 11:351
Equilateral triangular ducts
  fully developed flow in, SUP:223–225
  hydrodynamically developing flow in,
    SUP:240–242
  with rounded corners, SUP:225–226
  thermally developing flow in, SUP:242–243
Equilibrium
  across a curved surface, 9:192
  Kelvin equation and, 10:89
  between liquid and its vapor, 1:198ff
  in a plasma, see Plasma

Equilibrium film pressure, 9:227
Equilibrium flows, 2:252ff; 8:8
Equilibrium flow singularities, see Singularities,
  associated with equilibrium flows
Equilibrium freezing, 28:262
Equilibrium phase change, 31:113
Equilibrium phase diagram, 22:188; 28:233, 235,
  239
Equilibrium radiation, 5:258
  thermodynamics of, 5:261
Equilibrium segregation coefficient, Czochralski
  process, 30:331
Equilibrium size distribution, in homogeneous
  nucleation, 14:243–301
Equilibrium superheat, 16:64–66
Equilibrium vapor pressure, 10:87
Equivalent property of state
  heat capacity, 22:201, 209, 296
  latent heat, 22:202
  phase front, 22:202
  thermal conductivity, 22:201
Equivalent radial distance, 30:290
Equivalent thermal conductivity, 5:83
Ergun correlation, 19:101, 110; 25:151, 204, 223
Error function, 5:12; 24:169, 173
Erythrocytes, high temperature injury, 22:322
Escoa fintube, 14:198
ESPROSE.m field model, steam explosion,
  29:180, 182, 195–201
Established flow, SUP:5
Ethane-ethylene, film boiling, 16:134, 137
Ethanol
  effect of heating-surface
    orientation on frequency of liquid–wall
      contacts, 18:287
    orientation on mean liquid–wall contact
      time, 18:286
    roughness on frequency and liquid–wall
      contact time, 18:285
  frequency of liquid–wall contacts, 18:282
  heat flow rates, effect of heater shape and
    material, 16:191
  hystograms of liquid–wall contact time dis-
    tribution, 18:277
  mean liquid–heating surface contact time,
    18:278
  mean relative liquid–heating surface contact
    time, 18:284
  mean vapor phase–heating surface contact
    time, 18:280

Ethanol (*continued*)
nucleate pool boiling, data available, **16**:170
Ethanol-benzene
A₀, **16**:119
boiling, effect of subcooling, **16**:111–113
boiling site density, **16**:77–80
bubble departure, diameter, **16**:95
equilibrium superheat, effect of various physical properties, **16**:69
film boiling, **16**:138
heat transfer coefficients, **16**:107–108
prediction, **16**:127
Ethanol-butanol, bubble departure
diameter, **16**:95–96
frequency, **16**:96, 102
Ethanol-cyclohexane
A₀, **16**:119
convective boiling, **16**:144
nucleate boiling, onset, **16**:72
Ethanol-isopropanol, bubble growth, experimental data, **16**:91–92
Ethanol-water
A₀, **16**:119
activation superheat, **16**:72–74
boiling, effect of subcooling, **16**:111–113
boiling site density, **16**:77–80
effect of contact angle, **16**:77
bubble departure
diameter, **16**:95–96, 100–101
frequency, **16**:96, 102
bubble growth, experimental data, **16**:90, 91–92
contact angle, effect on boiling inception, **16**:70, 71
convective boiling, **16**:144, 146
deactivation superheat, **16**:72–74
equilibrium superheat, effect of physical properties, **16**:68–69
heat transfer coefficient, prediction, **16**:124–127
onset of nucleate boiling, **16**:71
peak nucleate heat flux, **16**:127–129
slope of saturation curve, **16**:67
ETHICCA effect, **29**:153
Ethyl alcohol in forced convection film boiling, **5**:96, 97
Ethylene glycol-water, convective boiling, **16**:144
Eucken correction, **3**:279ff
Euler equations, **24**:241, 242
Eulerian point of view, **3**:184

Euler–Lagrange equations, **8**:27; **24**:40, 42, 47, 96
complex heat and mass transfer problems, **19**:196
for steady-state heat conduction processes, **19**:198, 199
for Stefan problems, **19**:40
variational description of heat transfer processes, **19**:207–209
Euler–Mascheroni constant, **26**:21
Euler number, **8**:150; **SUP**:40
Euler's transformation, **8**:15
Eutectic system, **28**:233
Evaporation, **16**:148; **22**:275
compound drop, **26**:48–52
and condensation
effect of compared with passive conduction, **31**:95
estimate of term of, **31**:5
convective, **16**:148–150
convective droplet, **26**:58–59, 70–71
in cryogenic systems, **5**:351
from free surface, **1**:125ff, 131ff
importance of, **31**:5
mass rate of, **31**:29, 91
microscopic description, **21**:57
molten metal drop, **26**:83
moving liquid droplet
electric field, **26**:83–84
gaseous environment, **26**:6, 31–32, 52–83
Reynolds number, **26**:59–77
shape deformation, **26**:79–81
multicomponent droplet, **26**:66, 77–79
slurry fuel droplet, **26**:81–82
stagnant medium, **26**:54–57
in thermosyphon, **9**:87
Evaporation coefficient, **9**:218; **21**:73–74
Evaporative heat exchangers, **31**:271, 319
Exact solutions, **SUP**:61–62
Exchange factors, **23**:166–168
calculation of, **31**:411
Exit temperature
cross-flow heat exchanger, **26**:290
parallel-stream heat exchanger, **26**:273–274
Exit velocity, transfer coefficients and, **13**:35
Exodus method, **5**:45
EXO-FITS experiments, **29**:147
Exothermic regeneration stage, **9**:114
Expansion, bed, **25**:152, 155, 166, 190, 193, 201, 236

Expansion cooling, in shear flow, **13**:229
Expansion parameter, **1**:35, 39
Expansion rate
  ratio of to system volumetric elasticity, **31**:117
    and system elasticity, **31**:118–119
Experimental technique, **9**:136
Explicit Euler scheme, **24**:244
Exploding bubble technique, **10**:105
Exploding drop method, in nucleation studies,
    10:100–101
Explosions, **29**:129–131
  detonation limits, **29**:100–102
  detonation velocity, **29**:62, 63
  hydrogen combustion
    combustion parameters, **29**:62–65
    detonation, **29**:62, 90–102
    fast flames, **29**:86–90
    nuclear reactor safety, **29**:59–61, 77, 102–
      123
    subscale models, **29**:118–121
    thermohydraulic codes, **29**:113–138
  physical explosions, **29**:129
  spontaneous, **29**:158
  steam explosions, **29**:132, 158–168
    described, **29**:132–135
    ESPROSE.m field model, **29**:180, 182,
      195–201
    history, **29**:135–136
    industrial accidents, **29**:129–136
    integral explosions assessments, **29**:192–
      195
    nuclear reactor safety, **29**:129, 131, 136,
      147
    premixing, **29**:148–152
    propagation, **29**:180–182
    triggering, **29**:132, 158–168
  stratified, **29**:182–185
  thermal-chemical explosions, **29**:187–192
  vapor explosions, **29**:129
Exponential axial wall heat flux, **SUP**:21, 28 29
  in circular duct, **SUP**:83–85, 135–136
  in concentric annular ducts, **SUP**:296–297
  for longitudinal flow over circular cylinders,
    SUP:364
  in parallel plates, **SUP**:159, 188
Exponential growth rate, **9**:339
Exponential integral, **1**:5; **8**:250
Exponential Kernel approximations, **8**:251, 252
Exponential repulsive potential, **3**:262, 290, 295
Extended film model theory, **26**:69

Extended surface plate fin heat exchanger,
    26:227, 228
Extended surfaces, *see* Fins
External condensing flows, in porous media,
    30:141–148
External fins, heat exchangers, **30**:216
External flow in boiling, **1**:245ff
  analysis, **1**:254ff
  experimental data, **1**:256ff
  flow patterns, **1**:247ff
External flows, **5**:365
  flow across cylinders, **5**:365
External heat and mass transfer experiments on
    porous bodies, **1**:124ff, 153ff
  with drying of moist solids, **1**:128ff, 139ff
  with evaporation from free surface, **1**:125ff,
    131ff
  with porous cooling, **1**:130f, 144ff
  with sublimation, **1**:153ff
External problems, **8**:161
External receivers, for central receiver systems,
    18:30–34
Extinction coefficient, **9**:377
Extinction index, **9**:363
Extrinsic constitutive laws, **31**:106
Extrinsic radiation similarity parameters, **3**:217ff
Extrusion, **28**:146–151
  chemical reaction and conversion, **28**:218–220
  flow in dies, **28**:201–212
  moisture transport, **28**:213–218
  screw extrusion of polymers, **28**:145–151,
    220–226
    combined heat and mass transfer, **28**:213–
      220
    material characterization, **28**:152–155
    single-screw extruder, **28**:147, 149, 155–
      187
    twin-screw extruder, **28**:151, 187–201,
      225–226
Eye, structure, **22**:249
Eyring model, shear viscosity and, **15**:79–80
Eyring-Powell equation, **2**:362, 370, 378

F

Fabrics
  ignition tests for, **10**:240–241, 244, 273, 279–
    280
  protective, **4**:129ff

FACET (software), **30**:339
Factor, particle shape, **25**:183
FAI/ASCo study, **29**:323
FAI/DCH studies, **29**:220, 221, 222, 225, 226, 230–231
　FAI/DCH-1, **29**:316, 317
　FAI/DCH-3, **29**:318
Falkner–Skan flows, **2**:142, 167f
　axisymmetric jet stagnation region, **26**:111, 120, 122
Falkner–Skan parameter, **4**:323
Falling ball viscometer, **15**:72–75
　and shear rate for CMC solutions, **15**:74
Falling spheres, terminal velocity of, **15**:73
Falling supported beam, direct contact condensation on, **15**:237–239
False-starting methods, errors, **19**:53–54
Fanning friction coefficients, for polyethylene oxide solutions, **15**:95
Fanning friction factor, **SUP**:39–40; **15**:80; **19**:255; *see also* Apparent friction factors; *and appropriate ducts in the text for locating tabular values*
　degradation effect on, **15**:98
　dimensionless heat transfer coefficient and, **15**:117
　solvent effect on, **15**:96–97
Fan power, for cooling towers, **12**:13–14, 33, 41–45
Fan spray sheet, direct condensation on, **15**:234–235
Faraday's constant, **12**:245
Faraday's law of induction, **1**:275
Farming, **31**:406–407, 424
FARO experiments, **29**:147
Fast-bubble regime, **19**:98
Fauske and Associates, Inc. (FAI) studies direct containment heating, **29**:219, 221, 222, 225
FDM, *see* Finite difference method
FEM, *see* Finite element method
Fermat's principle, **6**:145; **30**:262
Fibers
　coarse, **25**:293
　fine, **25**:293
Fiber wick heat pipe, vs. nonwicked transistor, **20**:291–293
Fick principle, **22**:25
Fick's first law of diffusion, **12**:204
Fick's law, **1**:283; **2**:46; **3**:111; **31**:8, 9, 38, 60
Fick's relation, **2**:114f

Field
　alternating magnetic, **25**:179
　applied magnetic, **25**:152
　　effect on fluidization, **25**:152, 153, 157, 166, 169, 200, 232
　axially directed magnetic, **25**:169, 177
　colinear to direction of fluid flow, **25**:153, 178, 179, 232, 234
　constrict magnetic, **25**:158, 161, 162
　electromagnetic, **25**:160
　external electric, **25**:153
　　effect on fluidization, **25**:153
　external magnetic, **25**:151, 165, 200, 201, 203, 205, 224, 225
　impulsed magnetic, **25**:166
　longitudinal magnetic, **25**:162, 164, 201, 232
　moderate magnetic, **25**:223, 229
　nonuniform magnetic, **25**:153, 162, 178, 203
　pulsating magnetic, **25**:166, 167
　rotary magnetic, **25**:166
　running magnetic, **25**:166
　strong magnetic, **25**:223, 229
　transverse magnetic, **25**:153, 162, 163, 164, 178, 201, 232
　uniform magnetic, **25**:169, 171, 200, 203, 205, 220, 237
　weak magnetic, **25**:223, 229
Filled polymeric systems, thermal conductivity, **18**:227–230
Filling, vessel, **25**:130
Film boiling, **1**:198f; **5**:55, 56; **7**:78; **11**:2; **23**:9, 117–120; *see also* Film-boiling heat transfer; Pool boiling; Rod flow film boiling
　in annular spaces, **5**:74
　convective heat transfer coefficient, **23**:118–119
　critical wavelength in, **5**:434
　of cryogenics, **5**:428
　from cylinders, **5**:436
　dominant wavelength in, **5**:116
　effect of gravity on, **5**:435
　effect of increased gravitational field on, **5**:73
　effect on subcooling on, **5**:91
　in external flow, **11**:124–144
　from flat plates, **5**:437
　flow patterns in, **5**:75
　forced flow in channels, **11**:144–187
　Grashof number in, **11**:118, 210, 244–246, 251
　heat exchange and, **11**:51

of helium I, **5**:431
of helium II, **5**:497
of helium, **5**:429
on horizontal cylinders, **11**:56
from horizontal plate, **11**:90–91
of hydrogen, **5**:430; **11**:187
influence of radiation on, **5**:428
interface in, **11**:52
investigations, **23**:109–110
of liquid droplets, **5**:90
liquid hydrogen, **5**:82
liquid metal, **5**:74
of methane, **5**:439
methanol, **5**:58, 59
and minimum film boiling heat flux
    steady-state heat transfer in helium I and,
        17:147–148
    transient heat transfer in helium I
        onset of, **17**:151
        recovery and, **17**:151–153
minimum heat flux in, **5**:109
mist flow and, **11**:144, 164–170
most probable wavelength in, **5**:116
of nitrogen, **5**:429
of oxygen, **5**:430
quenching, **23**:118–119
research needs, **23**:123
rod flow and, **11**:144
single-phase analysis, **30**:148–149
slug flow and, **11**:144, 162–164
from small wires, **5**:431
from sphere, **5**:429
turbulent analysis of, **5**:77
turbulent friction factor correlation, **5**:79
vapor film thickness in, **11**:60
vs. nucleate boiling, **14**:131
Film boiling from horizontal cylinders
    critical diameter, **5**:68
    effect of subcooling on, **5**:96
    forced convection, **5**:95
    freon, **5**:67, 113
    influence of diameter on heat transfer coeffi-
        cient, **5**:66
    isopropanol, **5**:67
    natural convection, **5**:95
Film boiling from horizontal surfaces, **5**:82
    bubble spacing in, **5**:86
    freon-II on aluminum surface, **5**:88
    heat transfer coefficient for, **5**:83
    heat transfer coefficient in, **5**:88

instability, **5**:83
most probable wavelength, **5**:80
physical model, **5**:85
vapor film thickness, **5**:88
vapor velocity, **5**:85
water on aluminum surface, **5**:88
Film boiling from vertical cylinders, **5**:76
Film boiling from vertical surfaces, **5**:74
    experimental data for, **5**:81
Film boiling heat flux, minimum, **14**:132
Film-boiling heat transfer, **11**:51–187
    around cylinder, **11**:134, 139, 143–144
    average heat transfer coefficient in, **11**:86
    calculations in, **11**:78–84
    dimensionless temperature profiles in, **11**:83–
        84, 136
    dimensionless velocity profiles in, **11**:78–81,
        136
    experimental investigation of, **11**:105–124
    forced flow boiling in channels, **11**:144–187
    Froude number and, **11**:135, 140–141
    geometry and orientation of heated surface in,
        11:105, 110–157
    for horizontal cylinders and spheres, **11**:108
    in horizontal tube, **11**:159–162
    laminar flow with unstable film interface in,
        11:84–95
    Prandtl number and, **11**:76
    pressure, gravity, and subcooling in, **11**:105,
        110–117, 133
    rod flow in vertical tube, **11**:151–159
    in spheres, **11**:96, 107
    subcooling and, **11**:133
    temperature difference and, **11**:133
    theoretical study of, in laminar flow,
        11:62–84
    turbulent vapor flow and, **11**:95–105
    turbulent viscosity distribution in, **11**:102
    unstable film interface in, **11**:84–95
    vapor film thickness measurements in, **11**:60,
        119
    on vertical surfaces, **11**:120
    from vertical tubes, **11**:107–108, 151–159
Film boiling on cylinders
    benzene, **5**:64
    carbon tetrachloride, **5**:64
    ethyl alcohol, **5**:64
    nitrogen, **5**:64
    n-pentane, **5**:64
    water, **5**:64, 65

Film boiling on horizontal tubes
  general correlation, **5**:70
  isopropanol, **5**:70
Film boiling on horizontal wires, **5**:73
  near critical pressure, **5**:73
Film boiling on spheres, **5**:91
  liquid helium, **5**:91
  liquid nitrogen, **5**:91
Film boiling regimes, **31**:265
Film condensation, **21**:59–61, 104–115
  heat transfer coefficient, **21**:105
  multicomponent vapor, **21**:107–115
    forced convection condensation, **21**:113–114
    heat-transfer coefficient, **21**:108–109
    Kármán–Pohlhausen method, **21**:108–110
    Minkowycz–Sparrow theory, **21**:109–110
    natural convection condensation, **21**:114–115
    nonazeotropic mixture, **21**:110–112
    noncondensable gas effect, **21**:107–108
    thermodynamic cycle, **21**:110–111
  Nusselt number, **21**:105–106
  single component vapor, **21**:105–107
Film cooled wall temperature, **7**:363
Film cooling, **3**:80ff; **5**:199; **7**:321–379
  analysis, **7**:330–342
  applications, **7**:358
  blowing rate parameter, **7**:331, 339
  correlations, **7**:334–337
  defined, **7**:322
  effectiveness, **7**:324, 326–333
    high-speed flow, **7**:327, 329
    impermeable wall concentration, **7**:329, 330, 355
    incompressible flow, **7**:327
    isoenergetics, **7**:328
  effect of injection angle on, **7**:358, 373
  experimental studies, **7**:342–359
  free stream turbulence effect, **7**:354
  full-coverage, **16**:243, 310–319
    mixing length model, **16**:318
  heat sink model, **7**:330, 337
  heat transfer measurements, **7**:360
  high speed flow, **7**:327, 340, 361
  incompressible flow, **7**:326, 338, 351
  influence of boundary layer thickness, **7**:359
  injection
    of air into air, **7**:351
    through discrete holes, **7**:341

isoenergetic, **7**:328
large temperature difference effects, **7**:358
mass transfer analogy, **7**:329, 359
measurements, **7**:371
porous injection in, **7**:353, 355
slot geometry effects on, **7**:351, 356
stagnation region, **7**:368
surface effects on, **7**:334
three-dimensional, **7**:341, 369
two-dimensional, **7**:323, 338, 340, 351, 361
  compressible flow, **7**:340, 361
  incompressible flow, **7**:351
variable free stream velocity effects in, **7**:338, 354
Film cooling on horizontal cylinder
  heat transfer coefficient, **5**:66
  influence of radiation on heat transfer coefficient, **5**:66
Film-deep (rivulet) regime, **31**:293
Film formation, **28**:340–341
Film-fracture mechanism, **9**:206
Film heating, **7**:342, 351
Film length, slug flow calculations, **20**:98
Film pool boiling, **11**:55–124; *see also* Film boiling; Film-boiling heat transfer
  high speed movies of, **11**:55–56
  mechanism of, **11**:55–62
Film theory, **3**:63
Film zone, geometry of, **20**:85–87
Filonenko formula, **25**:48, 76
Filter, magnetically stabilized bed, **25**:233
Filtering
  fluid, **25**:234; *see also* Spatial decomposition
Filtering mechanism, **9**:332
Fin effectiveness, **2**:435f
Fin effectiveness factor, **19**:147, 154
Finite-amplitude convection, heat- and mass-transfer, **20**:327
Finite annulus length effect, induced rotational flow, **21**:165
Finite difference, **22**:178, 209, 238, 269
  spreadsheet solution, **22**:180
Finite-difference formulations, **31**:106
  alternatives to, **31**:107
Finite-difference method, **9**:99; **19**:42–43; **25**:47; *see also* Hydrodynamically developing flow; Thermally developing flow
  for fully developed flow, **SUP**:64–65
  for simultaneously developing flow, **SUP**:77
  stability of, **5**:341

transport phenomena in Czochralski crystal, **30**:347, 350
Finite-difference methodology, **20**:229–230
Finite-difference modeling, twin-screw extruder, **28**:195–197
Finite element, **22**:191, 200, 296, 333
Finite-element geometric modeling approach, **31**:421
Finite-element mesh, **31**:336
Finite-element method (FEM), transport phenomena in Czochralski crystal, **30**:347–350
Finite-element modeling, twin-screw extruder, **28**:193–195
Finite height strips, potential flow convection, **20**:362–363
Finite volume, **24**:195, 196
Finite volume model (FVM), transport phenomena in Czochralski crystal, **30**:347, 350
Finned heat pipes, indirect cooling techniques, **20**:280–281
Finned surfaces, **31**:161, 276–277
Finned tubes
    geometry of, **15**:23
    in heat transfer augmentation, **15**:22–23
Finning, **31**:159, 312–313
    transverse, **31**:259–261
    in annular channels, **31**:245–249
Fins, **8**:32; **9**:3, 76
    axial, **19**:58
    bundles of tubes without, **14**:226–228
    efficiency, **19**:152
    heat exchangers, **30**:216
    horizontal, 60 degree V-thread, **19**:151
    on horizontal and slanted tubes, **14**:172–207
    immersed in gas-fluidized beds, heat transfer investigation summary, **19**:147–150
    phase-change processes around cylinders and, **19**:57–58
    radiant interaction between, **2**:440ff
    radiating-conducting, **2**:400, 432ff
    relative efficiency of heat transfer, **19**:147
    single plate-type, **2**:433ff, 437
    single tubes with, **14**:192–195
    single tubes without, **14**:212–226, 228–231
    slanted single tubes without, **14**:204–206
    tube bundles with, **14**:195–203
    tube bundles without, **14**:178–192
Fin-tube radiator, **2**:433f, 436ff
    annular, **2**:443
    interactions, **2**:443f

Fire
    classification of, **10**:223–225
    convection above line type of, **10**:250–251
    densometric analysis of, **10**:229
    enclosed, **10**:258–260
    firebrand spotting in, **10**:260–262
    fuel load evaluation in, **10**:262–266
    heat and mass transfer in, **10**:221–222
    hot air plume in, **10**:245–260
    ignition in, **10**:235–245
    losses from, **10**:220
    mass spectrometry and gas chromatography in, **10**:228–229
    natural convection in, **10**:250–252, 258
    other materials than wood in, **10**:269
    phenomena of, **10**:221–223
    reaction rates in, **10**:232–233
    spontaneous heating in, **10**:236–237
    spread of, **10**:260–275
    sprinkler system in, **10**:245
    thermal explosion theory in, **10**:238
    tolerance levels in, **10**:242–243
    wildland, **10**:222
    wood chemistry in, **10**:225–226
Firebrands
    burning characteristics and terminal velocities of, **10**:263–264
    spotting of, **10**:262–262
Fire department vehicles, **10**:276
Fire fighter, equipment for, **10**:275–277
Fire insurance, **10**:220
Fire plume, **10**:245–260; *see also* Plume
Fire protection, pyrolysis testing and, **10**:233–235
Fire research
    building codes and, **10**:277–279
    defined, **10**:220–221
    exhaust hoods and, **10**:272
    fabric testing methods in, **10**:273
    fire fighter and, **10**:275–280
    flow visualization in, **10**:273
    fuels in, **10**:272–273
    heat and mass transfer in, **10**:219–280
    instrumentation in, **10**:272–280
    product testing in, **10**:279–280
    pyrolysis products and, **10**:280
    tall buildings and, **10**:278
    temperature measurement in, **10**:274
    velocity measurement in, **10**:274–275
    wind tunnels in, **10**:272

Fire retardants, **10**:234
Fire-safety oriented ignition tests, **10**:238–243
Fire spread
  fuel characteristics in, **10**:262–270
  large-scale studies of, **10**:269–271
  rates of, **10**:267–269
  theories of, **10**:266–267
Fire station locations, **10**:275–276
Fire storm, **10**:223–225
  plume in, **10**:246
First Law of Thermodynamics, **1**:276; **21**:240–241
First normal force difference, degradation effect on, **15**:86
First-order upwind scheme, **24**:226, 235
FITS experiments, **29**:147
Fixed-bed regenerator
  "boxed" process, **20**:139–149
  dimensionless parameters, **20**:133–135
  equivalence with rotary bed regenerator, **20**:136–137
  mathematical representation, **20**:133
Fixed beds, **25**:155; *see also* Packed beds
Fixed-identity insulation system, entropy generation minimization in, **15**:39–43
Flame
  absorption coefficient, **27**:124–125
  boiler furnace, **27**:158, 161, 164–167
  gas reformer furnace, **27**:169–170
  jet engine combustion chamber, **27**:177, 180–181
Flame heights, temperature profiles in open fires, **17**:205–216
Flame quenching, entropy production, **21**:261–263
Flames
  adiabatic flame temperature, **29**:62, 63
  buoyancy-driven diffusion flames, **30**:66, 67
  diffusion flame, **29**:62
  diffusion flames
    forced flame, **30**:52–65
    natural flame, **30**:65–78
  energy balance in open fires and, **17**:203–204
  fast flames, **29**:86–90
  flame speed, **29**:75, 112
  speckle photography
    axisymmetric candle flames, **30**:296–299
    gaseous flame, **30**:299–302
  temperature measurement, laser speckle photography, **30**:293–302, 304, 306

turbulence, **29**:73, 74, 81
Flame sheet, **2**:256ff, 262f
Flame Sherwood number, **30**:67
Flame speed, **29**:75, 112
Flammability limits, **29**:65–71
  direct containment heating, **29**:303
  nuclear accidents and, **29**:105–106
  thermohydraulic codes, **29**:116
Flammability tests and ratings, **10**:240–241, 245; *see also* Ignition
Flash photolysis, use of, in velocity measurements, **8**:298
Flat channels, heat transfer enhancement in, **31**:259–263
Flat heat pipes, indirect cooling techniques, **20**:278–280
Flatness factor, **8**:237–332, 340
Flat-pack modules, forced convection, **20**:236–237
Flat plate, *see also* Turbulence, free-stream; Turbulent boundary layers
  flow of transformer oil over, **8**:115
  glycerine flow over, **8**:115
  heat transfer in laminar flow, **8**:115
  influence of heating or cooling on heat transfer, **8**:115
  influence of Prandtl number on heat transfer, **8**:115
  isothermal, **9**:283
  velocity and temperature measurements for turbulent flow over, **8**:305
  vertical, **9**:283
Flat-plate boundary-layer heat transfer and solutions, **1**:317ff, 323ff, 331
  compressible flow, **1**:325ff
    assessment of results, **1**:329
    hysteresis effect, **1**:328f
    remarks on assumptions, **1**:329f
  incompressible flow, **1**:323ff
Flat-plate collectors
  aspect ratio, **18**:12–15
  convective loss suppression, **18**:15–17
  geometries and orientation for natural convection losses, **18**:9
  natural convection, **18**:11–19
  thermosiphon unit, **18**:5
Flat plate flows
  chemically reacting boundary layers, **2**:167f, 171, 173, 174ff, 183, 184f, 192, 213ff, 227ff, 231ff, 243ff

initial relaxation near leading edge, **2**:219ff
non-Newtonian heat transfer, **2**:390
Flat-plate heat pipe, **20**:287–289
Flight distance, **27**:52–53, 96
Floating elements technique, **3**:51f
Floating random walk, **5**:45
Flocculation, **25**:231
Florida aquifer, liquid waste disposal in, **14**:40
Flow
    accelerated fluid, **25**:77
    acceleration, **25**:83, 97, 98
    alloy solidification, **28**:270–288, 318
    back, **9**:243
    barrel chemical vapor deposition reactors,
        28:378
    base, **9**:326
    boundary layer-driven, in solar buildings,
        18:68–72
    bulk density-driven, in solar buildings, **18**:67–
        68
    in closed stoves
        patterns of, **17**:248–254
        resistance to, **17**:244–248
    cone, **9**:296
    creeping, **25**:292, 297, 298
    cross, **25**:252, 283, 286, 292, 295, 296
    decelerated fluid, **25**:77
    deceleration, **25**:83, 84, 95
    with drag reduction, **12**:77–110
    extruder dies, **28**:201–212, 226
    fluid, see Fluid flow
    free surface, **25**:296
    fully developed, see Fully developed flow
    gas, fluidized bed, **25**:152
    gas, see Gas flow
    heat carrier, dissociating, **25**:73
    hydrodynamically developing, see Hydrody-
        namically developing flow
    hydrodynamic structure of, **31**:162,
        165–166
    incompressible, **25**:299
    induction, **9**:242
    instability of laminar, **9**:321
    isothermal, **25**:268
    laminar, **25**:286, 292, 300
        fully developed, **25**:299
    liquid, **25**:152
    models of, **12**:79–85
    natural, **9**:335
    non-Newtonian material, **28**:146, 152

rheological fluid, **25**:73
separation, **9**:335, 342
simultaneously developing, see Simulta-
    neously developing flow
tapered channel, **28**:176
thermally developed, **12**:89–91
thermally developing, see Thermally develop-
    ing flow
transients, **9**:310
transition, **9**:321
with viscous dissipation, **3**:234ff
Flow average temperature, **SUP**:45
"Flow behavior indices", **2**:363
Flow distribution, in loop systems, **31**:126–
    133
Flow-field, periodic, **24**:304
Flow impedance, **31**:124
Flow inertia, **31**:124, 133, 148
Flow intermittence, in transition region, **31**:188–
    189
Flow model development, **9**:145
Flow modeling, two-phase, **31**:152
Flow past marginal stability, **30**:50
Flow pattern boundaries in annular-dispersed
    flows, **1**:359ff
Flow patterns, **5**:75
    in boiling, **1**:240f, 247ff
    of two-phase flow, **1**:355ff, 357ff
Flow-rate pulsations, as heat transfer enhance-
    ment, **31**:181–182
Flow rates, **25**:92, 96
    gas, variable speed, **25**:84
    gas atomization of liquid metals, **28**:19–20
    heat carrier, **25**:16
    hot gas, periodic, **25**:60
    mass, **25**:8
    phasic, **31**:130–133
Flow regimes, **24**:120
    in tube banks, **8**:107
Flow region, zones in, **31**:173–174
Flow resistance, **31**:133–136
    distribution of in the loop, **31**:136
Flow structure, liquid jet impingement, **26**:136–
    147
Flow swirling, **31**:159, 161
Flow systems, **31**:147–149
Flow turbulators, **31**:159–160, 161, 165, 245
    heat transfer and, **31**:177
    separated flow regions as means for goal-
        directed, **31**:166–178

Flow turbulators (*continued*)
  trends in heat transfer variations with discrete,
    31:185–187
Flow visualization, 9:24, 343
  with dye injection, 9:51
  light interruption technique, 9:139
  use of fish flakes for, 9:51
Flow work
  definition of, **SUP**:13
  effect of, **SUP**:80–82, 144, 157, 176, 184
Fluctuation, 25:160
  amplitude of density, 25:164
  bed voidage, 25:171
  frequency, 25:163, 165, 219, 230, 231
  local density, 25:160, 164
  pressure drop, 25:163, 170, 219
Flue gas, 9:114
Fluence rate, of radiation, 22:281
Fluid axial heat conduction, **SUP**:53, 56–57
  in circular ducts, **SUP**:79, 81, 83, 111–118,
    132–135
  in concentric annular ducts, **SUP**:309, 314–
    315
  in parallel plates, **SUP**:156, 174–176, 177,
    183
  in rectangular ducts, **SUP**:216
  thermal boundary condition for, **SUP**:17, 111–
    112, 132–133
Fluid bed heat exchanger, 30:236–237
Fluid bulk mean temperature, **SUP**:45
Fluid compressibility, 24:312–314
Fluid consistency index, 24:108
Fluid dynamics
  extrusion die, 28:201–212, 226
  spray deposition, 28:1–3
    impact region, 28:23–69
    spray region, 28:3–22
Fluid element
  acceleration and deceleration of, 12:199
  mass-average velocity and, 12:198
Fluid flow, *see also appropriate ducts*
  with boiling, 1:196ff
  comparisons of solutions for, **SUP**:394, 397–
    399
  dimensionless groups for, **SUP**:39–44
  fluidized beds, 23:232
  fluid properties and heat transfer, 18:107–108
  generalized solutions for, **SUP**:44
  hindered settling, 23:261
  packed beds, 23:192

periodically constricted tubes, 23:219
physical quantities for, **SUP**:38–39
restricted channel, 18:97–99
  average velocity, 18:99
  blackage ratio, 18:97–98
  reference velocity, 18:99
single tube, 18:89–99
  critical flow concepts, 18:91
  drag of the tube, 18:95–97
  ideal fluid, 18:89–90
  separation bubble, 18:91–92
  separation of boundary layer, 18:90
  stability of boundary layer, 18:92
  Strouhal number, 18:93
  supercritical flow regime, 18:91, 93
  types of flow regimes, 18:90–91
  velocity distribution around the tube,
    18:94–95
  in the wake, 18:92–94
three-phase systems, 23:253
tube in a bank
  bank arrangement and geometry, 18:99–100
  flow pattern changes for in-line banks,
    18:103
  flow regimes and Reynolds numbers,
    18:101
  friction drag, 18:106
  mean velocity in free cross section, 18:105
  pressure coefficient, 18:101
  pressure drag, 18:106
  staggered banks, 18:100
  surface roughness effects on drag, 18:106
  total bank drag, 18:101
  transverse pitch effect on velocity distribu-
    tion, 18:104–105
  velocity distribution, 18:101–106
Fluid friction irreversibilities, heat transfer and,
  15:29–32
Fluidization
  bubbling, 25:152, 167
  defined, 10:168
  incipient, 25:152, 158, 184
  initiation of, 10:169–170
  liquid–solid particulate, 25:166
  mechanism, 25:163
  particulate, 25:152, 168
  quality, 25:152, 163, 188
Fluidization, bed
  bed voidage at minimum fluidization, 19:105–
    108

group I systems, **19**:112
large particle systems, **19**:99
minimum fluidization velocity, **19**:108–111
particle classification system, **19**:100–105
quality of, **19**:99, 115
Fluidization velocity, minimum, **19**:99, 108–111
Fluidized-bed heat exchanger, **31**:462–463
Fluidized beds, **23**:187
    axial dispersion, **23**:257
    bed expansion, **23**:241, 253
    behavior of, **10**:169–171
    coating for, **10**:211–213
    conduction through emulsion layer of,
        10:193–194
    defined, **10**:168
    emulsion layer in, **10**:194–195
    emulsion packet in, **10**:186–188, 196
    entrance effects in, **10**:206
    experimental heat transfer results in, **10**:202–
        206
    flat plates in, **10**:208
    flow field near heat exchange surface of,
        10:194–195
    flow properties in, **10**:169–171
    fluidization conditions in, **10**:204–205
    fluid properties in, **10**:198
    gas holdup, **23**:255
    gas or liquid, **10**:168
    geometric variables in, **10**:205–206
    granular solid material properties in, **10**:203
    heat balance in, **10**:212
    heat transfer between surface and, **10**:180–213
    heat transfer coefficients for, **10**:177–178
    heat transfer in, **10**:167–213; **23**:251, 258, 261
    height and length of, **10**:205
    immersed bodies in, **10**:206–207
    immersed tubes in, **10**:209–210
    liquid, **25**:152
    liquid holdup, **23**:257
    magnetic, behavior, **25**:151, 152, 194, 201,
        206, 220, 223, 228, 237
    mass transfer, **23**:251, 258, 261
    minimum fluidization in, **10**:203–204
    minimum fluidization velocity, **23**:233, 239,
        241
    multistage, **25**:235
    non-Newtonian effects, **23**:232
    random surface renewal model of, **10**:196
    small cylinders in, **10**:208
    spheres in, **10**:207–208

temperature and pressure in, **10**:206
three-phase systems, **23**:253
tube bundles in, **10**:210
tubes in, **10**:209–210
uniform surface renewal model of, **10**:197
Fluidized heat particle, heat removal by,
    14:153
Fluidized heat transfer, principal mode of con-
    version to, **14**:126
Fluidized systems, liquid–solid, **10**:179–180
Fluid mean axial velocity, **SUP**:38
Fluid mechanics
    direct-contact transfer studies
        drop deformation, **26**:12–13, 18–19
        electric field effect, **26**:13–19
        inertialess translation, **26**:4
        Reynolds number translation, **26**:5–11, 16–
            18
        surface-viscous effect, **26**:13–16
        surfactants, effect on drop motion, **26**:11–
            12
        weakly inertial translation, **26**:5
    heat and mass transfer relationships in, **15**:60–
        61
    of non-Newtonian fluids in circular pipe flow,
        15:86–98
    of suspensions, **9**:134
Fluid phase, point equations for, **31**:24
Fluid properties
    effect on fluidization, **25**:152
        density, **25**:152
        viscosity, **25**:152
    influence of
        on heat transfer, **8**:112
        in heat transfer to liquids, **8**:128
    temperature-dependent, **12**:91
Fluids, *see also* Liquid; Newtonian fluids; Non-
    Newtonian fluids; Nucleate pool boiling
    Bingham plastic, **25**:255
    classification of, **15**:61–62
    dilatant, **25**:255
    elastic, **25**:260
    exceptional, heat flow rates, **16**:160–161
    $F(p_r)$ dependent on, **16**:199, 224–225
    heat flow
        with molecular weight, **16**:160
        with pressure, **16**:159
    heat transfer between solid particles and,
        10:171–180
    heat transfer rate

Fluids (*continued*)
    alternatives to molecular weight, **16**:183,
        197–198, 224
    correlation with molecular weight, trends in,
        16:181–182
    variation with molecular weight, **16**:160,
        179–182, 193–197, 222–224
    heat transfer rate parameters relating to
        reduced pressure and saturation-reduced
        temperature, **16**:183–184
    hydrodynamically developing rectangular duct
        flow solutions, **19**:268, 270–271
    incompressible, steady flow with negligible
        dissipation, **19**:250–251
    Maxwell, **25**:260
    Oldroyd, **25**:260, 278, 292
    one-dimensional motion, **19**:6–7
    particle systems, subgroup IIA, **19**:113
    Powell–Eyring, **25**:260, 281, 283
    properties
        effects on boiling, **16**:175–184
        effects on free-convection heat transfer,
            16:334
        effects on heat flow rates, **16**:192–200, 219–
            225
        in heat flow analysis, **16**:162
        interdependence, **16**:162, 163, 165–166
        temperature-dependent effect, **16**:244
        temperatures for evaluation, **16**:334
        variation, in heat transfer analysis, **16**:168
        variations between, **16**:175–176
    pseudo-plastic, **25**:255, 260, 279, 281, 282
    rheopectic, **25**:256
    Rivlin–Erickson, **25**:259
    second order, **25**:28
    shear-thickening, **25**:256
    shear-thinning, **25**:255, 256
    structure and evaluation, **16**:179–181
    thermodynamic properties, variation with
        molecular weight, **16**:176–179
    thermophysical properties, **21**:2–3
    thixotropic, **25**:258
    time-dependent, **19**:249; **25**:255, 256
    time-independent, **25**:255, 256
    visco-elastic, **19**:249; **25**:255, 256, 260, 312
    visco-plastic, **25**:255, 256
    viscous, **25**:255, 277
        model, **25**:5
Fluid-to-fluid modeling, of CHF
    dimensional analysis and, **17**:30–31

    empirical rule and, **17**:30
Fluid vibration, heat transfer enhancement by,
    30:242–243
FLUTE system, **29**:153
Flux, **24**:195, 203, 234, 235, 238, 239, 240
    by ordinary diffusion, **12**:204
    predicted net, **31**:383–384
Flux condition, as gas–solid interface,
    31:59–60
Flux divergence
    mass balance and, **31**:114–121
    uncertainty in, **31**:397
    values of, **31**:349
    variance in, **31**:360–364
Foil thickness, thermal contact conductance,
    20:266
Food extrusion, **28**:220, 223
Food(s)
    actual cooking data for, **17**:283–284
    specific heats of, **17**:277
Foodstuffs, thermal conductivity, **18**:211–213
Forced convection, **8**:22; **SUP**:6–7; **20**:220–243;
        24:121–126; **30**:19–20; *see also* Convective
        heat transfer
    agitated vessel, **7**:122
    artificially waved liquid layer, **7**:133
    concentric rotating annulus, **7**:128
    convective heat transfer and, **11**:212–243
    critical region, **7**:25–55
    cross flow, **7**:116
    cylinder with crossflow in, **11**:212–239
    in Czochralski process, **30**:316
    discrete flush heat sources–channel flow,
        20:220–226
    falling liquid film, **7**:132
    freezing, **19**:73–78
    gases, liquids, and viscous oils, **30**:20–24
    general methods of solutions for laminar flow,
        SUP:61–77
    internal boiling and, **30**:136–141
    jet flow, **7**:135
    liquid metals, **30**:24
    melting, **19**:73–74, 78
    packed beds, **7**:134
    phase-change problems, **19**:3–4
    single-phase, jet impingement boiling, **23**:7
    temperature loading in, **11**:238
    three-dimensional package arrays, **20**:232–243
        actual circuit cards, **20**:242–243
        fully populated arrays, **20**:232–238

missing modules or height differences, 20:238–241
tube flow, 7:103
two-dimensional protruding elements, 20:227–232
Forced convection boiling, 5:403, 440
of liquid hydrogen, 5:403, 449
of liquid nitrogen, 5:449
maximum heat flux in, 5:449
temperature fluctuation in, 11:43–45
Forced convection condensation, 21:113–114
Forced convection film boiling, 5:56, 91
benzene, 5:96, 97
of benzene in horizontal tubes, 5:99
carbon tetrachloride, 5:96, 98
comparison of constant wall temperature and constant heat flux, 5:99
effect of radiation on, 5:94
ethyl alcohol, 5:96, 97
on hemisphere cylinder, 5:99
hexane, 5:96, 98
on horizontal cylinders, 5:92
in internal flows, 5:99
in nuclear reactors, 5:99
of water in horizontal tubes, 5:99
Forced convection flows, boundary layer approximations in, 15:178–179
Forced convection heat transfer in neighborhood of critical point, 5:355
Forced convection transition boiling of water in annulus, 5:104
Forced fields, SUP:6
Forced flame
laminar, 30:52–60
turbulent, 30:60–65
Forced flow
dryout heat flux and, 29:50
thermal microscales, 30:9–13
Forced flow film boiling
in channels, 11:144–187
in horizontal tube, 11:159–162
mechanisms of, 11:144–151
mist flow film boiling and, 11:164–170
slug flow of, 11:162–164
in vertical tube, 11:151–159
wall temperature in, 11:145–146
Forced plume, definition, 16:3
Force field, effect on transition boiling heat transfer, 18:302–303
Force–momentum principle applied to film

boiling on cylinders, 5:61
Force of electrical origin, 14:109
Forces
bond, 25:163
buoyancy, 25:230
cohesion, 25:172, 188
downward, 25:200
drag, 25:229, 230
driving, 25:175
interparticle cohesive, 25:188
interparticle magnetic, 25:152, 157, 165, 172, 181, 189, 191, 195, 202, 204, 219, 220, 224, 229, 230, 232, 239
inward, 25:169
magnetic interfacial, 25:109
strong cohesive, 25:222
surface electrostatic, 25:166
Forchheimer coefficient, 24:112
Forchheimer's constant, heat- and mass-transfer, 20:317
Forchheimer's equation, 14:4
Forchheimer term, 24:112
Forest fires, 10:220; see also Fire
Forgó heat exchanger, 12:50
Forgó tower, 12:59
Format of published papers, SUP:416–419
presentation of solutions, SUP:417–419
problem statement, SUP:416–417
solution techniques, SUP:417
Form loss coefficient, 31:122
Form losses, 31:134
Form parameters, 8:319, 324
Forward approaches, for ray-tracing procedures, 31:344–346
Forward pathlength methods, 31:373, 422
parallelization of, 31:413–421
performance of, 31:348–364
Forward spacing steps, 31:257
resistance of channels with turbulators shaped as, 31:258
turbulators shaped as, 31:253
Fossil energy reserves, 19:97
Fouling
calculation of influence of on decrease of heat transfer coefficient, 31;320
control of, 31:294–304
experimental studies of effect of, 31:296–299
mechanisms of, 31:435–446
monitors for, 31:451
physical devices to control, 31:452–453

Fouling beds (*continued*)
  prediction of rate of, **31**:467–469
  research and development on, **31**:434
Fouling, heat transfer enhancement, **30**:197–198,
    246–247
  displaced enhancement device, **30**:227–230
  extended surfaces on heat exchangers,
      30:216–217, 218
  fluid bed heat exchanger, **30**:237, 238
  with heat transfer area enlargement, **30**:209–
      213
  heat transfer intensification, **30**:201–208
  HiTran® radial mixing elements, **30**:227–228
  injection, **30**:244
  porous layer enhancement, **30**:232–233
  resonating pulse reactor, **30**:243–244
  rotating surfaces, **30**:241–242
  rough surfaces on heat exchangers, **30**:217,
      219–227
  sensible heat exchangers, **30**:213–216
  Spirelf® system, **30**:228–230
  stirring, **30**:239
  structured surfaces, **30**:234–235
  suction, **30**:244
  surface scraping, **30**:240
  vibration, **30**:242–243
Fouling thermal resistance, **31**:298, 302–304
Four-equation drift flux model, **31**:109–111
Fourier–Biot law, **11**:321–322, 352, 360, 362
Fourier equation, **25**:6
Fourier number, **1**:175; **9**:313; **15**:246
  coalesced bubble region, **20**:54
  Fo, **25**:17
  for gas film, **19**:143
  narrow space nucleate boiling, **20**:49
  in shear flow, **13**:221, 262
Fourier's law, **2**:434; **5**:135; **25**:12; **31**:23, 42, 61,
    65
  heat conduction in structure liquids and,
      18:163–164
Fourier transform, **8**:50
  heat- and mass-transfer, **20**:327
Four-stream heat exchanger, **26**:226
Fractional moisture saturation, **13**:143, 196
"Fracture" of a fluid, **2**:363
Fragmentation
  hydrodynamic, **29**:168–175
  melt fragmentation, **29**:270–274
  thermal, **29**:175–176
Fra Mauro crater, moon, **10**:5

Fraunhofer–Fresnel diffraction, **6**:184
Freckles, alloys, **28**:243–244
Fredholm integral equation, **8**:35
Fredholm integrodifferential equation, **19**:16
Free boundary flows, **9**:274
Free–bound radiation in a plasma, **4**:235ff
Free convection, **8**:11, 19, 53; **9**:32, 274; *see also*
    Convective heat transfer
  critical region, **7**:55–66
  cylinder, **7**:101
  electronic packages, **20**:243
  in electrostatic fields, **1**:286f
    in MHD problems, **1**:286ff
    thermal instability, **1**:287
  horizontal wires, **21**:42–44
  influence of, on rotating systems, **5**:141
  laminar heat transfer by, **11**:267–295
  nucleate boiling and, **20**:2–4
  plate
    horizontal, **7**:100
    vertical, **7**:98
  in reduced gravity, *see* Reduced gravity
  sphere, **7**:103
  turbulent heat transfer by, **11**:295–307
  vertical surfaces, **21**:37–42
    laminar flow, **21**:37–39
    turbulent flow, **21**:39–42
Free convection boundary layer, **11**:269
Free convection effects, **8**:127
Free convection heat transfer, **11**:306–307; *see
    also* Convective heat transfer
Free convection on vertical surfaces, **4**:1ff
  laminar flow, **4**:4ff
    circular cylinder, **4**:17ff
    flat plate, **4**:4ff
      differential equation method, **4**:5ff
      experimental work, **4**:11ff
      integral equation method, **4**:8ff
    Grashof number, very low, **4**:43ff
    nonsteady conditions, **4**:33ff
      experimental work, **4**:41ff
      negligible thermal capacity, **4**:33ff
      significant thermal capacity, **4**:39ff
    physical properties effects, **4**:28ff
      large temperature differences, **4**:29f
      near-critical conditions, **4**:30ff
      viscous dissipation, **4**:32f
    radiative effects, **4**:28
    surface temperature effects, **4**:22ff
      differential equation method, **4**:22ff

experimental work, **4**:26
integral equation method, **4**:24ff
vibration, effects of, **4**:47ff
laminar flow instability, **4**:51ff
experimental work, **4**:51f
theoretical work, **4**:52f
turbulence, **4**:50ff
experimental work, **4**:58ff
theoretical work, **4**:54ff
Free energy of formation, **9**:197
Free–free radiation in a plasma, **4**:237f
Free–free transitions, **5**:304
Free jet impingement
single-phase conditions, **20**:254–257
thermal cooling of electronic components,
20:251–252
Free-module conduction, *see* Conduction
Free molecule flows, **7**:164, 169, 184, 187
Free-rise velocity, **20**:107–108
Free-stream acceleration effects, **7**:18
Free-stream turbulence, *see also* Turbulence
influence of, on heat transfer in tube banks,
8:134
Free-stream turbulence level, influence of, on
heat transfer, **8**:122, 123
Free-stream velocity, **20**:376–377
Free-surface flow, defined, **13**:206
Free-surface liquid jet, **26**:106–107, 205
axisymmetric jet, **26**:109–119
stagnation zone, **26**:137–141
turbulent flow, **26**:153–154, 157
experimental studies, **26**:135–170
heat/mass transfer, **26**:147–162
hydraulic jump, **26**:114, 201–203, 205
local heat transfer, **26**:159–160, 167–169
planar jet, **26**:107–108, 119–123
flow regions, **26**:146–147, 167–170
stagnation zone, **26**:147, 162–167
radial flow region, **26**:141–144, 156–160
splattering, **26**:195–198
stagnation zone, **26**:137–141, 147, 162–167
theory, **26**:107–123
Free turbulence, **8**:291
Free volume model, thermal conductivity of
polymeric liquids, **18**:214–215
Freeze drying, **5**:325
Freezing, **8**:49; *see also* Solidification
conduction-controlled, **19**:58
conduction-dominated
approximate integral methods, **19**:34–39

asymptotic expansions, **19**:28–33
coordinate transformations, **19**:16–24
historical development, **19**:4–5
integrodifferential equations, **19**:14–16
other methods, **19**:39–44
power series expansions, **19**:12–14
similarity transformation, **19**:6–12
weak formulations and enthalpy method,
19:24–28
forced convection, **19**:73–78
assumptions, **19**:74
Graetz solution, **19**:74–75
ice band structure, **19**:75–77
of pipes convectively cooled from outside,
19:75–77
inside vertical cylinders, natural convection
and, **19**:62
inward, of spherical masses of liquid PCM,
19:31
natural convection, **19**:47–49, 60
fin effects on cylindrical geometries, **19**:57–
62
in rectangular geometries, **19**:67–68
in nature, **19**:1–2
one-dimensional, finite-difference method,
19:42
one-phase, power-series solution, **19**:20–21
process, parts of, **19**:30
in rectangular geometries, **19**:67–68
two-phase, natural convection effects, **19**:49
Freezing kinetics coefficient, **28**:31
Freon-II, film boiling of, **5**:88
Freon-12, transition boiling, **18**:260–263
Freon-113
frequency of liquid–wall contacts, **18**:281
mean liquid–heating surface contact time,
18:278
mean relative liquid–heating surface contact
time, **18**:283
mean vapor phase–heating surface contact
time, **18**:279
transition boiling, experimental studies,
18:260–263
Freon-114, transition boiling, **18**:260–263
Freons
nucleate boiling formulas, **20**:18–19
values for, **20**:22–23
Frequency, dominant, **9**:334
Frequency factor, unidirectional regenerator operation, **20**:150

Friction, **19**:248
    factor, **25**:295–299, 303, 306, 312; *see also*
        Coefficient
    frozen bed, **25**:219
    interparticle, **25**:181
    mutual, in helium II, **17**:91–97
    resistance, **25**:48
    wall shear (term), **31**:134
Friction coefficient, **24**:11, 14, 17
Friction drag
    fluid flow past a single tube, **18**:95–96
    fluid flow past a tube in a bank, **18**:106
Friction factor–Reynolds number product, *see*
        also Fanning friction factor; *and appropri-*
        ate ducts in question for locating tabular
        values
    comparisons for some ducts, **SUP**:392–399
    effective in a heat exchanger, **SUP**:406–407
    for fully developed flow, **SUP**:40
Friction factors, **2**:382; **5**:363; **9**:175; **21**:11–12,
        15; *see also* Fanning friction factor
    Bingham plastics, **23**:202
    Carreau fluids, **23**:211
    dimensionless groups to fluids, **19**:254–257
    Ellis model, **23**:199
    Grashof number/Reynolds number ratio,
        15:201–202
    packed beds, **23**:196
    power law fluids, **23**:196
    slug flow calculations, **20**:98, 114–115
    solvent effects on, **15**:96–97
    in turbulent film boiling, **5**:79
    turbulent heat transfer under cooling, **21**:36
    for various concentrations, **15**:94
    for viscoelastic fluids, **15**:90–96
Friction velocity, **8**:319
Fritz–Ende relation, **4**:178f
Froessling number, **3**:19f, 23f; **26**:152, 206
Front temperature, propagating, **24**:294–295
Front velocity shear, propagating, **24**:294
Frost, **5**:469; **19**:81
    thermal conductivity of, **5**:470
Frost formation, **5**:496
    in cryogenic systems, **5**:469
    inside of tubes, **5**:469
    on vertical surfaces, **5**:469
Froude number, **4**:53; **5**:467; **11**:135, 140–141;
        16:6, 51; **19**:99; **26**:152, 201–202, 206
    acoustic, **24**:312–313
    CHF in helium and, **17**:324, 326

of cryogens, **5**:468
    determination, **16**:17
    range, in entrainment modeling, **16**:51–52
Frozen flow, **2**:111
FTS experiments, **29**:165
Fuelbed
    energy balance in open fires, **17**:202–203
    radiation from, heat transfer to pan and,
        17:216–220
Fuel coolant interactions, **29**:129, 195
Fuel interfacial area transport equation,
        ESPROSE.m field model, **29**:198
Fuel rods, **SUP**:354
Fuels
    economy, theoretical, **17**:275–283
    ignition of, **10**:235
Fuel spray
    slurry fuel droplet, vaporization, **26**:81–82
    stagnant medium, evaporation, **26**:54–57
    vaporization, **26**:53–54, 83–84
Fully developed flow, *see also appropriate ducts*
    comparison of solutions for, **SUP**:392–397
    comparison of some methods for, **SUP**:276
    definition of, **SUP**:7
    hydrodynamically, **SUP**:5
    single-screw extruder, **28**:160–161
    solution methods for, **SUP**:61–68
    summary index of solutions for, **SUP**:387–388
Fully developed heat transfer condition, **3**:133
Fully developed temperature field, **13**:237
    in shear flow, **13**:258–259
FUMO code, direct containment heating, **29**:234
Function
    orthogonal, **24**:81–83
    trial, **24**:47, 50, 60, 66, 71, 79–84
Fundamental band, **8**:231
    in molecular gas band radiation,
        12:144–145
Fundamental boundary conditions, **SUP**:31–34;
        see also Concentric annular ducts; Eccentric
        annular ducts
Furnaces
    boiler furnaces, **27**:158–167
    gas-fired furnaces, **27**:196–199
    gas reformers, **27**:167–173
    radiative–convective heat transfer, **27**:128,
        129, 144, 164–166
    with throughflow and heat-generating region,
        27:154–157
Fused salts, transport coefficients for, **12**:238

Future research, areas of, **SUP**:419–420
FVM, *see* Finite volume model

G

Galerkin coefficients, **26**:18
Galerkin–Kantorowich variational method,
   SUP:75, 103, 121, 142, 151, 219, 252
Galerkin method, **1**:103ff; **26**:17
   heat- and mass-transfer
      natural convection, **20**:324–325
      property variations, **20**:326
Galerkin–Zhukhovitskii variational method, **9**:76
Gallium antimonide, properties, **30**:321
Gallium arsenide
   chemical vapor deposition, **28**:373–375, 385,
      390
   crystal growth, **30**:361, 414
   properties, **30**:321, 322, 346
   transition temperatures to transparency, **30**:340
Gallium drops, fragmentation, **29**:172–173
Gallium nitride, properties, **30**:321
Gallium phosphide, properties, **30**:321
Gas absorption
   radiative heat transfer analysis, **27**:93–95
   scattering and, **27**:92–99
   simulation, **27**:50–56
Gas absorption coefficient, **27**:13–14, 52, 64, 95,
      161
   in a flame, **27**:124, 125
   monochromatic, **27**:183
   nonuniform, **27**:81–85
   uniform, **27**:62–81
Gas absorptivity, in molecular gas band radia-
      tion, **12**:131
Gas blowthrough, **29**:248–249, 253–256
Gas bubble
   condensation in immiscible liquid, **26**:47–48
   entrainment requirements, **21**:289
Gas bubble mixing, **31**:159, 161
Gas chromatography, **10**:228–229
Gas compression work, **SUP**:13
Gas convection
   in fluidized bed, **14**:158–160
   fluid particle systems subgroup IIA, **19**:113
Gas core nuclear propulsion system, **5**:36
Gas distributor plates, **19**:178–180
Gas elements
   emission, **27**:56–59

heat balance equations, **27**:47
simulating radiative heat transfer absorption,
   27:50–56, 62–66, 71–74, 80–85
Gaseous flames, speckle photography, **30**:299–
   302
Gaseous radiation, physical processes in, **5**:263
Gases
   DCH debris, interaction, **29**:284–291
   in fluidized bed, **10**:167–168
   forced convection, **30**:20–24
   heat transfer in, **14**:107
   heat transfer microscales, **21**:267–268
   incompressible, energy generated by viscous
      dissipation, **21**:252–253
   noncondensable, condensation heat transfer
      effect, **21**:107–108
   spectral behavior of, **31**:388–390
   stagnant, entropy production, **21**:249–254
Gas-filled cavities, nucleation and, **10**:120–121
Gas film, steady state conduction across,
   10:181–188
Gas-fired furnaces, **27**:196–199
Gas flows, **8**:94
   heat transfer, **18**:115–118
Gas fluidization, **19**:97
Gas-fluidized beds, **10**:168
   gas convection, **14**:158–160
   heat transfer between unused tubes and,
      14:149–328
   radiation and, **14**:154–158
Gas holdup, three-phase systems, **23**:255
Gas injection, **3**:74f
Gas jet, **26**:106, 205; *see also* Array of round
      nozzles; Array of slot nozzles
   heat and mass transfer related to, **13**:1–58
   impinging, *see* Impinging flow
Gas–liquid heat exchanger, *see also* Heat
      exchangers
   bath temperature and, **15**:34–35
Gas–liquid interface, **31**:26–28, 55–56, 66–68
   mass balance equation, **20**:87
   Maxwell stress, **21**:127–128
Gas–liquid slug flow, velocity profiles, **20**:112–
   113
Gas–liquid systems, hydrodynamic fragmenta-
      tion, **29**:169–170
Gas–liquid two-phase flow simulation, model
      selection for, **31**:108
Gas mass velocity, **9**:121
Gas mixture, heat capacity of, **13**:163

gas–particle mixture radiation
  attenuation constant, **27**:17
  scattering, **27**:93–99
  scattering albedo, **27**:17–18
  scattering phase function, **27**:18
Gas phase
  arbitrary curve in, **13**:171
  closure for, **31**:57–59, 65
  convective transport in, **13**:169–175
  heat transfer process in, **31**:46–48
  mass transport in, **13**:165–175
  moisture in, **13**:142
  point equations for, **31**:24–25
  volume-averaged thermal energy equations
    for, **31**:54
  volume-averaged transport equations for,
    31:30–39
Gas-phase continuity equation, **13**:147; **31**:46, 71
  for water, **31**:31–39
Gas-phase diffusion, **31**:70–71, 75, 77, 78
  closure problem with, **31**:68–72
  passive problem with, **31**:80
Gas-phase diffusion equation, **13**:166–169
Gas-phase mass transfer, **31**:85–89
Gas-phase species continuity equation, **13**:149
Gas-phase thermal equation, **31**:47
Gas-phase transport, **31**:37
Gas-phase velocity field, **13**:170
Gas pores, alloy solidification, **28**:244–245, 327
Gas radiation, **27**:13; *see also* Molecular gas
    band radiation
  absorption coefficient, **27**:13–14, 52, 64, 95,
    161
  in a flame, **27**:124, 125
  monochromatic, **27**:183
  nonuniform, **27**:81–85
  uniform, **27**:62–81
  directional absorptivity, **27**:15
  directional emissivity, **27**:14
  gas volume and solid walls
    heat balance equations, **27**:46–49
    Monte Carlo simulation, **27**:49–85
    READ method, **27**:86–90, 107–146
  gray gas, **27**:13
  from isothermal gas volume, **27**:15–16
Gas reformer furnace, **27**:167–173
Gas–solid fluidization
  advantages, **25**:151
  temperature distribution, **25**:151
Gas–solid heat transfer, **10**:175

Gas–solid mixing, **25**:151
Gas temperature, blackbody intensity for, **12**:130
Gas-to-gas film cooling, **7**:324, 351
Gas transmissivity, in molecular gas band radia-
    tion, **12**:131
Gas turbine cooling, **5**:199
Gas turbine engines, heat flux measurements,
    23:344–346
gas–wall system
  heat balance equations, **27**:46–49
  Monte Carlo simulations
    emission from gas volume, **27**:56–59
    emission from solid walls, **27**:59–61
    gas absorption, **27**:56–59
    reflection and absorption by solid walls,
      27:61–85
  READ method, **27**:86–90, 107–146
Gaunt factor, **5**:303
Gaussian beam profile, **22**:274
Gaussian distribution, **8**:335
GEA air-cooled condenser system, **12**:3, 14
Generalized Newtonian fluids, **13**:211; **28**:152–
    153
  fluidized beds, **23**:232
  hindered settling, **23**:261
  packed beds, **23**:193–194
Generalized solutions
  for conjugated problem, **SUP**:59–60
  for conventional convection problem,
    SUP:57–59
  for velocity problem, **SUP**:44
General methods for solutions, *see* Methods for
    solutions
General transport equation, **12**:231–244
  dilute solutions and, **12**:234–237
  eddy diffusivity in, **12**:237–239
  for fused salts, **12**:238
  mass system and, **12**:231–232
  mixed system and, **12**:233–234
  molar system and, **12**:232
  number of transport coefficients in, **12**:237–
    241
Generation of heat internal to a slab, **1**:60f, 90ff
Generator (pump) coefficient, **1**:280
GENMIX computer program, **13**:72–78
  for air boundary layer above lake, **13**:81
  for Ekman layer problem, **13**:94–95
  for hyperbolic and partially parabolic layers,
    13:106–110
  for lake surface boundary layer, **13**:84

for radial pool, **13**:111–112
in THIRBLE problems, **13**:69, 72
for unsteady layers, **13**:112
Geometric albedo, lunar, **10**:13–14
Geometric conservation law, **24**:206
Geometric modeling, **31**:336–364
effects of, **31**:358–364
Geometric-optics scattering, **23**:147–151
Geometric scale factor, **4**:171ff
Geometries
steady-state heat transfer in helium II
peak heat flow values, **17**:115–118
recovery heat flux and film boiling heat
transfer, **17**:118–119
transient heat transfer in helium II and,
17:123–124
Geometry
annulus, **25**:379–380
backward facing step, **25**:324, 354, 358, 391–
394
complex, **25**:388–396, 398
corrugated duct, **25**:392, 394–395
impinging jet, **25**:394–396
nonperiodic, **25**:405–406
obstacle, **25**:394
plane channel, **25**:324–325, 348, 350, 354,
357, 378–388
transverse rib, **25**:388–391
Geophysical applications, **8**:66
Geophysical phase-change problems, **19**:82
Geophysical phenomena, **5**:131
Geopressured geothermal systems, heat transfer
in, **14**:88–91
Geothermal energy recovery processes, direct
contact condensation in, **15**:272–273
Geothermal power, **9**:63
Geothermal reservoir
combined free and forced convection in, **14**:49
isothermal heating, **14**:20
liquid-dominated axisymmetric type, **14**:87
need for, **14**:2
oscillating solution for, **14**:45
transient pressure drawdown in, **14**:87
withdrawal and injection of fluids in, **14**:83–84
Geothermal systems
classification of, **14**:2
convection heat transfer in, **14**:3–15
geopressured, **14**:88–91
heat transfer in, **14**:1–99
hot water, *see* Hot water geothermal systems

lumped parameter analysis in, **14**:91–96
multiphase flow, numerical simulations,
30:140–141
physical modeling in, **14**:77
two-phase flow, **30**:118–119
two-phase heat transfer processes in, **14**:88
water-steam type, **14**:77–78
Geothermal wellbore, heat transfer in, **14**:98–99
Geothermal wells, drawdown and buildup data
for, **14**:80–81
Germanium crystals
by Czochralski process, **30**:313
properties near melting point, **30**:346
transition temperatures to transparency,
30:340
Gibbs equation, **11**:10
Gibbs free energy function, **10**:90; **12**:261;
14:288–291; **21**:244
Gibbs potential, **16**:67
Gibbs thermodynamic relation, **21**:248, 271
Gladstone–Dale equation, **6**:276, 287
Glass
heat transfer in, **11**:326
transient temperature distribution in, **11**:410
Glass beads porous layer heat transfer using,
14:71–72
Glass lubrication, **24**:26–29
Glass melting tanks, heat transfer in, **11**:420
Glass microspheres, **9**:133
Glass particles, **9**:119
Global balance equations, for thermohydraulic
systems, **31**:145–152
Global coupling, between mass and momentum
balances, **31**:120
Global force balance, slug pressure drop, **20**:92–
95
Global heat transfer, in Czochralski process,
30:403–406
Globally shared memory, **31**:404
Global momentum
balance slug flow modeling, **20**:114
slug pressure drop, **20**:94–95
Global phase change, scaling of, **31**:150–151
Global pressure, multiphase mixture, **30**:108
Global scaling criteria, for integral systems,
31:149–152
GMF systems, **27**:161, 205
Gold, laser vaporization, **28**:129–131
Gold foil, thermal contact conductance, **20**:264–
265

Gold-plated surfaces, dropwise condensation, 21:103
Goodman's solution
    applied to semiinfinite slab, 1:57ff
    to slab of finite thickness, 1:62ff
Gouy–Stodola theorem, 15:25, 35, 47, 51
Governing differential equation, 31:70, 71
Governing point equations, in mass/energy transport, 13:128–133
Gradient, applied magnetic field, 25:182
Gradient of the average, 31:13
Graetz method, 19:282
Graetz number, 2:368ff; 13:221, 228, 262; SUP:50, 101; 15:102, 233–234; 19:255
    defined, 13:234
Graetz–Nusselt problem, SUP:100, 169
Graetz problem, 3:107; 8:22; SUP:99–105; see also Graetz–Nusselt problem
    extensions of, SUP:109–120
    with finite viscous dissipation, SUP:109–111
    with fluid axial heat conduction, SUP:111–118
    for non-Newtonian fluids, SUP:109
Graetz series, SUP:100
Graphite, see Surface combustion of
Grashof number, 1:24, 40, 280; 2:372; 3:2; 4:3ff, 151; 5:375; 7:15, 64; 9:288, 313, 324; 11:118, 200, 244–246, 251; 14:117; SUP:410; 15:157, 167, 192, 239; 16:331–334; 19:61, 62, 256; 21:280; 24:216, 217; 25:267
    defined, 15:148, 154
    local, 9:324
    natural convection cooling, 20:203–204
    nucleate boiling, 20:2–4
    temperature-gradient, 9:303
Grashof number/Reynolds number ratio, 15:198
    friction factor and, 15:201–202
    Nusselt number and, 15:204
Grates, open fires and, 17:211–214, 229, 231–232
Gravitational oscillations, 5:129
Gravitational potential, bed, 25:224
Gravitation effects, 5:457
    on boiling of freon, 5:113, 462
    on boiling of liquid nitrogen, 5:460
    in liquid phase, 13:191
    on nucleate and film boiling, 5:459
Gravity, multiphase flow and, 30:93–94
Gravity override, 30:93, 154

Gray diffuse enclosure, 2:401, 407ff
Gray diffuse surface, 5:17
Gray gas, 5:28; 8:259; 27:13
    myth of, 12:116–119
    vs. nongray gas, 12:118
Gray gas approximation, see Radiation–convection
Gray material
    radiative heat transfer in, 11:395–400
    transient temperature distribution in, 11:408
Gray medium, 1:2
Gray surfaces, 2:401; 27:11, 12, 61, 147
Greases, thermal conductivity, 18:211–213
Green's formula, 19:239
Green's function, 8:53; 19:14, 15
Grid
    adaptive, 24:213
    composite, 24:217
    multigrid, 24:254
    staggered, 24:194, 207, 210
Grid size, effects of, 31:358–359
Griffith number, in steady shear flow, 13:234
Grösste Nutzarbeit, defined, 15:8
Groundwater contamination, by nonaqueous phase liquids, 30:174–183
Groundwater remediation
    interfacial chemical nonequilibrium, 30:111
    multiphase flow effect, 30:152
Group III–V crystals, growth by Czochralski process, 30:319–325
Gr/Re, see Grashof number/Reynolds number ratio
Gukhman number, 1:124, 133ff, 140f
Gulf General Atomic, 12:7

H

Hadley–Apennine lunar region, 10:6
Hadley regime, 5:217
Hadley Rille, 10:2
Hagen–Rubens law, 23:143
Hagen–Rubens relation, 9:363
Hall current, 1:272
Hall effect, 1:272
Hall parameter, 4:308
Hampson heat exchanger, 26:226, 227, 252, 254
Hankel transform, 22:258
HARDCORE model, direct containment heating, 29:280

Hardness ratio, thermal contact conductance, 20:266

Harmonic equation, 8:42

Harmonic oscillator, 5:269; 8:232
  vibrational nonequilibrium in, 8:248

Harsh environments, heat pipe thermal control, 20:297

Harten's TVD scheme, 24:240, 241, 242, 243, 244, 245, 246

Hartmann number, 1:281, 288

Hartmann problem, 1:300ff
  basic equations, 1:300ff
  flow characteristics, 1:304f
  heat transfer characteristics, 1:305f
    convection, 1:306
    heat transfer, 1:309f
    internal heat distribution, 1:306ff
    induced magnetic field, 1:302ff

Hartree flow, 8:21

Hawaii, magma energy in, 14:97

H boundary condition, 19:295

H1 boundary condition
  definition, 19:253, 254
  dimensionless groups, 19:255
  Newtonian fluids
    simultaneously developing flows, 19:300, 301
    thermal entrance length, 19:298–299
  thermally developed laminar flow through rectangular channels, 19:278, 279–280

H2 boundary condition
  definition, 19:253, 254
  dimensionless groups, 19:255
  thermally developed laminar flow through rectangular channels, 19:278, 280

Heat, 22:4, 5
  boiling, 25:259
  capacity, 22:296; 25:312
  convective, 25:273, 311
    forced, 25:270, 271
    free, 25:271, 279, 286
  eddy diffusivity and, 15:124, 130
  exchangers, 25:252
  fluid mechanics and, 15:60–61
  flux, 25:111, 286, 290
    density, 25:6, 8, 19, 27, 61, 62, 112
  history, 22:1
  hot-gas–cold-tube, 25:97
  internal, generation, 22:158
  inter-phase, 25:272

laminar natural convection, 25:281

leakage, 25:103, 111

model, see Model

overflow, 25:111

of reaction, 25:152

release, 25:33, 43

transfer, 22:1–18
  conjugate unsteady, 25:31
  convective turbulent, 25:18
  countercurrent, see Countercurrent heat transfer
  high temperature, 22:222
  low temperature
    without phase change, 22:158
    with phase change, 22:183
  whole body, 22:134
  unsteady, 25:43, 291
  gas-to-liquid, 25:72

Heat, transfer to pan
  calculated efficiencies, 17:223–226
  in closed stoves, 17:254–263
  convective transfer, 17:220–223
  measurement of, 17:226–227
  radiation from fuelbed, 17:216–217
    emission of soot particles, 17:218
    emissivity of gases, 17:218
    radiation from flames, 17:218
    total emissivity of flames, 17:219–220

Heat and mass flows, calculations of, 1:171ff

Heat- and mass-transfer, see also External heat and mass transfer experiments on porous bodies; Internal heat and mass transfer in porous bodies
  in capillary-porous bodies, 1:123ff
  with filtration, 1:173
  natural convection
    horizontal direction, 20:330–343
      boundary-layer flows, 20:332–336
      convection onset, 20:330–332
      enclosed porous layers, 20:336–341
      transient approach to equilibrium, 20:341–343
    horizontal line source, 20:346–347
    physical model, 20:316–318
    point sources, 20:343–346
    vertical direction, 20:318–330
      finite-amplitude convection, 20:327
      high Rayleigh number convection, 20:329–330
      nonlinear initial profiles, 20:323–325

Heat- and mass-transfer (*continued*)
    onset of convection, **20**:318–323
    soret diffusion, **20**:328–329
Heat- and mass-transfer problems
    analytic transformations by computer, **19**:230–233
    estimation of approximations using convolution-tube function, **19**:233–241
    quantitative analytical methods, **19**:192–194
    solution methods, **19**:194–197
        approximate analytical, **19**:194
        physical and analog modeling, **19**:194
    temperature field determination in infinite plate, **19**:210–223
        one-dimensional, **19**:223–226
    temperature field of complex-shaped bodies, **19**:226–230
    unsteady-state, variational solution methods, **19**:197–202
    variation description with variable thermophysical characteristics, **19**:203–210
Heat balance
    equations, **27**:46–49, 97–98
    in fluidized beds, **10**:212–213
    Monte Carlo method, **27**:151–154
    in woodburning stoves, **17**:289–290
        examples of, **17**:290–293
        techniques for measurement of, **17**:293–296
Heat-balance integral, **1**:53; **19**:34, 195
Heat capacity
    in drying process, **13**:155
    of suspensions, **9**:114
Heat capacity function, **19**:147
Heat capacity rate ratio, three-fluid heat exchanger, **26**:268
Heat capacity ratio, direct containment heating, **29**:235–236
Heat carrier pulsations, **31**:162
Heat conduction, **8**:23, 29, 45, 48, 66, 67, 77; **24**:39; **31**:75–76, 77–79
    axial, *see also* Conjugated problems
        in wall, **SUP**:30–31
        within fluid, *see* Fluid axial heat conduction
    basic principles, **18**:162–164
    in bilayer sphere, **15**:285
    classical theory of, **11**:321
    in cylinders, **24**:58, 66
    equation, hyperbolic, **24**:42
    finite difference method, **5**:340
    Fourier–Biot law of, **11**:321–322, 352, 360, 362

Fourier law of, **10**:39
    of a highly rarefied gas, **2**:276ff
    linear, **24**:41, 42, 49, 53, 60, 67, 79, 90
    nonlinear, **24**:43–55, 71, 74, 76, 88, 94
    in plates, **24**:58, 74, 76, 78, 84, 88, 90
    in prismatic bar, **8**:49
    in slab with variable properties, **5**:337
    solution by Wiener–Hopf method, **8**:49, 50
    in spheres, **24**:59, 71
    transient, **24**:39, 41
    two-phase system, **23**:372
    unsteady, **8**:57
        through single particle, **10**:185
Heat conduction equation, phase change, **28**:78–79
Heat conduction parameter, peripheral, **SUP**:28
Heat conduction problem
    with initial temperature distribution, **1**:67f
    nonlinear, **19**:193–194
Heat conduction term, **31**:42, 47
Heat conductivity, **4**:233
    of rocks and soils, **8**:45
Heat diffusion, nonlinear, **31**:138
Heated horizontal surfaces, free convection about, **14**:63
Heated plasma, entropy production, **21**:252–253
Heated ribbon gage, **23**:336–337
Heated surfaces, porous layer adjacent to, **14**:56
Heated vertical pipes, flow stability of non-Newtonian fluids in, **15**:205–206
Heater
    effects, on nucleate pool boiling heat transfer, **16**:187–192, 212–219
    material
        effect on heat flow rate in nucleate pool boiling, **16**:169
        effects on heat flow rates, **16**:190, 218
        roughness, effects on heat flow rates, **16**:187–188, 213–217
    shape
        effect on heat flow rate in nucleate pool boiling, **16**:169
        effects on heat flow rates, **16**:188–190, 217
    shape and material, effects on heat flow rates, **16**:190–192, 218–219
    types, **16**:170
Heat exchange, film boiling and, **11**:51
Heat exchanger design, entropy generation minimization in, **15**:25–38
Heat exchanger duct, lowest entropy generation rate in, **15**:33

Heat exchanger fins, **9**:3, 76
Heat exchanger geometry, for minimum irreversibility, **15**:29–33
Heat Exchanger Institute (HEI), **31**:465–467
Heat exchangers, **8**:144, 150, 157; *see also* Compact heat exchangers; Liquid metal heat transfer; *and specific heat exchangers*
  base tube bank, **12**:49
  counterflow, **15**:26, 28
  cryogenics, **26**:220, 222, 252, 305
  design conditions of, **31**:450–452
  for dry cooling tower, **12**:39–52
  effectiveness, **26**:290–304
  evaporative, **31**:271
  Forgó type, **12**:49–50
  gas-liquid, **15**:34–35
  heat transfer, **26**:219
  irreversibility of, **15**:29, 52
  manufacturing methods of, **31**:162, 166
  natural draft for, **12**:57–62
  optimization of, **15**:33
  optimum cooling and spatial positioning of, **15**:44
  for oxygen, **5**:364
  plate-finned, **31**:177
  plate fin type, **12**:45–48
  reducing mass and size of, **31**:159
  for Saturn rockets, **5**:363
  shallow, **12**:44
  solutions for, **12**:56–61
  spine fin type, **12**:48
  with zero pressure drop, **15**:27–29
Heat exchanger surface, optimum duct geometry and, **15**:32–33
Heat-exchanger system
  "boxed" regenerators
    cyclic steady state, **20**:147–148
    Nusselt assumptions and governing equations, **20**:141–144
    overall balance equation, **20**:144–146
    overall heat-transfer coefficient, **20**:148–149
    weak and strong periods, **20**:140–141
Heat exchange surface, **10**:184–185
Heat flow, **21**:241–242
Heat flux, **21**:248; **23**:285–286; **27**:25; *see also* Critical heat flux
  in chemical mixtures, **2**:115f
  circuit integration, **20**:181
  coalesced bubble region, **20**:49, 52
  convective, **21**:260

dimensionless, **SUP**:49
dropwise condensation, **21**:87
equation, **24**:42, 45, 90, 94
equation for, **10**:47
in evaporation, **16**:148
Fourier's law and, **10**:46
heat-transfer coefficient and, **20**:77–78
local heat flux, **26**:34
low liquid level nucleate boiling, **20**:56–57
in MHD, **1**:277, 283
narrow space nucleate boiling
  clearance and pressure, **20**:47–79
  emission frequency, **20**:43–45
  isolated bubble region, **20**:40–41
nucleate boiling, **20**:5–8
prescribed, **24**:302
radiative, *see* Radiative heat flux
turbulent, **7**:22, 52
wall
  nonuniform, **3**:131ff
  uniform, **3**:127ff, 138
wall heat flux, **27**:98, 143, 152, 154, 157
wall jet zone, **26**:114–118
Heat flux field vector, **19**:197
Heat flux gage, **23**:283–284; *see also* Layered gage
  plug-type, **23**:311
  thin-film, **23**:293–294, 315–321, 337
    calibration, **23**:320
    data analysis, **23**:318
    heated, **23**:338–340
    semi-infinite geometry, **23**:317
Heat-flux heating, constant, **19**:56
Heat flux measurements, **23**:279–353
  active heating, **23**:326–342
    constant heat flux, **23**:326–331
    constant surface temperature, **23**:331–342
  calibration, **23**:342–343
  categories, **23**:287
  combustions, **23**:352–353
  convection boundary layer, **23**:284–285
  in experimental transition boiling studies, **18**:257
  heat flux gage, *see* Heat flux gage
  high-temperature, **23**:351–352
  issues, **23**:281–287
  overview, **23**:279–281
  spatially distributed measurements, application, **23**:349–351
  spatial temperature difference, **23**:287–307
    circular foil gage, **23**:300–306

Heat flux measurements (*continued*)
  in-depth temperature, **23**:298–299
  layered gage, **23**:288–298
  radiometer, **23**:306–307
  wire-wound gage, **23**:299–300
  surface energy balance, **23**:281–282
  surface thermal boundary condition, **23**:282–283
  temperature change with time, **23**:307–326
    coaxial thermocouple, **23**:313–314
    null-point calorimeter, **23**:311–313
    optical method, **23**:321–326
    slug calorimeter, **23**:307–311
    thin-film method, **23**:315–321
    thin-skin method, **23**:314–315
  terminology, **23**:281
  time-resolved measurements, **23**:343–349
    basic research, **23**:346–349
    gas turbine engines, **23**:344–346
    reciprocating engines, **23**:343–344
    turbulence at stagnation point, **23**:347–348
Heat flux measurement techniques, **2**:55f
Heat flux microsensor, **23**:296–297
Heat generation by electromagnetic fields, **1**:349
Heating
  frictional drag and, **18**:96
  isothermal, **19**:36
  linear, thermal conductivity measurement in structured liquids, **18**:176
  liquid, around liquid–heating surface contact, **18**:250–252
  methods, in experimental transition boiling studies, **18**:254–257
  spontaneous, **10**:236–237
Heating parameter, **26**:82
Heating surface
  liquid contacts, *see* Liquid–heating surface contacts
  material, effect on transition boiling heat transfer, **18**:293–295
  nonisothermality, effect on transition boiling heat transfer, **18**:301–302
  orientation effects
    on transition boiling heat transfer, **18**:303
    on transition boiling mechanisms, **18**:282–285
  roughness effects
    on frequency and liquid–wall contact time with ethanol boiling, **18**:285
    on transition boiling, **18**:273–274

  on transition boiling heat transfer, **18**:290–292
  sections, physical mechanisms acting at, **18**:253–254
  temperature oscillations
    with mean subcooling, **18**:275
    with saturated liquid boiling and small subcooling, **18**:275
    with strongly subcooled liquid boiling, **18**:276
  thermal activity, effect on transition boiling heat transfer, **18**:296–298
  wettability effects on transition boiling heat transfer, **18**:264–265, 292–293
  wetted part, zones in transition boiling region, **18**:243
Heat kicker, indirect cooling techniques, **20**:277–278
Heat loss coefficients, overall
  active solar collectors, **18**:9–10
  compound parabolic collectors, **18**:25
  thermal network for, **18**:8
Heat of vaporization, **31**:56, 64
Heat pipes, **7**:219; **9**:3
  ammonia filled, **7**:320a
  applications, **7**:273
  control, **7**:276
  cryogenic, **7**:320a
  definition of, **7**:220
  description of, **7**:220
  electrohydrodynamic, **14**:139–140
  flexible type, **7**:320b
  fluids for, **7**:235
  functions of, **7**:224
  indirect cooling techniques
    bellows concept, **20**:275–277
    cylindrical structure, **20**:274
    finned heat pipes, **20**:280–281
    flat pipes, **20**:278–280
    flat-plate heat pipe, **20**:287–289
    heat kicker, **20**:277–278
    high-capacity semiconductors, **20**:278
    micro-heat-pipe concept, **20**:283–285
    multiple array cooling, **20**:285–286
    printed wiring boards, **20**:286–287
    self-regulating evaporative-conductive link, **20**:283
    single- and double-pipe configurations, **20**:281–283
    thermal switch, **20**:274–275

trapezoidal configuration, **20**:283–284
lithium-filled, **7**:320c
magnetic field effects on, **7**:320b
material test for, **7**:235–249
  compatibility of components, **7**:246
  life tests of components, **7**:246
  wicks, **7**:236
  working fluid, **7**:235
mercury-filled, **7**:320a
operating characteristics of, **7**:249–273
  basic studies, **7**:270
  heat transfer limit investigations, **7**:250
phenomenology, **7**:220
rotating type, **7**:223
sodium-filled, **7**:320c
surface tension effects, **7**:320c
surveys, **7**:234, 320d
theory, **7**:278, 320d
thermal control, **20**:272–299
  direct cooling, **20**:289–295
  indirect cooling, **20**:273–289
  operation schematic, **20**:272
  system-level control, **20**:295–297
threaded wall-artery wick, **7**:320d
transient behavior of, **7**:320e
types of, **7**:220
vibrational environment effect on, **7**:320b
wicks of, **7**:221, 236
Heat radiation, from gases, **12**:116; *see also*
  Molecular gas band radiation
Heat rejection
  methods of, **12**:10–15
  radiation in, **12**:12
Heat removal, by fluidized particle absorption,
  14:153
Heat removal rate, nuclear reactors, **29**:1, 2
Heatron Thermek spined tubes, **14**:198
Heat shields, **8**:11
Heat sinks
  lakes and oceans as, **12**:10
  for power plants, **12**:4–29
  in thermal power generation, **12**:2
Heat sink temperature, power generation cost
  and, **12**:16–17
Heat transfer, **25**:151, 252, 258, 259, 261, 274–
  278, 281–285, 288, 289, 291, 296, 308–
  312; *see also* Convection; Convective heat
  transfer; Energy transfer; Film boiling heat
  transfer; Free convection heat transfer; Heat
  conduction; Radiative heat transfer; Transi-

ent heat transfer; *and appropriate ducts*
age–density distribution function and, **10**:189–
  190
alloy solidification, **28**:238
analysis of, **12**:88–100
in annular-dispersed flows, **1**:402ff
in annular turbulator-provided tubes, calculation of, **31**:313–320
augmented nucleate boiling, surface configuration, **20**:70–77
availability in, **15**:7–8
average atmosphere temperature in, **12**:25
between bed and immersed surface, **25**:151
between bed and wall, **25**:151
blowing parameter, **16**:250–251
in blunt bodies, **1**:331ff, 347f
on bodies of revolution, **5**:137
between boiler and boiler tubes, **14**:150–158
in boiling, **11**:2
boiling, **14**:127–139
Boltzmann transformation, **30**:120
bubbling bed model of, **10**:176–177
by conduction in structured liquids, *see*
  Thermal conductivity, of structured liquids
by convection in structured liquids, **18**:167
chemical vapor deposition, **28**:402–403
closed system in, **15**:9
coeffcient
  convective, **16**:144
  nucleate boiling, prediction, **16**:117–127
coefficient of, **31**:208, 257
comparisons of solutions, **SUP**:394–397, 399–
  405
comparison with experimental data, **12**:100–
  109
condensation, **31**:275–294
conduction vs. radiation in, **11**:320–324
conductive vs. viscous effects in, **15**:12–13
constant heat flux mode of, **12**:78, 100
constant wall temperature mode of, **12**:78, 100
convection, **SUP**:6
convective, **11**:199–251; **25**:4; *see also* Convective heat transfer
coupling with fluid friction irreversibilities,
  15:29
from cylinders in crossflow, **11**:231–232
in Czochralski system, **30**:338–344, 403–406
degradation effect on, **15**:118
dimensionless groups for, **SUP**:46–57

Heat transfer (*continued*)
　　direct containment heating, **29**:215–330
　　downflow with side mixing in, **10**:189
　　dryout heat flux, **29**:34–53
　　during crystal formation, **11**:411
　　and early flow turbulization, **31**:177
　　eddy diffusivities in, **12**:89
　　effects of roughness and transpiration,
　　　　16:300–301
　　EHD-coupled condensation, **14**:137
　　electric fields and, **14**:107
　　electrohydrodynamically coupled, **14**:124, 142
　　electrohydrodynamically induced, **14**:107–143
　　enhanced, **24**:306; *see also* Heat transfer
　　　　enhancement
　　and entrance region in constant flux mode,
　　　　12:95–98
　　and entrance region in constant temperature
　　　　mode, **12**:98–100
　　entropy production, **21**:259–260
　　experimental correlations in, **10**:173–175
　　experimental measurement techniques in,
　　　　10:172–173
　　experimental studies of, **12**:103–104
　　external walls in, **10**:184–185
　　extrusion of non-Newtonian materials,
　　　　28:145–151, 220–226
　　　　chemical reaction and conversion, **28**:218–
　　　　　　220
　　　　combined heat and mass transfer, **28**:213–
　　　　　　220
　　　　material characterization, **28**:152–155
　　　　moisture transport, **28**:213–218
　　　　single-screw extruder, **28**:147, 149, 155–
　　　　　　187
　　　　twin-screw extruder, **28**:151, 187–201,
　　　　　　225–226
　　in film boiling, at Liedenfrost point, **16**:133
　　in fire research, **10**:219–280
　　first studies of, **SUP**:2
　　in flows with drag reduction, **12**:77–110
　　in fluidized beds, **10**:167–213
　　fluidized beds, **23**:251
　　between fluidized beds and surface, **10**:180–
　　　　213
　　gas convective, **14**:158–160
　　in gases, **14**:107
　　gas–solid, **10**:175
　　generalized correlation, **20**:11–13
　　generalized solutions for, **SUP**:57–60

　　in geopressured geothermal systems, **14**:88–91
　　in geothermal wellbore, **14**:98–99
　　in glass melting tanks, **11**:420
　　global heat transfer in Czochralski process,
　　　　30:403–406
　　heat exchanger, **26**:219
　　heat exchanger irreversibility figure in, **15**:29–
　　　　33, 52
　　in helium I
　　　　high heat flux, **17**:137–138
　　　　　　steady-state heat transfer, **17**:138–148
　　　　　　transient heat transfer, **17**:148–153
　　　　low heat flux, **17**:124–126
　　　　　　instability and critical Rayleigh numbers,
　　　　　　　　17:129–130
　　　　　　observations on heat transport in cryo-
　　　　　　　　genic Bénard cells, **17**:130–137
　　　　　　thermophysical properties of helium I
　　　　　　　　and helium gas, **17**:126–129
　　in helium II
　　　　high heat flux, **17**:103
　　　　　　steady-state heat transfer, **17**:104–119
　　　　　　transient heat transfer, **17**:119–124
　　　　low heat flux
　　　　　　dissipation in helium II: quantized vor-
　　　　　　　　tices, **17**:85–91
　　　　　　experimental background of superfluid-
　　　　　　　　ity: two-fluid theory, **17**:75–85
　　　　　　mutual friction in helium II, **17**:91–97
　　　　　　sources of data for helium II, **17**:102–103
　　　　　　summary of properties of quantum tur-
　　　　　　　　bulence, **17**:97–101
　　in hot-water geothermal systems, **14**:15–77
　　hydraulic resistance and, **11**:170–187
　　hydrogen combustion, **29**:59–123
　　impinging gas jets in, **13**:1–58
　　interfacial thermal equilibrium, **30**:110
　　irreversibility in, **15**:4–10, 29–33
　　laminar boundary layer thickness and, **10**:183
　　laminar flow and, **15**:98–104; **30**:50–51
　　laminar thermal convection and, **15**:151–165
　　in laminated media, **15**:286–287
　　laser speckle photography, **30**:267–276
　　liquid-encapsulated Czochralski process,
　　　　30:338
　　liquid jet, **26**:189–193
　　　　array, **26**:171–179, 175–177
　　　　free-surface jet, **26**:122, 147–162, 162–170
　　　　jet inclination, **26**:182–189
　　　　jet splattering, **26**:196–198

planar jet, **26**:122, 162–170, 192–195
stagnation zone, **26**:110–113, 128, 129
submerged jet, **26**:124–126, 131–135
wall roughness, **26**:193–195
in liquids, **14**:107–143
lost energy in, **15**:4–10
low liquid level nucleate boiling, **20**:55–58
  in liquid film, **20**:58–63
in lunar surface layer, **10**:72–79
measurement of, in turbulent flows, **8**:293
mechanism for, **10**:181–199
melt–water interactions, **29**:129–201
in MHD couette flow, **1**:321f
microgravity, **30**:43–51
microscales, **21**:263–268
  convection, **30**:19–33
  kinetic scales, **21**:263–265
  Kolmogorov microscales, **21**:265
  mean transport, **21**:264–265
  $Pr \rightarrow 0$, **21**:266–267
  $Pr \geq 1$, **21**:267–268
  $Pr \rightarrow \infty$, **21**:267
  Taylor scale, **21**:264
  thermal scales, **21**:265–268
  two-phase film, **30**:33–43
modeling
  continuum approach, **28**:65–66
  discrete approach, **28**:66–68
molten pools, **29**:3–33
moving liquid droplet
  condensation, **26**:32–47, 83–84
  droplet interaction, **26**:84–89
  droplet oscillation, **26**:44
  phase change at drop surface, **26**:19–32
  vaporization, **26**:48–84
multicomponent, **16**:147
multiphase flow in porous media, **30**:93–95,
  186–187
  capillary forces, **30**:93, 95, 96–98
  gravity and, **30**:93–94
  interfacial tension, **30**:96
  models, **30**:98–110
  multicomponent systems, **30**:162–186
  non-Darcian effects, **30**:111–112
  nonequilibrium effects, **30**:110–111
  phase saturation, **30**:95–96
  relative permeability, **30**:98, 99
  two-phase, single-component systems,
    30:112–162
  viscous forces, **30**:93, 94

natural convection cooling, vertical plates and
  channels, **20**:185–187
natural convection in rectangular cavities,
  8:179
near critical point, **7**:1–86
  boiling, **7**:74–82
    film, **7**:78
    nucleate, **7**:76
    pseudo, **7**:81
  equation of motion and energy, **7**:15–25
    boundary layer flow, **7**:17
      buoyancy effects, **7**:17
      dissipation, **7**:18
      free stream acceleration effects, **7**:18
    channel flow, **7**:19
      acceleration effects, **7**:21
      buoyancy effects, **7**:20
      dissipation, **7**:21
      turbulent shear stress and heat flux, **7**:22
      effect of variable properties on, **7**:22
  forced and free convection combined, **7**:66–
    74
    experimental results, **7**:67
    heat transfer deteriorations, **7**:68
      buoyancy effects on shear stress dis-
        tribution, **7**:69
      influence of wall heat flux, **7**:74
      local deterioration in heat transfer
        coefficient, **7**:71
      shear stress distribution effects on
        turbulence, **7**:70
  forced convection, **7**:25–55
    correlation of experimental data, **7**:43–51
      acceleration, buoyancy, and dissipation
        effects, **7**:44
      existing correlations, **7**:48
      large temperature differences, **7**:47
      limiting form of correlations, **7**:46–48
      small temperature differences, **7**:46
    experimental data, **7**:31–42
      gaps in experimental data, **7**:38
      local heat transfer coefficient, **7**:35
    experimental measurements, **7**:26–31
    presentation of data in terms of
      dimensionless groups, **7**:30
      heat transfer coefficient, **7**:29
      local conditions, **7**:27
    semiempirical theories, **7**:51–54
  free convection, **7**:55–66
    experimental results, **7**:55

Heat transfer (*continued*)
    temperature differences, large and
        small, **7**:56, 58
    theoretical methods and correlations,
        7:63
        basic correlations, **7**:64
        theoretical correlations, **7**:65
    physical properties near critical point, **7**:3–
        15
    molecular structure, **7**:8
    property variation effects on heat transfer,
        7:10
        effects of temperature difference, **7**:11
        limit as temperature difference tends to
            zero, **7**:12, 14, 15
    thermodynamic properties, **7**:3
        compressibility and velocity of sound, **7**:7
        heat capacity, **7**:6
        law of corresponding states, **7**:5
        van der Waals' model, **7**:3
        transport properties, **7**:8
    in noncircular ducts, **8**:44
    in non-Newtonian fluids, **15**:64–66, 98–126
    nucleate pool boiling, **16**:105–117
    in other geothermal systems, **14**:97–98
    packed beds, **23**:229
    from packet to wall, **10**:191
    particle motion and, **10**:182
    perfluoromethylcyclohexane and, **14**:125
    phase change
        enthalpy method, **28**:76–79
        interface tracking method, **28**:76, 79
    phonon conduction in, **10**:43, 45
    phonon in, **11**:353–354
    physical quantities for, **SUP**:45–46
    Prandtl number and, **12**:102
    pressure and, **11**:105, 110–117
    problems of, **31**:334
    pulsed combustion, **30**:83–85
    pulsed-laser-induced phase transformations
        melting, **28**:75–109, 135
        sputtering, **28**:109–123
        vaporization, **28**:123–135, 136
    radiative, *see* Radiative heat transfer
    in rarefied gases, **7**:163–218
        accommodation coefficients, **7**:165
        external flows, **7**:187
            free molecule flow, **7**:187
            temperature jump (slip) regime, **7**:192
            transition regime (M 1), **7**:193, 194

        cones, **7**:206
        cylinders, **7**:200
        flat plate, sharp leading edge, **7**:202
        spheres, **7**:202
        stagnation point, blunt body, **7**:196
    gas at rest, **7**:169
        free molecule conditions, **7**:169
        temperature jump approximation, **7**:170
        transition regime, **7**:170
            concentric cylinders, **7**:178
            parallel plates, **7**:171
        internal flows, **7**:183
        free molecule flow, **7**:184
        temperature jump (slip) regime, **7**:184
    in rivers, bays, lakes, and estuaries, *see*
        THIRBLE
    second-law analysis in, **15**:1–53
    in semitransparent solids, **11**:317–433
    in separated regions, **8**:120
    simplified formula for, **20**:22–23
    single tube
        determination, dimensionless relation,
            18:107
        local, **18**:110–122
            in gas flow, **18**:115–118
            theoretical calculations, **18**:110–115
            in viscous-fluid flow, **18**:118–112
        mean, **18**:122–129
            calculations, **18**:127–129
            at high Reynolds numbers, **18**:124–127
            at low Reynolds numbers, **18**:123–124
            power equation for generalization of ex-
                perimental data, **18**:107
            reference temperature and, **18**:108–110
            restricted channel, **18**:129–131
    slug flow, **20**:84
    between solid particles and fluid,
        10:171–180
    in solids, **31**:143–145
    spray deposition, **28**:1–3
        impact region, **28**:23–69
        spray region, **28**:3–22
    steady state measurement techniques in,
        10:172–173
    steam explosions, **29**:129–201
    steam-to-water, **12**:11
    in supercritical flow of hydrogen, **5**:368
    in supersonic nozzles, **16**:245–247
    temperature difference in, **12**:10–11
    theoretical models of, **10**:176

thermal communication, **26**:219
three-phase systems, **23**:261
through constant-velocity boundary layer,
    effects of transpiration, **16**:274–280
through emulsion layer, **10**:193–194
through "packets" of particles, **10**:186
transient, **11**:375–390
tube in a bank
    crossflow, calculation, **18**:154–156
    determination, dimensionless relation,
        18:107
    local transfer, **18**:131–136
    mean transfer, **18**:136–148
        of closely spaced banks, **18**:145–147
        in critical flow regime, **18**:142–145
        at low Reynolds numbers,
            18:138–139
        in mixed-flow regime, **18**:139–142
        variation along a bank, **18**:137–138
    power equation for generalization of ex-
        perimental data, **18**:107
    Prandtl number and, **18**:107–108
    reference temperature and, **18**:108–110
turbulent, **11**:295–306
turbulent-boundary-layer, **16**:243–244; *see*
    also Boundary layer
    effect of acceleration, **16**:282–288
    research at Stanford, **16**:241–365
    on smooth, flat plate, **16**:244–245
    for unheated starting length, **16**:245–246
turbulent flow, **26**:162–170; **30**:51, 82–83; *see*
    also Turbulent flow; Turbulent heat
    transfer
unsteady, **25**:4
unsteady state techniques in, **10**:173
velocity profile and, **15**:91
versus gas velocity, **21**:317
viscoelasticity influence on, **15**:212
water evaporation in, **12**:13
work transfer and, **15**:6n
Zabrodsky microbreak model of, **10**:176
Heat transfer analysis, rheology, **15**:70–86
Heat transfer augmentation
    entropy generation and, **15**:21–25
    entropy generation number and, **15**:24
    through finned tubes, **15**:22–23
Heat transfer capacity function, **19**:153–154
Heat-transfer coefficient in film boiling on
    cylinders, **5**:63, 66
Heat-transfer coefficient in turbulent boundary

layers, **2**:33ff, 45, 59ff
Heat-transfer coefficients, **SUP**:45; **25**:164; *see*
    also Adiabatic heat-transfer coefficient;
    Coefficient; Dropwise condensation
in annular-dispersed flows
    above CHF, **1**:427ff
    below CHF, **1**:423ff
    in two component mixtures, **1**:433ff
"boxed" regenerator design, **20**:148–149
bubbling, **21**:320–321, 329
    versus superficial gas velocity, **21**:324–325
calculation, **19**:151–152
circular cavities, **29**:17–18
coalesced bubble region, **20**:49–50
computation, **19**:161
condensation, **21**:108–109
controlling factors, **19**:146
convective, film boiling, **23**:118–119
determined using specklegram data, **30**:269–
    271
entrainment, **21**:315
film condensation, **21**:105
between fluidized bed and immersed surface,
    19:163–166
for fluidized beds, **10**:177
forced convection
    flat-pack modules, **20**:236–237
    high-heat-flux source, **20**:225–226
    missing module effect, **20**:238–239
    three-dimensional arrays, **20**:232–238
    turbulent flow models, **20**:238
    two-dimensional protruding elements,
        20:227–228
friction factor and, **15**:116–117
general dimensionless structure, **21**:320
heat pipe cooling techniques
    grooved design, **20**:278
    vapor temperature, **20**:277–278
Heat-transfer to and from immersed surfaces,
    19:118–119
heat transfer to and from immersed surfaces
    small horizontal tubes and fluidized beds of
        small particles, **19**:119–122
    smooth surfaces and fluidized beds of large
        particles, **19**:135–142
    smooth vertical tubes and fluidized beds of
        small particles, **19**:122–135
    high temperature, **19**:169–172
    for horizontal and slanted tubes, **14**:160–
        172

Heat-transfer to and from immersed surfaces
  (*continued*)
  horizontal bubbling surface, versus superficial
    gas velocity, **21**:330–331
  horizontal copper tube, **19**:150–151
  hydrodynamic behavior, bed, **25**:165, 195
  immersed surface to bed, **25**:164
  in impinging flow, **13**:6–7
  impinging liquid jets, **23**:1
  inclination angles, results of analysis, **20**:35–
    36
  in-line tube bundles
    horizontal, **19**:157–159
    vertical, **19**:158–159
  integral mean, **13**:13–26
  interfacial, **21**:70–71, 280
    versus superficial gas velocity, **21**:283, 285
  interlayer, **21**:315
  jet impingement cooling, **20**:251–252
  local instantaneous, **10**:187
  low liquid level nucleate boiling, **20**:56–57
    surface configurations, **20**:61–62
  maximum, **19**:114–118
  narrow space nucleate boiling
    coalesced bubble region, **20**:43–49
    isolated bubble region, **20**:40–41
    pressure and, **20**:46–47
  natural convection
    immersion cooling, **20**:213–217
    irregular surfaces, **20**:208–210
    two-dimensional protrusion, **20**:211–212
    vertical surfaces, **20**:205–206, 205–209
      isothermal channel wall, **20**:208–209
  nondimensional correlations for, **10**:202
  nucleate boiling
    liquid levels, **20**:7–8
    pressure factor, **20**:17–18
    surface conditions, **20**:2
    temperature difference, **20**:6–7
    thermal boundary layer thickness, **20**:9–10
  in nucleation, **10**:150–151
  outlet flow conditions vs. array of nozzles for,
    13:27–41
  for packed fluidized beds, **14**:232–233
  particle to fluid, **25**:177
  pool height effect, **21**:331–332
  Prandtl number effect, **21**:66–67
  pressure dependence, **20**:20–21
  radiative, **19**:163–166
  rough vs. smooth surface, **19**:146

  staggered tube bundles
    horizontal, **19**:157–158
    vertical, **19**:158–159
  steady-state conduction and lateral mixing,
    19:138
  stratified configuration, **21**:312
  surface, **21**:94
  surface inclination, **20**:27–28
  surface renewal, **21**:315
  swirling jets and, **13**:41
  total, **19**:16, 174–175
  tube roughness of size and, **14**:230–231
  turbulence promoters and, **13**:41
  turbulent free convection, **21**:40–41
  vapor–liquid interface, **21**:70
  for vertical tubes, **14**:212–231
  wall, **21**:323
  wall to bed, **25**:177
  wire-mesh grids and, **13**:41–43
Heat-transfer controlled collapse, of stagnant
  bubble, **15**:243
Heat-transfer distribution, lateral heat transfer
  effect and, **15**:43
Heat transfer enhancement
  additives, **30**:235
  in annular channels
    with an inner grooved tube, **31**:241–245
    with one-sided protrusion-groove turbula-
      tors, **31**:253–259
  area enlargement and, **30**:199, 209–213
  boiling, **30**:264–274
  by artificial flow turbulization, **31**:267–268
  in channels, recommendations on choice of
    means for, **31**:311–313
  choice of in straight channels and around tube
    bundles, **31**:182–185
  choice of method of, **31**:161–187
  compound enhancement, **30**:244–246
  with condensation of vapor mixtures on
    vertical surfaces, **31**:287–292
  with condensation vapor mixture on horizontal
    tubes, **31**:293–294
  described, **30**:197, 199–200, 246–247
  displaced enhancement devices, **30**:227–230
  due to transverse finning in annular channels,
    31:245–249
  due to transverse finning in tube bundles in
    longitudinal flow, **31**:249–253
  efficiency of
    evaluating, **31**:305–311

as function of diaphragm height and pitch, **31**:192–193

electrically heated test surfaces, **30**:213–216

extended surfaces (fins), **30**:199, 216–217, 218

in flat and triangular channels, **31**:259–263

fluid bed heat exchanger, **30**:236–237, 238

at fouling on tube surfaces, **31**:294–304

influence of Prandtl number on, **31**:209–212

injection and, **30**:244

intensification, **30**:199–208

in liquid flow in tubes, **31**:201–203

methods of convective, **31**:159

porous layer enhancement, **30**:232–233

resonating pulse reactor, **30**:243–244

rotating surfaces and, **30**:241–242

rotating twisted tape devices, **30**:230

rough surfaces, **30**:217, 219–227

rough surfaces and additives, **30**:246

sensible heat exchangers, **30**:213–216

stirring and, **30**:237, 239

structured surfaces, **30**:234–235

suction and, **30**:244

with supercritical hydrocarbon flow in tubes, **31**:228–232

surface scraping and, **30**:239–240

surface tension devices, **30**:230–236

swirl flow and additives, **30**:245–246

swirl flow devices, **30**:230

treated surfaces, **30**:232–233

in tube bundles in longitudinal flow and annular channels, **31**:232–259

in tubes, **31**:187–232

in turbulent flow, theoretical methods of predicting, **31**:196–197

with vapor condensation, **31**:278–287, 292–293

vibration and, **30**:242–243

Heat transfer enhancement devices, **31**:453

Heat transfer enhancement methods

analysis of different, **31**:178–182

conditions governing choice of, **31**:162–166

Heat-transfer-enhancing fences, forced convection, **20**:240–241

Heat transfer fluxes, **19**:16

Heat transfer *j* factor, **15**:126; **19**:256

Heat transfer measurements in turbulent boundary layers, **2**:59ff

Heat transfer models

downward-facing surface, **20**:28–36

in drag reduction, **12**:86–88

Heat transfer–momentum relation, Reynolds–Prandtl analogy in, **12**:109

Heat transfer number, *see* Incremental heat transfer number

Heat transfer process

in gas phase, **31**:46–48

in liquid phase, **31**:43–46, 64

in solid phase, **31**:41–43

variation description with variable thermophysical characteristics, **19**:203–210

Heat transfer rates, **19**:179

fluid withdrawal and, **14**:50

variables in, **10**:180–181

Heat transfer surface

development of, **31**:245

methods of calculating effective, **31**:304–320

Heat transfer surface rotation, **31**:161

Heat transfer systems, nucleation and, **10**:86–87

Heat transfer theory, dropwise condensation, **21**:92–100

Heat transfer tunnels, Stanford, **16**:254–255

concave wall rig, **16**:266–268

curvature rig (HMT-4), **16**:264–266

full-coverage film-cooling rig (HMT-3), **16**:261–264

mixed convection tunnel, **16**:268–270

rough plate rig (HMT-2), **16**:260–262

smooth plate rig (HMT-1), **16**:255–262, 270

Heat transfer unit, height and, **15**:265

Heatup, **24**:284

Heaviside step function, **20**:159–160

Heavy-liquid density, bubble-induced entrainment effect, **21**:303–304

Heavy-liquid viscosity, bubble-induced entrainment effect, **21**:305–307

HECTR models, **29**:108, 109, 111

Height differences, forced convection, **20**:238–241

Heisenberg principle, **5**:279

Heisler charts, **22**:161–166

Helical electrodes, **21**:130–131

Helical flow geometry, **13**:228

Helium I

critical heat flux in, experimental data, **17**:321–337

properties of, **17**:126–127

Helium II, **5**:480

boiling heat flux in, **5**:490

Helium II (*continued*)
  film boiling in, **5**:497
  heat transfer to, **5**:485
  Kapitza resistance in, **5**:484
  thermal conductivity of, **5**:481
Helium
  properties of, **17**:127–129
  relative change in heat transfer coefficient,
    21:16
Helium liquid, heat transfer in
  anomalous problems and, **17**:68
  basic considerations, **17**:67–68
  high heat flux results, **17**:73–74
  influence of solid characteristics, **17**:68–70
  numerical values, **17**:70–71
  present directions, **17**:71–73
  summary, **17**:74–75
Helium–turbine cycle, **12**:7
Helmholtz critical wavelength, **23**:97
Helmholtz free energy, **5**:261
Helmholtz instability, **5**:107, 416
  in boiling, **1**:234ff
Hemicellulose, fire and, **10**:226
Hemispherical isotropy, **23**:136
n-Heptane
  droplet deformation, **26**:80
  droplet vaporization, **26**:72–73
Heptane-methylcyclohexane, A_0, **16**:119
Heptane–superposed layers, conversion, **29**:32–
    33
Herschel–Bulkley fluid, **24**:109, 118
Heterogeneous chemical reaction, *see* Chemical
    reaction
Heterogeneous porous medium, **23**:373
Hewlett–Packard minicomputer, **10**:19
n-Hexadecane, vaporization, **26**:67, 70–71, 78–
    79
Hexagonal duct, **SUP**:262–263
  longitudinal thin fins within, **SUP**:370
n-Hexane, vaporization, **26**:56–57, 59, 67, 70
Hexane in forced convection film boiling, **5**:96,
    98
n-Hexane-*n*-octane, bubble growth, experimental
    data, **16**:93
n-Hexane-toluene, film boiling, **16**:138
High field effects in MHD, **1**:272f
High Flux® surface, heat transfer enhancement
    with, **30**:232
High heat regions, surface inclination, **20**:27–28
High-pressure liquid-encapsulated Czochralski

system (HPLEC), **30**:323–325, 351, 361
  convection in, **30**:412–423
  modeling, **30**:424
High-pressure melt ejection (HPME), **29**:215,
    216, 247
High-Rayleigh number convection, **20**:329–330
High-speed photography, of film boiling, **11**:55–
    59
High-temperature heat transfer probes, **19**:181
  design, **19**:166–169
  types, **19**:163–166
Hill's spherical vortex, moving liquid droplet,
    26:4, 7, 9, 10
Hindered settling, *see also* Sedimentation; Set-
    tling
  non-Newtonian effects, **23**:261
Hi Tran® radial mixing elements, heat transfer
    enhancement, **30**:227–228
Hittorf transference number, **12**:246
Hohlraum (cavity), **12**:120
  blackbody radiosity and, **12**:124
  in thermodynamic equilibrium, **12**:127
Hole ablation, direct containment heating,
    29:249–253
Holography, **6**:200
Holtz model, of trapped inert gas, **10**:159–160
Homogeneous chemical reaction, *see* Chemical
    reaction
Homogeneous elements, **31**:374–375
Homogeneous fluid, stratifying process of,
    24:293–295
Homogeneous model correlation in annular-
    dispersed flows, **1**:369f
Homogeneous nucleation, **1**:214f;
    14:281–339
  classical theory in, **14**:288–305
  classical theory modifications in,
    14:305–313
  description of problem in, **14**:285–288
  diffusion cloud chamber experiments in,
    14:325–330
  embryo size distribution in, **14**:293–298
  formation of embryos and nuclei in, **14**:288–
    293
  liquid superheats in, **10**:111
  Lothe–Pound theory in, **14**:309–311, 337–339
  model of, **14**:260–262
  piston cloud chamber experiments in, **14**:318–
    325
  shock tube experiments in, **14**:335–337

steady-state nucleation rate in, **14**:298–305
steady-state solutions vs. experiments in, **14**:317–337
steady-state time lag in, **14**:313–317
supersonic nozzle experiments in, **14**:330–334
superstructure in, **14**:282–284
Homogeneous nucleation temperature, **29**:139
Honeycombs, **18**:15–17
Horizontal chemical vapor deposition reactors, **28**:346, 365–375
   gallium arsenide, **28**:373–375
   MOCVD, **28**:373–375
   silicon, **28**:365–373
Horizontal circuit boards, natural convection, **20**:220
Horizontal circular cylinders, **24**:3–5, 8, 31–32, 126, 151
   flow patterns for natural convection inside, **8**:212
   governing equations for natural convection inside, **8**:197
   influence of initial conditions on natural convection inside, **8**:223
   influence of thermal boundary conditions on natural convection inside, **8**:216, 217
   natural convection experiments using silicone oil, **8**:211
   natural convection inside, **8**:196
   streamline patterns for natural convection inside, **8**:206–208, 216–219, 220, 221
   temperature profile for natural convection inside, **8**:204, 205, 211, 212, 217, 219, 220, 222
   velocity profiles for natural convection inside, **8**:204, 205, 209, 211, 217–219, 221, 222
Horizontal cylinders
   boiling about, **1**:247ff
   film boiling on, **5**:57
   natural convection in, **8**:164, 165, 167, 169, 172; **11**:200–210
Horizontal heated surfaces, mixed convection about, **14**:63–65
Horizontal isothermal cylinder, thermal convection heat transfer rates for, **15**:170
Horizontal layer, **24**:175
Horizontal line source, heat- and mass-transfer, **20**:346–347
Horizontal low-fin tube, **21**:120–123
Horizontal plate, **24**:137

Horizontal plate geometries, melting and freezing in, **19**:45
Horizontal porous layer
   convection in, **14**:76–77
   free convection in, **14**:71–73
   mixed convection in, **14**:76–77
Horizontal surfaces, **9**:296
   in enclosures in solar buildings, **18**:57–63
Horizontal tubes
   local heat transfer coefficients for, **14**:160–172
   packed-fluidized beds and, **14**:232–234
   superimposed free convection in, **SUP**:411–413
   thermal convection in, **15**:196–197
   total heat transfer coefficients for, **14**:172–207
Horizontal wires, **25**:260, 280
   free convection, **21**:42–44
Hot-film sensors, **8**:293
Hot-plate technique, thermal conductivity of structured liquids, **18**:168–174
Hot tears, **28**:237
Hot-wall chemical vapor deposition reactor, **28**:344, 345, 399–408
Hot water, injection of along wells or vertical fissures, **14**:60–61
Hot water geothermal systems
   boundary layer simplifications in, **14**:55–71
   experimental investigations in, **14**:71–77
   heat transfer in, **14**:15–77
   linear stability theory in, **14**:17–29
   numerical solutions in, **14**:34–54
   perturbation and finite amplitude analysis in, **14**:29–34
Hot-wire anemometer bridges, **23**:340–341
Hot-wire anemometers, **8**:292, 293
   calibration procedures in, **8**:294
   influence of nonisothermal flow on, **8**:294
   influence of turbulence intensity on, **8**:295
   influence of wire length on, **8**:294
   statistical analysis of signals from, **8**:336
Hot wire–cold wire method, **8**:310
Hot wire probes
   aging of, **8**:312
   calibration of, **8**:312, 313
   construction of, **8**:311
   cross-wire probes, **8**:296
   electronic circuit of, **8**:311
   sensitivity of, to flow direction, **8**:296
   time constant of, **8**:296
   wall effects on, **8**:296

Hot-wire/thermal conductivity probe technique, **18**:175–176

Howarth's transformation, **1**:32; **3**:225

HPLEC, *see* High-pressure liquid-encapsulated Czochralski system

HPME/DCH study, **29**:247, 270–271

Huka Falls, New Zealand, **14**:52–54

Huygens' principle, **6**:146, 169

Hybrid scheme, **24**:195, 226

"Hycam" camera, in temperature field measurements, **11**:22

Hydraulic axial distance, dimensionless, **19**:255

Hydraulic design, methods of appropriate, **31**:304–320

Hydraulic diameter, **SUP**:422

Hydraulic drag of banks
  calculation methods, **18**:148–150
  mean drag, **18**:150–153
    critical flow, **18**:151–153
    laminar flow, **18**:150
    mixed flow, **18**:150–151
  proposals for calculation of, **18**:153–154

Hydraulic jump
  free-surface liquid jet, **26**:114, 201–203, 205
  internal, **24**:300–301

Hydraulic loss, **25**:10, 74

"Hydraulic radius", **2**:428

Hydraulic resistance, **8**:150
  in annular turbulator-provided tubes, **31**:313–320
  calculation of, in tube banks, **8**:154
  heat transfer and, **11**:170–187
  increasing, **31**:163
  influence of Reynolds number on, **8**:151
  influence of tube pitch on, **8**:152
  in in-line tube banks, **8**:153
  in staggered tube banks, **8**:152, 153
  in tube banks, **8**:151

Hydraulic resistance coefficient, **31**:193–196
  caused by wall friction, **31**:204
  of different tubes, **31**:215–216
  influence of Reynolds number on, **31**:203–204
  plotted versus Reynolds number, **31**:205
  and protrusion profile, **31**:214
  ratio of, **31**:163
  versus diaphragm height, **31**:222–225
  for water as function of the Reynolds number, **31**:216–217

Hydraulic resistance law, **31**:226

Hydraulic resistance ratio, **31**:246

Hydrocarbon gases, heat flow rates, **16**:193, 194, 222, 223

Hydrocarbons
  moving liquid droplet
    drop deformation, **26**:79–80
    vaporization, **26**:56–57, 59, 67, 70, 72, 78–79
  woodburning stoves and, **17**:300–301

Hydrodynamic ablation, **28**:109–110

Hydrodynamically developed flow, **SUP**:5; *see also appropriate ducts in question*

Hydrodynamically developing flow, **SUP**:6
  axially matched solutions in, **SUP**:68–69, 85, 160–161
  in circular ducts, **SUP**:85–97
  in circular sector ducts, **SUP**:266–267
  in concentric annular ducts, **SUP**:297–301
  in eccentric annular ducts, **SUP**:333–336
  experimental results, **SUP**:96–97
  finite difference solutions for, **SUP**:73, 88–89, 163–164, 211–213, 242, 298–299, 333–336
  finite element solutions for, **SUP**:213
  integral solutions for, **SUP**:69–70, 74, 85–87, 160
  linearized solutions for, **SUP**:70–73, 74, 87, 163, 210, 240–242, 297–298
  in parallel plates, **SUP**:160–167
  in rectangular ducts, **SUP**:209–213
  solution methods for, **SUP**:68–74
  solutions by boundary layer type, idealizations, **SUP**:68–73
  solutions of Navier–Stokes equations for, **SUP**:73–74, 89–97, 164–167, 300
  in triangular ducts, **SUP**:240–242

Hydrodynamic behavior, submerged jet impingement cooling, **20**:252–253

Hydrodynamic entrance length, **SUP**:6, 41–42; **15**:88; *see also appropriate ducts*
  for circular ducts, **SUP**:98–99
  for concentric annular ducts, **SUP**:288–289
  for parallel plates, **SUP**:168–169
  for rectangular ducts, **SUP**:198–199

Hydrodynamic entrance region, **SUP**:6

Hydrodynamic fragmentation, **29**:168–175

Hydrodynamic heat transfer theory, **31**:186

Hydrodynamic instability, **5**:117
  multistep model of CHF and, **17**:14–19
  one-step model of CHF and, **17**:11–14

Hydrodynamic operator, 3:185
Hydrodynamic parameters
  forced convection, cubical element array,
    20:233–234
  slug flow modeling, 20:114
  liquid film, 20:88–92
Hydrodynamics
  jet impingement, 23:2–6
  liquid jet impingement, 26:136–147, 157
Hydrodynamic thermocapillary model (IHTCM),
    Czochralski process, 30:350
Hydrodynamic velocity, 12:204
Hydrogen
  convective heat transfer to, 5:360
  film boiling of, 11:187
  between infinite parallel plates, 5:35
  spectral absorption coefficient, 5:35
  in supercritical turbulent flow, 5:368
Hydrogen-bubble technique, 8:297
  velocity profiles, 20:112–113
Hydrogen combustion
  combustion parameters, 29:62–65
    autoignition temperature, 29:84–86
    burning velocity, 29:75–78
    combustion completeness, 29:71–75
    flammability limits, 29:65–71
    ignition sensitivity, 29:82–84
    turbulent burning rates, 29:78–82
  detonation, 29:62, 90–102
  diffusion flame, 29:62
  during direct containment heating, 29:297–
    308
  fast flames, 29:86–90
  nuclear reactor safety, 29:59–61, 77, 102–123
    deflagration, 29:108–118
    pressure transients from combustion,
      29:107
  premixed combustion, 29:62
  subscale models, 29:118–121
  thermohydraulic codes, 29:113–118
Hydrogen event containment transient response
  (HECTR) model, 29:108, 109, 111
Hydrogen flame, 6:326
Hydrogen peroxide, 31:449
Hydrophobic oils, steam condensation on,
  15:230–231
Hydrophobic surface, 31:276
Hydrosphere, as heat transfer medium, 12:10
Hydrostatic pressure, quasi-equilibrium slug-
  ging, 20:125

Hydroxyethylcellulose (Natrosol), 19:329
Hyperbolic equations, 8:53
Hyperbolic layer, in two-dimensional floating
  layers, 13:106–109
Hypereutectic solution, 28:241
Hypersonic flow, 1:331ff
Hypersonic nozzles, 8:11
Hyperthermia, 22:13, 222, 235, 289, 360–361
Hypoeutectic alloys, 28:239, 262
Hypothermia, 22:360–361
Hysteresis, nucleate boiling, 23:49
Hysteresis effects
  in annular-dispersed flows, 1:424f
  in MHD, 1:328f
  in nucleate boiling, 1:204ff
Hysteresis loop, 25:179

I

Ice condenser containments, direct containment
  heating, 29:328–329
Ice melting and friction, 24:7, 29–34
Ideal fluids, 8:95
Ideal gas law, 31:48–50
IDEMO model, steam explosions, 29:180
IFCI code, 29:151
Ignition, 29:188; see also Thermal ignition
  droplet, 26:60–61
  fire and, 10:235–245
  radiation induced, 10:243
  in wildland fires, 10:244–245
Ignition energy, 29:83
Ignition sensitivity, 29:82–84
Ignition temperatures, for various materials,
  10:239
Ignition tests
  fire-safety oriented, 10:238–243
  for various fabrics, 10:240–241
IHTCM model, see Hydrodynamic thermocapil-
  lary model
Illumination intensity, 6:151
Immersed bodies
  in fluidized bed, 10:206–207
  nondimensional correlations for, 10:201
Immersed surfaces, 25:253, 256
  in gas-fluidized bed, heat transfer rate, factors,
    19:178
  heat transfer to and from, 19:111–142
    heat transfer coefficient, 19:118–142

Immersed surfaces (*continued*)
   small horizontal tubes and fluidized beds of
      small particles, **19**:119–122
   smooth vertical tubes and fluidized beds of
      small particles, **19**:122–135
      maximum heat transfer coefficient,
         19:114–118
   smooth surfaces and fluidized beds of
      large particles, **19**:135–142
   high temperature heat transfer
      experimental techniques of measurement of
        total and radiative heat transfer coeffi-
        cients, **19**:163–166
      heat transfer fluxes and coefficients,
        19:161–163
      models, **19**:173–178
      probe design, **19**:166–169
   mechanistic theory of heat transfer, **19**:142–
      145
   rough and finned, heat transfer from, **19**:145–
      154
Immersed tubes, gas fluidized beds and, **14**:149–
   238
Immersion cooling
   natural convection, **20**:213–217
   thermal control of electronic components,
      20:243–251
Immiscible fluids, single bubbles in, **15**:245–246
Immiscible interfaces, condensation on, **15**:229–
   231
Impact pressure in annular-dispersed flows,
   1:389f
Impedance, **31**:133–136
   matrices of flow, **31**:147–149
Impinging flow
   acceleration pressure drop in, **13**:31
   arrays of round nozzles in, **13**:46–49, 51
   arrays of slot nozzles in, **13**:49–51
   on concave surfaces, **13**:43
   contour lines in, **13**:50–51
   heat transfer coefficient in, **13**:6–7
   high-pressure arrays of nozzles in, **13**:52–58
   hydrodynamics of, **13**:2–5
   integral mean transfer coefficients in, **13**:13–
      26
   jet length in, **13**:4
   mass transfer coefficient in, **13**:6–7
   mean heat and mass transfer coefficient
      correlations for, **13**:27–41
   nozzle array in, **13**:12–13
   nozzle-to-plate distance in, **13**:9

numbers of transfer units in, **13**:12
and optional spatial arrangement of nozzles,
   13:45–52
outlet flow conditions vs. transfer coefficients
   for array of nozzles in, **13**:27–41
Sherwood number in, **13**:9–12
for single nozzles, **13**:8–18
stagnation flow in, **13**:4–5
swirling jets in, **13**:41
turbulence promoters in, **13**:41
variations in coefficients for, **13**:8–13
velocity field of, **13**:2–3
wall jet flow and, **13**:5
wire-mesh grids and, **13**:41
Impinging-jet reactors, **28**:345, 346, 394–396
Implicit scheme, **24**:195
Incident radiant flux, **2**:405
Incident radiation, **1**:6
Incipience model, modified, **23**:16
Incipient boiling temperature, **11**:3–4
Incipient ignition, *see* Thermal ignition
Inclination effects, on thermosyphon, **9**:65
Inclined surfaces, **9**:296
Incompressible flow, **24**:218, 219
   defined, **13**:208
Incompressible fluid, as viscous fluid, **15**:62
Incompressible gas, energy generated by viscous
   dissipation, **21**:252–253
Incompressible liquid, differential energy equations for, **SUP**:13
Incremental heat transfer number, **SUP**:51–52,
   420
   for circular duct, **SUP**:103, 105, 127, 128
   for parallel plates, **SUP**:171, 172, 181, 182
Incremental pressure drop number, **SUP**:42–43;
   19:255; *see also other appropriate ducts in*
   the text
   for circular duct, **SUP**:97–98
   for concentric annular ducts, **SUP**:287–288
   for parallel plates, **SUP**:168
   for rectangular ducts, **SUP**:198–199
   for triangular ducts, **SUP**:226, 230, 237, 238
Incremental solidification, **28**:22
Independent number of components, **12**:200–201
INDFL variable, **27**:127, 129, 205
INDFUL variable, **27**:129, 205
INDGW variable, **27**:125, 161, 205
Indirect cooling techniques
   heat pipes, **20**:273–289
   thermal control of electronic components,
      20:182–183

Indium antimonide, properties, **30**:321
Indium arsenide, properties, **30**:321
Indium bismide, properties, **30**:321
Indium foils, enhancement of, thermal contact conductance, **20**:264
Indium nitride, properties, **30**:321
Indium phosphide
  crystal growth, **30**:323, 324, 360
  crystal quality, **30**:323
  Czochralski process, **30**:414–417
  properties, **30**:321, 346
  transition temperatures to transparency, **30**:340
INDNT variables, **27**:126, 161, 205
INDRDP variable, **27**:129, 205
Induced magnetic field, **1**:302ff, 319ff
Induced rotational flow, **21**:141–142
  boundary layers, **21**:142
  heat transfer, **21**:162–163
    mixed-mode flow, **21**:168–170
    natural convection, **21**:167–168
    zero axial flow, **21**:162–168
  isothermal flow
    secondary flow, **21**:146–147
    stability limit, **21**:149
  mass transfer, **21**:170–174
    wiper blades effect, **21**:174–176
  stability, transition, and flow regimes, **21**:143–144
    axial flow effect, **21**:156–159
    Couette flow, **21**:144–147, 150–151
    eccentricity effect, **21**:159–160
    finite annulus length effect, **21**:156
    isothermal flow, **21**:144–160
    nonisothermal flow, **21**:160–161
    Taylor vortex flow, **21**:147–152
    torque transport, **21**:161–162
    turbulent flow, **21**:153–155
    wavy vortex flow, **21**:152–153
  thermal stability, **21**:142
  transition to turbulence, **21**:154–155
  transport characteristics, **21**:142
Induction stirring, alloy solidification, **28**:317–318
Industrial accidents, *see also* Nuclear reactor safety
  steam explosions, **29**:129–136
Industrial gas-fired furnaces, **27**:196–199
INDWBC variable, **27**:126–127, 161, 163, 205
Inelastic fluids, **24**:106
  thermal convection in, **15**:208–211

turbulent thermal convection in external flows of, **15**:189–192
Inelastic solution, **25**:308
Inert gas, in nucleation model, **10**:159–160
Inertia
  convective, **24**:115
  flow, **31**:147–149
  porous, **24**:112
Inertia factor, **21**:28–29
Inertial subrange, **25**:335
Inertia matrix, **31**:129–130
Inert liquids, natural convection cooling, **20**:184–185
Infinite cylindrical medium, radiative flux for, **11**:344–350
Infinite harmonic temperature wave, approximation of, **14**:253
Infinite horizontal plates, natural convection between, **8**:165
Infinite long pointed needle, potential flow convection, **20**:368–369
Infinitely long concentric cylinder containing gray gas, **5**:31
Infinite parallel plates separated by gray gas, **5**:28
Infinite slab geometry
  full nonlinear radiation term, **19**:33
  linearized radiation condition, **19**:33
  saturated liquid, **19**:37
Infinity condition, **SUP**:105
Inflow boundary, liquid metal droplet, **28**:6
Information technology, **31**:463–464
Infrared emission, **8**:230
Infrared radiating gas, temperature profiles in, **8**:254
Infrared radiation, **8**:229
Infrared spectrum, **8**:231
Inhomogeneous medium, model of, **14**:262–263
Inhomogeneous properties, **31**:373–387
Initial conditions, **25**:357, 387; **31**:106
Initial relaxation near the leading edge of flat plates, and supersonic wedges and cones, **2**:219ff
Initial temperature difference, for dry cooling towers, **12**:66–70
Injection, **8**:12, 40; **25**:84, 98
  effect, **25**:83
  gas
    axisymmetric, **25**:137
    isothermal, **25**:131, 132
  heat transfer enhancement by, **30**:244

Injection (*continued*)
  nonisothermal jet, **25**:132
  parameter, **25**:84
Injection and discharge points, **31**:122
Injection parameter, *see* Blowing parameter
Injectors, **2**:77f
Injury
  atlas, **22**:300
  erythocyte, high temperature, **22**:322
  function, quantitative, **22**:233, 314, 329
  thermal, **22**:11, 16
Inlet temperature, three-fluid heat exchanger,
  26:267, 268
Inlet velocity profile, the effect of, **SUP**:109,
  142, 146, 164, 190
In-line tube banks, **8**:136, 137, 139–141, 143,
  145, 146
  hydraulic resistance in, **8**:151, 153
  pressure drop coefficients, **8**:154
Inner expansion, **8**:7
Inner limit similarity equations, **4**:344f
  solutions, **4**:441
Inner product, in linear diffusion, **15**:302–304
Instability
  general aspects of, **9**:339
  hydrodynamic, **9**:322
  laminar, **9**:322
  thermal, **9**:322
Insulation, **4**:87
  absorption and scattering in, **9**:369
  cryogenic types of, **9**:354
  effect of conduction in, **9**:356, 358
  entropy generation rate in, **15**:40
  evacuated multilayer, **9**:354
  evacuated porous, **9**:354
  high vacuum, **5**:343; **9**:354
  low temperature insulation, **5**:341
  multiple layer, **5**:343
  multishield type of, **9**:350
  physical properties of, **9**:366
  powder, **5**:345
  radiation shields as, **15**:47–48
  rigid foam, **5**:345
  spectral transmission of, **9**:367
  super-, **9**:355
  super insulation, **5**:329
  thermal diffusivities of, **9**:355
Insulation design, "constant q" type, **15**:41
Integral equations, **2**:418ff; **8**:34
  of boundary layer, **3**:37ff

energy equation, **3**:38
momentum equation, **3**:37
von Kármán's integral equation, **3**:37
for radiative transfer in an enclosure, **3**:193ff
solutions of, **2**:420ff
Integral mean transfer coefficients
  for array of round nozzles, **13**:18–22
  for array of round orifices, **13**:22
  for array of slot nozzles, **13**:22–26
  equations for single nozzles and, **13**:15–18
  impact angle and, **13**:45
  in impinging jet flow, **13**:13–26
  single slot nozzle in, **13**:18
Integral methods, **8**:116, 163, 287; **25**:270;
  31:113–145; *see also* Dorodnitsyn integral
  method; Pohlhausen integral method
  application of, **31**:107
  approximate, **19**:34–39
  advantage, **19**:34
  disadvantages, **19**:39
  for nonlinear problems, **1**:52ff
Integrals, evaluation of, by Monte Carlo methods, **5**:46
Integral solutions, **SUP**:85–87, 160, 192, 194
Integrated band intensities, **5**:295
  for CH_4, **5**:295
  for CO, **5**:295
  for CO_2, **5**:295
  for H_2O, **5**:295
  for NH_3, **5**:295
  for NO, **5**:295
  for N_2O, **5**:295
Integrating sphere, theory of, **10**:16
Integration in a polygonal region, **19**:231–232
Integrodifferential equations, **2**:440; **19**:14–16
Integument, **22**:13
Intensity
  magnetic field, **25**:154, 162, 171, 178, 179,
    181, 183, 185, 189, 191, 206, 239
    nonuniform, **25**:200
    varying, **25**:184
  of radiation, **1**:3
Interaction methods, **2**:422f
Interaction particle and magnetic field, **25**:155
Interdendritic liquid, **28**:232, 233, 239, 251, 326
Interface-transfer problem, **13**:64
Interfacial area concentration, large, **31**:105
Interfacial chemical nonequilibrium, **30**:111
Interfacial drag, in porous media, **29**:42–47
Interfacial forces, balance between, **21**:62–63

Interfacial friction factor, hydrodynamics, slug flow, **20**:89–90

Interfacial heat transfer, dimensionless correlation of data, **21**:285–286

Interfacial instability, severe slugging, **20**:116

Interfacial shear, hydrodynamics, slug flow, **20**:90

Interfacial temperature, **9**:216

Interfacial tension
  bubble-induced entrainment effect, **21**:306
  multiphase flow and, **30**:96

Interfacial tension–liquid density ratio, entrainment onset, **21**:297–298

Interfacial thermal equilibrium, **30**:110

Interferograms, **9**:315; **24**:300

Interferometer (Fabry–Perot), **9**:141

Interferometer adjustment, **6**:211

Interferometer interference contrast, **6**:237

Interferometry, **8**:176, 178, 179

Interior core, inviscid, **24**:282–284

Intermediate friction factors, *see also* Friction factor
  measurement of, **15**:93–94

Intermediate heat transfer, in once-through system, **15**:112

Intermediate regime
  between parallel plates, **2**:300ff
  between unspecified surfaces, **2**:303f

Intermeshing region, twin-screw extruder, **28**:190–192, 195–197

Internal boiling
  and forced convection, **30**:136–141
  natural convection and, **30**:122–136

Internal emissivity, in molecular gas band radiation, **12**:157

Internal energy, **2**:281
  and the accommodations coefficients, **2**:337ff

Internal flow in boiling, **1**:240ff
  analysis, **1**:242ff
  critical heat flux trends, **1**:241f
  experimental data, **1**:244f
  flow patterns, **1**:240f

Internal flows, **8**:162
  effect of viscous dissipation in, **8**:162
  influence of body forces on, **8**:162
  thermal convection in, **15**:195–206
  thermal instabilities in, **8**:162

Internal friction, **SUP**:13

Internal heat and mass transfer in porous bodies, **1**:164ff

boundary conditions, **1**:173f
  dimensionless variables, **1**:174ff
  experiments, **1**:177ff
  solutions, **1**:176
  theoretical development, **1**:165ff

Internal heat generation, **22**:158
  and non-Newtonian heat transfer, **2**:376ff, 379ff

Internally finned tubes, **SUP**:366

Internal problems, **8**:161

Internal radiative transfer, one-dimensional melting and freezing, **19**:42

Internal thermal energy generation, **SUP**:56
  in fully developed flow, **SUP**:82, 158, 224–225, 249, 262, 277–278
  in thermally developing flow, **SUP**:109, 129, 176

Internal transference number, **12**:246

Interphase heat transfer
  liquid fluidized beds, **23**:251
  packed beds, **23**:229
  three-phase systems, **23**:261

Interphase mass transfer, **9**:218
  liquid–solid beds, **23**:258
  packed beds, **23**:229
  three-phase beds, **23**:259

Interphase mass transfer theory, **21**:67–72
  Maxwell velocity distribution function, **21**:6
  net condensation rate, **21**:68–70

Interphase transfer laws for mass, momentum, and energy, **31**:109

Intersection location, of rays, determining, **31**:342–343

Interval-based integration scheme, **31**:392

Intracellular ice, **22**:294

Intrinsic average, **31**:11–13

Intrinsic average mass fraction, **31**:32

Intrinsic constitutive laws, **31**:106

Intrinsic radiation similarity parameters, **3**:213ff

Inverse Abel transform, **30**:295

Inverse power potential, **3**:262, 290, 295

Inverse transformation method, **27**:51, 60

Inviscid asymptotes, **9**:335

Inviscid flows, **8**:13

Inviscid theory, Taylor bubble translational velocity, **20**:101–102

Inviscid theory strength, **26**:10

Ionic components, and equations of change, **12**:200–201

Ion implantation, **31**:461

Ion slip, 1:273
Ion transport
    diffusion and convection in, 12:196
    migration in, 12:196
Iron–carbon alloy, solidification, 28:288
Irradiation
    light, 22:249
    thermal, 22:238
Irregular surfaces, natural convection cooling, 20:209–213
Irreversibility
    defined, 15:7
    of heat exchangers, 15:4–10, 29–33, 52
Irreversibility minimization, applications of, 15:51
Irrotational flow entry, SUP:90
ISG variable, 27:102–103
Isobaric, 25:10
Isoelectric point, 12:259–260
Isoenergetic temperature, 7:327, 328
Isolated bubble region narrow space nucleate boiling, 20:36, 40–41
Isomorphous system, 28:233
Isopropanol–water
    A₀, 16:119
    bubble growth, experimental data, 16:90
Isopropyl alcohol, film boiling of, 11:58
Isosceles triangular ducts, SUP:227–235
    fluid flow in, SUP:227–232
    friction factors for, SUP:230–231, 241
    heat transfer in, SUP:232–235
    hydrodynamically developing flow in, SUP:240–242
    with inscribed circular cores, SUP:349
    Nusselt numbers for, SUP:230–231, 234–235
    thermally developing flow in, SUP:244–246
Isothermal cylinder, potential flow convection, 20:359–360
Isothermal flow, induced rotational flow, 21:144–160
    axial flow effect, 21:156–159
    Couette flow, 21:144–147, 150–151
    eccentricity effect, 21:159–160
    finite annulus length effect, 21:155–156
    Taylor vortex flow, 21:147–152
    turbulent flow, 21:153–155
    wavy vortex flow, 21:152–153
Isothermal gas radiation, 12:162–174; see also Molecular gas band radiation
    cylinder band absorptance in, 12:166–167
    for gas within one wall, 12:167–169
    for gas within two walls, 12:169–174
    script-F interchange factor in, 12:168–174
    slab band absorption in, 12:163–165
    sphere band absorption in, 12:165–166
    well-stirred model in, 12:162–163
Isothermal hydraulic resistance coefficients, ratios of, 31:195–196
Isothermal surfaces, 2:401
    potential flow convection
        conic, 20:367–368
        elliptical cylinders, 20:361–362
        finite height strips, 20:362–363
        flat plate streamline conduction, 20:357–358
        infinitely long pointed needle, 20:368–369
        planar motion, 20:355–356
        spherical, 20:369–370
            oblate axisymmetric potential flow, 20:373
            prolate axisymmetric potential flow, 20:372
            thin disk of finite diameter, 20:373–374
            transverse curvature, 20:370–371
        spherical cap, 20:374
        wedge, 20:363–364
Isotherm migration method, 19:40–41
Isotopic disturbance or turbulence
    homogeneous medium and, 14:260–262
    model of, 14:260–262
Isotropic pressure of radiation, 21:247
Isotropic scattering, 21:255–256; 27:157; 31:364–373
Isotropic systems, 31:87
Isotropic turbulence, velocity fluctuation dissipation rate, 21:186
Ispra
    α-mode failure experiments, 29:147
    KROTOS facility, 29:178–179, 182
ISW variable, 27:102–103
Iteration, 8:28, 35
IVA3 code, 29:151
IW variable, 27:109, 205

J

Jackson's theory, 2:343ff
Jacobian ray weighting, 31:362–364

Jacobi-elliptic integral, **8**:47
Jacob number, **15**:246; **30**:34
JAERI experiments, **29**:165
Jakob number, **1**:218f; **16**:86
Jamin interferometer, **6**:197
Jelling
  curve, **25**:173
  point, **25**:173
Jet-ambient mechanisms, **16**:3
Jet combustion, hydrogen, during direct containment heating, **29**:299–302
Jet engines, combustion chambers, **27**:173–181
Jet flow, heat and mass transfer in, **13**:1–58; *see also* Impinging flow
Jet impingement, *see also* Liquid jet impingement
  heat transfer, **23**:328–331
  hydrodynamics, **23**:2–6
  inviscid pressure and velocity, distributions, **23**:4–5
  jet configurations, **23**:2–3
  local saturation conditions, **23**:5
Jet impingement boiling, **23**:1–123; *see also* Critical heat flux; Nucleate boiling; Transition boiling
  film boiling, **23**:9, 117–120
  high-density integrated circuits, **20**:257–258
  nucleate boiling, **23**:7–8
  research needs, **23**:120–123
  single-phase forced convection, **23**:7
  study summaries, **20**:254
  system-specific effects, **23**:9
  thermal cooling of electronic components, **20**:251–258
    free jet impingement, **20**:251–252
    submerged jet impingement, **20**:252–258
  transition boiling, **23**:8–9
  vs. pool boiling, **20**:252–253
Jct inclination, liquid jet impingement, **26**:182–189
Jet pulsation, liquid jet impingement, **26**:198–201
Jets, *see also* Impinging flow; Momentum jet
  axial, **25**:131, 132
  behavior studies, **16**:53
  buoyant, **9**:339
    classes, **16**:13
    equation of state, for modeling, **16**:19–20
    experimental studies, **16**:13–16

  characteristics, **16**:2–3
  circular, **25**:131
  concentration profiles, **16**:5
  decay, models, **16**:2, 32–33
  definition, **16**:3
  effects of ambient stratifications, **16**:42–51
  entrainment modeling, for range of Froude number values, **16**:30–36
  far field, **16**:4
  flow establishment, general formulation, **16**:23–25
  flow regimes, **16**:3–4
  gas, *see* Gas jet
  impinging, **25**:136
  near field, **16**:4
  nonbuoyant, **9**:339
  physical dimensions, coordinate system for, **16**:4–6
  round, *see* Round jet
  steady axisymmetrical, **13**:65–66
  submerged, **25**:131
  swirling, **13**:41
  temperature difference decay, entrainment modeling, **16**:32–33
  temperature profile, **16**:5
  trajectory
    calculation in unstratified quiescent ambients, **16**:25–30
    coordinate system for, **16**:4–6
    modeling, **16**:2, 34–35
    prediction, **16**:6–13
    in stratified quiescent ambients, **16**:43–46
  turbulent buoyant, diffusion, **16**:1–58
  in unstratified flowing ambients
    comparative calculations, **16**:36–42
    pure initial coflow, **16**:39–42
  in unstratified quiescent ambients, comparative calculations, **16**:25–36
  vapor, Zuber hydrodynamic instability model, **16**:130
  velocity decay, modeling, **16**:32–33
  width, prediction, **16**:6–13
  zone of established flow, **16**:4
    dimensionless variables, **16**:53–54
    governing equations, **16**:8, 21
  zone of flow establishment, **16**:3–4
Jet splattering, liquid jet impingement, **26**:195–198
Jet velocities, very high, CHF and, **17**:21–22
J-factor, **15**:126; **19**:256; **21**:171; **25**:260

Johns Hopkins University, **15**:84
Joint probability density distribution, **8**:337, 341
Joulean dissipation of energy by diffusion, **21**:272
Joule heating, **14**:127
  tissue, **22**:243
Joule's law, **1**:275
Jump conditions
  species mass for air and water, **31**:37
  for volume-averaged quantities, **31**:28
"Jumping of the latent heat peak", **19**:27
Just noticeable difference (JND), **4**:121

K

Kantorovich method, **19**:196–197, 224, 242
Kapitza resistance, **5**:484
  heat transfer in liquid helium and
    anomalous problem, **17**:68
  basic considerations, **17**:67–68
  high heat flux results, **17**:73–74
  influence of solid characteristics, **17**:68–70
  numerical values, **17**:70–71
  present directions, **17**:71–73
  summary, **17**:74–75
Kármán integral method, **25**:288
Kármán–Nikuradse expression, **19**:325
Kármán–Nikuradse–Martinelli three-layer model, **19**:323
Kármán–Pohlhausen integral technique, **26**:114, 118
Kármán–Pohlhausen method, **SUP**:69, 160, 192, 194; **19**:34; **21**:109–110
  for laminar boundary layer, **11**:68
Kellog catalyst-regenerator bed, **19**:134–135
Kelvin effect, **31**:98
Kelvin equation, **10**:87–92
Kelvin–Helmholtz instability, **30**:19
Kern–Seaton correlation, **31**:443
Kinematic energy dissipation rate, nearly homogeneous turbulence, **21**:194
Kinematic gas viscosity, **19**:99
Kinematic reference time, **31**:147
Kinetic energy
  during laser ablation, **28**:112–116
  of molecular rotation, **2**:277
  of molecular translation, **2**:276f
  of molecular vibration, **2**:277

strong nearly homogeneous turbulence, **21**:207–208
  turbulent, **2**:72ff
Kinetic energy correction factor, **SUP**:43; *see also appropriate ducts in the text*
Kinetic microscales, **30**:5–9
Kinetic theory, **9**:218
  moving liquid droplet, **26**:31–32
King's law, **8**:294
Kirchhoff's laws, **1**:3; **2**:402; **3**:182f; **5**:18, 262; **11**:321; **12**:128; **27**:11–12
Kirpichev number, **1**:175; **24**:80
KIVA-DCH model, direct containment heating, **29**:233
Klinger continuum model
  comparison with other models, **22**:45, 46, 58–59
  governing equations, **22**:37
  Green's function solution, **22**:37–40
  mean tissue temperature
    effective thermal conductivity, **22**:44
    multiple models, **22**:40–41
    vascular geometry variations, **22**:42
Knudsen condensation coefficient, **5**:353
Knudsen layer, **28**:111, 125
Knudsen number, **1**:6; **2**:273, 275; **3**:2; **7**:164; **9**:357; **21**:72–73; **28**:351
Knudsen's method, **2**:304, 317ff
Kogan-multistage flash evaporator process, **15**:271
Kolmogorov microscales, **9**:115; **21**:265
Kolmogorov–Obukhov scales, **25**:325
Kolmogorov scales, **21**:268, 269; **30**:7, 18, 48, 62, 74
  radiation-affected thermal, **21**:269–270, 273
  viscous oils, **21**:267
Kolmogorov spectrum, **25**:335, 344
Konovalov's first rule, **16**:68, 82
Konsetov model, **21**:279
Konsetov turbulent heat transfer model, **21**:331–332
Kossovich number, **1**:175
Kozeny–Carman equation, **25**:151
Kozeny–Carmen equation, **28**:266
Kozeny constant, **25**:293, 313
  modified, **25**:297, 313
Kozicki Reynolds number, **19**:331
Kramers–Unsöld formula, **5**:303
Kreiss difference scheme, **14**:51
Kreith-Summerfield correction, **2**:387

Krischer's method, **1**:146ff
Kronecker delta, heat- and mass-transfer, **20**:324–325
KROTOS experiments, **29**:147, 165, 178–180, 182
Kubelka–Munk theory, light absorption and scattering, **22**:278
Kutateladze number, CHF in helium and, **17**:324–325, 334

L

Labile atoms, **3**:269ff
Labuntzov–Kryukov theory, **21**:72–73, 79
Ladenberg–Reiche function, **5**:283
Lagendijk model, **22**:87–88
Lagrange–Burmann expansions, interfacial, **19**:32
Lagrange equations, **19**:198
Lagrange stress function, **19**:258
Lagrangian derivatives, elimination of, **31**:114
Lagrangian function (Lagrangian), **24**:40, 41, 42, 43
Lagrangian integral time scale, **9**:149
Lagrangian thermodynamics, **8**:23
Laguerre polynomials, **8**:69
Lakes, as heat sinks, **12**:10
Lakes and estuaries, heat and mass transfer in, *see* THIRBLE
Lake surface
    air boundary layer above, **13**:79–82
    boundary conditions for, **13**:83–84
    boundary layer in water at, **13**:82–85
    constitutive equation for forces acting on, **13**:184–192
    continuous, **13**:186
    convective transport in, **13**:175–192
    GENMIX computer and, **13**:84
    geometry and physics of, **13**:82–83
    gravitational effects in, **13**:191
    hydrostatic equilibrium and, **13**:186
    moisture in, **13**:142
    quasi-steady state transport in, **13**:190
"λ point", **5**:330
Lambert's cosine law, **2**:402f, 405; **27**:9
Laminalization, flow core, **25**:95
Laminar boundary flow, entropy generation surface for, **15**:15

Laminar boundary layer, heat transfer in, **11**:62–68; *see also* Boiling boundary layer; Boundary layer equations
Laminar burning velocity, **29**:75–77, 78–79
Laminar convection, Rantz–Marshall correlation, **28**:8, 11
Laminar convection problems, approximate solutions to, **11**:265–310
Laminar flame
    forced flame, **30**:52–60
    natural flame, **30**:65–72
    quenched, **21**:261–262
Laminar flow, *see also appropriate ducts for solutions*
    approximate analysis for, **11**:267–272
    in a circular tube, **8**:22
    definition of, **SUP**:5
    early studies of, **SUP**:2, 78–79
    experimental heat transfer studies, **19**:304–305
        Newtonian fluid, **19**:306–316
        non-Newtonian fluid, **19**:316–322
    external, **2**:389ff
    forced convection problems, **19**:75–77
    forced flow in round pipes, **21**:6
    heat transfer and, **15**:98–104; **30**:50–51
    heat transfer from vapor inside film with stable interface, **11**:62–84
    hydrodynamically developed
        friction factor
            Newtonian fluids, **19**:261–262
            non-Newtonian fluids, **19**:262–268
        velocity profile
            Newtonian fluids, **19**:257–259
            non-Newtonian fluids, **19**:259–261
    hydrodynamically developing, **SUP**:6; **19**:268
        hydrodynamic entrance length, **19**:274–276
        incremental pressure drop, **19**:272–274
        velocity profiles, **19**:269–271
    hydrodynamically fully developed, **SUP**:5
    inertial terms in, **11**:308
    instability of, **9**:321
    internal, **2**:388
    mathematical model, **28**:3–10
    mean drag of banks, **18**:150
    microgravity, **30**:46
    in packed beds, **23**:194
    simultaneously developing, **SUP**:7; **19**:300
        Newtonian fluid, **19**:300, 301
        non-Newtonian fluid, **19**:301–306
    summary index of solutions for, **SUP**:386–388

Laminar flow (*continued*)
　　thermally developed
　　　Newtonian fluid, **19**:277–280
　　　non-Newtonian fluid, **19**:280–289
　　thermally developing
　　　Newtonian fluid, **19**:289–291
　　　non-Newtonian fluid, **19**:291–297
　　　thermal entrance length, **19**:297–300
　　thermally fully developed, **SUP**:7
　　transition of, **9**:321
　　transition to turbulent, **11**:301–303
　　two-phase, **30**:37–39
　　with unstable film interface, **11**:84–85
　　vertical surface, supercritical pressures, **21**:37–
　　　39
　　viscosity measurements and, **15**:70–77
Laminar free convection, **11**:274–275; *see also*
　　Convective heat transfer; Laminar heat
　　transfer
　　from nonuniform temperature difference,
　　　11:77
　　for vertical flat plate, **11**:272–274
Laminar heat transfer, *see also* Laminar free
　　convection
　　by free convection, **11**:267–272
　　from circular cylinder, **11**:277–280
　　empirical correlations for, **15**:101
　　experimental results of, **15**:102–103
　　fully developed, **15**:98–99
　　to non-Newtonian fluids in rectangular pas-
　　　sages, **19**:248–249
　　from sphere, **11**:280–281
　　in thermal entrance region, **15**:100–104
Laminarization, **8**:320
Laminar jet, **26**:152–153
　　axisymmetric jet
　　　heat flow, **26**:156–157
　　　radical flow region, **26**:114–119
　　　stagnation zone, **26**:109–113, 123, 127–130
　　planar jet
　　　parallel flow region, **26**:122–123
　　　stagnation zone, **26**:119–122
Laminar mixed convection
　　in external flows of non-Newtonian fluids,
　　　15:184–188
　　in inelastic fluids, **15**:184–187
　　for viscoelastic fluids, **15**:187–188
Laminar momentum, along flat plate, **15**:14–16
Laminar natural convection
　　irregular surfaces, **20**:209–213
　　Taylor bubble velocity, **20**:105

Laminar pipe flow, entropy generation profiles
　　in, **15**:14
Laminar sublayer, **3**:43; **9**:123
Laminar thermal convection
　　in external flows of viscoelastic fluids,
　　　15:178–184
　　for horizontal cylinder, **15**:170–172
　　for slender vertical cone, **15**:172–173
　　Sutterby model of, **15**:166
　　from three-dimensional axisymmetric surface,
　　　15:174–176
　　transient, **15**:177–178
　　variable temperature case in, **15**:161–162
Laminar thermal convection from vertical flat
　　plate
　　constant heat flux case in, **15**:167
　　constant temperature case in, **15**:168–169
　　to other time-dependent models, **15**:166–169
Laminar thermal convection heat transfer
　　constant heat flux case in, **15**:156–157
　　exact solutions in, **15**:153–157
　　to power law fluid from other geometrical
　　　surfaces, **15**:169–176
　　to second-order fluid, **15**:179
　　to two-dimensional surface, **15**:169–173
　　from vertical flat plate to power law fluid,
　　　15:151–165
Laminar thermal convection in power law fluid
　　approximate integral solutions in, **15**:157–162
　　constant heat flux case in, **15**:163–165
　　constant temperature case in, **15**:158–161
　　theoretical predictions vs. experimental data
　　　in, **15**:162–165
Laminar–turbulent flow solutions, **11**:303–306
Laminar–turbulent transition, for constant flux
　　heating of liquids in vertical pipes, **15**:196
Laminated media
　　heat transfer in, **15**:286–287
　　linear diffusion in, **15**:283–321
　　materials requirements in, **15**:284
Landau's theory, **2**:344
Langer's transformation, **8**:21
Langhaar velocity profile, **SUP**:70, 87, 138, 145,
　　152, 193
Langrangian function, **19**:198–199
Laplace equation, **5**:43; **10**:87–92; **16**:64
　　in finite difference form, **5**:43
　　pressure difference, **21**:116
Laplace–Kelvin equation, **10**:87, 92, 95
Laplace method, **2**:244f
Laplace transform, **8**:38, 50

balanced-symmetric regenerators, **20**:173–175
unidirectional regenerator operation, **20**:151–152
Laplace transformation, **19**:16, 195; **25**:100, 108
Large eddy, **25**:324
  simulation
    applications, **25**:370–396
    literature, **25**:332–333
    prospects, **25**:333, 396–401, 405–407
  turnover time, **25**:325, 329, 389, 400
Large path length limit, **8**:253, 257, 264
Large scale equations, **25**:338
Laser ablation, **28**:109–110
  kinetic energies, time-of-flight, **28**:112–116
  surface topography, **28**:101–109
Laser–Doppler method, **8**:298, 342
Laser–Doppler velocimetry (LDV), liquid jet
    impingement, **26**:137, 142, 158
Laser–Doppler velometer, **9**:136
Laser-induced thermal chemical vapor deposition, **28**:342–343
Lasers, **22**:10
  heating, popcorn effect, **22**:321
  high pulsed power, **22**:319
  irradiation, **22**:248
  pulsed laser melting, **28**:75–109, 135
  pulsed laser sputtering, **28**:109–123, 136
  pulsed laser vaporization, **28**:125–135, 136
  steam formation, **22**:253
  temperature, blood flow, **22**:266
Laser speckle photography
  heat and mass transfer, **30**:267–306
  operating principles, **30**:256–267
Latent heat, **9**:190; **22**:183, 188
  equivalent localized, **22**:202
  pattern, nonlinear, **22**:192
  of phase change, large, **31**:105
  of vaporization, **2**:121
Latent-heat transport, heat transfer models, **20**:28, 31–33
Law of large numbers, **5**:12
Law of mass action, **2**:123f
Law of the Wake, **16**:307
Law of the wall, **16**:309, 346
Laws of thermodynamics, **21**:240–243
Lax–Wendroff scheme, **24**:242
Layered gage, **23**:288–298
  advantages, **23**:291, 298
  Heat flux microsensor, **23**:296–297
  pattern, **23**:295
  resistance temperature devices, **23**:288

sensitivity, **23**:291–292
  thermal resistance thickness, **23**:292–293
  thermocouples, **23**:288–290
  thermopile, **23**:291
  transient response, **23**:292
Lead, **9**:122
Leading edge, **25**:298
Leading edge effect, **9**:280, 311
  propagation of, **9**:321
Lead–tin alloy, **28**:236, 239
  macrosegregation, **28**:291–292, 295–304, 317
  solidification, **28**:288–304, 309, 322
Least squares fit, **22**:368, 385–386
Least squares methods
  approximate solution by, **SUP**:66
  continuous, **SUP**:66
  discrete, **SUP**:66
  point-matching, **SUP**:66
Leaving radiant flux, **2**:405
LeChatelier rule, **29**:69
LEC process, *see* Liquid-encapsulated
    Czochralski process
Leddineg instability, **31**:148
Lee's approximations, **2**:201f, 237
Le Fevre–Rose theory, **21**:92–96
Leibnitz rule, **3**:202
  three-dimensional representation of, **31**:13
Leidenfrost boiling, **9**:83, 87
Leidenfrost phenomenon, **5**:56, 90
Leidenfrost point, **16**:133
  film boiling at, **16**:139–142
Length-scale constraints, **31**:16–23
  associated with local form, **31**:19–21
  decomposition of, **31**:15
  disparate, **31**:16
Lennard–Jones potential function, **3**:262f, 291, 295
Lenticular nucleus, free energy of formation of, **15**:230
LES, *see* Large eddy, simulation
LETOT, *see* Large eddy, turnover time
Lévêque approach, **2**:366, 371ff
Lévêque equation, **2**:371, 375
Lévêque method, **SUP**:100, 115, 121
Lévêque series, **SUP**:106
Lévêque solution, **SUP**:105–107, 171–173
Lévêque-type solutions, **SUP**:216, 242, 251, 267, 271–272, 365
  extended, **SUP**:126–127, 182, 302–312
Lever law, alloy solidification, **28**:262, 263
Levitation, bed, **25**:189, 231

Lewis number, 1:283; 2:47f, 117; 3:280; 4:270f, 320
  frozen, 4:275
  heat- and mass-transfer
    enclosed porous layer, 20:339–341
    high Rayleigh number convection, 20:328–329
    horizontal direction, 20:332
    point source, 20:344–345
    time-dependent concentration, 20:342–343
    vertical direction, 20:320–322
Liebmann method, SUP:64, 216, 237
Light, 27:3–7
Light absorption and scattering, 22:249, 278, 279
Lighthill gas, 2:133
Light-liquid density, bubble-induced entrainment effect, 21:304, 306
Light-liquid viscosity, bubble-induced entrainment effect, 21:305, 307
Light-water reactors, heat transfer, 29:2–3, 131
Limited flight path tests, direct containment heating, 29:226–227
Linear diffusion
  double-contact resistance in, 15:291–292, 319–320
  equation for, 15:287–288
  existence of solution in, 15:297–306
  existence theorem and, 15:298–299
  homogenization of internal and external boundary conditions in, 15:293–297
  interface conditions in, 15:288–292
  linear contact resistance in, 15:289, 315
  n-layer limiting cases in, 15:308–310
  perfect thermal contact in, 15:289, 309–210, 314
  specialization to two-layer slab with double contact resistance, 15:310–319
  Sturm-Liouville problem in, 15:299–302
  unified n-layer model in, 15:292, 297–298
  uniqueness of solution in, 15:306–308
Linear heat conduction problem, 19:193
Linear heating technique, thermal conductivity measurement in structured liquids, 18:176
Linear heat transfer, space-time relationship in, 15:287
Linearization methods, SUP:70–73
Linearized solutions
  Langhaar's, SUP:70–71, 87, 163, 210, 240, 242, 297
    stretched coordinate, SUP:71–72, 87, 163, 210, 241, 242, 298
    Targ's, SUP:71, 87, 298
Linearly varying wall temperature, SUP:119, 219, 252, 316
Linear stability analysis
  heat- and mass-transfer, 20:319
  linear behavior and, 15:207
Linear stability theory, 9:323; 14:17–29
Line broadening, 8:231
Line fire
  convection above, 10:250–251
  mathematical model of, 10:254
  plume instability in, 10:253–254
Line-focusing collectors, 18:19–27, 75
  compound parabolic, 18:20–25
  natural convection, 18:19–27
Line heat-source method, 10:48–49
Line integral method, see Contour integral method
Line intensity, 8:232
  variation of, with wavenumber, 8:233
Line radiation, 5:277
  collision broadening, 5:278
  doppler broadening, 5:279
  line broadening, 5:277
  in a plasma, 4:234f
Lines, magnetic field, 25:154, 190, 229, 237
Line source, 9:287
Line source flow method, 3:274
Liouville's differential equation, 8:20
Liquefied natural gas, 9:405
  steam explosions, 29:130
Liquid cooling, discrete thermal sources, 20:213–217
Liquid crystals
  constant heat flux method, 23:328–331
  heat flux measurements, 23:323
Liquid deficient region, narrow space nucleate boiling, 20:39
Liquid drop, equilibrium vapor pressure of, 10:87
Liquid droplets
  drag and stability, 21:333–342
  drag coefficient, 21:334–337
Liquid drop mixing, 31:161
Liquid-drop model, 9:199
Liquid-encapsulated Czochralski process (LEC process)
  described, 30:322

Group III–V compounds, **30**:424
heat transfer in, **30**:338, 424
phase transformations, **30**:335, 337
simulation studies, **30**:348–349
Liquid fallback, severe slugging, **20**:118–119
Liquid film characteristics
hydrodynamics, slug flow, **20**:88–92
low liquid level nucleate boiling, **20**:58–63
thickness, heat transfer models, **20**:33–35
Liquid film cooling, **7**:324
Liquid film flow rate, **1**:388
Liquid film thickness measurements, **1**:384ff
Liquid-fluidized bed, **10**:168; **25**:152; *see also*
Fluidized bed
Liquid–gas multicomponent systems
boiling of multicomponent liquids, **30**:173–
174
convection in unsaturated media, **30**:167–168
drying of porous media, **30**:168–172
flow and heat transfer, **30**:167–184
nonaqueous phase liquids, **30**:174–183
Liquid–heating surface contacts
liquid heating around, **18**:250–252
mean contact time
correlation of experimental data, **18**:288
with ethanol boiling, **18**:278
with Freon-113 boiling, **18**:277
with water boiling, **18**:279
mean relative contact time
with ethanol boiling, **18**:284
with Freon-113 boiling, **18**:283
with water boiling, **18**:284
processes, **18**:248–253
hydrodynamic instability of vapor–liquid
interface, **18**:249–250
liquid heating around liquid–wall contact,
18:250–252
vapor bubble growth and evanescence,
18:252–253
Liquid helium, **5**:91, 329
film boiling of, **5**:91
natural convection, **5**:375
Liquid holdup, **1**:381ff, 388
Liquid hydrogen, **5**:326
forced convection boiling, **5**:403
natural convection in, **5**:375
thermodynamic properties of, **5**:327
Liquid jet, **26**:106–107; *see also specific liquid
jets*
array, **26**:170–182, 205

cooling, **26**:105, 170, 205
heat transfer, **26**:189–193
hydraulic jump, **26**:114, 201–203, 205
jet inclination, **26**:182–189
jet pulsation, **26**:198–201
mass transfer, **26**:124–126, 147–162, 171–174
modified impingement surface, **26**:189–193
motion of impingement surface, **26**:203–204
stagnation zone, velocity gradient, **26**:109–
110, 120
wall roughness, **26**:193–195
Liquid jet impingement
arrays, **26**:170–182
flow structure, **26**:136–147
free-surface jet, **26**:109–123, 135–170
heat transfer
arrays, **26**:171–179
jet inclination, **26**:182–189
jet pulsation, **26**:198–201
liquid jets, **26**:124–127, 131–135, 147–162
modified impingement surface, **26**:189–193
motion of impingement surface, **26**:203–
204
wall roughness, **26**:193–195
hydraulic jump, **26**:114, 201–203, 205
hydrodynamics, **26**:136–147, 157
jet pulsation, **26**:198–201
jet splattering, **26**:195–198
laser–Doppler velocimetry (LDV), **26**:137,
142, 156
metals, **26**:204
Nusselt number, **26**:206
submerged jet, **26**:123–135
theory, **26**:107–123
transport, factors affecting, **26**:182–204
Liquid layers
immiscible, entrainment onset, *see* Entrainment
initially stratified, *see* Bubble-induced
entrainment
Liquid–liquid interface
models for heat transfer, **21**:279–281
Nusselt number, **21**:280, 285
Liquid–liquid systems, hydrodynamic fragmentation, **29**:170–175
Liquid metal, **6**:437
Liquid metal convection, **20**:376–379
Liquid metal heat transfer in ducts, **3**:101ff
boiling, **3**:109f
condensation, **3**:108ff

Liquid metal heat transfer in ducts (*continued*)
   film, **3**:108f
   high velocity, **3**:109
   ordinary, **3**:109
  effectiveness coefficient, **3**:147, 149
   fully developed, **3**:147
  effects of
   axial conduction, **3**:106f
   properties, temperature dependent, **3**:104, 108
   Reynolds number, small, **3**:104, 106f
   thermal convection, **3**:107f
   wall heat flux
    cosine, **3**:135f
    nonuniform, **3**:131ff
    uniform, **3**:127ff, 138
  forced convection, laminar flow, **3**:106ff
  fully developed heat transfer condition, **3**:133
  in
   annular spaces, **3**:127ff, 131ff, 136ff
    eccentric, **3**:106
    externally heated, **3**:129ff
    internally heated, **3**:127ff
   ducts
    specified heat flux, **3**:126ff
    symmetrical, **3**:105
    unsymmetrical, **3**:105f
   heat exchangers, **3**:140ff
    concurrent flow, **3**:154ff
    countercurrent flow, **3**:165ff
    double-pipe, **3**:141ff
    effectiveness, **3**:140, 159
    efficiency, **3**:140, 149, 151f, 157, 169
  magnetohydrodynamics, **3**:110
  Nusselt number, **3**:129, 130f, 133f, 136, 138, 153, 159ff, 170
   local, **3**:163ff
  turbulent convection, **3**:111ff
   axial heat diffusion, **3**:116ff
   eddy diffusivities, **3**:123ff, 171
   plug flow idealization, **3**:119ff
Liquid metals, **8**:70; **9**:32, 33
  film boiling of, **5**:74
  forced convection, **30**:24
  heat transfer microscales, **21**:266–267
  nucleation of, **10**:157–162
Liquid metal sprays
  droplet formation, **28**:18–21
  splat quenching, **28**:23–54
Liquid mixture, *see also specific mixture*

  azeotrope, **16**:63
   linear mixing law, **16**:124
   phase equilibrium diagram, **16**:62–63
   work of bubble formation, **16**:67
  binary
   boiling site density, **16**:75–77
   bubble growth, **16**:82–90
   bubble growth experiments, **16**:90–94
   condensation, **16**:147
   convective boiling, **16**:144
   ideal, **16**:62
   slope of vapor pressure curve, **16**:65–67
  boiling
   effect of pressure, **16**:112–116
   from enhanced surfaces, research needs, **16**:150
  contact angle
   effect on boiling inception, **16**:68–70
   effect on boiling site density, **16**:77
  convective boiling, **16**:142–143
  film boiling, **16**:133–138
   effect of relative volatility, **16**:137
   temperature and composition profiles, **16**:135
   vs. pure fluid, **16**:135–136
  multicomponent
   boiling, **16**:59–156
   research needs, **16**:150–151
  nucleate boiling, **16**:142
  subcooled boiling, **16**:110–113
  ternary
   boiling, **16**:116–117
   research needs, **16**:151
Liquid nitrogen
  film boiling of, **5**:91
  natural convection, **5**:375
Liquid oxygen, thermodynamic properties of, **5**:327
Liquid periods, heat transfer models, **20**:29–31
Liquid phase
  closure for, **31**:57, 63–64
  heat transfer process in, **31**:43–46
  volume-averaged thermal energy equations for, **31**:53
  volume-averaged transport equations for, **31**:28–30
Liquid-phase continuity equation for water, **31**:24
Liquid-phase epitaxy, **28**:341
Liquid-phase transport, **31**:37

Liquids
  complex, classification, **18**:164–165
  ebullition and cavitation in, **10**:92–95
  flow of, over single tubes, **8**:128
  forced convection, **30**:20–24
  heat transfer in, **8**:128
  heat transfer microscales, **21**:267–268
  inertia force, effect on bubble departure, **16**:97–99
  low thermal conductivity, **10**:148–157
  multicomponent, boiling of, **30**:173–174
  nonpolymeric, *see* Nonpolymeric liquids
  polymeric, *see* Polymeric liquids
  sheared structured, thermal conductivity measurement, **18**:176–178
  single-component, work of bubble formation, **16**:67
  stratified immiscible, heat transfer by gas bubbling across interface, **21**:279–286
  superheated, *see* Superheated liquids
  temperature distribution in, **10**:136–139
  temperature measurement, speckle photography, **30**:302–304
Liquid slug
  bubble velocity, **20**:107–108
  length and slug flow, **20**:95–97
  liquid holdup in, **20**:109–110
  zone
    geometry, **20**:85–86
    slug pressure drop, **20**:92–95
Liquid–solid fluidized beds, **23**:231
  axial dispersion, **23**:250
  bed expansion, **23**:241
  heat transfer, **23**:251
  mass transfer, **23**:251
  minimum fluidization velocity, **23**:233
Liquid–solid fluidized systems, **10**:179–180
Liquid–solid interface, **31**:26, 54–55, 65–66
Liquid stimulators, **31**:276
Liquid subcooling, dryout heat flux and, **29**:49
Liquid superheat, **11**:3–9
  in boiling, **11**:9–17
  Boltzmann principle in, **11**:6
  bubble growth in, **11**:11–17
  Döring–Volmer theory in, **11**:7, 9
  heating surface and, **11**:40
  incipient boiling temperature in, **11**:3–4
  maximum, **11**:4–9
  in pool boiling, **11**:30–45

Volmer theory and, **11**:7–9
Liquid superheat fluctuation, statistical characteristics of, **11**:30–45
Liquid–vapor interface, *see also* Vapor–liquid interface
  curvature of, **10**:122–125
Liquid–vapor interface phenomena, **9**:215
Liquid viscosity-controlled regime, **23**:95
Liquid volume fraction, **1**:365, 381f, 390, 392
Liquid–wall contacts
  frequency
    effect of heating-surface orientation with ethanol boiling, **18**:287
    with ethanol boiling, **18**:282
    with Freon-113 boiling, **18**:281
    with water boiling, **18**:282–283
  hystograms of distribution with ethanol boiling, **18**:277
  mean, effect of heating-surface orientation with ethanol boiling, **18**:286
  mean relative, at point q_{max} under different subcoolings, **18**:276
LMTD method, *see* Log-mean temperature difference method
Loading ratio, **9**:116
Local chemical equilibrium, **2**:46
Local entropy generation, in convective heat transfer, **15**:11–25
Local entropy generation rate, nonmonotonic dependence of, **15**:18
Local form
  length-scale constraints associated with, **31**:19–21
  transformation of nonlocal form to, **31**:18–19
Local heat flux, **26**:34
Local heat transfer
  array, **26**:179
  to flat plates, **8**:116
  forward stagnation point, **8**:118
  influence of wall temperature distribution on, **8**:126
  jet inclination, **26**:184–186
  jet pulsation, **26**:198–201
  liquid jets
    axisymmetric jet, **26**:182
    free-surface jet, **26**:159–160, 167–169
    submerged jet, **26**:130–133
  to single tubes, **8**:116
  variation in, **15**:175
  to wedge-shaped bodies, **8**:117

Local heat transfer coefficients, *see also* Heat transfer; Heat transfer coefficients
  slanted and horizontal tubes, **14**:160–172
  for vertical tubes, **14**:207–212
Local liquid holdup, quasi-equilibrium slugging, **20**:125–126
Locally homogeneous flow (LHF) model, **26**:54
Locally similar solution, **2**:145ff
  for catalytic reaction around a hypersonic blunt body, **2**:157ff
Local mass equilibrium, **31**:52–53
Local potential, **24**:39–40
Local similarity, **4**:333ff
Local thermal equilibrium, **4**:246ff; **23**:370, 426, 451–455; **31**:53–56
Local thermodynamic equilibrium, **11**:321; **12**:127
  narrow-band properties in, **12**:134
  principle of, **31**:50
  radiative decay of temperature fluctuations and perturbations for, **14**:251–255, 259–263
Lockart-Martinelli correlation, **1**:370f, 374, 379, 384
Log-mean temperature difference method, three-fluid heat exchanger design, **26**:306–309, 320–321
Loitsyanskiy's invariant, **21**:198
Long hot wire method, **3**:274f
Longitudinal flow
  heat transfer enhancement in tube bundles in, **31**:232–259
  transverse fins in tube bundles in, **31**:249–253
Longitudinal rolls, **9**:345
Longitudinal thin fins, V-shaped, **SUP**:374–375
Longitudinal thin fins, **SUP**:366–384
  from opposite walls within rectangular ducts, **SUP**:372–374
  within a circular duct, **SUP**:367–370
  within square or hexagonal ducts, **SUP**:370–372
Longitudinal triangular fins, within a circular duct, **SUP**:375–378
Long regenerator
  recuperative operation, **20**:155, 157
  unidirectional operation
    temperature distribution–unbalanced regenerator, **20**:161–170
    three-dimensional plot, **20**:169–170
    vanishing transfer potential, **20**:158–170

Long Valley (Calif.), hot springs discharge at, **14**:49
Loop flow impedance, **31**:126
Loop segment, conveniently selected primary, **31**:127
Loop system
  flow distribution in, **31**:126–133
  stability of, **31**:149
Lorentz broadening, **5**:278
Lorentz forces, **19**:80
  alloy solidification, **28**:247, 316, 317
Lorentz line profile, **8**:232
  influence of pressure on, **8**:232
  for strong overlapping lines, **8**:242
Lorentz–Lorenz equation, **6**:284
Lorentz profile, **5**:282
Lorentz transformation, **1**:274
Los Alamos Scientific Laboratory, **14**:97
Loschmidt, method of, **3**:283f
Loss-of-coolant accident, nuclear accidents, **29**:59
Loss-of-flow accident, nuclear accidents, **29**:1–2
Lost available work
  defined, **15**:7
  reference heat reservoir and, **15**:9
Lost energy, in heat transfer, **15**:4–10
Lost heat, **21**:240
Lothe–Pound nucleation rate equation, **14**:309–310, 337–339
Lothe–Pound theory, **14**:337–339
Low electromagnetic work, entropy production, **21**:270–272
Low energy surfaces, **9**:229
Lower flammability limit, **29**:303
Low-gravity environment, **24**:312
Low heat regions, surface inclination, **20**:27–28
Low liquid level nucleate boiling, **20**:55–63
  heat-transfer characteristics, **20**:55–58
  in liquid film, **20**:58–63
Low pressure chemical vapor deposition (LPCVD), **28**:344, 399–408
Low thermal conductivity liquids, **10**:148–162
Low-yield processes, **21**:141
Lubrication, **25**:298
  glass softening, **24**:26–29
  melting, **24**:1–38
Luikov number, **1**:175
Lumped models, **28**:10, 11
Lumped-parameter analyses, **14**:91–96
Lumped-parameter modeling, **31**:138

Luna missions, **10**:1, 10, 41, 53, 58
Lunar basalt, *see also* Moon
  specific heat of, **10**:60, 62
  thermal conductivity of, **10**:44–45, 57
  thermal diffusivity of, **10**:64
Lunar breccia, specific heat of, **10**:61
Lunar fines, *see also* Lunar materials
  average composition of, **10**:25
  core–tube density and, **10**:55–56
  line–heat–source method for, **10**:49
  in mare regions, **10**:42
  properties of, **10**:23–24
  simulated, **10**:50
  solar albedo for, **10**:28
  specific heat of, **10**:63–64
  spectral directional reflectance for, **10**:23–29
  thermal conductivity of, **10**:45–59
  thermal diffusivity of, **10**:68–70
Lunar materials, *see also* Lunar fines; Lunar
    rocks
  backscattering from, **10**:20–21
  bidirectional reflectance apparatus for, **10**:17
  bidirectional reflectance results for, **10**:19–21
  directional reflectance apparatus for, **10**:16–17
  directional reflectance results for, **10**:21–30
  directional spectral emittance for, **10**:15–16
  experimental results for, **10**:19–34
  "fairy castle structure" of, **10**:10
  fines properties in, **10**:23–24
  heat transfer within lunar surface layer in,
    10:39–80
  line heat-source method for, **10**:48–49
  measuring techniques for, **10**:16–19
  oldest, **10**:6–7
  specific heat of, **10**:39–40
  spectral emittance of, **10**:18–19, 30–34
  thermal conductivity of, **10**:39–40
  thermal diffusivity of, **10**:40
  thermal parameter of, **10**:40
  thermophysical properties of, **10**:1–77
Lunar olivine dolerite, specific heat of, **10**:61
Lunar reflectances, defined, **10**:13
Lunar rock chips, directional spectral reflectance
    of, **10**:22
Lunar rocks
  clinopyrosene, **10**:47
  specific heat of, **10**:59–63
  thermal conductivity of, **10**:42–45
  thermal diffusivity of, **10**:64–68
Lunar Sample Analysis Program, **10**:3

Lunar samples, numbering of, **10**:4
Lunar surface
  temperatures of, **10**:76
  variable vs. constant properties for, **10**:77
Lunar surface layer
  heat transfer in, **10**:39–80
  thermal parameter of, **10**:70–72
Lykoudis number, **1**:292
Lyman Series, **5**:275
Lyon integral, **6**:519

M

MAAP code, direct containment heating, **29**:232,
  234, 247–248, 320
MAAP3B model, **29**:247, 322
MAAP 4 code, **29**:247
Mach number, **3**:2; **24**:193, 201, 203, 204, 231,
  262
Mach–Zehnder interferometer, **8**:167, 273
Macrobalance equations, three-fluid heat ex-
  changer, **26**:263
Macroscopic scale, *see* Darcy scale
Macroscopic system, boundary of, **31**:28
Macrosegregation
  alloys, **28**:243, 251, 266, 327
    double diffusive convection, **28**:240–241
    electromagnetic stirring, **28**:320–325, 327
    inhibition, **28**:308–310, 316
    semitransparent analog alloys, **28**:272
    shrinkage and buoyancy, **28**:306, 308
    tin–lead alloys, **28**:291–292, 295–304, 317
Madejski analysis, in boiling nucleation, **10**:152–
  156
MAGG, *see* Multizone adaptive grid generation
MAGICO experiments, **29**:153–154
Magma, energy extraction from, **14**:97
Magnesium–magnesium junctions, enhancement
  of, **20**:264
Magnetically saturated, **25**:196
Magnetically stabilized, **25**:167, 201
Magnetically stabilized fluidized, **25**:171, 172,
  176, 180
Magnetically unstabilized, **25**:168
Magnetic fields, **25**:152
  alloy solidification, **28**:246, 316–319
  colinear, **25**:153, 167
  Czochralski process and, **30**:318, 390–391,
    422–423

Magnetic fields (*continued*)
  effect on bed behavior, **25**:167
  homogeneity, **25**:154
  longitudinal, **25**:153
  nonuniformity, **25**:167
  in plasmas, *see* Plasma
  transverse, **25**:167
  uniform, **25**:154, 167
Magnetic fluid conditioning, **31**:458–459
Magnetic flux, **25**:233
  distributor downcomer, **25**:235
  flowmeter for solids, **25**:235
  fluidized bed, behavior, **25**:151, 152, 194, 201, 206, 220, 223, 228, 237
  moment, **25**:179
  valve for solids, **25**:234
Magnetic forces used for producing reduced gravity effects, **4**:150f
Magnetic-interaction parameter, **1**:280
Magnetism, residual, **25**:167
Magnetization, **25**:165
  bed, **25**:169, 175, 204, 239
  moderate, **25**:179
  particles, **25**:167, 175, 184, 202
  remnant, **25**:179
  strong, **25**:173
  weak, **25**:173
Magnetohydrodynamic model, Czochralski process, **30**:333–334, 351
Magnetohydrodynamics, **3**:110; **25**:155; *see also* MHD
Magnetostriction, **1**:273f
Manned Spacecraft Center, Houston, **10**:42
Marangoni convection, in Czochralski process, **30**:316, 384
Marangoni effect, **16**:132
Marangoni flows, **10**:156
Marangoni number, **30**:43, 49
Mare Imbrium, spectral emittance from, **10**:11
Mare Tranquilitatis, **10**:4, 8–10
Marginal distribution function, **5**:9
Markov chain, **5**:4
Mars
  atmosphere of, **14**:265, 275
  planetary boundary layer of, **14**:266
  radiative decay for, **14**:256
  thermosphere of, **14**:266
Martinelli–Nelson correlation, **1**:372ff, 382
Martinelli two-phase parameter, **5**:373
Mascarenhas thermodynamical theory, **14**:119

Mass
  closure for, **31**:56–60
  conservation, **14**:8–10
  interphase transfer laws for, **31**:109
  volume-averaged equation of, **31**:6–7
Mass absorption coefficient, **12**:126
Mass action law, *see* Law of mass action
Mass average velocity, **31**:7
  defined, **12**:204
  for dilute solutions, **12**:234
  for fluid volume element, **12**:198
Mass balances
  and flux divergence equations, **31**:114–121
  slug flow, **20**:86–87
Mass burning rate, droplet, **26**:64
Mass conservation, **24**:222
  multiphase flow, **30**:100–101, 104
Mass conservation equation, laser vaporization, **28**:127, 129
Mass diffusion, ultrashallow p+-junction formation, **28**:96–101
Mass diffusion coefficient, **2**:46ff
Mass diffusive flux, **31**:35–36
Mass/energy transport equations, **13**:126–153
  boundary conditions in, **13**:133–137
  governing point equations and, **13**:128–133
  volume-averaged, **13**:137–153
Mass entrainment, rate, **21**:278
Mass equilibrium, local, **31**:4
Mass extinction coefficient, **3**:181
Mass flow rate, local distribution of, **31**:119–121
Mass flux, **25**:310
  in nozzle, **2**:58, 102
Mass fraction
  of species, **2**:114, 133
  weighted average heat capacity, **13**:155
Mass fraction–mass density relations, **31**:34–35
Mass production rate, **2**:153
Mass rate of formation, **2**:113f
Mass sources in porous bodies, **1**:169ff
Mass spectrometry, fire and, **10**:228–229
Mass system
  general transport equation and, **12**:231–232
  migration and, **12**:221–225
  in ordinary diffusion, **12**:204–209
  vs. molar system, **12**:211–212
Mass transfer, **5**:144; **22**:2, 4, 5, 13–16; **25**:252, 260, 262, 267–272, 275–281, 283, 284, 285, 288, 290, 291, 312; *see also* Boundary layer

convective, **25**:271, 272
in convective boiling, **16**:144, 145
in cryogenic systems, **5**:348
diffusion model of, **31**:36–38
eddy diffusivities of, **15**:130
electrochemical, measurement, convective loss
        of cavity receivers and, **18**:44–45
extrusion of non-Newtonian materials,
        28:145–151, 220–226
    chemical reaction and conversion, **28**:218–
        220
    combined heat and mass transfer, **28**:213–
        220
    material characterization, **28**:152–155
    moisture transport, **28**:213–218
    single-screw extruder, **28**:147, 149, 155–187
    twin-screw extruder, **28**:151, 187–201,
        225–226
in fire research, **10**:219–280
forced, **25**:278
free, **25**:279
induced rotational flow, **21**:170–174
influence of, on heat transfer, **5**:144
interphase, **25**:271, 272
laser speckle photography, **30**:267–276
liquid jet
    array, **26**:171–174
    free-surface jet, **26**:147–162
    submerged jet, **26**:124–126
liquid-phase process of, **31**:64
liquid–solid beds, **23**:251
microscales, **30**:51
    diffusion flame, **30**:52–78
    pulsed combustion, **30**:78–75
moving liquid droplet
    condensation, **26**:32–48, 83–84
    droplet interaction, **26**:84–89
    no phase change at drop surface, **26**:19–31
    phase change at drop surface, **26**:31–32
    vaporization, **26**:48–84
multicomponent, **16**:147
nonuniformity of in impinging flow, **13**:38
packed beds, **23**:229
pulsed-laser-induced phase transformations
    melting, **28**:75–109, 135
    sputtering, **28**:109–123
    vaporization, **28**:123–135, 136
rheology and heat related to, **15**:60–61
in rivers, bays, lakes, and estuaries, *see*
        THIRBLE

from rotating disk, **5**:167
from rotating surfaces, **5**:158
in thermal stratification, **5**:397
three-phase beds, **23**:258
in turbulent pipe flow, **15**:128–129
unsteady, **25**:291
Mass transfer analogy, film cooling, **7**:329, 359
Mass transfer coefficient, **3**:14
    and array of slot nozzles, **13**:42
    arrays of nozzles vs. outlet flow conditions for,
        13:27–41
    in impinging flow, **13**:6–7
    integral mean, *see* Integral mean transfer
        coefficients
    swirling jets and, **13**:41
    trip wires and, **13**:42
    turbulence promoters and, **13**:41
    vs. nozzle-to-plate distance, **13**:39
Mass transfer measurements, **7**:94
    electrochemical method, **7**:87
        forced convection, **7**:103
        free convection, **7**:98
Mass transfer model in boiling, **1**:231f
Mass transfer parameter, *see* Blowing parameter
Mass transport
    chemical vapor deposition, **28**:349, 352
    convective transport in, **13**:169–175
    electrical current by, **12**:244–247
    in gas phase, **13**:165–175
Mass velocity, CHF in helium and, **17**:323–324,
        325, 327, 328, 329, 330–332, 333, 336
MASTRAPP, *see* Multizone adaptive scheme for
        transport and phase change processes
Matched asymptotic expansions, **8**:7, 214
Matched solutions, **SUP**:85, 160–161
Matching, **8**:10
Material thermal properties, heat transfer coeffi-
        cient effect, **21**:86–90
Mathematical methods, **8**:1
Mathematical model
    cross-flow solution, **26**:281–290
    multifluid heat exchanger, **26**:234
    parallel-stream solution, **26**:271–281
    three-fluid heat exchanger, **26**:231–237, 256–
        269
Matrix algebra, **8**:62
Matrix solver, **24**:195, 246
Mats, **25**:292
Matter, conservation of, **13**:62–63
Maximum drag reduction asymptote, **12**:80

Maximum heat flux
  comparison with correlation, **5**:423
  in saturated liquid nitrogen, **5**:422
Maximum heat transfer coefficient, **19**:153
Maximum liquid superheat, experimental determination of, **11**:7–9; *see also* Liquid superheat
Maximum useful work, **15**:8
Maxwell–Boltzmann distribution, **4**:245
Maxwell equations, **1**:274
  electromagnetics, **21**:270
Maxwell fluid, **24**:109
Maxwellian distribution, **2**:278, 285, 317, 323
Maxwellian velocity distribution, **5**:39
Maxwell's law, **2**:278
Maxwell stress, gas–liquid interface, **21**:127–128
Maxwell velocity distribution function, **21**:68
MBE, *see* Molecular-beam epitaxy
MC3D code, **29**:151
Mean absorption coefficient, **5**:312
  Planck mean, **5**:313
  Rosseland mean, **5**:313
Mean axial velocity, **SUP**:38
Mean beam length, **5**:310
Mean free path, **2**:273; **9**:357; **31**:365
  of phonons and electrons, **9**:361
Mean heat transfer, **8**:126, 138
  in tube banks, **8**:138
Mean heat transfer coefficients, calculation of, **8**:127
Mean line width, **8**:243
  temperature effect on, **8**:243
Mean radiation pressure, **3**:190
Mean spectral emissivity, **5**:291, 292
  for Elsasser model, **5**:291
  for statistical model, **5**:291
Mean transfer coefficients
  integral, *see* Integral mean transfer coefficients
  reduction of by unfavorable outlet conditions, **13**:40
Mean wall temperature, **SUP**:45
Measurement, **22**:364
  thermal property, **22**:187, 342
Mechanical degradation, in viscoelastic fluids, **15**:85–86
Mechanical disequilibrium, **31**:111
Mechanical dispersion, heat- and mass-transfer, **20**:325–327
Mechanical energy balance, **21**:242
Mechanical energy integral equation, **SUP**:10, 70

Mechanical stress, caused by rapid laser heating, **22**:321
Medium, saturated bed, **25**:196
  magnetically nonuniform, **25**:197
Meksyn's method, **8**:14
MELCOR code, direct containment heating, **29**:218, 232, 234, 245–247, 320
Mellin transformation, **8**:37–40
Melt fragmentation, **29**:270–274
Melting, **4**:33; **8**:30, 31, 34, 75; **24**:1–38
  back, due to reradiating surface, **19**:43
  close-contact, *see* Close-contact melting
  conduction-dominated
    approximate integral methods, **19**:34–39
    asymptotic expansions, **19**:28–33
    coordinate transformations, **19**:16–24
    historical development, **19**:4–5
    integrodifferential equations, **19**:14–16
    other methods, **19**:39–44
    power series expansions, **19**:12–14
    similarity transformation, **19**:6–12
    weak formulations and enthalpy method, **19**:24–28
  conduction stage, **19**:51
  convection stage, **19**:52
  flow visualization, **19**:3
  forced convection, **19**:73–74, 78
  of ice layer by radiant head, **19**:38
  ice melting and friction, **24**:7, 29–34
  importance in nature, **19**:1–2
  with internal radiative effects, **19**:43–44
  isothermal, **19**:23, 35
  natural convection, **19**:44–47
    complications, **19**:68–73
    fin effects on cylindrical geometries, **19**:57–62
    phase-change around cylinders, **19**:49–57
    in rectangular geometries, **19**:62–67
  one-phase
    isotherm migration method, **19**:40–41
    in semi-infinite solid, approximate integral solution, **19**:35–36
  problems, classes of, **19**:16
  pulse-laser-induced, **28**:75–109, 135
  in rectangular geometries, **19**:62–67
  of solid in semi-infinite plane, **19**:6
  transition stage, **19**:51
Melting cylindrical tube, **8**:6
Melting range, **19**:61
Melting slab, **8**:4

Melt rate
  density ratio and, **19**:11
  dimensionless time and, **19**:20
  reduction by supercooling, **19**:10–11
Melt region thickness, **19**:9, 10
Melt velocity, magnitude, **19**:7
Melt volume, **19**:55
Melt–water interactions, explosive, *see* Steam
  explosions
Membrane
  barrier to growth of ice, **22**:294
  cell, **22**:291
Mercury
  heat transfer, in rotating systems, **5**:207
  sphere rotating in, **5**:165
Mercury drops, fragmentation, **29**:171–172, 174
Meridional circulation, **24**:282–283
Mesomicroscale, **30**:11, 13, 63
Message passing, **31**:404
Metabolism, **22**:4
Metal alloys, solidification, **28**:288–308, 316,
  327
Metal casting industry, steam explosions, **29**:130
Metal drops, conduction freezing, **28**:17–18
Metallic coatings
  oxide formation, **20**:270–271
  thermal contact conductance enhancement,
    20:263, 266–270
Metallic foils, thermal contact conductance,
  20:263–266
Metal–organic chemical vapor deposition
  (MOCVD), **28**:342, 373–375
  barrel reactor, **28**:385
  impinging-jet reactor, **28**:395
  rotating-vertical-pedestal reactor, **28**:346,
    389
Metal powders, atomization, **28**:18–19
Metals
  liquid, boiling heat transfer coefficient, **16**:161
  liquid jet impingement, **26**:204
  metal alloys, solidification, **28**:288–308, 316,
    327
  moving liquid droplet, evaporation, **26**:83
  pulsed-layer sputtering, **28**:109–112
    thermal and electronic effects, **28**:116–123
    time-of-flight measurements, **28**:112–116,
      136
Metal sprays, *see* Liquid metal sprays
Metal vapor
  condensation coefficient, **21**:73–75

dropwise condensation, **21**:92
Meteorology, **5**:208; *see also* Planetary boundary
  layer
Methane
  radiation–conduction interaction in, **8**:266,
    267, 272
  radiative transfer in, **8**:259, 261
Methanol
  film boiling of, **5**:58, 59
  natural convection boiling at, **5**:103
Methanol-amyl alcohol, A_0, **16**:119
Methanol-benzene, A_0, **16**:119
Methanol-ethanol, slope of saturation curve,
  16:67
Methanol-ethanol-water, boiling, **16**:116
Methanol-water
  bubble departure diameter, **16**:95
  convective boiling, **16**:144
  equilibrium contact angle on graphite,
    16:70
  slope of saturation curve, **16**:67
Méthode flash, **31**:2
Method of characteristics, **8**:53, 54
Method of closure, **31**:2
Method of discrete exchange factor (DEF),
  30:339–340
Method of intermediate limits, **8**:13
Method of multiple scales, **8**:12
Method of steepest descent, **8**:253
Method of strained coordinates, **8**:4
Method of volume averaging, **31**:2
Methods for solutions
  for fully developed flows, **SUP**:61–68
  general, **SUP**:61–77
  for hydrodynamically developing flows,
    SUP:68–74
  for thermally developing flows, **SUP**:74–77
Methylethyl ketone-toluene, A_0, **16**:119
Methylethyl ketone-water, A_0, **16**:119
Metzner–Friend equation, **15**:105
MFM, *see* Multiphase flow model
MFTF program, **29**:147
MHD channel flow, **8**:56
MHD channel flow heat transfer, **1**:299ff, 316f;
  see also Hartmann Problem
MHD equations, **1**:275ff
MHD free convection problems
  in horizontal cylinder, **1**:297f
  on vertical parallel plates, **1**:295ff
  on vertical plate, **1**:290ff

MHD free convection problems (*continued*)
  with constant magnetic field, 1:292f
  with variable magnetic field, 1:293f
  in verticle pipe, 1:298
MHD-limitations of classic theory, 1:271ff
Microbalance equations, three-fluid heat exchanger, 26:258–263
Microbial-induced corrosion (MIC), 31:445
Microclimate, 22:8
Microcontinuum model, heat conductivity of blood, 18:210–211
Microelectronics industry, Czochralski process for crystal growth, 30:313–319
Microfins, 14:203
Microfoil heat flux gage, 23:293
Microgravity, 30:43–51
Micro-heat-pipe concept, 20:283–285
  trapezoidal configuration, 20:283–285
Microscale properties, 9:115
Microscales
  heat transfer
    convection, 30:19–33
    microgravity, 30:43–51
    two-phase film, 30:33–43
  kinetic scales, 30:5–9
  laminar flow, 30:46
  mass transfer, 30:51
    diffusion flame, 30:52–78
    pulsed combustion, 30:78–85
  origin of, 30:4–5
  thermal scales, 30:9–19
  turbulent flow, 30:46–49
Microscopic inertia coefficient, multiphase flow, 30:111, 112
Microscopic scale, *see* Closure scale; Microscales
Microsegregation
  alloy solidification, 28:262
  Czochralski process, 30:332
Microthermocouple probe, in temperature field determination, 11:19–20
Microvascular perfusion, blood, 22:228, 276
Mie scattering, 5:37
Mie theory, 23:141–142, 145
Migration, *see also* Diffusion
  in electrochemical transport, 12:221–231
  electrostatic potential and, 12:225–226
  mass system in, 12:221–225
  mobility measurement in, 12:229–231
  mobility units in, 12:228–229

molar system and, 12:226–228
Migration potential, 12:258–259
Milionshchikov, equation, 25:76
Milne–Eddington approximation, 3:204ff; 27:23
Milne integral equation, 3:203
MIMD (multiple instruction multiple data stream) architecture, 31:402–403
Minimum fluidization velocity
  definition, 23:233
  prediction
    Bingham plastics, 23:239, 241
    Carreau fluids, 23:239, 241
    power law, 23:234
Minimum heat flux, effect of pressure and subcooling with nitrogen, 5:425
Minimum heat transfer asymptote
  thermal entrance lengths corresponding to, 15:125
  in turbulent flow, 15:111–112
Minimum heat transfer coefficient, 5:68
Minimum spark ignition energy, 29:82–83
Minimum transport velocity, 9:132
Minkowycz–Sparrow theory, 21:109–110
Miscellaneous
  doubly connected ducts, SUP:352–353
  singly connected ducts, SUP:278–279
  solution methods for fully developed flow, SUP:67–68
Missing modules, forced convection, 20:238–241
Mist flow, 1:356, 358
  film boiling, 11:164–170
MISTRAL code, 29:151
Mitigation
  alternative methods of, 31:458–462
  methods of, 31:446–469
MIXA experiments, 29:155, 156, 165
Mixed convection, 9:75; 24:144–153
  about vertical impermeable surfaces, 14:61–63
  in vertical tubes, 15:215
Mixed flow
  heat transfer, 18:139–142
  mean drag of banks, 18:150–151
Mixed force convection, electronic packages, 20:243
Mixed mode flow, induced rotational flow, 21:168–170
Mixed system, general transport equation and, 12:233–234

Mixing, **9**:43, 44, 57; *see also* Axial dispersion; Péclet number
  liquid–solid fluidized beds, **23**:250
  packed beds, **23**:229
  solids, **25**:167
  three-phase systems, **23**:257
Mixing cup temperature, **SUP**:45
Mixing length, **8**:288
Mixing zone, presence of, **31**:171–172
Mixture energy balance, **31**:111–112
  integral formulation of, **31**:138–139
Mixture energy conservation equation, **31**:136–143
Mixture flow rates, **31**:126–130
Mixture model of local balance equations, **31**:109
Mixture momentum balance, **31**:112
Mixture pressure, multiphase mixture, **30**:108
Mixture velocity, Taylor bubbles, **20**:104–105
MMM, *see* Multiphase mixture model
Mobility, **9**:146
  bed, **25**:161
  electrochemical, **12**:222
  measurement of, **12**:229–231
Mobility units, migration and, **12**:228–229
MOCVD, *see* Metal–organic chemical vapor deposition
Model
  Baish, *see* Baish models
  capillary bundle, **25**:274, 304, 306, 310, 313, 314
  Carreau, **25**:297, 298
  cell, **25**:292
  Charny, *see* Charny models
  Chato, *see* Chato model
  Chen and Holmes continuum, *see* Chen and Holmes continuum model
  converging-diverging cross-section, **25**:300, 304, 305
  cylindrical channel, **25**:275
  Keller and Seiler, *see* Countercurrent heat transfer
  Klinger continuum, *see* Klinger continuum model
  Lagendijk, *see* Lagendijk model
  Mitchell and Myers, *see* Countercurrent heat transfer
  multiple, *see* Klinger continuum model
  parallel-plate channel, **25**:298, 300–304, 307, 310, 313, 314

penetration, **25**:272
Pennes bioheat equation, *see* Pennes bioheat equation
simplified bioheat equation, *see* Weinbaum–Jiji models
submerged object, **25**:274
superposition, *see* Chato model
thermal conductance, *see* Pennes bioheat equation
three layers, *see* Weinbaum–Jiji models
two phase, *see* Weinbaum–Jiji models
Weinbaum–Jiji, *see* Weinbaum–Jiji models
Wissler, *see* Wissler models
Wulff continuum, *see* Wulff continuum model
zero-vorticity, **25**:293, 298
zero-vorticity cell, **25**:295
Modeling
  alloy solidification, **28**:249–250
    dual scale model, **28**:263
    history, **28**:250–252
    micro- and macromodels, **28**:261–264
    multidomain model, **28**:251–252
    multiphase model, **28**:264
    single-domain models, **28**:252, 253–261, 326
    submodels, **28**:264–269
  approaches of, **31**:373–379
  chemical vapor deposition
    artificial neural network models, **28**:408–413
    deposition, **28**:403–406
    heat transfer model, **28**:402–403
  direct containment heating, **29**:232–234
    CONTAIN code, **29**:232, 234, 241–245, 275, 283, 301, 311, 320–321, 328–329
    convection-limited containment heating (CLCH) model, **29**:241, 242, 243, 301, 311, 320
    MAAP code, **29**:232, 234, 247–248, 320
    MELCOR code, **29**:232, 234, 245–247, 320
    single-cell equilibrium (SCE) model, **29**:232, 235–238
    two-cell equilibrium (TCE) model, **29**:238–240, 301, 311
  dryout heat flux
    mechanistic model, **29**:42–47
    one-dimensional models, **29**:36–42
  hydrogen combustion, subscale models, **29**:118–121
  probability-based, **31**:390–394

Modeling (*continued*)
  pulsed laser melting, **28**:76–80
  pulsed laser vaporization, **28**:123–135, 136
  screw extrusion, **28**:149–151
    single-screw extruder, **28**:155–176
    twin-screw extrusion, **28**:190–197
  spray deposition
    convective cooling, **28**:13–14
    heat transfer, **28**:65–68
    liquid metal droplet impact, **28**:68–69
    radiative cooling, **28**:13–14
    recalescence, **28**:15–17
    solidification, **28**:17–18
    splat quenching, **28**:35–54
  steam explosions
    CHYMES model, **29**:150, 151, 152, 155, 156
    CULDESAC model, **29**:180, 186
    droplet capture model, **29**:140
    ESPROSE.m field model, **29**:180, 182, 195–201
    IDEMO model, **29**:180
    parametric, **29**:135–137
    PM-ALPHA, **29**:151, 153–154, 156, 195
    premixing, **29**:148–152
    propagation, **29**:180–182
    spontaneous nucleation, **29**:137, 138–140
    thermal detonation, **29**:137, 140–147
    THERMAL model, **29**:151
    thermodynamic, **29**:135–137
    vapor film collapse, **29**:160–163
  thermal, **22**:7–9, 11, 13, 16
  thermal–chemical explosions, **29**:188–190
Modified Oseen linearization, **8**:193
Modified permeability, **24**:110
Modified power-law model, **28**:153
Modified Rayleigh number, **5**:393
Modular Accident Analysis Program (MAAP) code, **29**:232, 234, 247–248, 320
Module-averaged heat transfer, **26**:175–177
Modulus, magnetic, **25**:196
Moiré method, **SUP**:63
"Moist body", **1**:123f
Moisture
  in liquids and gas phases, **13**:142
  porous media and, **13**:120
Moisture content
  defined, **13**:142
  as function of space and time, **13**:127
Moisture distribution, in sand, **13**:122–123

Moisture saturation
  defined, **13**:143
  fractional, **13**:143, 196
Moisture transport, screw extrusion, **28**:151, 213–218
Moisture transport equation, **31**:2, 95–99
Molal activity coefficient, **12**:262
Molar activity coefficient, **12**:262
Molar-average velocity, for dilute solutions, **12**:234
Molar equation of continuity, **12**:209
Molar ordinary diffusion, **12**:209
Molar system
  general transport equation and, **12**:232
  migration and, **12**:226–228
  vs. mass system, **12**:211–212
Molding, **25**:292
Molecular-beam epitaxy (MBE), **28**:340
Molecular diffusivity, **25**:291
Molecular gas, "gray"nature of, **12**:117
Molecular gas band radiation, **12**:115–190; *see also* Nonisothermal gas radiation; Thermal radiation
  absorption bands in, **12**:132
  asymmetric and symmetric bands in, **12**:142
  band absorption in, **12**:155, 158
  band head in, **12**:140, 154
  band shapes in, **12**:141–142
  channel emittance vs. wall layer transmittance in, **12**:185
  cold wall layer transmittance in, **12**:186–187
  in engineering, **12**:116
  engineering properties in, **12**:155–162
  equivalent line model in, **12**:147
  experimental observations of, **12**:128–133
  exponential wide band model parameters in, **12**:148–149
  fundamental band in, **12**:144–145
  gas absorption bands in, **12**:141–143
  gray-gas myth in, **12**:116–119
  hot band in, **12**:146
  internal emissivity in, **12**:157
  isothermal gas radiation in, **12**:162–174
  nonisothermal gas radiation in, **12**:174–188
  optical collision diameters in, **12**:136
  overlapping bands in, **12**:153
  properties in, **12**:118–120
  script-F transfer function in, **12**:168–174, 189
  slab band absorption in, **12**:163–165
  thermal radiation properties in, **12**:128–162

total absorptivity in, **12**:157
total emissivity in, **12**:156
Molecular heat capacity, **2**:298
Molecular heat conduction coefficient, **2**:297
Molecular transfer ratio, effective transmittance
  and, **12**:184–185
Molecular weight
  effect on fluid heat flow, *see* Fluids, heat
    transfer rate
  effect on thermal conductivity of polymeric
    liquids, **18**:218–220
Molecules, activation energy for, **10**:92, 112
Molten core–concrete interaction, **21**:277
Molten fuel coolant interactions, **29**:129
Molten metal droplet, evaporation, **26**:83
Molten polymers, *see also* Shear flow; Steady
    shear flow
  Biot number and, **13**:207, 222–225, 238–239,
    243, 262
  elongation flow in, **13**:260–262
  heat transfer in, **13**:206
  nearly shear flow in, **13**:261
  Poiseuille flow in, **13**:218–219
  rheological properties of, **13**:206–207, 230–
    231
  shear flow in, **13**:212–259
  steady shear flow in, **13**:227–250
  thermal properties of, **13**:209–210
  viscosity curve for, **13**:215
  viscous dissipation in shear flows of, **13**:205–
    264
Molten pool heat transfer, **29**:3–4
  circular cavities, **29**:17–20
  convection in superposed layers, **29**:31–33
  rectangular cavities, **29**:4–17
  spherical cavities, **29**:20–27
  transient convection, **29**:27–31
Molten resin, **25**:292
Moment coefficient, **5**:143
  spinning cone, **5**:142
Moments
  method of, **1**:101ff
  second-order, **25**:7
  third-order, **25**:7
Momentum, **31**:27
  conservation of, **13**:62–63; **14**:10–12
  coupling of to mass and energy balances,
    31:107
  eddy diffusivities of, **15**:130–132
  heat vs. mass transfer in, **15**:129–130

interphase transfer laws for, **31**:109
volume-averaged equation of, **31**:7–9
Momentum accommodation coefficient, **7**:166
Momentum balance, **25**:299
  in annular-dispersed flows, **1**:366ff
  decoupling from mass and energy equations at
    local level, **31**:113
  equation, **25**:258, 259
  global, **31**:121–126, 147
  heat- and mass-transfer, natural convection,
    20:317–318
  of loop system, **31**:126
  severe slugging, **20**:120–121
  for single loop, **31**:125–126
  slug pressure drop, **20**:94–95
  for system component, **31**:121–124
  transfer, **25**:252, 259
Momentum conservation
  Czochralski process, **30**:327
  multiphase flow, **30**:100, 104–105
Momentum conservation equation
  chemical vapor deposition, **28**:353, 359–360,
    361–362
  laser vaporization, **28**:128, 129
Momentum equations, **2**:8f, 18f, 112, 140, 147,
    171, 197, 214, 227, 236, 377; **3**:178; **8**:8,
    27, 43, 67; **SUP**:7–9; **19**:250, 251, 257
  ESPROSE.m field model, **29**:196–197
  moving liquid droplet, inertialess system, **26**:4
  for turbulent boundary layers, **8**:288
Momentum flux, **2**:18; **12**:198
  change in, **31**:123
Momentum flux correction factor, **SUP**:43; *see
    also appropriate ducts in the text*
Momentum integral equation, **SUP**:10
Momentum jet
  buoyant, **16**:1–2
    downstream radial growth, modeling, **16**:35
  circular buoyant
    algebraic models, **16**:6
    differential models, **16**:6–13
    modeling schemes, **16**:6–13
  definition, **16**:3
  downstream radial growth, modeling, **16**:35
  entrainment modeling, **16**:7–13
    governing equations, **16**:8
  Gaussian velocity profiles, **16**:5
  modeling
    in flowing ambient media, **16**:10–13
    in quiescent ambient media, **16**:9–10

Momentum thickness, **4**:329; *see also* Boundary layer thickness
Momentum transfer, **12**:259–260
Monochromatic absorption coefficient, **27**:183
Monochromatic emissivity, **27**:11
Monomer unit type, effect on thermal conductivity of polymeric liquids, **18**:218
Monte Carlo method, **5**:1, 2
  advantages of
    in solution of radiation problems, **5**:38
    in treating conduction problems, **5**:45
  algorithms for, **31**:343
  applied to conduction problems, **5**:40
  Beer's law, **27**:53–56
  definition of, **5**:2
  disadvantages of, **5**:37
  energy method, **27**:86, 87
  evaluation of error, **5**:11
  evaluation of integrals by, **5**:46
  heat balance, **27**:151–154
  industrial applications
    boiler furnaces, **27**:158–167
    circulating fluidized bed boiler, **27**:187–193
    gas reformer furnaces, **27**:167–173
    jet engine combustion chambers, **27**:173–181
    nongray gas layer, **27**:181–187
    three-dimensional systems, **27**:193–199
  parallelization of, **31**:407–413
  performance comparison of, **31**:371–373
  photons, **27**:56–57
  radiative heat transfer
    emission, **27**:56–61
    gas absorption, **27**:50–56
    reflection and absorption by solid walls, **27**:61–85
  RAT1 program, **27**:64–75, 80, 83, 88, 122, 203
  RAT2 program, **27**:76–84, 88, 109, 122, 203
  READ method, **27**:46, 86–91, 107–157
  used for determining equation of state for gases, **5**:48
Monte Carlo ray tracing approach, **31**:334, 388
Monte Carlo simulation, radiative heat transfer, **23**:161–166
Moon, *see also* Lunar (*adj.*)
  absorption band on, **10**:7
  albedo of, **10**:7, 13–14
  backscattering from, **10**:7
  bidirectional spectral reflectance of, **10**:8, 15

blackbody radiation from, **10**:11, 15–16
bond albedo of, **10**:13
directional reflectance of, **10**:14–15
fines on, *see* Lunar fines
Fra Mauro area on, **10**:5, 9
infrared radiation from, **10**:40
landings on, **10**:1
Le Monnier area on, **10**:9
Ocean of Storms on, **10**:5
Oceanus Procellarum of, **10**:6
oldest materials from, **10**:6–7
origin of, **10**:1
Plato crater on, **10**:9
reflected light phase distribution function of, **10**:9
remote sensing data for, **10**:7–11
Sea of Cold on, **10**:8
Sea of Moisture on, **10**:8–9
Sea of Serenity on, **10**:8
Sea of Showers on, **10**:11
Sea of Tranquility on, **10**:4–6, 8–10
specific heat of, **10**:39–41
surface composition of, **10**:40–41
surface polarization of, **10**:7
surface temperature of, **10**:13
temperature variation on, **10**:77
thermal conductivity of, **10**:39–59
thermal radiation measurements for, **10**:12–35
thermophysical property reference values for, **10**:79–80
Tycho crater on, **10**:8
variable surface temperature comparisons for, **10**:13
Moon rocks, *see* Lunar rocks
Moon-shaped ducts, **SUP**:272–273
Morse potential, **2**:346
Motion
  equations of, **12**:197–200
  particles, **25**:154, 191, 197, 204
Moving boundary, **22**:184, 204
Moving boundary measurements, **12**:246
Moving liquid droplet
  condensation, **26**:32–43
    compound drop, **26**:47–48
    droplet shape deformation, **26**:44
    electric field, **26**:17, 83–84
    gaseous environment, **26**:6, 31–32
    spherically symmetric, **26**:31–32
    spray of drops, **26**:44–47
    surfactants, **26**:43–44

direct-contact transfer, **26**:1–3
  fluid mechanics, **26**:3–19
    drop deformation, **26**:12–13,
      18–19
    electric field effects, **26**:13–19
    inertialess translation, **26**:4
    Reynolds number translation, **26**:5–11,
      16–18
    surface-viscous effects, **26**:13–16
    surfactants, effect on drop motion,
      26:11–12
    weakly inertial translation, **26**:5
  heat and mass transport, **26**:19–89
    condensation, **26**:44
    evaporation, **26**:79–81
drag force
  electric field, **26**:17
  gaseous environment, **26**:7–9
  inertialess system, **26**:4
  Reynolds number, **26**:9, 10, 34
droplet array, **26**:84
droplet group, **26**:84
droplet oscillation, **26**:44–49
drop motion, **26**:11–19, 43–44
electric field effect, **26**:13–19, 26–31, 83–84
evaporation
  electric field, **26**:83–84
  gaseous environment, **26**:6, 31–32, 52–83
  Reynolds number, **26**:59–77
  shape deformation, **26**:79–81
heat/mass transfer, **26**:19–89
Hill's spherical vortex, **26**:4, 7, 9, 10
hydrocarbons
  drop deformation, **26**:79–80
  vaporization, **26**:56–57, 59, 67, 70, 72, 78–
    79
kinetic theory, **26**:31–32
metals, evaporation, **26**:83
Nusselt number, **26**:21–23
surfactants, **26**:11–12, 43–44
vortex, **26**:10, 11
Moving solid particles, unsteady state conduction
  to, **14**:151–154
Multibubble systems, **15**:248–260
Multichannel systems, **9**:74
Multicomponent condensation, **15**:232–233
Multicomponent droplet, vaporization, **26**:66,
  77–79
Multidimensional flow, dryout heat flux and,
  29:50–52

Multidimensional numerical modeling, **30**:155–
  161
Multidirectional solidification, alloys, **28**:282–
  285
Multidomain models, alloy solidification,
  28:251–252
Multifluid bubbling pool, **21**:277–343
  bubble-induced entrainment, **21**:299–311
  configurations, **21**:311
  drag and instability of liquid droplets settling
    in continuous fluid, **21**:333–342
    background, **21**:333–337
    experiment, **21**:337–338
    liquid droplets, **21**:334–337
    results, **21**:338–341
    solid spheres, **21**:334
  entrainment onset, *see* Entrainment
  gas-sparged, **21**:278
  heat transfer by gas bubbling across interface,
    21:279–286
    apparatus and procedure, **21**:281–283
    Blottner's interfacial relationship, **21**:279–
      280
    experimental data, **21**:283–286
    models, **21**:279–281
  heat transfer from horizontal bubbling surface
    to overlying pool, **21**:327–333
    background, **21**:327–329
    experiment, **21**:329–330
    results, **21**:330–332
  heat transfer from liquid pool to vertical wall,
    21:318–327
    experiment, **21**:321–322
    heat transfer measurements, **21**:323–325
    model development, **21**:325–327
    physical phenomena, **21**:318–319
    previous studies, **21**:319–321
    void fraction measurements, **21**:322–323
  heat transfer with entrainment across interface,
    21:311–318
  rising gas bubbles, **21**:278
  stratified configuration, heat transfer, **21**:312
Multifluid heat exchanger, **26**:220–222, 225–227
  design, **26**:304
  dimensionless groups, **26**:237
  literature review, **26**:252, 254–255
  mathematical model, **26**:234
Multi-geometry passages, heat exchanger with,
  SUP:405–410
Multigrid, **24**:254

Multigrid method, **25**:383
Multilevel model, *see* Subgrid, model, multilevel
Multi-particle system, **25**:252
Multiphase flow, in porous media, **30**:93–95, 98–100, 186–187
  capillary forces, **30**:93, 94, 96–98
  gravity and, **30**:93–94
  interfacial tension, **30**:96
  multicomponent systems, **30**:162–186
  non-Darcian effects, **30**:111–112
  nonequilibrium effects, **30**:110–111
  phase saturation, **30**:95–96
  relative permeability, **30**:98, 99
  two-phase, single component systems, **30**:112–162
  viscous forces, **30**:93, 94
Multiphase flow model (MFM), **30**:98–102, 186
  comparison with other multiphase flow models, **30**:107–110
  conservation equations, **30**:100–101
  drawbacks, **30**:101–102
  drying of porous material, **30**:170, 171
  two-phase analysis, **30**:127, 161
Multiphase mixture model (MMM), **30**:103–108, 185, 186
  comparison with other multiphase flow models, **30**:107–110
  conservation equations, **30**:105–107
  drying of porous material, **30**:170, 172
  external boiling, **30**:149–151
  two-phase analysis, **30**:143, 144–146, 149–151, 161, 162
  two-phase boiling convection, **30**:127–134, 139–140
Multiphase transport equations, **31**:1, 2
Multiply connected ducts
  analytical results, **SUP**:354–365
  definition of, **SUP**:17
  thermal boundary conditions for, **SUP**:36
Multiregion analysis, **SUP**:354
Multispecies diffusion model, **31**:110
Multistage flash evaporator water desalination, **15**:271–272
Multistage fluidized bed, **25**:235
Multitrain systems, in bubble columns, **15**:255–260
Multizone adaptive curvilinear finite volume scheme, **30**:351, 352–357
Multizone adaptive grid generation (MAGG), **30**:352–354

Multizone adaptive scheme for transport and phase change processes (MASTRAPP), **30**:356–357, 361
Mushy zone, **22**:190, 201, 212
  alloy solidification, **28**:232–233, 239–242, 245–246, 247, 251
  modeling, **28**:251, 253
  semitransparent analog alloys, **28**:270

                                    N

Nahme number, in steady shear flow, **13**:234, 242–243, 262
Naphthalene, **5**:158; **25**:177
NAPLs, *see* Nonaqueous phase liquids
Napthalene sublimation technique, **20**:237–238
Narmco epoxy-fiberglass structure, **15**:46
Narrow-band thermal radiation, **12**:133–138
Narrow space nucleate boiling, **20**:36–55
  bubble behavior, **20**:36–40
  clearance factor, **20**:37–39
  coalesced bubble region, **20**:42–49
  heat transfer characteristics, **20**:40–49
  isolated bubble region, **20**:40–41
  thermal cooling of electronic components, **20**:248
National Aeronautics and Space Administration, **10**:4, 42, 277
National Bureau of Standards, **10**:276, 279–280
National Power Survey, **12**:2
Natrosol (hydroxyethylcellulose), **19**:329
Natural convection, **5**:375; **8**:11, 12, 15, 161; **9**:274; **24**:127–143, 214
  boundary-layer simplifications of, **9**:280
  buoyancy forces and, **30**:384–385
  in a closed-end tube, **8**:162, 163
  complications, **19**:68–73
  in confined channels, **8**:162
  cooling of electronic components, **20**:183–220
  in cylinders, **5**:395
  in cylindrical geometries, fins in phase-change processes, **19**:57–62
  in Czochralski process, **30**:315, 362
  discrete sources–in enclosures, **20**:217–220
  discrete sources–liquid cooling, **20**:213–217
  discrete thermal sources–vertical surfaces, **20**:202–209
  external, **9**:274
  flat-plate collectors, **18**:11–19

freezing, **19**:60
  in rectangular geometries, **19**:67–68
of gaseous nitrogen, **5**:375
gas vs. particle, **19**:100
global effects, **19**:49
governing equations for, **9**:275
heat- and mass-transfer
  horizontal direction, **20**:330–343
    boundary layer flows, **20**:332–336
    convection onset, **20**:330–332
    enclosed porous layers, **20**:336–341
    line source, **20**:346–347
    transient approach to equilibrium,
      20:341–343
  physical model, **20**:316–318
  point source, **20**:343–346
  vertical direction, **20**:318–330
    convection onset, **20**:318–323
    finite-amplitude convection, **20**:327
    high Rayleigh number convection,
      20:329–330
    nonlinear initial profiles, **20**:323–325
    sores diffusion, **20**:328–329
from heated horizontal plates, **5**:377
horizontal continuous casting process, **19**:60–61
from horizontal cylinders, **11**:200–210
for horizontal cylinders, **5**:375
between infinite parallel horizontal plates,
  8:165
initial studies, **19**:44–49
inside horizontal cylinders, **8**:167, 169, 196
in interior of a horizontal cylinder, **8**:164
internal, **9**:274
internal boiling in porous media, **30**:122–134
irregular surfaces, **20**:209–213
laser speckle photography, **30**:267–268
  converging channel flows, **30**:281–283
  heat transfer coefficient, **30**:269–271
  isothermal single vertical wall, **30**:272–274
  isothermal vertical channel flows, **30**:274–278
  one-dimensional refraction, **30**:268–269
  upward-facing isothermal surfaces, **30**:278–280
line-focusing collectors, **18**:19–27
  compound parabolic collectors, **18**:20–25
  single-axis-tracking parabolic collectors,
    18:26–27
of liquid helium, **5**:375

of liquid nitrogen, **5**:375
melting
  internal, of horizontal cylinders, **19**:59–60
  inward, of vertical cylinders, **19**:61
  of PCM inside horizontal tube, **19**:58–59
  in rectangular geometries, **19**:63–67
Nusselt number for, **5**:375; **9**:288
in partially filmed container, **5**:395
plumes, **9**:322
point-focusing systems, **18**:27–46
  cavity receivers for central receiver systems,
    18:35–41
  cavity receivers for parabolic dish systems,
    18:41–46
  external receivers for central receiver systems, **18**:30–34
Rayleigh–Bénard problem, **30**:25–33
in rectangular cavities, **8**:166, 168, 169, 175
in rotating cylinders, **8**:164
rotation and, **30**:367–369, 385–386
similarity solutions, **9**:282
in solar buildings, **18**:46–75
  building envelope, **18**:47–49
  double-pane windows, **18**:49
  multiple-building zones, **18**:67–75
    boundary layer-driven flow, **18**:68–72
    bulk density-driven flow, **18**:67–68
    Trombe walls, **18**:72–75
  single-building zones, **18**:49–66
    surface roughness effects, **18**:63–66
    triangular spaces, **18**:50–51
    vertical and horizontal surfaces in enclosures, **18**:57–63
    vertical surfaces in enclosures,
      18:51–57
  single-pane windows, **18**:48–49
for sphere, **5**:375
in spherical cavities, **8**:164
surface tension and, **30**:365–366, 384–385
transients, **9**:310
between vertical isothermal plates, **8**:167
vertical plate, **30**:25
vertical plates and channels, **20**:185–202
  convection–radiation interaction, **20**:201–202
  optimum spacing, **20**:202
  staggered arrays, **20**:198–201
  vs. actual circuit boards, **20**:195–198
Natural convection boiling of methanol, **5**:103
Natural convection condensation, **21**:114–115

Natural convection film boiling, 5:56
    from horizontal cylinders, 5:57
Natural draft, for heat exchanger–cooling tower
    combination, 12:57–59
Natural flame
    laminar, 30:65–72
    turbulent, 30:73–78
Navier–Stokes equations, 5:133; 15:72; 24:196,
    197, 218, 219
    first approximate solution of, SUP:165
    first complete solution of, SUP:166
    generalized solutions for, SUP:44
    in impinging flow, 13:4
    induced rotational flow, 21:144–146
    methods of solutions, SUP:73–74
    solutions for, SUP:73–74, 89–97, 164–167,
        300
NBS/NRC Steam Tables, 31:115, 116
n-Butane, nucleate pool boiling curve, 16:106
n-Butanol-water, convective boiling, 16:144
Near-equilibrium flows, 2:251ff
Nearly homogeneous scalar field, second-order
    model, 21:199–202, 233–234
Nearly homogeneous turbulence, second-order
    model, 21:193–202
Nearly homogeneous velocity field
    model formation, 21:213–214
    second-order model, 21:194–199, 232–233
Near-wall
    condition, 25:19
    damping, 25:351, 358–350, 403
        Van Driest function, 25:358–360
Nernst–Einstein equation, 12:229, 237, 266–267
Net emission, 3:188
Net spectral flux, 12:123
Net surface flux, variance in, 31:360–364
Network thermodynamics, 22:287
Neumann condition, 31:80
Neural networks, chemical vapor deposition
    modeling, 28:408–413
Neutron cross section, 9:117
Neutron diffusion, 5:37
Nevada Test Site (NTS) tests, 29:299
Newtonian, 25:252, 255, 260, 267–269, 279–
    283, 285, 288, 292–298, 304, 306, 307, 310
    generalized, 25:255
Newtonian fluids, 2:359f; 19:256; 28:152
    boundary conditions, 19:252–253
    constitutive equations, 19:251–252
    flow in rectangular duct, 19:314–316
    future research, 19:343–346

generalized, 13:211
governing equations
    conservative, 19:250–251
    constitutive, 19:251–252
    dimensionless groups and generalized
        solutions, 19:253–257
heat transfer studies
    in circular tubes, 19:304–305
    in rectangular geometries, 19:306–316
hydrodynamically developed laminar flow
    friction factor, 19:261–262
    velocity profile, 19:257–259
hydrodynamically developing laminar flow
    hydrodynamic entrance length, 19:274–275
    incremental pressure drop, 19:272
    velocity profiles, 19:269
laminar heat transfer in, 15:100
noncircular geometry applications, 19:248
primary normal stress difference for, 15:178
simultaneously developing flows, 19:300, 301
thermal convection in, 15:146–149
thermal entrance length, 19:297–299
thermally developed flow in rectangular duct,
    19:277–280
thermally developing flow in rectangular duct,
    19:289–290
turbulent channel flow, 19:322–323
    fluid mechanics, 19:323–325
turbulent heat transfer through rectangular
    ducts, 19:334–339
velocity distributions, 19:269
Newtonian models, generalized, 15:78–80
Newtonian solution in MHD, 1:334f
Newton–Raphson method, slug flow calcula-
    tions, 20:98
Newton–Raphson scheme, 4:332
Newton's second law, 12:204
Newton's third law, 12:212
New Zealand
    Broadlands geothermal field in, 14:95
    Wairakei geothermal field in, 14:52–54, 84–
        85
NGM variable, 27:124, 206
ngprnt function, 27:129
n-Heptane, 14:125
    limiting superheat of, 10:104
n-Hexane, limiting superheat of, 10:104
Nickel–copper alloys, 28:233, 236
Nield's analysis, heat- and mass-transfer, 20:318–
    319
Nitrogen, transition boiling, 18:260–263

Nitrogen-argon
  boiling
    heat transfer, **16**:109–110
    incipience, **16**:72–73
  bubble departure
    diameter, **16**:95–97, 98
    frequency, **16**:102
  bubble growth
    experimental data, **16**:92
    variation with vapor-liquid composition and growth time, **16**:92–93
  heat transfer coefficient, prediction, **16**:127
  inertia force, effect on bubble departure, **16**:97–99
Nitrogen-methane, heat transfer coefficient, prediction, **16**:127
Nitrogen-oxygen, heat transfer coefficient, prediction, **16**:127
Noble metal-plated surfaces, dropwise condensation, **21**:103
Noise
  electrical, **22**:406
  magnetic, **22**:406
  Nuclear detonation, light source for burns, **22**:249
  Nucleation
    of gas bubbles from solution, **22**:293
    of vapor by laser irradiation, **22**:321
    $1/f$, **22**:405–406, 424, 429–433
    shot, **22**:405, 424, 429–433
    white, **22**:404–405, 424, 429–433
Nomenclature, **19**:82–83, 181–184, 242, 346–349
Nonaqueous phase liquids (NAPLs), groundwater contamination by, **30**:174–183
Nonazeotropic multicomponent mixture, **21**:110–112
Non-Boussinesq effects, **24**:313–315
Noncircular cross section, **9**:36
Noncircular ducts
  elliptical channels, **8**:44
  triangular channels, **8**:44
Noncondensables, **9**:231, 264
  direct condensation in presence of, **15**:239–242
Non-Darcian effects, **24**:111–118
  multiphase flow, **30**:111–112
Nondimensional correlating equations
  nucleate boiling, **20**:2–18
    evaluation of constants, **20**:9–11

Nonemitting medium, two-flux approximation, **23**:137
Nonequilibrium, **8**:250
Nonequilibrium flows of high energy gases with dissociation and recombination, **2**:218ff
Nonequilibrium radiation, **3**:183
Nonequilibrium source function, **8**:249
Nongray gas, behavior of, **12**:118
Nongray gas layer, **27**:181–187
Nongray nondiffuse surface, **5**:17
Nonhomogeneous nonequilibrium boiling, profile function for, **31**:141
Nonintermeshing twin-screw extruders, **28**:188
Nonisothermal flows, **8**:304
  induced rotational flow, **21**:160–161
Nonisothermal gas radiation, **12**:174–188
  band location in, **12**:182
  Curtis–Godson narrow-band scaling in, **12**:176–177
  equation-of-transfer solution for, **12**:174–176
  radiation-to-molecular transport parameter in, **12**:183
  script-F transfer function for, **12**:189
  wall–layer transmissivity in, **12**:178–188
  wide-band scaling in, **12**:177–178
Nonisothermal vertical plate, heat transfer from, **11**:294–295
Non-Isothermicity, **25**:98
Nonisothermity effect, under artificial turbulization conditions, **31**:225–228
Nonlinear effects, **9**:336, 341
Nonlinear initial profiles, heat- and mass-transfer, **20**:323–325
Nonlinear integrodifferential equations, **5**:34
Nonlinear problems, **8**:40
Nonlocal form, transformation to local form, **31**:18–19
Nonlocal thermodynamic equilibrium, radiative theory of temperature perturbations for, **14**:255–259
Nonmonotonic dependence, of local entropy generation rate, **15**:18
Non-Newtonian, **25**:253, 255, 260–264, 268, 270, 272, 282, 288
Non-Newtonian behavior, **9**:296; **15**:61
Non-Newtonian effects
  liquid–solid fluidized beds, **23**:232
  packed beds, **23**:192
  sedimentation, **23**:261
  three-phase beds, **23**:253

Non-Newtonian fluids, **4**:33; **19**:256; **23**:189;
    24:106; *see also* Viscoelastic fluids
    asymptotic friction factor behavior of, **15**:88–
        92
    boundary conditions, **19**:252–253
    in circular pipe flow, **15**:59–132
    classification, **15**:61–63; **19**:249–250
    constitutive equations, **19**:251–252
    density of, **15**:63–66
    in drag and heat transfer studies, **15**:64
    drag-reducing viscoelastic
        turbulent channel flow, **19**:328–334
        turbulent heat transfer through rectangular
            ducts, **19**:342–343
    empirical turbulent heat transfer correlations
        for, **15**:107
    engineering applications, **19**:248
    external flows of, **15**:151–178
    in falling ball viscometer, **15**:72–73
    flow stability in, in heated vertical pipes,
        15:205–206
    fluid mechanics, **2**:389f; **15**:86–88
        in entrance region, **2**:390
            between parallel flat plates, **2**:390
            in round tubes, **2**:390f
            transition to turbulence, **2**:381ff
        friction factor in, **15**:90
    future research, **19**:343–346
    general stability considerations with respect to,
        15:207–208
    governing equations
        conservative equations, **19**:250–251
        constitutive, **19**:251–252
        dimensionless groups and generalized so-
            lutions, **19**:253–257
    heat transfer, **2**:358ff, 389ff; **15**:64–66, 98–
        126
        from cylinders, **2**:390f
        in ducts, **2**:366ff
        from flat plate, **2**:390
        forced convection, **2**:390
        with internal heat generation, **2**:376ff, 379ff
        natural convection, **2**:390
        nucleate boiling rates, **2**:392
        between parallel flat plates, **2**:374f, 380
        in round tubes, **2**:366ff, 380
            transition to turbulence, **2**:381ff
        with viscous dissipation, **2**:376ff
    heat transfer studies
        in circular tubes, **19**:304–305

    in rectangular geometry, **19**:316–322
    hydrodynamically developed laminar flow,
        friction factor, **19**:262–268
    hydrodynamically developing laminar flow
        hydrodynamic entrance length, **19**:274,
            275–276
        incremental pressure drop, **19**:272–274
        velocity profiles, **19**:269
    hydrodynamic entrance length for, **15**:88
    laminar heat transfer in, **15**:100–104
    laminar mixed convection in external flows of,
        15:184–188
    laminar thermal convection in, **15**:151–178
    mechanical degradation of, **15**:85–86
    physical properties of, **15**:63–64
    Prandtl number and, **15**:66–67; **19**:300
    predictions, deviation of, **19**:139–140
    purely viscous, **19**:249, 251–252
        nonelastic, **19**:249
        turbulent channel flow, **19**:325–328
        turbulent heat transfer through rectangular
            ducts, **19**:339–441
    Reynolds number and, **15**:66–67
    rheological classification of, **15**:150
    shear thinning and thickening characteristics
        of, **15**:168
    simultaneously developing flows, **19**:301–306
    solute effects in, **15**:83–85
    solvent chemistry effects in, **15**:79–83
    specific heat and, **15**:63–66
    thermal conductivity in, **15**:63–66
    thermal convection in, **15**:143–223
    thermal entrance length, **19**:297–300
    thermally developed laminar flow through
        rectangular ducts, **19**:280–289
    thermally developing flows in rectangular
        channels, **19**:290–297
    thermal properties, **2**:364ff
        heat capacity, **2**:365
        thermal conductivity, **2**:354f
    turbulent flow
        external, **2**:392
        internal, **2**:388f
    turbulent thermal convection in external flows
        of, **15**:180–195
Non-Newtonian materials, **28**:146
    material characterization, **28**:152–155
    screw extrusion, **28**:146–151
        combined heat and mass transfer, **28**:213–
            220

flow in dies, **28**:201–212
single-screw extruder, **28**:15–187, 147, 149
twin-screw extruder, **28**:151, 187–201,
    225–226
Nonorthogonal boundary, radiative heat transfer,
    27:99–104
Nonorthogonal curvilinear coordinates, **24**:196
Nonoverlapping line limit, **8**:236
Nonpolar gases, **3**:261ff, 278f
Nonpolymeric liquids, thermal conductivity
    biological liquids, **18**:209–211
    greases and foodstuffs, **18**:211–213
    suspensions, flowing media, **18**:200–209
        Brownian motion and, **18**:206–207
        deformed drops/particles, **18**:205–206
        generalized analytical framework, **18**:201–
            203
        spherical drops/particles, **18**:203–204
        theoretical and experimental comparisons,
            18:207–209
    suspensions, stagnant media, **18**:178–200
        basic equations for generalized analytical
            framework, **18**:191–193
        bounds on, **18**:197–200
        conventional approach, **18**:179–190
        expressions for $(k_e)_{ij}$ for different inclusion
            shapes, **18**:193–197
        generalized analysis, **18**:190–197
        linear heat flow model, **18**:180
        nonlinear heat flow model, **18**:181–182
Nonround channels with narrow comers,
    31:165–166
Nonscattering medium, **1**:3
Non-similar solutions of conservation of species
    equation *see* Conservation of species equa-
    tion
Nonstationary heat transfer
    determined by external conditions, **21**:44–46
    thermoacoustic perturbations, **21**:46–47
Nonuniform fluid inlet temperature, **SUP**:109,
    128–129
Nonuniformity, bed particles, **25**:220
Nonuniform temperature difference, laminar free
    convection from, **11**:275–277
Nonviscometric flow, **13**:260–262
Normal component, of stress boundary condi-
    tion, **31**:40
Normal heat transfer regime, **31**:229
No-slip boundary condition, **SUP**:8
Nozzle, **25**:127

Nozzle configurations *see* Air experiments
Nozzle design, atomization, **28**:10, 18
Nozzle exit velocity
    for array of slot nozzles, **13**:25
    blower rating and, **13**:45
Nozzle heat flux, **2**:87ff
Nozzle inlet, *see* Boundary layer
Nozzles
    high-performance arrays of, **13**:52–58
    optimal spatial arrangements of, **13**:45–52
    slot, *see* Array of slot nozzles
Nozzle-to-plate distance
    in impinging flow, **13**:36, 45
    mass transfer coefficient and, **13**:39
n-Pentane
    limiting superheat in, **10**:104
    pool boiling of, **5**:105
    transition boiling, **18**:260–263
NRAY variable, **27**:53, 62–63, 66, 88–89, 96–97,
    127, 206
NRC DCH studies, **29**:216–219, 220
NTS studies, **29**:299
NTU, **26**:222, 256, 267, 274
Nuclear industry, **31**:106
Nuclear power plant
    dry cooling for, **12**:3
    heat rejection for, **12**:9
Nuclear power reactors, accidents, **21**:277
Nuclear reactor cooling, **9**:114
Nuclear reactors, **9**:76; *see also* Nuclear reactor
    safety; Reactor coolant system; Reactor
    pressure vessel
    boiling water reactors, **29**:219, 324–328
    breeder reactors, **29**:1–2, 9
    containment studies, **29**:216, 217
    convection, **29**:53
    direct containment heating, **29**:215–216
    dryout heat flux, **29**:34–35, 52–53
        forced flow and, **29**:50
        liquid subcooling, **29**:49
        mechanistic model, **29**:42–47
        multidimensional flows, **29**:50–52
        one-dimensional models, **29**:35–42
        overlying liquid layer depth, **29**:48–49
        system pressure, **29**:49–50
        unconsolidated porous media and, **29**:47–48
    hydrogen combustion, *see* Hydrogen combus-
        tion
    light-water reactors, **29**:2–3, 131
    molten pool heat transfer, **29**:3–4, 53

Nuclear reactors (*continued*)
  circular cavities, **29**:17–20
  convection in superposed layers, **29**:31–33
  rectangular cavities, **29**:4–17
  spherical cavities, **29**:20–27
  transient convection, **29**:27–31
Nuclear reactor safety
  accident management strategy, **29**:3
  α-mode failure, **29**:147, 149–151, 193–195
  containment failure, **29**:216
  direct containment heating, **29**:215–216, 235–
    236
  fuel coolant interactions, **29**:129, 195
  hydrogen-air combustion, **29**:59–61, 77, 102–
    123
    deflagration, **29**:108–118
    pressure transients from combustion,
      29:107
  light water reactors, **29**:131
  loss-of-coolant accident, **29**:59
  loss-of-flow accidents, **29**:1–2
  reactor coolant system blowdown, **29**:235,
    237, 238, 248, 256, 257, 310
  steam explosions, **29**:129, 131, 136, 147, 193–
    195
  TMI-2 incident, **29**:1, 2, 59–60, 235
  transient-over-power accidents, **29**:1, 2
Nuclear Regulatory Commission (NRC), direct
    containment heating studies, **29**:216–219,
    220
Nucleate boiling, **1**:193ff, 198f, 226f; **5**:412,
    440; **7**:76; **9**:84; **11**:2; **23**:7–8, 10–53
  augmentation, **20**:63–77
    prepared surface heat characteristics, **20**:70–
      77
    vapor nucleus stability, **20**:64–70
  of benzene, **5**:443
  bubble behavior, **20**:2
  characteristics, **20**:70–77
  circular, free-surface jets, **23**:18–19
  comparison of results for hydrogen, **5**:411
  of cryogens, **5**:440
  of cyclohexane, **5**:443
  direct cooling with heat pipes, **20**:290–291
  effect of acceleration on, **5**:420
  effects of gravity on, **5**:459
  effects of pressure and subcooling on, **5**:419
  electric field and, **14**:112
  FC-72 and FC-87 mixture, **23**:18–19
  generalized Stanton numbers for, **5**:409

  heater geometry, **23**:53–54
  heat transfer coefficient
    organic fluids, **20**:1–2
    surface conditions, **20**:1–2
    thermal property similarities, **20**:18–23
  of heptane, **5**:443
  hysteresis, **23**:49
  incipience, **23**:17–18
    temperature excursions, **23**:46–48
  investigations, **23**:12–15
  jet impingement vs. pool boiling, **20**:252–253
  jet suction, **23**:53, 56–57
  of liquid hydrogen, **5**:82
  local measurements along impingement sur-
    face, **23**:42–45
  low liquid levels, **20**:55–63
    heat transfer characteristics, **20**:55–58
    liquid film heat transfer characteristics,
      20:58–63
  maximum heat flux, **5**:416, 420
  of methane, **5**:417
  of methanol, **5**:443
  minimum heat flux, **5**:416, 421
  modified incipience model, **23**:16
  moving impingement surface, **23**:50–51
  multiple-jet impingement, **23**:35–41
    circular water jets, **23**:36–38
    nine-jet configuration, **23**:39
    saturated water and R-113, **23**:40–41
  narrow space, **20**:36–55
    bubble behavior, **20**:36–40
    heat transfer characteristics, **20**:40–49
    model characteristics, **20**:49–55
  of neon, **5**:415
  of nitrogen, **5**:412
  nondimensional correlating equation, **20**:2–18
    constant evaluation, **20**:9–11
    elementary process formulation, **20**:5–8
    free convection, **20**:2–4
    generalized correlation of heat transfer,
      20:11–13
    nucleation factor, **20**:13–15
    pressure factor, **20**:15–18
  nozzle geometry, **23**:51–53
  onset, **23**:11, 16–20, 121
  of oxygen, **5**:413
  of pentane, **5**:443
  prediction of, **5**:441
  research needs, **23**:121–122
  of saturated liquids, **5**:443

single-jet impingement, **23**:21–35
  correlations, **23**:35, 39
  F98 surfactants, **23**:23–24
  fluid properties, **23**:26–27
  free-surface, circular jets, **23**:21–28
  free-surface, planar jets, **23**:28–29
  jet velocity effects, **23**:21–25, 28–32
  local film thickness, **23**:24
  nozzle/heater dimensions, **23**:27, 33–34
  nozzle-to-surface spacing, **23**:28, 34–35
  saturated liquid nitrogen, **23**:32
  saturated water and R-113, **23**:26–27, 30–
    31
  subcooling effects, **23**:26, 29, 33
  submerged, confined, and plunging jets,
    23:29–35
  surface orientation/impingement angle,
    23:27, 29
  trends, **23**:35–38
  wall superheat, **23**:23, 25, 28
stagnation point temperature, **23**:44
steady-state heat transfer in helium I
  above homogeneous nucleation tempera-
    ture, **17**:142–145
  below homogeneous nucleation tempera-
    ture, **17**:140–142
of subcooled liquids, **5**:440
subcoolings, **23**:16
surface conditions, **23**:49–50
surface configuration, **20**:23–36
  boiling curves and bubble behavior, **20**:23–
    27
  heat transfer model–downward-facing sur-
    face, **20**:28–36
  latent-heat transport, **20**:31–33
  liquid and vapor periods, **20**:33
  liquid film thickness, **20**:33–35
  results analysis, **20**:35–36
  sensible heat transport, **20**:29–31
  surface inclination, **20**:27–28
surface temperature distributions, **23**:10–11
temperature excursions, **23**:46–48
thermal control of electronic components,
  20:243–258
  heater widths, **20**:246
  heat transfer characteristics, **20**:250–251
  high flux spaces, **20**:249–250
  narrow space, **20**:248
  surface roughness, **20**:246–247
thermal shock, **23**:51

transient heat transfer in helium I and, **17**:150–
  151
  heat flux conditions, **17**:145–146
  peak conditions, **17**:145–147
  temperature difference, **17**:147
transition from single-phase convection, **23**:17
under zero gravity, **5**:466
wall jets, **23**:45–46
of water, **5**:443
Nucleate pool boiling, *see also* Fluid
  critical heat flux, **16**:127
    experimental investigations, **16**:127–130
  curves, **16**:106–107
  effect of third component, **16**:116–117
  energy flows, **16**:230–231
  future studies, **16**:150
  heat flow rates in, **16**:157–239
    advanced statistical techniques, **16**:168
    burnout, **16**:201–202, 225–229
    comparison of correlations, **16**:159–160
    conclusions, **16**:206–210
    correlation of raw data, **16**:167–168
    data accuracy, **16**:171–175
    data analysis, **16**:166–168
    data analysis results, **16**:185–203
    data available, **16**:169–175
    data distribution, **16**:169–170
    data input, **16**:167
    data reduction, **16**:161
    data reproducibility, **16**:171–175
    dimensional analysis, **16**:168
    effects of fluid properties, **16**:192–200,
      219–225
    effects of heater, **16**:187–192, 212–219
    effects of operating conditions, **16**:186–187,
      210–212
    empirical correlation, **16**:203–205
    existing correlations, **16**:158–161
    fluid property effects, research prospects,
      16:199–200
    $F(p_r)$ for refrigerants, **16**:187, 212
    $F(p_r)$ from data sets with wide range of
      pressure, **16**:186–187, 211–212
    further data needed, **16**:202
    German (Karlsruhe) data, **16**:171–173
    information available, **16**:170
    Japanese (Kyushu) data, **16**:173–175
    natural convection, **16**:201–202, 229
    practical applications of analysis, **16**:164–
      165

Nucleate pool boiling (*continued*)
  reformulation of correlations, **16**:161–164
  research applications of analysis, **16**:165–166
  scope of data, **16**:169–170
  simplification of correlations, **16**:163–164
  theoretical correlation, **16**:205–206
  unexplained differences between experimenters, **16**:200–201, 225
  heat transfer, **16**:105–117
  areas, **16**:120–121
  effect of composition, **16**:114
  effect of pressure, **16**:112–116
  heat transfer coeffcient, **16**:61
  decrease, physical explanation for, **16**:107–110
  as function of composition, **16**:123
  ideal, **16**:123–124
  prediction, **16**:117–127
  predictive methods, **16**:118–123
  ultimate prediction method, **16**:123
  variation with composition, **16**:106, 110, 112–113, 114
  heat transfer rates in, *see* Nucleate pool boiling, heat flow rates in
  local bubble point temperature, **16**:122–123
  local saturation temperature, **16**:122
  of mixtures, effect of subcooling, **16**:110–113
  peak nucleate heat flux, **16**:127–133
  physical phenomena and mechanics, **16**:130–133
  wall superheat, **16**:118–120
  ideal, **16**:123–124
  variation with composition, **16**:106
Nucleation, **9**:89, 183
  activated molecules in, **10**:92–94
  activation energy for, **10**:112
  active cavities size range in, **10**:127–134
  analysis and experiment in, **10**:129–133
  Bergles–Rohsenow graphical procedure and, **10**:157–158
  boiling, **10**:85–162
  boiling superheat in, **10**:148–162
  boiling surface characterization in, **10**:133–134
  and bubble dynamics, **1**:198ff
  bubble growth in, **10**:100–105
  in bulk phase, **9**:199
  cavity activation in, **10**:132–134
  cavity stability in, **10**:134–148

  constant surface temperature profiles in, **10**:130
  criteria in, **10**:128–129
  cycle in, **10**:136–138
  dimensionless superheat vs. dimensionless temperature gradient in, **10**:156
  equilibrium cluster radius in, **10**:96–97
  experimental findings in, **10**:146–148
  experimental observations in, **10**:99–107
  exploding drop method in, **10**:100–101
  explosive, **10**:87, 105, 107–108
  gas-filled cavities in, **10**:120–121
  heat flow densities in, **10**:147
  heat transfer coefficient in, **10**:150–151
  heterogeneous, **10**:111–117
  homogeneous, **10**:92–95, 111
  homogeneous, *see* Homogeneous nucleation
  liquid metals in, **10**:157–162
  low thermal conductivity liquids in, **10**:148–157
  Madejski model in, **10**:154–156
  Marangoni flows in, **10**:156
  nuclei appearance in, **10**:94–95
  from plane surfaces, **10**:111–113
  from preexisting gas or vapor phase, **10**:117–127
  process of, **9**:182
  pulsed heating methods in, **10**:106
  rate of, **9**:202
  ring bubble technique in, **10**:103
  of solid phase, **9**:203
  sounds of, **10**:87, 105
  from spherical projections and cavities, **10**:113–117
  superheat limits and, **10**:95–111
  vapor embryo and, **10**:112–113
  vapor-filled cavities and, **10**:122–127
Nucleation centers, **1**:209
Nucleation factor correlation equations, **20**:13–15
  estimated values, **20**:12, 14
  heat transfer formula, **20**:22–23
  surface factor, **20**:15
  values, **20**:11–12
Nucleation temperature, **28**:12–13
Nuclei, formation of in homogeneous nucleation, **14**:284–285
Null-point calorimeter, **23**:311–313
Numbers of transfer units, in impinging flow, **13**:12
Numerical method

advection terms, discretization-scheme
  Arakawa, **25**:371
  central, **25**:385, 402
  QUICK, **25**:385, 396, 402
computer code, **25**:329–331, 384, 389, 399
for hot water geothermal systems, **14**:34–54
pressure velocity coupling scheme
  PISO, PISOC, **25**:385, 402
  Poisson equation, **25**:371
  SIMPLE, SIMPLER, **25**:385, 390, 402
spatial discretization technique
  finite difference, **25**:338, 401–402
  finite volume, **25**:338, 401–402
  resolution requirements, **25**:323, 326–329
  spectral expansion, **25**:334, 383, 401–40
time-stepping scheme, **25**:402, 406–407
  Adams–Bashfort, **25**:371, 396, 402
  Crank–Nicolson, **25**:402
  leapfrog, **25**:402
Numerical modeling
  multidimensional, **30**:155–161
  of three-phase systems, **30**:185
Numerical simulations, boiling in a forced flow
  through porous media, **30**:140–141
Numerical solutions
  linearized transient flow, **24**:288–291
  nonlinear transient flow, **24**:294
  steady-state convection in, **14**:43
  three-dimensional flows, **24**:310–311
Numerical stability, **8**:172
Numerical studies, in permeable ocean, **14**:40–41
NUREG-1150, **29**:217, 324–325, 329
NUREG/CR-6075, **29**:218
NUREG/CR-6109, **29**:219
NUREG/CR-6338, **29**:219
Nusselt equation, condensation heat transfer,
  21:66–67
Nusselt number, **2**:29, 368ff; **3**:2ff, 129, 130f,
  133f, 136, 138, 153, 159ff, 170; **4**:3ff,
  154ff, 270; **7**:30, 65, 185; **11**:54, 72–75,
  93–94, 100–101, 105, 118–121, 128–131,
  160–161, 208–211, 216, 229–230, 240,
  248–250, 277–279, 291, 304–307, 310;
  12:87; **14**:34, 58, 72, 117, 138, 197–199,
  201–204, 235; **15**:18, 60, 102, 106, 121,
  123, 148, 160–162, 183, 191, 194–195,
  198, 203, 233; **16**:334; **21**:22, 260, 273;
  25:12, 16, 27, 29, 30, 34, 53, 63, 82, 116,
  122, 123, 133, 135, 136, 177, 204, 259,
  260, 263, 266–273, 280–283, 296, 313;

  28:8; *see also* Heat transfer rate;
  appropriate ducts, types of flow and thermal
  boundary conditions in the text for
  numerical values
  analytical predictions of, **15**:127
  in annular-dispersed flows, **1**:429
    MHD, **1**:289
    porous body problems, **1**:124, 133ff,
      139ff
    radiation with convection and conduction,
      1:24, 34, 36ff, 40ff
  approximate expression for, **15**:167
  in approximate pseudosimilar analysis,
    15:194
  aqueous polyacrylamide solutions in rectan-
    gular duct, **19**:320
  asymptotic or limiting, **SUP**:58
  augmented nucleate boiling surface config-
    uration, **20**:73–74
  average, **25**:263, 286, 287
  average numerical solution, **19**:65, 66, 67
  axial variation, **21**:165
  Biot number and, **13**:243
  coalesced bubble region, **20**:53–54
  comparisons of, **SUP**:392–397, 399–405
  convective, **19**:144, 145
  correlation, **SUP**:57–58
  definition of, **SUP**:47–49
  deviations, **19**:131–132
  effective in a heat exchanger, **SUP**:408–409
  elasticity and, **15**:211
  film condensation, **21**:105–106
  for flow over rotating disk and cone, **5**:157
  forced convection
    cubical element array, **20**:233, 235, 237
    flat-pack modules, **20**:236–237
    flush heat sources, **20**:222–223
      computed local number, **20**:223–224
    heat-transfer results–high-heat-flux source,
      20:225–226
    measured (square source) and predicted
      (two-dimensional) source, **20**:224–225
    napthalene sublimation technique, **20**:237–
      238
    small rib spacing, **20**:228–229
    two-dimensional protruding elements,
      20:227–228
  free convection, **21**:40, 43–44
  heat- and mass-transfer
    enclosed porous layer, **20**:340–341

Nusselt number (*continued*)
    finite-amplitude convection, **20**:327
    high Rayleigh number convection, **20**:328–329
    heat transfer relations, **21**:319
    in impinging flow, **13**:7
    induced rotational flow, **21**:162–163
    influence of aspect ratio on, **8**:184
    in laminar convection heat transfer to power-law fluids, **15**:153
    for laminar flow of power law fluids in rectangular channels, **19**:280–289
    in laminar thermal convection constant temperature case, **15**:162
    liquid jet impingement, **26**:206
        arrays, **26**:174–175, 177, 179–181
        free-surface jet, **26**:115–117
        jet configuration, **26**:183–187
        stagnation zone, **26**:112–113, 121, 151–152, 154–156, 162, 166–167, 194–195
        submerged jet, **26**:123, 127–129, 131–133
    liquid–liquid interface, **21**:280, 285
    local, **3**:163ff; **25**:282, 283, 286–288
    maximum, **19**:102, 116, 159
    and mixed convection, **16**:337–340
    mixed-mode flow, **21**:168–169
    module-average, **26**:174, 175, 177
    molten pool heat transfer
        circular cavities, **29**:17–20
        rectangular cavities, **29**:4–17
        spherical cavities, **29**:21–33
    moving liquid droplet, **26**:21–23
        compound drop, **26**:49–52
        deformed droplets, **26**:80
        droplet interactions, **26**:84, 86
        multicomponent drop, **26**:79
    multifluid bubbling pool, **21**:326
    narrow space nucleate boiling, **20**:49–50
    natural convection, **9**:288
        in closed-end tubes, **8**:164
        correlating equation development, **20**:382–383
        enclosure sources, **20**:218–220
        immersion cooling, **20**:217
        inside square cavities, **8**:194–196
        irregular surfaces, **20**:210–211
        isothermal sphere, **20**:371
        liquid metal convection, **20**:378–379
        in rectangular cavities, **8**:179–183, 195, 196
        various shapes, **20**:374–376

    vertical channels, **20**:187–191, 203–205, 207–208
        asymptotic limiting relations, **20**:191, 193, 195
        nucleate boiling, **20**:4
        wedges and spheroids, **20**:375
    nomenclature scheme for, **SUP**:19, 34
    normalized, **21**:33–34, 45
    overall mean, **SUP**:48
    porous layers and, **14**:68
    Rayleigh number and, **14**:42
    regenerator theory, **20**:138–139
        governing equations, **20**:141–144
    relationship with Stanton number, **SUP**:53
    on rotating bodies of revolution, **5**:139
    for rotating cylinder, **5**:166
    as shear flow parameter, **13**:262
    for shrouded rotating disk, **5**:181
    for sphere, **15**:174–175
    in steady shear flow, **13**:239–241
    submerged jet impingement systems, **20**:255–256
    temperature difference and, **13**:242
    total interlayer heat transfer model, **21**:315–316
    transient, **26**:25–26
    for two-phase flow, **5**:446, 447
    unidirectional regenerator operation numerical solution, **20**:150–151
    unsteady, **25**:20
    variation of with Grashof/Reynolds ratio, **15**:200
    versus Rayleigh number, **15**:210, 212–213
    versus speeds of rotation ratio, **21**:166–167
    versus Taylor number, **21**:163–164
    vertical plates and channels vs. actual printed circuit boards, **20**:195–196, 198
    viscous dissipation and, **13**:241–242
    zero axial flow, **21**:163–164
Nusselt number ratio, **19**:340; **31**:163, 246, 256
Nusselt–Reynolds type equation, **2**:32ff
Nusselt's IV model, regenerator design equivalence, **20**:136–137
Nusselt's analysis, **3**:108f
Nusselt theory, **21**:105
    of filmwise condensation, **5**:57
NWM variable, **27**:124, 206
NXTGAS subroutine, **27**:109
Nylon, **25**:6
    melt, **25**:268

O

Oberbeck equations, **4**:5
Oblate (planetary) spheroid, potential flow convection, **20**:373
Oboukhov–Corrsin scale, **30**:12, 13, 16, 48
Ocean of Storms, **10**:5
Oceans, as heat sinks, **12**:10
Oceanus Procellarum, **10**:6
Octane, compound droplet, vaporization, **26**:78
Ocular injury, due to intense light, **22**:248
Ohmic heating, **1**:306f
Ohm's law, **1**:275
Oil recovery, multiphase flow effects on, **30**:93–94, 152–153, 154, 159
Oils, viscous, forced convection, **30**:20–24
Oldroyd fluid, **15**:180; **24**:109
Olivine dolerite, lunar, **10**:59–61
OM experiments, **29**:165
OMVPE, *see* Organometallic vapor-phase epitaxy
Once-through cooling, **12**:12
  climatic factors in, **12**:24
  heat transfer mechanism in, **12**:13
  water for, **12**:20–21
Once-through system, intermediate heat transfer in, **15**:112–114
One-dimensional flow
  multiphase, **30**:164–167
  two-phase, **30**:114–121
One-dimensional heat transfer, models, **30**:209–213
One-dimensional two-phase flow, integral methods for, **31**:143
One-dimensional two-phase models, of geothermal systems, **14**:78–82
One-dimensional unsteady vertical distribution models
  auxiliary relations in, **13**:91–92
  boundary conditions in, **13**:92
  differential equations for, **13**:89–91
  Ekman layer in, **13**:93–95
  mathematical formulation in, **13**:89–93
  type of problem in, **13**:89
  warm-water column cooling and, **13**:95–98
One-equation model, **23**:369, 399
  comparison with experimental values, **23**:441
  comparison with the two-equation model, **23**:449, 455

effective thermal conductivity tensor, **23**:399, 453
  mapping vectors, **23**:404
"One-shot" method for measuring density in annular-dispersed flows, **1**:398f
Onsager's reciprocity relations, **8**:25
Open bath test chamber, thermal cooling of electronic components, **20**:250–251
Open boundary treatment, **24**:219
Open thermodynamic systems, for second-law analysis, **15**:5–8
Operating temperatures, active solar collectors, **18**:3
Operative temperature, **4**:83
Optical cross-correlation, **9**:138
Optically thick approximation, **1**:6ff, 13ff, 15, 19f; **3**:208f
Optically thick limit, **8**:252, 253, 264
Optically thick radiation, **8**:257, 258
Optically thin approximation, **1**:6, 8f, 13f, 20f; **3**:207f
Optically thin limit, **8**:235, 252, 254, 256, 263, 265
  influence of thermodynamic nonequilibrium on, **8**:252, 258
Optical method, **23**:321–326
  liquid crystals, **23**:322–324
  phase-change paint, **23**:322
  thermographic phosphors, **23**:324–325
Optical sensing technique, **9**:145
Optical spatial arrangement, of nozzles in impinging flow, **13**:45–52
Optical speckle, defined, **30**:255
Optical spectra, dispersion analysis of, **11**:357
Optical thickness, **1**:4, 6; **3**:200; **5**:29; **8**:12; **9**:377
  effect of, **31**:349–353
Optimum thickness, thermal contact conductance, **20**:271
Order of magnitude analysis, **24**:115
Order of magnitude estimates, **31**:82–85
Ordinary diffusion
  defined, **12**:203–204
  diffusivities in, **12**:207–208, 213–219
  diffusivity measurement in, **12**:220–221
  in electrochemical transport, **12**:203–221
  Fick's law and, **12**:204
  flux by, **12**:204
  mass system and, **12**:204–209
  mixed systems in, **12**:219–224

Ordinary diffusion (*continued*)
  molar system in, **12**:209–219
  Stefan–Maxwell equations in, **12**:208–209, 213–219
Organic vapors, dropwise condensation, **21**:91–92
Organometallic vapor-phase epitaxy (OMVPE), **28**:342
Orientation, particle, **25**:197, 237
Orientation of heating surface, effect on transition boiling heat transfer, **18**:303
Ornstein and Van Wyck's method, **2**:304
Orr–Sommerfeld, **9**:323
Oscillating disk method, **3**:256f, 260
Oscillating wires, **31**:453
Oscillations, internal-gravity, **24**:288–291, 300–301
Oscillators
  equation of motion for, **11**:356
  rate of change of vibration energy in system of, **8**:247
Oscillatory approach
  in linearized model, **24**:290–292
  in sidewall-heated cavity, **24**:297–302
Oscillatory convection, **24**:174
Osculation plane, **6**:141
Oseen constants
  influence of, on temperature profiles, **8**:208
  for natural convection inside horizontal circular cylinders, **8**:203
Oseen method, **8**:170, 172, 199, 214
Osmotic stress, **22**:212, 294
Ostrogradsky–Gauss equation, **19**:201
Ostwald–de Waele model, **15**:88, 151; **19**:251–252
Ostwald–de Waele power-law fluid, **24**:108
Outer expansions, **8**:7
Outer limit similarity equations, **4**:341ff
  solution, **4**:435ff
Outflow boundary, liquid metal droplet, **28**:6
Outlet flow conditions, mean transfer coefficients and, **13**:40
Ovens, **9**:3
Overshoot, **9**:313, 314
  temperature, **9**:320
Overshoot phenomenon
  in close-contact melting, **19**:72
  during transition phase, **19**:65
  heat transfer rates, **19**:46, 47
Overstability, **5**:205

Overstability Rayleigh number, **14**:29; *see also* Rayleigh number
Overstability wave number, **14**:29
Overtone bands, **8**:231
Oxidation, DCH debris, **29**:282–284
Oxide crystals
  defects in, **30**:319
  growth by Czochralski process, **30**:319
Oxide formation, thermal contact conductance enhancement, metallic coatings, **20**:270–271
Ozone, **31**:449

P

p^+-junction formation, heat transfer and mass transfer, **28**:96–101
Pacific ocean, northern, temperature, salinity, and density, **16**:46–48
Packed-bed condenser, direct-contact, **15**:265–266
Packed beds
  axial dispersion, **23**:229
  capillary models, **23**:195
    Blake–Kozeny, **23**:195, 197
    Blake model, **23**:195, 197
    Kozeny–Carman, **23**:195, 197
  cell models, **23**:206
    free surface, **23**:207, 209–210
    zero vorticity, **23**:207, 209–210
  dilute polymer solutions, **23**:222
  drag coefficient, **23**:207
  flow, laminar, **23**:194
  friction factor, **23**:196
  heat transfer, **23**:229
  mass transfer, **23**:229
  mixing, **23**:229
  non-Newtonian fluid flow, **23**:192
  nonspherical particles, **23**:215
  polymer adsorption, **23**:226–227
  polymer retention, **23**:226
  pressure drop, **23**:196
  slip effects, **23**:228
  submerged object model, **23**:206
  tortuosity, **23**:195–196
  viscoelastic effects, **23**:206–207
  voidage profiles, **23**:216
  wall effects, **23**:216
  wall–polymer molecule interactions, **23**:194, 206, 225

Packed fluidized beds, **14**:231–238
  horizontal tubes in, **14**:232–233
  vertical tubes and, **14**:234–235
Packed spheres, **27**:193–195
Packet model, **19**:128
Packing density, **25**:298
PAHR, *see* Postaccident heat removal
Pametrada project, **9**:65, 93
Pancake chemical vapor deposition reactors,
  28:346, 386–389
Panradiometer, **4**:71ff
Pan size, stove efficiency and, **17**:267–268
Paper industry
  explosions in, **10**:107
  steam explosions, **29**:130
Parabolic dish systems
  cavity receivers, **18**:41–46
  electrochemical technique for mass transfer
    measurement, **18**:44–45
Parabolic equations, **8**:13, 57
Paraffin, natural convection to, inside rectangular
  cavities, **8**:184
Parahydrogen, **5**:330
Parallelepipeds, as modeling elements, **31**:359–
  360
Parallel flow region, **26**:105
  free-surface planar jet, **26**:147, 167–170
  laminar planar jet, **26**:122–123
Parallelism, of Monte Carlo ray tracing algorithms, **31**:407–413
Parallel plate channel, **3**:242ff
Parallel plates, **SUP**:153–195
  arbitrary axial wall temperature distribution
    for, **SUP**:177–179
  arbitrary heat flux distribution for,
    SUP:185
  axially matched solutions for, **SUP**:160–161
  conjugated problems for, **SUP**:188–189
  constant heat fluxes at each wall in, **SUP**:157–
    158, 179–184, 192–193
  constant temperature at one wall, heat flux at
    other wall in, **SUP**:158
  constant temperatures at each wall in,
    SUP:155–157, 169–177, 190–191
  constant wall temperature and wall heat flux
    for, **SUP**:154–158
  energy content of fluid in Graetz–Nusselt
    problem for, **SUP**:174–175
  exponential axial wall heat flux in, **SUP**:159,
    188

finite wall thermal resistance in, **SUP**:158–
  159, 185–187
fluid axial heat conduction in, **SUP**:156, 174–
  176, 177, 183
friction factors for, **SUP**:153, 163, 167–168
fully developed flow in, **SUP**:153–159
fundamental solutions for, **SUP**:154, 177–179,
  185, 193–195
Graetz–Nusselt problem for, **SUP**:169–177
hydrodynamically developing flow in,
  SUP:160–169
hydrodynamic entrance length for, **SUP**:168–
  169
integral solutions for, **SUP**:160, 192, 194
linearized solutions for, **SUP**:163
numerical solutions for, **SUP**:163–164
Nusselt numbers for, **SUP**:154–159, 172–174,
  181–182, 187, 190–191, 193–194
simultaneously developing flow in, **SUP**:189–
  195
solutions of Navier–Stokes equations for,
  SUP:164–167
specified wall heat flux distribution for,
  SUP:179–185
specified wall temperature distribution for,
  SUP:169–179, 190–191
thermal entrance length for, **SUP**:171, 181,
  193
thermally developing and hydrodynamically
  developed flow in, **SUP**:169–189
viscous dissipation, effect of, **SUP**:156–158,
  176–178, 183–184
Parallel rotating disks, **5**:182
  comparison of theory and experiment, **5**:188
  inlet configurations for source flow, **5**:196
  laminar analysis of, **5**:183
  with laminar source flow, **5**:182
  Nusselt number for, **5**:184
  with source in center, **5**:193
  Turbulent flow, **5**:191
  with turbulent source flow, **5**:191
Parallel-stream heat exchanger, **26**:227–229
  boundary conditions, **26**:264
  design, **26**:309–315
  energy equations, **26**:258–260, 265–266
  exit temperature, **26**:273–274
  literature review, **26**:237–247
  mathematical model, solution, **26**:271–281
  multifluid exchanger, **26**:254
  temperature effectiveness, **26**:300–302

Parallel-walled cavities, *see* Cavities
Parallel-walled passages, *see* Passages
Parametric mapping function, 31:336
Parametric modeling, steam explosions, 29:135–137
Parl's calculation method, 20:152–153
PARSEC model, direct containment heating, 29:233
Partially parabolic layer, in two-dimensional floating layers, 13:109–111
Partially stabilized bed, 25:193, 220, 231
Partial mass enthalpy, 13:150
Partial molal Gibbs energy function, 12:261
Particle attachment mechanism, 31:442
Particle diameter, dryout heat flux and, 29:34–35
Particle drag law, 9:155
Particle dynamics, 9:123
Particle–fluid interaction, 9:132
Particle and heat transfer surface contact, for smooth v. rough surfaces, 19:152–153
Particle image velocimetry (PIV), 30:255
Particle–particle interaction, 9:123
Particle properties, effect on fluidization, 25:152, 194
    buoyant weight, 25:152
    density, 25:152
    dielectricity, 25:153
    flowability, 25:171
    magnetic
        fluid permeability, 25:154, 160, 197, 203
        susceptibility, 25:186
    shape, 25:152, 203, 229
    size, 25:152, 167, 169, 176, 203, 209
        distribution, 25:152
Particle Reynolds number, 9:124
Particles, 25:151–154, 275
    cast-iron, 25:155
    classification, 19:100, 180
        based on laminar flow, 19:102–103
    coarser and heavier, 19:134
    cobalt, 25:233
    critical size of, 9:125
    cylindrical, 25:275
    ferrite, 25:197
    ferromagnetic, 25:160, 164
    flux, 9:175
    group 1, 19:103
    group III, 19:104–105
    group IIA, 19:103–104
    group IIB, 19:104

iron, 25:164
    large, 19:180
        heat transfer data, summary, 19:137
        immersed in fluidized beds, heat transfer correlations for smooth surfaces, 19:135–142
        maximum heat transfer coefficients, 19:116–117
    magnetic, 25:155, 166, 205
    magnetized, 25:166
    mass flux and density measurements of, 9:143
    mild steel, 25:198
    model, 25:160
    molecular-sieve, 25:232
    nickel catalyst, 25:169
    nonmagnetic, 25:153
    non-spherical, 25:292
    number density of, 9:123
    particulate slurries, 25:253, 292
    permanently magnetized, 25:167
    size
        heat transfer coefficients, at high temperatures, 19:169–171
        maximum heat transfer coefficient and, 19:115–116
    small, 19:104, 134, 180
        groups I and IIA, 19:112
        heat transfer between immersed surface and fluidized bed, 19:127
        maximum heat transfer coefficient and, 19:115–116, 117
    solid spherical glass, 9:122
    spectral behavior of, 31:390
    spherical, 25:292
    steel shot, 25:166, 169
    subgroup IIA, 19:113–114
    subgroup IIB, 19:114
    suspension, 25:255, 256
    system, 25:256, 273
    unmagnetized, 25:166
Particles in suspension, thermal conductivity
    deformed, 18:205–206
    spherical, 18:203–204
Particle slip, 9:150
Particle velocities, 9:137, 175
    influence of particle size on, 9:122
    measurement of, 9:137
    terminal, 9:119
Particulate fouling, 31:441–443
Particulates, presence of, 31:441

Partition function, **2**:125f, 133
Paschen Series, **5**:275
Passages
  circular tube, **2**:432
  parallel-walled, **2**:432
  radiant transmission characteristics, **2**:430ff
  tapered gap, **2**:430f
  tapered tube, **2**:430f
Passive–active decomposition, **31**:80–82
Passive conduction, compared with effect of
    evaporation and condensation, **31**:95
Passive effective diffusivity tensor, **31**:5
Path length attenuation technique, **31**:377
Path length-based methods, **31**:346, 366–369
PCVD, *see* Photo-chemical vapor deposition
Peaceman–Rachford procedure, **22**:269
Peak nucleate heat flux, **14**:132
Péclet number, **3**:107; **4**:296, 301; **5**:220;
    SUP:52–53, 58; **15**:70, 246; **19**:74, 256;
    24:120; **25**:3, 269, 313; *see also* Fluid axial
    heat conduction
  cell, **24**:195, 226, 228
  flame, **21**:262
  fluidized beds, **23**:241–242
  forced convection, **20**:222–223
  packed beds, **23**:229
PECVD, *see* Plasma enhanced chemical vapor
    deposition
Pellets, **25**:252, 278
Peltier effect, **22**:388
Penetration depth, **8**:24; **9**:124
  of light incident on tissue, **22**:282
Penetration distance, **1**:53, 55, 84
Penetration length, **1**:6; **3**:215
Penndorf extension, **23**:146–147
Pennes bioheat equation
  analysis by Shitzer and Klinger, **22**:28–29
  experimental measurements
    basis for mathematical model, **22**:23
    description, **22**:22
  governing equations, **22**:25, 58, 75, 76, 114
  history, **22**:19
  model deficiencies, **22**:26, 29–30, 30–32, 34–
    36, 45, 50–51, 58, 96–97, 106
  perfusion heat source term, **22**:26, 54, 74, 106,
    113, 120, 124, 126
  theoretical formulation
    boundary conditions, **22**:27
    comparison with experiments, **22**:27
    effect of blood flow, **22**:27–28

thermal equilibration parameter, **22**:25, 52, 74
PEO, *see* Polyethylene oxide solution
Perelman's method, **8**:37
Perfect gas, differential energy equation for,
    SUP:13
Perfluoromethylcyclohexane, **14**:125
Periodic (cosine) profile, heat- and mass-transfer,
    20:324–325
Periodic flow theory, regenerator theory, **20**:138–
    139
Peritectic reactions, **28**:233
Perkin–Elmer spectrometer, **10**:17–18
Permafrost, **9**:3, 83
Permeability, **24**:110; **25**:154, 230
  alloy solidification, **28**:264, 276
  bed effective, **25**:157, 239
  cell membrane, **22**:292, 294
  force space, **25**:169, 198, 239
  magnetic field, **25**:154, 184
  magnetic fluid, **25**:154, 160, 197, 203
  multiphase flow, **30**:98
  temperature dependence, **22**:293
Permeability tensors, **31**:40
Permeable surfaces, *see* Turbulent boundary
    layers
Perturbation methods, **8**:3; **9**:39
  coordinate perturbation, **8**:3
  parameter perturbation, **8**:3
  regular perturbation, **8**:3
  singular perturbation, **8**:3
Perturbations, **19**:30, 33, 60, 196
  method of solution, **22**:306
Perturbation solution methods, **2**:220ff
Perturbation solutions, **8**:67
  potential flow convection, **20**:371
Perturbation techniques, **2**:253ff
Perturbation theory, **19**:28–29
pH, steady shear velocity and, **15**:83
Phase
  bubble, **25**:161
  definition of, **31**:105
  discrete, **25**:161
  emulsion, **25**:169, 170
Phase and velocity distribution in annular-
    dispersed flows, **1**:381ff
Phase change, **8**:24, 29, 75; *see also* Freezing;
    Melting
  convection and, **19**:3
  electromagnetic effects, **19**:80–81
  forced convection effects, **19**:3–4

Phase change (*continued*)
 frost, **19**:81
 geophysical processes, **19**:82
 heat ablation, **19**:81
 in heat conduction problems
  with ablation, **1**:75ff, 82
  with given surface heat flux, **1**:74f
  with step in surface temperature, **1**:72ff
 instability-induced convection, **19**:79–80
 literature review, **19**:2
 in manufacturing processes, **19**:2
 model simplification, **19**:4
 multidimensional problems, **19**:3
 one-dimensional, **19**:3, 4
  driven by constant temperature boundary
   conditions, **19**:9
  with isothermal boundary conditions, **19**:5
 physical nature, **19**:9
 in porous media, **19**:78–79
 sublimation, **19**:81–82
 two-phase, **19**:4–5
Phase-change paint, **23**:322
Phase-change process, **22**:187
Phase-change transformations
 pulsed laser melting, **28**:75–109, 135
 pulsed laser sputtering, **28**:109–123
 pulsed laser vaporization, **28**:123–135, 136
Phase diagram, **22**:188, 209
Phase equilibrium diagram, **16**:62–63
 bubble point line, **16**:62
 dew point line, **16**:62
Phase equilibrium thermodynamics, **16**:61
Phase indicator function, **23**:377
Phase interface, **22**:190
Phase saturation, multiphase flow and, **30**:95–96
Phase transformations, radiative effects in,
 11:411–413
Phase transition enhancement, **31**:266
Phasic energy balances, **31**:112
Phonon conduction
 heat flux and, **10**:46
 heat transfer and, **10**:45
 temperature and, **10**:43
Phonon conductivity, **11**:353–354
 from semitransparent solids, **11**:359–362
Phonons, **9**:361
Photo-chemical vapor deposition (PCVD),
 28:342
Photodissociation, **5**:263
Photoionization, **5**:263

Photoluminescent volumetric imaging (PVI),
 phase saturation measurement by, **30**:96
Photon bundle, **5**:14
Photon distribution function, **5**:257
Photon idea, **3**:183
Photon–molecule interaction, wavenumbers in,
 12:139
Photons
 absorption and emission of, **12**:126
 blackbody radiation and, **12**:125
 mean free path of, **11**:322
 Monte Carlo method, **27**:45–46
 ray of, **12**:126
 straight-line travel of, **11**:320
 in thermodynamic equilibrium radiation,
  12:120–121
 vibration–rotation line and, **12**:139
 wavenumbers of, **12**:139
Physical explosions, **29**:129
Physical properties, critical region, **7**:3
Physical vapor deposition, **28**:339–340
Physiconeural model, chemical vapor deposition,
 28:413
Pigment epithelium, **22**:265, 273
Pipeline, **25**:98
 complex geometry, **25**:111
Pipelining, **31**:401, 407
Piston cloud chamber experiments, **14**:318–
 325
Pitzer eccentric factor, **16**:160, 183
PIV, *see* Particle image velocimetry
Planar impingement, potential flow convection,
 20:364–365
Planar jet, **26**:107
 array, **26**:171, 182
 flow structure, **26**:144–147
 free-surface liquid jet, **26**:107–108, 119–123
 heat transfer, **26**:122, 162–170, 192–195
 jet inclination, **26**:186–189
 submerged jet, **26**:135
Planar motion
 conduction along streamlines, **20**:357
 general formulation, **20**:354
 negligible conduction along streamlines,
  20:355–356
Planar-stagnation-flow chemical vapor deposition reactors, **28**:397–399
Planck averaged absorption coefficient, **11**:405–
 406
Planck blackbody radiosity, **12**:124, 133

Planck constant, **2**:345; **3**:183; **9**:362; **12**:120; **22**:421

Planck distribution, **5**:6

Planck function, **4**:239, 245f; **11**:326–327

Planck internal mean absorption coefficient, **12**:156

Planck mean absorption coefficient, **5**:28, 314
  for ammonia, **5**:315
  for carbon dioxide, **5**:315
  for carbon monoxide, **5**:315
  for methane, **5**:315
  for nitric oxide, **5**:315
  for nitrous oxide, **5**:315
  for sulphur dioxide, **5**:315
  for water vapor, **5**:315

Planck mean coefficient, **8**:259

Planck mean emission coefficient, **12**:156

Planck number, **21**:258, 262

Planck's law, **3**:183; **5**:259; **9**:362; **11**:321; **27**:7–8

Planck's spectral energy distribution, **5**:13

Plane liquid sheet, direct condensation on, **15**:233–234

Plane parallel layer, radiative transfer in, **11**:333–338

Planetary boundary layer, **25**:370–378, 401

Plasma, **4**:229ff
  characteristic properties, **4**:231ff
    blackbody radiation, **4**:238ff
    bremsstrahlung, **4**:237f
    line radiation, **4**:234f
    recombination radiation, **4**:235ff
  composition, **4**:256ff
  heated, entropy production, **21**:252–253
  heat transfer, **4**:270ff
    in absence of an externally applied electric or magnetic field, **4**:270ff
    basic transport equations, **4**:272ff
    laminar boundary layer equations, **4**:276ff
    results of reentry studies, **4**:278ff
    in presence of a magnetic field, **4**:304ff
    electrically insulating surface, **4**:307ff
    electrode heat transfer, **4**:305ff
    in presence of an electric current, **4**:282ff
    electrically conducting surface, **4**:289ff
    electrically insulating surface, **4**:282ff
      constricted arc, **4**:283ff
      experimental studies, **4**:292ff
        cathode axis parallel to a plane anode, **4**:295ff
        cylinder geometry with annular anode, **4**:302ff
      free-burning arcs with plane anodes, **4**:293ff
  influence of on heat transfer, **4**:231ff
  plasma–wall boundaries, **4**:252ff
  sheaths, **4**:252ff
  steady, dense plasmas, **4**:242ff
  temperature increase of plasma, **4**:262ff
    reduction of heat conductivity by a magnetic field, **4**:267ff
    reduction of wall heat fluxes by transpiration cooling, **4**:269f
  thermodynamic equilibrium, **4**:244ff
    local thermal equilibrium, **4**:246ff
    excitation equilibrium, **4**:249
    ionization equilibrium, **4**:250ff
    kinetic equilibrium, **4**:248f
    perfect thermodynamic equilibrium, **4**:245f
  thermodynamic properties, **4**:256ff
  transport properties, **4**:260ff

Plasma arc, **22**:10

Plasma deposition, **28**:53

Plasma enhanced chemical vapor deposition (PECVD), **28**:342

Plasmon decay, **28**:120–123

Plastic coating, **31**:461

Plastic films, reflectance of, **9**:388

Plastics, U.S. consumption of, 1963-79, **15**:60

Plate, **25**:257, 258, 262, 286–291, 312
  flat, **25**:252, 254, 286, 289
  horizontal, **25**:259, 261, 286, 288
  isothermal, **25**:287
  moving, **25**:289, 290
  stationary, **25**:259
  vertical, **25**:259, 261, 286, 288

Plate-fin heat exchanger, **26**:227, 228

Platinum wires, film boiling on, **5**:66

PLK method, **8**:3

Plug flow, **1**:358

Plug flow idealization, **3**:119ff

Pluglike flow, **9**:122

Plume, **9**:282, 297
  axisymmetric, **9**:339
  axisymmetric flow, **9**:301
  axisymmetric swirling turbulent, **10**:247–250
  behavior studies, **16**:53
  boundary conditions of, **9**:299
  convected energy, **9**:299
  definition, **16**:3

Plume (*continued*)
  of dry cooling towers, **12**:36
  enclosed fire and, **10**:258–260
  end effects of, **9**:301
  experimental studies of, **10**:256–260
  in fire studies, **10**:245–246
  forced, **9**:297
  instability in, **9**:335
  line source, **9**:299
  multiple fire whirl formation in, **10**:257–258
  stability of in line fire, **10**:253–255
  temperatures of, **9**:300
  thermal, **9**:300
  unperturbed, **9**:337
  of wet cooling towers, **12**:22–23
Plume rise, from horizontal line source, **14**:59–60
PM-ALPHA model, steam explosions, **29**:150, 151, 153–154, 156, 195
Pohlhausen integral method, **2**:179ff, 227f, 234ff
Pohlhausen solution, **SUP**:143, 195
Pohlhausen transformation, **4**:22
Point, transition, **25**:195
Point equations, governing, **31**:23–28
Point-focusing systems, **18**:27–29, 75–76
  cavity receivers, **18**:35–41
    for central receiver systems, **18**:35–41
    for parabolic dish systems, **18**:41–46
  external receivers for central receiver systems, **18**:30–34
Point-matching methods, **SUP**:65–66
Point source, heat- and mass-transfer, **20**:343–346
Point source flow method, **3**:285f
Poiseuille flow, **8**:29; **13**:218–222; *see also* Geometry, plane channel
  converging or diverging, **13**:261
  defined, **13**:221
Poisson's equation, **12**:247; **21**:189
Polar gases, **3**:265ff, 281, 298
Polarization effects, **3**:183
Polar symmetry, heat conduction problems involving, **1**:64, 70f, 79
Pollution
  by woodburning stoves, **17**:300–301
  quantitative prediction of, **13**:62
  thermal or chemical, **13**:62
Polyacrylamide, **25**:278
Polyacrylamide oxide, **15**:102
Polyacrylamide-PAM (Separan), **19**:329

Polyacrylamide solutions, **15**:64
  friction factors in rectangular duct, **19**:332
  pH vs. shear velocity in, **15**:83
  solvent effects on turbulent heat transfer for, **15**:116
  steady shear viscosity vs. average shear rate for, **15**:75
  time-dependent terminal velocity for, **15**:74
  turbulent friction factors, **19**:330–331
  turbulent heat transfer results for, **15**:114
Polyatomic fluid, condensation coefficient, **21**:79–80
Polyatomic molecules, **8**:231
Polydimethylsiloxane, thermal conductivity, **18**:218–222
Polyethylene oxide, **19**:329; **25**:281, 286
  drag reduction in, **12**:77
Polyethylene oxide solution, **15**:64, 102, 167; *see also* Polyox solution
  drag-reducing, **15**:183
  Fanning friction coefficients for, **15**:95
Polygonal ducts, *see* Regular polygonal ducts
Polymer coating, dropwise condensation, **21**:103–104
Polymeric film, development of, **31**:435–436
Polymeric liquids, thermal conductivity
  multicomponent systems
    filled polymeric systems, **18**:227–230
    polymer solutions, **18**:223–227
      altered free-volume state model, **18**:226–227
      experimental observations, **18**:223–225
  single-component systems, **18**:213–222
    molecular characteristics and, **18**:218–220
    strain-rate effect, **18**:220–222
    temperature effect, **18**:213–218
      experimental observations, **18**:217–218
      modified Bridgman equation, **18**:216–217
      WLF correlation, **18**:214–216
Polymeric regime, **12**:80
Polymer melt, **25**:292
  extensibility, **25**:308
  solution, **25**:259, 264, 267, 276–278, 281, 286
Polymer retention in packed beds, **23**:226–227
Polymers
  flow in dies, **28**:201–212
  melting, **28**:224
  molten, *see* Molten polymers
  screw extrusion, **28**:146–151

combined heat and mass transfer, **28**:213–220

material characterization, **28**:152–155

single-screw extruder, **28**:147, 149, 155–187

twin-screw extruder, **28**:151, 187–201, 225–226

thermal degradation of, **12**:78

Polymer solutions

degradation effects in, **15**:115–120

viscosity of, **15**:60

Polyox solution, **15**:64, 84, 88; **25**:259

asymptotic friction in, **15**:94

degradation effects in, **15**:118

Nusselt number for, **15**:102

in tap water, **15**:86–87

Polyox WSR-FRA, **15**:175

Polyphosphate, **31**:449

Polysilicon films, excimer laser melting, **28**:81–96

Ponderomotive coefficient, **1**:280f

Ponderomotive force, **1**:273f

Pool boiling, **1**:233ff; **5**:401; *see also* Film-boiling heat transfer

binary mixture, **30**:173–174

CHF in, **17**:5–7

experiments in, **1**:238ff

geometry and orientation of heated surface in, **11**:110–117

of helium, **5**:406

of hydrogen, **5**:405

liquid superheat fluctuation in, **11**:30–45

maximum heat flux in, **5**:402

minimum heat flux in, **5**:402

of nitrogen, **5**:404, 408

of oxygen, **5**:407

in reduced gravity, *see* Reduced gravity

saturated, of liquid metal and nonmetal fluid at very low pressures, **17**:22–23

temperature fluctuation in, **11**:38–43

Pool boiling curve, **16**:105–106

Pool fire

laminar, **30**:65–72

turbulent, **30**:73–78

Popcorn effect, during laser heating, **22**:321

Population growth, waste heat cooling and, **12**:2

Porosity, **24**:110; **25**:298, 314; *see also* Voidage

fluidized beds, **23**:241–242

packed beds, **23**:216

Porous bodies, heat and mass transfer in, **1**:123ff

Porous cooling, **1**:130f, 144ff

Porous inertia effects, **24**:112

Porous layer heat transfer enhancement devices, **30**:232–234

Porous layers

convective heat transfer in, **14**:67–68

experimental investigations in, **14**:71–72

horizontal, **14**:76

inclined, **14**:73–74

vertical, **14**:74–76

Porous media, **8**:29; **25**:292, 305, 308; *see also* Drying process; Radiative heat transfer

capillary action in, **13**:121

characteristic time for flow in, **13**:170

drying of, **30**:168–172

drying process in, **13**:119–120

dryout heat flux in, **29**:47–48

heat, mass, and momentum transfer in, **13**:119–200

interfacial drag in, **29**:42–47

liquid phase flow for two-fluid system in, **13**:176

mass and energy transport in, **13**:126–153

multiphase flow, **30**:93–95, 186–187

capillary forces, **30**:93, 94, 96–98

gravity and, **30**:93–94

interfacial tension, **30**:96

models, **30**:98–110

non-Darcian effects, **30**:111–112

nonequilibrium effects, **30**:110–111

phase saturation, **30**:95–96

relative permeability, **30**:98, 99

viscous forces, **30**:93, 94

multiphase multicomponent flow, **30**:102–164, 185–186

liquid-gas systems, **30**:167–184

one-dimensional, **30**:164–167

three-phase systems, **30**:184–185

one-dimensional two-phase flow in, **30**:114–121

quantitative effects in, **13**:191

temperature gradients in, **13**:191–192

two-phase, single-component system multiphase flow, **30**:112–113, 161–162

external boiling flows, **30**:148–152

external condensing flows, **30**:141–148

internal boiling and forced convection, **30**:136–141

internal boiling and natural convection, **30**:122–136

Porous media (*continued*)
  one-dimensional, **30**:114–121
  steam injection, **30**:152–161
  unsaturated, convection in, **30**:167–168
Porous-medium systems
  heat- and mass-transfer
    natural convection, **20**:325–327
    physical model, **20**:316–318
Porous nesting, hot water geothermal effects in, **14**:37
Porous plate, **9**:294
Porous surfaces, augmented nucleate boiling, **20**:72–77
Posnov number, **1**:175
Postaccident heat removal (PAHR), nuclear accidents, **29**:1, 9
Postburn cooling, **22**:235, 240
Post-dryout heat transfer, in a porous bed, **30**:135
Potential, in electrochemical systems, **12**:196
Potential core, **26**:106
Potential flow convection
  axisymmetric flow
    general formulation, **20**:365–367
    planar impingement, **20**:369
  circular cylinder, **20**:359–361
    isothermal cylinder, **20**:359–360
    uniformly heated cylinder, **20**:360–361
  cone, **20**:367–368
  derivations and solutions, **20**:354–357
    planar motion, **20**:354
    planar motion–streamline conduction, **20**:355–357
  elliptical cylinder, **20**:361–362
    finite height strip–front face, **20**:362–363
    isothermal surface, **20**:361–362
    uniformly heated surface, **20**:362
  flat plate, **20**:357–359
    isothermal plate–streamline conduction, **20**:357–358
    uniformly heated plate–streamline conduction, **20**:358–359
  infinitely long pointed needle, **20**:368–369
  mean Nusselt number for various shapes, **20**:374–376
  summary of solutions, **20**:374–376
  wedges and spheroids, **20**:375
  planar flow, normal plane impingement, **20**:364–365
  research background, **20**:353–354

solution applicability, **20**:376–382
  correlating equation development, **20**:382–383
  liquid metals convection, **20**:376–379
  rising bubble convection, **20**:379–382
  spherical, **20**:369–370
    axially symmetric convection, **20**:372–374
    isothermal cap, **20**:374
    transverse curvature, **20**:370–371
    wedge, **20**:363–364
Potential-flow nozzle, **2**:11
Potential function
  Buckingham, **3**:263, 291, 295
  exponential repulsive, **3**:262, 290, 295
  inverse power, **3**:262, 290, 295
  Lennard–Jones, **3**:262f, 291, 295
Powder classification
  bed fluidization and, **19**:99–105
  groups IIB and III, **19**:136, 138
  Nusselt number prediction, **19**:180
Powder classification scheme, **25**:152
Powell–Eyring model, **19**:333
Power cycle, rejection temperature of, **12**:5
Power generation
  cooling system vs. cost in, **12**:15
  cost of, **12**:16
Power industry
  dry cooling in, **12**:3
  waste heat removal in, **12**:1–2
Power law, **2**:362; **25**:257–264, 267–271, 273–279, 281, 283–286, 286, 290, 291, 298, 299, 306, 310, 312
  consistency index, **25**:313
  flow-behavior index, **25**:313
  fluid, **24**:108
  general consistency index, **25**:313
  index, **24**:108
  modified general consistency index, **25**:313
  Newtonian model and, **15**:80
Power-law dimensional formulas, for generalizing transition boiling heat transfer data, **18**:307
Power-law fluid flow
  liquid–solid fluidized beds, **23**:236
  packed beds, **23**:197
  sedimentation, **23**:262
  three-phase systems, **23**:253
Power-law fluids
  constant heat flux in, **15**:200–202

laminar thermal convection in, **15**:151–165, 169–176
thermal instability in, **15**:210
Power-law index, **15**:102, 190
Power-law model, **19**:251–252; **28**:153
Powerline, high voltage as source of burns, **22**:245
Power parks, **12**:9, 24
  heat concentration in, **12**:28
  storm-making potential of, **12**:29
Power plants, pollution from, **13**:62; *see also* THIRBLE
Power requirements for cookstoves, **17**:296
Power semiconductor coolers, **20**:280–281
Power series, **8**:13
Power spectral density, slug velocity, **20**:113
Poynting theorem, **21**:270
Prandtl–Eyring, **25**:257, 258
Prandtl number, **2**:15, 23, 117, 372, 385, 386f; **3**:2, 4, 40; **4**:3ff, 151, 271, 320; **9**:288; **11**:76, 126–129, 155, 200, 212, 269–271, 278–279, 283, 288, 300–302, 305–306; **12**:88, 92, 102; **14**:117; **SUP**:52, 58; **15**:12, 60, 70, 104, 121, 158, 176, 181, 185, 192–193; **16**:308; **24**:231; **25**:264, 265, 270, 280, 281, 283, 313; **28**:8, 288; *see also* Simultaneously developing flow
  boundary condition, **19**:256
  CHF in helium and, **17**:326
  defined, **15**:68, 148, 154
  effective, **12**:8
  effects of, **24**:291
  equilibrium, **4**:275f
  frozen, **2**:48; **4**:275f
  heat transfer coefficient effect, **21**:66–67
  heat transfer of tubes in crossflow and, **18**:107–108
  in impinging flow, **13**:7
  influence of, **31**:192, 208–213
    on heat transfer to tubes and tube banks, **8**:112
  influence on heat transfer from rotating surfaces, **5**:155
  limiting values of, **9**:334
  melting of anisotropic metallic PCM, **19**:63
  molecular, **21**:225
  multifluid bubbling pool, **21**:325–326
  nucleate boiling, **20**:4
    narrow space, **20**:49–50
  potential flow convection, **20**:353
  correlating equation development, **20**:382–383
  liquid metal convection, **20**:376, 378–379
  pure and saline water, **18**:345–348
  seawater
    comparison at 1 bar pressure for various temperatures and pressures, **18**:349
    temperature and pressure effects, **18**:347
  simultaneously developing flows
    Newtonian fluids, **19**:300, 301
    non-Newtonian fluids, **19**:301–306
  submerged jet, **26**:129–130
  submerged jet impingement systems, **20**:255–256
  turbulence, **21**:228
  turbulent, **12**:86; **25**:18, 26; **31**:186
    definition, **16**:349
    and roughness, **16**:310
    variation with boundary layer and flow, **16**:349–355
  use of, **15**:66–70
  water
    at various temperatures and pressures, **18**:349
    at various temperatures and salinities, **18**:358
  zero axial flow, **21**:164–165
Prandtl number dependence, **19**:52
Prandtl number method, **3**:275f
Preburnout region, narrow space nucleate boiling, **20**:43–45
Precipitation fouling, heat transfer enhancement, **30**:201–206, 208
Precipitator, electrostatic, **25**:234
Precooling cryogenic equipment, **9**:82
Predicted net flux, **31**:383–384
Preimpingement jet, **26**:136–137, 144–147, 165
Premagnetized bed, **25**:201
Premature truncation, **31**:356–357
Premixed fuel, combustion of injected, *see* Combustion, of injected premixed fuel
Premixing, steam explosions, **29**:132, 147–148
  experiments, **29**:152–158
  models, **29**:148–152
Prepressurization, **16**:72, 74, 75
Pressure, **25**:323, 339; **31**:147
  augmented nucleate boiling, **20**:68–70
  CHF in helium and, **17**:323–324, 330
  correction, **24**:200
  film-boiling heat transfer and, **11**:105, 110–117

Pressure (*continued*)
  heat-transfer coefficient and, **20**:77–78
  influence of on accommodation coefficient,
      see Accommodation coefficient
  melting, **24**:29–34
  modified, **25**:339
  narrow space nucleate boiling
      clearance and heat flux, **20**:47–79
      heat-transfer coefficient, **20**:46–47
  nucleate boiling formulas, **20**:15–18, 20–21
  pressure-based formulation, **24**:194
  pressure drop number, **24**:10
  pressure–velocity coupling, **24**:195
  static, **25**:323
  total, **25**:323
Pressure broadening, **8**:231
  in thermal radiation, **12**:136
Pressure change, **31**:118
Pressure coefficient, **8**:107
  tube in a bank, **18**:101
Pressure defect, **SUP**:42
Pressure discontinuity, **31**:122
Pressure distribution on tube surface, ideal fluid
      flow and, **18**:89–90
Pressure drag
  fluid flow in a single tube, **18**:96
  fluid flow past a tube in a bank, **18**:106
Pressure drop, **25**:151, 152, 160, 200, 205, 296,
      302–308
  across the bed, **25**:154, 167, 168, 169, 180,
      188, 193, 194, 200, 201, 210, 218, 223,
      238
  in annular-dispersed flows, **1**:368ff, 373f, 379
      in vertical conduits, **1**:376ff
  in condensation, **9**:258
  fluctuation, **25**:154, 219, 220
  liquid–solid fluidized beds, **23**:241
  loss, **25**:208
  packed beds, **23**:194
  particle shape effect, **23**:215
  in pipes, **5**:361
      of helium at high wall temperatures,
          5:363
  shearthinning fluids, **23**:194
  slug flow, **20**:92–95, 114
  Taylor bubbles, gas pocket propagation,
      20:102–103
  in two-phase flow, **5**:449
  viscoelastic effects, **23**:217
  wall effects, **23**:216

Pressure drop coefficients, in tube banks,
      8:154
Pressure effects
  and temperature effects
      enthalpy of water, **18**:343
      entropy change for seawater, **18**:345
      Prandtl number for seawater, **18**:347
      saline expansion coefficient of seawater,
          18:336
      Schmidt number of seawater, **18**:351
      specific heat of seawater, **18**:338
      thermal expansion coefficient of seawater,
          18:332
      transition boiling heat transfer, **18**:306
Pressure field, **9**:296
Pressure gradient flows, **8**:305
Pressure oscillations, effects of, **2**:82ff, 94
Pressure path length, **8**:235
Pressure pulsations, as heat transfer enhance-
      ment, **31**:181–182
Pressure term, **9**:279, 293
Pressurized discharge of cryogens, **5**:379
  comparison of theory and experiment, **5**:387
Pressurized systems, and system mechanical
      compliance, **31**:146–147
Pressurized water reactors (PWR), direct con-
      tainment heating, **29**:219
Primary normal stress difference, for Newtonian
      fluid, **15**:178
Principle of corresponding states, **7**:5
Principle of local thermal equilibrium, **31**:43
Printed wiring board (PWB)
  forced convection, **20**:221
  heat pipe cooling techniques, **20**:286–287
Prismatic bars, conduction in, **8**:49
Prisms, **25**:252
Probabilistic risk analysis (PRA), direct con-
      tainment heating, **29**:217
Probability-based modeling, **31**:390–394
Probability density distribution, **8**:296, 337, 338
  of temperature, **8**:325, 326, 329
  of velocity, **8**:325, 326, 339, 342
Probability density function, **5**:7
  slug velocity, **20**:113
Probability distribution, **5**:5
  comparison of, for temperature and velocity
      measurements, **8**:332, 333
Probability integral, **2**:286
Probe
  boundary layer, **2**:54f

heat transfer, **25**:164, 166, 191, 210, 238
Process
  molecular-sort adsorption, **25**:232
  particulate removed, **25**:232
Process control, alloy solidification, **28**:235, 237, 238, 308–326, 327–328
Profile estimates, **31**:138–139, 143
Prolate (ovary) spheroid, potential flow convection, **20**:372
Promoters, **9**:210, 231
Propagation, steam explosions, **29**:132, 134–135, 168, 185–187
  experiments, **29**:177–180
  hydrodynamic fragmentation, **29**:168–175
  modeling, **29**:180–182
  stratification, **29**:182–185
  thermal fragmentation, **29**:175–176
Propane, nucleate pool boiling curve, **16**:106
Propane-*n*-butane, nucleate pool boiling curves, **16**:106
Propane-water, A_0, **16**:119
n-Propanol-water, equilibrium contact angle on graphite, **16**:70
Properties
  temperature dependent effects of, **3**:104, 108
  thermophysical, **22**:9, 10
Property dependences, treatment of realistic, **31**:364–400
Property ratio method, in heat transfer analysis, **12**:91
Property variations, effects of, **2**:39ff, 96
Protoslugs, defined, **20**:85–86
PRTDAT subroutine, **27**:122, 123, 127, 162
Pseudo-binary system, **31**:8
Pseudo boiling, **7**:87
Pseudo-boiling phenomenon, **5**:355
Pseudodropwise condensation, **21**:130
Pseudo film boiling, **5**:372
Pseudoplastic fluids, **2**:359, 363, 372; **24**:107
  with yield stress, **24**:107
Pseudoplasticity index, **24**:108
  Nusselt number–Rayleigh number variation for, **15**:210
  parameter variations with, **15**:190
Pseudoplastic material, **28**:152
Pseudoplastics, **19**:249
Pseudopolimarized bed, **25**:154, 157
Pseudo random numbers, **5**:10
Published papers, format of, **SUP**:416–419
Pulsation characteristics, in channel cross sec-

tions, **31**:170–171
Pulsation velocity components, along channel axis, **31**:168
Pulse combustion, **30**:78
  heat transfer, **30**:83–85
  microscales, **30**:80–81
  skin friction and heat transfer, **30**:82–83
  unsteady turbulence, **30**:78–80
Pulsed-electric discharge, **31**:461
Pulsed laser melting, **28**:75–76, 135
  experimental verification of melting, **28**:80–81
    conductance, **28**:81–87
    pyrometry, **28**:87–96
  heat transport, thermal modeling, **28**:76–80
  topography formation, **28**:101–109
  ultrashallow junction formation, **28**:96–101
Pulsed laser sputtering
  metals, **28**:103, 106, 109–112
    thermal and electronic effects, **28**:116–123
    time-of-flight measurements, **28**:112–116, 136
Pulsed laser vaporization, computational modeling, **28**:123–135, 136
Pulse heating technique, vaporization temperatures in, **10**:108
Pulselike inputs, *see* Boundary conditions, in heat conduction problems
Pumping of cryogenics, **5**:378
  influence on vehicle thrust, **5**:380
  sloshing and splashing problem, **5**:379
Purely viscous fluids, **24**:106
Purely viscous material, **28**:152
PVI, *see* Photoluminescent volumetric imaging
PWR, *see* Pressurized water reactors
Pyrolysis, **10**:226–235
  experimental studies of, **10**:227–229
  fire protection and, **10**:233–235
  interpretation of results in, **10**:229–231
  mathematical models of, **10**:231–233
  products of, **10**:280
  of wood, **17**:184–188
Pyrometers, in radiative heat transfer measurement, **11**:424–425
Pyrometry, semiconductor melting, **28**:87–96

Q

QG variable, **27**:161, 207
QND variable, **27**:27, 207

Quadrature, **23**:139
Quadrilateral ducts, **SUP**:260–262
Qualitative map, **25**:173
Quality in annular-dispersed flows, **1**:408f
quality, woodburning cookstove efficiency and, **17**:266–267
  used in domestic sector, **17**:161
  weight-loss experiments, in open fires, **17**:228–236
  wood
    charcoal, **17**:180–181
    chemical nature of, **17**:175–176
    physical properties and, **17**:176–180
    populations involved in scarcity or deficit situations and, **17**:162
Quanta, definition, **27**:3, 4
Quantum theory, **27**:4–5
Quantum turbulence, in helium II, properties of, **17**:97–101
Quasi-equilibrium severe slugging, **20**:124–128
Quasi-neutral, **1**:271
Quasi-stationary approximation, **19**:29
Quasi-stationary assumption, **19**:28
Quasi-stationary flows, **6**:388
Quasi-steady assumption, **19**:28
Quasi-steady closure problem, **31**:69
Quenching, **5**:90
Quenching regime, **29**:87
QUEOS experiments, **29**:155
QUEST study, **29**:324
QUICK scheme, **24**:239
Quiescent beds, **19**:127
QW values, **27**:163, 201

R

Rabinowitch–Mooney equation, **15**:71
Rabinowitch–Mooney factor, **25**:299
Radial flow region
  free-surface jet, **26**:141–144, 156–160
  submerged jet, **26**:129–134
Radial momentum equation, liquid metal droplet, **28**:3
RADIAN program, **27**:28–29, 46, 203
  absorbing–emitting gas, **27**:107–109, 122–130
  boiler furnaces, analysis, **27**:158–167
  output, **27**:130–133
  program listing, **27**:110–122

RADIAN1 program, **27**:20
Radiant conductivity, **23**:166–168
Radiant energy density, **3**:179f
Radiant energy flux, heat flux and, **10**:47
Radiant energy flux vector, **3**:189
Radiant flux boundary condition, **SUP**:24–25
  for circular duct, **SUP**:81–82, 123–124
Radiant heat, gain by pans in presence of walls, **17**:241–244
Radiant heat flux, **1**:4
Radiant heating, ablation of semitransparent solids with, **11**:388–390
Radiant heat load, **4**:83
Radiant heat transfer rate, **12**:123
RADIANW program, **27**:203
  output, **27**:143–144
  program listing, **27**:134–142
  surfaces separated by nonparticipating gas, **27**:130, 132, 145–146
Radiating-conducting fins, *see* Fins
Radiation, **8**:11, 12, 34, 66; **22**:281; *see also* gas–particle mixture radiation; Gas radiation; Molecular gas band radiation; Radiative–convective system; Radiative heat transfer; Thermal radiation
  application of Duhamel's principle to, **8**:78
  blackbody, **27**:7–9
  at cryogenic temperatures, **5**:473
  definition, **27**:3, 7
  effects on phase change, **19**:37–38
  emissive power, **27**:10
  energy transfer through, **11**:322–324
  and free convection, experimental results, **1**:45ff
  from gases, **12**:116
  in gas fluidized bed and immersed tube transfer, **14**:155–158
  in heat rejection, **12**:12
  heat transfer in packet, high-temperature heat transfer model, **19**:173
  heat transfer to pan and, **17**:216–220
  ideal surface in, **12**:11
  ignition and, **10**:243
  Kirchhoffs law, **27**:11–12
  Lambert's cosine law, **27**:9
  natural convection cooling, **20**:201–202
  Planck's law, **27**:7–8
  and rarefaction interaction, **7**:177
  scattering, **27**:17
    anisotropic, **27**:146–154

isotropic, **27**:157
radiative–convective system, **27**:92–99
solid surfaces, **27**:10–12
spectral intensity and, **11**:324
spectroradiometric curves, **27**:7–9
Stefan–Boltzmann law, **27**:8–9
thermodynamic equilibrium, **12**:120–122
Wein's displacement law, **27**:8
Radiation absorption, *see* Absorption
Radiation-affected turbulence, entropy production, **21**:268–270
Radiation characteristics, of semitransparent solids, **11**:365–375
Radiation–conduction interaction, **8**:264
  in carbon dioxide, **8**:266, 267
  in carbon monoxide, **8**:266, 267
  comparison between optically thin limit and large path length limit, **8**:267
  in methane, **8**:266, 267
  in water vapor, **8**:266, 267
Radiation–convection heat transfer, **3**:176ff
  similarity parameters, **3**:213ff
    extrinsic, **3**:217ff
    intrinsic, **3**:213ff
  theoretical considerations, **3**:177ff
    conservation of
      energy, **3**:178
      mass, momentum, and energy for a radiating fluid, **3**:190ff
        continuity equation, **3**:191
        energy equation, **3**:192
        momentum equation, **3**:191f
      momentum, **3**:178
      momentum for radiative transport, **3**:189ff
      radiant energy, **3**:187f
        monochromatic, **3**:188
    equation of transport, **3**:183ff, 199ff, 203ff
      approximation
        discrete coordinate method, **3**:206f
        Milne–Eddington approximation, **3**:104f
        Schuster–Schwarzschild approximation, **3**:203f
        spherical harmonics method, **3**:206
      formal solutions, **3**:199ff
    integral equation for radiation transfer in an enclosure, **3**:193ff
    Milne integral equation, **3**:203

types of flows
  boundary layer, **3**:226ff
    without dissipation, **3**:231ff
    with viscous dissipation, **3**:234ff
  channel, **3**:236ff
    one-dimensional, **3**:237ff
    two-dimensional, **3**:242ff
      cylindrical channel, **3**:242ff
      parallel plate channel, **3**:242ff
  couette, **3**:219ff
    gray approximation, **3**:221ff
      constant properties, **3**:221ff
      variable properties, **3**:225f
Radiation effects, **4**:67ff
  in nucleate boiling, **1**:214f
Radiation emission, *see* Emission
Radiation energy absorption distribution method, *see* READ method
Radiation flux, **3**:179, 207ff
Radiation flux vector, **3**:179
Radiation heat transfer, *see also* Radiative heat transfer
  with an absorbing medium, **1**:2ff
    by combined conduction and radiation, **1**:17ff
    by combined convection and radiation, **1**:23ff
    by pure radiation, **1**:12ff
  in semitransparent solids, **11**:390–411
  between surfaces, **2**:400ff
Radiation heat transfer cavities, *see* Cavities
Radiation heat transfer passages, *see* Passages
Radiation intensity
  attenuation, **27**:148–149
  source function, **27**:21–23
  thermal radiation, **27**:9–10, 13–14
Radiation parameter, **SUP**:24
Radiation pressure tensor, **3**:189f
Radiation shields, multiple optimally cooled, **15**:47–48
Radiation technique for measuring density in annular-dispersed flows, **1**:394ff
Radiation transfer equations, **11**:325–327
Radiative conductivity, **1**:26
  for semitransparent solids, **11**:362–365
Radiative–convective system, **27**:45–46
  analysis
    energy method, **27**:86, 87
    READ method, **27**:46, 86–90, 107–146
    heat balance equations, **27**:46–49

Radiative–convective system (*continued*)
  heat transfer in furnaces, **27**:128, 129, 144,
    164–166
  Monte Carlo simulation, **27**:49–85
  scattering by particles, **27**:93–99
Radiative cooling, modeling, **28**:13–14
Radiative decay
  realistic ways of calculating, **14**:254–258
  of temperature perturbations for local ther-
    modynamic equilibrium, **14**:251–255
  of turbulent temperature fluctuations for,
    14:259–268
Radiative effects
  in melting and solidification of semitranspar-
    ent crystals, **11**:413–419
  at stagnation point, **1**:345f
Radiative energy
  absorption-reflection characteristics, **27**:61–85
  computation, **27**:86, 87
  gas layer, **27**:50–53
  gas–wall system, **27**:47–48, 60
  Monte Carlo simulations, **27**:50
  READ method, **27**:46, 86–90
    absorbing–emitting gas, **27**:146–157
    nonparticipatimg gas, **27**:130–146
    RADIAN, **27**:107–133
Radiative energy transfer, **11**:320
Radiative equilibrium, **8**:250, 277
Radiative exchange factors, **5**:20
Radiative flux
  in energy transfer, **11**:329–352
  equation for, **11**:336
  general analysis of plane parallel layer in,
    11:333–338
  for infinite cylindrical medium, **11**:344–350
  optically thin approximation of, **11**:331–333
  for plane layer with optically smooth inter-
    faces, **11**:338–343
  for semiinfinite medium, **11**:343–344
  in spherical shell, **11**:350–352
Radiative flux equations, **8**:250, 252
Radiative flux vector, **8**:246
Radiative heat flux, **8**:279; **21**:246
  comparison of, for several gases, **8**:280
  versus absorption and scattering, **21**:256–257
  versus emission and absorption, **21**:255–256
Radiative heat flux equation, divergence, **27**:20
Radiative heat flux vector, **27**:20, 25, 98
Radiative heat transfer
  absorbing–emitting gas

    RADIAN, **27**:107–133
    READ, **27**:146–157
  cylindrical coordinate system, **27**:99–102
  directional and boundary effects in, **11**:400–
    403
  gas volume and solid walls, **27**:46–49
  in gray materials, **11**:395–400
  heat balance equations, **27**:46–49
  industrial applications
    boiler furnaces, **27**:158–167
    circulating fluidized bed boiler, **27**:187–193
    gas reformer furnaces, **27**:167–173
    jet engine combustion chambers, **27**:173–
      181
    nongray gas layer, **27**:181–187
    three-dimensional systems, **27**:193–199
  Monte Carlo method
    emission, **27**:56–61
    gas absorption, **27**:50–56
    READ method, **27**:46, 84–91, 107–157
    reflection and absorption by solid walls,
      27:61–85
  nonorthogonal boundary, **27**:99–104
  nonparticipatimg gas, **27**:133–146
  optical methods in, **11**:422–424
  in plane parallel layer, **11**:333–338
  pyrometers in, **11**:424–425
  scattering, **27**:92–99
  spectral effects on, **11**:403–407
  spectral thermal remote sensing in, **11**:425–
    431
  surfaces separated by nonparticipating gas,
    27:130–146
  temperature measurement in, **11**:420–431
  temperature sensors in, **11**:421–422
Radiative heat transfer equations, **27**:19–23
Radiative heat transfer, porous media, **23**:133–
  183
  continuum treatment, **23**:134–136
  discrete-ordinates approximation, **23**:139–141
  modeling dependent scattering, **23**:168–179
    bed properties, **23**:176–177
    dependence-included discrete-ordinates
      method, **23**:173–179
    opaque spheres, **23**:171–172
    Percus–Yevick model, **23**:169
    scaling, **23**:171–173
    single particle properties, **23**:174–175
  noncontinuum treatment, **23**:161–166
    emitting particles, **23**:165–166

opaque particles, **23**:162–164
semitransparent particles, **23**:164–165
phase function, **23**:135
radiant conductivity, **23**:166–168
radiative properties, **23**:152–161
  independent scattering limits, **23**:154–156
  opaque spheres, **23**:160–161
  particle interactions, **23**:153–156
  spectral scattering coefficient, **23**:152–153
  transmittance, **23**:156–159
  volumetric size distribution function, **23**:153
single particle properties, **23**:141–151
  geometric- or ray-optics scattering, **23**:147–151
  Hagen–Rubens law, **23**:143
  Mie theory, **23**:141–142, 145
  Penndorf extension, **23**:146–147
  prediction comparisons, **23**:141–147
  Raleigh theory, **23**:142–143
  Snell law, **23**:147, 149
solid conductivity effect, **23**:179–181
two-flux approximations, **23**:136–139
Radiative internal energy, **21**:245
Radiative lifetime, **8**:249
Radiatively participating gas, energy equation for, **8**:246
Radiative Péclet number, **1**:26
Radiative stress, **21**:245–247
Radiative theory, for nonlocal thermodynamic equilibrium, **14**:255–259
Radiative transfer, **8**:255
  in carbon dioxide, **8**:256, 259, 260
  in methane, **8**:256, 259, 261
  Monte Carlo simulation of, **31**:334–336
  on motions in planetary atmosphere, **14**:249–279
  in water vapor, **8**:256, 259, 261
Radiative transfer equations, **8**:70
  one-dimensional, **23**:177
Radical flow region, laminar axisymmetric jet, **26**:114–119
Radicals, **3**:269ff
Radioactive waste repositories, convection in unsaturated media, **30**:167–168
Radio frequency energy injection, **31**:459
Radiometer, **23**:306–307
Radiometric apparatus, **2**:317f
Radiosity, **1**:4, 6; **2**:405; **27**:34–35
Radio waves, **27**:3

Radius cylinder, **25**:313
  hydraulic, **25**:299, 301, 314
  sphere, **25**:313
Radius of cavities in nucleate boiling, **1**:209ff
Radon transform, **30**:295
Random numbers, **5**:7, 9
  generation of, **5**:10
RANDOM subroutine, **27**:53, 109–110
Rankine cycle, **12**:4
  vs. Brayton cycle, **12**:7
Rankine–Hugoniot equation, **19**:25
Rantz–Marshall correlation, **26**:53, 70; **28**:8, 11
RAN variable, **27**:53, 207
Rapid phase transitions, **29**:129
Rarefaction and radiation interaction, **7**:177
Rarefied gas dynamics, **5**:38
  analogy to radiation heat transfer, **5**:38
Rarefied gases
  external flows, **7**:187
  gas at rest, **7**:169
  heat transfer, **7**:163
    by heat conduction, **2**:291ff
    between coaxial cylinders, **2**:282ff, 292f
    from a flat plate, **2**:284
    between infinite parallel plates, **5**:39
    from a moving plate, **2**:279
    between Parallel plates, **2**:280ff, 292
    between unspecified surfaces, **2**:284
  internal flows, **7**:183
  stagnation region, **7**:196
  transition regime, **7**:170, 193, 194
Raschig rings, Aroclor flow rates and, **15**:269
RAT1 program, **27**:64–75, 80, 83, 88, 203
  flowchart, **27**:72
  output, **27**:73–75
  program listing, **27**:67–71
RAT2 program, **27**:81–84, 88, 109, 203
  flowchart, **27**:82
  output, **27**:83–84
  program listing, **27**:76–80
Rate of production of species, **2**:120, 135f
Rate point transfer, **25**:214, 231, 238
  solids discharge, **25**:172, 173
Rate processes, **22**:290, 314
Ratio, kinetic to magnetic energy, **25**:196
  magnetic energy density to particle energy density, **25**:224
  magnetic to gravitational energy, **25**:172
  magnetic to kinetic energy, **25**:172
  particle to discharge tube diameter, **25**:172

Rational activity coefficient, **12**:262
Rayleigh–Bénard problem, buoyancy-driven
   flow, **30**:25–33
Rayleigh equation of number, **11**:14
Rayleigh function, dissipative, **25**:6
Rayleigh inviscid criterion for rotational instability, **21**:157
Rayleigh–Jeans formula, **5**:261
Rayleigh number, **4**:3ff, 151f; **5**:375; **8**:161, 166;
   9:281; **11**:201, 209–211, 245–246, 250,
   279, 282, 291, 301–303, 305, 309–310;
   14:18–20, 71, 138; **SUP**:410, **15**:206, 210;
   19:52, 55, 256, 304; **24**:5, 120, 215; **25**:133,
   279
   cell aspect ratio and, **14**:47
   critical, **14**:20–24; **24**:176–178
      heat transfer in helium I and, **17**:129–130
   critical value of, **8**:165
   in deformable geothermal reservoir, **14**:51
   heat- and mass-transfer
      finite-amplitude convection, **20**:327
      high convection model, **20**:329–330
      horizontal direction, **20**:332
      point source, **20**:343–346
      time-dependent concentration,
         20:342–343
      vertical direction, **20**:320–321
   isotherms and, **14**:38–40
   modified, **4**:151ff
   molten pool heat transfer
      circular cavities, **29**:17–18, 20
      rectangular cavities, **29**:4–9, 12–17
      spherical cavities, **29**:21, 24–30
   natural convection
      enclosure sources, **20**:219–220
      two-dimensional protrusion, **20**:211–212
   Nusselt number and, **14**:42
   power series and, **14**:34
   salinity, **14**:27
   secondary flow fluctuations and, **14**:75
   steady-state streamlines and isotherms in
      relation to, **14**:48
   turbulent flow, **21**:39
   in two-layer system, **14**:36
   two-phase, **30**:129, 144
   unicellular motion in flow at, **14**:44
Rayleigh problem
   compressible fluid, **24**:314–316
   incompressible fluid, **24**:315
Rayleigh–Ritz method, **2**:425
Rayleigh scattering, **5**:37

Rayleigh theory, **23**:142–143
Ray number, effect of, **31**:354–355
Ray-optics scattering, **23**:147–151
Rays
   parallelize by, **31**:412–413
   redirection of, **31**:366–367
   selecting launch locations of, **31**:340
   starting direction of, **31**:341–342
Ray splitting, **31**:367–369
Ray tracing, **31**:336–364
   forward approaches for, **31**:344–346
Reactants conservation equation, *see* Conservation of reactants equation
Reaction coordinate, **2**:127f
Reaction rate, **2**:126f, 134
Reactions, *see also* Boundary layer; Catalytic
   surface reactions
   of injected combustible gases, **2**:207ff
   at surface, *see* Surface reaction
Reactor coolant system (RCS)
   blowdown, **29**:235, 237, 238, 248, 256, 257,
      310
   direct containment heating, **29**:215, 235
Reactor pressure vessel (RPV)
   blowdown, **29**:235, 237, 238, 248, 256, 257,
      310
   coejected RPV water, **29**:322–323
   direct containment heating, **29**:215, 216
   discharge phenomena, **29**:248–256
   gas blowthrough, **29**:248–249, 253–256
Reactors, *see* Nuclear reactors; Nuclear reactor
   safety
READC, **27**:109
READ method, **27**:46, 86–91
   absorbing–emitting gas, **27**:146–157
   nonparticipating gas, **27**:130–148
   RADIAN, **27**:107–133
READ values, **27**:72–73, 84–85, 88, 89, 109
   cylindrical coordinate system, **27**:100
   scattering phenomena, **27**:96–97
   stopping printout, **27**:129–130
Recalescence, **28**:13, 15–17
Recalescence temperature, **28**:17
Reciprocating engines, heat flux measurements,
   23:343–344
Reciprocity
   relation, **2**:408
   theorem for cavities, **2**:426f
Recombination, *see also* Dissociation-recombination
   arbitrary, **2**:46

Recombination radiation in a plasma, **4**:235ff
Recombination rate coefficient, **2**:197
Recovery factor, **7**:189
  adiabatic, **2**:15
Recovery temperature, **7**:189, 326, 327
Recrystallization, **22**:196
Rectangular cavities, **8**:166, 174
  boundary layer regime in, **8**:177
  centerline temperatures for free convection in, **8**:178
  conduction regime in, **8**:177
  convection, **29**:4–17
  dimensionless parameters for free convection in, **8**:176
  flow oscillations for natural convection in, **8**:189
  influence of horizontal partitions on free convection in, **8**:182
  Nusselt numbers for natural convection in, **8**:195, 196
  secondary flows for natural convection in, **8**:189
  streamline patterns for natural convection in, **8**:189
  temperature profile for natural convection in, **8**:176
  tertiary flows for natural convection in, **8**:190, 191
  transition regime in, **8**:177
  velocity profiles for natural convection in, **8**:176
Rectangular ducts, **8**:70; **SUP**:196–222
  finite difference solutions for, **SUP**:211–213
  finite element solution for, **SUP**:213
  friction factors for, **SUP**:199–200, 212
  fully developed flow in, **SUP**:196–209
    fluid flow, **SUP**:196–202
    heat transfer, **SUP**:202–209
  hydrodynamically developing flow for, **SUP**:209–213
  linearized solutions for, **SUP**:210
  linearly varying wall temperature for, **SUP**:219
  longitudinal thin fins within, **SUP**:372–373
  Nusselt numbers for, **SUP**:200, 202–206, 209, 214–216, 218, 220–221
  with semicircular short sides, **SUP**:273–275
  shape factor for, **SUP**:201
  simultaneously developing flow in, **SUP**:219–222
  specified axial wall heat flux for, **SUP**:203–

  209, 217–219
  specified wall temperature for, **SUP**:202–203, 213–217
  thermal entrance lengths for, **SUP**:217
  thermally developing and hydrodynamically developed flow for, **SUP**:213–219
Rectangular geometries
  freezing in, **19**:67–68
  future research, **19**:343–345
  heat transfer studies, laminar flow, **19**:304–305
    Newtonian fluids, **19**:306–316
    non-Newtonian fluids, **19**:316–322
  heat transfer to Newtonian and non-Newtonian fluids, **19**:247–249
  governing equations, **19**:250–257
  hydrodynamically developed laminar flow
    friction factor
      Newtonian fluids, **19**:261–262
      non-Newtonian fluids, **19**:262–268
    velocity profile
      Newtonian fluids, **19**:257–259
      non-Newtonian fluids, **19**:259–261
  hydrodynamically developing laminar flow, **19**:268, 270–271
    hydrodynamic entrance length, **19**:274–277
    incremental pressure drop
      Newtonian fluids, **19**:272
      non-Newtonian fluids, **19**:272–274
    velocity profiles, **19**:269
  melting in, **19**:62–67
  simultaneously developing laminar flow
    Newtonian fluids, **19**:300
    non-Newtonian fluids, **19**:301–306
  thermally developed laminar flow
    Newtonian fluids, **19**:277–280
    non-Newtonian fluids, **19**:280–289
  thermally developing laminar flow
    Newtonian fluids, **19**:289–291
    non-Newtonian fluids, **19**:291–297
    thermal entrance length, **19**:297–300
  turbulent flow, **19**:322–323
    fluid mechanics
      Newtonian fluids, **19**:323–325
      non-Newtonian fluids, **19**:325–334
    heat transfer
      Newtonian fluids, **19**:334–339
      non-Newtonian fluids, **19**:339–343
Rectangular-groove cavities, *see* Cavities
Rectangular slot, natural convection in, **8**:168, 169

Recuperative operation, unidirectional regenerators, **20**:154–158
Reduced enthalpy of evaporation, **7**:6
Reduced gravity, **4**:144ff
  combustion, **4**:215ff, 221
    candle flame, **4**:215
    fuel droplets, **4**:215ff
    solid fuels, **4**:217f
  condensation, forced flow, **4**:209ff, 220f
    flow behavior, **4**:210f
    noncondensable gas, **4**:214
    pressure drop, **4**:211ff
    vapor–liquid interface, **4**:213f
  condensation without forced flow, **4**:206ff, 220
    laminar film condensation on a vertical surface, **4**:207f
    laminar-to-turbulent transition, **4**:208f
    transient time to establish laminar condensate film, **4**:209
  experimental production of, **4**:146ff
    airplane trajectory, **4**:150
    drop tower, **4**:146ff
    magnetic forces, **4**:150f
    rockets, **4**:150
    satellites, **4**:150
  forced convection boiling, **4**:200f
    designs involving substitute body forced, **4**:205f
    two-phase heat transfer, **4**:201ff
  free convection, elimination, **4**:145
  free convection in, **4**:151ff, 218f
    fluid flow, **4**:152ff
      boundary layer theory, **4**:152f
      boundary layer transition, **4**:153
      Rayleigh numbers encountered in low-gravity, **4**:153f
      threshold of convective motion, **4**:152
    heat transfer, **4**:154ff
    transient development times of boundary layers, **4**:156ff
  gravity as an independent parameter, **4**:144f
  pool boiling, **4**:158ff, 219f
    bubble dynamics in saturated nucleate boiling, **4**:174ff
      diameter at departure, **4**:181f
        experimental results, **4**:181f
        theoretical relations, **4**:181
    bubble dynamics in subcooled boiling, **4**:194ff

    bubble growth, **4**:195f
    forces acting on bubbles, **4**:196ff
    critical heat flux, **4**:166ff
      experimental behavior, **4**:167ff
      theory, **4**:167
    film, boiling, **4**:173f
      experimental results, **4**:173f
      theoretical relations, **4**:173
    forces acting during growth, **4**:182ff
    growth rates, **4**:177ff
      experimental results, **4**:178ff
      higher heat flux effects, **4**:191ff
      nucleation cycle and coalescence, **4**:175ff
      rise of detached bubbles, **4**:190f
      theoretical relations, **4**:178
    minimum heat flux between transition boiling and film boiling, **4**:171f
    nucleate pool boiling, **4**:158ff
      experimental results, **4**:159ff
      theory, **4**:158f
      transition region for pool boiling, **4**:171
  space applications, **4**:144
Reduced isotherms, **7**:6
Reentrant cavity, augmented nucleate boiling, **20**:67–70
Reentry studies, **4**:278ff
Reference electrode, in electrostatic potential measurement, **12**:225
Reference temperature, **8**:95, 113
  effect of fluid physical properties on heat transfer and, **18**:108–110
  use of, for heat transfer to single tubes and tube banks, **8**:113
Reference velocity, fluid flow in a restricted channel and, **18**:99
Refined vacuum evaporation, **28**:340
Reflectance, **2**:405
  defined, **10**:13
  directional, **10**:14–15
Reflection, by solid walls, **27**:61–85
Reflective power, **27**:11
Reflective shields, **9**:382, 385
Reflectivity, definition, **27**:11
Refractive index, **9**:363
  photon travel and, **11**:320
Refractory surfaces, **5**:21
Refrigerants
  nucleate pool boiling, data available, **16**:169
  nucleate pool boiling heat flow rates

effect of pressure, **16**:187, 212
  under similar conditions, **16**:193–194, 220–222
Refrigeration plant, Carnot efficiency of, **15**:45
Refrigerator power requirement, cold space environment and, **15**:46–47
Regenerator theory
  "boxed" process, **20**:139–149
    cyclic steady state (equilibrium), **20**:147–148
    governing equations–Nusselt's assumptions, **20**:141–144
    overall balance equation, **20**:144–146
    overall heat-transfer coefficient, **20**:148–149
    weak and strong periods, **20**:140–141
  classical model assumptions, **20**:138–139
  cyclic equilibrium and thermal effectiveness, **20**:135–136
  operation classification, **20**:137–138
  regenerator parameters, **20**:133–135
  rotary system–fixed bed equivalents, **20**:136–137
  unidirectional operation, **20**:149–176
    aperiodic or recuperative operation, **20**:154–158
    balanced-symmetric regenerator, **20**:173–175
    completely unbalanced regenerators, **20**:170–173
    general case solution, **20**:151–154
    long regenerator, **20**:158–170
    unbalanced-symmetric regenerator, **20**:174, 176–177
Regime, **24**:120
  bubbling bed, **25**:192
  fixed bed, **25**:218, 220, 223, 229
  fixed-bed high field intensity, **25**:205
  frozen-bed, **25**:231, 237
  magnetically stabilized bed, **25**:194, 220, 230
  partially stabilized bed, **25**:220, 230, 237
  solidified bed, **25**:234
  stabilized bed, **25**:205, 219, 229, 237
Region, magnetic field
  high, **25**:204
  low, **25**:204
  medium, **25**:204
  moderate, **25**:188, 205
  strong, **25**:189, 205, 214

weak, **25**:188, 205
Regression analysis, low liquid level nucleate boiling, **20**:56–58
Regular perturbation, **8**:3
Regular polygonal ducts, **SUP**:262–263
  with central circular cores, **SUP**:346–349
  within circular ducts, **SUP**:350–352
Regular regime, **8**:67
Reichardt formula, **25**:19, 27, 75
  from turbulent flow momentum transfer, **15**:105
"Reiner-Rivlin" fluid, **2**:363
Reistad effectiveness, defined, **15**:29
Relaminarization, **9**:321, 334
Relative laws of friction and heat transfer, **3**:45ff, 48ff
Relative permeability
  multiphase flow, **30**:98
  in porous media, **29**:42–47
Relative transference number, **12**:246
Relaxation, *see* Dissociative relaxation; Initial relaxation
Relaxation times, **2**:298; **9**:126; **24**:41, 109
  in thermodynamic equilibrium, **12**:127
Remanence, **25**:179
Remelting, alloy solidification, **28**:262, 263
Remote sensing, lunar data from, **10**:7–11
Research, basic, heat flux measurements, **23**:346–349
Reservoir cavities, liquid–vapor interface in, **10**:124–125
Reservoir management, multiphase flow, numerical simulations, **30**:140–141
Residence-time distribution, screw extrusion, **28**:176–180
Residual, *see* Subgrid
Residual magnetism, **25**:160
Resistance law, **31**:257
  similarity region of, **31**:204–205
Resistance temperature coefficients, **23**:289
Resistance temperature devices, **22**:363; **23**:288, 319
Resistance thermometers, **8**:299
Resonance, **24**:307
Resonating pulse reactor, **30**:243
Response, rheological, **25**:204
Retardation time, **24**:109
Retina, **22**:251, 255, 265, 270
Reverse emission path method, **31**:347–348, 373, 422

Reverse emission path method (*continued*)
  parallelization of, **31**:413–421
  performance of, **31**:348–364
Reverse ray tracing method, **31**:379, 398
Reverse simulations, **31**:346–348
Reversing-flow shuttle brushes, **31**:452
Review articles, **22**:157, 289
Reynolds analogy, **2**:23f, 47, 96, 98f; **4**:346ff;
  31:196
  Colburn form, **2**:23
  in MHD couette flow, **1**:322
  von Kármán form, **2**:16, 23, 98f
Reynolds number, x, **16**:244
Reynolds number, **1**:280; **2**:4, 381f; **3**:2ff, 5,
  37ff, 57f, 71f; **4**:54ff, 208, 345; **9**:340;
  11:99, 127, 131, 154, 156, 172, 200, 212,
  216–217, 244–247, 251; **12**:102, 105, 109,
  186; **14**:5, 198; **SUP**:52; **15**:16, 22, 32, 60,
  70, 88–90, 97, 109–110, 121, 124, 195,
  205, 239; **24**:120, 215, 219, 231, 239;
  25:151, 157, 158, 176, 177, 237, 260, 268,
  279, 281, 283, 284, 286, 292, 294–296,
  298, 299, 307, 308, 312–314, 323–330;
  28:8; **31**:295, 307; *see also* Dimensionless
  axial distance; Navier–Stokes equations
  acoustic, **24**:312–316
  Archimedes number and, **19**:102, 107, 110
  for array of round nozzles, **13**:20
  for array of slot nozzles, **13**:25
  and artificial turbulators, **31**:189–191
  augmented nucleate boiling, surface config-
    uration, **20**:74
  axial, **21**:169
  based on apparent viscosity, **19**:256
  based on superficial gas velocity, **21**:280
  based on the Kolmogorov scale, **30**:8
  for beds of large particles, **19**:143
  bubble, **21**:309
  bubble population density, **20**:7–8
  critical, **3**:6; **21**:337, 339; **31**:188–189
    versus critical Weber number, **21**:340–
      341
    versus drag coefficient, **21**:338–339
  defined, **15**:68, 148
  development of turbulent friction factor for
    Newtonian fluids flowing in rectangular
    ducts, **19**:324–325
  dimensionless groups, **19**:99
  droplet drag as function of, **21**:336
  enthalpy-thickness, **16**:247

  and effects of convex curvature, **16**:321–
    324
  and Stanton number, *see* Stanton number
  experimental vs. predictions at different Ar-
    chimedes number, **19**:110, 111
  Fanning friction factor and, **15**:80
  forced convection
    small rib spacing, **20**:228–229
    surface fences (turbulators), **20**:239–240
    three-dimensional arrays, **20**:23
  gas–solid system, **19**:106
  generalized, **19**:256
  heat transfer *j* factor and, **15**:126
  high performance arrays of nozzles and, **13**:57
  hydraulic resistance coefficient
    as function of, **31**:195
    plotted versus, **31**:205
    for water as function of, **31**:216–217
  on hydraulic resistance coefficient in tubes,
    31:203–204
  ice band structure and, **19**:76
  in impinging flow, **13**:3, 7
  induced rotational flow, **21**:153–154
  influence of, **31**:197–208
    on flow around cylinders, **8**:97
  Kozicki generalized, **19**:256
  at large temperature differential, **19**:304
  liquid droplets, **21**:335
  magnetic, **1**:282
  maximum nucleate boiling heat flux in helium
    and, **17**:321–322, 324, 325
  and mixed convection, **16**:336–340
  momentum-thickness, effects of acceleration,
    16:289–291
  moving-liquid droplet
    condensation processes, **26**:32
    direct-contact transfer fluid dynamic,
      studies, **26**:5–11, 16–18
    evaporation, **26**:59–77
  Newtonian, **19**:256
  rotational, **21**:151
  roughness, **16**:302–305, 344, 346
    variation of *B* with, **16**:309–310
  in shear flow, **13**:230
  for single nozzles, **13**:15–16
  for single slot nozzles, **13**:18
  slug flow calculations, **20**:98
  small, **3**:104, 106f
  solid spheres, **21**:334
  in stagnation flow, **13**:5

submerged jet impingement systems, **20**:255–256

thermal boundary layer and, **12**:97

turbulent, **21**:209

use of, **15**:66–70

Weissenberg number and, **19**:331–332

Weissenberg numbers for, **19**:333

Reynolds number effects, **8**:121

Reynolds number/Grashof number ratio, **15**:198

Reynolds stress, **8**:288; **9**:152

Rheocasting, electromagnetic, **28**:318

Rheological constitutive equation, **13**:211–212

Rheological fluid flow, **25**:73

Rheological properties, **23**:189; **25**:312

bulk solutions, **23**:226

of fluids, **2**:358ff

in situ packed beds, **23**:226

Rheology

fluid mechanics and mass transfer related to, **15**:60–61

for heat transfer analysis, **15**:70–86

magnetically stabilized bed, **25**:174

Rheopectic fluids, **2**:359f; **24**:107

flow curves for, **15**:63

Rhombic ducts, **SUP**:259–261

Ribbed-duct geometry, forced convection, **20**:229–231

Richards equation, **30**:103, 109–110

Riemann zeta function, **3**:134

Rig A/B program, **29**:147

Right triangular ducts

fluid flow in, **SUP**:233–236

heat transfer in, **SUP**:236–237

hydrodynamically developing flow in, **SUP**:241–242

thermally developing flow in, **SUP**:244

Rigid-rotor, harmonic-oscillator model, **8**:240

Rigid-rotor approximations, **8**:233

Rising bubble convection, **20**:379–382

bubble mobility, **20**:379–380

bubble shape, **20**:379–381

Rising bubble technique, in nucleation studies, **10**:103

Risk-orientated accident analysis methodology (ROAAM), **29**:193–194

Ritz method, **8**:28; **19**:196

Rivers and bays, heat and mass transfer in, **13**:61–114; *see also* THIRBLE

ROAAM methodology, **29**:193–194

Roast's law, **16**:63

Roberts' experiments, **2**:312f, 325f, 328, 343, 348f

Rocket nozzles, **5**:327

Rockets, their use in producing reduced gravity, **4**:150

Rocket thrust-chamber measurements, **2**:74ff

Rocks, lunar, *see* Lunar rocks; Terrestrial rocks

Rod bundle, *see* Circular cylinders

Rod-climbing effect, *see* Weissenberg effect

Rod flow–film boiling

in horizontal tube, **11**:159–164

hydraulic resistance for, **11**:182

in vertical tube, **11**:151–159

Rods, heat conduction in, **1**:66ff

Rogowski electrodes, **14**:118

Rolled tubes, **31**:276–277

Rolling contact melting, **24**:21–26

dimensionless group, **24**:23

Rolling depth, **31**:238–240, 243, 311

Rolling pitch, **31**:235–236

Rosenzweig's matching technique, **8**:70

Rossby number, **5**:209

in plume studies, **10**:248

Rosseland absorption coefficient, **5**:316

for carbon dioxide, **5**:317

for carbon monoxide, **5**:318

for water vapor, **5**:317

Rosseland approximation, **1**:7f; **3**:209f, 216; **8**:264

Rosseland limit, **8**:252, 253

Rosseland mean absorption coefficient, **10**:47

Rosseland mean extinction coefficient, **3**:211

Rosseland radiative conductivity, **11**:362–364

Rotary-matrix exchanger, mathematical representation, **20**:133

Rotary-matrix regenerators, "boxed" process, **20**:139–149

Rotary regenerators, **5**:244

dimensionless parameters, **20**:134–135

equivalence with fixed-bed regenerator, **20**:136–137

Rotating annulus

without axial flow, **5**:208

with axial flow, **5**:236

Rotating bodies of revolution in forced flow field, **5**:166

Rotating concentric cylinders, **5**:207

without axial flow, **5**:208

with axial flow, **5**:236

comparison of experiment and theory, **5**:226

Rotating concentric cylinders (*continued*)
　flow configurations in, **5**:212
　rate of heat transfer in, **5**:216
Rotating cones, **5**:135, 140
　in axisymmetric flow field, **5**:170
　in forced flow field, **5**:167
　local friction coefficient, **5**:143
　in mixed flow, **5**:149
　transition on, **5**:171
　in turbulent flow, **5**:149
　turbulent flow regime, **5**:149
Rotating cylinders, **5**:166; **6**:317; **8**:164; **25**:259,
　　291
　cone, **25**:291
　disk, **25**:261
　flow regimes on, **5**:174
　Nusselt number on, **5**:166
　sphere, **25**:291
　surfaces, **25**:291
　in uniform flow stream, **5**:172
Rotating disks, **5**:129, 130, 133, 140; **8**:22
　in axisymmetric flow field, **5**:168
　convective heat transfer, **5**:129
　effect on surface roughness on flow over,
　　5:156
　effects of shrouds, **5**:175
　experiments, **5**:144
　flow regimes, **5**:175
　in forced flow field, **5**:167
　local Nusselt number, **5**:140
　in mixed flow, **5**:149
　moment coefficient for, **5**:153
　rotating between parallel walls, **5**:179
　stability of flow on, **5**:156, 168
　transient heat transfer, **5**:145, 148
　in turbulent flow, **5**:149
　turbulent flow regime, **5**:149
Rotating flow, **24**:286
Rotating fluids heated from below, **5**:203
　critical Rayleigh number, **5**:205
　mercury, **5**:206
　overstability, **5**:205
Rotating heat pipe, **7**:223
Rotating isothermal body, **5**:161
　Nusselt number for, **5**:162
Rotating sphere, **5**:159
　boundary layer thickness on, **5**:161
　in liquid mercury, **5**:165
　Nusselt number for, **5**:164, 174
　separation on, **5**:160

　in uniform flow stream, **5**:172
　velocity distribution on, **5**:160
Rotating systems, **5**:129
　Grashof number, **5**:142
　Reynolds number, **5**:142
　stability of, **5**:130
Rotating test rig, **9**:38
Rotating tubes, **5**:207
　without axial flow, **5**:208
　with axial flow, **5**:241
　flow regimes in, **5**:242
　heat transfer to, **5**:243
Rotating twisted tape, in heat exchange
　　enhancement devices, **30**:230
Rotating-vertical-pedestal MOCVD reactor,
　　28:346, 389
Rotating viscometer, **15**:75–76
Rotating wires, **31**:453
Rotation
　in Czochralski process, **30**:364–365, 385–390
　natural convection and, **30**:367–369, 385–386
　surface tension and, **30**:369
Rotational energy and the accommodation coef-
　　ficient, **2**:338
Rotational flow, induced, *see* Induced rotational
　　flow
Rotational quantum number, **8**:232
Rotational transitions, **8**:231
Roughness, **8**:103
　effect on viscous sublayer, **16**:344, 346
　effects on boundary layer, prediction, exam-
　　ples, **16**:360–361
　effects on turbulent boundary layer, **16**:300–
　　311
　sand grain, **16**:301, 309
　surface, *see* Surface roughness effects
Roughness effects, **9**:79
Rough surfaces, **31**:159, 161
　heat transfer enhancement and fouling,
　　30:217, 219–227, 246
　melting of, **24**:15–18
Rough tubes, turbulence in, **31**:196
Round disks, **8**:24
Round jets
　boundary conditions for, **13**:74
　expected results for, **13**:75
　geometry and physics of, **13**:73–74
　importance of, **13**:74
　method adaptation for, **13**:75
　in surrounding stream, **13**:73–75

Round nozzles, vs. slotted, **13**:36
Round pipes, *see* Supercritical pressures
ROVER Program, **5**:327
Row-reduced echelon matrix, **31**:128
Rows of cylinders
  in-line, **25**:308
  staggered, **25**:308
RPV, *see* Reactor pressure vessel
Rugeley dry cooling tower, **12**:43, 51, 59
Runge–Kutta integration scheme, **SUP**:135, 178
Runge–Kutta method, **2**:435; **4**:331f; **5**:133
Russian roulette, **31**:357–358
Rybczynski–Hadamard solution, **26**:4, 11, 27
Rydberg Series, **5**:274

S

Saffman's invariant, **21**:197
Saha–Eggert equation, **4**:238
Saline expansion coefficient
  calculation, **18**:359
  seawater, temperature and pressure effects,
    18:336
  water
    temperature and salinity effects, **18**:337
    at various temperatures and salinities,
      18:354
Salinity and temperature effects
  enthalpy of water, **18**:344
  entropy change for water, **18**:346
  Prandtl number for water, **18**:348
  saline expansion coefficient of water, **18**:337
  specific heat of water, **18**:339
  thermal expansion coefficient of water, **18**:333
Salinity Rayleigh number, **14**:27
Salt deposition
  and carry-away, **31**:295
  with cold water flow past annular turbulator-
    equipped tubes, **31**:294–296
  on inner surface of annular diaphragm-
    equipped and helical tubes, **31**:301–304
  on outer surface of annular groove-provided
    tubes, **31**:299–301
Salt fingers, **28**:244
Salts, **31**:439
Sandia National Laboratories (SNL) studies
  α-mode failure, **29**:147, 151
  direct containment heating, **29**:216–217, 219,
    220–223, 225

melt–water interactions, **29**:164, 165, 190
SNL/ANL study, **29**:228
SNL/CES studies
  SNL/CES-1, **29**:231, 322
  SNL/CES-2, **29**:322
  SNL/CES-3, **29**:231, 322
SNL/CE studies, **29**:221, 222, 231
SNL COREXIT facility, **29**:224, 225
SNL CTTF facility, **29**:224
SNL/DCH studies, **29**:216, 220–222, 225–
  226, 247, 273, 302
SNL/HIPS studies
  SNL/HIPS-4W, **29**:323
  SNL/HIPS-6W, **29**:323
  SNL/HIPS-8C, **29**:276
  SNL/HIPS-9W, **29**:323
  SNL/HIPS-1OS, **29**:277
SNL/IET studies, **29**:220–223, 231, 244, 246,
  281, 289, 310, 313, 318–319
  SNL/IET-1R, **29**:288
  SNL/IET-3, **29**:288, 306, 309
  SNL/IET-4, **29**:302, 306, 309
  SNL/IET-5, **29**:228, 229, 230, 300, 302
  SNL/IET-6, **29**:302, 306, 309
  SNL/IET-7, **29**:302, 306, 309, 316, 319
  SNL/IET-8A, **29**:229
  SNL/IET-8B, **29**:229, 300, 313–315, 316,
    319, 321, 323
  SNL/IET-9, **29**:230, 276
  SNL/IET-10, **29**:230, 279
  SNL/IET-11, **29**:230, 276, 302, 303, 306
  SNL/IET-12, **29**:230
SNL/LFP studies, **29**:221, 222, 226–227, 281,
  293
SNL/TDS studies, **29**:220, 221, 222, 226
SNL/WC studies, **29**:220–222, 227, 293
  SNL/WC-1, **29**:228
  SNL/WC-2, **29**:228
  SNL/WC-3, **29**:228
Sand layer, water distribution in, **13**:122–123
Sandwich-type heat pipe, cooling techniques,
  20:285–286
SASM, **29**:218, 220, 228, 288
Satellites, their use in producing reduced gravity,
  4:150
Saturated boiling, **1**:198; **5**:444
  of liquid hydrogen, **5**:445
Saturated bulk liquid, CHF in external flow of
  behaviors of vapor and liquid in high heat flux
    in nucleate boiling, **17**:3–5

Saturated bulk liquid (*continued*)
  CHF in pool, intermediate, and forced flow
    boiling, **17**:5–9
  mechanism of CHF, **17**:9–19
Saturated liquid, boiling curve, **23**:6
Saturation, magnetic, **25**:186
  particle classification, **25**:227, 238
Saturation temperature, **16**:62
Scale analysis, **24**:9–13, 115
Scale of turbulence, **3**:3f, 11
Scale ratio parameter, **21**:186
Scales
  length, **24**:299
  time, **24**:285
  velocity, **24**:298
Scaling, **31**:145, 151, 439–441
Scaling arguments, **24**:298–302
Scattering, **5**:36, 264; **27**:17; **31**:344
  anisotropic, **27**:146–154; **31**:364–373
  direction of, **31**:369–371
  isotropic, **27**:157; **31**:364–373
  of light, **22**:278
    anisotropy in tissue, **22**:283
    coefficient, measured in tissue, **22**:279
  quasi-isotropic, **23**:136–139
  radiative–convective system, **27**:92–99
Scattering albedo, **27**:17–18, 95, 152–154, 190
Scattering coefficient, **3**:180f; **9**:369; **27**:149–
  150
  volumetric, **3**:181
Scattering mean beam length, **31**:366
Scattering phase functions, **31**:370–371
  gas–particle mixture, **27**:18
Scattering properties, spectral variation in,
  31:388
SCDAP/RELAP5, direct containment studies,
  29:218
SCE model, direct containment heating, **29**:232,
  233, 235–238
Scheil assumption, **28**:262
Scheme, regime delineation, **25**:223
Schiller velocity profile, **SUP**:41, 69, 192, 194
Schleiermacher's method, **2**:304ff
Schlichting velocity profile, **SUP**:160–161, 192
Schliere, **6**:135, 144, 150, 274, 320
Schlieren enthalpy, **6**:275
Schlieren interference, **6**:202
Schlieren lens, **6**:155, 184, 187
Schlieren optics, in convection effects, **14**:119
Schmidt–Boelter gage, **23**:299–300

Schmidt number, **2**:117; **4**:16, 320; **15**:240;
  25:184, 268, 271, 314
  influence of, **9**:308
  mass transfer, **21**:171
  for pure and saline water, **18**:348–351
  seawater
    at atmospheric pressure for various tem-
      peratures and salinities, **18**:350
    temperature and pressure effects, **18**:351
    at various temperatures and pressures,
      18:359
Schrage's theory, modification, **21**:72–73
Schuster number, *see* Albedo
Schuster–Schwarzschild approximation, **3**:203f;
  23:136–139; **27**:23–31
Schvab–Zeldovich property, **30**:53, 60, 61, 65,
  70, 73, 88
Schwartz–Christoffel transformation, **8**:45, 46
  for natural convection inside rectangular
    cavities, **8**:189
  for natural convection inside rectangular
    channels, **8**:193
Schwarz–Neumann alternating method, **SUP**:63,
  273, 352, 353
Sclera, **22**:265
Scraped surface heat exchangers, **30**:239–240
Screw extrusion, modeling, **28**:149–151
Script-F transfer fraction
  for isothermal gas radiation, **12**:168–174
  for nonisothermal gas radiation, **12**:189
Seas, lunar, *see* Moon
Seawater, *see also* Water
  density, **18**:330–332
  diffusivity, **18**:328–330
  entropy, **18**:345
  Prandtl number
    comparison at 1 bar pressure for various
      temperatures and pressures, **18**:349
    temperature and pressure effects, **18**:347
  saline expansion coefficient, temperature and
    pressure effects, **18**:336
  Schmidt number
    at atmospheric pressure for various tem-
      peratures and salinities, **18**:350
    temperature and pressure effects, **18**:351
    at various temperatures and pressures,
      18:359
  specific heat, **18**:337–343
    comparison of change at various tempera-
      tures and pressures, **18**:341

temperature and pressure effects, **18**:338
temperature and pressure effects on entropy, **18**:345
thermal conductivity, **18**:327–328
thermal expansion
  at atmospheric pressure and various temperatures, **18**:336
  at various temperatures and pressures, **18**:335
thermal expansion coefficient, temperature and pressure effects, **18**:332
viscosity, **18**:327
Secondary refrigerant cycle, water freezing in, **15**:271
Second-law analysis
  dilute polymer solution as, **15**:179
  in heat transfer and thermal design/research, **15**:1–53
  open thermodynamic systems in, **15**:5–8
Second Law of Thermodynamics, **21**:240–243
Second-order model
  nearly homogeneous scalar field, **21**:199–202, 233–234
  nearly homogeneous turbulence, **21**:193–202
    equation for vorticity tensor function, **21**:195–196
    vorticity scaler function, **21**:196
  nearly homogeneous velocity field, **21**:194–199, 232–233
Second-order upwind scheme, **24**:227, 238
Sectorial duct, truncated, **SUP**:269
Sedimentation, *see also* Hindered settling; Settling
  concentrated suspensions, **23**:261
  hindered settling velocity, **23**:261
    power law fluids, **23**:261
    viscoelastic fluids, **23**:261
Sedimentation potential, **12**:258–259
Seebeck effect, **22**:363, 388
Seeding, **31**:462
Segmented plate heaters, **23**:333
Self-absorption, **27**:48, 58, 96–97, 131
Self-emitted radiation, **4**:234
Self-similar flows, **2**:142
Self-similar solutions of species conservation equation, *see* Conservation of species equation
Self-wiping twin-screw extruders, **28**:189, 192
Semiconductor chip
  heat pipe cooling, **20**:272–274

vs. conventionally cooled chips, **20**:290–293
Semiconductors
  chemical vapor deposition, **28**:341–342
  Czochralski process for crystal growth, **30**:313–319
  impurity redistribution in, **15**:287
  pulsed laser melting, **28**:80–96
Semifluidized bed, **25**:190
Semiinfinite medium, radiative flux from, **11**:343–344
Semiinfinite solid, **8**:24
  transient heat transfer in, **11**:376–381
Semitransparent analog alloys, **28**:270–288
Semitransparent crystals, radiative effects on
  melting and solidification of, **11**:413–419
  directional reflectance of, **11**:365–371
  radiative effects on phase transformation in, **11**:411–413
Semitransparent materials
  phase changes, **19**:37–38
  solidification and melting of, **11**:411–420
Semitransparent solids, **2**:400; *see also* Radiative flux
  ablation of with radiant heating, **11**:388–390
  absorptance of, **11**:365–373
  combined conduction and radiative heat transfer in, **11**:390–411
  conduction in, **11**:390–411
  directional absorptances and transmittances for, **11**:369–370
  directional and boundary effects on heat transfer in, **11**:400–403
  directional and hemispherical emittance in, **11**:373–375
  effective thermal conductances for, **11**:359–362
  examples of, **11**:318
  experimental absorption spectra for, **11**:358–359
  Fresnel reflection of, **11**:366
  gray media and, **11**:395–400
  heat transfer in, **11**:317–433
  hemispherical reflectance, absorptance, and transmittance of, **11**:371–375
  irradiation of, **11**:385–388
  nongray radiative transfer in, **11**:432
  optical properties of, **11**:355–359
  phonon conductivity for, **11**:360–362
  physical properties of, **11**:352–365
  radiation characteristics of, **11**:365–375

Semitransparent solids (*continued*)
  radiation heat transfer in, **11**:390–411
  radiative conductivity for, **11**:363–365
  radiative transfer temperature measurement
      for, **11**:420–431
  recovered vs. predicted temperature profiles
      for, **11**:429–431
  spectral effects in, **11**:403–407
  spectral regions of, **11**:395
  specular reflection in, **11**:402
  temperature distribution in, **11**:425–431
  thermal conductivity in, **11**:352–354
  and transient conduction–radiation in partici-
      pating media, **11**:407–411
  transient heat transfer in, **11**:375–390
  transmittance of, **11**:365–373
Sensible heat exchange, **31**:64
Sensible heat exchangers, fouling tests, **30**:213–
      216
Sensible heat transport, **20**:28–31
Sensible heat units, for energy storage, **15**:34–38
Separan ^{c}AP–273^c, **25**:308
Separan solution, **15**:64–65, 82, 116; *see also*
      Polyacrylamide solution
  degradation effects on, **15**:118
  NaOH additives in, **15**:96–97
  turbulent flow in, **15**:112
Separated flows
  acoustic radiation, **6**:73
  boundary layer solutions, **6**:101, 107, 110, 116
  breakaway separation, **6**:6
  Chapman–Korst model, **6**:7, 17, 22
  critical point, **6**:34, 56, 112
  downstream facing step, **6**:4, 33, 41, 114
  eddy viscosity, **6**:16, 120
  free interaction, **6**:84
  invisid solutions, **6**:2
  lip shocks, **6**:19, 50, 55
  low density flows, **6**:38
  notch induced separation, **6**:5, 60, 68, 75
  ramp induced separation, **6**:4, 73
  recompression, **6**:27, 34
  recovery temperature, **6**:94
  reviews articles, **6**:6
  shock-induced separation, **6**:4, 5
  slow flows, **6**:2, 61
  stability, **6**:36
  supersonic flows, **6**:19
  transition, **6**:35, 38, 77, 87, 92
  transpiration, **6**:80

  unsteady heat transfer, **6**:479
  upstream facing step, **6**:4, 5, 6, 67, 83, 84, 89,
      95
  wedge flows, **6**:37, 51, 52, 65
Separation, **8**:96; **9**:343
  biochemical, **25**:235
  flow, **9**:342
  heat transfer at, **8**:120
  influence of Reynolds number on, **8**:98
  point, **25**:283
  solids–solids, **25**:187, 234
  in tube banks, **8**:110
Separation point, **8**:97
Sequential volume searching and traversing,
      31:342
Series inversion, **8**:14
Series truncation, **8**:13
Settling, *see also* Hindered settling; Sedimenta-
      tion
  concentrated suspensions, **23**:261
  in shearthinning media, **23**:261
  velocity, **23**:261
Settling length, **SUP**:41
Severe Accident Scaling Methodology (SASM),
      29:218, 220, 228
Severe slugging
  Boe's criterion, **20**:121–122
  cycle for, **20**:116–121
  quasi-equilibrium, **20**:124–128
  stability criterion, **20**:122–124
  transient phenomena, **20**:115–116
Shape factor, **SUP**:201, 228; **24**:43, 62, 65
Shared memory, **31**:403–404
Shear
  rate, **25**:255, 269, 314
  shear thickening fluid, **24**:107
  shear thinning fluid, **24**:107
  stress, **25**:75, 255, 301
      wall, **25**:14, 299, 301, 307, 314
  stress tensor, **24**:108
  thickening, **25**:255, 263, 280
  thinning, **25**:255, 264, 269–271, 281, 288, 308
Shear anisotropy, dimensionless parameter,
      21:203
Shear direction, **13**:213
Sheared structured liquids, thermal conductivity
      measurement, **18**:176–178
Shear flow, **13**:212–259
  activation energy and, **13**:216, 218
  with closed stream lines, **13**:250–259

Couette flow in, **13**:219–222
dimensionless parameters in, **13**:252–253
dimensionless variables in, **13**:231–233
elongation flow and, **13**:260–262
expansion cooling in, **13**:229
fully developed temperature field in, **13**:258–259
heat transfer studies in, **13**:220
kinematically developed velocity in, **13**:253
master curve in, **13**:217
nearly steady, **13**:219
open or closed stream lines in, **13**:219–222
Poiseuille flow in, **13**:219–222
steady, *see* Steady shear flow
stream lines in, **13**:219–222
thermal boundary condition in, **13**:222–227
unsteady, **13**:256–258
viscous dissipation in, **13**:205–264
wall thermal capacity in, **13**:225–227
Shear flow program, universal numerical, **13**:235–238
Shear layer, **9**:322
Shear stress, **3**:49f; **28**:152
nonlinear dependence on shear rate, **19**:251
slug flow, **20**:89
turbulent, **7**:22, 55, 70
Shear stress continuity, liquid metal droplet, **28**:6
Shear stress distribution, **8**:303
Shear stress measurements, electrochemical method, **7**:136
Shear surfaces, defined, **13**:212
Shear thickening fluids, **2**:359
Shear thinning effects
liquid–solid fluidized beds, **23**:232
bed expansion, **23**:241, 253
minimum fluidization velocity, **23**:234, 239, 241
packed beds, **23**:194
settling, **23**:261
three-phase systems, **23**:253
Shear thinning fluids, **2**:359, 363, 372; **23**:189–190
Shear thinning material, **28**:152
Shear velocity
degradation effect on, **15**:119
solvent effects on, **15**:81
Shear viscosity, **13**:214–219
Sheaths, plasma, **4**:252ff
Shell-and-tube multifluid heat exchanger, **26**:225, 227

Shell side, **25**:252, 292
Shepherd's map, **29**:104
Sherwood number, **3**:14f; **9**:304; **25**:260, 267, 271–273, 288, 314; **26**:22, 79, 80, 84
for array of round nozzles, **13**:20, 37
for array of slot nozzles, **13**:25–26, 34
for disks and cones rotating in air, **5**:158
forced convection, interbarrier spacing modules, **20**:240–241
heat- and mass-transfer
enclosed porous layer, **20**:340–341
finite-amplitude convection, **20**:327
high Rayleigh number convection, **20**:329
in impinging flow, **13**:7–12
liquid–solid fluidized beds, **23**:253
mass transfer, **21**:172–174
mean integral for high performance arrays, **13**:57
packed beds, **23**:230
for shrouded rotating disk, **5**:181
for single nozzles, **13**:15–16
variation in for arrays of slot nozzles, **13**:34
wiper blades effect, **21**:175
Shock, **24**:231, 242
Shock tube experiments, **14**:335–337
Shock tube method, **3**:275
Shock tubes, **2**:187f; **24**:242
autoignition temperature, **29**:85
Shock wave, *see* Surface reaction
Shock wave–boundary layer interaction, **8**:11
Showerhead injection, **2**:78f
Shrinkage, alloy solidification, **28**:305–306
Shrouded rotating disk, **5**:175
convection in, without source flow, **5**:177
Nusselt number, **5**:181
Sherwood number, **5**:181
with sink flow, **5**:199
transition on, **5**:182
with veil cooling, **5**:200
Shvets' method, **1**:83ff
Side-chilled mold, solidification, **28**:235, 237
Sidewall
partially-conducting, **24**:295–296
perfectly conducting, **24**:281–294
Sieder–Tate correction, **2**:372f, 375, 387
SIGMA facility, **29**:174
Signal to noise, **22**:369, 424, 430
Silicon
chemical vapor deposition, **28**:365–373, 382–385, 386

Silicon (*continued*)
  polysilicon film melting, **28**:81–96
  surface topography after laser ablation, **28**:101–109
  ultrashallow doping profiles, **28**:96–101
Silicon crystals
  by Czochralski process, **30**:313–314, 315, 317–319, 364
  properties near melting point, **30**:346
  transition temperatures to transparency, **30**:340
Silicone oil
  experiments with, in horizontal circular cylinders, **8**:211
  natural convection to, in rectangular cavities, **8**:184
Silicone oil/water, entrainment heat transfer data, **21**:316
Silicon–water superposed layers, convection, **29**:32–33
SIMD (single instruction multiple data stream) architecture, **31**:402–403
Similarity, **9**:284; **25**:290
  for laminar boundary layers, **4**:319ff
    conditions for, **4**:322ff
    equations for solution, **4**:324ff
    large values of the pressure–gradient parameter, **4**:341ff
      inner limit equations, **4**:344f
      outer limit equations, **4**:341ff
    local similarity, **4**:333ff
      asymptotic expansion of equations, **4**:335ff
      determination of f, by successive approximations, **4**:337ff
    numerical integration procedure, **4**:331ff
    numerical results, **4**:330f
    range of solutions and parameters, **4**:330
    special classes of reduced equations, **4**:327ff
      computation of boundary layer properties, **4**:328
      no mass transfer, **4**:327f
      Prandtl number equal to unity, **4**:328
      pressure–gradient parameter equal to zero, **4**:328
      two-dimensional flow, **4**:327
      viscosity proportional to temperature, **4**:328
  local-time, **25**:17

for mass diffusion, **9**:285
for thermal transport, **9**:285
variable, **9**:283
Similarity flows, **9**:9, 16, 17, 20, 21
Similarity law, **31**:226
Similarity regime, **8**:163
Similarity solutions, for laminar thermal convection heat transfer, **15**:152–153
Similarity theory, **8**:94
Similarity variable, **1**:35, 39; **2**:146
Similar solutions, **4**:356ff, 389ff, 401ff
  for a power–law viscosity relation, **4**:370ff
  for a Sutherland viscosity–temperature relation, **4**:426ff
Similiarity transformations, **19**:6–12
SIMPLE algorithm, **13**:102; **24**:194, 195, 201, 202
Simple fluid, **15**:211
SIMPLER algorithm, **28**:195, 255, 382
Simpson's rule, **2**:422
Simulations
  BEER program, **27**:53–55
  Czochralski system
    flow and temperature, **30**:395–402
    liquid-encapsulated process, **30**:348–349
    transport phenomena, **30**:347–352
  numerical
    boiling in a forced flow through porous media, **30**:140–141
    multiphase flow in geothermal systems, **30**:140–141
  radiative heat transfer, **27**:62–85
    gas absorption, **27**:50–56
    radiation emission from gas volume, **27**:56–59
    radiation emission from solid walls, **27**:59–61
    reflection and absorption by solid walls, **27**:61–85
  RAT1 program, **27**:64–75, 80, 83, 88
  RAT2 program, **27**:76–84, 88, 109
Simultaneous heat and mass transfer, **8**:62
Simultaneously developing flow, **SUP**:7, 16, 34
  in circular ducts, **SUP**:138–152
  in circular duct with a twisted tape, **SUP**:384
  comparisons of solutions for, **SUP**:403–405
  in concentric annular ducts, **SUP**:319–321
  with constant wall heat flux, **SUP**:144–149, 192–193, 219–222, 242–245

with constant wall temperature, **SUP**:138–144, 190–191, 219–222, 242–245, 384
with constant wall-to-fluid temperature difference, **SUP**:152
for eccentric annular ducts, **SUP**:337–340
with finite wall thermal resistance, **SUP**:149–151
for fundamental solutions, **SUP**:193–195, 319–321, 337–340
generalized solution for, **SUP**:58
for parallel plates, **SUP**:189–195
for rectangular ducts, **SUP**:219–222
solution methods for, **SUP**:77
in triangular ducts, **SUP**:242–245
Sine ducts, **SUP**:253–356
Single-axis-tracking parabolic trough collectors
geometries and orientation for natural convection losses, **18**:9
natural convection, **18**:26–27
Single band gas, **8**:258
Single bubbles
in immiscible fluids, **15**:245
in motion, **15**:244–248
noncondensables and, **15**:247
Single-bubble trains, **15**:248–254
Single-cell equilibrium (SCE) model, direct containment heating, **29**:232, 233, 235–238
Single-domain models
alloy solidification, **28**:250–252, 326
dual-scale model, **28**:263
micro- and macromodels, **28**:261–264
mixture models, **28**:253–256
multiphase model, **28**:264
submodels, **28**:264–269
two-phase models, **28**:256–261
Single exposure speckle photography, **30**:291–293, 304
Single laminar bubble train, **15**:255
Single nozzles, integral mean transfer coefficients for, **13**:15–18
Single-phase heat exchangers, **31**:319
Single-phase impingement heat transfer, **20**:254–255
Single plate-type fin, *see* Fins
Single region analysis, **SUP**:354
Single-screw extruder
experimental results, **28**:184–187
heat transfer, **28**:147, 149
mixing characteristics, **28**:180–184
modeling, **28**:155–176

axial formulation, **28**:170–175
developed flow, **28**:161–164
fully developed flow, **28**:160–161
three-dimensional transport, **28**:164–170
two-dimensional transport, **28**:156–160
residence-time distribution, **28**:176–180
tapered screw, **28**:176, 225
Single slot nozzle, integral mean transfer coefficient for, **13**:18
Single stagnant bubbles, **15**:242–244
Single tubes, **8**:94
boundary layer on, **8**:96
combined free and forced convections, **8**:127
constant surface temperature, **8**:119
cross flow of air in, **8**:113
cross flow of oil in, **8**:113
cross flow of water in, **8**:113
distribution of skin friction around, **8**:102
drag coefficient for flow around, **8**:102
experimental velocity distribution around, **8**:99
flow pattern around, **8**:95
gas flows around, **8**:94
heat transfer to, **8**:94, 112, 116
heat transfer to air flow over, **8**:124, 127, 129
heat transfer to oil flow over, **8**:124, 129
influence of blockage ratio in flow around, **8**:103
influence of blockage ratio on heat transfer to, **8**:132
influence of heating or cooling on heat transfer to, **8**:129
local heat transfer on, **8**:116, 119
local Nusselt numbers on, **8**:116
mean heat transfer to, **8**:126
pressure distribution around, **8**:96
in pressure of other tubes, **14**:210–212
separation point on, **8**:97, 100, 101
turbulent boundary layer on, **8**:97
wake regime, **8**:98
Singly connected ducts, **SUP**:8, 62
definition of, **SUP**:17
thermal boundary conditions for, **SUP**:17–31
Singularities, **8**:3
associated with equilibrium flows, **2**:252ff
Singular perturbation, **8**:3
Singular perturbation method, **2**:255ff
Sintered surfaces
augmented nucleate boiling, **20**:70–71
optimum thickness, **20**:72–73
Six-equation two-fluid model, **31**:108

Skewness factor, **8**:329–332, 340
Skin, **22**:222–226
  properties of, **4**:116ff
Skin friction, turbulent flow, **30**:82–83
Skin-friction coefficient, **2**:14, 15f, 20ff, 39, 98f; **3**:14, 31
  adiabatic, **2**:21
  Coles', **2**:21, 98f
  diabetic, **2**:20ff
  low speed, **2**:21
Skin-friction gage, **23**:337–338
Slab
  heat conduction in
    for semiinfinite slab, **1**:56ff, 68ff, 80f, 83f, 86ff
    for slab of finite thickness, **1**:61ff, 71, 93ff
  semitransparent, **11**:416–419
Slab band absorption, in isothermal gas radiation, **12**:163–165
Slanted single tubes
  without fins, **14**:204–206
  with fins, **14**:206–207
Slanted tubes
  heat transfer coefficients for, **14**:172–207
  local heat transfer coefficients for, **14**:160–172
  in packed fluidized beds, **14**:233–234
Slats, horizontal and vertical, **18**:17
Slender body flows, **2**:246
Slender vertical cone, laminar thermal convection from, **15**:172–173
Slip, **8**:95
Slip effects, packed beds, **23**:228
Slip flow regime, **28**:351
Slip ratio in annular-dispersed flows, **1**:364ff, 390
Slip regimes, **9**:358
Slip velocity (thermal creep), **7**:185
Slot nozzles
  single, **13**:15–18
  vs. round, **13**:36
Slow burn, hydrogen combustion, **29**:108–114
Slug bed, **25**:219
Slug calorimeter, **23**:307–311
Slug film boiling regime, **31**:265, 269–271
Slug flow, **1**:358, 359ff; **8**:54; **SUP**:16, 64, 75, 186, 220
  defined, **20**:83–84
  geometry, **20**:85–86
  hydrodynamics, **20**:84
  Nusselt numbers for, **19**:290, 292–295

separator movement, **20**:116–117
severe slugging, **20**:115–129
  Boe's criterion, **20**:121–122
  cycle dynamics, **20**:116–121
  quasi-equilibrium, **20**:124–128
  stability criterion, **20**:122–124
  transient phenomena, **20**:115–116
of solids over surface, **10**:187
steady, **20**:84
steady-state modeling, **20**:85–115
  auxiliary relations, **20**:95–97
  average void fraction, **20**:88
  calculation procedures, **20**:97–100
  dispersed bubble velocities, **20**:107–108
  liquid film hydrodynamics, **20**:88–92
  liquid holdup, **20**:109–110
  mass balances, **20**:86–87
  pressure drop, **20**:92–95
  slug length and frequency, **20**:110–113
  Taylor bubble translational velocities, **20**:100–107
terrain-induced, **20**:84
transient slugging, **20**:84, 115–116
two-phase, **20**:83–129
Slug flow film boiling, subcooling and, **11**:183
Slug frequency and slug length, **20**:110–113
Slugging, **25**:180, 181
Slumped bed, **25**:210
Slurry fuel droplet, vaporization, **26**:81–82
Slurry region, **28**:269
Smagorinsky
  constant, **25**:341–343, 385
  model, *see* Subgrid, model, Smagorinsky
Small aspect ratio ducts, **SUP**:280–283
  approximate solution methods for, **SUP**:67
Smelt dissolving tank operations, **10**:107
Smelt explosions, **29**:130
SMIN variable, **27**:81–82, 207
Smoke filaments, **9**:344
Snell's law, **6**:144; **11**:341; **23**:147, 149
Sodium carbonate analog alloy, **28**:270, 287
Sodium chloride solution, diffusivity at atmospheric pressure for various salinities, **18**:329
Sodium hydroxide, in Separan solution, **15**:96–97
Soil remediation, by soil vapor extraction, **30**:175–177
Soil vapor extraction (SVE), **30**:94, 111, 175–177

Solar buildings
  building envelope, **18**:47–49
  double-pane windows, **18**:15, 49
  enclosure correction studies with vertical and
    horizontal heat fluxes, summary, **18**:58–
    59
  heating and cooling
    vertical and horizontal surfaces in enclo-
      sures, **18**:57–63
    vertical surfaces in enclosures, **18**:51–57
  multiple building zones, **18**:67–75
    boundary layer driven flow, **18**:68–72
    bulk density driven flow, **18**:67–68
    interzone natural convection, summary,
      18:70
  natural convection, **18**:46–75, 76–77
  single-building zones, **18**:49–66
    surface roughness effects, **18**:63–66
    triangular spaces, **18**:50–51
  single-pane windows, **18**:48–49
  thermal design characteristics, **18**:10
  thermal models, **18**:46–47
  types, **18**:5–6
Solar collectors
  active, **18**:1–4, 9–10
  efficiency and overall heat-loss coefficients,
    18:6–10
  flat-plate, **18**:11–19
  instantaneous thermal efficiency, **18**:7
  line-focusing, **18**:19–27
  operating temperatures, **18**:3
  overall heat loss coefficients, **18**:9–10
  passive, **18**:5–6
  point-focusing systems, **18**:27–46
  single-axis-tracking parabolic trough collec-
    tors, **18**:26–27
  types, **18**:2
Solar power plants, **16**:331, 335
Solar radiation measurements, **4**:77ff
Solar reflectance, for lunar fines, **10**:28–30
Solenoid, **25**:154, 157, 162, 164, 166
Solid conductivity, radiative heat transfer,
  23:179–181
Solid–gas interface, **31**:26, 55, 66
Solidification, **8**:35; **22**:183
  alloys, **28**:231–238, 326–328; *see also*
    Macrosegregation
    from bottom wall, **28**:277–282
    continuum model, **28**:270–288, 311
    mathematical models, **28**:249–269

metal alloys, **28**:288–308, 316, 327
  multidirectional solidification, **28**:282–285
  physical phenomena, **28**:238–249
  process control strategies, **28**:235, 237, 238,
    308–326, 327–328
  semitransparent analog alloys, **28**:270–288
  from sidewall of rectangular cavity, **28**:270–
    277
  unidirectional solidification, **28**:277–282
  zero gravity, **28**:309
  of binary eutectic-forming solutions, **19**:79
  of a liquid in a square channel, **1**:115ff
  spray deposition
    incremental solidification, **28**:22
    liquid metal droplet, **28**:11–18
    modeling, **28**:17–18
    splat solidification, **28**:22
Solidification curve, **25**:173
Solidification point, **25**:173
Solid–liquid interface
  alloys, **28**:232
  augmented nucleate boiling, **20**:64–67
  determination, **19**:8
  instability, **19**:73–78
  position, **19**:7, 36
  suction or blowing effect along, **19**:6
Solid nucleation, **28**:12, 15
Solid particles
  heat transfer between fluid and, **10**:171–180
  mixing, **31**:161
Solid phase
  closure for, **31**:56, 61–63
  heat transfer process in, **31**:41–43
  point equations for, **31**:23–24
  volume-averaged thermal energy equations
    for, **31**:53
  volume-averaged transport equations for,
    31:28
Solid-phase epitaxy, **28**:341
Solids
  heat transfer in, **31**:143–145
  slug flow of over surface, **10**:187
  temperature distribution in, **10**:144
Solid spheres, drag coefficient, **21**:334
Solid surfaces, **25**:311; *see also* Wall elements
  Kirchhoff's law, **27**:11–12
  thermal radiation, **27**:10–12
  wettability, **21**:61
Solid–vapor interface, augmented nucleate boiling, **20**:64–67

Solutal buoyancy, **28**:239, 251
Solute
    concentration, **22**:195, 212
    redistribution during solidification, **22**:293
Solute dispersion, heat- and mass-transfer,
    20:325
Solute redistribution model, **28**:251
Solutions, polymeric
    concentration and temperature effects on
        thermal conductivity, **18**:223–225
    thermal conductivity, **18**:223–227
Solvent chemistry effects, in non-Newtonian
    fluids, **15**:79–83
Solvent velocity, for dilute solutions, **12**:234
Sonic field, heat transfer in, **31**:182
Soret diffusion, heat- and mass-transfer, **20**:328–
    329
Soret term, **28**:356
Sound, effects of, on heat transfer, **3**:28ff
Sound velocity, **7**:8
Source-and-sink laws, **13**:63
Source function, radiation intensity, **27**:21–23
Sources, **8**:15
    distributed, **9**:294
Source term, **24**:200, 201, 202, 206, 208, 209,
    239, 241, 244, 254, 255
Spacers, **9**:382
Spalding's correlation, **26**:53
Sparrow's velocity profile, **SUP**:71, 87, 139, 145
Spatial average temperature, **31**:54
Spatial averaging theorem, **31**:13–14
Spatial decomposition, **25**:333–339
    by filtering, **25**:333–335, 337
    by volume averaging, **25**:337–339
Spatial deviation, average of, **31**:21–22, 33
Spatial deviation mass fraction, **31**:35
    gradient of, **31**:33
Spatial deviation temperature, **31**:50, 51
Spatial deviation transport equation, **31**:58
Spatial deviation variables, **31**:15
Spatial region, parallelize by, **31**:410–412
Special concentration equation, chemical vapor
    deposition, **28**:354, 360, 362
Special form, of averaging theorem, **31**:19
Species conservation equation, *see* Conservation
    of species equation
Species continuity equations, **31**:9
Species densities, **31**:9
Species equation, **8**:8, 27
Species momentum equations, **31**:9

Species production rate, *see* Rate of production
    of species
Species velocities, **31**:9
Specific heat, **2**:277; **5**:335
    apparent, **22**:190
    calculation, **18**:360
    at cryogenic temperatures, **5**:335
    of lunar fines, **10**:63–64
    of lunar materials, **10**:39–40, 59–63
    of metals at low temperatures, **5**:335
    seawater, **18**:337–343
        change at various temperatures and pres-
            sures, **18**:341–342
        temperature and pressure effects, **18**:338
    two-phase, **30**:36
    water, pure, **18**:337–343
        change at 0° for various pressures, **18**:340
        change at 10° for various pressures, **18**:340
        change at 20° for various pressures, **18**:341
        changes for various temperatures and sali-
            nities, **18**:355
Specific heat method, **19**:27
Specific intensity, **5**:255
Specified wall heat flux distribution, *see also*
    Arbitrary heat flux distribution
    in circular ducts, **SUP**:82–83, 124–136
    for concentric annular ducts, **SUP**:292–296,
        312–318
    for parallel plates, **SUP**:179–185
    for rectangular ducts, **SUP**:203–209, 217–219
Specified wall temperature distribution, *see also*
    Arbitrary axial wall temperature distribution
    in circular ducts, **SUP**:99–120
    for concentric annular ducts, **SUP**:292–296,
        312–318
    for parallel plates, **SUP**:169–179, 190–191
    for rectangular ducts, **SUP**:202–203, 213–317
Specklegram
    full-view interrogation of, **30**:264–265
    point-by-point interrogation of, **30**:261–264
Speckle interferometry, **30**:265–267, 295
Speckle photography, **30**:255
    flame temperature measurement, **30**:293, 304,
        306
        axisymmetric candle flames, **30**:296–299
        premixed gaseous flames, **30**:299–302
        tomographic reconstruction of temperature
            field, **30**:287, 288, 293–296, 306
    full-view interrogation, **30**:264–265
    liquid temperature measurement, **30**:302–304

natural convection problems, **30**:267–268
  converging channel flows, **30**:281–283
  heat transfer coefficient, **30**:269–271
  isothermal single vertical wall, **30**:272–274
  isothermal vertical channel flows, **30**:274–278
  one-dimensional refraction, **30**:268–269
  upward-facing isothermal surfaces, **30**:278–280
  point-by-point interrogation, **30**:261–264
  principles, **30**:256–261
  single-exposure speckle photography, **30**:291–293, 304
  speckle interferometry, **30**:265–267, 295
  turbulent flow with density fluctuation, **30**:283–285
   fluctuating density and temperature fields, **30**:289–291
   laminar and inhomogeneous density fields, **30**:285–289
   single-exposure speckle photography, **30**:291–293, 304
Speckle shearing interferometry, **30**:266, 267
Spectral absorption coefficient, **8**:231; **31**:389
  hydrogen, **5**:35
Spectral absorption efficiency, **23**:141
Spectral absorptivity, **5**:18
Spectral band, parallelize by, **31**:409–410
Spectral band absorptance, **8**:235, 240, 241
Spectral behavior
  of gas, **31**:388–390
  of particles, **31**:390
Spectral directional reflectance
  angle of illumination and, **10**:28–30
  bulk density and, **10**:26, 27
  for lunar fines, **10**:23–29
Spectral effects, in radiative heat transfer, **11**:403–407
Spectral emission measurements, **2**:317
Spectral emittance
  for lunar materials, **10**:31–34
  vs. bulk density, **10**:32
Spectral intensity, **3**:178f
  energy radiation and, **11**:324
Spectral lines, **8**:231
  half-width of, **8**:232
  spacing of, **8**:233
  strong overlapping lines, **8**:236
  variation of intensity with rotational quantum number, **8**:232

variation with pressure and temperature, **8**:232
Spectrally dependent properties, **31**:388–400, 423
Spectral normal reflectance, **9**:388
Spectral radiant energy density, **5**:255
Spectral radiant flux, **12**:123
Spectral radiant intensity, **12**:123
Spectral radiative flux, **8**:250
  expression for, **11**:336
Spectral scattering coefficient, **23**:152–153
Spectral scattering efficiency, **23**:141
Spectral thermal remote sensing, for semitransparent solids, **11**:425–431
Spectroradiometric curves, thermal radiation, **27**:7–9
Specular-diffuse surfaces, **2**:445ff
  nonplanar, **2**:449f
Specularly-reflecting surfaces, **2**:400, 426f, 445ff
Speedup, **31**:408
Sphere band absorption, in isothermal gas radiation, **12**:165–166
Sphere in uniform flow, **3**:4, 10
Spheres, **8**:12; **24**:5, 7, 126, 151; **25**:252, 254, 257, 259, 262, 263, 267–279, 291, 312
  close-contact melting inside, **19**:70–71
  electrically heated copper, **25**:264
  freezing, **19**:31
  heat conduction in, **15**:285
  heat transfer to, **1**:343f
  inward solidification, **19**:60
  isothermal, **25**:264
  laminar heat transfer for, **11**:280–281
  one-phase problems with convective and radiative heating, **19**:32–33
  potential flow convection, **20**:369–370
Spherical cap, potential flow convection, **20**:374
Spherical cavities, *see also* Cavities
  convection, **29**:20–27
Spherical droplet, inertialess terminal velocity, **26**:4
Spherical harmonics method, the, **3**:206
Spherical projections and cavities, nucleation from, **10**:113–117
Spherical shell, radiation flux in, **11**:350–352
Spherical surface, mechanical equilibrium in, **15**:260
Spherical symmetry, heat conductor problems involving, **1**:64
Spheroidal state, **5**:56
Spheroids, potential flow convection, **20**:375

Spinning bodies of revolution, **5**:132

Spinning cone, influence of free convection, **5**:141

Spin-up, **24**:286

Spiral-shaped strips, **9**:117

Spirelf® system, heat transfer enhancement, **30**:228–230

SPIT-18,19 experiments, **29**:217

Spitzer's formula, **4**:264

Splat solidification, **28**:22

SPMD (single program multiple data) approach, **31**:406

   individual-ray, **31**:413–421

Sponge-ball cleaning, **31**:452, 454, 456

Spontaneous explosions, **29**:158

Spontaneous heating, materials subject to, **10**:236–237

Spontaneous nucleation model, steam explosion, **29**:137, 138–140

Spontaneous nucleation temperature, **29**:139

Spray

   condensation, **26**:44–47

   droplet deformation, **26**:79–80

   droplet interaction, **26**:84

   droplet transport, **26**:87–89

   evaporation, **26**:53–54

   spray deposition, **28**:18–21

Spray column condensers, direct contact condensation in, **15**:262–264

Spray deposition, **28**:1–3

   impact region, **28**:22–23

      solidification of multiple liquid metal droplets and sprays, **28**:54–69

      splat cooling of single liquid metal droplet, **28**:23–54

   spray region

      convective cooling of single liquid metal droplet, **28**:3–11

      in-flight solidification of single liquid metal droplet, **28**:11–18

      sprays, **28**:18–22

Spray flow, **1**:358

Spread factor, **28**:24

Spreading coefficient, **9**:189, 229

Sputtering, pulse-laser-induced, **28**:103, 106, 109–123

Square array, **26**:171, 181

   heat transfer, **26**:174, 175

Square cavities

   natural convection inside, **8**:188

Nusselt numbers for natural convection in, **8**:194, 195

Square ducts, *see also* Rectangular ducts

   longitudinal fins within, **SUP**:370–372

Square residual

   of energy equation, **24**:94

   initial, **24**:48, 59, 60, 69, 72, 81

SRN, *see* Round nozzles

SSN, *see* Single slot nozzle

Stability, **9**:52, 55, 91, 340; **24**:174

   condition for, **9**:291

   equations of, **9**:323

   experimental values of, **9**:328

   of flow over rotating surfaces, **5**:156

   of laminar flows, **9**:307

   limits of, **9**:325

   of a liquid-vapor interface in pool boiling, **1**:234ff

   neutral, **9**:324, 327

   plane, **9**:340

   of rotating systems, **5**:130

Stability criterion

   severe slugging, **20**:122–124

      classical vs. quasi-equilibrium operations, **20**:128

Stability limits, predicted values of, **9**:324

Stability loss, of laminar flow in triangular channels, **31**:262

Stabilization

   hydrodynamic

      flow, **25**:64, 65

      section, **25**:124

Stabilized bed, **25**:182, 206

   flow, thermal hydrodynamically, **25**:121

Stabilizing surface, **9**:342

Staggered grid, **24**:194, 207, 210

Staggered tube banks, **8**:135, 136, 139, 140, 142, 144, 146

   heat transfer in, **8**:116

   hydraulic resistance in, **8**:151–153

   influence of Reynolds number in, **8**:110

   pressure drop coefficients, **8**:155

   separation points in, **8**:110

Staggered vertical plate/card arrays

   heat dissipation ratio, **20**:199–200

   natural convection cooling, **20**:198–201

   temperature reduction, **20**:199–200

Stagnant bubble heat transfer-controlled collapse of, **15**:243

Stagnant-cap model, **26**:12

Stagnant film theory, **16**:278
Stagnant gas, entropy production, **21**:249–254
Stagnation
  point, **25**:283
  region, **25**:279
Stagnation flow
  jet length and, **13**:4
  for single nozzles, **13**:11
Stagnation point, **8**:7, 19
  chemical reaction effects, **1**:364f
  vs. jet axis/surface point, **13**:44
Stagnation-point heat transfer
  with applied magnetic field, **1**:332ff
    enthalpy gradient, **1**:339ff
      stagnation heating, **1**:341ff
        heat transfer to spheres, **1**:343f
          to cylinders, **1**:344f
      velocity gradient, **1**:334ff
        comparison of solutions, **1**:335ff
        Newtonian flow, **1**:334f
        similarity solutions, **1**:335
  rarefied gases, **7**:196
Stagnation pressure, **2**:53
Stagnation region flows, **2**:195ff
Stagnation region heat flux, **2**:200f
Stagnation regions, **31**:229
Stagnation temperature, **2**:53; **7**:189
Stagnation zone, **26**:105, 108, 205
  liquid jets
    free-surface jet, **26**:137–141, 147, 162–167
    laminar jet, **26**:109–113, 119–122, 123, 127–130
    submerged axisymmetric jet, **26**:123, 127–130
    velocity gradient, **26**:109–110, 120
  transport, **26**:108
Standard deviation, **5**:12
Stanford University, **14**:97
  Thermosciences Laboratory
    apparatus and techniques, **16**:254–270
    data reduction, **16**:270–271
    turbulent-boundary-layer heat transfer research, **16**:242–243
Stanton number, **2**:14, 15f, 20, 22ff, 27, 29, 39, 96, 99, 386; **3**:40f, 77, 90; **7**:189; **8**:306; **12**:87–88, 98–99, 104; **SUP**:53; **15**:123; **19**:341; **21**:14–15; **25**:26, 99, 103, 113; **31**:248
  distribution downstream of steps in wall temperature, **16**:245–246

drag reduction and, **12**:97
effect of humidity on, **16**:271
effects of acceleration, **16**:286–287
increase in, due to turbulization, **8**:306
measured values of, **12**:106–107
  in pressure gradient flows, **8**:306, 307, 319
relationship of enthalpy-thickness Reynolds number, **16**:249, 280–281
response, to step increase in blowing, **16**:278, 279
rough surface, **16**:302–306
thermogravity effect, **21**:19–20
variation
  with x-Reynolds number at constant velocity, **16**:274–275
  with blowing and suction, at constant x-Reynolds number, **16**:274–276
  with blowing parameter, **16**:276, 277
  during strong acceleration, **16**:246–247
  with effects of curvature, **16**:320–327
  with enthalpy-thickness Reynold's number, at various accelerations, **16**:283–286
  with enthalpy-thickness Reynolds number for mild deceleration with transpiration, **16**:298
  with enthalpy-thickness Reynolds number for strong deceleration with transpiration, **16**:298–299
  in full coverage film cooling, **16**:312–317
  versus buoyancy parameter, **21**:17–18
STAR-CD (software), **30**:351
Stark broadening, **5**:278, 281
Star-shaped ducts, **SUP**:276
State
  bubbling, **25**:203
  fixed bed, **25**:203
  fluidized, **25**:154, 166
  fully developed fluidized, **25**:154
  lockup, **25**:232
  packed, **25**:204
  pseudopolimerized, **25**:154, 203
  quiescent, **25**:193
  semistabilized, **25**:189
  stabilized, **25**:189, 195, 204
  stable ("liquid") fluidized bed, **25**:170, 193
  transition, **25**:204
  unfluidized ("solid") bed, **25**:170
  unstabilized, **25**:204
  unstable ("vapor") fluidized bed, **25**:170, 193, 195

State parameter, **25**:84
Static pressure, **2**:15
  distributions, **2**:57ff
Stationary current distribution, **14**:120
Statistical analysis, **8**:335
  of hot wire signals, **8**:336
Statistical models, **5**:291; **8**:290
Steady axisymmetrical jets, in THIRBLE classification, **13**:65–66
Steady multiphase flows, **30**:165–166
Steady shear flow
  defined, **13**:218
  dimensionless parameters in, **13**:233–235
  experimental studies in, **13**:248–250
  fully developed temperature field in, **13**:237
  heat transfer studies in, **13**:220
  Nusselt number in, **13**:239–241
  with open stream lines, **13**:227–250
  shear viscosity and, **13**:214–219
  velocity field in, **13**:235
Steady shear velocity, degradation effect on,
  15:119
Steady-state flow field, heat- and mass-transfer,
  20:345–346
Steady-state slug flow modeling, **20**:85–115
  auxiliary relations, **20**:95–97
  calculation procedures, **20**:97–100
  dispersed bubble velocities, **20**:107–108
  liquid film hydrodynamics, **20**:88–92
  liquid holdup, **20**:109–110
  mass balances, **20**:86–87
  pressure drop, **20**:92–95
  slug length and frequency, **20**:110–113
  Taylor bubble translational velocities, **20**:100–
    107
  void fractions, **20**:88
Steady two-dimensional layer model
  elliptic case in, **13**:104–106
  mathematical formulation of, **13**:99–100
  parabolic and hyperbolic cases in, **13**:100–106
  for two-dimensional floating layers, **13**:98–
    106
Steam
  high-pressure, heat transfer coefficient, **21**:90
  low-pressure, heat transfer coefficient, **21**:83
  near-atmospheric pressure, heat transfer coef-
    ficient, **21**:82–83
Steam bubbles, heat transfer to water, **15**:254
Steam condensation, direct, **15**:268–269
Steam explosions, **29**:129–131

  described, **29**:132–135
  history, **29**:135–136
  integral explosion assessments, **29**:192–195
  modeling
    droplet capture model, **29**:140
    ESPROSE.m field model, **29**:180, 182,
      195–201
    parametric, **29**:135–137
    premixing, **29**:148–152
    propagation, **29**:180–182
    spontaneous nucleation, **29**:137, 138–140
    thermal detonation, **29**:137, 140–147
    thermodynamic, **29**:135–137
  nuclear reactor safety, **29**:129, 131, 136, 147,
    193–195
  premixing, **29**:132, 147–158
  propagation, **29**:132, 134–135, 168, 185–187
    experiments, **29**:177–178
    hydrodynamic fragmentation, **29**:168–175
    stratification, **29**:182–185
    thermal fragmentation, **29**:175–176
    thermal–chemical explosions, **29**:187–192
    triggering, **29**:132, 158–159, 167–168
    integral experiments, **29**:164–166
    prevention, **29**:166–167
    single droplet experiments, **29**:163–164
    vapor film collapse, **29**:132, 159–163
Steam formation, by laser, **22**:253
Steam injection, in porous media, **30**:152–161
Steam mass flow ratio, **15**:264
Steam override, gravity effects in, **30**:93, 154
Steepest descent, **8**:14
Stefan–Boltzmann constant, **5**:261; **9**:363;
  12:124
Stefan–Boltzmann convection, **11**:200
Stefan–Boltzmann law, **9**:362; **19**:193; **21**:246;
  22:421; **27**:8–9
Stefan condition, **19**:192
Stefan constant, **2**:321
Stefan diffusion tube problem, **13**:169
Stefan–Maxwell equations, **12**:236, 243, 269,
  271; **31**:8
  diffusivity types in, **12**:213–219
  generalization of, **12**:267–268
  for migration, **12**:225
  in ordinary diffusion, **12**:208–209
Stefan–Neumann solution, **19**:8, 20, 29–30, 32
Stefan number, **22**:203; **24**:4
  constant-flux, **19**:12
  definition, density change and, **19**:11

modified, **22**:185, 206
natural-convection-dominated freezing and,
    19:60
viscous heating, **24**:14
Stefan problems
  Biot's variational method, **19**:39–40
  boundary conditions, constant heat flux and
    convection, **19**:12
  boundary element method or boundary
    integral method, **19**:44
  brute-force numerical solutions, **19**:42
  characteristics, **19**:8–9
  complex variable methods, **19**:39
  coordinate transformations, **19**:16–24
  definition, **19**:5
  exact close-form solutions, **19**:11–12
  finite-difference method, **19**:42
  inverse method, **19**:41–42
  isotherm migration method, **19**:40–41
  multidimensional, natural convection in,
    19:49–50
  nonlinearity, **19**:17–18
  one-dimensional, **19**:21–22
    transform methods, **19**:16
    two-phase, **19**:6–7, 24–28
  one-phase, **19**:12–13
  power-series expansions, **19**:12–14
  reduction to integrodifferential equations,
    19:14–16
  solution methods, **19**:9
  two-phase, one-dimensional, **19**:6–7
  weak solutions, **19**:24–28
    principal advantages, **19**:25–26
Step-function profile, heat- and mass-transfer,
    20:324–325
Stephan method, **3**:284f
Step in surface temperature, *see* Boundary
    conditions, in heat conduction problems
Sticking probability, **30**:206
Stirring, heat transfer enhancement by, **30**:237,
    239
Stirring force, nucleate boiling, **20**:3
Stockmayer function, **3**:266
Stoichiometric equation, **2**:124
Stoichiometric relations, and equations of
    change, **12**:201
Stokesean fluid, momentum balance, **21**:247–
    248
Stokes flow, *see* Creeping flow
Stokes' law, **15**:72

bubble velocity, **20**:107–108
Stokes' second problem, **24**:306
Stokes stream and potential functions, **20**:365–
    366
Stokes stream function, **9**:302
Stokes' theorem, **2**:415, 418
Stove design
  metal, use in, **17**:297
  mud, as building material, **17**:296–297
Stoves
  carbon monoxide emission by, **17**:301–302
  closed
    control, efficiency and, **17**:284–289
    convective heat transfer to pan in, **17**:254–
      263
    flow-through stoves, **17**:244–254
    indoor air quality and, **17**:299–307
    single-pan, advantages of, **17**:298–299
    as system for cooking, **17**:169–172
    wall effects and, **17**:236–244
Straight channels, heat transfer enhancements
    method in, **31**:182–185
Strained coordinates method, **19**:31
Strain model, **25**:340
Strain-rate effect, on thermal conductivity of
    polymeric liquids, **18**:220–222
Strain-rate tensor, **25**:339
Stratification, **9**:131, 132, 345
  adiabatic, **9**:290
  of cryogens, **5**:388
    analysis of, **5**:390
    cause of, **5**:389
    thermal stratification layer, **5**:389
  effect of, **9**:278
  hydrogen jet combustion, **29**:302–303
  stable, **9**:290
Stratification temperature distribution, **24**:302
Stratified explosions, **29**:182–185
Stratified liquids, heat transfer with entrainment,
    21:311–318
  bubble-induced entrainment
    onset, **21**:313
    rate, **21**:313–314
  heat transfer, versus gas velocity, **21**:317
  overall interlayer heat transfer with entrain-
    ment, **21**:314–317
Stratifying process
  from non-stratification, **24**:293–295
  from pre-existing stratification, **24**:280–293
Streaking camera, **9**:137

Streaks, *see* Viscous sublayer, streaks
Stream function, **1**:35, 39; **2**:139, 146, 170, 187f;
    4:320; **13**:99
    generalized, **9**:283
Stream-function equation, liquid metal droplet,
    28:5
Streamline conduction, potential flow convection
    flat isothermal plate, **20**:357–358
    flat uniformly heated plate, **20**:358–359
    planar motion, **20**:355–357
Streamline flow, **SUP**:5
Streamlines
    for natural convection inside horizontal circu-
        lar cylinders, **8**:206–208, 211, 216–222
    for natural convection inside rectangular
        cavities, **8**:189
    in shear flow, **13**:221
Streamwise velocity component, RMS values of,
    in turbulent flows, **8**:302
Stress
    driving, **25**:175
    repose, **25**:181
    shear, wall, **25**:14, 299, 301, 307, 314
    yield, **25**:174, 204
Stress boundary condition
    normal component of, **31**:40
    tangential component of, **31**:40
Stress differences, normal, **19**:249
Stress jump condition, **31**:27
Stress tensor, **2**:358
Stretched coordinates, **8**:12
Strong nearly homogeneous scalar field, evolu-
    tion in
    strong degenerating isotropic velocity field,
        21:217–220
    strong evolving homogeneous turbulence,
        21:216–224
    strong evolving nearly homogeneous velocity
        field, **21**:220–224
Strong nearly homogeneous velocity field, evo-
    lution, **21**:204–209
Strong nonoverlapping line limit, **8**:239, 242
Strouhal number, **8**:98, 99; **18**:93
    dependence of, on Reynolds number, **8**:99
    jet pulsation, **26**:199
Structure
    bed, **25**:165, 237
    function, **25**:345
    magnetically stabilized fluidized bed, **25**:177
    polymerized bed, **25**:229

pseudopolymeric, **25**:163, 165, 197
pseudopolymerized bed, **25**:154, 200, 201
Structured surfaces, heat transfer enhancement
    with, **30**:234–235
Sturm–Liouville equations, **8**:68; **12**:96, 98
Sturm–Liouville operator, **15**:303
Sturm–Liouville problem, **3**:125, 128ff, 140,
    154ff; **SUP**:100, 112; **15**:285, 299–300
Sturm–Liouville theory, **15**:284, 320
Subchannel analysis, **SUP**:358–360
Subcooled boiling, **1**:191f, 198
Subcooled liquid drops, **9**:225
Subcooled vapor, **7**:4
Subcooling
    convective cooling and, **19**:66
    effect on transition boiling heat transfer,
        18:304–306
    effects on CHF, **17**:23–24
    film-boiling heat transfer and, **11**:133
    incipient boiling temperature and, **10**:149
    melting inside rectangular cavity, **19**:67
    melting rate and, **19**:54, 63
    natural-convection heat transfer, **19**:50
Subcooling number, **19**:8, 59
Subcritical flow regime, **8**:136, 142
Subcritical pressure, **7**:2
Subcritical temperature, **7**:8
Subgrid
    diffusivity, **25**:346
    energy, **25**:339–340, 343, 350, 374
    model, **25**:339–356, 403–405
        dynamic, **25**:343, 351–354, 396, 404
        gradient diffusion, **25**:339–346
        higher order, **25**:346–348
        multilevel, **25**:349–354
        nonconventional, **25**:354–356
        scale similarity, **25**:349–354
        Smagorinsky, **25**:340–343, 371, 403–405
        structure function, **25**:345
    Prandtl number, **25**:346, 361–362, 374
    scalar flux, **25**:339, 346
    stress, **25**:339, 363
    terms, **25**:335–339, 363
    viscosity, **25**:339–341, 343, 345
"Sublayer", **2**:386, 387
Sublimation, **1**:153ff; **8**:19; **19**:81
Submarine volcanism, steam explosions, **29**:130
Submerged jet impingement, thermal cooling of
    electronic components, **20**:252–258
Submerged liquid jet, **26**:106, 205

axisymmetric jet, **26**:123–135
    turbulent flow, **26**:132–133
cooling, **26**:205
experimental studies, **26**:123–135
heat transfer, **26**:124–126, 131–135
mass transfer, **26**:124–126
planar jet, **26**:135
Prandtl number, **26**:129–130
radial flow region, **26**:129–134
Substitute kernel method, **8**:34
Substrate material, **9**:233
Subsurface remediation, multiphase flow effects,
    30:93, 94, 111, 152
Successive–approximation method, **25**:31, 49, 112
Suction, **5**:144; **9**:294; *see also* Transpiration
    definition, **16**:272
    heat transfer enhancement by, **30**:244
Suction effect, **25**:83
    gas, **25**:84
Suit
    diving, **22**:8
    space, **22**:8
Sulfur-reducing bacteria (SRB), **31**:445
Supercomputers, **25**:330
Superconducting magnets, support for, **15**:43–44
Superconducting systems, refrigerator power
    requirement in, **15**:46
Superconductivity, **5**:326
    large-scale applications of, **15**:43
Supercooling, melt rate reduction, **19**:10–11
Supercritical conditions, **7**:2
Supercritical flow regime, **8**:97; **18**:91, 93
Supercritical fluid, **7**:2
Supercritical heat transfer, **7**:2
Supercritical hydrocarbon flow in tubes, heat
    transfer enhancement with, **31**:228–232
Supercritical pressures, **7**:2; **21**:1; *see also* Free
    convection
    boundary-layer approximation, **21**:3
    degraded heat transfer, **21**:47–48
    forced flow in round pipes
        hydrodynamic entry region, **21**:5–6
        laminar flow, **21**:5–6
        turbulent flow, **21**:6–17
        turbulent mixed convection, **21**:17–32
        turbulent transfer at varying heat flux,
            21:32–34
        turbulent transfer under cooling, **21**:34–37
    general premises and approaches to problem
        solution, **21**:3–5

heat transfer augmentation, **21**:47–48
nonstationary heat transfer, **21**:44–47
thermophysical properties of fluid, **21**:2–3
turbulent energy balance equation, **21**:4
Supercritical region, **25**:84
Supercritical temperature, **7**:8
Superficial averages, **31**:11–13
Superficial average velocity, **31**:31
Superficial entrainment flux, **21**:315
Superficial mass velocity, **1**:365
Superflatness factor, **8**:329, 330, 332
Superfluid behavior, **5**:329
    specific heat, **5**:331
    thermal conductivity, **5**:330
Superfluidity, **5**:481
    experimental background of, two-fluid theory,
        17:75–85
Superheat, *see also* Boiling liquid superheat;
        Liquid superheat
    dimensionless, in Madejski nucleation model,
        10:156
    emergent, **10**:122, 126
    minimum boiling, **10**:148–162
    nonuniform, **10**:127–128
Superheated gaseous stream, partial condensa-
        tion of, **15**:264
Superheated liquid drops, **10**:99
Superheated liquids, **7**:4
    equilibrium cluster radii for, **10**:97
    in nucleate boiling, **1**:205ff
    theoretical vs. experimental values for, **10**:99
Superheating, **22**:208
    density anomaly and, **19**:68–69
    natural-convection dominated freezing and,
        19:60
    Survival map, **22**:297
    Survival signature, **22**:295
    Sweat, **22**:277
Superheat limits
    classical treatment in, **10**:95–99
    expression for, **10**:97–98
    studies of, **10**:99–100
Superheat limit temperature, **29**:139
Superimposed free convection
    in compact heat exchangers, **SUP**:414
    in horizontal circular tubes, **SUP**:411–413
    influence of, **SUP**:410–414
    in vertical circular tubes, **SUP**:413–414
Superinsulation, **5**:329
    mean thermal conductivity of, **5**:329

Superposed layers, convection, **29**:31–33, 53
Superposition methods, **SUP**:18, 34–35, 119, 136, 316–317
Superposition principle, **2**:417; **21**:33
Supersaturation pressure ratio, **9**:194
Superskewness factor, **8**:329, 330, 332
Supersonic flow, **8**:53
Supersonic nozzle experiments, **14**:330–335
Supporting electrolyte, in electrochemical transport, **12**:241
Suppression
  bubble, **25**:168
  convective losses in flat-plate collectors, **18**:15–17
Surface area, **25**:164
Surface combustion of graphite, **2**:152ff
Surface condensation, **21**:58
Surface condenser, **21**:55
Surface configuration
  augmented nucleate boiling heat-transfer characteristics, **20**:70–77
  experimental apparatus, **20**:24
  heat flux and boiling curve, **20**:24
  low liquid level nucleate boiling, **20**:59–60
  natural convection cooling, **20**:209–213
  nucleate boiling
    boiling curves and bubble behavior, **20**:23–27
    downward-facing surface model, **20**:28–36
    heat transfer, **20**:2
    inclination angles, **20**:28–29
    liquid and vapor profiles, **20**:33–35
    liquid film thickness, **20**:33–35
    pressure dependence, **20**:20–21
    surface inclination, **20**:27–28
    thermal cooling of electronic components, **20**:246–247
  thermal contact conductance, metal foil enhancement, **20**:265–266
Surface curvature
  concave, **16**:326–331
  convex, **16**:319–327
  effect on boundary layer, **16**:319–331
Surface curves, nucleate boiling and, **20**:23–36
Surface emittance, **8**:262
  influence on radiative transfer in gases, **8**:262, 263
Surface factor and nucleation factor, **20**:15
Surface fences (turbulators), **20**:239–240
Surface flux values, **31**:396

Surface force, **31**:9
Surface free energy, **9**:185
Surface heat flux, **2**:120f, 160, 205, 239
Surface heat transfer coefficient, **21**:94
Surface inclination
  boiling heat transfer, **20**:24–25
    angle variations, **20**:25–27
  heat-transfer coefficient, **20**:27–28
Surface latent heat, **9**:187
Surface normals, **31**:339
Surface radiosity, **8**:250, 264
Surface reactions, **2**:135f, 138ff
  behind strong moving shock, **2**:185ff
  catalytic, *see* Catalytic surface reactions
  and heat transfer from high temperature dissociated air, **2**:161ff
Surface reaction solutions, **2**:172ff
Surface recombination, **2**:246ff
Surface roughness, **8**:145; **19**:147, 150, 180
Surface roughness effects
  on boiling, **1**:202ff
  on drag and flow pattern in fluid flow, **18**:97, 106
  heating surface, on transition boiling heat transfer, **18**:290–292
  on natural convection in solar buildings, **18**:63–66
Surfaces
  bed, **25**:165
  concave bed, **25**:188
  enhanced, **31**:453–457
  flat, **9**:343
  fouling control on, **31**:294–304
  hear transfer probe, **25**:164, 165, 192, 206, 220
  horizontal, **9**:339, 343
  hydrophobic, **31**:276
  immersed, **25**:190, 205
  to provide enhancement of heat transfer, **31**:275
  slightly inclined, **9**:339
  treatment of, **31**:461–462
Surface science, fundamental equations of, **10**:87–92
Surface temperature, in dry cooling, **12**:26–27
Surface tension, **9**:184
  critical value of, **9**:229
  in Czochralski process, **30**:316, 364, 365–367, 369, 384–390
  effects in annular-dispersed flows, **1**:380

natural convection and, **30**:365–366, 384–385, 386
rotation and, **30**:369
Surface tension-controlled regime, **23**:95
Surface tension force, condensation heat transfer enhancement, *see* Condensation heat transfer
Surface tension heat exchange enhancement devices, **30**:230–232
Surface topography, laser ablation, **28**:101–109
Surface variables in boiling, **1**:202ff
Surface vibration, **31**:161
Surface-viscous effect, drop motion, **26**:13
Surfactants, moving liquid droplet, **26**:11–12, 43–44
Surveyor missions, **10**:1, 7, 10, 41, 73, 78
Suspension flow, **9**:128, 171, 175
Suspensions, density profiles of, **9**:157
Suspensions, thermal conductivity
  flowing media, **18**:200–209
    Brownian motion and, **18**:206–207
    deformed drops/particles, **18**:205–206
    generalized analytical framework, **18**:201–203
    spherical drops/particles, **18**:203–204
    theoretical and experimental data, **18**:207–209
  stagnant media, **18**:178–200
    basic equations, **18**:191–193
    bounds on, **18**:197–200
    conventional approach, **18**:179–190
    effective thermal conductivity, **18**:179
    first-order expressions for different inclusion shapes, **18**:193–194
    generalized analysis, **18**:190–197
    inclusions of arbitrary shapes, **18**:197
    linear heat flow model, theoretical expressions, **18**:180
    multidimensional heterogeneous media, **18**:188
    nonlinear heat flow model, theoretical expressions, **18**:181–182
    particle shape effects, **18**:183
    size distribution effects, **18**:186–188
Sutherland constant, **4**:326
Sutherland viscosity law, **4**:326
Sutterby model, of laminar thermal convection, **15**:166
SVE, *see* Soil vapor extraction
Swirlers, **31**:159

Swirl flow heat exchange enhancement devices, **30**:230, 245–246
Swirling flow, **9**:128
Swirling jets, in heat and mass transfer, **13**:41
SWTI-5 study, **29**:316
SWTI-6 study, **29**:316
SWTI-12 study, **29**:316
Symmetric-balanced regenerators, **20**:137–138
  recuperative operation, **20**:155–156
  unidirectional operation, **20**:150
Symmetric-unbalanced regenerators, **20**:137–138
  recuperative operation, **20**:157–158
  unidirectional operation, **20**:150
Synchronization, proper, **31**:403
Synthetic boiling, electrolysis as, **14**:136
System
  catalytic, **25**:152
  cyclic fixed-bed, **25**:232
  fluidized bed, **25**:164
  moving settled bed, **25**:232
  noncatalytic, **25**:152
System admittance matrix, **31**:128
System component, momentum balance for, **31**:121–124
System elasticity, expansion rate and, **31**:118–119
System-level thermal control, heat pipes, **20**:295–297
  harsh environments, **20**:297
  sealed cabinet electronic equipment, **20**:296–297
System mechanical compliance, pressurized systems and, **31**:146–147
System scaling criteria, **31**:150
System volumetric compliance, **31**:118
System volumetric elasticity, ratio of net expansion or contraction rate to, **31**:117

T

Tall buildings, fire safety in, **10**:278
Tangential component, of stress boundary condition, **31**:40
Tangential corotating twin-screw extruder, **28**:147, 193, 194
Tapered gap passages, *see* Passages
Tapered screw extruders, **28**:176, 225
Tapered tube passages, *see* Passages

Tap water, chemical analysis of, **15**:80–82
Tarset, **29**:166
Taylor bubbles
  defined, **20**:83, 85–86
  severe slugging, blowout, **20**:120–121
  translational velocities, **20**:100–107
Taylor instability, **5**:68, 88, 114, 425
  in boiling, **1**:234ff
  for rotating fluids heated from below, **5**:205
Taylor number
  critical
    appearance of vortices, **21**:150–151
    dependence on axial Reynolds number,
      21:158
    induced rotational flow, **21**:148, 151–152
    nonisothermal induced rotational flow,
      21:160
    variation with eccentricity ratio, **21**:159
    wavy vortex flow, **21**:153
  induced rotational flow, **21**:157–158
  versus Nusselt number, **21**:163–165
Taylor scale, **21**:264, 267–269; **30**:6, 18, 61, 74, 75
  liquid metals, **21**:266
  viscous oils, **21**:267
Taylor series, **2**:423, 424; **8**:34; **25**:15, 17
Taylor series expansion, **19**:9; **31**:19
Taylor vortex
  critical speeds for onset, **21**:156
  induced rotational flow, **21**:146–152
T boundary condition
  dimensionless groups, **19**:253, 254, 255
  Newtonian fluids
    simultaneously developing flows, **19**:300,
      301
    thermal entrance length, **19**:298–299
  non-Newtonian fluid, thermally developed,
    19:295
  thermally developed laminar flow through
    rectangular channels, **19**:277
TCE model, direct containment heating, **29**:218,
  219, 233, 238–240, 301, 311, 320
TEACH code, **24**:194, 195
Technische Arbeitsfähigkeit, **15**:8
Technology, gas–solid, fluidization, **25**:151
  magnetically stabilized bed, **25**:232
Temperature, **2**:277; **3**:182; **22**:4, 5, 13; **25**:269,
  290; *see also* Wall temperature
  average, **31**:50
  of blood, **22**:5
  characteristic, **25**:15

damping, wave, **25**:12
dimensionless, **19**:255
distribution, **24**:47
  initial, **24**:58, 64
    constant, **24**:63, 66
    general, **24**:43, 48, 51
    parabolic, **24**:51, 60
field, **25**:269
free-stream, **16**:334
  as function of heat flux, **20**:184–185
  as function of space and time, **13**:127
  gradients, natural convection, **20**:347–348
  heat transfer probe surface, **25**:214
  influence of under artificial flow turbulization
    conditions, **31**:225–228
intermediate-reference, **16**:334
low liquid level nucleate boiling, **20**:59–60
mean-mass, heat carrier, **25**:8
measurement, **22**:9, 10, 14, 16
  accuracy, **22**:370–371
  correlation to, due to conduction, **8**:317
  flame temperature, **30**:293–302, 304, 306
  fundamental limitation, **22**:408–409
  laser therapy, **22**:361, 387, 403–404
  liquid temperature, **30**:302–304
  noise, **22**:404–408, 424, 429–433
  precision, **22**:372
  in radiative transfer, **11**:420–431
  reproducibility, **22**:372–373
  resolution, **22**:371, 423–424
  safety, **22**:373
  simultaneous with velocity, **8**:309, 310
  spatial temperature gradients, **22**:399–402
  spatial temperature response, **22**:398–399
  thermistor-based, **22**:409–413
  thermocouple-based, **22**:413–416
  in turbulent flows, **8**:299, 316, 318
nucleate boiling formulas, **20**:18–23
profile, **25**:264, 265
  in bed, **25**:214
regulation, **22**:360
spatial deviation, **31**:50, 51
surface, **24**:63, 66, 73
thermal control of electronic components,
  20:182–183
transducer
  classification, **22**:362–364
  impedance, **22**:369–370
  linearity, **22**:367, 375, 389
  sensitivity, **22**:368, 375, 389

specificity, **22**:369
time constant, **22**:369, 395–398
unidirectional operation, long regenerators,
  20:160–161
wall, distribution, **25**:6
Temperature boundary condition, exponential
  variation in, **9**:289
Temperature continuity, **8**:253
Temperature correlation, dimensionless,
  SUP:58–59
Temperature cross, parallel-stream exchanger,
  26:274, 276–278
Temperature dependence
  of thermal conductivity, **7**:9
  of viscosity, **7**:9
Temperature-dependent fluid properties, **12**:90–
  94
  future research in, **SUP**:420
  for gases, **SUP**:415–416
  influence of, **SUP**:414–416
  for liquids, **SUP**:416
Temperature differential number, **15**:12
Temperature distribution
  cross-flow heat exchanger, **26**:281–286
  for free convection in rectangular cavities,
    8:166
  parallel-stream heat exchanger, **26**:271–273
Temperature effectiveness, three-fluid heat exchanger, **26**:297–304
Temperature effects
  and pressure effects
    enthalpy of water, **18**:343
    entropy change for seawater, **18**:345
    Prandtl number for seawater, **18**:347
    saline expansion coefficient of seawater,
      18:336
    Schmidt number of seawater, **18**:351
    specific heat of seawater, **18**:338
  and salinity effects
    enthalpy of water, **18**:344
    entropy change for water, **18**:346
    Prandtl number for water, **18**:348
    saline expansion coefficient of water,
      18:337
    specific heat of water, **18**:339
    thermal expansion coefficient of water,
      18:333
  thermal conductivity of polymeric solutions,
    18:223–225
  thermal conductivity of single-component

polymeric liquids,
  18:213–218
  experimental observations, **18**:217–218
  modified Bridgman equation, **18**:216–217
  WLF correlation, **18**:214–216
  thermal expansion coefficient of seawater,
    18:332
Temperature field
  bubble growth and, **11**:15–16, 18–30
  of complex-shaped bodies, **19**:226–230
  fully developed, **13**:237, 258–259
  maximum, **11**:29–30
Temperature field determination
  comparison of results in, **11**:27–30
  dynamic calibration in, **11**:20–21
  experimental apparatus and procedure in,
    11:21–27
  method in, **11**:18–19
  microthermocouple probe in, **11**:19–20
Temperature flatness factor, **8**:334
Temperature fluctuation
  in forced convection boiling, **11**:43–45
  in pool boiling, **11**:38–43
  in turbulent boundary layer on a flat plate,
    8:305
  for turbulent flows with pressure gradient,
    8:325
  in two-phase boiling boundary layer, **11**:32–38
Temperature gradient, age of earth and, **15**:286
Temperature interaction zone (TIZ), **29**:163
Temperature jump, **9**:217; **SUP**:119
  approximation, **7**:170
  distance, **2**:275, 292, 296, 299
  measurements, **2**:293f
  regime, **2**:291f; **7**:184, 192
  theory, **2**:296ff
Temperature overshoot, immersion cooling techniques, **20**:245–246
Temperature perturbations, radiative decay from,
  14:251–259
Temperature problem, **SUP**:10–16
  categories, **SUP**:10
  complete solution for, **SUP**:419
Temperature profiles
  comparison of, with velocity profiles, **8**:334
  comparison of various methods for determin-
    ing values of, in turbulent flow, **8**:334
  for free convection between vertical isothermal
    plates, **8**:167
  for free convection in rectangular slot, **8**:168

Temperature profiles (*continued*)
  for free convection inside horizontal cylinders,
    8:167
  for natural convection in enclosed cavities,
    8:171
  for natural convection in rectangular cavities,
    8:176, 182–185, 191, 192
  for natural convection inside horizontal circu-
    lar cylinders, **8**:170, 204, 205, 211, 212,
    217, 219–222
  for radiative equilibrium in gases, **8**:279
  for turbulent flows with pressure gradient,
    8:322, 324
  within an infrared radiating gas, **8**:254
Temperature pulsations
  heating surface, **18**:270–273
  and vapor content on heating surface, mea-
    surement, **18**:259
Temperature sensors, in radiative heat transfer,
    11:421–422
Temperature skewness factor, **8**:333
Temperature slip, **8**:257
Temperature spectra, **8**:305
Temporary bubbling, **25**:230
Terminal equilibrium level, slug flow, **20**:97–98
Terminal velocity, intertialess, spherical drop,
    26:4
Termination cutoff level, **31**:356–358
Terrain-induced slugging, *see* Severe slugging
Terrestrial fines, thermal conductivity of, **10**:50
Terrestrial rock powders, thermal conductivity of,
    10:52
Terrestrial rocks, thermal properties of, **10**:43–44
Tertiary flows, **8**:190
  for natural convection inside rectangular
    cavities, **8**:190, 191
Test, solids, discharge, **25**:173
Test level filter, **25**:351
Tetrahedrons, as modeling elements, **31**:359–360
TEXAS code, **29**:151
TFM program, **27**:27–29, 203
Thallium bismide, properties, **30**:321
Theory of distributions, **23**:378
  convolution product, **23**:379
Thermal
  behavior, bed, **25**:165
  conductance models
    contribution of blood flow, **22**:20
    effective tissue conductance, **22**:20, 82
    inadequacies, **22**:21
  conductivity, **22**:237

cross-section mean, **25**:9
  insulated material, **25**:112
  turbulent, **25**:49
design, **25**:2
diffusivity, **25**:7, 61
dose, **22**:361
energy, **25**:268
imaging
  calibration, **22**:424–428
  detectors, **22**:422–424
  digital filtering, **22**:428–433
  medical, **22**:421–422
  noise equivalent temperature difference,
    22:424
  Stefan–Boltzmann law, **22**:421
inertia, effect, **25**:81, 98
properties, **22**:381, 394
  measurement, **22**:187, 342
regulation, **22**:7, 16
stress, **22**:304
  azimuthal and radial, **22**:307
  fracturing, **22**:177
  in vitrification, **22**:312
unsteadiness, **25**:23, 100
  parameter, **25**:64
Thermal ablation, **28**:110
Thermal accommodation, incomplete, **7**:172
Thermal accommodation coefficient, **7**:165;
    9:358
Thermal axial distance, dimensionless, **19**:255
Thermal boundary conditions, **SUP**:16–36; *see
    also* Boundary conditions
  comparisons of, **SUP**:389–392
  defined, **SUP**:16
  for doubly connected ducts, **SUP**:31–36
  finite wall axial heat conduction, **SUP**:30–31
  five fundamental kinds of, **SUP**:31–33
  general classification of, **SUP**:18
  H, *H*1 boundary condition, **SUP**:20, 25–26
    for circular ducts, **SUP**:82–83, 124–135,
      144–149
    for concentric annular ducts, **SUP**:294, 315
    for eccentric annular ducts, **SUP**:328–329
    for elliptical ducts, **SUP**:249–251
    for parallel plates, **SUP**:157–158, 179–184,
      192–193
    for rectangular ducts, **SUP**:203–206, 217–
      222
    for triangular ducts, **SUP**:224–226, 232,
      235–239, 242–243, 245
  *H*2 boundary condition, **SUP**:20, 25–27

for rectangular ducts, **SUP**:205, 207, 216, 219, 222

for triangular ducts, **SUP**:225–226, 230–232, 236–237

*H*4 boundary condition, **SUP**:21, 25, 27–28

for rectangular ducts, **SUP**:208–209

*H*5 boundary condition, *see* Exponential axial wall heat flux

idealizations for wall thermal conductivity, **SUP**:24

influence of, on natural convection inside cavities, **8**:172

for multiply connected ducts, **SUP**:36

nomenclature used for, **SUP**:19, 33–34

nonuniform conditions in, **9**:285

power-law variation of, **9**:286, 296

in shear flow, **13**:222–227

for singly connected ducts, **SUP**:17–31

summary table of, **SUP**:20–21

superposition methods and, **SUP**:34–35

T boundary condition, **SUP**:20, 22–23

for circular ducts, **SUP**:79–80, 99–118, 138–144

for concentric annular ducts, **SUP**:292, 314

for elliptical ducts, **SUP**:249–251

for parallel plates, **SUP**:155–157, 169–177, 190–191

for rectangular ducts, **SUP**:202–203, 213–217, 219–222

for triangular ducts, **SUP**:224, 233–234, 238, 242–243, 245

*T*3 boundary condition, **SUP**:20, 23–24; *see also* Wall thermal resistance

*T*4 boundary condition, **SUP**:20, 24–25; *see also* Radiant flux boundary condition

Δ*t* boundary condition, **SUP**:21, 29, 152

Thermal boundary layers, **6**:180; **8**:14

in boiling, **1**:231

concept of, **11**:11

influence of Prandtl number on, **8**:35

instabilities in, **14**:69–71

maximum temperature in, **11**:26

Reynolds number and, **12**:97

stripping mechanism, **16**:109, 120–121

temperature fluctuation in, **11**:32–38

thickness, **11**:12–13

nucleate boiling, **20**:9–10

and two-phase boiling boundary layer, **11**:32–38

variables in, **11**:31

Thermal capacity

effects of, **9**:310, 314

parameter, **9**:313

surface, **9**:327

Thermal–chemical explosions, **29**:187–192

Thermal–chemical vapor deposition, **28**:342

Thermal communication, **26**:219

Thermal conduction, mechanism of, **2**:274

Thermal conductivity, **2**:274; **7**:8–13; **15**:63–66; *see also* Transport properties of dilute gases

apparent values of, **9**:353

in convection heat transfer, **11**:286

effective, **13**:140, 158–164

effect of pressure and void size on, **9**:371

of evacuated powders, **5**:346

free convection, **21**:39–40

of frost, **5**:470

generalized coordinates, **24**:47, 73

heat flux and, **10**:46–47

indirect methods for determining, **9**:404

of low temperature solids, **9**:360

of lunar fines, **10**:45–59

of lunar materials, **10**:39–40, 42–59

measurement of, **10**:48

measuring effective in a partially saturated porous medium, **31**:2

of metals at low temperatures, **5**:334

near critical region, **7**:13

near pseudocritical temperature, **21**:43

parameters for, **10**:48

penetration depth, **24**:75, 79

pressure effects in, **10**:58

seawater, **18**:327–328

of selective foams, **5**:346

of semitransparent solids, **11**:352–354, 359–362

of structured liquids

classification of complex liquids, **18**:164–166

experimental techniques, **18**:166–168

basic philosophy, **18**:169

coaxial cylinder method, **18**:172

enclosed apparatus, **18**:173–174

hot-plate technique, **18**:168–172

hot-wire/thermal conductivity probe technique, **18**:175–176

linear heating technique, **18**:176

for sheared liquids, **18**:176–178

steady methods, **18**:168–174

transient methods, **18**:174–176

heat conduction principles, **18**:162–164

nonpolymeric liquids

Thermal conductivity (*continued*)
   biological liquids, **18**:209–211
   greases and foodstuffs, **18**:211–213
   suspensions, flowing media,
     18:200–209
   suspensions, stagnant media,
     18:178–200
   surface, dropwise condensation, **21**:88–89
   temperature dependence, **7**:9; **10**:51
   for terrestrial fines, **10**:50
Thermal constriction resistance, **9**:359
Thermal contact conductance
   defined, **20**:259–260
   enhancement of, **20**:262–271
     conductive greases, **20**:263–264
     metallic foils, **20**:263–266
     mold compound and substrate-spreader inter-
       face, **20**:262–263
Thermal contact resistance
   chip-bond and bond-aluminum interfaces,
     20:260–261
   chip design, **20**:259–261
   diebond material thickness, **20**:260–261
   electronic component cooling, **20**:258–271
Thermal control–electronic components
   boiling, **20**:243–244
   forced convection, **20**:220–243
     discrete flush heat sources–channel flow,
       20:220–226
     three-dimensional package arrays, **20**:232–
       243
       actual circuit card arrays, **20**:242–243
       fully populated arrays, **20**:232–238
       missing modules or height differences,
         20:238–241
     two-dimensional protruding elements chan-
       nel flow, **20**:227–232
   heat pipes, **20**:272–299
     direct cooling, **20**:289–295
     indirect cooling, **20**:273–289
     system-level thermal control, **20**:295–297
   immersion cooling, **20**:244–251
   jet impingement cooling, **20**:251–258
   natural convection, **20**:183–220
     discrete sources–enclosures, **20**:217–220
     discrete sources–liquid cooling, **20**:213–217
     discrete thermal sources–vertical surfaces,
       20:202–209
     irregular surfaces, **20**:209–213
     vertical plates and channels, **20**:185–202

     interacting convection and radiation,
       20:201–202
     optimum spacing, **20**:202
     staggered plate or card arrays, **20**:198–
       201
     vs. actual electronic circuit boards,
       20:195–198
   theoretical background, **20**:181–183
   thermal contact resistance, **20**:258–271
     conductance enhancement, **20**:262–271
Thermal convection
   effects of, **3**:107f
   governing equations in, **15**:145–150
   in horizontal layer of non-Newtonian fluid,
     15:206–213
   in horizontal tubes, **15**:196–197
   in internal flows, **15**:195–206
   laminar, *see* Laminar thermal convection
   in non-Newtonian fluids, **15**:143–223
   rheological conditions in, **15**:148–150
   turbulent, *see* Turbulent thermal convection
   in vertical tubes, **15**:197–206
Thermal convection experiments, difficulty of,
   15:215
Thermal coupling effect, **9**:324
Thermal creep (slip velocity), **7**:185
Thermal design
   methods of appropriate, **31**:304–320
   second-law analysis in, **15**:1–53
   of solar buildings, **18**:10
Thermal detonation model, steam explosion,
   29:137, 140–147
Thermal diffusion, **6**:344
Thermal diffusivity, **5**:41; **7**:14
   of lunar fines, **10**:68–70
   of lunar materials, **10**:40
   of lunar rocks, **10**:64–68
   measurement of, **9**:404
   ratio of, subcooling and, **19**:11
Thermal diffusivity measurements, **6**:332
Thermal disequilibrium, **31**:111
Thermal dispersion, natural convection, **20**:325
Thermal effectiveness, regenerator theory,
   20:135–136
Thermal efficiency
   active collectors, **18**:6–10
   instantaneous, **18**:7
   rejection temperature and, **12**:5
Thermal energy closure problem, **31**:72–74
   passive, **31**:81–82

Thermal energy conservation, augmented nucleate boiling and, **20**:78–79
Thermal energy equation, **31**:91
total, **13**:154–158
Thermal energy source number, **SUP**:56
Thermal entrance length, **SUP**:7, 50–51
for circular ducts, **SUP**:105, 115, 126, 133, 143, 146
for concentric annular ducts, **SUP**:318–319
for eccentric annular ducts, **SUP**:339
future research in, **SUP**:420
minimum heat transfer asymptote and, **15**:125
for parallel plates, **SUP**:171, 181, 193
for rectangular ducts, **SUP**:217
turbulent flow and, **15**:108
Thermal entrance problem, **SUP**:7, 15–16
Thermal entrance region, **5**:370; **9**:115
laminar heat transfer in, **15**:100
Thermal equilibrium
liquid in, **31**:116
local, **31**:4
principle of local, **31**:43
vapor in, **31**:115–116
Thermal expansion
seawater
at atmospheric pressure and various temperatures, **18**:336
at various temperatures and pressures, **18**:335
water
at atmospheric pressure for various temperatures and salinities, **18**:335
at various temperatures and pressures, **18**:334
Thermal expansion coefficient
calculation, **18**:359
water
pure and saline, **18**:332–337
at various temperatures and salinities, **18**:353
Thermal explosion theory, fire and, **10**:238
Thermal fragmentation, **29**:175–176
Thermal ignition
incipient, **2**:216ff
problems, **2**:213ff
Thermal instability, **8**:162, 165, 169, 212
in free convection MHD problems, **1**:287
for natural convection inside horizontal circular cylinders, **8**:223
in power law fluids, **15**:210

Thermal insulation systems, **15**:38–51
Thermal intermittency, **30**:11, 12, 17
Thermally developed flow, **SUP**:7; *see also* Fully developed flow; Laminar flow
analysis of, **12**:89–91
temperature problem for, **SUP**:13–15
Thermally developing flow, **SUP**:7, 15–16; *see also* Conjugated problems; Simultaneously developing flow
in annular sector duct, **SUP**:271–272
areas of future research, **SUP**:420
categories, **SUP**:15–16
in circular duct, **SUP**:99–138
in circular sector duct, **SUP**:267
in circular segment ducts, **SUP**:267
comparisons of solutions for, **SUP**:400–403
in concentric annular ducts, **SUP**:301–319
conformal mapping method for, **SUP**:75
in eccentric annular ducts, **SUP**:336–337
in elliptical ducts, **SUP**:251–252
finite difference solutions for, **SUP**:76, 102–103, 138, 139, 142, 145, 146, 148, 190, 192, 214–215, 219, 222, 268, 320
finite element method for, **SUP**:76
for longitudinal flow over circular cylinders, **SUP**:365
Monte Carlo method for, **SUP**:76, 171
in parallel plates, **SUP**:169–189
in rectangular ducts, **SUP**:213–219
in semicircular duct, **SUP**:267–269
separation of variables methods for, **SUP**:74
simplified energy equation for, **SUP**:75
temperature problem for, **SUP**:15–16
in triangular ducts, **SUP**:242–246
variational methods for, **SUP**:75
Thermal management, **31**:432
Thermal microscales, **30**:9–13
buoyancy-driven flow, **30**:13–19
forced flow, **30**:9–13
Thermal nonuniformity, **31**:265
Thermal parameter
of lunar materials, **10**:40, 70–72
of simulated lunar fines, **10**:68
Thermal penetration depth, **31**:144
Thermal pollution, from power parks, **12**:29
Thermal potential, **8**:23
Thermal power cycle
heat sinks and, **12**:4–5
heat transfer problem in, **12**:4–10

Thermal power plant, energy balance in, **12**:24–25

Thermal properties
    temperature dependent, **1**:32, 68ff, 90ff, 101f
    thermodynamic similarity, **20**:18–23

Thermal radiation, **12**:120–128; **27**:9–14; *see also entries under Radiation and Radiative*; Molecular gas band radiation
    Elsasser model of, **12**:134
    in gases, **5**:255
    Goody curve in, **12**:137
    Goody model of, **12**:134–135
    narrow-band properties of, **12**:133–138
    pressure broadening in, **12**:136
    radiant intensity and flux in, **12**:123–126
    vibrational quantum numbers in, **12**:144
    wide-band properties in, **12**:138–155

Thermal radiation heat transfer, molecular gas band radiation and, **12**:116

Thermal radiation measurements, lunar, **10**:12–35

Thermal radiation properties, **5**:253

Thermal radiation transfer, **5**:13
    probabilistic model of, **5**:13

Thermal resistance, **21**:87–88

Thermal stability, induced rotational flow, **21**:142

Thermal state variable, **31**:141

Thermal thickness, **4**:329

Thermal triode damping, **9**:83

Thermal variability factors (TVF), **4**:107f

Thermal wall resistance parameter, **SUP**:23, 55–56

Thermic conductivity probes, **18**:175–176

THERMIR experiments, **29**:147

Thermistor
    Boltzmann factor, **22**:362–363
    calibration, **22**:385–386, 411
    dissipation constant, **22**:383–384, 409–410
    electrical leakage, **22**:393–394
    impedance loading, **22**:407–408
    linearization, **22**:376–383
    spatial temperature gradients, **22**:399–402
    spatial temperature response, **22**:398–399
    surface measurements, **22**:403
    time constant, **22**:395–398

Thermistor radiometer, **4**:69f

Thermoacoustic perturbations, nonstationary heat transfer, **21**:46–47

Thermocapillary-driven flow, **30**:43; *see also* Microgravity

Thermocouple plug fabrication, **2**:55, 77

Thermocouples, **23**:288–290
    calibration, **22**:389–390
    characteristics, **22**:389–390
    coaxial, **23**:313, 314
    construction, **22**:386–387
    differential, **23**:336
    electrical leakage, **22**:393–394
    error analysis, **22**:390–393
    impedance loading, **22**:407–408
    instrumentation, **22**:389, 414–416
    Peltier effect, **22**:388
    reference, **22**:392–393, 414–416
    Seebeck effect, **22**:363, 388
    spatial temperature gradients, **22**:399–402
    spatial temperature response, **22**:398–399
    surface measurements, **22**:403
    thin-skin, **23**:314–316
    time constant, **22**:395–398
    vapor content and temperature pulsations on heating surface, **18**:259

Thermodiffusion ratio, **28**:356

Thermo/diffusocapillary convection, alloy solidification, **28**:287

Thermodynamic approach
    advantage of, **12**:270–271
    chemical potential and, **12**:263–268
    concepts and definitions in, **12**:261–263
    diffusivity and, **12**:270–271
    driving force in, **12**:263–268
    electrochemical potentials in, **12**:268–270
    Nernst–Einstein relation in, **12**:266–267
    practical vs. thermodynamic model and, **12**:270–271
    Stefan–Maxwell equation in, **12**:267–268

Thermodynamic coupling, temperature and solute concentration, **22**:291

Thermodynamic driving force, **12**:263–268

Thermodynamic equilibrium
    at a curved interface, **1**:198ff
    departures from
        in cryogenic systems, **5**:353
        due to radiation, **8**:246
    hohlraum and, **12**:127
    local, **11**:321; **12**:127
    in a plasma, *see* Plasma
    pumping rate in, **12**:127
    radiation in, **12**:120–122
    relaxation times in, **12**:127

Thermodynamic modeling, steam explosions,

29:135–137
Thermodynamic nonequilibrium
  effects of, on conduction–radiation interaction,
    8:277
  in gaseous radiation, **8**:273
  for radiation in carbon monoxide, **8**:274–276
  for radiation in diatomic gases, **8**:275
  for radiation in single band gases, **8**:274,
    275
Thermodynamic properties
  critical region, **7**:3
  of cryogenic fluids, **5**:327
Thermodynamic relations, **31**:10–11, 48–53
  in drying process, **13**:164–165
Thermodynamics, **8**:23; *see also* Second law
  first and second laws of, **15**:2
  second law of, **12**:4
Thermodynamic state, of two-phase flow,
    31:137–138
Thermodynamic system entropy generation re-
    duction in, **15**:10
Thermoelectric sensitivity, **23**:290
Thermoelectromagnetics, system fixed in stag-
    nant media, **21**:271–272
Thermoexcel, augmented nucleate boiling,
    20:74–77
Thermographic phosphors, heat flux measure-
    ments, **23**:324–325
Thermography, **22**:10, 11
Thermohaline convection, **14**:26–28
  heat- and mass-transfer
    porous model, **20**:326
    Soret diffusion, **20**:328–329
    stability chart, **20**:322
Thermohydraulic codes, hydrogen combustion,
    29:113–118
Thermomechanical models, Czochralski process,
    30:357–362
Thermometer, **22**:2, 3
Thermophoretic effects, **31**:442
Thermophysical characteristics, variational de-
    scription of heat transfer process, **19**:203–
    210
Thermophysical properties, **31**:229
  fluid, **21**:2–3
  reference values for lunar studies of, **10**:79–
    80
Thermopile
  circular foil gage, **23**:304
  layered gage, **23**:291

thin-film, **23**:296
Thermosolutal convection, **28**:239–243
  freckle formation, **28**:243–244
  gravity effects on, **30**:94
  macrosegregation, **28**:243
  magnetic damping, **28**:317
Thermosyphons, **5**:244; **9**:1, 3
  analytical systems, **9**:39
  applications of, **9**:3, 83
  circular open liquid metal systems, **9**:33
  classification of, **9**:3, 4
  closed loop, **9**:72, 93, 97, 104
  closed systems, **9**:49, 55, 57, 93
  condensing, **9**:97
  constant wall heat flux in, **9**:20, 35
  constant wall temperature in, **9**:102
  effects of rotation of, **9**:38, 65, 73
  evaporating, **9**:97
  exchange mechanism in, **9**:43
  favorable pressure gradients in, **9**:28
  filling of, **9**:86
  governing equations for, **9**:99
  inclination effects in, **9**:37, 39, 40
  inclined, **9**:67
  inclined open systems, **9**:38
  influence of filling on, **9**:78, 81
  influence of leading edge on, **9**:97
  instability in, **9**:34
  liquid metal, **9**:101
  liquid metals in, **9**:33
  maximum heat flux in, **9**:85
  mixed convection systems in, **9**:3, 75
  noncircular, **9**:36
  noncircular closed systems, **9**:46
  noncircular cross sections in, **9**:36
  noncircular open, **9**:36
  nuclear boiling in, **9**:84
  open, **9**:92
  optimal Prandtl number in, **9**:46
  perturbation methods for, **9**:39
  pressurization effects in, **9**:40
  rotating liquid metal, **9**:71
  rotating two-phase, NaK, **9**:80
  roughness effects in, **9**:79
  semiclosed, **9**:66
  stagnant region in, **9**:22, 49
  stagnation phenomena in, **9**:24
  two-phase, **9**:77, 93
  two-phase and critical states in, **9**:3
  two-phase flow in, **9**:78, 84

Thickness of film
  slug flow calculations, **20**:98
  slug pressure drop, **20**:95
Thin disk of finite diameter, axisymmetric
    potential flow, **20**:373–374
Thin film
  evaporation, direct cooling with heat pipes,
    20:293–294
  narrow space nucleate boiling, bubble beha-
    vior, **20**:37–38
Thin-film surface thermometer, **21**:89
Thin-fin assumption, **2**:434
Thin fins, *see* Longitudinal thin fins
THIRBLE (Transfer of Heat in Rivers, Bays,
    Lakes and Estuaries), **13**:61–114
  classification in, **13**:65–69
  fluid motion in, **13**:63
  general case of, **13**:68
  GENMIX and, **13**:69, 72
  interface-transfer problem in, **13**:64
  jet mixing phenomena in, **13**:65–66
  laws and models for, **13**:62–63
  one-dimension unsteady vertical distribution
    models in, **13**:89–98
  scientific components of, **13**:62–63
  steady axisymmetrical jets in, **13**:65–66
  subjects related to, **13**:63
  three-dimensional steady jets in, **13**:67
  two-dimensional floating layers in, **13**:98–113
  two-dimensional parabolic phenomena in,
    13:70–72
  two-dimensional steady boundary layers in,
    13:66, 79–88
  two-dimensional steady jets and plumes in,
    13:73–79
  warm-water layer problems in, **13**:95–98
  zero-dimensional (stirred-tank) problems in,
    13:65
THIRMAL model, steam explosions, **29**:151
Thixotropic fluids, **2**:359f; **24**:107
  flow curves for, **15**:63
Three-dimensional axisymmetric surface, lami-
    nar thermal convection from, **15**:174–176
Three-dimensionalities, **24**:310–311
Three-dimensional package arrays
  actual circuit cards, **20**:242–243
  forced convection, **20**:232–243
  fully populated arrays, **20**:232–238
  missing modules or height differences,
    20:238–241

Three-dimensional processes, in THIRBLE
    classification, **13**:67–68
Three-dimensional systems
  industrial gas-fired furnaces, **27**:196–199
  packed spheres, **27**:193–195
Three-dimensional transport, single-screw extru-
    der, **28**:164–170
Three-equation model, **31**:108
Three-fluid heat exchanger, **26**:219–223, 319–
    321
  design, **26**:304–321
    log-mean temperature difference method,
      26:306–309, 320–321
  dimensionless groups, **26**:234–237
  dimensionless inlet temperature, **26**:268
  effectiveness, **26**:290–304
  energy equations, **26**:258–264
  extended surface plate fin exchanger, **26**:227,
    228
  heat capacity rate ratio, **26**:268
  inlet temperature, **26**:267, 268
  literature review, **26**:237–253, 255–256
  macrobalance equations, **26**:263
  mathematical model, **26**:231–237, 256–269
    solution, **26**:269–290
  microbalance equations, **26**:258–263
  temperature cross, **26**:274, 276–278
Three Mile Island, Unit 2 (TMI2), nuclear
    accident, **29**:1, 3, 59–60, 235
Three-phase fluidized beds, **23**:253
  axial dispersion, **23**:257
  bed expansion, **23**:253
  gas holdup, **23**:255
  heat transfer, **23**:261
  liquid holdup, **23**:257
  mass transfer, **23**:259
Three-phase multicomponent systems, in porous
    media, **30**:184–185
Three-phase system, **31**:84
Threshold, injury, **22**:316
Throttle valve, **25**:133, 134, 136
Thyristor
  heat-pipe cooling of, **20**:281, 283
  thermal control, **20**:181
Time
  dimensionless, **19**:20, 21; **25**:113
  filtration, **25**:233
  integral dimensionless, **25**:99
  relaxation, **24**:41, 109
  residence, hear carrier, **25**:106

retardation, **24**:109
stabilization, **25**:115
Time constant, **22**:279, 369, 395–398
    analysis for radiation, **22**:285–287
    thermal, **22**:283
Time-dependent fluids, **15**:61; **24**:106, 107
Time derivative of the volume average, **31**:15
Time-independent fluids, **24**:106, 107
Time-of-flight measurements, kinetic energies
    during laser ablation, **28**:112–116
Time rate of change of volume-averaged pres-
    sure, **31**:117–118
Time scales
    diffusive, **24**:285
    heatup, **24**:285
    sidewall boundary layer, **24**:285
Tin-coated stainless-steel, thermal contact con-
    ductance enhancement, **20**:268–270
Tin droplets, fragmentation, **29**:174
Tin–lead alloy, **28**:236, 239
    macrosegregation, **28**:291–292, 295–304, 317
    solidification, **28**:288–304, 309, 322
Tin–nickel alloy, thermal contact conductance
    enhancement, **20**:268
Tissue
    absorption of light, **22**:279
    burn, **22**:228
    coagulation, **22**:277, 319
    combustion, **22**:321
    convective cooling, **22**:267, 276
    denaturation, **22**:280
    desiccation, **22**:319
    Joule heating, **22**:243
TIZ, *see* Temperature interaction zone
Toluene, soil contamination by, **30**:181–182
Tomographic reconstruction, speckle photogra-
    phy, **30**:287, 288, 293–296, 306
Tom's effect, **12**:77; *see also* Drag reduction
Topology, of body in electrical field, **22**:243
Torque transport, induced rotational flow,
    21:161–162
Tortuosity channel, **25**:183, 201
Tortuosity effect, **31**:36
Tortuosity factor, **24**:110; **25**:298, 308, 314
    packed beds, **23**:195–196
Total absorption coefficient, **27**:149
Total band absorptance, **8**:234, 235, 241, 243
    general correlations for, **8**:244, 245
    limiting forms of, **8**:239
    limiting solution of, by rigid-rotor, harmonic-

        oscillator band model, **8**:242, 243
    limiting values of, **8**:235, 236
    square root limit of, **8**:239
Total band absorption, for Elsasser band model,
    8:238, 239, 240
Total band intensity, **8**:233
Total emissivity, **5**:306
    of air, **5**:309
    of carbon dioxide, **5**:308
    of carbon monoxide, **5**:308
    of hydrogen, **5**:308
    of metals at low temperatures, **5**:474
    of vacuum deposited metallic coatings, **5**:475
    of water vapor, **5**:307
    of water vapor–carbon dioxide mixtures,
        5:308
Total enthalpy, **2**:116, 207
Total enthalpy transport equation, **31**:10
Total extinction coefficient, **9**:368
Total heat transfer in film boiling from a
    cylinder, **5**:63
Total hemispherical emissivity, **5**:14
Total interlayer heat transfer model, **21**:315–316
Total mass density, **31**:7
Total momentum equation, **31**:8
Total package thermal resistance, direct cooling
    with heat pipes, **20**:293–294
Total radiant energy, **3**:181f
    from isothermal gas volume, **27**:15–16
Total radiant energy density, **3**:188
Total thermal energy equation, **13**:154–158
Total Variation Diminishing (TVD) schemes,
    24:240, 241, 242, 243, 244, 245, 246
Trailing function, **8**:25
Transducers, **8**:292, 293
Transference numbers, **12**:246
Transfer equations, specular moments, **21**:246
Transformation coordinate, **19**:16–24
Transformation of boundary layer equations into
    incompressible form, *see* Boundary layer
    equations
Transient condensation of cryogenic fluids,
    5:354
Transient conductance, semiconductor melting,
    28:81–87
Transient conduction–radiation, in semitranspar-
    ent solids and participating media, **11**:407–
    411
Transient convection, nuclear reactors,
    29:27–31

Transient equilibrium approach, heat- and mass-transfer, **20**:341–343
Transient flow, **24**:167
Transient flow field, heat- and mass-transfer, **20**:344–346
Transient flow response, **9**:311
Transient heat exchanger problem one fluid, **1**:64f
Transient heat transfer
   in gray material heated by diffuse incident flux, **11**:381
   in gray solid heated by collimated flux, **11**:377–379
   and irradiation of semitransparent solids, **11**:385–388
   in nongray solid heated by collimated flux, **11**:379–381
   in semiinfinite solid, **11**:376–381
   in semitransparent plate, **11**:381–385
   in semitransparent solids, **11**:375–390
Transient internal flows, **5**:390
Transient laminar thermal convection, from flat vertical plate to Bingham plastic fluid, **15**:177–178
Transient multiphase, multicomponent flows, **30**:166–167
Transient-over-power accidents, nuclear accidents, **29**:1, 2
Transient radiation problems, **5**:36
Transient response, **9**:314
Transients, **9**:320
   delay of amplification in, **9**:334
   vigorous, **9**:334
Transition, **8**:97, 141; **9**:10, 17, 19, 27, 28, 42, 321
   natural, **9**:332
Transitional flow, **25**:392, 394, 406
Transition boiling, **5**:55, 56, 102; **23**:8–9, 108–117
   array of circular water jets, **23**:112–113
   boiling curve, **23**:111
   bubble pattern in, **5**:113
   circular, confined jet, **23**:117
   departure from film boiling, **18**:242
   departure from nucleate boiling, **18**:242
   effect of additives in, **5**:109
   effect of agitation on, **5**:109
   experimental methods
      heat flux measurement, **18**:257
      heating methods, **18**:254–257

for physical mechanisms, **18**:258
      wall temperature measurement, **18**:257–258
   experimental studies, **18**:260–263
   free-surface
      circular jet, **23**:112–113
      planar jets, **23**:116
   heating surface orientation effects, **18**:282–285
   heating surface roughness effects, **18**:273–274
   on horizontal cylinders, **5**:113
   from horizontal surfaces, **5**:111
   influence of pressure and subcooling on, **5**:428
   investigations, **23**:109–110
   mathematical models
      based on processes causing liquid–superheated wall contacts, **18**:247–253
      heat removal due to unsteady heat conduction, **18**:244–246
      liquid film drying, **18**:247
      physical mechanisms at heating surface sections, **18**:253–254
   mechanism, **18**:243–244
      acoustic methods, **18**:265–267
      heating surface temperature pulsations, **18**:270–273
      steam quality in wall region, **18**:259, 264
      vapor content probes, **18**:267–270
      visual observation, photography, and optical methods, **18**:258
   planar, free-surface jet, **23**:108
   research needs, **23**:123
   subcooling, **23**:108, 112
   wetted region size, **23**:113
   wetting temperature, **23**:115
   zones on wetted part of heating surface, **18**:243
Transition boiling heat transfer
   data
      empirical methods of generalization, **18**:307–311
      semiempirical methods of generalization, **18**:311–318
   effects of
      ambient conditions, **18**:302–306
      contamination, **18**:295–296
      heating surface
         characteristics, **18**:290–298
         materials, **18**:293–295
         nonisothermality, **18**:301–302
         roughness, **18**:290–292

thermal activity, **18**:296–298
wettability, **18**:264–265, 292
liquid subcooling, **18**:304–306
pressure, **18**:306
process unsteadiness, **18**:298–301
Transition flow, **5**:365
Transition from annular-dispersed to slug flow,
1:363ff
from slug to annular-dispersed flow, **1**:359ff
Transition regime, for natural convection inside
rectangular cavities, **8**:181
Transitions
bound–bound, **5**:263
bound–free, **5**:263
free–free, **5**:263
Translational velocity
slip parameters, **20**:105–106
slug flow, **20**:95–97
Taylor bubbles, **20**:100–107
liquid slug dispersion, **20**:107–108
Translation region, twin-screw extruder, **28**:190–
192
Transmissive power, **27**:11
Transmissivity, **27**:11
Transmittance, in semitransparent solids,
11:366–371
Transmitted light imaging, phase saturation
measurement by, **30**:96
Transparent or optically thin approximation, **5**:30
Transpiration, **8**:32
effect on boundary layer, **16**:272–282
Transpiration cooling, **4**:269f; **7**:322, 323
Transpiration radiometer, **23**:306–307
Transport
laws of, **13**:63
mass, *see* Mass transport
Transport coefficients, for supporting electrolytes, **12**:241
Transport equations, **27**:21
conjugate problem, **26**:22
general, **12**:231–244
in ordinary diffusion, **12**:204
solutions
two flux method, **27**:23–31
zone method, **27**:31–42
Transport phenomena, *see also* Electrochemical
transport
alloy solidification, **28**:238–248, 268–269,
326
magnetic field, **28**:318–319

semitransparent analog alloys, **28**:270–288
chemical vapor deposition, **28**:346–353, 362–
365
in the Czochralski process, **30**:315–318, 325,
326, 403
convection, **30**:362–423
melt flow, **30**:362–370
models, **30**:326–362
simulations, **30**:347–352
direct containment heating, **29**:275–281
screw extrusion of polymers, **28**:149–151,
213–226
chemical reaction and conversion, **28**:218–
220
in dies, **28**:205–212
moisture transport, **28**:213–218
single-screw extrusion, **28**:147, 149, 155–
187
tapered screw extruders, **28**:176, 225
twin-screw extrusion, **28**:151, 187–201,
225–226
spray deposition, **28**:1–3, 22
impact region, **28**:23–69
spray region, **28**:3–22
Transport problems, defining approximating
profiles in, **31**:138
Transport properties, **2**:36ff
critical region, **7**:3
of cryogenic fluids, **5**:328
Transport properties of dilute gases, **3**:253ff
concentration diffusivity, **3**:283ff, 287ff
experimental data, **3**:287ff
experimental techniques, **3**:283ff
dissociated gases, methods for, **3**:286f
Loschmidt, method of, **3**:283f
point source flow method, **3**:285f
Stefan, method of, **3**:284f
theory, **3**:290ff
binary diffusion coefficients, **3**:290ff,
296
from data on binary mixtures viscos-
ities, **3**:296ff
for potential functions of
Buckingham, **3**:291
exponential repulsion, **3**:290
inverse power repulsion, **3**:290
Lennard–Jones, **3**:291
dissociated gases, **3**:297f
multicomponent diffusion, **3**:299
polar gases, **3**:298

Transport properties of dilute gases (*continued*)
   potential parameters from
      diffusion, **3**:291ff
      viscosity, **3**:293, 295
         using potential functions of
            Buckingham, **3**:295
            exponential repulsion, **3**:295
            inverse power repulsion, **3**:295
            Lennard–Jones, **3**:295
   thermal conductivity, **3**:272ff
      experimental data, **3**:276ff
      experimental technique, **3**:272ff
         concentric cylinder method, **3**:273
         dissociated gases, methods for, **3**:276
         line source flow method, **3**:274
         long hot wire method, **3**:274f
         Prandtl number method, **3**:275f
         shock tube method, **3**:275
      theory, **3**:278ff
         mixtures, **3**:282f
            inert, **3**:282
            reacting, **3**:282f
         pure gases, **3**:278f
            monatomic, **3**:278
            polar, **3**:281
            polyatomic nonpolar, **3**:278f
  viscosity
      experimental data, **3**:259ff
      experimental technique, **3**:255ff
         capillary flow method, **3**:255f, 260
         dissociated gases, methods for, **3**:257f
         oscillating disk method, **3**:256f, 260
      theory, **3**:261ff
         Chapman–Enshog theory, **3**:261
            using potential functions of
               Buckingham, **3**:263
               exponential repulsion, **3**:262
               inverse power repulsion, **3**:262
               Lennard–Jones, **3**:262f
         labile atoms and radicals, **3**:269ff
         mixtures, **3**:267ff
         pure gases, **3**:261ff
            nonpolar, **3**:261ff
            polar, **3**:265ff
Transport theorem, **14**:8
  general, **31**:14–15
Transverse curvature, **9**:303
  potential flow convection, **20**:370–371
Transverse fins, in tube bundles in longitudinal
    flow, **31**:249–253

Transverse flow, heat transfer enhancement in,
  31:263–264
Transverse grooves, as turbulators, **31**:181
Transverse pitch, effect on velocity distribution,
  18:104–105
Trapezoidal ducts, **SUP**:256–259
Trapezoidal rule, **2**:422
  unidirectional regenerator operation, **20**:154
Trapezoid heat pipe configuration, **20**:283–284
Trapped gases, vapor bubbles and, **10**:119
"Traversing" method for measuring density in a
  conduit, **1**:398f
Triangular channels, **8**:44
  heat transfer enhancement in, **31**:259–263
Triangular ducts, **SUP**:223–246
  arbitrary, **SUP**:237–240
  equilateral, **SUP**:223–225, 240–243
    with rounded corners, **SUP**:225–226
  fully developed flow in, **SUP**:223–240
  hydrodynamically developing flow in,
    SUP:240–242
  isosceles, **SUP**:227–235, 240–242, 244–246
  Nusselt numbers for, **SUP**:224–226, 230–231,
    234–239, 243, 245
  right-angled, **SUP**:233–237, 241–242, 244
  thermally developing flow in, **SUP**:242–246
Triangular spaces, in solar buildings, **18**:50–51
Tribology, **24**:1–38
Trichloroethylene, soil contamination by, **30**:182
Trickle bed reactors, convection in, **30**:168
Triggering, steam explosion, **29**:132, 158–159,
  167–168
  integral experiments, **29**:164–166
  prevention, **29**:166–167
  single droplet experiments, **29**:163–164
  vapor film collapse, **29**:132, 159–163
Triple correlations, **8**:289
  of temperature and velocity components,
    8:305
Triple-pipe three-fluid heat exchanger, **26**:227,
  229
Triple point
  gallium, **22**:364
  rubidium, **22**:364
  succinonitrile, **22**:364
  water, **22**:364
Tripping wires, **3**:16f, 25
Trip-wire experiments, **13**:41
Trombe walls, natural convection, **18**:72–75
True transference number, **12**:246

Truncated sectorial duct, **SUP**:269
Tube banks
  calculation of hydraulic resistance in, **8**:154
  closely spaced geometries, **8**:147
  combined free and forced convection, **8**:140
  comparison of available heat transfer results
    in, **8**:148
  comparison of heat transfer to staggered and
    in-line tube banks, **8**:142, 148
  comparison with single tubes, **8**:134
  critical flow in, **8**:152
  drag in, **8**:106, 111
  effect of roughness of heat transfer in, **8**:145
  efficiency of, **8**:148
  flow of gases in, **8**:95
  flow of viscous liquids in, **8**:95
  flow pattern in, **8**:105
  flow regimes in, **8**:107
  heat transfer in, **8**:94, 112, 141
  heat transfer to
    at high Reynolds numbers, **8**:145
    at low Reynolds numbers, **8**:140
    at subcritical or mixed flow conditions,
      8:142
  heat transfer to viscous liquids in, **8**:144
  hydraulic resistance in, **8**:94, 150
  hydraulic resistance of, **8**:150, 151
  influence of number of rows on heat transfer,
    8:156
  influence of pitch on heat transfer in, **8**:111,
    135, 142, 144, 146
  influence of Reynolds number in, **8**:106
  influence of tube pitch, **8**:108
  influence of turbulence intensity on heat
    transfer in, **8**:134
  in-line arrangement, **8**:105
  local heat transfer in, **8**:133, 135, 137, 138
  mean heat transfer in, **8**:138
  Nusselt number calculations for gas flow
    through, **8**:156
  optimal arrangement of, **8**:157
  physical properties, influence of, **8**:95
  potential flow through, **8**:111
  pressure coefficient in, **8**:107, 108
  relative efficiency of in-line and staggered
    tube banks, **8**:157
  staggered arrangement, **8**:105
  transition in, **8**:141, 152
  velocity distribution in, **8**:107
Tube bundles, **31**:263–264

annular-grooved in longitudinal flow, **31**:315–
    318
heat transfer enhancement in, **31**:182–185,
    232–259
inclined, **31**:277–278
in longitudinal flow, transverse fins in,
    31:249–253
Tube flow, **8**:22, 29
Tube length-mean heat transfer, **31**:191–193
Tube rows, **3**:7f
Tubes, *see also* Cylinders; Fins; Single tubes;
    Slanted single tubes; Slanted tubes; Vertical
    tubes
  in a bank
    crossflow, heat transfer calculations,
      18:154–156
    drag, **18**:106
    general, flow pattern, **18**:99–101
    velocity distribution, **18**:101–106
  bundles
    heat transfer from, **19**:154–161
    horizontal, **19**:159, 160, 181
    relative pitch, **19**:160
  diameter, **19**:126–129
  finned
    horizontal, 60 degree V-thread, **19**:151
    immersed in gas-fluidized beds, heat trans-
      fer investigation summary, **19**:147–150
    relative efficiency of heat transfer, **19**:147
  in fluidized beds, **10**:209–210
  helical, bundle, **25**:114
  horizontal
    heat transfer coefficients, **19**:156
    rough, heat transfer rates between fluidized
      bed, **19**:146–147
    v. staggered tube bundles, **19**:157
  horizontal bundles, **19**:181
    in-line, **19**:159
    staggered, **19**:160
  horizontal smooth cylindrical, **19**:180
  natural convection in, **8**:163
  pitch, **19**:159
  steel, **25**:53, 104
  vertical
    heat transfer to and from immersed surface,
      19:129–135
    rough, immersed in fluidized beds, **19**:147
    smooth cylindrical, **19**:180
Tubes, single
  drag, **18**:95–97

Tubes, single (*continued*)
  heat transfer, **18**:110–131
    in gas flow, **18**:115–118
    local transfer, theoretical calculations, **18**:110–115
    mean transfer, **18**:122–129
    mean transfer calculations, **18**:127–129
    restricted channels, **18**:129–131
    viscous-fluid flow, **18**:118–122
  ideal fluid flow, **18**:89–99
  local heat transfer, **18**:110–122
    in gas flow, **18**:115–118
    theoretical calculations, **18**:110–115
    in viscous-fluid flow, **18**:118–122
  real fluids, general flow pattern, **18**:90–92
  restricted-channel flow, **18**:97–99
  velocity distribution around, **18**:94–95
Tubular Heat Exchanger Manufacturers Association (TEMA), **31**:464–467
Tubular three-fluid heat exchanger, **26**:224–227
Tumor, **22**:290, 300
Turbine, blade, **25**:127
Turbine blade cooling, **9**:97, 104
Turbine cooling, **9**:91
Turbo-B® structured surface, heat transfer enhancement with, **30**:234
Turbulator height, **31**:312
Turbulator pitch, **31**:312
Turbulators, **9**:118; **31**:246
  choosing optimal parameters of, **31**:319–320
  equidistant smoothly outlined, **31**:180
  far-spaced, **31**:196–197
  forced convection, **20**:238–241
  increasing height of, **31**:208–209
  influence of artificial in forcing growth of critical Reynolds number, **31**:189–191
  influence of relative position of, **31**:172
  influence of shape of, **31**:213–219
  interaction of artificial with varying Prandtl number, **31**:208
  method of making, **31**:214–219
  pitch of, **31**:202
  shape of, **31**:202
  smoothly outlined, **31**:178
  transverse equidistant protrusions used as, **31**:183
  transverse grooves as, **31**:181, 184–185
Turbulence, **8**:285; **9**:16, 34
  alloy solidification, **28**:318–325, 327
  effects of acceleration on, **2**:67f, 72ff

  energy diffusion in, **8**:289
  energy dissipation in, **8**:289
  energy generation in, **8**:289
  experimental methods in, **8**:292
  flame propagation and, **29**:73, 74, 81
  flammability limits, **29**:106
  fluid flow in single tubes and, **18**:97
  free-stream, effect of on heat transfer, **3**:1ff, 8f, 28
    blunt body, **3**:5
    cylinder in cross flow, **3**:4ff, 16ff
      flow patterns for
        critical flow, **3**:5f
        subcritical flow, **3**:5f
        supercritical flow, **3**:5f
      measurement near stagnation point, **3**:19ff
    flat plate in parallel flow, **3**:13f
      with
        favorable pressure gradient, **3**:24ff
        moderate, favorable pressure gradient, **3**:23f
        zero pressure gradient, **3**:21ff
    flat plate in perpendicular flow, **3**:16
    local curvature effects, **3**:2
    sphere in uniform flow, **3**:4, 10
    tube rows, **3**:7f
    viscosity effects, **3**:30f
  generation, **9**:22; **25**:23; **31**:168
    distance variations, **31**:168–169
    mechanism, **25**:21
  in grids, **8**:291
  influence of, on heat transfer, **8**:290
  kinetic energy of, **8**:288, 289
  length scales in, **8**:289
  models, **25**:331, 340, 346–347, 389, 401
    algebraic Reynolds stress, **25**:347
    k–ε, **25**:340, 389, 401
    Reynolds stress transport, **25**:346
  near walls, **8**:291
  onset of, **9**:20
  parameters, asymptotic, **21**:212
  pulsation generation, **25**:41
  radiation-affected, entropy production, **21**:268–270
  radiative decay and, **14**:263–268
  and rotation, **25**:381, 384
  scales, **25**:325
  secondary flow, **25**:403
  spectrum, **25**:335–336, 341, 344

statistical model of, **8**:290

as a statistical phenomenon, **8**:286

theory, **25**:341, 344, 345, 355

transition to, **25**:356, 392, 394

wave approach to study of, **8**:344

Turbulence–energy equation, **13**:91

Turbulence–energy production, **8**:303

Turbulence intensities, **3**:3ff; **8**:122, 138

comparison of, for isothermal and nonisothermal turbulent flows, **8**:323

influence of

on heat transfer, **8**:138, 140

in hot-wire anemometry, **8**:295

measurement of

in pressure gradient flow, **8**:308, 309

for turbulent channel flow, **8**:321

Turbulence intensity distributions, experimental
measurements of, **8**:302

Turbulence level, influence of, on flow pattern,
8:103

Turbulence models, for Czochralski flows,
30:329

Turbulence promoters, **9**:117

in heat and mass transfer, **13**:41

Turbulent

energy, **25**:23

kinetic, **25**:75

flow, unsteady, **25**:7

structure, **25**:19, 21, 61, 72, 74

heat flux, **25**:7

Prandtl number, **25**:18, 26

pulsation intensity, **25**:75

stress, **25**:7, 23

viscosity, **25**:75

relative, **25**:80

Turbulent boundary layers, **3**:34ff; **8**:35

compressible on flat plate, **3**:52ff

drag, **3**:52ff

heat transfer, **3**:55ff

cross flow effects, **3**:63f

in entrance region of a tube, **3**:41f

gas injection, **3**:74f

heat transfer and, **11**:12

permeable surfaces, **3**:62ff

tube, **3**:77ff

shear stress, **3**:45, 49f

Turbulent boundary-layer theory, **19**:145

Turbulent burning rates, **29**:78–82

Turbulent burning velocity, **29**:78–79

Turbulent bursts, **9**:334

Turbulent convection, *see* Liquid metal heat
transfer

Turbulent decay, **2**:90ff

Turbulent flame

forced flame, **30**:60–65

natural flame, **30**:73–78

Turbulent flow characteristics, **2**:384f

Turbulent flows

analysis of structure of, **31**:205–208

analytic study of, **15**:120–126

comparison of nonisothermal and isothermal
conditions, **8**:323

critical region, **7**:22

dimensional approach, **30**:1–3

fluid mechanics

Newtonian fluids, **19**:323–325

non-Newtonian fluids, **19**:325–334

free-surface jet, **26**:204–205

axisymmetric, **26**:153–154, 157

planar, **26**:146, 162–170

frictional drag of, **12**:77

heat flux, **7**:22, 52

heat transfer, **11**:295–296; **26**:162–170; **30**:51,
82–83

Newtonian fluids, **19**:322–323, 334–339

non-Newtonian fluids, **19**:339–343

heat transfer measurements in, **15**:113

historical development of, **15**:104–108

induced rotational flow, **21**:153–155

and intermediate heat transfer in once-through
system, **15**:112–114

for isothermal, incompressible, zeropressure,
gradient conditions, **8**:299

laser speckle photography, **30**:283–293

in MHD, **1**:350f

microgravity, **30**:46–49

microscales, **30**:4–5

heat transfer, **30**:19–51

kinetic scales, **30**:5–9

mass transfer, **30**:51–85

thermal scales, **30**:9–19

minimum heat transfer asymptote in, **15**:111–
112

momentum equations for, **8**:288

for nonisothermal conditions, **8**:305

packed beds, **23**:212

prediction methods for, **8**:287

recent experimental studies in, **15**:108–120

on rotating disk, **5**:15

shear stress, **7**:22, 55, 70

Turbulent flows (*continued*)
  simultaneous measurement of velocity and
      temperature in, **8**:310
  skin friction, **30**:82–83
  statistical analysis of, **8**:335
  submerged axisymmetric jet, **26**:132–133
  supercritical pressures, **21**:6–17
    boundaries in absence of Archimedes forces
        effects, **21**:10
    Boussinesq relation, **21**:7
    degraded heat transfer regime, **21**:11–12
    eddy viscosity, **21**:13
    friction factor, **21**:11–12, 15
    models, **21**:12–13
    Nusselt number, **21**:16
    relations for turbulent transport coefficients,
        21:7–8
    relative change in heat transfer coefficient,
        21:16
    Stanton number, **21**:14–15
    turbulent momentum transport coefficient,
        21:8
    upstream flow history effects, **21**:13–14
  Taylor bubble velocity, **20**:105
  transition from laminar flow, **11**:301–303
  two-phase, **30**:39–43
  unsteady, **30**:78–80
  velocity and temperature profiles in, **8**:334
  vertical surface, supercritical pressures, **21**:39–
      42
Turbulent flow transition region, heat transfer
    enhancement in, **31**:187–196
Turbulent fluctuations, **8**:287
Turbulent free convection problems, approximate
    solutions to, **11**:265–310
Turbulent heat flux, **8**:290
Turbulent heat transfer
  by free convection, **11**:295–306
  in degrading polymer solutions, **15**:115–120
  for drag-reducing viscoelastic fluids, **15**:112
  at heat flux varying lengthwise, **21**:32–34
  solvent and solute effects in, **15**:114–115
  in thermal entrance length, **15**:108–111
  under cooling, **21**:34–37
    friction factor, **21**:36
    velocity profiles, **21**:35
Turbulent heat transfer rates, **2**:385f
Turbulent–laminar flow solutions, **11**:303–306
Turbulent mixed convection
  balance equation of turbulent stresses, **21**:24

  buoyancy parameters, **21**:24–25
  hydraulic drag, **21**:28
  inertia factor, **21**:28–29
  supercritical pressures, **21**:17–32
    coefficient of turbulent viscosity, **21**:23
    density fluctuations, **21**:23
    horizontal pipes, **21**:30–32
    Nusselt number, **21**:22
    parameter $\tilde{k}$, **21**:20
    Stanton number versus buoyancy para-
        meters, **21**:17–18
    thermogravity effect, **21**:19–20
    velocity profiles, **21**:22
    vertical pipes, **21**:17–30
    wall temperature distributions, **21**:21
  threshold acceleration parameter, **21**:29
  upward water flow, **21**:25–26
  velocity profiles, **21**:28
Turbulent motions, water–air interface and, **13**:64
Turbulent nearly homogeneous scalar field,
    evolution, **21**:214–231
Turbulent pipe flow, mass transfer in, **15**:128–
    129
Turbulent Prandtl number, **8**:290
Turbulent pulsations, energy of, **31**:168
Turbulent stresses, balance equation, **21**:24
Turbulent thermal convection
  in external flows of non-Newtonian fluids,
      15:188–195
  for inelastic fluids, **15**:189–192
  relevant stress components for, **15**:180
Turbulent transport process modeling, **21**:185–
    232
  anisotropy degree, **21**:187
  correcting tensors, **21**:193
  fundamental parameters of turbulence, **21**:192
  governing equations, **21**:188–190
  nearly homogeneous two-point correlations for
      closely spaced points, **21**:191–193
  problem of evolution, **21**:202–231
    "mandolin" and "toaster", **21**:218–220
    nearly homogeneous velocity field model
        formation, **21**:213–214
    strong nearly homogeneous scalar field, *see*
        Strong nearly homogeneous scalar
        field
    strong nearly homogeneous velocity field,
        21:204–209
    turbulent nearly homogeneous scalar field,
        21:214–231

turbulent nearly homogeneous velocity field, **21**:203–214

weak nearly homogeneous scalar field, *see* Weak nearly homogeneous scalar field

weak nearly homogeneous velocity field, **21**:209–213

second-order model, nearly homogeneous turbulence, **21**:193–202

Turbulent vapor flow, in film-boiling heat transfer studies, **11**:95–105

Turbulent viscosity, **8**:288

Turbulization, **8**:305

artificial with weakly developed turbulent structure, **31**:189–191

heat transfer augmentation due to, **31**:212–213

heat transfer enhancement by artificial flow, **31**:267–268

of a vapor phase, **31**:267

Turning moment

bodies of revolution, **5**:137

influence of mass transfer, **5**:145

Twin-screw extruder, **28**:187–190

experimental results, **28**:197–201

heat transfer, **28**:151, 225–226

modeling, **28**:190–197

finite-difference approach, **28**:195–197

finite-element approach, **28**:193–195

translating and intermeshing regions, **28**:190–192

Twisted tapes, within circular ducts, **SUP**:366, 379–384

developing flows, **SUP**:384

fully developed flow in, **SUP**:379–384

fluid flow, **SUP**:379–381

heat transfer, **SUP**:381–384

Two-cell equilibrium (TCE) model, direct containment heating, **29**:218, 219, 233, 238–240, 301, 311, 320

Two-cell kinetic model, direct containment heating, **29**:233

Two-dimensional floating layers

hyperbolic layer in, **13**:106–109

partially parabolic layer in, **13**:109–111

radial pool and, **13**:111–112

steady nearly radial flow and, **13**:112

steady two-dimensional layer model in, **13**:98–106

in THIRBLE classification, **13**:98–113

unsteady layers in, **13**:112–113

Two-dimensional method, for hot water and

vapor-dominated geothermal reservoirs, **14**:83

Two-dimensional model, heat- and mass-transfer, **20**:329–330

Two-dimensional parabolic phenomena

auxiliary relations in, **13**:71–72

boundary conditions in, **13**:70–71

differential equation in, **13**:70

mathematical characteristics of, **13**:70–72

predictions for, **13**:72

in THIRBLE, **13**:70–72

Two-dimensional steady boundary layers, in THIRBLE classification, **13**:66

Two-dimensional steady boundary layers adjacent to phase interfaces

air boundary layer above a lake, **13**:79–82

boundary layer in water at lake surface, **13**:82–85

combined air–water layer for, **13**:85–88

Two-dimensional steady jet phenomena

predictions for, **13**:75

round jet in surrounding stream, **13**:73–75

in THIRBLE, **13**:73–79

and vertically rising warm-water plume in stratified surroundings, **13**:76–79

Two-equation model, **23**:370–371

arrays of cylinders, **23**:422

boundary conditions, **23**:430

closure, **23**:386

closure problems, **23**:390, 400, 404, 407

numerical methods, **23**:415

closure variables, **23**:390

comparison with one-equation model, **23**:445, 449

conductive cross-coefficient, **23**:397, 423

deviations, representation, **23**:389

dimensionless variables, **23**:431

finite-difference solution, **23**:434

Fourier series solution, **23**:433

heat transfer coefficient, **23**:396, 424

interfacial flux, **23**:385

nodular system, **23**:443, 456

numerical experiments, **23**:435

stratified systems, **23**:411, 447

symmetric unit cells, **23**:395, 398, 404, 410

transport coefficients, **23**:371, 396, 410

volume-averaged equations, **23**:395

Two-equation turbulence model, **24**:211, 264

Two-fluid heat exchanger, **26**:219–220, 231

effectiveness, **26**:290–291

Two-fluid interface, transfer problem in, 13:64

Two-fluid model, 31:109

Two-flux approximation, radiative heat transfer, 23:136–139

Two-flux model, 9:379

Two-layer slab, in linear diffusion region, 15:310–319

Two-phase binary systems, 5:348

Two-phase boiling boundary layer, temperature fluctuation in, 11:32–38

Two-phase boundary layer approximations, 30:143

Two-phase films
  microscales, 30:33–34
    dimensionless number, 30:34–37
    laminar two-phase flow, 30:37–39
    turbulent two-phase flow, 30:39–43

Two-phase flow, 5:446; 9:78, 84
  cost of analyzing transient phenomena of, 31:107
  coupling permeability tensors for, 31:1
  and drift-flux parameters, 31:109–111
  flow patterns in, 5:452
  importance of, 31:105–106
  methods of analysis of, 31:106–108
  Nusselt numbers for, 5:446, 447
  one-dimensional, 30:114–121
  physics of, 31:106
  pressure drop in, 5:449
  selection of models for, 31:108–113
  systems for, 31:74
  thermodynamic state of, 31:137–138
  variables of importance, 1:356f

Two-phase Rayleigh number, 30:129, 144

Two-phase, single-component systems
  multiphase flow, 30:112–114, 161–162
    external boiling flows, 30:148–152
    external condensing flows, 30:141–148
    internal boiling and forced convection, 30:136–141
    internal boiling and natural convection, 30:122–136
    one-dimensional, 30:114–121
    steam injection, 30:152–161

Two-phase slug flow characteristics, 20:83–84

Two-phase specific heat, 30:36

Two-phase theory, 25:152

Two-point correlations, closely spaced points, 21:191–193

U

UC High Flux, augmented nucleate boiling, 20:75–77

UFT, see Unsaturated flow theory

Ultrashallow doping profiles, 28:96–101

Ultrasound treatment, 31:459–461

Ultraviolet treatments, 31:459–461

Unbalanced regenerators, temperature distribution, 20:161–170

Unbalanced-symmetric regenerator
  unidirectional operation, 20:174, 176–177
    effectiveness plot, 20:176–177

Unbalanced unidirectional regenerator, 20:170–173
  dimensionless parameters, 20:171
  effectiveness charts, 20:170–172

Unbounded liquid interfaces, direct condensation in, 15:233–242

Undercooling, 22:176

Unicellular motion, 8:185

Unidirectional flow, 13:213

Unidirectional regenerator operation, 20:149–176
  aperiodic or recuperative operation, 20:154–158
  balanced-symmetric regenerator, 20:173–175
  completely unbalanced regenerator, 20:170–173
  long regenerator–vanishing transfer potential, 20:158–170
  unbalanced-symmetric regenerator, 20:174, 176–177

Unidirectional solidification, alloys, 28:277–282

UNIDRE double precision function, 20:154

Unified n-layer horizontal model
  and linear diffusion in laminated media, 15:292–293, 297–298
  solution for, 15:304–306

Uniform heated surfaces, potential flow convection, 20:358–359
  conic, 20:368
  cylinder, 20:360–361
  elliptical cylinders, 20:362
  finite height strips, 20:363
  infinitely long pointed needle, 20:369
  planar motion, 20:356
  spherical, 20:370
  wedge, 20:364

Uniform heat flux surface condition, 9:325

Uniform heat sources, in radiating gases, **8**:254, 255
Uniform retardation model, **26**:12
Universal numerical shear flow program, **13**:234–238
Universal velocity law for suspension, **9**:116
Universal velocity profile, **5**:369
  effect of property variations on, **5**:369
Unresolved, *see* Subgrid
Unsaturated flow theory (UFT), **30**:102
  comparison with other multiphase flow models, **30**:107–110
  drying of porous material, **30**:169
  two-phase analysis, **30**:143, 145
Unsaturated porous media, convection in, **30**:167–168
Unstable fluidized bed, **25**:170
Unstable wavelength, dependence on applied electric field intensity, **21**:128–129
Unsteadiness, **25**:81
  hydrodynamic, **25**:74, 127
  parameter, **25**:47
  thermal, **25**:45, 127
    parameter, **25**:74
  velocity, **25**:95
Unsteady-state heat and mass transfer problems, variational solution methods, **19**:197–202
Unsteady wall layer heating, **25**:18
Unstretched coordinates, **8**:12
Upgraded heat transfer regime, **31**:229
Upwind, first-order upwind scheme, **24**:226, 235
Utilization factor, unidirectional operation, **20**:161–164

V

Valles Caldera, New Mexico, **14**:97
Valves
  magnetic, **25**:234
  throttle, **25**:133, 134, 136
Vanishing transfer potential, unidirectional operation, **20**:158–170
Vapor
  multicomponent film condensation, **21**:107–115
  physical explosions from, **10**:107–108
  single component film condensation, **21**:105–107
Vapor binding, **5**:99

Vapor blockage, **9**:88
Vapor bubbles, *see also* Bubble
  equilibrium vapor pressure and, **10**:87
  formation of, **10**:119
  growth and evanescence, **18**:252–253
Vapor chambers, heat pipe, **7**:223
Vapor condensation
  heat transfer enhancement with, **31**:278–287
  mixture on horizontal tubes, **31**:293–294
  with vapor–air mixture on vertical tubes, **31**:292–293
  with vapor mixtures on vertical surfaces, **31**:287–292
Vapor content
  probes, **18**:267–270
  and temperature pulsations on heating surface, measurement, **18**:259
Vapor dome, low liquid level nucleate boiling, **20**:61–63
Vapor embryo
  in conical cavity, **10**:116
  in nucleation, **10**:112–113
Vapor explosions, **29**:129
Vapor-filled cavities, nucleation and, **10**:122–127
Vapor film collapse
  experiments, **29**:132, 159–163
  modeling, **29**:160–163
Vapor film thicknesses, measurements of, **11**:60, 119
Vapor formation, explosive, **10**:87
Vapor–gas Schmidt number, **15**:240
Vapor generation rate, **31**:113
  nonequilibrium, **31**:141–142
Vaporization, **22**:280
  benzene, **26**:78
  compound drop, **26**:48–52
  convective droplet, **26**:58–59, 70–71
  n-decane, **26**:67, 78–79
  drag force, **26**:50
  droplet spray, **26**:87–89
  enthalpy of, **13**:157
  fuel spray, **26**:53–54, 83–84
  n-heptane droplet, **26**:72–73
  n-hexadecane, **26**:67, 70–71, 78–79
  n-hexane, **26**:56–57, 59, 67, 70
  moving liquid droplet
    electric field, **26**:83–84
    gaseous environment, **26**:52–54
    Reynolds number, **26**:59–77

Vaporization (*continued*)
  shape deformation, **26**:79–81
  multicomponent droplet, **26**:66, 77–79
  octane, **26**:78
  pulse-laser-induced, **28**:123–135, 136
  slurry fuel droplet, **26**:81–82
  stagnant medium, **26**:54–57
Vaporization temperature, pulse heating technique in, **10**:108
Vaporized mass diffusion fraction, **16**:85
Vapor–liquid interface, *see also* Condensation heat transfer
  augmented nucleate boiling, **20**:64–67
  electrohydrodynamic instability, condensation enhancement, **21**:126–131
  heat transfer coefficient, **21**:70
  hydrodynamic instability, **18**:248–250
  shape, **21**:117
  temperature at, **21**:66–67, 74–75
  transport at, **21**:65–60
    condensation coefficient determination, **21**:73–80
    interphase mass transfer theory, **21**:67–72
    modification of Schrage's theory, **21**:72–73
    Nusselt equation, **21**:66–67
Vapor–liquid phase equilibria, **16**:61–63
Vapor mass balance, integral formulation of, **31**:138–139
Vapor mass conservation equation, **31**:136–143
Vapor nucleus, **16**:64
  conical cavity, **20**:64–67
  pressure and stability, **20**:68–70
  reentrant cavity, **20**:67–70
  stability, **20**:64–70
  variation in vapor pressure, **20**:65–67
Vapor periods, heat transfer models, **20**:29–31
Vapor phase, nucleation from, **10**:117–127
Vapor-phase density of water, intrinsic average, **31**:4
Vapor phase–heating surface mean contact time correlation of experimental data, **18**:289
  with ethanol boiling, **18**:280
  with Freon-113 boiling, **18**:279
  with water boiling, **18**:281
Vapor-phase solutions, dilute, **31**:99
Vapor production rate in film boiling from a cylinder, **5**:63
Vapor slug, heat flux into, **11**:163–164
Vapor space condenser, direct cooling with heat pipes, **20**:289–290

Vapor temperature, dimensionless, **10**:130
Vapor-to-surface temperature differences, dropwise condensation, **21**:91
Vapor trapping in nucleate boiling, **1**:206ff
Vapor void distribution, profile for, **31**:139–141
Variable surface temperature, **8**:126
  rotating cone, **5**:135
  spinning bodies of revolution, **5**:133
Variable wall temperature, **8**:35, 163, 164
Variance, **5**:12
Variational calculus, **8**:32
Variational methods, **8**:22; **SUP**:66–67, 75; *see also* Galerkin–Kantorowich variational method
  based on local potentials, **8**:26
Variational principle, **25**:272
  Bateman's, **24**:40
  Biot's, **24**:2
  solution, **24**:39
  with vanishing parameter, **24**:40, 46
    canonical form, **24**:42–43, 45, 90
    general form, **24**:40, 46, 48
    for general hyperbolic equation, **24**:42
    for linear heat conduction problems, **24**:41
    modified form, **24**:42
    for nonlinear problems, **24**:44
Vector, particle magnetization, **25**:224
Vector machines, **31**:401–402
Vee-groove cavities, *see* Cavities
Vehicle thrust, influence of pressurization system on, **5**:380
Veil cooling, **5**:200
  experimental setup for, **5**:202
Velocity, **25**:269
  bubbling, **25**:166, 195
  characteristic, **2**:35
  comparison of various methods for determining values of, **8**:328, 329
  critical gas, **25**:158
  dimensionless, **19**:255
  dimensionless angular, **25**:314
  elastic waste, **25**:199
  fluid, **25**:152
  fluidizing, **25**:167, 203
  free-stream, **25**:289
  gas, **25**:151, 153, 161, 164
  gradient, **25**:302
  incipient gas, **25**:168, 170
  intermediate, **25**:176
  interstial, **25**:314

fluid, **25**:155, 175, 184, 190, 230
mass-bubbling, **25**:162
maximum, **25**:314
mean, **25**:314
  flowrate, **25**:8
  mean-mass, **25**:82
measurements of, **9**:137, 142
  by laser–Doppler method, **8**:298
  by visual techniques, **8**:297
  in gas–solid flows, **9**:137, 142, 159, 175
  near a wall in turbulent flows, **8**:314, 316
  simultaneous with measurement of tem-
    perature, **8**:309, 310
  in turbulent flows, **8**:292
  in turbulent pressure gradient flows, **8**:320
minimum
  bubbling, **25**:158
  fluidization, **25**:154, 162, 181
particle, **25**:204
power-law, **25**:13
  point, **25**:314
profile, **25**:48, 264, 302
pulsation, rms, **25**:78
superficial
  gas, **25**:161, 169, 180, 238
  mass flow, **25**:309, 313
tangential, **25**:271
terminal, **25**:152, 163
transition, **25**:168, 170, 179, 181, 184, 195
voidage wave, **25**:199
Velocity correction, **24**:200
Velocity distribution
  flow past a single tube, **18**:94–95
  flow past a tube in a bank, **18**:101–106
Velocity fields
  averaged in channel cross sections, **31**:170–
    171
  at entrance, in steady shear flow, **13**:235
  natural convection, **20**:316–317
  potential flow convection
    liquid metal convection, **20**:376–377
    rising bubble convection, **20**:379–382
Velocity fluctuation dissipation rate, isotropic
  turbulence, **21**:186
Velocity gradient, liquid jet stagnation zone,
  26:109–110, 120
Velocity measurements, electrochemical method,
  7:144
Velocity of sound, **7**:8
Velocity overshoots, **SUP**:91–92

Velocity problem, **SUP**:7–10
  complete solution for, **SUP**:419
  generalized solutions for, **SUP**:44
  for hydrodynamically developed flow, **SUP**:7–
    8
  for hydrodynamically developing flow,
    SUP:8–10
Velocity profiles, 3:43f; *see also* Bodoia velocity
    profile; Langhaar velocity profile; Schiller
    velocity profile; Schlichting velocity pro-
    file; Sparrow's velocity profile; Velocity
    overshoots; Wang and Longwell velocity
    profile
  comparison of, for isothermal and noni-
    sothermal turbulent flows, **8**:323
  comparison of various methods for determin-
    ing values of, in turbulent flows, **8**:334
  experimental measurements of, in pressure
    gradient flows, **8**:307
  for free convection inside horizontal cylinders,
    8:167
  heat transfer and, **15**:91
  liquid slugs, **20**:111–113
  logarithmic, 3:44
  measurement of, in turbulent pipe flow, **8**:303
  for natural convection in enclosed cavities,
    8:171
  for natural convection in rectangular
    cavities, **8**:176, 184–186, 191, 192
  for natural convection inside circular cylin-
    ders, **8**:170, 204, 205
  for natural convection inside horizontal circu-
    lar cylinders, **8**:209, 211, 217–219, 221,
    222
  for nonisothermal turbulent flows with pres-
    sure gradient, **8**:322
  universal, 3:43f, 171
Velocity spectra, **8**:346
Venus
  atmosphere of, **14**:265–267, 275
  radiative decay for, **14**:256
  thermosphere of, **14**:266
Vertical condenser tube, optimum fin geometry,
  21:117–119
Vertical cylinders, natural convection from,
  11:213–215
Vertical-distribution models, one-dimensional
  unsteady, **13**:89–98
Vertical fissures, injection of hot water along,
  14:60–61

Vertical flat plate(s), **24**:122, 127, 139, 142, 144, 167
convection heat transfer between, **11**:290–295, 308–310
laminar free convection in, **11**:272–274
Vertically eccentric spheres, convective heat transfer between, **11**:288–290
Vertically rising warm-water plume, **13**:76–79
boundary conditions for, **13**:77
expected results for, **13**:77–78
geometry and physics for, **13**:76
importance of, **13**:77
method adaptation for, **13**:77
Vertical needles, **9**:303
Vertical parallel plates, **8**:162
natural convection between, **8**:165, 166
Vertical permeable surface, free convection about, **14**:56–59
Vertical plates and channels, natural convection cooling, **20**:185–202
buoyancy-induced flow and heat transfer, **20**:185–187
card-on-board configuration, **20**:185, 187
comparison with electronic circuit boards, **20**:195–198
composite relations, **20**:194–195
convection–radiation interaction, **20**:201–202
experimental prototypes, **20**:196–197
isothermal wall flow rates, **20**:191–192
Nusselt number, **20**:189–191
optimum spacing, **20**:194–195, 202
Rayleigh number, **20**:189–191
staggered arrays, **20**:198–201
symmetric–asymmetric nomenclature, **20**:187–188
uniform flux flow rates, **20**:191–192
uniform heat flux (UHF) boundary conditions, **20**:188, 191
uniform wall temperature (UWT) boundary conditions, **20**:188, 191
velocity and temperature fields, **20**:189–190
Vertical surfaces, mixed convection about, **14**:61–63
Vertical tubes
CHF of forced flow in, **17**:24–25
empirical correlation of CHF data of water, **17**:27–30
inlet subcooling enthalpy, relationship to CHF, **17**:25–27
constant heat flux in, **15**:198–202

local heat transfer coefficients for, **14**:207–212
in packed fluidized beds, **14**:234–235
superimposed free convection in, **SUP**:413–414
thermal convection in, **15**:197–206
total heat transfer coefficients for, **14**:212–231
Vibration, **8**:99
heat transfer enhancement by, **30**:242–243
of heat transfer surface, **31**:277
Vibrational constant, in photon interactions, **12**:139
Vibrational energy and the accommodation coefficient, **2**:338
Vibrational nonequilibrium, **8**:248, 250
Vibrational relaxation time, **8**:247, 248
Vibrational rotational bands, **5**:272
of CO_2, **5**:273
of CO, **5**:273
of water vapor, **5**:273
Vibrational transitions, **8**:231
Vibration–rotation bands, **8**:230, 231
Virk's maximum drag reduction asymptote, **15**:87, 90
Virk's minimum drag asymptote, **19**:330
Viscasil-300M, **28**:154–155, 187, 188
Viscoelastic effects, **25**:257, 281, 304, 305, 308
media, **25**:272
packed beds, **23**:217
sedimentation, **23**:262
solution, **25**:278, 308
Viscoelastic flow, in rectangular channel, **19**:320
Viscoelastic fluids, **2**:359ff; **23**:189, 191; **24**:107; see also Non-Newtonian fluids
capillary tube results for, **15**:94–96
characteristic time of, **15**:77–79
defined, **15**:61–62
drag-reducing phenomenon in, **15**:69
eddy diffusivities for, **15**:130–132
heat and mass transfer analogy for, **15**:126–132
intermediate friction factors in, **15**:92–96
laminar mixed convection in external flows of, **15**:187–188
laminar thermal convection heat transfer, **15**:180
laminar thermal convection in external flows, **15**:178–184
mechanical degradation in, **15**:85–86
in steady shear flow, **15**:60
thermal convection in, **15**:143–223

turbulent heat transfer performance in, **15**:104

Viscoelasticity, **25**:259–261, 268, 269, 278, 279, 284, 286, 308

Viscoelasticity number, **2**:381

Viscoelastic material, **28**:152

Viscoinelastic material, **28**:152

Viscometer

    capillary tube, **15**:70–72, 113

    Couette, **15**:76

    falling ball, **15**:72–75

    rotating, **15**:75–76

Viscometric flow, **13**:212–259; *see also* Shear flow

Viscosity, **2**:15, 38f; **7**:8–10; **25**:30, 269; *see also* Transport properties of dilute gases

    apparent, **25**:253, 255

    coefficient, volume, **25**:5

    dependence on shear rate, **19**:251

    effective, **24**:118, 175; **25**:253, 255, 257, 260, 275, 276, 279, 283, 300, 314

    fluid, **25**:155

    fluidizing gas, **25**:167, 177

    influence of, on heat transfer to tubes and tube banks, **8**:114

    kinematic, **25**:155

    magnetostabilized bed, apparent, **25**:175

    Newtonian, **25**:314

    Newton's law of, **13**:211

    of polymer solution, **15**:60

    pressure dependence of, **13**:218

    Reynolds number and, **15**:60

    seawater, **18**:327

    stress-shear, **25**:281, 298, 312

    temperature dependence of, **7**:9

    turbulent, **25**:75

    Viscosity measurements, laminar flows and, **15**:70–77

    zero-shear, **25**:281, 298, 312

Viscosity effects, **3**:30f

Viscosity variations, heat- and mass-transfer, **20**:326–327

Viscous dissipation, **8**:162; **9**:279, 292; **SUP**:13; **25**:268

    cardioid duct, **SUP**:277–278

    circular duct, **SUP**:80, 82, 109–111, 129–132, 142, 146, 148–149

    concentric annular ducts, **SUP**:294

    elliptical ducts, **SUP**:249

    equilateral triangular duct, **SUP**:224

    Nusselt number and, **13**:240–241

    parallel plates, **SUP**:156–158, 176–178, 183–184

    polygonal ducts, **SUP**:263

Viscous drag tensors, **31**:40

Viscous flow, **SUP**:5

Viscous-fluid flow, heat transfer, **18**:118–122

Viscous forces, multiphase flow, **30**:93, 94

Viscous heating melting, **24**:13–15

Viscous layer, **3**:43f

Viscous oils

    forced convection, **30**:20–24

    heat transfer microscales, **21**:267–268

Viscous sublayer, **25**:327, 381–287, 400

    streaks, **25**:381–382, 384, 385, 387

Visual techniques, **8**:297

    hydrogen bubble technique, **8**:297

    use of

        in velocity measurements, **8**:297

        for velocity profile measurements, **8**:303

Vitrification, **22**:176, 213, 304

VLSI–ULSI (very large scale integration–ultra large scale integration) technology, **30**:314

Voidage, **25**:165, 196, 299

    around heat transfer surface, **25**:165, 239

    bed, **25**:167, 170, 174, 185, 197, 201, 202, 205, 223, 224, 229, 230, 233, 237

    uniform, **25**:198

Voidage profiles, packed beds, **23**:216

Void fraction, **1**:364ff; **9**:114; **25**:314

    for forced convection boiling, **5**:102

    liquid slug zone, **20**:109–110

    measurements, **21**:322–323

    slug units, **20**:88

    surface inclination, heat transport models, **20**:32–33

Voight function, **5**:281

Volatiles, combustion in open fires, **17**:206, 234–236

Volcanism, steam explosions, **29**:130

Volmer's rate equation, **29**:138

Volmer theory, in liquid superheat, **11**:7–9

Volterra integral equation, **8**:38

    representation of surface reaction, **2**:172ff

Volterra integrodifferential equation, **19**:16

Volume-averaged equations

    in mass/energy transport, **13**:137–153

    phase average in, **13**:138

Volume-averaged flux, **31**:58

Volume-averaged mass diffusive flux, **31**:34

    representation for local, **31**:33

Volume-averaged pressure, time rate of change of, **31**:117–118
Volume-averaged temperature, **31**:144
Volume-averaged thermal energy equations, **31**:53–54
Volume-averaged transport equations, **31**:12–13, 28–39
  for energy, **31**:41–48
  for momentum, **31**:39–41
Volume average of a time derivative, **31**:14–15
Volume averaging, **23**:378; **31**:11–23
  averaging theorem, **23**:379
  cellular average, **23**:410
  disordered porous media, **23**:384
  drying, **30**:169
  geometrical theorem, **23**:382
  length scale constraint, **23**:382, 385
  spatial deviation, **23**:381
  spatially periodic system, **23**:384
Volume change, rate of, **31**:146–147
Volume condensation, **21**:58
Volume displacement, rate of, **31**:146–147
Volume dissipation function, **8**:23
Volume goodness factor, **SUP**:393, 395–397
Volumetric absorption coefficient, **5**:257; **8**:232
  variation of, over entire band, **8**:234
Volumetric averaging, in heat conservation equation derivation, **14**:6–14
Volumetric combustion, direct containment heating, **29**:307–308
Volumetric concentration expansion coefficient, **20**:318
Volumetric emission coefficient, **5**:257
Volumetric expansion coefficient, **25**:23, 25
Volumetric extinction coefficient, **3**:181
Volumetric flow rate, local distribution of, **31**:119–121
Volumetric flux divergence equation, **31**:111
Volumetric heat capacity, **9**:353
Volumetric size distribution function, **23**:153
Volumetric thermal-expansion coefficient, **20**:318
von Kármán form of the Reynolds analogy, *see* Reynolds analogy
von Kármán-Polhausen equations, **2**:390
von Kármán's friction law, **3**:44
Von-Misses transformation, **8**:74
Vortex flow, wavy, **21**:152–153
Vortex layer, **9**:342
Vortex shedding, **8**:98

Vortex structure, **31**:174–176
Vortex zones, **31**:186
  key feature of, **31**:167
Vortices, **8**:96, 97; **16**:326–329
  liquid droplet, **26**:10, 11
  quantized, in helium II, **17**:85–91
  shedding of, **8**:99
  structure development, **25**:22
  vibrations due to, **8**:99
Vorticity, **8**:165, 166, 170; **25**:340, 343
Vorticity distribution, **31**:171
Vorticity equation, liquid metal droplet, **28**:5
Vorticity scalar function, equation, **21**:196

W

Waiora aquifer, New Zealand, **14**:52
Wairakei geothermal field, New Zealand, **14**:52–54
  production and subsidence history at, **14**:84–85
  transient pressure declines at, **14**:92
Wake column, **21**:291
Wakes, **8**:98
  flow, **18**:92–94
  influence of Reynolds number on, **8**:98
  symmetrical regime in, **8**:98
  velocity fluctuations in, **8**:98
Wake volume, **21**:291
Wall axial heat conduction boundary condition, **SUP**:30–31
Wall boundary conditions, *see* Boundary conditions, wall
Wall boundary layer, in impinging flow, **13**:5
Wall curvature, effects, **16**:243
Wall effects, **8**:314
  in classic MHD theory, **1**:271f
  in closed stoves
    heat loss through walls, **17**:237–241
    radiant heat gain by pans, **17**:241–244
  measurements of, **8**:315
  packed beds, **23**:216
  on probability density distributions, **8**:326
  in turbulence measurements, **8**:296, 297
Wall elements, **27**:10–12, 124, 147–148
  heat balance equations, **27**:47
  radiative energy, **27**:48
  simulating radiative heat transfer absorption, **27**:74–75, 80

emission from solid walls, **27**:59–61
  reflection and absorption, **27**:61–62
Wall friction, **31**:123
  hydraulic resistance coefficient caused by,
    31:204
Wall function, **25**:365–366
Wall heat flux, **27**:98, 143, 152, 154, 157
Wall heat flux distribution, *see* Arbitrary heat
    flux distribution; Specified wall heat flux
    distribution
Wall heating, unsteady, **25**:23
Wall inclination, heat- and mass-transfer, **20**:336
Wallis correlation, hydrodynamics, slug flow,
    20:90
Wall jets, **23**:45–46
  critical heat flux, **23**:101–103
Wall jet zone, **26**:105, 107
  heat flux, **26**:114–118
  temperature, **26**:118
  transport, **26**:108–109
Wall–layer transmissivity, in nonisothermal gas
    radiation, **12**:178–188
Wall layer transmittance
  cold, **12**:186–187
  vs. channel emittance, **12**:185
Wall peripheral mean temperature, **SUP**:45
Wall–polymer molecule interaction, packed beds,
    23:194, 206, 225
Walls
  catalytic, *see* Catalytic wall
  nondimensional correlations for in heat trans-
    fer, **10**:199–200
  thermal capacity of, in shear flow, **13**:225–
    227
Wall shear stress, **SUP**:38–39
  fluctuations in, **8**:293
Wall superheat, in boiling nucleation, **10**:154
Wall surface temperature, in bubble formation,
    10:142
Wall temperature, **16**:334
  distribution, **21**:33–34
    turbulent mixed convection, **21**:21
  in forced flow film boiling, **11**:145–146
  standard deviations in determination, **21**:76
Wall temperature measurement in experimental
    transition boiling studies, **18**:257–258
Wall thermal conductivity, idealizations of,
    SUP:24, 390
Wall thermal resistance boundary condition,
    SUP:20–21, 23–24, 27

for circular duct, **SUP**:80–81, 120–123, 149–
    152
for concentric annular ducts, **SUP**:318–319
for elliptical ducts, **SUP**:250–251
for parallel plates, **SUP**:158–159, 185–187
for sine ducts, **SUP**:255–256
Wall thermal resistance parameter, **SUP**:23, 55–
    56
Wall turbulence, **8**:304
  experimental measurements of, **8**:300, 301
Wandbindung, **6**:156, 159, 258
Wang and Longwell velocity profile, **SUP**:142,
    166
Warming time, as function of size of tubers,
    17:168
Warm-water column, cooling of, **13**:95–98
Warm-water plume, vertically rising, **13**:76–79
WASH-1400 study, **29**:193
Waste heat
  atmosphere and, **12**:2
  efficient use of, **17**:297–298
  rejection of in thermal cycle, **12**:1–2
Waste heat cooling, importance of, **12**:2
Water
  access to, **31**:431
  ambient conditions, jet behavior in, **16**:46–51
  boiling
    frequency of liquid–wall contacts, **18**:283
    mean liquid–heating surface contact time,
      18:279
    mean relative liquid–heating surface contact
      time, **18**:284
    mean vapor phase–heating surface contact
      time, **18**:281
  comprehensive chemical analysis of, **15**:80–82
  condensation/evaporation coefficient, **21**:73–
    74
  density anomaly, **19**:68
  diffusivity, **18**:328–330
  direct containment heating and, **29**:312–324
  enthalpy
    changes for various temperatures and sali-
      nities, **18**:356
    temperature and salinity effects, **18**:344
  entropy
    changes for various temperatures and sali-
      nities, **18**:357
    temperature and salinity effects, **18**:346
  film boiling in annulus, **5**:100
  film boiling of, **5**:88; *see also* Film boiling

Water (*continued*)
  forced convection boiling of, **5**:99
  gas-phase continuity equation for, **31**:31–39
  heat flow rate, **16**:160–161
    effect of heater shape and material, **16**:191, 218
  as heat transfer medium, **12**:10
  laminar flow in rectangular channels, **19**:313–314
  nucleate pool boiling
    data available, **16**:169
    heat transfer rates data differences between experimenters, **16**:200–201, 225, 226–227
  for once-through cooling, **12**:20–21
  physical properties, **21**:2–3
  Prandtl number
    temperature and salinity effects, **18**:348
    at various temperatures and salinities, **18**:358
  pure
    density, **18**:330–332
    enthalpy, **18**:343–344
    entropy, **18**:345
    Prandtl number, comparison at various temperatures and pressures, **18**:349
    specific heat
      change at 0° for various pressures, **18**:340
      change at 10° for various pressures, **18**:340
      change at 20° for various pressures, **18**:341
    temperature and pressure effects on enthalpy, **18**:343
    temperature and pressure effects on entropy, **18**:346
    thermal expansion at various temperatures and pressures, **18**:334
  saline, density, Gebhart-Mollendorf relation for density of, **16**:54–55
  saline expansion coefficient
    temperature and salinity effects, **18**:337
    at various temperatures and salinities, **18**:354
  salinity, in formulation of jet flow establishment, **16**:23–25
  seawater, *see* Seawater
  specific heat
    changes for various temperatures and salinities, **18**:355
    temperature and salinity effects, **18**:339
  steam bubble heat transfer to, **15**:254
  thermal conductivity, **18**:327–328
  thermal expansion at atmospheric pressure for various temperatures and salinities, **18**:335
  thermal expansion coefficient
    temperature and salinity effects, **18**:333
    at various temperatures and salinities, **18**:353
  transition boiling
    in annulus, **5**:100
    experimental studies, **18**:260–263
  viscosity, **18**:327
  wave number, **5**:83
  for wet towers, **12**:21–22
  Wien's displacement law, **5**:259
Water–air interface, turbulent motions at, **13**:64
water-boiling tests, on open fires, **17**:228
water-cooled circular foil gage, **23**:305
Water–ethylene glycol, bubble growth, experimental data, **16**:90–91
Water-evaporation method, cookstove efficiency and, **17**:265–266, 268–274
Water fouling, **31**:444
water–glycerine, bubble departure, diameter, **16**:95–97
water–glycerol, A_0, **16**:119
water–glycol, A_0, **16**:119
water–methyl ethyl ketone
  boiling site densities, experimental data, **16**:75–77
  film boiling, **16**:138
water–pyridine, A_0, **16**:119
water–steam two-phase geothermal systems
  discharge enthalpy for, **14**:95
  experimental investigations in, **14**:87–88
  heat transfer in, **14**:77–88
  multidimensional models of, **14**:82–87
  one-dimensional models of, **14**:78–82
Water stream, dimensionless interfacial temperature and temperature gradients along, **15**:241
Water surface boiling, heat transfer enhancement at, **31**:271–274
Water vapor
  band absorption for, **12**:150
  radiation–conduction interaction in, **8**:266, 267, 270, 272
  radiative transfer in, **8**:256, 259, 260
  rotation bands in, **8**:231

total emissivity of, **12**:161

Watt, James, **21**:55

Wave mechanics, **27**:3, 5–6

Wave number, **8**:232; **9**:326; **SUP**:179; **24**:214, 228, 251; **25**:333–336, 357

  critical, **24**:176

  implementing with CDF, **31**:391–394

Wave propagation, prediction of kinematic and thermal, **31**:120

Wave velocity, **27**:4

Wavy flow, **1**:358

Wavy gas–liquid interface, **10**:88

Wavy vortex

  critical speeds for onset, **21**:156

  induced rotational flow, **21**:152–153

WC-1 study, **29**:316, 318

WC-2 study, **29**:316, 318, 321

Weak nearly homogeneous scalar field

  evolution in weak degenerating isotropic velocity field, **21**:226–229

  evolution in weak evolving homogeneous turbulence, **21**:224–231

  evolution in weak nearly homogeneous velocity field, **21**:229–231

Weak nearly homogeneous velocity field, evolution, **21**:209–213

Weber number, **1**:363; **26**:36, 202; **31**:288

  critical, **21**:336–337, 339–340; **23**:69

  liquid droplets, **21**:335

  versus critical Reynolds number, **21**:340–341

Wedge flows, **8**:35

Wedges, **9**:296

  potential flow convection, **20**:363–364

    Nusselt number, **20**:375

  supersonic, initial relation near leading edge of, **2**:219ff

  supersonic flow about, **2**:167f, 171, 173, 174ff, 183, 184f

Wedge-shaped bodies, **8**:117

Wedge-shaped duct, **SUP**:264

Weight, bed, apparent, **25**:200

Weighted residuals method, **1**:97ff; **8**:32; **SUP**:65; **19**:195

Weinbaum–Jiji models

  criticism of Pennes, **22**:19, 96–97

  early two phase model

    blood phase

      continuity law, **22**:99–100

      energy balance equations, **22**:100

      scaling law, **22**:98

    governing equation, **22**:102

schematic view, **22**:96–97

simplified bioheat equation

  applicability

    asymbiotic expansion, **22**:139

    Baish superposition model, **22**:138

    blood flow dependence, **22**:140

    first order analysis, **22**:137–140

    unequal size vessel pairs

      asymptotic analysis, **22**:142–143

      effective conductivity, **22**:144

      experimental observations, **22**:141

      experimental verification, **22**:145–146

      governing equations, **22**:141–142

  derivation, **22**:124–128

  effective thermal conductivity, **22**:45, 128–129

  experimental observations, **22**:130

  implementation into peripheral tissue model

    cutaneous layer, **22**:133–134

    deep tissue layer, **22**:132

    intermediate tissue layer, **22**:133

    results, **22**:135–137

  justification, **22**:123–124

  preliminary calculations, **22**:124

  second order asymptotic analysis

    applicability of analysis, **22**:149

    closure condition, **22**:148

    conclusions, **22**:150

    first order equations, **22**:148

    second order equations, **22**:149

    zero order equations, **22**:148

  thermal equilibration length, **22**:129–130

skin layer, anatomical observations, **22**:102

three layer model

  analytical solution, **22**:114–118

  conclusions, **22**:123

  cutaneous layer

    blood flow effects, **22**:107

    difference with earlier model, **22**:107

    governing equations, **22**:113–114

  deep tissue layer

    assumptions, **22**:105, 110

    governing equations, **22**:71, 105, 108–110

    nondimensionalization, **22**:111

    thermal equilibration lengths, **22**:106

  intermediate tissue layer

    description, **22**:107

    governing equations, **22**:112–113

  parameter estimation, **22**:118–120

  physical description, **22**:103

Weinbaum–Jiji models (*continued*)
    schematic view, **22**:104
    simulation results, **22**:120–123
    tissue phase
        energy balance equation, **22**:101
        schematic view, **22**:101
Wein's displacement law, **27**:8
Weissenberg effect, **15**:60, 63; **19**:249
Weissenberg–Modney relation, **2**:388
Weissenberg number, **15**:77, 181, 183; **19**:256–257, 333, 334; **24**:142, 153
Weissenberg rheogoniometer, **15**:73, 76, 113, 117, 179
Welander–Keller mechanism, **14**:44
Welding, **19**:80–81
Wet cavity tests, direct containment heating, **29**:227–228
Wet cooling towers
    cost of, **12**:3
    noise and appearance of, **12**:22
    plumes and drift in, **12**:22–23
    vs. dry, **12**:4, 18
    water consumption in, **12**:21–22
Wet–dry hybrid cooling system, **12**:23
Wettability, condensation, **21**:61–65
Wettability of heating surface, effect on transition boiling heat transfer, **18**:264–265, 292–293
wgprnt function, **27**:129
Whole body heat transfer, **22**:134
Wick design
    direct cooling with heat pipes, **20**:290–293
    fluid–vapor transport and heat transfer, **20**:292–293
Wick inclination, heat pipe, **7**:221, 236, 320b
Wicks, thermal conductivity, **7**:320b
Wide-band scaling, in nonisothermal gas radiation, **12**:177–178
Wiener–Hopf method, **8**:49
Wien's displacement law, **9**:362
Wildland fires
    fuel analysis in, **10**:262–266
    ignition in, **10**:244
    vortices in, **10**:246
Windows
    double-pane, **18**:15
    single-pane, **18**:48–49
Wind tunnel for external heat and mass transfer experiments, **1**:125ff
Wind-tunnel measurements, incremental transfer in, **11**:236–238

Winfrith Technology Centre
    α-mode failure experiments, **29**:147
    melt–water interactions, **29**:155, 165
Wiper blades effect, induced rotational flow, **21**:174–176
Wire mesh, **31**:453
Wire-mesh grids, in heat and mass transfer, **13**:41–43
Wires, **8**:127
Wire-wound gage, **23**:299–300
Wissler models
    applicability of Pennes model, **22**:150, 152
    criticism of Weinbaum–Jiji models, **22**:137, 140
    vascular heat transfer, **22**:71
    whole body heat transfer, **22**:134
Wissler transform, **22**:262
WKBJ method, **8**:19; **SUP**:171, 301, 310
Wolverine Korodense® tube, fouling, **30**:219
Wood
    burning, moisture content and, **17**:232–234
    chemical nature of, **10**:225–226; **17**:175–176
    combustion of, **17**:188–189
        example solutions for solid-phase combustion, **17**:196–200
        as fuel, **17**:175–181
        mathematic modeling, **17**:189–196
        phenomenological description of, **17**:181–189
    combustion systems, two modes of operation, **17**:182–184
    moisture distribution during drying of, **13**:125
    open fires and
        characteristics of heat transfer to a pan, **17**:216–227
        experimental results, **17**:227–236
        flame heights and temperature profile, **17**:205–216
        physical properties of, **17**:176–180
        proportion as fuel in total energy consumption, **17**:161
        thermodynamic considerations, **17**:200–204
    pyrolysis of, **10**:228–231
Woodburning cookstoves, *see also* Stoves
    example heat balances for, **17**:290–293
    performance estimations and designs
        critique of efficiency measurements, **17**:263–275
        design considerations and stove optimization, **17**:296–299

efficiency and fuel economy, **17**:275–284
    heat balances, **17**:289–296
    stove control, **17**:284–289
  processes and system, **17**:165–166
    cooking, **17**:166–169
    resources and economics, **17**:172–175
    stove, **17**:169–172
Work of adhesion, **9**:227
Work of separation, **9**:191
Work transfer, **15**:6
Worm devices, **31**:159, 161
Worsøe–Schmidt perturbation, *see* Lévêque solution; Lévêque-type solutions
Wrede–Harteck gauge, **3**:258
Wulff continuum model
  analytical solution
    comparison with Pennes, **22**:35, 76
    parameterization, **22**:34
  criticism of Pennes, **22**:30–32, 36, 151
  derivation of alternate equation
    anistropic blood flow, **22**:32–33, 37, 46, 49, 55, 101
    complete blood–tissue equilibration, **22**:33–34, 48, 58
WUMT experiments, **29**:165

X

X-rays, **27**:3
X-ray techniques in density measurement, **1**:394ff

Y

YAG, properties near melting point, **30**:346
Yang, method of, **1**:106ff
Yawed cylinders, forced convection from, **11**:239–243
Yield stress, **25**:174, 175, 181, 204

in magnetically stabilized bed, **25**:174
Young's equation, **21**:62
Young's fringes, speckle photography, **30**:261

Z

Zabrodsky microbreak heat transfer model, **10**:176
Zeldovich factor, **14**:304
Zener's theory, **2**:341ff
Zero axial flow, induced rotational flow, **21**:163–168
Zero-defect, **25**:283
Zero-dimensional (stirred-tank) problems, in THIRBLE classification, **13**:65
Zero discharge, **31**:431, 433
Zero gravity, **5**:131
  alloy solidification, **28**:309
Zero pressure drop, in heat exchangers, **15**:27–28
Zero pressure gradient, in steady shear flow, **13**:244
Zero-shear, viscosity, **25**:281, 298, 312
Zero-vorticity
  cell model, **25**:295
  model, **25**:293, 298
Zero wall shear stress, **13**:246
Zeta potential, **12**:253–255
Zhukovsky number, **6**:383
Zinc selenide, properties, **30**:321
Zinc telluride, properties, **30**:321
ZM program, **27**:39–42, 203
Zone
  stabilized adsorption, **25**:232
  stagnant, **25**:161
Zone method, **27**:31–42
Z transform, **22**:417
Zuber–Findlay distribution parameter, **31**:110
Zwanzig's theory, **2**:349f

Appendix
Tables of Contents, Volumes 1–31

Volume 1

R. D. Cess, The Interaction of Thermal Radiation with Conduction and Convection Heat Transfer

Theodore R. Goodman, Application of Integral Methods to Transient Nonlinear Heat Transfer

A. V. Luikov, Heat and Mass Transfer in Capillary-Porous Bodies

G. Leppert and C. C. Pitts, Boiling

Mary F. Romig, The Influence of Electric and Magnetic Fields on Heat Transfer to Electrically Conducting Fluids

Mario Silvestri, Fluid Mechanics and Heat Transfer of Two-Phase Annular-Dispersed Flow

Volume 2

D. R. Bartz, Turbulent Boundary-Layer Heat Transfer from Rapidly Accelerating Flow of Rocket Combustion Gases and of Heated Air

Paul M. Chung, Chemically Reacting Nonequilibrium Boundary Layers

F. M. Devienne, Low Density Heat Transfer

A. B. Metzner, Heat Transfer in Non-Newtonian Fluids

E. M. Sparrow, Radiation Heat Transfer between Surfaces

Volume 3

J. Kestin, The Effect of Free-Stream Turbulence on Heat Transfer Rates

A. I. Leont'ev, Heat and Mass Transfer in Turbulent Boundary Layers

Ralph P. Stein, Liquid Metal Heat Transfer

R. Viskanta, Radiation Transfer and Interaction of Convection with Radiation Heat Transfer

A. A. Westenberg, A Critical Survey of the Major Methods for Measuring and Calculating Dilute Gas Transport Properties

Volume 4

A. J. Ede, Advances in Free Convection

Alice M. Stoll, Heat Transfer in Biotechnology

Robert Siegel, Effects of Reduced Gravity on Heat Transfer

E. R. G. Eckert and E. Pfender, Advances in Plasma Heat Transfer

C. Forbes Dewey, Jr., and Joseph F. Gross, Exact Similar Solutions of the Laminar Boundary-Layer Equations

Volume 5

John R. Howell, Application of Monte Carlo to Heat Transfer Problems
Duane P. Jordan, Film and Transition Boiling
Frank Kreith, Convection Heat Transfer in Rotating Systems
C. L. Tien, Thermal Radiation Properties of Gases
John A. Clark, Cryogenic Heat Transfer

Volume 6

A. F. Charwat, Supersonic Flows with Imbedded Separated Regions
W. Hauf and U. Grigull, Optical Methods in Heat Transfer
E. K. Kalinin and G. A. Dreitser, Unsteady Convective Heat Transfer and Hydrodynamics in Channels
B. S. Petukhov, Heat Transfer and Friction in Turbulent Pipe Flow with Variable Physical Properties

Volume 7

W. B. Hall, Heat Transfer near the Critical Point
T. Mizushina, The Electrochemical Method in Transport Phenomena
George S. Springer, Heat Transfer in Rarefied Gases
E. R. F. Winter and W. O. Barsch, The Heat Pipe
Richard J. Goldstein, Film Cooling

Volume 8

I. J. Kumar, Recent Mathematical Methods in Heat Transfer
A. Žukauskas, Heat Transfer from Tubes in Crossflow
Simon Ostrach, Natural Convection in Enclosures
R. D. Cess and S. N. Tiwari, Infrared Radiative Energy Transfer in Gases
Z. Zarić, Wall Turbulence Studies

Volume 9

D. Japikse, Advances in Thermosyphon Technology
Creighton A. Depew and Ted J. Kramer, Heat Transfer to Flowing Gas–Solid Mixtures
Herman Merte, Jr., Condensation Heat Transfer
B. Gebhart, Natural Convection Flows and Stability
C. L. Tien and G. R. Cunnington, Cryogenic Insulation Heat Transfer

Volume 10

Richard C. Birkebak, Thermophysical Properties of Lunar Materials: Part I: Thermal Radiation Properties of Lunar Materials from the Apollo Missions
Clifford J. Cremers, Thermophysical Properties of Lunar Media: Part II: Heat Transfer within the Lunar Surface Layer
Robert Cole, Boiling Nucleation

Chaim Gutfinger and Nesim Abuaf, Heat Transfer in Fluidized Beds
S. L. Lee and J. M. Hellman, Heat and Mass Transfer in Fire Research

Volume 11

N. H. Afgan, Boiling Liquid Superheat
E. K. Kalinin, I. I. Berlin, and V. V. Kostyuk, Film-Boiling Heat Transfer
Vincent T. Morgan, The Overall Convective Heat Transfer from Smooth Circular Cylinders
G. D. Raithby and K. G. T. Hollands, A General Method of Obtaining Approximate
    Solutions to Laminar and Turbulent Free Convection Problems
R. Viskanta and E. E.Anderson, Heat Transfer in Semitransparent Solids

Volume 12

F. K. Moore, Dry Cooling Towers
Yona Dimant and Michael Poreh, Heat Transfer in Flows with Drag Reduction
D. K. Edwards, Molecular Gas Band Radiation
Aharon S. Roy, A Perspective on Electrochemical Transport Phenomena

Volume 13

Holger Martin, Heat and Mass Transfer between Impinging Gas Jets and Solid Surfaces
D. Brian Spalding, Heat and Mass Transfer in Rivers, Bays, Lakes, and Estuaries
Stephen Whitaker, Simultaneous Heat, Mass, and Momentum Transfer in Porous Media: A
    Theory of Drying
Horst H. Winter, Viscous Dissipation in Shear Flows of Molten Polymers

Volume 14

Ping Cheng, Heat Transfer in Geothermal Systems
T. B. Jones, Electrohydrodynamically Enhanced Heat Transfer in Liquids—A Review
S. C. Saxena, N. S. Grewal, J. D. Gabor, S. S. Zabrodsky, and D. M. Galershtein, Heat
    Transfer between a Gas Fluidized Bed and Immersed Tubes
G. M. Shved, Influence of Radiative Transfer on Certain Types of Motions in Planetary
    Atmospheres
George S. Springer, Homogeneous Nucleation

Supplement 1

R. K. Shah and A. L. London, Laminar Flow Forced Convection in Ducts

Volume 15

Adrian Bejan, Second-Law Analysis in Heat Transfer and Thermal Design
Y. I. Cho and James P. Hartnett, Non-Newtonian Fluids in Circular Pipe Flow

A. V. Shenoy and R. A. Mashelkar, Thermal Convection in Non-Newtonian Fluids

Samuel Sideman and David Moalem-Maron, Direct Contact Condensation

Patricia E. Wirth and Ervin Y. Rodin, A Unified Theory of Linear Diffusion in Laminated Media

Volume 16

Benjamin Gebhart, David S. Hilder, and Matthew Kelleher, The Diffusion of Turbulent Buoyant Jets

John R. Thome and Richard A. W. Shock, Boiling of Multicomponent Liquid Mixtures

M. G. Cooper, Heat Flow Rates in Saturated Nucleate Pool Boiling—A Wide-Ranging Examination Using Reduced Properties

R. J. Moffat and W. M. Kays, A Review of Turbulent-Boundary-Layer Heat Transfer Research at Stanford, 1958–1983

Volume 17

Y. Katto, Critical Heat Flux

John M. Pfotenhauer and Russell J. Donnelly, Heat Transfer in Liquid Helium

K. Krishna Prasad, E. Sangen, and P. Visser, Woodburning Cookstoves

M. S. Sohal, Critical Heat Flux in Flow Boiling of Helium I

Volume 18

Ren Anderson and Frank Kreith, Natural Convection in Active and Passive Solar Thermal Systems

A. Žukauskas, Heat Transfer from Tubes in Crossflow

A. Dutta and R. A. Mashelkar, Thermal Conductivity of Structured Liquids

E. K. Kalinin, I. I. Berlin, and V. V. Kostiouk, Transition Boiling Heat Transfer

David J. Kukulka, Benjamin Gebhart, and Joseph C. Mollendorf, Thermodynamic and Transport Properties of Pure and Saline Water

Volume 19

L. S. Yao and J. Prusa, Melting and Freezing

S. C. Saxena, Heat Transfer between Immersed Surfaces and Gas-Fluidized Beds

N. M. Tsirelman, Variational Solutions of Complex Heat and Mass Transfer Problems

James P. Hartnett and Milivoje Kostic, Heat Transfer to Newtonian and Non-Newtonian Fluids in Rectangular Ducts

Volume 20

Kaneyasu Nishikawa and Yasunobu Fujita, Nucleate Boiling Heat Transfer and Its Augmentation

Yehuda Taitel and Dvora Barnea, Two-Phase Slug Flow

Branislav S. Baclic and Peter J. Heggs, Unified Regenerator Theory and Reexamination of the Unidirectional Regenerator Performance

G. P. Peterson and Alfonso Ortega, Thermal Control of Electronic Equipment and Devices

Osvair V. Trevisan and Adrian Bejan, Combined Heat and Mass Transfer by Natural Convection in a Porous Medium

Stephen R. Galante and Stuart W. Churchill, Applicability of Solutions for Convection in Potential Flow

Volume 21

A. F. Polyakov, Heat Transfer under Supercritical Pressures

Ichiro Tanasawa, Advances in Condensation Heat Transfer

David Moalem Maron and Shimon Cohen, Hydrodynamics and Heat/Mass Transfer near Rotating Surfaces

B. A. Kolovandin, Modeling the Dynamics of Turbulent Transport Processes

Vedat S. Arpaci, Radiative Entropy Production—Heat Lost to Entropy

George Alanson Greene, Heat, Mass, and Momentum Transfer in a Multifluid Bubbling Pool

Volume 22 **Bioengineering Heat Transfer**

John C. Chato, A View of the History of Heat Transfer in Bioengineering

Caleb K. Charny, Mathematical Models of Bioheat Transfer

Kenneth R. Diller, Modeling of Bioheat Transfer Processes at High and Low Temperatures

Jonathan W. Valvano, Temperature Measurements

Volume 23

D. H. Wolf, Frank P. Incropera, and Raymond Viskanta, Jet Impingement Boiling

Massoud Kaviany and B. P. Singh, Radiative Heat Transfer in Porous Media

R. P. Chhabra, Fluid Flow, Heat, and Mass Transfer in Non-Newtonian Fluids: Multiphase Systems

Thomas E. Diller, Advances in Heat Flux Measurements

Michel Quintard and Stephen Whitaker, One- and Two-Equation Models for Transient Diffusion Processes in Two-Phase Systems

Volume 24

Adrian Bejan, Contact Melting Heat Transfer and Lubrication

B. D. Vujanović and B. S. Bač lić, Variational Solutions of Transient Heat Conduction Through Bodies of Finite Length

A. V. Shenoy, Non-Newtonian Fluid Heat Transfer in Porous Media

Wei Shyy, Elements of Pressure-Based Computational Algorithms for Complex Fluid Flow and Heat Transfer

Jae Min Hyun, Unsteady Buoyant Convection in an Enclosure

Volume 25

E. K. Kalinin and G. A. Dreitser, Unsteady Convective Heat Transfer in Channels

S. C. Saxena, V. L. Ganzha, S. H. Rahman, and A. F. Dolidovich, Heat Transfer and Relevant Characteristics of Magnetofluidized Beds

U. K. Ghosh, S. N. Upadhyay, and R. P. Chhabra, Heat and Mass Transfer from Immersed Bodies to Non-Newtonian Fluids

Michele Ciofalo, Large-Eddy Simulation: A Critical Survey of Models and Applications

Volume 26

Portonovo S. Ayyaswamy, Direct-Contact Transfer Processes with Moving Liquid Droplets

B. W. Webb and C.-F. Ma, Single-Phase Liquid Jet Impingement Heat Transfer

D. P. Sekulić and R. K. Shah, Thermal Design Theory of Three-Fluid Heat Exchangers

Volume 27

Wen-Jei Yang, Hiroshi Taniguchi, and Kazuhiko Kudo, Radiative Heat Transfer by the Monte Carlo Method

Volume 28   **Transport Phenomena in Materials Processing**

Dimos Poulikakos and John M. Waldvogel, Heat Transfer and Fluid Dynamics in the Process of Spray Deposition

Costas P. Grigoropoulos, Ted D. Bennett, Jeng-Rong Ho, Xianfan Xu, and Xiang Zhang, Heat and Mass Transfer in Pulsed-Laser-Induced Phase Transformations

Yogesh Jaluria, Heat and Mass Transfer in the Extrusion of Non-Newtonian Materials

Patrick J. Prescott and Frank P. Incropera, Convection Heat and Mass Transfer in Alloy Solidification

Roop L. Mahajan, Transport Phenomena in Chemical Vapor-Deposition Systems

Volume 29   **Heat Transfer in Nuclear Reactor Safety**

V. K. Dhir, Heat Transfer from Heat-Generating Pools and Particulate Beds

John H. S. Lee and Marshall Berman, Hydrogen Combustion and Its Application to Nuclear Reactor Safety

D. F. Fletcher and T. G. Theofanous, Heat Transfer and Fluid Dynamic Aspects of Explosive Melt–Water Interactions

Martin M. Pilch, Michael D. Allen, and David C. Williams, Heat Transfer During Direct Containment Heating

Volume 30

Vedat S. Arpaci, Microscales of Turbulent Heat and Mass Transfer

C. Y. Wang and P. Cheng, Multiphase Flow and Heat Transfer in Porous Media

Euan F. C. Somerscales and Arthur E. Bergles, Enhancement of Heat Transfer and Fouling Mitigation